ASTRONOMY

AT PLAY IN THE COSMOS

SECOND EDITION

ASTRONOMY

AT PLAY IN THE COSMOS

SECOND EDITION

ADAM FRANK

University of Rochester

with contributions by

JEFF BARY

Colgate University

CAROL LATTA

Rochester Academy of Science

W. W. NORTON & COMPANY

Independent Publishers Since 1923

Norton & Company has been independent since its founding in 1923, when William
and Mary D. Herter Norton first published lectures delivered at the People's In-
education division of New York City's Cooper Union. The firm soon expanded
the Institute, publishing books by celebrated academics from America
road. By midcentury, the two major pillars of Norton's publishing program—trade books
and college texts—were firmly established. In the 1950s, the Norton family transferred control of the company to its employees, and today—with a staff of five hundred and hundreds
of trade, college, and professional titles published each year—W. W. Norton & Company
stands as the largest and oldest publishing house owned wholly by its employees.

Editor: Erik Fahlgren
Project Editor: Michael Fauver
Editorial Assistant: Sara Bonacum
Managing Editor, College: Marian Johnson
Managing Editor, College Digital Media: Kim Yi
Production Manager: Eric Pier-Hocking
Media Editor: Rob Bellinger
Associate Media Editor: Arielle Holstein
Media Project Editor: Marcus Van Harpen
Media Editorial Assistant: Jasvir Singh
Ebook Production Manager: Danielle Lehmann
Marketing Manager, Astronomy: Katie Sweeney
Design Director: Rubina Yeh
Designer: DeMarinis Design LLC
Photo Editor: Travis Carr
Director of College Permissions: Megan Schindel
Permissions Manager: Bethany Salminen
Copyeditor: Heather Whirlow Cammarn
Proofreader: Jennifer Greenstein
Composition: Graphic World
Illustration Studio: Graphic World
Manufacturing: Transcontinental Interglobe, Inc.

Permission to use copyrighted material is included on page C-1.

ISBN 978-0-393-93522-6

Library of Congress Cataloging-in-Publication Data

Names: Frank, Adam, 1962- author. | Bary, Jeff, author. | Latta, Carol
    (Carol E.), author.
Title: Astronomy : at play in the cosmos / Adam Frank, University of
    Rochester ; with contributions by Jeff Bary, Colgate University, Carol
    Latta, Rochester Academy of Science.
Description: Second edition. | New York : W. W. Norton & Company, [2020] |
    Includes index.
Identifiers: LCCN 2019045124 | ISBN 9780393673999 (paperback) |
ISBN
    9780393419955 (epub)
Subjects: LCSH: Astronomy--Textbooks. | LCGFT: Textbooks.
Classification: LCC QB45.2 .F73 2020 | DDC 520--dc23
LC record available at https://lccn.loc.gov/2019045124

W. W. Norton & Company, Inc., 500 Fifth Avenue, New York, NY 10110-0017

wwnorton.com

W. W. Norton & Company Ltd., Castle House, 75/76 Wells Street, London W1T 3QT

1 2 3 4 5 6 7 8 9 0

TO MY PARENTS INGRID FRANK AND GEORGE RICHARDSON.
THANK YOU FOR TEACHING ME THAT WONDER IS THE
HIGHEST VIRTUE AND PURSUING WONDER
THE HIGHEST ASPIRATION.

BRIEF CONTENTS

CONTENTS

1

Getting Started
SCIENCE, ASTRONOMY, AND BEING HUMAN 1

2

A Universe Made, a Universe Discovered
THE NIGHT SKY AND THE DAWN OF ASTRONOMY 21

PREFACE

When teaching introductory astronomy to nonscience majors, I realized that most of my students were not reading their textbook. These students, many of whom are humanities and social science majors, definitely liked to read. They just don't like to read textbooks. This is the main reason I wanted to write a book that has everything a text needs but *doesn't read like a textbook*.

The most important distinction between *Astronomy: At Play in the Cosmos* and other textbooks is this book's use of more than one voice to explain science and share the excitement of discovery. I began a second career as a popular-science writer during physics graduate school. While working on my PhD thesis I also began writing for small magazines such as *Exploratorium Quarterly*; later I made regular contributions to national publications such as *Discover* and *Astronomy* magazines. The training I received, from some of the best science writers and editors in the country, included the importance of "narrative drive" in writing to nonscientists about science. The stories of science, and the science itself, is easier for nonscientists to access when it's part of a broader human narrative: difficulties faced and overcome, people willing to endure long hours in service of an inner drive to know—these make science interesting and memorable to any reader.

With that experience in mind, I started this project by conducting interviews with a diverse group of working scientists and used their voices to help me tell the story of the cosmos. Quotes from the interviews, with scientists who include Crystal Brogan, Miki Nakajima, David Jewitt, Margaret Geller, and Eric Wilcots, are woven into the narrative and provide the all-important human element. Brief profiles of these researchers appear in each chapter, and lengthier versions are included in Appendix 8. The quotes and profiles give students a face and story to accompany the science they're learning. This approach also fulfills one of the cardinal rules of good science writing: let someone else tell the story. For student readers faced with a dizzying array of facts and details to digest, this device can mean the difference between falling asleep and falling in love with a topic.

In this book I have tried to use the fundamental skills I learned as a magazine writer—such as effective transitions and the proper balance of detail and narrative—to make a book that students *will* read. The Second Edition includes **Anatomy of a Discovery** figures, developed by Jeff Bary of Colgate University, which provide students with a visual representation of how a chapter-relevant discovery was made and the people who made it. Discoveries such as how planetary systems form, the evolution of the Hertzsprung-Russell diagram, and ripples in space-time are illustrated in a way that can help students see the process of science and the people who participate in that process.

By using interviews and stories from real astronomers, I also had the chance to show students how diverse this science can be. STEM fields (science, technology, engineering, and math) can thrive only if they draw from, and support, as wide a pool of talent as possible. The scientists in the Second Edition represent people with many kinds of stories and backgrounds and will, hopefully, let students see there is no one "kind" of astronomer (except the really excited and passionate kind).

Although *Astronomy: At Play in the Cosmos* was written to be read, my more than 30 years of teaching introductory astronomy taught me that a textbook also serves as a study tool and reference. To this end, I developed several in-chapter features that make the Second Edition a key resource for all students:

- Each chapter begins with a set of learning goals keyed to that chapter's sections. I have tried to provide goals that span Bloom's taxonomy and require that a student not only knows the basics but also can explain the concepts to a classmate.

CRYSTAL BROGAN

"When I was much younger, I wanted to be an archaeologist," says Crystal Brogan. "I loved *Raiders of the Lost Ark*, and I thought finding clues and assembling ideas about how stuff happens in the past was really exciting." Eventually she concluded that all the biggest mysteries on that front had been solved and decided that assembling clues about the Universe would be just as exciting.

One of Brogan's most exciting moments in science came when she and her colleagues used the ALMA telescope to get one of the highest-resolution views ever of a young planet-forming disk, HL Tau. "It was the gaps that were really amazing," she says of the image. "The gaps were a very telling sign of planet formation."

MARGARET GELLER

Margaret Geller is a trailblazer. You can tell that from the awards she's received, including the MacArthur Fellowship (the "Genius" award)—in part for her willingness to take on a seemingly impossible project at the frontiers of her field. Born in Ithaca, New York, she got a bachelor's degree in physics at UC Berkeley. After getting a PhD in physical cosmology, Geller headed to the Harvard-Smithsonian Center for Astrophysics—where she did groundbreaking research on the Universe's large-scale structure.

Geller keeps a keen eye on how science works in the rest of society. She has collaborated on films and TV shows and in 2004 wrote the essay "The Black Ribbon," which explored how poverty keeps many bright, talented students from pursuing careers in science. As she says, "The differences between the population of physical scientists and the population of the United States—or of the world—are not subtle."

"The downside of all that pattern hunting is that sometimes we perceive patterns that don't exist or don't mean anything significant."

- Each chapter provides integrated Section Summaries so that students can quickly review the major concepts covered in each section.
- Checkpoint questions at the end of each section provide an opportunity for students to stop and assess their understanding before moving to the next section.
- Each chapter ends with an overarching Chapter Summary, organized by learning goals, so that students can review, in one or two paragraphs, the points covered in the learning goals.
- For easy reference, a running glossary in the margins provides definitions of terms that appear in boldface type in the text.

Mathematical literacy is an important part of scientific literacy. In astronomy, seeing how the math works often helps students better understand a concept or work through a problem. Recognizing that many students have different mathematical backgrounds (and anxieties), I treat most quantitative aspects of topics *separately*, in **Going Further** boxes, so that instructors can adjust the level of mathematics in their courses. In the **Going Further** boxes I explain each step so that students understand why and how the math is being applied.

At the end of each chapter, students will find three categories of questions and problems to help them check their understanding:

- **Narrow It Down** includes multiple-choice questions that range from fact-based questions to ranking exercises.
- **To the Point** poses open-ended, qualitative questions that can be used to spark a discussion in class.
- **Going Further** offers quantitative problems that support the **Going Further** boxes by asking students to use the skills they learned in those boxes.

As I wrote each chapter, I considered how the concepts in the book could be animated or simulated to make them even clearer. An NSF Career Award in 1997 allowed me to begin developing interactives for my classes. That experience showed how powerful digital tools can be for introductory astronomy, but only if they are clearly and cleanly integrated into the textbook and the course. For this textbook, Jeff Bary and I worked together not only to identify the concepts chosen for the nearly 50 interactive features but also to develop the storyboards that have gone into each of these online tools. Identified by a thumbnail image in the text and ebook, these simulation-based interactive features can be used in class, recitations, or labs and allow students to "play" with ideas first encountered in the text. These same "smart" graphics appear again in modified form as interactive assessment questions in Norton's online tutorial and homework system, Smartwork5.

The 18 chapters of *Astronomy: At Play in the Cosmos* are organized with the understanding that certain topics (extraterrestrial life, black holes, the Big Bang, etc.) are more effective at holding student interest than others. This organization is designed to develop a cadence whereby these popular topics come at the end of each section, enabling instructors to reignite students' excitement. The treatment of life in the Universe comes at the end of the planet unit rather than in the book's final chapter (which some classes never reach because of time constraints). A chapter on black holes concludes the stellar evolution unit, and cosmology concludes the unit on galaxies and large-scale structure. Each of these chapters presents a broad and engaging view that speaks both to the current state of the science and to the questions that students bring to the topics.

The Second Edition has too many changes to list in one place. Here are some of the highlights:

- Each chapter now includes one **Anatomy of a Discovery** figure that illustrates, in a simple yet visually appealing style, the people and the processes involved in making an important discovery relating to the chapter.

INTERACTIVE:
Anatomy of the Sun

ANATOMY OF A DISCOVERY

Are there ripples in space-time?

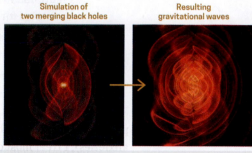

Simulation of two merging black holes

Resulting gravitational waves

theory

Gravitational lensing: test of general relativity

In 1915, **Albert Einstein** published his general theory of relativity. It revolutionized our understanding of the interactions of massive bodies with the fabric of space and time. In it he predicted that the path of even a massless wave of light is altered when passing near a massive object.

hypothesis

Einstein's theory also predicts the generation of gravitational radiation by objects accelerating through space-time. The merger of very massive, dense objects such as neutron stars or black holes is predicted to radiate brief bursts of the Universe's most energetic gravitational waves (ripples in space-time), which may be detectable on Earth.

confirmed by observation

Using the Arecibo radio telescope, **Russell Hulse** and **Joseph Taylor** discovered the first binary pulsar in 1974. They noticed the system's orbital period was decreasing, as though the two stars were losing energy via gravitational waves—the mechanism Einstein had predicted 60 years earlier. Although seen as a confirmation of one prediction of Einstein's theory, this evidence of gravitational waves was indirect.

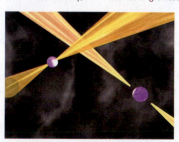

Artist's conception of binary pulsar

LIGO's Hanford Observatory

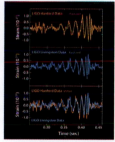

Detecting gravitational waves

confirmed by experiment

In 2015, an international team including **Manuela Campanelli** used the Laser Interferometer Gravitational-Wave Observatory (LIGO) to make the first direct detection of gravitational waves from the merger of two massive black holes. The "ripples" LIGO measured have amplitudes on the order of one-thousandth of the diameter of an atomic nucleus!

- I interviewed six new scientists for the Second Edition: Lucianne Walkowicz in Chapter 1, Crystal Brogan in Chapter 5, Miki Nakajima in Chapter 6, Didier Saumon in Chapter 11, Meg Urry in Chapter 16, and Margaret Geller in Chapter 17.

- Every figure was evaluated, and based on user and reviewer feedback, many illustrations were tweaked to ensure their clarity, efficacy, and accessibility. Examples of new or significantly revised figures include the following:

- An increased emphasis on helping students visualize motions of the sky with new figures illustrating the celestial sphere (Figure 2.5), meridians and the zenith (Figure 2.7), and the small-angle formula (Figure 2.10)
- New photos in Figures 7.24 and 7.25, showing the evidence of water on Mars
- Figure 13.27 on light curves of Type II and Type Ia supernovas, which has been annotated for further clarity
- A new illustration (Figure 14.22) to illustrate the accompanying new material on gravitational waves
- Figure 17.13, new to the Second Edition, which helps students visualize the gravitational attraction of the Great Attractor

- Astronomical images are now accompanied by icons that indicate the part of the electromagnetic spectrum at which they were photographed and by scale bars to facilitate comparisons.

- I added Checkpoint questions at the end of numbered sections to allow students to check that they are ready to proceed.
- Throughout the text and **Going Further** boxes, key equations have been clarified via the addition of verbal descriptions and interim steps.
- *At Play in the Cosmos* videogame hints are now referenced in the margin where relevant.
- Key new topics have been added throughout the Second Edition, including the history of the Universe, starting with the Big Bang (Chapter 1); the importance of "naked-eye" observations (Chapter 2); bound/unbound orbits, free-fall, and weightlessness (Chapter 3); charge-coupled devices, adaptive and active optics, and light pollution (Chapter 4); Pluto and other trans-Neptunian objects, Kirkwood gaps, Centaurs, and categories of meteorites (Chapter 5); seafloor spreading, the link between tidal interaction and the size of the Moon's orbit, and the link between the Moon and life on Earth (Chapter 6); the role of conservation of momentum in gas giant formation (Chapter 8); Kepler-186f (Chapter 9); luminosity class (Chapter 11); cosmic recycling, hypernovas, the link between stellar evolution and the abundance of elements, and kilonovas (Chapter 13); the principle of equivalence, gamma-ray bursts and hypernovas, and gravitational waves (Chapter 14); Vesto Slipher (Chapter 16); and the Higgs particle, the Planck era, the GUTS era, the electroweak era, Georges Lemaître, and Olbers's paradox (Chapter 18).
- Many topics have been updated to reflect our current understanding of the science, including climate change (Chapter 1); exoplanet research, comet research, the comet water hypothesis, and dwarf planets (Chapter 5); changes in Earth's surface temperature and the formation of the Moon (Chapter 6); the geologic status of the surfaces of Mercury and Venus and the evidence of water on Mars (Chapter 7); Jupiter's atmosphere and Ganymede (Chapter 8); habitability considerations, the search for life on other planets and moons, and Search for Extraterrestrial Intelligence (SETI) signals (Chapter 9); jets and outflows (Chapter 12); globular clusters in the Milky Way's stellar halo, the structure of the galactic center, and stellar-mass black holes at the galactic center (Chapter 15); active galactic nuclei (AGNs) and the unified model of AGNs (Chapter 16); and X-ray halos, redshift surveys, studies of large-scale structure, and models of cold dark matter and hierarchical formation (Chapter 17).

In my introductory astronomy course, I have two overarching goals: scientific literacy and a love of science (especially astronomy). I believe it's imperative for my students, and all citizens, to have at least a basic understanding of science and physical concepts, es-

AT PLAY IN THE COSMOS THE VIDEOGAME

Navigate the dangers of a star-forming accretion disk in Mission 2 and recover an artifact of unknown but dangerous origin.

"It's very difficult to smash a planet once you've made it."

pecially with "science denial" becoming a prevalent part of social discourse. I want my students, our next generation, to have an understanding of how science works and to be both curious and critical about information presented to them. My great hope is that, throughout their lives, my students will turn up the volume or click the link when a science story is reported. This brings me to my second goal: I want my students to *love* science—but I'll be satisfied if they tell their roommates or relatives that their astronomy course was one of the best classes they took in college. I sincerely believe that we have the opportunity to make a difference when we teach this course and that the approach I've taken with this book will motivate more students to read, see some part of themselves in the diverse group of people who make discoveries, and share their excitement about science.

The teaching and learning program to accompany the Second Edition will continue to set the bar for this course in terms of quality of the material and innovation.

> "The Miller-Urey experiment happened back when the origin of life was viewed as a chemistry problem. . . . It's a software problem as much as a hardware problem."

At Play in the Cosmos: The Videogame

Backed by academic research on how students learn through play, we've developed and updated the videogame *At Play in the Cosmos* with the Gear Learning studio at the University of Wisconsin and the Learning Games Network, with authorial guidance from Adam Frank and Jeff Bary. The game challenges students to apply what they've learned and to learn more, by flying challenging missions and confronting problems in astronomy—such as finding habitable exoplanets. The game offers 22 missions, each one tied to specific learning objectives, plus an Exploration Mode in which students can visit and obtain data from any of the astrophysical objects in the game. Each mission asks students to apply the right tool to the job.

Instructors can have students play the game before or after class, in the classroom or the lab. Students can also play whenever they want outside of class.

The videogame reports student diagnostic data to a grade book, enabling instructors to assess students' engagement and progress. Smartwork5 now supports the videogame by providing instructors with questions and problems that will assess students' understanding of concepts after they have completed different missions.

Ancillaries for Students

Interactive Simulations

Jeff Bary, *Colgate University*

Nearly 50 easy-to-use and tablet-friendly interactive simulations enable students to play with physical relationships that are key to the study of astronomy. The simulations use art from both the textbook and the videogame. They are incorporated into the videogame, the ebook, and Smartwork5 online assessment. The simulations are also available for professors to use in the classroom.

smartw✷rk**5**

David Wood, *San Antonio College*
Douglas Shields, *University of Arkansas*
Violet Mager, *Penn State Wilkes-Barre*
Juan Cabanela, *Minnesota State University Moorhead*
William Dirienzo, *University of Wisconsin–Sheboygan*

Smartwork5 is Norton's online tutorial and homework system. Over 900 Smartwork5 questions support the Second Edition of *Astronomy: At Play in the Cosmos*—all with answer-specific feedback, hints, and ebook links. Questions include ranking and sorting tasks, selected end-of-chapter problems (both multiple-choice and algorithmic numeric entry), labeling exercises based on book art, guided inquiry activities based on the Interactive Simulations, and new postmission questions based on *At Play in the Cosmos:* The Videogame.

In addition, guided-inquiry "Process of Science" assignments help students apply the scientific method to important questions in astronomy, challenging them to think like scientists.

Smartwork5 offers template assignments but is also fully customizable. You can quickly find questions by filtering for specific learning objectives or problem type/series. You can also modify any system question or even create your own.

Smartwork5 provides rich diagnostic data on student performance, and it can integrate with campus Learning Management Systems (LMS).

Learning Astronomy by Doing Astronomy Workbook: Collaborative Lecture Activities, Second Edition

Stacy Palen, *Weber State University*

Ana Larson, *University of Washington*

Students learn best by doing. Devising, writing, testing, and revising suitable in-class activities that use real astronomical data, illuminate astronomical concepts, and ask probing questions that encourage students to confront misconceptions can be challenging and time-consuming. In this workbook, the authors draw on their experience teaching thousands of students in many different types of courses (large in-class, small in-class, hybrid, online, flipped, and so on) to present 36 field-tested activities that can be used in any classroom today. The activities have been designed to require no special software, materials, or equipment, and to take no more than 50 minutes to complete. Pre- and post-activity questions are now available in Smartwork5 so instructors can easily assess student understanding before and after each activity.

Instructor's materials, including PowerPoint versions of the pre- and post-activity class questions, are available at Norton's Instructor's Site.

Starry Night Planetarium Software (College Version) and Norton *Starry Night Workbook*

Steven Desch, *Guilford Technical Community College*

Michael Marks, *Bristol Community College*

Starry Night is a realistic, user-friendly planetarium simulation program designed to enable students in urban areas to perform observational activities on a computer screen. Norton's unique accompanying workbook offers observation assignments that guide students' virtual explorations and help them apply what they've learned from reading assignments in the text.

Ancillaries for Instructors

Instructor's Manual

Trace Tessier, *Quinnipiac University*

The Instructor's Manual is designed to help instructors prepare for lectures and exams. It contains chapter overviews, suggestions for using the text and the interactive simulations, and worked solutions for all in-chapter and end-of-chapter problems. The manual also includes notes for teaching with the corresponding *Learning Astronomy by Doing Astron-*

> "That means globular clusters give us a natural controlled experiment to test our stellar evolutionary tracks."

omy: Collaborative Lecture Activities and answers to the *Starry Night Workbook* exercises. In addition, the Instructor's Manual provides support for integrating the videogame into your course.

Test Bank

Steven Furlanetto, *University of California–Los Angeles*

The Test Bank assesses a common set of learning objectives, consistent with Smartwork5 online homework, and provides over 1,200 multiple-choice and short-answer questions. Questions are classified according to Bloom's taxonomy (remembering, understanding, applying, analyzing, and evaluating), difficulty level, and section. The Test Bank makes it easy to construct meaningful and diagnostic exams.

PowerPoint Lecture Slides

Matthew Gonderinger, *Wayne State University*

These ready-to-use lecture slides complement the text by providing summaries of central topics, integrated photographs and art, class questions (including questions that address common misconceptions), and links to the interactive simulations. These lecture slides are editable and are available in Microsoft PowerPoint format.

Norton also provides an update service—quarterly PowerPoint presentations on engaging new topics—that enables instructors to cover new developments in astronomy soon after they occur.

Norton Interactive Instructor's Guide (IIG)

This new and searchable online tool is designed to help instructors prepare for lectures in real time. It contains the following resources, each tagged by topic and chapter:

- Test Bank, in Examview, Word, and PDF formats
- Lecture Slides, in PowerPoint format
- All art and tables in JPEG and PowerPoint formats
- Interactive Simulations
- Instructor Notes
- Discussion Points
- Notes for teaching with the Anatomy of a Discovery figures and scientist interviews
- Notes for teaching with *Learning Astronomy by Doing Astronomy: Collaborative Lecture Activities*
- Solutions to all checkpoint and end-of-chapter problems
- Answers to the *Starry Night Workbook* exercises
- Support for integrating the videogame into your course

Norton Instructor's Resource Site

This web resource contains the following resources to download:

- Test Bank, in ExamView, Word, and PDF formats
- Instructor's Manual, in PDF format
- Lecture slides, in PowerPoint format
- All art and tables, in JPEG and PowerPoint formats
- Coursepacks, available in BlackBoard, Angel, Desire2Learn, Moodle, and Canvas formats

"Near a black hole, space-time is so distorted that your feet are being stretched toward the black hole much more strongly than your head."

> "For every kilogram of stuff like the kind you are made of, there are 10 kilograms of the mysterious dark stuff."

Acknowledgments

This book is the product of many years of work that could not have been completed without the combined effort of a team of highly creative professionals. I am forever indebted to Carol Latta for her tireless work and boundless energy for all things astronomical in helping me assemble the book's questions, summaries, images, and other features. Carol's intelligence and creativity were essential to the book. It was a great pleasure to work with Jeff Bary, a talented astronomer and skilled teacher, who had the idea for **Anatomy of a Discovery** figures and drove their authorship. Thanks to Anne DeMarinis for her creativity and attention to detail on everything when it comes to design, including the Second Edition interior and cover and her work with Jeff on the **Anatomy of a Discovery** figures. Karen Karlin's talent as a developmental editor is unparalleled, and I am deeply grateful for her work. Thanks too to Michael Fauver who coordinated the details on every page and, with Sara Bonacum, kept the trains running on time. Robert Bellinger brought an infectious sense of enthusiasm to his work in leading the development of the many digital components of the textbook, while Arielle Holstein helped produce the teaching and learning package. I appreciate the creativity and time that Katie Sweeney has devoted to marketing the book, as well as the videogame, to instructors everywhere. Finally, I am very grateful to have had Erik Fahlgren as the lead at W. W. Norton for this project. Erik's vision, dedication to excellence, perseverance, and good humor throughout the long work of creating and revising this book were essential to its completion. In general, the entire team at Norton was a pleasure to work with.

It's worth more than just a note that without the kindness and support of my wife, Alana Cahoon, this project would not have been as much fun, because nothing is as much fun without her.

I also want to thank the following instructors who provided feedback at the many stages of this book. I admire your dedication to your students and appreciate your help in making this book a better tool for learning (and loving!) science:

Second Edition Reviewers

Erik Aver, *Gonzaga University*
Merida Batiste, *Emory University*
Tim Beers, *University of Notre Dame*
Erin Bonning, *Emory University*
Gene Byrd, *University of Alabama*
Juan Cabanela, *Minnesota State University Moorhead*
Michael Chester, *Mansfield University*
Scott Cochran, *Central Michigan University*
Juliana Constantinescu, *University of Wisconsin–Whitewater*
James Cooney, *University of Central Florida*
Lindsay DeMarchi, *Northwestern University*
Michael Dunham, *State University of New York at Fredonia*
John Elliff, *Kirkwood Community College*
Jessica Ennis, *Augsburg University*
Duncan Farrah, *University of Hawai'i at Mānoa*
Jack Gabel, *Creighton University*
Christopher Gerardy, *University of North Carolina at Charlotte*
Matthew Gonderinger, *Wayne State University*
Erika Grundstrom, *Vanderbilt University*
Emily Hardegree-Ullman, *Colorado State University*

Philip Harrington, *Suffolk County Community College*
David Hedin, *Northern Illinois University*
Maureen Hintz, *Brigham Young University*
Kishor Kapale, *Western Illinois University*
Sean Kelly, *Santa Barbara City College*
Mahvand Khamesian, *Saginaw Valley State University*
Steven Kipp, *Minnesota State University, Mankato*
Davide Lazzati, *Oregon State University*
Joel Levine, *Saddleback College*
Norman Markworth, *Stephen F. Austin State University*
Neil McFadden, *University of New Mexico*
Brendon Mikula, *Indiana State University*
Greg Perugini, *Rowan College at Burlington County*
Bob Powell, *University of West Georgia*
Michael Rich, *University of California, Los Angeles*
Anna Sajina, *Tufts University*
Douglas Shields, *University of Arkansas*
Gina Sorci, *University of Louisiana at Lafayette*
David Stickler, *Allegany College of Maryland*
Miriam West, *Utah Valley University*
Gerry Williger, *University of Louisville*

First Edition Reviewers

William Bagnuolo, *Georgia State University*

Becky Baker, *Missouri State University*

Celso Batalha, *Evergreen Valley College*

Elisabeth Benchich, *University of North Carolina at Charlotte*

Jeffrey Bodart, *Chipola College*

David Bradford, *State University of New York College of Technology at Canton*

David Branning, *Trinity College*

Juan Cabanela, *Minnesota State University Moorhead*

Todd Carriero, *City College of San Francisco*

Clinton Case, *Truckee Meadows Community College*

Damian Christian, *California State University, Northridge*

Fred Ciesla, *University of Chicago*

Juliana Constantinescu, *University of Wisconsin–Whitewater*

James Cooney, *University of Central Florida*

Michael Corwin, *University of North Carolina at Charlotte*

Santo D'Agostino, *Brock University*

Kate Dellenbusch, *Bowling Green State University*

Alessandra Di Credico, *Northeastern University*

Jess Dowdy, *Abilene Christian University*

Donald Farnelli, *Rowan University*

Tim Farris, *Volunteer State Community College*

Andrew Fittingoff, *Laney College*

Ian Freedman, *Dutchess Community College*

Michael Frey, *Cypress College*

Thor Garber, *Pensacola State College*

Christopher Gay, *Santa Fe College*

Christopher Gerardy, *University of North Carolina at Charlotte*

Nancy Gerber, *San Francisco State University*

Clint Harper, *Moorpark College*

Paul Heckert, *Western Carolina University*

Michael Hood, *Mt. San Antonio College*

Olenka Hubickyj-Cabot, *San Jose State University*

Gerceida Jones, *New York University*

Robert D. Joseph, *University of Hawai'i*

Kishor Kapale, *Western Illinois University*

Julia Kregenow, *The Pennsylvania State University*

Monika Kress, *San Jose State University*

Danilo Marchesini, *Tufts University*

Brian Mazur, *University of Mississippi*

Suzanne Metlay, *Western Governors University*

Wouter Montfrooij, *University of Missouri*

Thomas Nelson, *University of Minnesota*

Christopher Palma, *The Pennsylvania State University*

Jon Pedicino, *College of the Redwoods*

Ylva Pihlström, *University of New Mexico*

John Pratte, *Arkansas State University*

Claude Pruneau, *Wayne State University*

Richard Rand, *University of New Mexico*

Michael Reid, *University of Toronto*

Ronald Revere, *Coastal Carolina University*

Andreas Riemann, *Western Washington University*

Lawrence Rudnick, *University of Minnesota*

Eric Schlegel, *The University of Texas at San Antonio*

Sara Schultz, *Minnesota State University Moorhead*

Kathy Shan, *University of Toledo*

John-David Smith, *University of Toledo*

Kelli Spangler, *Montgomery County Community College*

Edward Stander, *State University of New York College of Agriculture and Technology at Cobleskill*

Winfield Sylvester, *Queensborough Community College*

Brian Thomas, *Washburn University*

Bruce Twarog, *University of Kansas*

Geoffrey Tweedale, *University of Wisconsin–Milwaukee*

Christopher Tycner, *Central Michigan University*

Robert Tyson, *University of North Carolina at Charlotte*

Arun Venkatachar, *Ohio University*

Saeqa Vrtilek, *Northeastern University*

G. Scott Watson, *Syracuse University*

Paul Wiita, *Georgia State University*

Derek Wills, *The University of Texas at Austin*

Jie Zhang, *George Mason University*

First Edition Class Testers

Esteban Araya, *Western Illinois University*

Raymond Benge, *Tarrant County College*

Matthew Bobrowsky, *University of Maryland*

David Bradford, *State University of New York College of Technology at Canton*

Juan E. Cabanela, *Minnesota State University–Moorhead*

Todd Carriero, *City College of San Francisco*

Erik Christensen, *South Florida State College*

Aaron Clevenson, *Lone Star College–Montgomery*

James Dickinson, *Clackamas Community College*

Ethan Dolle, *Northern Arizona University*

Jess Dowdy, *Abilene Christian University*

Andrew Fittingoff, *Laney College*

Michael Frey, *Cypress College*

Jeffrey Gillis-Davis, *University of Hawaii*

Gerceida Jones, *New York University*

Kishor Kapale, *Western Illinois University*

Frank Pyzcki, *County College of Morris*

Ronald Revere, *Coastal Carolina University*

Brian Schuft, *North Carolina Agricultural and Technical State University*

Christine Staver, *County College of Morris*

ABOUT THE AUTHOR

Adam Frank is a physicist, astronomer, and writer. His research focuses on computational astrophysics with an emphasis on star formation and late stages of stellar evolution. His popular writing has focused on issues of science in its cultural context, including issues of science and religion and the role of technology in the human experience of time. He is a cofounder of National Public Radio's *13.7 Cosmos and Culture* blog.

ASTRONOMY

AT PLAY IN THE COSMOS

SECOND EDITION

"SOMETIMES . . . THE QUESTION THAT YOU BEGAN WITH WAS . . . NOT THE RIGHT QUESTION"

1

Getting Started
SCIENCE, ASTRONOMY, AND BEING HUMAN

Miles Apart and Years Between

1.1 **Newgrange, Ireland.** The entrance is barely wide enough for a grown person to pass through. Unlike the bright Irish summer day behind us, with its impossibly blue sky and iridescent green hills, the path ahead is dark, cool, and damp. The ceiling hangs low and the walls press in close. We thread single file down a narrow passageway, a corridor that was constructed 5,000 years ago, a full half millennium before the Egyptian pyramids were finished. The passage opens into a low circular chamber set into the stone. As my Irish astronomer colleagues and I circle the chamber to make room for our guide, the silence of five millennia settles on us like a layer of ash.

Until the day before I had never heard of Newgrange, a prehistoric monument 50 miles from Dublin (**Figure 1.1**). I was just an American astrophysicist in Ireland to collaborate with Tom Ray, an Irish astronomer who studies stars like the Sun. But Ray moonlights as an "archaeoastronomer" and is often asked to verify the astronomical orientations of archaeological sites. In a pub the night before, Ray suggested we visit Newgrange for a view into our collective astronomical past.

The guide waits as we settle down in the chamber, and begins. "Around fifty centuries ago, this great monument was built," he intones. "Without aid of wheels, the 26-ton slabs of granite were dragged across miles and set in place where you see them. We do not know what rituals were performed here. This, however, we do know: One morning each year on the winter solstice, the rising Sun aligns with the passageway you just traversed. On that morning, a beam of sunlight pierces the 24-meter-long hallway and falls in this central chamber. The light flares for a mere 14 minutes. Then it passes. The Sun rises above the hill, the alignment is broken, and the chamber remains dark for another year. We do not know much about the long-forgotten builders of this place, but the alignment of the hallway with the winter solstice tells us they were, to some degree, astronomers."

← The Milky Way, with its broad swath of countless stars, as seen from Andalusia, Spain. The dark areas are dust that partially obscures this spiral arm of our own galaxy. For as long as human beings have existed, the sky has been a source of inspiration and wonder.

Figure 1.1 Newgrange
The 5,000-year-old monument at Newgrange, Ireland, provides direct evidence of the astronomical interests and knowledge of its ancient builders.

Figure 1.2 Very Large Array
The Very Large Array (VLA) in New Mexico consists of 27 radio telescopes in a Y-shaped configuration. The data from all 27 antennas are combined electronically to give the resolution of a much larger telescope. The VLA detects radio waves coming from the regions around black holes and from clouds of gas where stars form.

INTERACTIVE:
Scale of the Universe

astronomy The study of celestial objects such as moons, planets, stars, nebulas, and galaxies.

Universe The totality of space, time, matter, and energy.

Socorro, New Mexico. The road disappears over the sand-colored hills and bright-blue sky. It's morning and the desert is already unnerving me. I've been driving for hours and have passed only two human outposts. Still, this seems like a fitting location for civilization's most advanced signal post to the stars. Lonely tufts of grass roll by as my rental car begins a long climb to a dry plateau. An hour later, I crest the final rise and reach the Very Large Array, or VLA (**Figure 1.2**).

Stretching 10 miles into the distance, they stand with their giant heads upturned, all silent, all facing the same patch of sky. The array is composed of 27 radio telescopes, each a 230-ton dish the size of an eight-story building. All the human effort and genius that went into their design serves a single purpose: to catch faint radio signals from the far ends of the Universe. The VLA's mission is to serve the astronomers who come here to convert pinpricks of electromagnetic radiation caught by the telescopes into a narrative of the Universe—stories that tell us where we are, how we got here, and where we're going.

Science and the Human Journey

Newgrange and the Very Large Array are separated by more than 7,000 kilometers (km) of Earth's surface and 5,000 years of human history. Still, despite all that space and all that time, they share an essentially human ambition. Each site stands as a stark reminder that human beings have always been fervent star watchers. The night sky has been the gateway to our greatest mysteries, igniting in us a seemingly innate and creative curiosity. This boundless inquisitiveness seems to be one of the great gifts that evolution bestowed on us. We do not have sharp claws. We do not have wings to fly. What we do have is an intelligence that takes the world in and then has the audacity to ask it questions.

Astronomy—the study of celestial objects such as moons, planets, stars, nebulas, and galaxies—is now a modern science, but science in its modern form is part of an old game we play as we try to understand and control the world around us (**Figure 1.3**). The game can be deadly serious when the answers we seek determine our survival—as they did for early farmers who saw how the timing of different constellations appearing in the night sky across the seasons could help them know when to begin planting. It can also be creative, and even playful, as it was for the 17th- and 18th-century astronomers who delighted in applying newly developed laws of physics to the movement of celestial objects.

From our very beginnings as humans hundreds of thousands of years ago, the nightly theater of stars, the Moon, and the planets has invited us to ask enduring questions of who we are, what we are, and how we got here. Our personal questions about life (What should I do with my time?), love (Whom should I marry?), and meaning (Why is life so hard sometimes?) are all, ultimately, framed against the background of our place in the **Universe**—the totality of space, time, matter, and energy. Astronomy asks big, exhilarating questions, and from its perspective almost every aspect of life and the Universe can be explored.

This book and the class that you're starting are about the modern story of astronomy. It is a story of planets, stars, galaxies, and the Universe as a whole. It is a story of origins and endings. It is a story of forces and their capacity to shape all that we see and experience.

Given humans' long history as explorers for whom asking questions defines our species, astronomy is also a gateway. It's a window that lets us see both where we came from and where we might be headed. It shows us what we know and what we don't know, and the play of ideas between those two domains. Like all other sciences, astronomy shows us how much we can learn about the world in which we find ourselves, if only we're willing to pay attention, ask questions, and be vigilant in our search for answers.

Figure 1.3 The Cosmological Perspective

Left: Earth "rising" from the perspective of the Moon, as seen by *Apollo* astronauts. Right: the Milky Way as seen from Earth. Middle: a market in a busy city. Everyday life is embedded in a cosmic context, but how often do we think about our place in the Universe?

Section Summary

- As ancient astronomical structures such as Newgrange show, human beings have always been star watchers, using their observations for both survival and inspiration in building our cultures.
- Astronomy, the study of celestial objects and the physical universe as a whole, helps us confront our individual existence in the context of the Universe.

CHECKPOINT **a.** Cite one example each of an ancient and a modern astronomical observation. **b.** Why is the comparison between ancient and modern observations significant to astronomy's role in human history and culture?

Very, Very Old and Really, Really Big

1.2 One of the most difficult aspects of learning the modern story of astronomy is the sheer size and age of the Universe. The numbers that astronomers use when accounting for astronomical distances and timescales can make your head spin. Somehow you have to get a feeling for what these numbers mean before you truly grasp what astronomy can tell you about the cosmos you live in.

Scientific Notation

How do scientists work with the extreme range of numbers that occur in astronomy? Consider the idea of *powers of 10*. To understand what scientists mean when they talk about powers of 10, let's use our imagination. Imagine what the world was like 10 years ago. That's not hard, because you were around then. Now imagine the world 100 years ago (**Figure 1.4**). That was just after World War I. You may have seen pictures from that time; there were still cars and telephones and even airplanes.

Next imagine the world 1,000 years ago. That was a couple of decades after 1000 CE, which was the time of medieval Europe, the height of the Islamic empires, and the start of the Song dynasty in China. Maybe you've seen a movie depicting those societies, but it's hard to imagine what day-to-day life was like so long ago. For example, famines were routine, most babies died early, and no human beings had traveled faster than 40 miles per hour (unless they had fallen off a cliff).

Figure 1.4 Time, History, and Powers of 10

Time, from the present backward, is shown in powers of 10 in years, along with some key historic events and attributes. There are four powers of 10 in years from the present to the end of the last ice age.

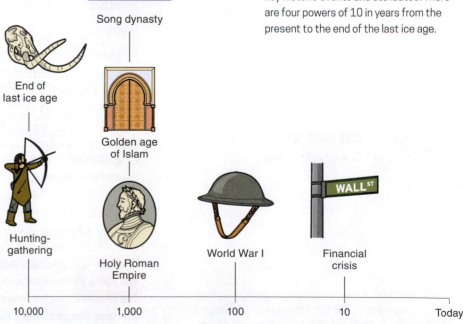

Song dynasty

End of last ice age

Golden age of Islam

Hunting-gathering

Holy Roman Empire

World War I

Financial crisis

10,000   1,000   100   10   Today

Time (years ago)

Let's keep going—imagine 10,000 years ago, just after the last ice age ended. Back then, human beings lived in nomadic tribes and had yet to invent agriculture. That was truly a different world (Figure 1.4).

In going from 10 years to 10,000 years ago, we stepped backward in powers of 10: first 10 years, then 100 years, then 1,000 years, and finally 10,000 years. In each jump we added another 0 after the 1. "Powers of 10" are so named because we can write 100 as the number 10 raised to the second power (that is, with an exponent of 2): $100 = 10^2$. Likewise, $1,000 = 10^3$ and $10,000 = 10^4$. Notice that $1 = 10^0$ and $10 = 10^1$. As our imagination experiment shows, thinking in terms of powers of 10 can be very powerful. By taking just four powers-of-10 steps in years, we went from the modern world all the way back to the time before the dawn of civilization.

The numbers involved in astronomy get much larger than 10,000. The age of Earth, for example, is about 5 billion years. With numbers of this magnitude, we have not only the challenge of grasping what the number means, but also the problem of writing it down. Scientists have a special way of writing very large and very small numbers, called **scientific notation**. In scientific notation, the number 5 billion is written as 5×10^9, which means 5 with nine zeros after it: 5,000,000,000. And 5 million, which is 1,000 times smaller than 5 billion, is written as 5×10^6, which means 5 with six zeros after it. In a similar fashion, the number 0.005 would be written as 5×10^{-3}; the negative exponent tells us how many places to the right of the decimal to shift the digit 5.

Sometimes astronomers are interested in only the basic scale of physical quantities in terms of powers of 10—what is called the **order of magnitude**. If we ask a question about distance but want only an order-of-magnitude answer, we don't care whether the distance is 150 meters versus 723 meters. Our order-of-magnitude answer would be just "hundreds of meters." When we want to know just the order of magnitude of a quantity, we do not need to specify the coefficient part of a number in scientific notation. We are interested only in its exponent. Dealing in orders of magnitude helps astronomers get a quick feel for the scales involved in a problem they're investigating. For example, is the object being studied the size of a person or the size of a star?

This ability to represent very large or very small numbers easily using scientific notation can help you digest the staggering range of scales that make up the modern story of astronomy. **Going Further 1.1** gives more details of how to work with scientific notation.

Section Summary

- The rules for working with numbers in scientific notation are straightforward and involve operations of both the coefficient and the exponent. Multiplying two numbers in scientific notation, for example, requires multiplying the coefficients and adding the exponents.
- Using scientific notation lets astronomers deal with very large and very small numbers. Considering just the order of magnitude of such numbers (the exponents in a number represented in scientific notation) allows them to get a quick grasp of the scale of the problem at hand.

CHECKPOINT   **a.** What is a power of 10? **b.** Describe how to divide two numbers expressed in scientific notation.

The Contents and Story of the Cosmos

1.3 Stories need actors. The actors in the modern story of astronomy are the contents of the Universe including planets, stars, and galaxies. It has taken astronomers and physicists about 400 years to develop an inventory of the kinds of things that fill the cosmos. In all likelihood, that inventory is not complete, and there is more to be discovered; that is what makes astronomy so exciting.

scientific notation A way to write very large or very small numbers using the form $a \times 10^b$, where the coefficient a is any real number and the exponent b is an integer.

order of magnitude The size of a physical quantity in powers of 10.

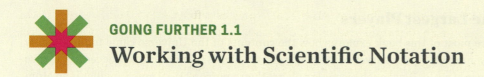

Working with Scientific Notation

Since most quantities in astronomy are expressed in scientific notation, it's important to know how to work with these kinds of numbers. When you write the number 245,000,000,000 (245 billion) in scientific notation, it looks like this:

$$2.45 \times 10^{11}$$

The number 2.45 is called the *coefficient*. The number 10 (without its exponent) is called the *base*. In scientific notation, the base is always 10. The number 11 above and to the right of the base is referred to as the *exponent* or the *power of 10*.

MULTIPLICATION. Let's start with multiplying two numbers in scientific notation. To do so, just multiply the two coefficients and add the exponents. For example,

$$(3.0 \times 10^7) \times (2.0 \times 10^5) = (3.0 \times 2.0) \times 10^{(7+5)}$$
$$= 6.0 \times 10^{12}$$

What do you do if the product of the two coefficients is greater than 10? Consider the following expression:

$$(4.0 \times 10^7) \times (3.0 \times 10^5) = (4.0 \times 3.0) \times 10^{(7+5)}$$
$$= 12 \times 10^{12}$$

There is nothing wrong with the result 12×10^{12}, but in general, the coefficients in numbers expressed in scientific notation should be less than 10 and greater than or equal to 1. To meet that requirement, in this example you shift the decimal place to the left and add one to the exponent:

$$12 \times 10^{12} = (1.2 \times 10^1) \times 10^{12} = 1.2 \times 10^{(1+12)}$$
$$= 1.2 \times 10^{13}$$

DIVISION. Dividing two numbers in scientific notation is similar. Divide the two coefficients and subtract the exponents. Here is an example:

$$(6.0 \times 10^4)/(2.0 \times 10^3) = (6.0/2.0) \times 10^{(4-3)} = 3.0 \times 10^1$$

What do you do if the quotient of the two coefficients is less than 10? Consider the following expression:

$$(2.0 \times 10^9)/(4.0 \times 10^5) = (2.0/4.0) \times 10^{(9-5)} = 0.5 \times 10^4$$

Again, the result 0.5×10^4 is not standard scientific notation. For the standard form, you shift the decimal place to the right and subtract one from the exponent:

$$0.5 \times 10^4 = (5.0 \times 10^{-1}) \times 10^4 = 5.0 \times 10^{(4-1)} = 5.0 \times 10^3$$

ADDITION AND SUBTRACTION. When adding or subtracting numbers written in scientific notation, you directly add or subtract the coefficients only if the exponent is the same for both numbers. This means that the equation

$$2.5 \times 10^3 - 1.5 \times 10^3$$

is the same as

$$2,500 - 1,500$$

which is equal to 1,000, or 1.0×10^3.

If you have to add or subtract two numbers with different exponents, you need to manipulate them to make their exponents the same. Say, for example, you have to add these two numbers:

$$(2.0 \times 10^3) + (1.0 \times 10^2)$$

This is just $2,000 + 100 = 2,100$. In scientific notation, 2,100 is written as 2.1×10^3. If you want to add these numbers in scientific notation, you need to represent the second number, 100, in thousands: 0.1×10^3 instead of 1.0×10^2. Now the equation becomes

$$(2.0 \times 10^3) + (0.1 \times 10^3) = 2.1 \times 10^3$$

Example: You're a space smuggler who stowed 150,000 packages of super-rare, high-grade Corporation coffee on your Sagan-class starship. You want to sell each package for 3,000 federation credits. Use scientific notion to calculate how much money you can make.

Answer: In scientific notation: $150,000 = 1.5 \times 10^5$ and $3,000 = 3.0 \times 10^3$. Multiplying the two numbers together, we find

$$(1.5 \times 10^5 \text{ packages}) \times (3.0 \times 10^3 \text{ credits/package})$$
$$= (1.5 \times 3.0) \times 10^{(5+3)} \text{ credits} = 4.5 \times 10^8 \text{ credits}$$

That's 450 million credits. You're rich!

Example: Aliens want to build a massive, planet-sized space station in two phases. The first phase will last 1 million years, and the second will last 100,000 years. How long will it take them?

Answer: 1 million $= 1.0 \times 10^6$ and $100,000 = 1.0 \times 10^5 = 0.1 \times 10^6$. So

$$(1.0 \times 10^6 \text{ years}) + (0.1 \times 10^6 \text{ years}) = (1.1 \times 10^6 \text{ years})$$

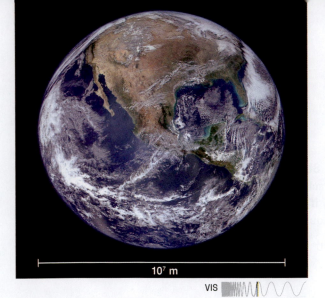

Figure 1.5 Size Scale of Planets
Earth. $L \approx 10^7$ meters.

The Largest Players

Since most of this book explores how the contents of the Universe came to be and how they interact, a quick survey of the major players in this cosmic drama is necessary to give a sense of scale. How much bigger is a planet than a person? How much smaller is an atom than a galaxy? Using your familiarity with powers of 10 and scientific notation, you can now get a first glimpse of the mind-boggling range of size scales that astronomy covers. In what follows, the letter L will denote the approximate size of something in meters. Let's start with something familiar and work our way up in scale.

YOU. Human beings are clearly one kind of "thing" in the Universe (even if we are unlikely to be universal). In terms of scale, the average human being is somewhere between 1 and 2 meters in height, so $L \approx 10^0$ meter. (The symbol $\approx$ means "approximately equal to.")

PLANETS. The first major players among cosmic structures are planets. **Planets** come in at least two basic types: either smaller spheres of rock like Earth ($L \approx 10^7$ meters; **Figure 1.5**) or larger spheres of gas (perhaps with a rocky core) like Jupiter ($L \approx 10^8$ meters). Notice that there are at least seven jumps in the power of 10 between the size of a human being and the size of a planet.

STARS. Planets are a type of **satellite**—they orbit a larger object, which is their host star. For the most part, planets always orbit stars—the next important astronomical actors in terms of size. **Stars** are balls of gas that produce energy by means of nuclear fusion in their cores. Our Sun is an average-sized star, at $L \approx 10^9$ meters (**Figure 1.6**), which is about 100 times larger than the Earth. Stars come in a wide range of sizes, from 10 times smaller to 1,000 times larger than the Sun.

PLANETARY SYSTEMS. A star and the planets that orbit it together compose a **planetary system**. Until the 1990s, astronomers did not know whether our system of the Sun and eight orbiting planets was the only one of its kind in the entire cosmos. Now we have enough evidence to infer that billions of other planetary systems exist in our galaxy. There are different ways to define the outer boundaries of a planetary system, but for now we can take $L \approx 10^{13}$ meters as the typical size (**Figure 1.7**). Notice that we have jumped 13 powers of 10 from the size of a human being, and we haven't even left our cosmic neighborhood yet.

GALAXIES. The distances between stars are immense. The Sun's nearest neighbor lies more than 1,000 times farther away than the orbit of Neptune, the outermost planet in our Solar System. That means the typical distance between stars is about $L \approx 10^{16}$ meters. But

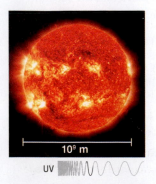

Figure 1.6 Size Scale of Stars
Our star, the Sun. $L \approx 10^9$ meters.

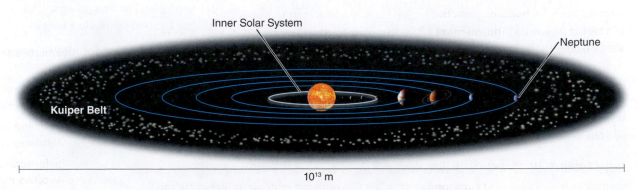

Figure 1.7 Size Scale of Planetary Systems
Our Solar System. $L \approx 10^{13}$ meters.

stars do not float around randomly in empty space. All stars are born, live, and die in vast "cities" of Suns called **galaxies**. Our galaxy, the Milky Way, contains 400 billion stars and stretches across $L \approx 10^{21}$ meters (**Figure 1.8**). Notice the jump in scale from the size of planetary systems to the size of galaxies. Stars (and their planetary systems) are the building blocks of galaxies, but they are hundreds of millions of times (eight powers of 10) smaller.

GALAXY CLUSTERS. In general, galaxies form together in groups and, on a larger scale, in **galaxy clusters**. A typical galaxy cluster can have 1,000 galaxies in it and stretch across $L \approx 10^{23}$ meters (**Figure 1.9**).

SUPERCLUSTERS. At the top end of the size scale of cosmic structures are so-called superclusters of galaxies. A **supercluster** is a collection of galaxy clusters, and these behemoths extend across $L \approx 10^{24}$ meters of space (**Figure 1.10**). That makes them the largest "things" in the cosmos. Superclusters are still in the process of forming, as the gravity in these structures continues to pull material together from one end of the cosmos to the other.

It's worth noting some of the units of size that will be used often in this book. The first, called an **astronomical unit**, is the average distance from the Sun to Earth $(1 \text{ AU} = 1.49 \times 10^{11}$ meters; see **Anatomy of a Discovery**). For larger distances, astronomers use the distance that light travels in a year—called, intuitively, the **light-year** $(1 \text{ ly} = 9.5 \times 10^{15}$ meters). Remember: A light-year is a unit of distance, not time. It can be difficult to wrap your mind around how far a light-year is. If you were to fly around Earth, you'd have covered about 40,000 km. One light-year, however, is about 10 *trillion* km. That means you would have to fly around Earth about 1 billion times to cover the same distance. In a modern jet plane, which takes a day to circumnavigate the globe, that would be a trip of 1 billion days (or 3 million years). That's how far a light-year is! For measuring even larger distances, though, astronomers often don't use the light-year but instead use another unit, called the **parsec (pc)**, which is about equal to 3.26 light-years.

The Smallest Players

From human beings to galaxy superclusters, we've been traveling upward in size scale. But if we move down in scale and investigate the microscopic world, what do we find in terms of the contents of the cosmos?

SINGLE-CELLED ORGANISMS. While single-celled organisms may or may not be a universal cosmic structure, they still represent one of the first important steps "down" in size relative to human beings. An amoeba, which is a relatively large single-celled organism, has a typical size of 0.2 millimeter (mm), or $L \approx 10^{-4}$ meter.

DUST. **Dust** particles are the smallest bits of solid matter in the Universe and are the starting point for building planets. The typical dust particle has a size scale of a millionth of a meter, or $L \approx 10^{-6}$ meter (**Figure 1.11**). These dust particles are composed of billions of atoms of elements such as carbon and silicon.

ATOMS. **Atoms** are the fundamental building blocks of all normal matter (non–dark matter) in the Universe. Atoms have a typical size scale of a tenth of a billionth of a meter, or $L \approx 10^{-10}$ meter (**Figure 1.12**). Each of the different elements in the Universe, from hydrogen to uranium, has a distinct kind of atom.

SUBATOMIC PARTICLES. Atoms are made of even smaller particles—electrons, protons, and neutrons—which are called **subatomic particles**. A proton, for example, is incredibly small, at a thousand-million-millionth of a meter, or $L \approx 10^{-15}$ meter (**Figure 1.13**). Some subatomic particles, such as the proton and neutron, are made of even smaller particles, called *quarks*.

10^{21} m

VIS

Figure 1.8 Size Scale of Galaxies
The Andromeda Galaxy, which is similar in size and shape to our Milky Way Galaxy. $L \approx 10^{21}$ meters.

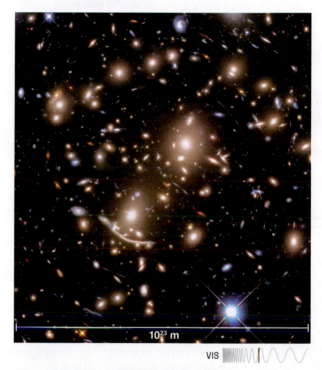

10^{23} m

VIS

Figure 1.9 Size Scale of Galaxy Clusters
Galaxy cluster. $L \approx 10^{23}$ meters.

astronomical unit (AU) The average distance from Earth to the Sun, 1.5×10^{11} meters; used as a standard of measure for Solar System objects.

light-year (ly) The distance that light travels in a year, 9.5×10^{15} meters; used as a standard of measure for large distances.

parsec (pc) A unit often used for very large distances, equal to approximately 3.26 light-years.

dust The smallest bits of solid matter in the Universe, composed of atoms such as carbon and silicon.

atom The fundamental building block of matter, consisting of a dense central nucleus surrounded by a cloud of negatively charged electrons.

subatomic particle Any particle smaller than an atom. Subatomic particles include protons, neutrons, and electrons.

How far away is the Sun?

observations

By the 17th century, astronomers had determined relative distances between planets in terms of the Earth-Sun distance. For instance, Saturn was known to be 9.5 times farther from the Sun than is Earth. Astronomers began devising observations that could reveal the Earth-Sun distance, and size of the Solar System, in absolute units.

predictions

Johannes Kepler, using his own tables of planetary motion, was the first to predict the 1631 transits of Mercury and Venus—the passing of planets in front of the Sun as seen from Earth. Kepler died in 1630, and no recorded observations of the 1631 transit of Venus exist.

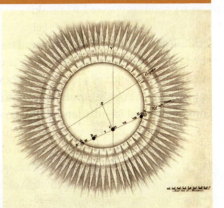

1761 Transit of Venus sketch (by Nicholas Ypey)

technique

Edmund Halley was first to recognize that careful observation of a Venus transit could provide an absolute measure of the distance to the Sun. He theorized that viewers at different latitudes, separated by known distances, would see slightly different transit paths and could measure the differences to get the distance to Venus. Then the Earth-Sun distance could be calculated. Halley encouraged astronomers to make these observations on June 6, 1761—a date he knew he would not live to see.

Jeremiah Horrocks then correctly predicted a Venus transit would occur in 1639. It was the last for another 122 years.

Horrocks observes projected image of 1639 transit of Venus

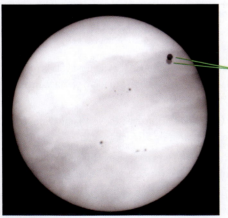

Transit of Venus, seen from two locations

Sun

observation

War, poor weather, and other factors kept most astronomers from viewing the 1761 transit, but expeditions were sent around the globe for the next one, in 1769. **Jean-Baptiste Chappe d'Auteroche** observed both—but died soon after the 1769 event. The average Earth-Sun distance wasn't accurately determined until the 1820s, when Johann Encke analyzed data from both transits. His result was just a few percent above today's accepted value, 1 astronomical unit (AU) = 1.49×10^{11} m.

Excerpt from Chappe's journal

We discuss all of these actors that make up the contents of the cosmos, from quarks to superclusters, in detail in the coming chapters. There are other important actors, such as electromagnetic fields and the newly discovered, very mysterious dark matter (a form of mass that interacts only weakly with the protons and neutrons that make up our bodies) and dark energy (a form of energy that is pushing the Universe apart). We will get to these in time.

Take a moment to reflect on the range of scales we have covered. When we were imagining what life was like in the past, just four jumps took us from our modern world to a nomadic existence. In thinking about the contents of the cosmos, we traveled from the largest cosmic scales (superclusters, at 10^{24} meters) through the size of human beings (10^0 meter) down to the size of a proton (10^{-15} meter), covering 39 powers of 10, or 39 orders of magnitude! The movement from the width of a proton (a million-billionth of a single meter) to the width of a supercluster (a million billion billion meters) seems beyond the imagination. Yet, using tools like scientific notation, we can begin to quantify and comprehend the almost infinite landscape across which modern astronomy ranges.

The Story of the Universe

Now that we've toured the Universe to see what's in it, we might want to know how all that "stuff" got out there in the first place. In other words, we've just seen what exists in space but haven't said anything about time. If space is the stage on which the drama of the Universe plays out, what is the story that's played out so far?

For astronomers, the history of the Universe begins about 13.8 billion (1.38×10^{10}) years ago with the Big Bang. The term *Big Bang* refers to the fact that, in its very early history (much less than a second old), our Universe was incredibly dense and hot. It was a kind of a dense, ultra-high-energy soup of radiation and matter particles that was already expanding. That is, space itself was stretching. Every point in space was moving away from every other point. As space expanded, the matter and radiation cooled (like steam blowing out of a teapot). One second after the Big Bang, for example, the temperature of the soup was 1 billion kelvins (abbreviated K; for comparison, room temperature is about 295 K). But by 1 minute after the Big Bang, the temperature dropped to about 10 million K, which is the same as the inside of the Sun. Three hundred thousand years later, the temperature had dropped to a few thousand kelvins, which is about the surface temperature of stars cooler than the Sun.

As temperatures dropped, gravity was able to draw matter in the Universe together. Eventually, and with a lot more physics, the stuff in these "clumps" of matter would go on

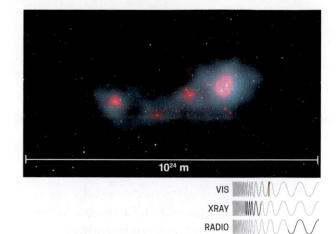

VIS
XRAY
RADIO

Figure 1.10 Size Scale of Superclusters
The core of the Shapley Supercluster. The blue colors show the distribution of matter using long wavelength light as detected by the Planck telescope. The red colors show x-rays from hot material between galaxies.

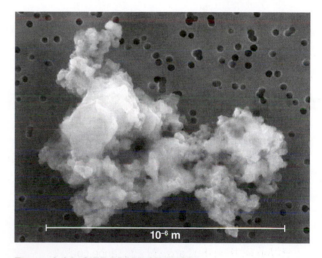

Figure 1.11 SIZE SCALE OF DUST
Dust particle. $L \approx 10^{-6}$ meter.

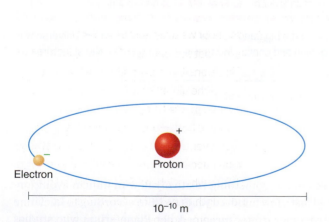

Electron
Proton

10^{-10} m

Figure 1.12 Size Scale of Atoms
Hydrogen atom (the Bohr model is shown). $L \approx 10^{-10}$ meter.

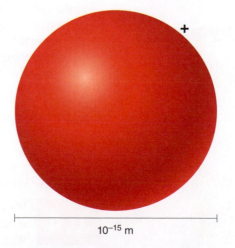

10^{-15} m

Figure 1.13 Size Scale of Subatomic Particles
Proton (which itself is composed of three quarks). $L \approx 10^{-15}$ meter.

Figure 1.14 Cosmic History Since the Big Bang
Schematic showing the history of the Universe, going back 13.8 billion years. The Universe began in a hot compressed state. As it expanded and cooled, structures formed that led to the cosmos as we see it now.

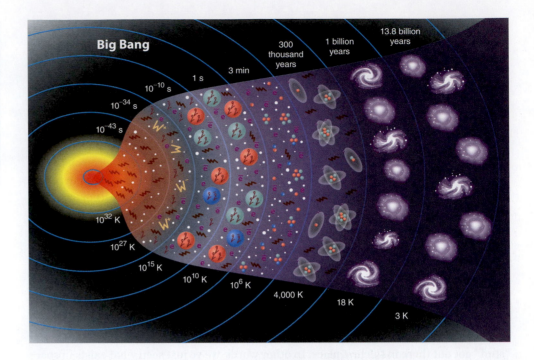

"We have always depended upon our observations of the natural world for survival."

to become all the structures we see in the Universe today, including planets and people (**Figure 1.14**). Today, fossil radiation left over from the Big Bang fills the Universe, which has cooled down to just a few kelvins above absolute zero (the lowest possible temperature).

As astonishing as this story and the range of scales it encompasses are, what's equally amazing is the fact that, even 150 years ago, this astronomical story could not have been told. Humans had yet to develop the tools to investigate these scales of size and time, so *their* stories were of a Universe much, much smaller (and younger) than the one we now know to exist. The expansion of *our vision* of the Universe and all the actors it contains is a story that is just as important to understand as the specifics of that astronomical vision. Science itself makes such a view possible. But what is science, where did it come from, and what does it require of us?

Section Summary

- A census of major cosmic structures, in order of increasing size, includes planets, stars, planetary systems, galaxies, galaxy clusters, and superclusters.
- All normal matter (non-dark matter) is made up of atoms, which are themselves made up of smaller, subatomic particles.
- The range in scale of all the objects in the cosmos covers 39 orders of magnitude, from $L \approx 10^{-15}$ meter to $L \approx 10^{24}$ meters.
- The story of the Universe begins with the Big Bang 13.8 billion years ago, when the Universe was much denser and hotter. Its expansion and cooling led to the formation of all the structures we see today.

CHECKPOINT   How much bigger is a galaxy than a star?

Why Science?

1.4   We may think of science as something relatively new in human evolution, something invented just a few hundred years ago. But according to Lucianne Walkowicz, an astronomer from Chicago's Adler Planetarium who studies planets that orbit other stars, that view is too narrow. Walkowicz was the Library of

Congress's Chair of Astrobiology, and she thinks broadly about astronomy and science as human endeavors. "Science is really the name that we give to a specific approach to learning about the world around us in a methodical way. Back before there was ever a notion of science, humans were observers of their environment—whether that means plants, animals, or weather." In that way, Walkowicz tells us, science is something deeply rooted in the way all human beings encounter the world. Part of that observation and learning is also deeply rooted in curiosity, which is a defining aspect of our species. Together, those features of human nature led to what is now called *science*.

Seeds of Science

"We have always depended upon our observations of the natural world for survival," says Walkowicz. "To my mind that's a very old, very fundamental human process that has allowed us to survive and to flourish." Because we have always used observations of the world to navigate it, our prehistoric ancestors were, in their own way, scientists. Even before there was such a thing as chemistry, people were already chemists because they learned, for example, how to extract dye from plants. And people were astronomers before there was astronomy, because as nomads and sailors they learned to navigate by the stars. All early cultures showed an intense interest and curiosity in the way the world worked. Using that curiosity, these early cultures tried to use what was around them to transform their world.

Observations are not just looking for random events. People have always been keenly interested in the parts of nature that fit into patterns. As Walkowicz puts it, "Human beings are very finely tuned to look for patterns. Look at a coat hook or a fire hydrant, and you might see that it looks like there's a face. So the patterns are not always meaningful, but because we humans depended for our survival on identifying patterns, we were always inclined to pay attention to them."

The origins of astronomy are deeply tied to this impulse. People found patterns in the night sky that could help them control their lives. Viewing the sky, early societies noticed the regular cycles of Sun, Moon, and constellations. Through trial and error over time, they could use the position of the rising Sun and the appearance of different constellations in the night sky across the seasons to determine when it was time to plant. The night sky imparted a sense of mystery and wonder, and paying attention to the sky led early people to a deeper mastery of the world around them.

Still, it took many thousands of years and the particular genius of many individuals and cultures before the pattern recognition of early societies developed into the practice of science that we recognize today. "The ability to represent ideas and relationships symbolically through math really allowed the transmission of those ideas much more easily and led to what we now call science," says Walkowicz.

The roots of this methodological approach to the world go back to ancient Greek cultures (500–300 BCE). Arab astronomers a thousand years later also played a key role in its development, as did Renaissance scholars in the 1400s and 1500s. In this way, slowly, people learned how to focus their curiosity into a specific way of asking questions.

What we really think of as modern science, however, grew out of beginnings forged in Europe 500 years ago. Now science and the technologies it produces dominate the world. Because you live in a scientific culture, you will live longer, enjoy better health, travel farther, experience more, and have access to a wider array of information than virtually every one of the ancestors whose genes you carry. Because of science and the technologies that can grow out of it, the world you inhabit also faces dangers and challenges that did not exist in the past. But what exactly is science, and what sets it apart from other human endeavors?

"Human beings are very finely tuned to look for patterns."

Figure 1.15 Scientific Method

The scientific method is the formal process by which science proceeds. It demands that scientists rigorously test hypotheses and discard or modify those that don't consistently predict results. In reality, the process is more complicated. Factors such as accidents and surprises can play an important role.

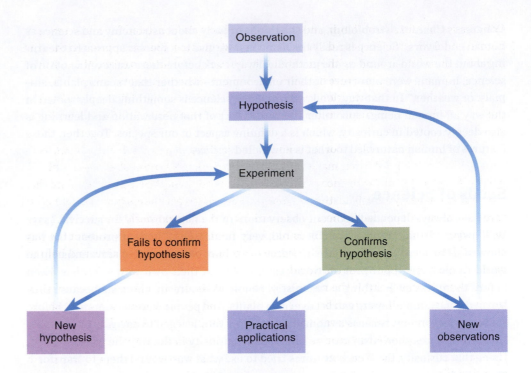

"Sometimes, you discover that the question that you began with was actually not the right question."

---

scientific method A body of techniques for acquiring new knowledge that includes systematic observation, experimentation, and theorizing.

hypothesis (pl. hypotheses) A proposed explanation of a phenomenon that must be rigorously tested by experiments.

theory An explanation for a phenomenon that has undergone extensive testing and is generally accepted to be accurate.

pseudoscience A claim, belief, or practice that is presented as scientific but does not adhere to a valid scientific method, lacks supporting evidence or plausibility, and/or cannot be reliably tested.

What Science Is and Is Not

"Often in formal education we get the view that science is a very clean and straightforward process," says Walkowicz. "The textbook view is that it involves knowing the right question to ask and then applying various very well-known techniques to testing that question. Then at the end you get an answer or somebody yells 'Eureka!'" But, as Walkowicz explains, the truth is much more interesting than that.

THE SCIENTIFIC METHOD. You may have already learned about the **scientific method**, which often begins when you make an observation about a particular phenomenon. Then you come up with a **hypothesis** to explain that phenomenon. Then, according to the standard account of the scientific method, you design an experiment to test that hypothesis. If the experiment fails, you have disproved your hypothesis. You have to go back to the drawing board, modify your hypothesis or invent a new one, and start again (**Figure 1.15**).

Take a concrete example: Imagine that you're watching leaves fall from a tree. You notice that some leaves tumble over and over as they fall. Others just drop straight down. That is your observation. Then you think, "I bet the leaves that tumble are wider than the ones that drop straight down because the wider ones present more air resistance." That is your hypothesis. So, you take some paper and cut out 10 leaf shapes ranging from very wide to very narrow and drop them from a second-story window, counting the number of times each one tumbles on the way down. That is your experiment. Finally, you look at the data you collected to see whether the width of the leaf really did affect how much it tumbled.

This story of the scientific method is neat and reasonable, but in real life things are usually a bit more complicated. "Science is often messier and more cyclic," says Walkowicz. "Sometimes, you discover that the question that you began with was actually not the right question. Your initial investigation shows you that you should have phrased the question differently or asked a different question entirely."

Many of the greatest discoveries in human history actually came from accidents in the lab, and only afterward did scientists test the findings through experiments. Penicillin, perhaps the most important medical breakthrough ever, was discovered by mistake. In

1928, biologist Alexander Fleming noticed that the bacteria he needed for certain experiments would not grow on petri dishes that had mold on them. Fleming reasoned that the mold was somehow inhibiting the growth of the bacteria, and he recognized that what he was observing needed more investigation. He dropped the experimental program he was planning and turned his attention to isolating the factor in the mold that was killing the bacteria. Fleming's moment of recognition led to the development of antibiotics for medical use.

Many typical descriptions of the scientific method do not account for accidents, intuition, or flashes of insight that make up some of science's most interesting stories of discovery. What really makes science work is a broader account of the scientific method that not only includes rigorously testing a hypothesis against experiments but also, and most importantly, requires a willingness to adapt or discard any hypothesis that doesn't stand up to experimental investigation. By using experiments to test hypotheses about the way the world works, scientists achieve a degree of certainty about their claims and help us gain reliable knowledge about the world.

One important point is the meaning of the word **theory** in science. Sometimes the terms *hypothesis* and *theory* can be used interchangeably. This occurs when scientists are speaking of a very specific observation or set of observations they wish to describe. We can think of cases like these as theories with a lowercase *t*, like a detective's "theory" about who committed a crime. This kind of theory can be overturned with just one or two counterexamples, meaning one or two experimental results that disprove the theory.

When scientists speak of Einstein's theory of relativity or the theory of evolution, they mean something different. We can think of these as theories with a capital *T*. They are well-developed bodies of knowledge that have been tested many times in many different ways over many decades or even over many centuries. A single new experimental result can't overturn the theory of relativity, because it has already been subjected to so much scrutiny. "It's very rare," says Walkowicz, "that you just totally blow a theory completely out of the water with a single new observation." Instead, that new result would itself be strongly scrutinized for mistakes in how the experiment was set up or analyzed. These "capital-*T* theories" are still subject to revision or to being overturned, but given their success over so much time, it takes considerable effort to show that they're wrong. The rare cases when a major theory is overturned are therefore considered scientific revolutions.

SCIENCE VERSUS PSEUDOSCIENCE. Sometimes it can be hard to tell what is science and what is not. People make claims about the world all the time, usually with the aim of getting you to part with your money. They will tell you that a vitamin supplement will give you 50 percent more energy or that a cure for acne works 75 percent of the time. It's important to know whether these kinds of claims are valid—to separate science from **pseudoscience**.

"The difference between science and pseudoscience goes back to how good humans are at perceiving patterns," says Walkowicz. "The downside of all that pattern hunting is that sometimes we perceive patterns that don't exist or don't mean anything significant." Walkowicz points to the often-claimed link between periods of a full Moon and intense activity at hospital emergency rooms. "A lot of times you will hear people say that 'Well, it was a full Moon last night. Everybody was acting so weird.'" But part of this link is just the fact that people notice the Moon when it's full, whereas they might not even notice the Moon when it's in a crescent phase. "So there is some observational bias," says Walkowicz, "in that you notice something (the full Moon) and then attribute something else you observe to it (activity at the ER). That ends up guiding your thinking about whether those two things are related." Despite what people think about the connection between ER activity and Moon phases, careful statistical studies show absolutely no relationship.

LUCIANNE WALKOWICZ

Lucianne Walkowicz likes to do science, but she likes to do a lot of things. Her journey to astronomy began in high school with new interests in physics and chemistry. "I wanted to find a way to combine them," Walkowicz says. A summertime program gave her the opportunity to explore the chemical reactions that occur in space, and from that point on Walkowicz was hooked. "I got connected to an astrochemist at New York University. I fell for astronomy after that!" Her day job is as an astronomer at the Adler Planetarium, where she does both science and science communication. "The best thing about working in a planetarium is definitely the questions people ask!" Walkowicz has also served in the unusual position of Library of Congress Chair of Astrobiology, where she's opened new conversations on how we view the possibilities of life in the Universe.

"The downside of all that pattern hunting is that sometimes we perceive patterns that don't exist or don't mean anything significant."

Figure 1.16 Astrological Zodiac
Astrology is a pseudoscience that purports to link constellation and planet positions to events in human lives.

"The point of science is it helps you remove some of the biases that we are often born with."

When it comes to pseudoscience, the false relationships can grow to an entire belief system. *Pseudoscience* can be defined as a belief or practice that people claim is scientific, or that is made to look scientific, but that has not been developed through the methodology and standards discussed in this chapter. In general, pseudoscientific claims lack the supporting evidence or the plausibility associated with ideas that are backed by scientific knowledge. In other words, the proponents of a pseudoscientific claim haven't tested their ideas, or their ideas have failed experimentally but the proponents don't want to give up their cherished beliefs.

One of the longest-lasting pseudosciences, astrology, predates the origins of modern science. **Astrology** is the belief that the position of the planets and constellations at the time of your birth affects the future course of your life. Much of the earliest history of astronomy is tightly woven with astrological forecasting. In recent years, many highly detailed, exacting statistical studies have been done to look for correspondences between astrological predictions and reality. These studies have always shown that astrology's predictions are no better than random guesses. As much as people want to believe otherwise, astrology simply has no predictive capability. The daily horoscopes that some of us read are, in general, so vague that you could read the horoscope for any of the astrological signs and find something related to your life (**Figure 1.16**).

What makes astrology a pseudoscience is its very explicit claim that a link exists between a set of phenomena (the exact positions of planets on the sky at the exact time you were born) and the future behavior of another set of phenomena (your life). Astrology appears to use precise data (planet positions) to make precise predictions (the tall, attractive stranger you will meet tonight). The problem is that these predictions do not turn out to be true most of the time—unlike, for example, astronomy's predictions of the motions of planets. Many practitioners of astrology will say that it doesn't make exact predictions; instead, it predicts trends in a person's life. Don't be fooled. This is really just a way of wiggling out of the uncomfortable position of making a completely wrong prediction. In a real science, if your predictions are wrong even 10 percent of the time (one out of every 10 predictions), it's time to look for a new hypothesis.

"Ideally," says Walkowicz, "science has a way of creating a methodology that records things in an unbiased way. The point of science is it helps you remove some of the biases that we are often born with." Ultimately, the ability of science to predict how the world behaves, and then to help humans build powerful technologies from those predictions, is the reason it has gained such wide acceptance over the last five centuries. What separates science from pseudoscience is the willingness of scientists, and of societies that accept science, to allow the world to speak for itself, as well as the open-mindedness to give up even our most cherished beliefs about the world if they turn out to be wrong.

Section Summary

- Science as we know it developed in stages, beginning with the ancient Greeks and continuing through Renaissance and later scholars, but its roots go further back to human beings' innate ability to find patterns in their experience.
- Scientific discovery usually results from rigorous testing of a hypothesis with experiments and requires a willingness to discard or adapt a hypothesis that doesn't pass these tests. Theories are well-determined bodies of knowledge that have been tested many times in many circumstances.
- A pseudoscience, such as astrology, makes claims that lack supporting evidence or plausibility.
- Science helps people remove biases in their search for patterns, predict how the world will behave, and build technologies from those predictions.

CHECKPOINT **a.** Give an example of two kinds of observations that many people think are linked but that are actually based in a bias. **b.** Explain how your example helps show the difference between science and pseudoscience.

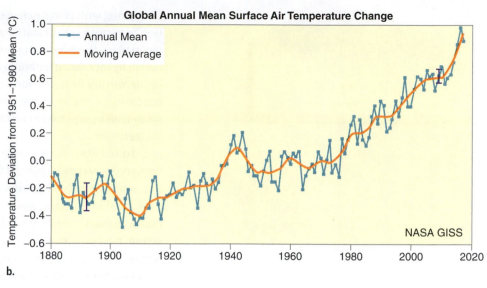

b.

Figure 1.17 Climate Change
Climate scientists agree that fossil fuel use has made Earth's climate unstable and that climate change poses a great danger to civilization. **a.** The Briksdal Glacier in Norway has receded in recent years. Many scientists attribute rapid glacial melting to global warming. **b.** Graph of the average global temperature going back 140 years. The graphed values compare the annual average temperatures to the mean value for 1951–1980, showing clear acceleration in the last few decades. Notice how the uncertainty (shown by dark blue vertical bars) gets smaller the closer we get to the present era.

Science, Politics, and the Human Prospect on a Changing Planet

1.5 Every human generation thinks it's special. Every generation thinks it's on the cusp of history, a time when the world will either enter a bright new era or begin a decline into something darker. Europeans in the 1700s were sure they were at the beginning of a new Age of Reason. Aztecs in the 1100s thought themselves entering an era of decline after a golden age of the past. Remarkably, the generations we belong to may be the first ones to be entirely correct in the belief that we are special. The reason for our uniqueness is inextricably tied to science.

Effects of Science and Technology

You were born into a world saturated with the fruits of science and its technologies. You were also born into a world challenged by science and its technologies. You already know some of the benefits of science: longer, healthier lives; fast, continent-spanning transportation; instant communication; the accessibility of all human knowledge through an Internet search. The challenges that science and technology bring us are also familiar to anyone who glances at the news.

Climate change tops the list of difficulties we face because of modern technologies. There is a consensus among scientists that, over the last 100 years, our rapidly expanding use of fossil fuels to power everything from transportation to increased crop yields has altered the state of Earth's climate (its long-term weather patterns; **Figure 1.17**). The consequences of climate change are likely to include increases in global temperatures, changes in rainfall patterns, changes in biological habitats, and rises in sea levels. How large these changes will be over the next century is unknown; predictions span from mild to overwhelming.

astrology The belief that there is a relationship between astronomical phenomena and events in the human world.

climate change A significant, long-term change in the statistical distribution of weather patterns.

Along with climate change, humanity faces challenges in the form of resource depletion in various manifestations. We are using up the things we need to live faster than we can replace them—and in some cases, we cannot replace them. Large-scale development fueled by increasing technological capacity has led to the exhaustion of many, if not most, of the world's fisheries, for example. Along with vanishing resources, society must contend with pollution of the resources we do have. Direct challenges from the products of science also continue to exist in the form of nuclear weapons and newer fears like biological terrorism.

In the 2002 book *Our Final Hour*, Sir Martin Rees, Astronomer Royal of England, reviewed a variety of threats facing the human race to measure our chances of survival. In a sobering calculation, Rees gave humanity 50/50 odds of making it, intact, to the next century. While many people would argue with Rees's odds, the idea that we face a critical epoch in human history has become widely accepted. That sobering reality, dependent so much on science and technology, is why you need to know what science is and how it works, regardless of your interest in any particular field of science.

As a member of a democratic society, you will be asked to make critical choices about the future. Almost all of those choices will involve science and technology in some way. From climate change to genetically modified crops to data security, in the years ahead, all of us will have many, many opportunities to vote on issues shaped by science and technology. You may express your opinions by casting votes in an election or by choosing which products you purchase. In some cases you may be faced with a parade of "experts" who will tell you what to believe about the science of climate change or the science of the new "green" car you're thinking of buying. By understanding how science works, how it reaches its conclusions, and how we can understand the limits of those conclusions, you will be able to make informed decisions about your own future. The dominance of science and technology in our lives brings us great benefits, but it also comes with some very real responsibilities. At this strange and critical moment in history, we neglect those responsibilities at our own very real peril.

We human beings have always shown great resourcefulness and a capacity to attend to and solve the problems that face us. Science has been part of that resourcefulness in the past and will continue to be in the future.

Astronomy and the Cosmic Perspective

If it is true that we're living in a critical historical moment driven by science and technology, then where should a person begin learning about science and how it works? With astronomy, of course!

Astronomy is one of the oldest sciences, dating far back into the millennia before written history. Astronomy also played a key role in the long march to the development of modern science, as new theories of the Solar System directly challenged the power structure of western Europe 500 years ago. But astronomy is also one of the newest sciences, pushing our understanding of distances and ages in ways that strain and invigorate our imaginations. At the same time, modern astronomy—with its stories of newly discovered planets orbiting far-flung stars, and galaxies colliding in 100-million-year-long slow-motion ballets—offers us more than just an understanding of how science works. It also offers a much-needed view of who we are, what we are, and how we got here.

In learning astronomy's stories—learning what we know about the Universe and how we have come to know it—you will gain something of great value: perspective. We may be small creatures in a very large cosmos, living short lives compared with those of stars and galaxies, but with the perspective we can gain from astronomy, we can come to understand our Universe. Not bad for a bunch of wingless, clawless mammals.

Section Summary

- We live in a critical historical moment driven by science and technology, which offer benefits and challenges for the current generation of human beings.
- Scientific consensus is that fossil fuel use is changing Earth's climate, and climate change creates the most challenging difficulties facing humanity. Other major challenges include resource depletion and the threats of nuclear weapons and biological terrorism.
- Astronomy, as one of the oldest and yet newest sciences, provides perspective on humans' place in the cosmos and is a starting point for learning about science and how it works.

CHECKPOINT **a.** Give an example of the way science and technology raise important issues that society must face. **b.** Give examples of how astronomy is both one of the oldest and newest sciences.

CHAPTER SUMMARY

1.1 Miles Apart and Years Between

The earliest humans sought to understand the Universe and their place in it. This endeavor continues today, manifested in such ways as modern science and, in particular, astronomy.

1.2 Very, Very Old and Really, Really Big

Scientific notation provides a concise way to use powers of 10 to write very large and very small numbers. Orders of magnitude are differences in powers of 10.

1.3 The Contents and Story of the Cosmos

A brief survey of the main categories of structures in the Universe and their sizes includes human beings (10^0 meter), planets (10^7–10^8 meters), stars (10^9 meters), planetary systems (10^{13} meters), distances between neighboring stars (10^{16} meters), galaxies (10^{21} meters), galaxy clusters (10^{23} meters), and galaxy superclusters (10^{24} meters). On smaller scales are single-celled organisms (10^{-4} meter), dust grains (10^{-6} meter), atoms (10^{-10} meter), and protons (10^{-15} meter). In astronomy, numbers with the same order of magnitude are often considered roughly equivalent. The story of the Universe begins with the Big Bang 13.8 billion years ago, when the Universe was much denser and hotter. Its expansion and cooling led to the formation of the structures we see today.

1.4 Why Science?

Science is a natural result of our native human curiosity and the ability to find patterns in our experience. Such discoveries typically have practical applications that enable humans to gain some control over (or understanding of) our environment. The modern scientific process asks questions and seeks answers that can be verified and repeated experimentally. Scientists formulate a hypothesis to explain a phenomenon and then design an experiment to test it. If the hypothesis is not disproved and discarded, further experiments may refine understanding of and strengthen support for it. Theories (such as the theory of relativity—think of them with a "capital T") are well-determined bodies of knowledge that have been tested many times in many circumstances. Pseudosciences, such as astrology, have not been rigorously tested to verify their claims to explain the causes and effects of phenomena. True scientific claims are predictive; that is, the expected outcome is achieved each time they are tested, given the statistical uncertainties.

1.5 Science, Politics, and the Human Prospect on a Changing Planet

Our world today faces many challenges, some created by science and the resulting technology, and some likely to be solved through that science and technology. Climate change is foremost among those challenges. Each citizen of the world must understand the process of science enough to comprehend the challenges and their potential solutions.

QUESTIONS AND PROBLEMS

Narrow It Down: Multiple-Choice Questions

1. Scientific notation is
   a. the use of Greek symbols to express scientific concepts.
   b. convenient shorthand for expressing very large or very small numbers.
   c. footnotes in scientific papers.
   d. the documentation of experimental results.
   e. formulas that express mathematical relationships.

2. Which of the following is written in standard scientific notation?
   a. 0.014×10^{28}
   b. 0.14×10^{27}
   c. 1.4×10^{26}
   d. 14.0×10^{25}
   e. 140×10^{24}

3. Which of the following is equivalent to 9,400,000?
   a. 9.4×10^5
   b. 9.4×10^6
   c. 9.4×10^7
   d. 0.94×10^8
   e. 0.94×10^9

4. A number is said to be an order of magnitude larger than another number if the first number is
   a. 2 times larger.
   b. 3 times larger.
   c. 5 times larger.
   d. 10 times larger.
   e. 100 times larger.

5. Consider a number (x) that is 4 orders of magnitude larger than another number (y). Which accurately describes the value of x compared to y?
   a. It is 4 times larger.
   b. It is 10 times larger.
   c. It is 1,000 times larger.
   d. It is 10,000 times larger.
   e. It is 400 times larger.

6. How many orders of magnitude larger is the Sun ($L_{Sun} \approx 10^9$ meters) than a terrestrial planet such as Earth ($L_{Earth} \approx 10^7$ meters)?
   a. 0   b. 1   c. 2   d. 5   e. 10

7. By how many orders of magnitude does the size of the gas giant planet Jupiter ($L_J \approx 10^8$ meters) exceed the size of a human being ($L_{HB} \approx 10^0$ meter)?
   a. 2   b. 4   c. 6   d. 8   e. 10

8. How many orders of magnitude are there between the diameter of the Sun ($L_{Sun} \approx 10^9$ meters) and the typical distance between stars ($L \approx 10^{16}$ meters)?
   a. 7   b. 10   c. 15   d. 20   e. 100

9. How many powers of 10 smaller than a human being ($L_{HB} \approx 10^0$ meter) is a typical grain of dust ($L_{DG} \approx 10^{-6}$ meter)?
   a. 2   b. 3   c. 6   d. 9   e. 15

10. The existence of ancient structures like Newgrange indicate that the humans who built them were aware of
    a. star motions.
    b. changing constellations.
    c. moon phases.
    d. related motions of Sun and Earth.
    e. existence of other planets.

11. Compared to the size of a galaxy cluster ($L \approx 10^{23}$ meters), a super-cluster ($L \approx 10^{24}$ meters) is _____. Choose all that apply.
    a. twice as large
    b. 1 order of magnitude larger
    c. 10 times as large
    d. 24 times larger
    e. 1 power of 10 larger

12. Which of the following describe(s) areas where informed citizens need to understand how science works? Choose all that apply.
    a. understanding the arguments for and against the need for human intervention in climate control
    b. weighing the positions for and against hydraulic fracking for obtaining natural gas
    c. discerning the accuracy of statements regarding the depletion of natural resources
    d. deciding whether children should be inoculated against various diseases
    e. deciding which album by Led Zeppelin was the band's best

13. How many powers of 10, in years, separate the end of the last ice age from today? (Hint: See Figure 1.4.)
    a. 10   b. 4   c. 3   d. 5   e. 2

14. Comparing the size of a human being ($L_{HB} \approx 10^0$ meter) to the size of an atom ($L \approx 10^{-10}$ meter),
    a. the human being is 9 times bigger.
    b. the human being is 10 times bigger.
    c. the atom is 9 orders of magnitude smaller.
    d. the human being is a billion times bigger.
    e. the atom is 10 powers of 10 smaller.

15. One of the earliest practical uses of astronomy was the timing of crop planting by
    a. lunar eclipses.
    b. the appearance of comets.
    c. the appearance of specific planets.
    d. the appearance of specific constellations.
    e. solar eclipses.

16. Which of the following characteristics of astrology is/are consistent with defining it as a pseudoscience? Choose all that apply.
    a. Its advocates take a perceived correlation and claim that it shows a causal relationship.
    b. Its proponents would not give up their beliefs, even if experiments showed that those beliefs were incorrect.
    c. Its advocates assert that there are underlying principles but fail to test their validity rigorously.
    d. Its hypotheses have not been supported by experimentation.
    e. Its predictions are so broad that they fit almost any possible outcome.

17. Which of the following are *not* characteristics of "capital T" scientific theories, such as Einstein's theory of relativity? Choose all that apply.
    a. They compose a large body of knowledge about a subject.
    b. They can be disproved with a single experiment.
    c. They have been tested many times.
    d. They cannot be overturned.
    e. When they are overturned, it's often considered a scientific revolution.

18. Which of the following is the correct order of actions that scientists take in implementing the simplest version of the scientific method?
    a. theory, observation, hypothesis, test, results
    b. hypothesis, test, results, retest, observation, potential theory
    c. observation, hypothesis, test, results, retest, results, potential theory
    d. theory, test, results, hypothesis, revised theory
    e. observation, theory, hypothesis, test, results, retest, results

19. True/False: Hospital emergency rooms are most active on nights with a full Moon.

20. True/False: Understanding the natural world through the application of scientific knowledge can give us a better understanding of our place in the Universe.

To the Point: Qualitative and Discussion Questions

21. Scientific knowledge is said to be simultaneously "tentative and reliable." Describe how this paradox is a strength of scientific knowledge, not a weakness.

22. The progress of modern science often is not linear. What aspects of modern science do not fit neatly into the scientific method?

23. Pseudoscience is often intended to sound and look a lot like science. In what ways do pseudoscientific understandings differ from scientific ones?

24. A recent study by the National Science Foundation found that fewer than 25 percent of Americans understand the true nature of scientific knowledge. How would you describe the process of acquiring scientific knowledge to someone in the other 75 percent?

25. What are the criteria for a successful scientific theory?

26. What are some of the ways science has changed the world? Do you see these as benefits or challenges?

27. Name several of the toughest challenges we face today as a result of technological advancement.

28. What types of questions ultimately cannot be answered through scientific investigation?

29. How do astronomy and astrology differ?

30. What are some practical applications for early astronomical knowledge of the cycles of the Sun, Moon, and constellations?

31. Solstice, such as that observed at Newgrange, relates to a definitive event and time in the changing of the seasons. What is the next solstice you will experience?

32. Recall what you have heard or read about climate change. Consider the source. What purpose or agenda might the speaker or writer have had? Cite examples that you judge to be based on science and others based on pseudoscience.

33. Humankind has been fascinated by the mysteries of the night sky since its earliest beginnings. Given our modern understanding of the Universe and our position within it, what thoughts, ideas, and questions come to mind when you look up at a starry sky?

34. Give an example of how a bias can appear in the observation of a pattern.

35. Do you agree or disagree with Martin Rees's statement that humanity has only 50/50 odds of surviving into the next century? Why or why not?

Going Further: Quantitative Questions

36. Humanity has existed in its culturally and anatomically modern form for approximately 200,000 years. Write this number in scientific notation.

37. How many orders of magnitude are there between the last ice age and 100 years ago?

38. Using the values given in the text, estimate how many times larger Jupiter is than Earth.

39. How many orders of magnitude are there between the sizes of a dust particle and a proton?

40. How many years would 1,500 generations represent, if each generation was 25 years? Give your answer in scientific notation.

41. Which is larger, the number of powers of 10 between the sizes of a planetary system and a galaxy, or those between an atom and a proton?

42. What proportion of the time since *Homo sapiens* appeared on Earth (about 300,000 years ago) have we been culturally and anatomically modern? Give your answer as a ratio. (Hint: See question 36.)

43. How many powers of 10, in years, separate you from the first *Homo sapiens*?

44. Write one million billion billion in scientific notation.

45. As of December 31, 2019, how many seconds have passed since the end of the year 1543, when Copernicus (see Chapter 3) published his Sun-centered model of the Universe? Give your answer in scientific notation. (Hint: There are 3.16×10^7 seconds in a year.)

"IT WAS LIKE A LIGHTBULB WENT ON, AND I KNEW, THIS IS IT FOR ME"

2

A Universe Made, a Universe Discovered

THE NIGHT SKY AND THE DAWN OF ASTRONOMY

An Old Obsession

2.1 Carol Latta never took astronomy graduate courses. Instead, Latta got an MBA and spent most of her career working for Xerox, then the world's largest maker of copy machines. "I was a product launch manager, among other things," she says. "It was a rewarding career, but it wasn't my passion."

Now Latta knows exactly where her passion lies—in the stars. On any given Friday night, you can find her, the current president of the Astronomy Section Rochester Academy of Science, out under the dark starlit skies of upstate New York with her beloved telescopes. "When I found astronomy, I was in my mid-50s," she says. "It was like a lightbulb went on, and I knew, 'This is it for me.' I've found that I like to do observing with binoculars or a telescope." (Many are available at the academy's 17-acre observatory site.) "But what I *love* is doing outreach, showing others how to find the spiral arms of a spectacular galaxy like M64, showing them the bright jewel of the globular cluster M13 or perhaps the glow of gas blown off a dying star as in the Helix Nebula" (**Figure 2.1**). "Besides observing and outreach, my other favorite activity is using my imagination to envision some of the cosmic events I read about. I can't think of anything else that comes close to this," says Latta, describing her love affair with astronomy. "Nothing else I have found has the same richness in it."

No one pays amateur astronomers such as Latta to spend hours building telescope domes, giving presentations to grade school kids, or holding public "star parties" where everyone gets a chance to look through a telescope. Amateur astronomers do what they do because they have become enchanted by the sky and its mystery. They are just the latest in a long line of men and women obsessed with the stars. It's a line that stretches back 50,000 years or more into the dim origins of human culture.

In this chapter we will explore the basic motions of Sun, Moon, stars, and planets visible to anyone who takes the time to look. We will also unpack some of humanity's

← The Milky Way as seen from temple ruins in Amman, Jordan.

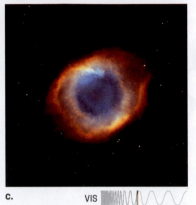

Figure 2.1 Deep-Sky Objects

Favorite targets for amateur astronomers' telescopes include galaxies, star clusters, and nebulas such as these, known as **a.** M64, **b.** M13, and **c.** the Helix Nebula (NGC 7293). Deep-sky objects lie far beyond our Solar System.

earliest attempts to understand the Universe and our place in it. Looking back is an essential step in understanding the modern story of astronomy and the development of the enormously powerful tool we call science. In these first attempts to make sense of the world, we can see the seeds of our own efforts today.

Dance of Night and Day: Basic Motions of the Sky

2.2 Most of us have never seen the night sky in its full splendor: the radical invention of artificial illumination more than 100 years ago erased the darkness of night for most of the world's inhabitants. If we encounter a dark sky, it's usually for just a few nights on a camping trip (**Figure 2.2**), and even then we generally don't have enough time to become familiar with the night sky and its motions. For 99.9 percent of human evolution, however, our ancestors lived in a world without artificial light and couldn't help but notice what happened on the sky.

"Every day the Sun looks like it's moving from east to west," says Carol Latta. "Every night the stars look like they are doing the same. The daily motion of the Sun and stars is the most fundamental astronomical fact and the most basic illusion." We now know that it's Earth that moves, not the sky. "It certainly seems like Earth is stationary," says Latta. "It doesn't feel like we are moving, so you can understand why the ancients thought it was the stars and the Sun that were in motion. But, of course, what's happening is simply that Earth is rotating." It was not until about the 1600s that the **rotation** of Earth—the planet's spin on its axis—was universally accepted. So it's the Earth that is spinning eastward. The story related here, the one we've now learned to be true, was hard won over thousands of years and reflects a triumph of science and human intelligence.

You don't need a powerful telescope to develop an accurate account of what happens on the sky. After all, "naked-eye" observations were all our ancestors had when they began charting the regular patterns that appeared in the motions of the Sun, Moon, planets, and

Figure 2.2 Light Pollution

a. Light pollution (shown in this satellite image) makes it impossible to see the skies as our ancestors saw them, but **b.** a camping trip to a remote location may provide a sense of the seemingly countless stars in the night sky.

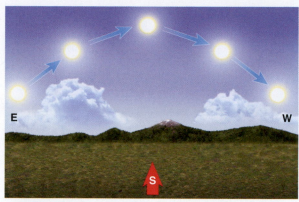

a. Daily path of the Sun

b. Nightly path of the stars

Figure 2.3 Daily Path of Sun and Stars
a. From northern latitudes, the Sun appears to rise in the east, reach its highest point in the south, and set in the west.
b. Stars also rise in the east, move across the sky, and set in the west. Here the constellation Orion, which includes the stars Betelgeuse and Rigel, follows that path each night.

stars. The most important aspect of making (and recording) observations is to have reference points that don't change each time you go out to do your observing. First, find a place you can return to each night (or day). Then, from your observation place, find a reference point on the horizon: it could be a mountain peak toward the west or a telephone pole in an eastward direction. Using these reference points, you can make observations such as where the Moon set relative to the mountain peak or where the Sun rose every day relative to the telephone pole. By repeating the observations over days, months, and even years, you can build a clear picture of the patterns of the Sun and Moon. By using simple instruments to measure the distance from one point on the sky to the other, you can make naked-eye observations such as how the planet Mars moves relative to specific stars night after night. It was in this way that early peoples built up a fundamental understanding of the basic motions of the sky.

INTERACTIVE:
Basic Motions of the Sky

Daily Motions and the Celestial Sphere

Each morning, at a time that varies as the seasons progress, the Sun appears to rise above the eastern **horizon** (the boundary between land and sky from an observer's perspective). As the day winds on, the Sun climbs until it reaches its highest **altitude**, or height above the horizon, at noon. It then descends toward the west, eventually setting by dropping below the western horizon. As the sky darkens, the stars appear as patterns of twinkling points of light whose positions relative to one another do not appear to change. As night progresses, stars also make a trek across the sky similar to the Sun's during the day, rising above the eastern horizon and setting into the western horizon (**Figure 2.3**).

THE CELESTIAL SPHERE. The exact path the Sun and stars take through the sky relative to the horizon in their east-to-west daily motion depends on where on Earth you're standing and what season it is. More specifically, the paths of the Sun and stars depend on your **latitude**—your distance north or south as measured from the equator. (Your **longitude** on Earth, which measures your position from east to west, referenced from a line running through Greenwich, England, does not affect the path of the Sun and stars on the sky.) Since Earth is a sphere, we can measure distances from one place to another on it in terms of an angle or **angular size**, given in degrees (°), minutes of arc (arcminutes, or ′), or seconds of arc (arcseconds , or ″). The North Pole has a latitude of 90° north; the South Pole has a latitude of 90° south. New York City's latitude is 40 degrees 43 minutes (40° 43′) north; and Sydney, Australia, has a latitude of 30° 51′ south.

Your latitude and the Sun's motion on the sky are related because of your *line of sight*, the line connecting your eyes to the Sun. Suppose you're standing somewhere on the equator (such as Quito, the capital of Ecuador) at noon. For you, at Quito, the Sun will take a path that brings it much higher on the sky at noon than it would be if you were farther

> "The daily motion of the Sun and stars is the most fundamental astronomical fact and the most basic illusion."

rotation The spin of an object around an internal axis.

horizon The horizontal circle defining the lower edge of the sky.

altitude The angular distance from a point on the sky to the point on the horizon directly below it.

latitude A measure of location specifying north–south position on Earth running from 0° at the equator to 90° north (North Pole) and 90° south (South Pole).

longitude A measure of location specifying east–west position on Earth.

angular size Angular distance measured on a sphere (such as Earth or the celestial sphere) in degrees, arcminutes, and arcseconds.

Figure 2.4 The Sun's Position at Different Latitudes

For most of the year, the Sun's position at noon as seen from **a.** Quito, on the equator, at latitude 0°, is much higher from the observer's southern horizon, S, than the Sun's position at noon as seen from **b.** Seattle, at latitude 47° north. a1 and a2 are the angle between the Sun's position in the sky and the horizon at noon. a2 is smaller for the person in Seattle, reflecting the fact that the Sun is much lower in the sky there.

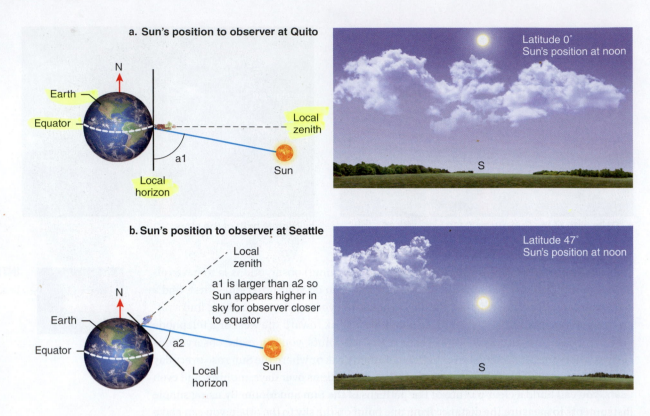

a. Sun's position to observer at Quito

b. Sun's position to observer at Seattle

a1 is larger than a2 so Sun appears higher in sky for observer closer to equator

Latitude 0°
Sun's position at noon

Latitude 47°
Sun's position at noon

celestial sphere An imaginary transparent sphere surrounding Earth, on which positions of stars, planets, and other celestial bodies are projected from extensions of Earth's coordinate system.

celestial equator The extension of Earth's equator onto the celestial sphere.

Polaris The star that is found very near Earth's north celestial pole (hence called the North Star) and around which the sky of the Northern Hemisphere appears to turn.

north celestial pole The extension of Earth's North Pole onto the celestial sphere.

south celestial pole The extension of Earth's South Pole onto the celestial sphere.

zenith The point on the sky directly overhead from an observer.

meridian A circular arc crossing the celestial sphere and passing through the local zenith and celestial pole.

declination The position of an object on the celestial sphere, measured similarly to latitude on Earth.

north, such as Seattle, Washington (**Figure 2.4**). Because Seattle is well north of Quito, the line of sight to the Sun in Seattle will always point more to the south than it will in Quito, and therefore the Sun will appear lower on the sky and closer to the southern horizon.

Before astronomers could understand the reason for the difference in the Sun's position at different latitudes, they needed a way to map the positions of objects such as the Sun and stars on the sky. To do so, they invented the concept of the celestial sphere. The **celestial sphere** is like an imaginary transparent globe that surrounds Earth with the positions of stars and planets painted on the inside. Earth's equator, poles, and lines of latitude and longitude extend out to this imaginary globe and act as reference points (**Figure 2.5**). If we take Earth's equator and extend it out into space, we create a line called the **celestial equator**. Then we can take the points that mark Earth's North and South poles and extend them out into space as well. Those points are called the celestial poles. Every day Earth rotates once on its axis. That means every day the Sun and stars appear to turn once around the celestial poles. "The star **Polaris** is very close to the **north celestial pole**," explains Latta, "so it looks like all the stars visible in the Northern Hemisphere turn around Polaris. That's why it's called the North Star. There is no bright star near the **south celestial pole**, so if you were in the Southern Hemisphere, you would see all the stars spinning around a relatively empty patch of sky at the southern celestial pole each night" (**Figure 2.6**). The celestial poles will always appear to you at an altitude above the horizon that is equal to the latitude of your position on the Earth.

"No matter where you are on Earth, only one-half of the celestial sphere is available to you at any given moment," explains Latta. Your location in latitude on Earth determines which part of the celestial sphere you can see (Figure 2.5). "Imagine extending the horizon around you out to infinity. You can only see above where the horizon line intersects the celestial sphere, and that determines which half of the sky you can see day or night. If you were standing exactly at the North Pole, you would see only the northern portion of the celestial sphere. If you were standing at the South Pole, you would only see the southern half of the celestial sphere."

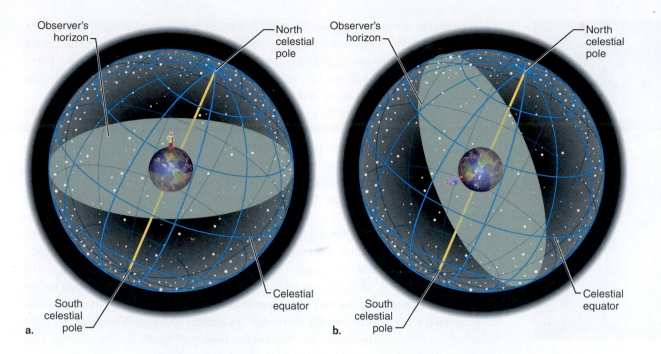

Figure 2.5
The Celestial Sphere
The visible portion of the celestial sphere depends on the observer's latitude but always includes one-half of the sky.

a.

b.

From where you're standing at any moment, the **zenith** is the point on the celestial sphere directly above your head. The **meridian** is an imaginary line that runs from the south point on the horizon, through the zenith, to the north point on the horizon (**Figure 2.7**). Thus, if you were standing at Earth's North Pole, your zenith would be the north celestial pole. If you were standing in a field outside of Chicago, your zenith and the north celestial pole would no longer overlap. How far apart would they be? Since Chicago is located at a latitude of about 42° north and the North Pole is at 90° north, the difference in angle between your zenith and the north celestial pole would be 90° − 42° = 48°.

What if you were in Quito, on the equator? Since Quito lies at 0°, the difference between the zenith and the north celestial pole is 90° − 0° = 90°. The north celestial pole would lie just at your northern horizon. The north–south position of an object on the celestial sphere is referred to as its **declination**, and it is measured similarly to latitude on Earth.

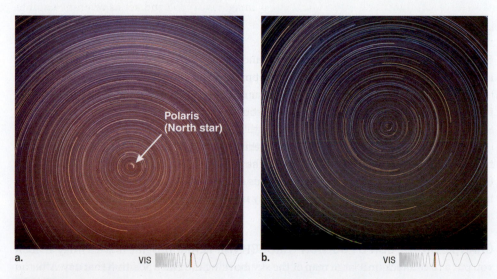

Figure 2.6 Star Trails
Earth's rotation creates the illusion that the Sun and stars move across our skies. In time-lapse photographs, the stars appear to create circular star trails around the celestial poles. **a.** Polaris lies nearly at the north celestial pole. **b.** There is no bright pole star at the south celestial pole.

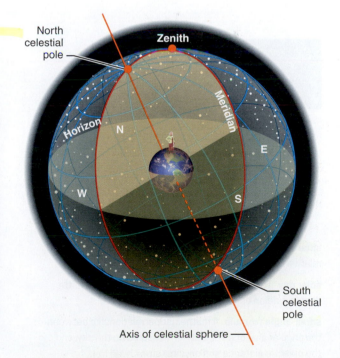

Figure 2.7 Meridian and Zenith
Meridians are imaginary circular lines drawn on the sky that pass through the north and south celestial poles. The zenith is the point directly above your head on the sky. There is always a meridian that passes through your zenith.

Figure 2.8 Angular Size

Angular size is measured as the number of degrees of arc that an object spans in the field of view of the observer. The Sun varies a little in apparent size because of Earth's orbital motion but generally appears to span 0.5°, about the width of a pinky finger held at arm's length.

INTERACTIVE:

Small-Angle Formula

---

sidereal day The time taken by Earth to rotate on its axis relative to the stars.

solar day The time between successive crossings of the same meridian on the sky by the Sun.

small-angle formula A mathematical relationship between the angular size of an object, as measured on the sky, and the object's distance and physical size.

arcminute A measure of angular size. There are 60 arcminutes in 1 degree.

analemma The figure-eight pattern formed by charting the daily position of the Sun at the same time of day over the course of 1 year.

The part of the celestial sphere that you can see at any moment also depends on the time of day, as Earth swings around in its daily rotation. At noon you can see the half of the celestial sphere with the Sun at its highest position. Twelve hours later, at midnight, Earth has swung around 180°, halfway through its daily rotation. Now you are facing directly away from the Sun and seeing the opposite half of the celestial sphere.

Using the celestial sphere, we can define a "day" in two ways. (1) A **sidereal day** is the time Earth takes to rotate on its axis relative to the stars—meaning the time it takes for a particular point on the celestial sphere (a star's location, for instance) to cross the same celestial meridian. Image a meridian as a line on the sky cutting through some marker on your horizon (like a telephone pole), going up to your zenith and reaching all the way to the other side of your horizon. As the night progresses, the stars will track across that meridian as they rise and set. So *a sidereal day is the time it takes a particular star to cross that meridian between one night and the next.* (2) A **solar day** is the time between successive crossings of the same celestial meridian on the sky by the Sun. Because Earth orbits the Sun, the Sun appears to move against the fixed background of stars across the celestial sphere. In other words, while the stars don't appear to move relative to each other on the sky, as the year progresses the Sun tracks a path through the positions of the fixed stars. That means the two measurements of the day are slightly different. The solar day is 24 "solar" hours (by definition), but the sidereal day is only 23 hours and 56 minutes long because of the Sun's movement relative to the fixed stars.

APPARENT SIZE AND THE SMALL-ANGLE FORMULA. Because the sky *appears* to us as an upturned bowl, a hemisphere, the distance between objects on the sky can also be measured in terms of angular size. The hemisphere of the sky extends through 180° of arc. The Moon and the Sun, as they appear from Earth, have a width of about 0.5° of arc, and your pinky finger held at arm's length also takes up about 0.5° (**Figure 2.8**). (For comparison, the monitor on a typical computer spans about 30° of arc in your field of view.)

While we're thinking about angles and the night sky, consider the relationship between how large something appears on the sky (which would be measured as an angle) and how large it *is* if you measure it with a ruler (sometimes called its linear size, measured in meters or kilometers). The relation between the two for astronomical applications is neatly expressed in something called the **small-angle formula**, discussed in **Going Further 2.1**.

Yearly Motions of the Sun

Imagine it's a hot summer day in a city in the northern United States. The Sun has been up since early morning, and it hangs high and oppressive on the sky.

Now imagine it's the middle of winter in the same city. The day is cold, short, and forlorn. People hustle across the windy street trying to fight off the chill. It's noon and the Sun hangs low on the southern half of the sky. Its slanting rays seem weak and ineffective.

These two paragraphs describe something you've probably noticed: measured at the same time of day, the Sun is higher on the sky during summer days than it is in winter. "Higher" means closer to the zenith. This difference shows that, along with the daily motion of the Sun and stars, there are yearly motions that affect the lengths of days (the amount of daylight) and seasons.

CHANGES IN THE SUN'S POSITION THROUGH THE YEAR. Imagine that each day at noon you put a thumbtack on a large map of the sky marking the Sun's position that day. After an entire year you would find that the thumbtacks marked out a thin figure eight on the sky (**Figure 2.9**). For someone living in the Northern Hemisphere, the thumbtack for December 21 would be very close to the lowest point of the figure eight, meaning that the Sun was closer to the southern horizon at noon on that day than on any other day of the year. The

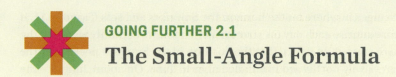

GOING FURTHER 2.1
The Small-Angle Formula

You've probably noticed that big things look small when they're far away and that small things look big when held right up to your eye. The ancient Greeks noticed this too and used geometry to discover how an object's apparent size, true size, and distance are related. When objects are far away, the relationship can be determined by the small-angle formula:

$$a = 206{,}265'' \times \frac{d}{D} = 206{,}265'' \times \frac{\text{object's true size}}{\text{object's distance from us}}$$

The small-angle formula requires the distance D and the physical size d to be expressed in the same units of length (meters or kilometers, for example). The term a refers to the object's *apparent size*—that is, how much of our field of view it appears to take up (**Figure 2.10**). We can measure this as an angle (see Figure 2.8) in units of degrees. For smaller apparent sizes, we subdivide each degree into 60 **arcminutes** (abbreviated 60′) and divide each arcminute into 60 *arcseconds* (abbreviated 60″); there are 3,600″ (60 × 60) in 1°. The small-angle

formula is usually expressed in terms of arcseconds. The value 206,265 comes from converting radians (an angular size unit) into arcseconds.

Let's apply this formula to some real astronomy. The Sun has a diameter, or linear size, of 1.39×10^6 kilometers (km). That's its true size, d. We orbit the Sun at a distance of 1.49×10^8 km. That's the distance, D. How large on the sky does the Sun appear to us? Using the small-angle formula, we calculate as follows:

$$a = 206{,}265'' \times \frac{1.39 \times 10^6 \text{ km}}{1.49 \times 10^8 \text{ km}} = 1{,}924''$$

We then convert this number into degrees:

$$a = 1{,}924'' \times \frac{1°}{3{,}600''} = 0.5°$$

The Sun takes up about one-half of a degree on the sky, so its apparent size is about the same as your pinky finger held up at arm's length.

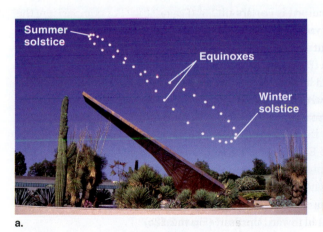

a.

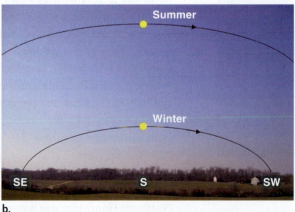

b.

INTERACTIVE:
Seasons

Figure 2.9 The Analemma

a. If you were to take a picture of the Sun's position every day at the same time, the Sun's location would trace out an analemma—a figure eight. Marking key events of the astronomical year on the analemma shows its relationship to the seasons and to Earth's motion around the Sun. **b.** The Sun's altitude on the sky is higher at noon in summer, lower in winter.

thumbtack placed 6 months later, on June 21, would be very close to the highest point on the figure eight. This was the day when the noontime Sun was closest to the zenith. On March 21 and September 21, the thumbtacks would be close to the middle of the figure eight. This cyclical motion of the Sun around the figure eight (called an **analemma**) each year was recognized as far back as two millennia ago and was the basis for creating sundials to track time.

The analemma is just one way to see how the Sun moves around the sky during a given year. You can also track solar motion by noticing where the Sun rises and sets each day. The path mapped out by the analemma—the Sun's noontime positions over the course

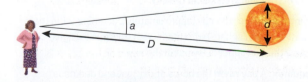

Figure 2.10 The Small-Angle Formula

The small-angle formula allows you to calculate the angular (apparent) size of an object on the sky if you know its distance from you and its physical size.

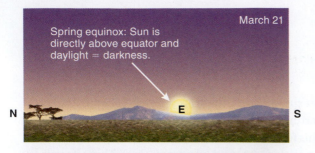

Spring equinox: Sun is directly above equator and daylight = darkness.

March 21

N   E   S

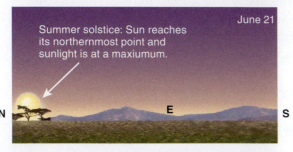

Summer solstice: Sun reaches its northernmost point and sunlight is at a maxiumum.

June 21

N   E   S

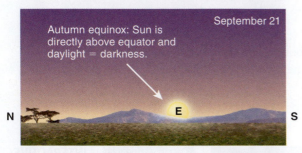

Autumn equinox: Sun is directly above equator and daylight = darkness.

September 21

N   E   S

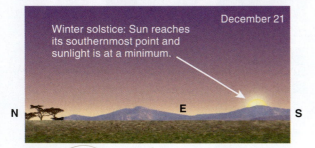

Winter solstice: Sun reaches its southernmost point and sunlight is at a minimum.

December 21

N   E   S

Figure 2.11 The Sun's Yearly Motions at Sunrise
The direction of sunrise on the solstices and equinoxes shows how the Sun's daily path across the sky changes during 1 year in the Northern Hemisphere.

summer solstice The day (approximately June 21 in the Northern Hemisphere) with the most hours of sunlight, when the Sun appears highest in the sky.

winter solstice The day (approximately December 21 in the Northern Hemisphere) with the fewest hours of daylight, when the Sun appears lowest in the sky.

equinox A day when the hours of daylight and darkness are equal. In the Northern Hemisphere, the autumn equinox occurs on approximately September 21; the spring equinox, approximately March 21.

orbit The cyclical path in space that one object makes around another object.

of a year—reflects changes in where on the horizon the Sun rises and sets (**Figure 2.11**). If you were awake before sunrise each day (as your farmer ancestors certainly were), in the Northern Hemisphere you would notice that the position of the Sun as it peeks over the horizon moves progressively northward from December to June. On about June 21—the **summer solstice**, the day of the year with the most sunlight—it stops its northward progression. From that point until about December 21—the **winter solstice**, the day with the least sunlight—the Sun rises progressively southward. The autumn and spring **equinoxes** (September 21 and March 21) mark the midpoint of this yearly journey. On these two days the amount of daylight and darkness are roughly equal.

"The length of the day (meaning the amount of daylight) changes. That's the one thing everyone notices about the seasons, even those who live in places where the temperature is fairly constant all year," says Latta. "The change in the length of the day and the changes in the path of the Sun through the sky are both a result of the yearly orbital journey of Earth around the Sun and the tilt of Earth's axis of rotation relative to that orbit."

EXPLAINING THE SUN'S YEARLY MOTIONS AND THE SEASONS. Earth's **orbit**, its cyclical journey around the Sun, defines the year, which lasts approximately 365¼ days. Recall from Chapter 1 that the average distance of Earth from the Sun is called an astronomical unit (AU), about 1.49×10^{11} m. The shape of an orbit is not perfectly circular but takes the form of a slightly squashed circle called an **ellipse**. An elliptical orbit means that Earth swings a little closer and then a little farther from the Sun as it completes its yearly orbit. "One of the most common misconceptions people have is that summers are hotter than winters because Earth is closer to the Sun in summer," says Latta. "Not only is that wrong, because Earth is actually closer to the Sun when the northern latitudes are in winter; it doesn't even explain why the length of the day changes or why the Sun is higher in the sky in summer. The real culprit is Earth's tilt."

If you love the seasons, then you have to be thankful for one of those lucky coincidences that our planet happens to rotate at an angle of 23.5° to a line that's perpendicular to the plane of its orbit around the Sun (**Figure 2.12a**). In other words, Earth's rotation (spin) axis is *tilted* with respect to its orbit. The angle between that axis and the perpendicular to the orbital plane is called the **obliquity** or **axial tilt**. (This is the same as the angle between the celestial equator and the orbital plane.) The axial tilt and its direction with respect to the fixed stars do not change as the year progresses; the North Star is the North Star in both winter and summer. "Since Earth's tilt always has the same orientation," says Latta, "the Sun appears to change its position on the sky as the year progresses. It will be low on the sky in northern winter when the Northern Hemisphere tilts away from the Sun. Six months later, the Sun will appear high in the summer sky (in the north) because the Northern Hemisphere is now tilted toward the Sun" (**Figure 2.12b**).

Earth's spin axis wobbles slightly, however, moving a little more than 1° per century. This phenomenon is known as **precession**. While the motion is small, the north celestial pole has moved enough that the Egyptians would not have recognized Polaris as the North Star.

The tilt of Earth's rotation axis, relative to its orbit around the Sun, is all you need to explain both the Sun's yearly motion through the sky and the seasons. For example, winter occurs in the Northern Hemisphere when the axis of rotation tilts away from the Sun, so the Sun's rays hit Earth at a shallow angle, spreading their energy over a larger area (**Figure 2.13a**). In summer, when Earth is tilted toward the Sun, the opposite effect takes place: the Sun's rays strike the surface at a steeper angle, concentrating more energy in a smaller surface area (**Figure 2.13b**).

Earth's tilt also explains the seasonally changing length of daylight. With no tilt, every day would experience the same amount of daylight. Because the rotation axis tilts, the Sun rises later and sets earlier in winter than in summer, making the winter day shorter. In summer, the Sun rises earlier and sets later, making the summer day longer. A longer day,

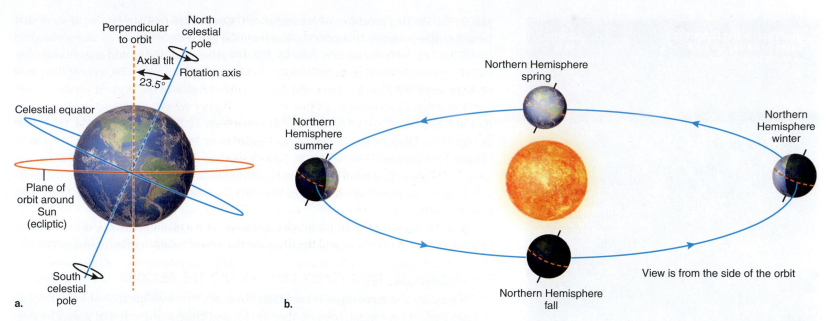

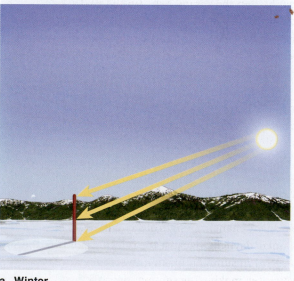

Figure 2.12 Earth's Tilt and the Seasons

a. Earth's axis of rotation tilts 23.5° to the perpendicular of its orbit around the Sun. The orientation of the tilt does not change as Earth orbits the Sun and thus provides all of our seasonal variations. **b.** As Earth orbits the Sun, the fixed orientation of its tilt axis means that when one hemisphere experiences summer, the opposite hemisphere experiences winter.

with sunlight striking the ground at a steeper angle, leads to warm summers. A shorter day, with sunlight striking the ground at a shallower angle, leads to cold winters. Near the equator, however, Earth's rotational tilt has less effect: the Sun always sits high on the sky at noon, the amount of daylight does not change dramatically, and therefore seasonal changes are minimal.

Yearly Motions of the Stars

Like the Sun, the stars also change their positions on the sky throughout the year. To our ancestors who lived their whole lives without artificial light, the drama of the nighttime sky was the only show in town. One of the most obvious changes over the course of the

> "One of the most common misconceptions people have is that summers are hotter than winters because Earth is closer to the Sun in summer."

ellipse The geometric shape of an oval, in which the sum of the distances from two points (foci) to every point on the ellipse is constant. A circle is a special type of ellipse.

obliquity or axial tilt The angle between a planet's rotation axis and a line perpendicular to the planet's orbital plane.

precession The rotation of a planet's spin axis, similar to that of a wobbling top.

a. Winter **b. Summer**

Figure 2.13 Angle of the Sun's Rays

The Sun's rays hit the ground more obliquely in winter and more directly in summer. Thus, the same amount of solar energy is spread over a larger area in winter, so the Sun's rays provide less heat. **a.** At noon in winter, the angle between a tall pole and the Sun's rays is large, so the pole casts a long shadow. **b.** At noon in summer, the Sun is almost overhead and its rays hit the ground at an angle closer to 90°, so the pole casts a short shadow.

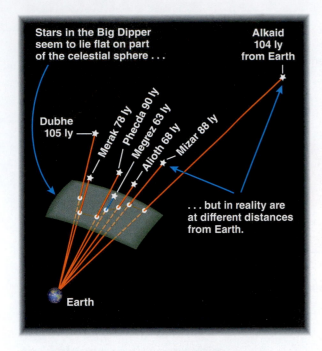

Figure 2.14 Stars in a Constellation

All of the stars in a constellation are at different distances from Earth, as this 3D map of the Big Dipper (in Ursa Major) shows.

Figure 2.15 Seasonal Constellations

As Earth moves from one side of its orbit in winter to the other side in summer, different constellations are visible on Earth's "night side." Some constellations appear in the sky primarily in summer months; others appear primarily in winter months. In the Northern Hemisphere, **a.** Sagittarius and Scorpius adorn the summer night sky. **b.** The constellations centered around Orion dominate the winter night sky.

year is the shifting positions of constellations. **Constellations** are groups of stars that form familiar patterns. In modern astronomy, the sky is divided into 88 zones based on constellations, and astronomers refer to objects by their position within one of these constellation zones ("the star-forming clouds in Taurus," for example). So, a constellation is both the stars that form a pattern and the designated region of the sky surrounding them.

Even though stars such as those in the Big Dipper are grouped together on the sky, they are, in general, not all at the same distance from us. For example, Alkaid, the star at the tip of the Dipper's handle, is 104 light-years away from Earth, but Merak, the star at the tip of the Dipper's ladle, is only 78 light-years away (**Figure 2.14**). The stars in a constellation will appear to have different levels of brightness. Brightness is an important characteristic of any object we view on the sky and, as we will see, distance is one factor that determines it.

For our ancestors, what mattered was that not all constellations are visible all year long. Some constellations can be seen only in the winter or only in the summer. The stories that early peoples imagined about each constellation often reflected the season when it appeared (**Figure 2.15**).

There are also **circumpolar constellations**, which are visible all year long. They do not rise and set each night, because they are located close to the celestial pole. The constellation Ursa Minor (the "Little Bear") includes the North Star, so it is one circumpolar constellation in the Northern Hemisphere.

Of particular importance are the constellations that appear along the path that the Sun charts on the sky. This may sound strange because no stars are visible when the Sun is up. But people made note of the position of the Sun on the sky at a certain time each day (noon, for example). By doing so, they found that the Sun makes a line across the background of the stars on the celestial sphere as the year progresses. That line is called the **ecliptic**, and the constellations of the **zodiac** lie in a band about 9° wide on either side of the ecliptic (**Figure 2.16**). As the year progresses, the Sun appears to march eastward along the ecliptic, and the constellations of the zodiac progressively become visible at night. (For followers of astrology, a person's date of birth indicates an astrological sign that corresponds to one of 12 constellations of the zodiac. Because of the way astronomers define constellations in terms of regions of the sky, there are actually 13 constellations in the zodiac: Ophiuchus is the 13th.)

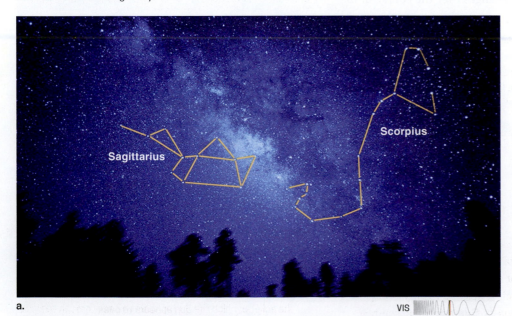

If we think of the year beginning in January (which is arbitrary), imagine Earth beginning on one side of its orbital track around the Sun. At night, the view of the sky faces the direction opposite the line linking Earth and the Sun. That view of the celestial sphere determines which stars (and which constellations) are visible. As Earth moves in its orbit around the Sun and the year progresses, the portion of the celestial sphere we face at night continually shifts. "It's this changing view of the celestial sphere as Earth swings around in its orbit that changes the constellations you can see," explains Latta. "Amateur astronomers like me get excited as the year unfolds because it means our favorite objects lying in different constellations start appearing in the night sky for us to observe."

For example, the constellation Taurus appears at night in the northern sky in winter. That means the night side of Earth faces in the direction of Taurus in January. In other words, the half of the celestial sphere that is visible at night in January includes Taurus. The constellation Sagittarius is visible during the half of the year stretching from spring through summer and into fall, so the night side of Earth faces in the direction of Sagittarius in June. In other words, the half of the celestial sphere that is visible at night in June includes Sagittarius. Why, then, is Sagittarius a winter astrological sign? The Sun was thought to be in the "house" of Sagittarius during specific winter months, because the position of the Sun on the celestial sphere in the winter crosses through the constellation Sagittarius.

Since Earth moves around the Sun, and not the other way around, it is Earth's motion that defines the plane of the ecliptic on the celestial sphere. The Sun appears to chart a course against the background of fixed stars because the line of sight from Earth to the Sun points to different background stars as the year rolls on and Earth spins around in its orbit.

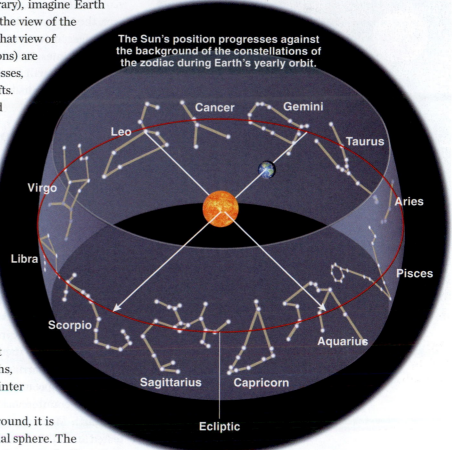

The Sun's position progresses against the background of the constellations of the zodiac during Earth's yearly orbit.

Figure 2.16 The Ecliptic and Zodiac
The ecliptic is the path that the Sun appears to follow through the sky over the course of a year. The zodiac represents the 12 constellations through which the ecliptic passes. (Astronomers include a 13th constellation, Ophiuchus, although it does not appear in most traditional versions of the zodiac.) As Earth moves through its orbit, its night side faces outward, so different constellations are visible at different times of year.

Section Summary

- The Sun and stars appear to move east to west because of Earth's rotation. The exact path on the sky depends on the observer's latitude: observers at the equator see the Sun reach higher above the horizon than those at other latitudes do.
- The celestial sphere is an imaginary sphere whose coordinates are extensions of Earth's poles, equator, and latitude/longitude lines.
- Because of Earth's orbital motion, the apparent position of the Sun against the background stars changes over the course of the year. The Sun's annual path against the background of the stars is the ecliptic.
- The Sun's daily position is higher (closer to the zenith) on the sky in summer than in winter. Seasons result from the 23.5° tilt of Earth's axis in relation to the line perpendicular to Earth's orbit. This affects the number of hours of daylight and how directly that light hits Earth.

CHECKPOINT Describe how the Sun's path through the sky changes during one year and how these changes are related to the cause of the seasons.

"It's this changing view of the celestial sphere as Earth swings around in its orbit that changes the constellations you can see."

Monthly Changes of the Moon

2.3 The change of the Moon's phases poses the same challenge of moving from what we see from the ground to the view of the Earth-Moon-Sun system from space. The monthly round of lunar phases made such an enormous impression on early cultures that the Moon became a principal actor in the mythologies of almost every culture. In later ages, it became a primary driver of astronomical science.

constellation A group of stars that form a pattern, and the designated region of the sky surrounding them.

circumpolar constellation A constellation that, from the viewer's perspective, never rises or sets.

ecliptic The path that the Sun appears to follow against the background stars, as defined by Earth's orbit around the Sun.

zodiac The 13 constellations that lie close to the plane of the ecliptic, which the Sun appears to pass through over the course of a year.

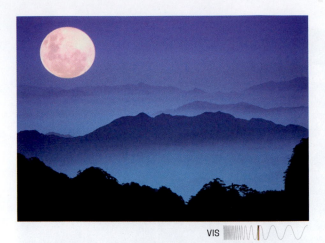

VIS

Figure 2.17 Ever-Changing View of the Moon
Orbiting Earth, the Moon presents an ever-changing appearance to viewers on Earth, although these changes repeat in cycles. Here, a full Moon rises above the horizon.

The Moon rises and sets each day, just as the Sun and stars do. Unlike the Sun, however, the Moon can be "up" during either the day or the night. This fact is directly related to changes in the Moon's appearance called **phases**. Half of the Moon is illuminated by the Sun at all times. The phases are defined by how much of the illuminated half of the Moon we see from Earth. We see the entire illuminated half during a full Moon (**Figure 2.17**), and none of the illuminated half during a new Moon.

A Scientific Model of the Moon's Phases

The explanation of the Moon's motion and cycles offers a wonderful example of a **scientific model**. In any science, including astronomy, a model is an idea or set of ideas that enables people to create testable explanations of how they think a particular aspect of the world works. The world presents us with phenomena such as the monthly cycle of the Moon and its phases. In response, we try to imagine a changing configuration of bodies such as Earth, Moon, and Sun that enables us to "recover" what we observe—meaning that the model accounts for our observations. A model often helps us understand how something we cannot see directly (the orbits of the Earth, Moon, and Sun) can create what we do see (the Moon's phases). A testable model makes predictions that we can verify by experiments or observations. When a model is testable, it can be incorporated into the methods of scientific investigation.

Some of our ancestors imagined the phases of the Moon as cycles of victory and defeat of two warring gods. We now know that the Moon is an approximately spherical body orbiting counterclockwise (as viewed by looking down at the Earth's North Pole) around a counterclockwise-spinning Earth on an approximately 30-day orbit and that the Earth-Moon system orbits the Sun once a year. We know this is true because we have tested and confirmed this model with everything from telescopes to space probes. Earlier in history, it was just a set of ideas that had to be tested. Let's see how this model of the Earth-Moon-Sun system explains what we see from the ground.

The Moon's phases occur as a direct result of its relative positions between the Sun and viewers on Earth. The Moon shines by reflecting sunlight; it emits no visible-wavelength-range light of its own. That means we can see only the side of the Moon that is directly exposed to sunlight. The half of the Moon that lies in shadow appears dark to us. To understand how this works, go into a dark room and place a flashlight on a table to represent the Sun. Face the flashlight and hold a tennis ball out to your side. The tennis ball denotes the Moon. You should see that one-half of the ball is in shadow and the other half is illuminated (and you see only one half of the ball, the side facing you). This configuration is exactly what happens with the Moon in its "quarter" phases, as we'll see shortly. Now hold the ball directly in front of you between your head and the flashlight. The illuminated side of the ball now faces away from you, and you can only see the shadowed side directly. This is exactly what happens with the Moon in its "new" phase, except that the dark side of the Moon cannot be seen at all when Earth, Moon, and Sun have the same orientation as your head, the ball, and the flashlight. Keep this model in mind as we now track through the Moon's phases in detail.

The Moon's Phases: What We See and Why

A *new Moon* occurs when the Moon is between Earth and the Sun—the point in its orbit when the Moon is closest to the Sun on the sky. The new Moon is easy to explain: it occurs when the Moon's orbital position is close to the line connecting Earth and the Sun (**Figure 2.18**). During a new Moon, the lunar face illuminated by the Sun points away from Earth and thus is essentially invisible to us.

As the Moon moves around Earth on its orbit, it appears progressively eastward of the Sun. This means that a few days after a new Moon, you might notice the Moon

phase (of the Moon) As seen by an observer on Earth, the appearance of the illuminated portion of the Moon, which changes cyclically as the Moon orbits Earth.

scientific model An idea or set of ideas used to create testable explanations.

synodic month The amount of time it takes the Moon to cycle through its phases.

sidereal month The amount of time it takes the Moon to make one complete orbit in relation to the background stars.

appearing in the late-afternoon sky as the Sun is sinking in the west. This phase is called *waxing crescent*.

The waxing crescent phase culminates in what is called the *first quarter* Moon. The first quarter phase, during which the Moon is at its highest altitude at sunset, occurs approximately 7 days after the new Moon. The illuminated part of the Moon always curves toward the Sun (since that's where the illumination is coming from). So from new Moon to first quarter Moon, the "horns" of the crescent point away from the Sun and toward the east.

The Moon continues its journey eastward relative to the Sun, rising later and later in the day. This is called the *waxing gibbous* phase. It culminates in the *full Moon*, which rises in the east just as the Sun is setting in the west. Occurring about 7 days after the first quarter Moon, the full Moon is at its highest point in the night sky at midnight (when the Sun is at its highest point on the other side of the world). The full Moon completes the first half of the lunar cycle, when the Moon has traveled halfway around its orbit and is on the opposite side of Earth from the Sun. The Moon appears full because the illuminated half of the Moon, the half facing the Sun, is now fully visible to viewers on Earth.

As the month continues, the Moon now begins to appear over the eastern horizon closer to midnight. This is called the *waning gibbous* phase and culminates in the *third quarter* Moon 7 days after the full Moon. The third quarter Moon rises at midnight and appears overhead when the Sun is rising. At this point, the Moon has completed three-quarters of its orbit and is on the other side of Earth from its position at first quarter.

The final quarter of the Moon's cycle occurs as the Moon moves back toward the line connecting Earth and the Sun. During this approximately 7-day period, called the *waning crescent* phase, more and more of the Moon's illuminated face becomes hidden from viewers on Earth. In the waning crescent phase (on the way to the next new Moon), the horns of the crescent once again point away from the Sun. When you see a sliver of the Moon on the sky as you walk to class in the morning, you're seeing it as a waning crescent.

It's worth noting that the Moon takes about 29.5 days to cycle through its phases. This period is called the **synodic month**. It is 2.2 days longer than the time it takes the Moon to make one orbit about Earth with respect to the background of distant stars—a period called the **sidereal month**. These two definitions of the month differ because

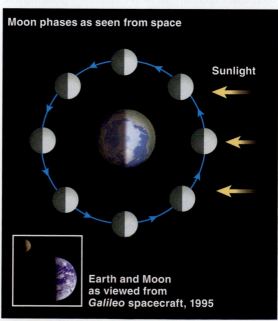

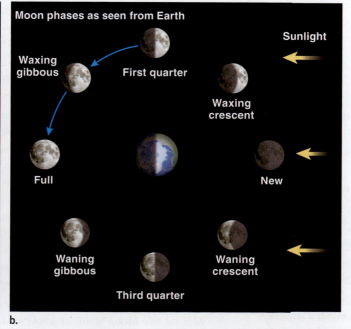

Figure 2.18 Phases of the Moon

a. The Sun always illuminates one-half of the Moon's surface, as well as Earth's. Seen from a point in space, the same portion of both Earth and Moon is illuminated. The Moon's phase indicates its location relative to Earth and the Sun. **b.** As the Moon orbits Earth, the proportion of the illuminated surface that can be seen from Earth changes from 0 percent (new Moon) to 100 percent (full Moon).

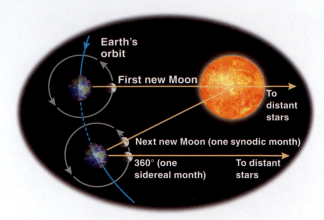

Figure 2.19 Synodic Month versus Sidereal Month
The synodic orbital period of the Moon (when it returns to the same position with regard to the Sun) is 2.2 days longer than its sidereal period (when it returns to the same position with regard to the distant stars) because Earth has moved in its orbit as well.

INTERACTIVE:

Lunar Phases

eclipse The passing of one celestial body through the shadow of another.

penumbra The outer region of a shadow cast by an extended object.

umbra The inner region of a shadow cast by an extended object.

total lunar eclipse An eclipse that occurs when the Moon passes through Earth's umbral shadow.

partial lunar eclipse An eclipse that occurs when the Moon passes partially through the Earth's umbral shadow.

total solar eclipse An eclipse that occurs when a region of Earth's surface passes under the Moon's umbral shadow and the Sun's disk is fully blocked.

while the Moon is completing one cycle of its orbit around Earth, the entire Earth-Moon system is moving through its own orbit about the Sun (**Figure 2.19**). After one synodic month, Earth has moved approximately one-twelfth of the way around the Sun. That means the Sun is located in a different part of the celestial sphere than it was at the beginning of the month. The difference between the synodic month and the shorter sidereal month is the time it takes the Moon to cross that extra portion of sky and catch up to the Sun.

Finally, notice that we always see the same "face" of the moon regardless of its phase. If you go back to our ball and flashlight model, imagine painting a smiley face on one side of the ball. So that the smiley face will always point toward you, the ball must be spinning relative to the flashlight as it goes around your head. If the ball didn't spin on its own axis, then the smiley face would always point in the same direction relative to the flashlight; sometimes you would see the face as the ball "orbits" your head, and sometimes you wouldn't. Since we always see the same face of the Moon, it too must be spinning on its axis as it orbits Earth. Most importantly, the time it takes to complete one spin rotation must be the same as the time it takes to complete one orbit around the Earth. Thus a day-night cycle on the Moon takes 1 month to complete. Astronomers call this synchronization of a body's orbit and its spin "tidal locking"; we'll explore it in more detail in Chapter 6.

Going Dark: Lunar and Solar Eclipses

It is perhaps one of the most remarkable circumstances in cosmic history, at least as far as human beings are concerned, that the diameter of the Sun is 400 times larger than the diameter of the Moon. Yet as seen from Earth, the disks of the Moon and the Sun have just about the same size on the sky (in other words, their angular diameters are about the same). That coincidence occurs because the Sun is not only farther away than the Moon (which means the Sun appears smaller on the sky), but exactly far enough away that their angular diameters nearly match. From that fluke of size and distance arises one of the most stunning of astronomical events: the solar eclipse. During an **eclipse**, one celestial body passes through the shadow of another. In this section we will explore the causes of solar eclipses and the related eclipses of the Moon (lunar eclipses).

Eclipses are all about shadows. When the light from an extended object such as the Sun hits the Moon or Earth, the shadow it casts has two parts: an extended **penumbra**, shaped like an open-ended cone; and a smaller **umbra**, shaped like the closing tip of a cone (**Figure 2.20a**).

A **total lunar eclipse** occurs when the Moon passes through the center of Earth's shadow, meaning that it passes first through Earth's penumbra and then through its umbra (see Figure 2.20a). Because this happens only when the Moon, Earth, and Sun lie along the same line, total lunar eclipses occur only during a full Moon. Since the Moon "shines" via reflected sunlight, during a total lunar eclipse the face of the Moon will first darken as it passes through the penumbra. When the Moon passes through the darker umbra, however, it does not disappear. Sunlight passing through Earth's atmosphere is redirected into the umbra and onto the face of the Moon. Blue light is more easily scattered by the atmosphere than red light, so during the umbral phase of a total lunar eclipse, the Moon takes on a dusky reddish glow (**Figure 2.20b**). The extended interval of the reddish hue depends on the conditions in Earth's atmosphere at the time of the eclipse. A typical total lunar eclipse lasts a few hours as the Moon traverses the entirety of Earth's shadow.

Why don't we see a lunar eclipse at every full Moon? The answer is that the Moon is not usually directly behind Earth. The plane of the Moon's orbit has a 5° tilt relative to the plane of Earth's orbit around the Sun (the plane of the ecliptic). So, during most full Moons, the Moon is either above or below Earth's shadow. Sometimes the Moon passes

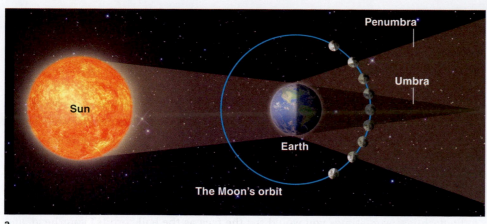

a.   b.   VIS

Figure 2.20 Total Lunar Eclipse

a. Lunar eclipses occur when the Moon passes into Earth's shadow. As it passes into the penumbra, the Moon darkens; as it passes into the umbra, Earth's shadow intensifies and grows across the face of the Moon until totality. As the Moon exits Earth's shadow, the steps reverse. **b.** A time-lapse series of a lunar eclipse, as Earth's shadow falls across the face of the Moon. The reddening that occurs at totality is caused by refracted sunlight passing through Earth's atmosphere and reflecting off the Moon.

only through the penumbra or only partially into the umbra. When the moon passes partially through the umbra we see a **partial lunar eclipse**.

A solar eclipse occurs when the Moon's shadow falls across a region of Earth. As **Figure 2.21** shows, even the penumbral lunar shadow is too small to cover all of Earth, and the umbra can cover only an even smaller area at any moment. Thus, unlike a lunar eclipse, which can be seen by everyone on the night side of Earth, a solar eclipse can be seen by only those living in the regions of Earth where the Moon's shadow falls.

Solar eclipses occur when the disk of the Moon blocks all or some of the disk of the Sun. This happens only during a new Moon, when the Moon lies directly along the line joining Earth and the Sun. **Total solar eclipses** occur only where the Moon's umbra falls on Earth, creating a spot approximately 270 km in diameter (though the spot size will vary from one eclipse to another). As Earth turns under this spot and the Moon moves in its orbit, this umbral spot moves across Earth in what is called the *path of totality*.

"Those people were priests who were, in a sense, the precursors of scientists and astronomers today."

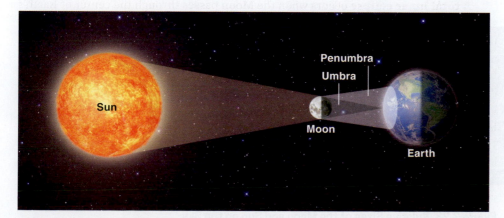

Figure 2.21 Total Solar Eclipse

Solar eclipses occur when the Moon's shadow falls across a region of Earth. Totality, lasting only a few minutes, is visible for observers in the umbral area, which is about 270 km in diameter. Partial eclipses are visible to those in the penumbral region.

VIS

Figure 2.22 The Sun's Atmosphere during an Eclipse
Because the Moon blocks the solar disk during a solar eclipse, the less bright solar atmosphere, called the corona, becomes visible around the darkened Moon. The red features are solar prominences erupting from the Sun's surface.

Because they are remarkable events for anyone who experiences them, total solar eclipses must have had a profound effect on early cultures. As the Moon begins to cross in front of the Sun, at first an arc-shaped "bite" appears to be taken out of the solar disk. As **totality**, or total concealment, is reached and the Moon completely obscures the Sun, the sky suddenly slips into darkness. Stars come out. Animals switch to nighttime behaviors. Streetlights switch on. As if this were not enough, behind the disk of the Moon, the Sun's usually obscured outer atmosphere, called the *solar corona*, becomes visible. Lower regions of the Sun's atmosphere, such as the chromosphere and eruptive prominences (explained in Chapter 10), may also be seen (**Figure 2.22**). Totality during a total solar eclipse is relatively short, lasting less than 8 minutes.

The Moon's orbit is not circular. Therefore, when the Moon is at **apogee** (its farthest distance from Earth), it appears smaller on the sky than when it is at **perigee** (the point closest to Earth; **Figure 2.23a**). Note that at apogee the Moon is about 406,000 km distant from the Earth, while at perigee it is 357,000 km distant. Earth's orbit is also not circular; its farthest and closest points from the Sun are called, respectively, **aphelion** and **perihelion**. That means the disk of the Sun also sometimes appears larger or smaller (**Figure 2.23b**). So sometimes, even in the umbral shadow, the disk of the Moon will not entirely block the disk of the Sun. When that happens, an **annular eclipse** occurs, in which a thin ring of Sun appears to surround the darkened disk of the Moon (**Figure 2.24**).

Partial solar eclipses occur in regions of Earth that fall under the Moon's penumbra. Usually regions within a few thousand miles of the umbra see a scalloped portion of the Sun blocked by the lunar disk.

Just as we don't see a lunar eclipse at every full Moon, solar eclipses don't occur at every new Moon. Again, the reason is the 5° tilt of the Moon's orbital plane relative to Earth's orbit around the Sun. Solar eclipses can occur only when the Moon's tilted orbit carries the Moon directly between Earth and the Sun. When does this alignment occur? Just as the axis of the spinning Earth points in the same direction in space as Earth orbits the Sun (creating the seasons), the tilt of the Moon's orbit maintains a fixed direction in space as Earth goes around the Sun. That means the location *where the Moon crosses the Earth-Sun orbital plane* during its monthly revolution changes throughout the year. Astronomers call the line connecting the locations of such plane crossings the Moon's **line of nodes** (**Figure 2.25**). Twice in its monthly orbit, the Moon touches the line of nodes: once when it crosses from above to below the plane of the Earth-Sun orbit, and once when it crosses from below to above. Only when the line of nodes aligns with either the full Moon or the new Moon can we see a lunar or solar eclipse. All of these factors make the

a.

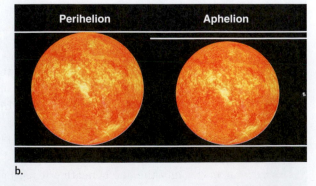

b.

Figure 2.23 Relative Sizes of the Moon and Sun on the Sky
Both the Moon and the Sun vary in their apparent angular size as seen from Earth, because the orbits of the Moon and Earth are not perfectly circular. **a.** *Perigee* and *apogee* refer to the closest and farthest distances between the Moon (or any object) and Earth. **b.** *Perihelion* and *aphelion* represent the closest and farthest distances between Earth (or any object) and the Sun.

Figure 2.24 Annular Eclipse
An annular eclipse appears when the angular size of the Moon is not as large as that of the Sun, because of the slightly elliptical orbits of the Moon and Earth.

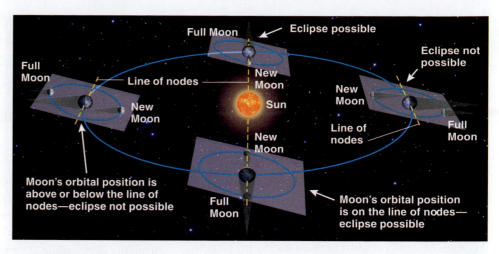

Figure 2.25 Line of Nodes
The Moon's orbit is tilted relative to the plane of Earth's orbit. The two orbits cross at the line of nodes. The position of the line of nodes relative to the line connecting Earth and the Sun rotates as Earth completes its orbit. Eclipses are possible only when full or new Moons align with the line of nodes.

occurrence of eclipses periodic, or cyclical—something ancient astronomers knew about 2,800 years ago.

Never, ever look directly at the Sun, even during a solar eclipse. Numerous guides available in print or on the Internet describe how to view a solar eclipse safely.

Section Summary

- As the Moon orbits Earth monthly, its phase changes because its position shifts relative to both Earth and the Sun. The phases cycle as follows: new Moon, waxing crescent, first quarter, waxing gibbous, full Moon, waning gibbous, third quarter, and waning crescent.
- Lunar eclipses occur when the Moon passes through Earth's shadow. Solar eclipses occur when a region of Earth's surface passes under the Moon's shadow. Both lunar and solar eclipses can be partial or total.
- The Moon's orbit is tilted by 5° with respect to Earth's orbit around the Sun, so eclipses occur only when Moon, Earth, and Sun are all in alignment, not at every full or new Moon.

CHECKPOINT Draw a diagram of the Moon, Earth, and Sun that illustrates why phases occur. Show how synchronization of the Moon's spin and its orbit allows the same side of the Moon to face toward Earth always.

Celestial Wanderers: The Motion of Planets

2.4 The fourth celestial motion to consider, the motion of the other planets in our Solar System, played a critical role in the history of astronomy and human thinking about our place in the Universe. Yet most citizens of the modern world know little about planetary motion. We are all familiar with the Sun, the Moon, and the stars, but few people can identify any of the planets on the sky, and fewer still can find and name them all.

Five planets (besides the one beneath our feet) can be seen with the naked eye: Mercury, Venus, Mars, Jupiter, and Saturn (**Figure 2.26**). In principle, it's easy to tell the difference between a star and a planet. "Planets look different," explains Carol Latta. "They don't twinkle. They also follow the ecliptic, just as the Sun and Moon do."

Stars are so far from Earth that they appear as nothing more than points of light on the sky. Turbulence in Earth's atmosphere bounces the light from the point-like stars,

totality The moment or duration of total concealment of the Sun or Moon during an eclipse.

apogee The distance of farthest approach of an object orbiting Earth.

perigee The distance of closest approach of an object orbiting Earth.

aphelion (pl. aphelia) The distance of farthest approach of an object orbiting the Sun.

perihelion (pl. perihelia) The distance of closest approach of an object orbiting the Sun.

annular eclipse A solar eclipse that occurs when the positions of Moon and Sun are such that the lunar disk does not fully block the solar disk.

partial solar eclipse An eclipse that occurs when a region of Earth's surface passes under the Moon's penumbral shadow.

line of nodes The line defined by the intersection of the Moon's orbital plane and Earth's orbital plane around the Sun.

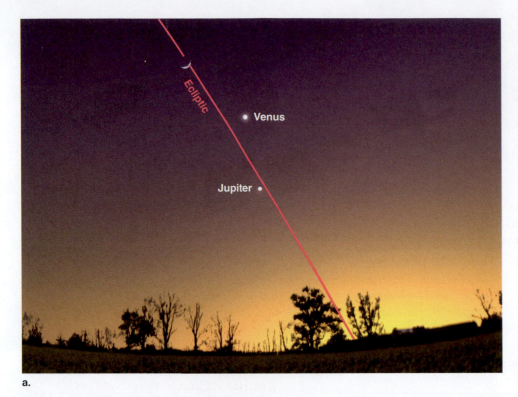

a.

b.

Figure 2.26 Planets Visible to the Naked Eye
Five planets are visible without a telescope; their locations, brightness, and lack of twinkling make them easily identifiable as planets, not stars. **a.** Since the orbits of all the planets and the Moon lie approximately in the same plane, these worlds are always found close to the ecliptic (Earth's orbital plane). Venus and Jupiter outshine any stars at their brightest, and their light is steady. **b.** Jupiter shines more brightly than a nearby star in the twilight sky.

scattering it from one moment to the next (as we will see in Chapter 4). Planets are so much closer to Earth than stars are that, in general, they appear as tiny disks instead of points on the sky. As individual rays of light from a planet pass through the atmosphere, they are scattered just as much as those from a star—but scattering for a point-like star moves the point very slightly (causing the familiar twinkling), whereas scattering a larger disk leaves you with a disk.

"Planet means 'wanderer' in ancient Greek, because wander across the sky is exactly what they do."

Basic Planetary Motion

"Planet means 'wanderer' in ancient Greek," says Latta, "because wander across the sky is exactly what they do. The planets move relative to the background stars. Stars don't change their position relative to each other on the sky—at least they don't do so in a way that can be seen with the naked eye. Planets do." To understand the motion of planets across the sky, says Latta, "imagine you went out every night at the same time and took a picture of the sky. If you compared the pictures over time, you would see the planets moving slowly eastward relative to the stars" (**Figure 2.27**). The eastward motion of the planets against the background stars (also called *prograde motion*) was something our ancestors noticed. They did not have artificial illumination to drown out the night sky or televisions, computers, or smartphones to distract them at night. Instead, they noticed what changed on the sky. So, night after night, month after month, they watched the planets' movement against the fixed stars.

As the planets march across the sky, they brighten and dim at different points in their motion. More importantly, they always appear close to the ecliptic. However, some planets can be seen anywhere along the ecliptic relative to the Sun, while other planets, such as

INTERACTIVE:
Retrograde Motion

retrograde motion Apparent motion of a planet in a direction opposite to its normal motion.

Time 1   Time 2   Time 3

Figure 2.27 Direct Planetary Motion
As the months pass, the planets move eastward against the background of the "fixed" stars, as Mars illustrates.

Mercury and Venus, can be seen only when they are close to the Sun, right before sunrise or right after sunset. Some planets move quickly across the starry field; some planets seem to crawl. Mars, for example, takes just a few months to cross a constellation, but Jupiter takes years to cover the same distance.

Retrograde Motion

As human cultures became more adept at charting the behavior of the planets, one facet of their motion became a puzzle that took more than 3,000 years to solve: a phenomenon called **retrograde motion**. "Let's say you are following Mars's motion," says Latta. "Night after night you will see it moving eastward against the background of the fixed stars. Then, all of a sudden, the eastward motion stops, and the planet reverses course to move westward. Then the westward motion stops, and the planet returns to an eastward path. The planet makes a kind of loop, and that loop is called retrograde motion" (**Figure 2.28**). These loops are easy to miss if you're not paying attention. "Mars's retrograde loop only covers a few degrees on the sky," says Latta. "That is only the width of a couple of Moons. You really have to look carefully over a long time to see it."

The appearance of the retrograde loops was an enigma for early astronomers. Why did the planets move like that? What model could explain the planetary motions and their strange retrograde loops? The previous sections introduced each basic celestial motion and then offered the modern explanation for that motion. In the case of planetary motion, we must wait a bit and let the story play itself out. Hidden within the mystery of planetary motion is a key to understanding the essential nature of science and how it came to play such a pivotal role in human culture.

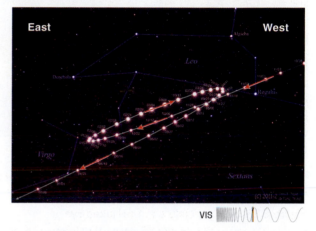

East   West

VIS

Figure 2.28 Retrograde Motion
The retrograde motion of a planet (here, Mars) occurs when it moves westward on the sky relative to the distant stars. Retrograde planetary motion baffled astronomers for centuries and became a major issue in developing Solar System models.

Section Summary

- Planets "wander" along the ecliptic relative to the fixed stars, moving eastward nightly but with different apparent speeds.
- Planets can be distinguished from stars because planets don't appear to twinkle.
- Retrograde motion occurs when planets stop their usual eastward motions against the fixed stars and move westward for a short period before returning to their eastward paths.

CHECKPOINT Describe the motion of planets against the fixed stars over the course of a year.

Stone and Myth: Astronomy Begins

2.5 The motions on the sky constitute the raw material of our astronomical heritage. Across the many thousands of years before written records (which begin about 2500 BCE), humans were just as filled with a sense of wonder by the Sun, Moon, stars, and planets as people are today. While the processes and methodology of science were still many millennia in the future, the basic need to make sense of the night sky led our ancestors to create explanations in the form of stories.

CAROL LATTA

It's hard not to catch Carol Latta's enthusiasm. After a career working in management at a major international corporation and raising two children, Latta found herself ready to take on something new. Returning to a local university, she rediscovered her enjoyment in learning science and then, unexpectedly, found a passion for astronomy. "I started out with math and science, then went to linguistics, and ended up working in business. It was only after I took my early retirement that I realized how much I missed math and science." Latta now conducts youth education through the Astronomy Section of the Rochester Academy of Science.

Figure 2.29 Astronomical Myths
Myth was one way that ancient cultures accounted for the phenomena they observed on the sky. In ancient Egyptian mythology, the falcon god, Horus, was said to be the sky. His right eye was believed to be the Sun and his left eye the Moon, and they traversed the sky when he flew across it.

> "As human cultures evolved, the sky itself became a stage for stories that related directly to human activities."

We often think of myths as false stories. But for the cultures that lived before history, **mythologies** were interconnected collections of stories that took the basic "data" of what was seen on the sky and transformed those observations into explanations (Figure 2.29).

Astronomical Myths and Astronomical Science

"The relationship between astronomy as a science and ancient mythologies is really fundamental," says Marcelo Gleiser, a cosmologist and professor at Dartmouth College. "At root they are both efforts to understand the Universe." Gleiser has studied everything from the physics of the early Universe to the origins of the first biologically active molecules. He has written widely on cosmology and its relation to myth. As Gleiser and others have shown, it is critical not to lose sight of the continuity between ancient mythology-based attempts to understand the world and the science practiced today by astronomers.

Myths often gave the sky special meaning and power. "As human cultures evolved," explains Gleiser, "the sky itself became a stage for stories that related directly to human activities. Constellations were attributed supernatural powers related to gods that ruled different aspects of life on Earth. There were gods of war in the sky and gods of love."

"Of course, the big difference between science and myth," says Gleiser, "is that myth focuses on using supernatural means of explanation. Science focuses on natural explanations using rationally based cause and effect. Still, the questions they both address and the motivations for asking those questions can have many similarities."

Megaliths and Early Astronomy

Along with telling mythological stories, early cultures were watching the sky and translating their understanding into concrete forms as well. Artifacts from the Americas, Egypt, the Middle East, India, China, and other regions provide evidence that many ancient peoples systematically observed astronomical phenomena. Across the world, there exist hundreds of giant stone monuments called **megaliths**, some of which date back 5,000 years or more (Figure 2.30a). Most importantly for our understanding of astronomy's history, at least some of these prehistoric monuments align with celestial cycles.

Newgrange, the megalith in Ireland shown in Chapter 1, aligns with the rising Sun during the winter solstice. The most famous megalith site is the enigmatic Stonehenge in England (Figure 2.30b), which consists of a series of circular rings. The outer circle is

Figure 2.30 Early "Observatories"
Many cultures left behind megaliths, such as **a.** these on the Isle of Lewis, Scotland, that were symbols of homage in some cases and served astronomical functions in others.
b. Stonehenge, in England, served religious purposes but also marked the motions of the sky, which are vital to agrarian cultures that must accurately gauge when to plant and harvest.

a.

b.

constructed from stone slabs weighing up to 50 metric tons; each had to be dragged from 40 km away. These stones are not set at just any location along the ring. Instead, their placement shows alignments with the motion of the Sun and the Moon. The main axis of the monument, for example, faces the horizon where the Sun rises on the summer solstice, the longest day of the year.

Stonehenge played a practical role in its society. "Stonehenge was built by people who were part of an agrarian society," says Gleiser. "By knowing how the skies were moving, you actually knew when to plant, you knew when to harvest, when the rains would come, when the cold would come. The builders of Stonehenge wanted to create the clock that would reproduce the motions of the sky." But to think of Stonehenge as only a clock is probably a mistake. "The location of burial mounds around the Stonehenge site also makes it clear that the site was used for religious purposes," says Gleiser. "By mimicking the motions of the skies, which were the realm of the gods, Stonehenge's creators were bringing that realm down to Earth."

Astronomy Culture and History

Although the evolution of humans can be traced back at least 6 million years, only in the relatively recent past did human cultures begin to develop the technologies that led to modern societies. Between 70,000 and 50,000 years ago, a flowering of consciousness occurred, in which toolmaking and the production of art exploded.

This earliest period of human culture occurred at the end of the **Paleolithic period**, when Earth's climate was locked in the deep freeze of an ice age. But about 10,000 BCE, at the beginning of the **Neolithic period**, conditions began to shift toward the warmer, wetter world we inhabit today.

The invention of farming 9,000 years ago changed everything. Agriculture allows surpluses to develop that need to be stored and protected. As a result, larger permanent settlements tend to grow and diversify. Neolithic-period villages gave way to the first true cities, such as Çatal Hüyük in eastern Turkey (7000 BCE) and Sumer in southern Iraq (5000 BCE).

Life in these cities depended on their inhabitants developing specialized skills as leatherworkers or metalsmiths. A priestly class also arose that looked after the supernatural aspects of life, which often meant attending to the night sky. Priests became the first true astronomers as they began keeping long-term records of lunar phases, lunar and solar eclipses, and planetary motions. The longevity of city-building civilizations meant that these records could be passed from one generation to the next—a critical step in the development of astronomy. The planet Saturn, for example, completes its motion against the fixed background of stars over the course of almost 30 years. Without long-term records, observers of the sky could not recognize that pattern.

One of the most important city-building civilizations for astronomy was the Babylonian Empire, centered in what is today southern Iraq. Babylon became the hub of a great civilization in 1700 BCE that lasted at least 1,000 years. It was the Babylonians who first began to maintain detailed records of celestial motions on small slabs of clay that were called cuneiform tablets (**Figure 2.31**). By cataloging both stars and constellations, the Babylonians developed schemes for predicting the risings and settings of the Sun and planets, as well as the length of the day.

Most cuneiform tablets, however, appear to be concerned more with astrology than with astronomy. But the Babylonians applied their astrological observations in new directions, including the use of mathematics to calculate variations in the length of day across the year. "The Babylonians were the first to have a preoccupation with mapping the motions of planets in the sky," says Gleiser. "The skies were a kind of open book to them that only certain people would be able to interpret. Those people were priests who were, in a sense, the precursors of scientists and astronomers today."

Figure 2.31 Cuneiform Tablets
The Babylonians recorded planetary and stellar motions on cuneiform tablets, showing that this civilization recognized the periodic nature of astronomical events. The Babylonians also used their observations to make astrological predictions for the king.

MARCELO GLEISER

The tragic loss of his mother when he was just a child left Marcelo Gleiser searching for answers even as a boy. Growing up in Rio de Janeiro, Gleiser found those answers in tales of the supernatural. "I was really fascinated with the possibilities that there might be greater intelligences living somewhere in the vastness of space," he says.

Gleiser earned his PhD in theoretical astrophysics from King's College London and eventually joined the faculty of Dartmouth College. One of the most popular science writers in Brazil, Gleiser calls science ". . . a very human creation and the product of the same curiosity that has moved our collective imagination for thousands of years."

mythology A collection of stories used to explain phenomena whose physical basis is not understood.

megalith A large stone used as a monument or part of a monument.

Paleolithic period The period extending from 2.6 million years ago (characterized by the earliest known use of stone tools) to 10,000 years ago.

Neolithic period The period extending from about 10,000 BCE to between 4500 and 2000 BCE, when farming was introduced.

Centuries later and 1,500 miles northwest of Babylon, philosophers in the city-states of Greece would explain the skies in the rational language of mathematics and natural cause and effect rather than divine intervention.

Section Summary

- Myth explains observations on the sky in terms of supernatural powers of various gods; science looks for natural explanations.
- Megaliths like Stonehenge reflect both astronomical and religious purposes of ancient peoples.
- With the development of farming and the ability to store food came the creation of specialization within societies. Priests often were trackers of the motions of celestial objects.
- The Babylonians were the first to create long-term records of the motions of planets.

CHECKPOINT How did the invention of agriculture change the relationship between human cultures and the sky?

The Greek Advancement of Science

2.6 "It all started with the Greeks," says Marcelo Gleiser. "They were the ones who discovered a new way of approaching nature. For the first time, rational arguments were brought to bear, rather than mythic or supernatural explanations. It was something very different and very new that began around 650 BCE."

> "It all started with the Greeks. They were the ones who discovered a new way of approaching nature."

Advances in Science

In cities dotting the Greek peninsula and the islands of the Aegean, a radical new culture arose during the last millennium before the common era. Thales of Miletus (624–547 BCE), a Greek citizen who lived in Turkey, was a forerunner in rejecting supernatural explanations of phenomena and claimed that the world could be understood through reason alone. In the centuries that followed, Greek philosophers took up Thales's approach and began constructing rational models to explain the natural world, including the astronomical motions of the Sun, Moon, and planets.

One of the most influential Greek mathematician-philosophers was Pythagoras (570–500 BCE), who developed a mathematical way of understanding music and harmony. Pythagoras and his followers (called the Pythagorean brotherhood) extended this approach and saw mathematics as a language for describing the underlying patterns they saw on Earth and on the sky.

Plato (428–347 BCE), another influential Greek philosopher, continued the Pythagorean enchantment with numbers and mathematics. Above his famous school in Athens (called the Academy), he had inscribed the words "Let None Who Is Ignorant of Geometry Enter." Plato argued that behind the appearances of the world lies a more perfect mathematical world that acts as a kind of blueprint for all we see. The world we experience is just a corrupted version of this ideal world of mathematical forms. As part of this doctrine of ideals, Plato held that all the celestial motions must be manifestations of a perfect underlying mathematical order. Plato famously asked his students to "save the appearances," meaning to develop a mathematical model of planetary motion that could account for what appears on the sky from the perspective of humans on Earth, such as the observed retrograde loops made by planets.

Aristotle and the Geocentric Model

Plato's most famous student was Aristotle (384–322 BCE), whose ideas dominated Western thinking about nature for more than 1,500 years. Aristotle was not a scientist in the modern sense of the word, because he never sought to validate his theories through rigorous experimentation. His mode of reasoning was to begin with a set of core beliefs that he

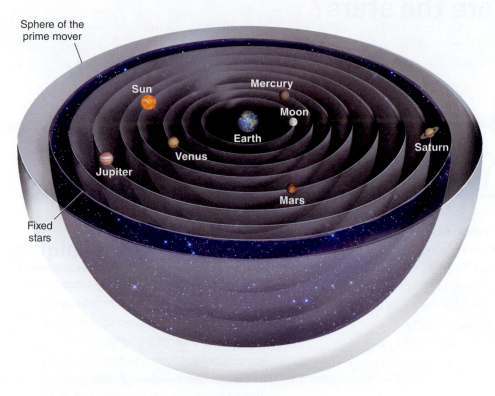

Sphere of the prime mover

Sun

Mercury

Moon

Earth

Venus

Saturn

Jupiter

Mars

Fixed stars

Figure 2.32 Aristotle's Geocentric Model of the Universe
A simplified rendering of Aristotle's model, which placed Earth at the center of the Universe. All other heavenly bodies were attached to crystalline concentric spheres rotating at different uniform speeds around Earth.

thought were self-evident (clearly true) and combine them with select observations. In this way, Aristotle attempted to deduce how the world worked.

In the arenas of physics and astronomy, Aristotle argued that the cosmos is divided into two domains. First there is Earth, which he claimed is spherical and the center of all the cosmos. Surrounding Earth is the celestial domain of the planets and stars. To account for celestial motions, Aristotle built on ideas proposed by Eudoxus (409–356 BCE), another of Plato's students, arguing that 55 nested crystalline spheres compose the heavens. Each sphere centers on Earth and rotates with a different speed (**Figure 2.32**). The Moon, Sun, planets, and stars are each attached to one of these spheres. The rotation of the crystalline spheres creates the appearance of celestial motion across the sky on Earth.

In Aristotle's model, the terrestrial, or "sublunar," domain is the realm of change and decay. The realm of the crystalline spheres, in contrast, is unchanging and perfect. Each crystalline sphere guides the planets through mathematically perfect circular motions at a perfectly constant, or uniform, speed. This emphasis on the perfection of the celestial realm, including the constant circular motion of the planets on their spheres, would play an important role in the evolution of astronomical thinking.

Many Greek astronomical thinkers accepted this kind of **geocentric model**, in which the Sun and planets orbit Earth. It is a tribute to the imagination, creativity, and ingenuity of Greek thinkers that Aristarchus (310–230 BCE), who lived after Aristotle, proposed a **heliocentric model**, in which Earth is just another planet orbiting the Sun (the center of the Universe). Aristarchus's model was not widely accepted, however. His opponents argued that if Earth really moves through space, then there should be a great wind due to the motion. It is worth noting that Aristarchus also proposed a method to estimate the distance to the Sun relative to the Earth–Moon distance.

geocentric model A model stating that Earth is the body around which all other Solar System objects orbit.

heliocentric model A model stating that the Sun is the body around which all other Solar System objects orbit.

How far away are the stars?

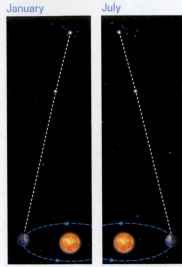

January   July

Parallax

observation

Ancient Greeks realized that if Earth orbits the Sun, as **Aristarchus** (3rd century BCE) suggested, the stars' relative positions would change over a year: the stellar parallax effect. Failing to observe parallax, however, most Greek philosophers concluded that Earth is fixed at the center of the Universe. They rejected the possibility that the stars are too distant for us to detect parallax.

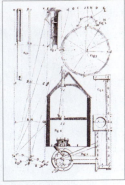

Robert Hooke's chimney telescope (ca. 1669)

techniques

The geocentric model was not challenged until the 16th century. **Galileo**, a firm believer in an Earth that moves, suggested observational strategies to search for parallax with an appropriate telescope—such as the chimney telescope Robert Hooke built. Unfortunately, Hooke's telescope was too crude to detect parallax.

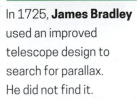

observations

In 1725, **James Bradley** used an improved telescope design to search for parallax. He did not find it.

In 1779, **William Herschel** tried another of Galileo's suggestions: to seek changes in the separation between paired stars as evidence of parallax. He too failed to measure it (but found that many star pairs are gravitationally bound systems).

confirmed by observation

In 1838, **Friedrich Bessel** measured the parallax of 61 Cygni, in the constellation Cygnus, as less than 0.000087 degree. He then estimated the distance to be 10.4 ly—near today's accepted value, 11.3 ly. Bessel thus showed that Earth orbits the Sun and that the stars are incredibly far. The Gaia mission measures parallaxes less than $\frac{1}{10,000}$ the size Bessel could. Gaia has calculated distances to more than 1.3 billion stars!

Bradley's zenith telescope (1727)

Herschel's reflecting telescope

Gaia space observatory (launched 2013)

Another argument against the heliocentric model was that if Earth orbits the Sun, then the changing view of the stars from one side of the orbit to the other would lead to changes in stellar position. This effect, called **parallax**, can be precisely defined as the apparent displacement of an observed object due to a change in the position of the observer. You can most easily see this effect by holding your finger up to your nose and closing only the left eye and then closing only the right eye. When you switch between views, notice that the positions of distant objects relative to your finger change depending on which eye is open. The Greeks reasoned that if Earth orbits the Sun, the position of nearby stars relative to the farther ones should shift as Earth moves from one side of its orbit to the other. Since they did not observe any parallax, they believed that Earth must be stationary (**Figure 2.33**). We now recognize that the stars are too distant for us to observe the parallax without a telescope, but the true distances to the stars were too large for the Greeks to imagine (see **Anatomy of a Discovery**).

Mathematical Advances

The reliance on mathematics and geometry enabled the Greeks to estimate the sizes of the Earth, Moon, Sun, and Solar System. Some of the estimates were wildly wrong; for example, the Greeks believed that the distance to the last crystalline sphere, which held the stars, was smaller than what we now know to be the size of the Solar System. Still, in some cases their reasoning was remarkably sophisticated. Eratosthenes (276–195 BCE) was a philosopher, an astronomer, and the head of the great library of Alexandria, Egypt. He used the length of midday shadows observed at two cities some 500 miles apart to infer Earth's size. Using only the record of shadow lengths and geometry, he deduced a radius of Earth that was off by only 14 percent of the actual value.

Greek astronomers built on and improved the observations of the Babylonians. One of the most famous Greek astronomers, Hipparchus (190–120 BCE), is credited with the development of trigonometry, which he may have used to construct the first reliable method for predicting solar eclipses. Hipparchus's most important contribution, however, was the development of the first comprehensive star catalog in the Western world. His catalog included a scheme for describing stellar brightness called the magnitude system, which remains in use today. Hipparchus's ideas and influence in astronomy were paramount in the ancient world and lasted for more than 300 years.

At the Center of Everything: The Ptolemaic Universe

Hipparchus's influence was eclipsed only by the appearance of the great Claudius Ptolemy (90–168 CE), whose efforts gave astronomers a new functional explanation of planetary motions (including retrograde loops). Ptolemy's most important legacy was the fulfillment of Plato's demand to create a geometrically based model of the Universe using perfect, uniform circular motion to recover the behavior seen in the night sky.

Like Aristotle's model, Ptolemy's model was geocentric: it placed Earth at the center of the Solar System. How can a model so wrong be considered a triumph? Ptolemy's model worked. It did a good job of predicting all the regular celestial motions, and prediction is a crucial feature of any model. So powerful and useful was Ptolemy's model that his book *Mathematical Syntaxis* (written in 150 CE) remained the standard textbook in astronomy for nearly 1,400 years. The Arab astronomers who carried science forward while western Europe was plunged into the Dark Ages called Ptolemy's book *Almagest*, or "The Greatest."

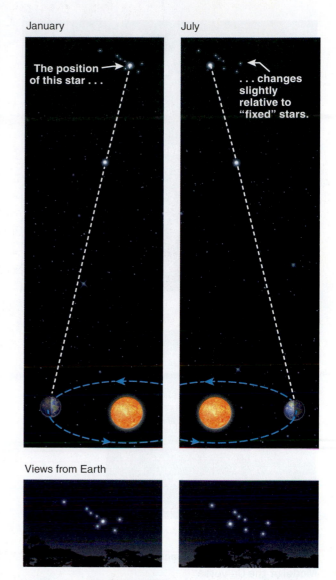

January   July

The position of this star . . .

. . . changes slightly relative to "fixed" stars.

Views from Earth

Figure 2.33 Stellar Parallax
Parallax occurs when the apparent position of a star changes as Earth moves from one side of the Sun to the other. (This effect is judged against "background" stars, whose positions do not noticeably change relative to one another). Since this effect could not be observed with the technology of the time, the ancient Greeks reasoned that Earth is stationary at the center of the Universe.

parallax A displacement or difference in the apparent position of an object viewed along two different lines of sight.

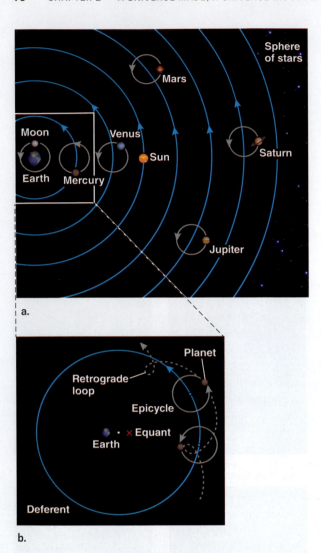

Figure 2.34 Ptolemy's Geocentric Model of the Universe

Ptolemy's model was based on the incorrect assumption that the Universe is geocentric. It included elaborate constructs such as **a.** epicycles, small circular paths on which planets move as they proceed on larger orbits, and **b.** an equant, a point displaced from the center of the planet's orbit. The planets move around a point midway between the Earth and the equant. Nonetheless, this model did an admirable job of predicting the observed motions of the Sun and planets.

epicycle A secondary orbit whose center point orbits Earth; devised by Ptolemy to explain retrograde motion.

equant A point displaced from the center of a planet's orbit around which the planet displays uniform motion.

Occam's razor A principle stating that among competing hypotheses, the simplest one should be selected.

How did Ptolemy manage such accuracy in predicting planetary and other celestial motions with a model that was completely wrong? Building upon the work of Eudoxus, Aristotle, and others, Ptolemy imagined the Sun, Moon, and planets all to be orbiting around Earth (**Figure 2.34a**). To account for retrograde motion and the variable speeds that planets manifest, Ptolemy's model included special mathematical devices. In his model, for example, the planets do not move directly in *deferents*—that is, circular orbits about Earth. Instead, each planet moves with uniform speed on a smaller circle called an **epicycle**, which creates the appearance of retrograde motions. The *center of the epicycle*, not each planet itself, then moves with uniform speed around Earth. When a planet's motion on an epicycle is in the same direction as the planet's orbital motion, it would appear to an Earth-based observer to move eastward on the sky. But when the planet loops around to the other side of the epicycle, it would appear to move westward on the sky. The combination of motions—the planet on its epicycle and the epicycle orbiting Earth—allowed Ptolemy's model to recover retrograde motion with uniform circular motion and to predict very accurately how planets move through the sky.

Ptolemy needed more than single epicycles to get his model to match observations. In some cases, smaller epicycles were required in addition to one main epicycle. Getting the variable speed of the planets in the night sky correct also required that Ptolemy assume Earth is slightly off center from the planetary orbits. He then had to adjust every planet's uniform motion relative to this new point, called the **equant** (**Figure 2.34b**).

A principle of modern science called **Occam's razor** states that explanations for natural phenomena should be as simple as possible when compared with other models. By today's understanding, Ptolemy's model, with its epicycles and the equant, clearly violates Occam's razor. It is important to recall, however, that astronomers of Ptolemy's time (and for the next 14 centuries) were under the sway of philosophical principles such as the demand for the heavens to be "perfect," including perfect circular motion. It took great intellectual courage (and some personal risk) to reject the geocentric model eventually.

Ptolemy can be celebrated as a culmination of the Greek genius, with its demand to know the world through reason and mathematics and its rejection of the supernatural. The Greek era changed human history and led directly to the flourishing science of our age. What will people say of our era 1,000 years from now? What will we leave behind that will be so lasting and of such consequence?

Section Summary

- Ancient Greeks approached the world from the perspective of reason and pursued mathematics as the key to understanding the sky's patterns. Plato believed that, unlike Earth, celestial objects and motion are perfect and their motions should be based on perfect mathematical order.
- Most Greek philosophers believed that Earth is the center of the cosmos and subscribed to Aristotle's geocentric model of the heavens, based on the existence of crystalline spheres rotating around Earth.
- Eratosthenes used the geometry of shadow lengths to measure Earth's radius fairly accurately. Hipparchus is credited with creating the first star catalog based on the brightness of stars.
- Ptolemy's geocentric model, though wrong, was quite successful at predicting motions of planets and dominated astronomy for more than 1,400 years. It required a complicated system of epicycles and equants to match the motions of the planets, including the explanation of retrograde motion.

CHECKPOINT Describe how Eratosthenes' use of the lengths of shadows to measure Earth's radius was an example of the ancient Greeks' emphasis on reason to explain the Universe.

CHAPTER SUMMARY

2.1 An Old Obsession

For thousands of years, human beings have looked to the sky and tried to explain the phenomena they witnessed there.

2.2 Dance of Night and Day: Basic Motions of the Sky

Without a telescope, accurate "naked-eye" observations can reveal patterns in the motion of objects on the sky. The Sun, Moon, stars, and planets appear to move from east to west each day (and night), with variations based on the observer's latitude on Earth. Earth's rotation creates this apparent motion. The celestial sphere is an imaginary transparent globe, surrounding Earth, on which the stars and planets appear to rest. Its equator and poles are extensions of Earth's equator and poles. An observer at any position on Earth can see only half of the celestial sphere. To Northern Hemisphere observers, the celestial sphere appears to turn around the star Polaris each day. The celestial poles appear at an altitude equal to the observer's latitude. Over the course of a year, the Sun's observed path changes, reaching higher altitudes in summer (with its highest point on the summer solstice) than in winter (with its lowest point on the winter solstice). Correspondingly, the Sun's rising and setting points move southward from the summer solstice to the winter solstice and then northward again. Seasons are the result of Earth's tilt, which makes for longer days and more direct solar radiation in summer. Although Earth's orbit is slightly elliptical, its varying distance from the Sun is not the cause of the seasons. A constellation is a recognizable group of stars and the area encompassing them. Which constellations are seen depends on the observer's location on Earth and Earth's position relative to the Sun. Only constellations on Earth's night side can be observed. Circumpolar constellations—close to one of the poles—are visible all year. The ecliptic is the Sun's apparent annual path through the sky.

2.3 Monthly Changes of the Moon

The Moon shows phases as a result of its 29.5-day orbit around Earth. The Moon reflects light from the Sun; the phase depends on how much of its illuminated half we can see. The full Moon occurs when the Moon is opposite the Sun and its entire face is illuminated by sunlight; the new Moon occurs when the Moon is between Earth and the Sun. In order, the phases are new Moon, waxing crescent, first quarter, waxing gibbous, full Moon, waning gibbous, third quarter, and waning crescent. From one night to the next, the Moon moves eastward against the background stars. The same hemisphere "face" of the Moon always points toward us because its spin period and orbital period are synchronized.

2.4 Celestial Wanderers: The Motion of Planets

Like the Moon, planets move eastward against the stars. However, they sometimes appear to be moving in the opposite direction—a phenomenon called retrograde motion.

2.5 Stone and Myth: Astronomy Begins

Before science provided testable models, people developed elaborate myths about the supernatural origins of celestial phenomena, but both myth and science seek to make sense of observations of the Universe. Early astronomers created megaliths such as Stonehenge to record and use motions of sky objects, and those monuments may have served ceremonial purposes. Knowing when to plant was an important practical use. The study of astronomy grew from the birth of civilizations, especially in Babylon. Record keeping enabled knowledge of events spanning many years.

2.6 The Greek Advancement of Science

The Greeks contributed to astronomy in many ways, such as the practice of mathematical modeling. The most noteworthy developments include Pythagoras's geometry, Aristotle's geocentric model of the Universe involving crystalline spheres, Eratosthenes' measurement of Earth's radius, and Hipparchus's invention of trigonometry and a star magnitude chart. Ptolemy built on their work to develop a detailed geocentric model of the Universe that was accepted for 1,400 years, despite being fundamentally wrong in placing Earth at the center. Ptolemy's model went a long way toward incorporating the retrograde motion of planets and was an important example of the concept of interpreting the world through reason.

QUESTIONS AND PROBLEMS

Narrow It Down: Multiple-Choice Questions

1. How does the altitude of the Sun at noon on the same day in the Northern Hemisphere's summer compare for two observers at latitudes 12° north and 54° north, respectively?
   a. It is the same for both observers because they are in the same hemisphere, experiencing summer.
   b. It cannot be determined without knowing their longitudes.
   c. It is 42° higher for the observer at 12° north because of the difference in latitude.
   d. It is 42° higher for the observer at 54° north because of the difference in latitude.
   e. The relative altitude of the Sun cannot be determined for the two locations without knowing the exact date.

2. Which observers on Earth can see Polaris on a clear night?
   a. all observers on Earth
   b. only observers above the Arctic Circle
   c. only observers in the Western Hemisphere
   d. only observers in the Southern Hemisphere
   e. only observers in the Northern Hemisphere

3. The small-angle formula relates the distance and diameter of an object to its observed angular size. From your understanding of the small-angle formula, or from personal observations of the world around you, comparing two spherical objects with different diameters, which of these statements is always true?
   a. At the same distance, the larger and smaller objects will appear the same size.
   b. At the same distance, the smaller object will appear larger than the other one.
   c. The smaller, farther object will appear smaller than the other one.
   d. The smaller, nearer object will appear larger than the other one.
   e. The larger, nearer object will appear smaller than the other one.

4. The Sun is highest on the sky at noon on
   a. the winter solstice.
   b. the spring equinox.
   c. the summer solstice.
   d. the autumn equinox.
   e. any day, because the Sun reaches the same altitude daily regardless of season.

5. Which statement about constellations is true?
   a. Any group of stars can be called a constellation.
   b. A constellation includes a group of stars within specific boundaries in the same region of the sky.
   c. All constellations visible in the Northern Hemisphere are circumpolar.
   d. All of the stars within a constellation are the same distance from Earth.
   e. All of the stars within a constellation are about the same brightness.

6. Which statement about constellations is true?
   a. Earth's orbital and rotational motions determine which constellations are visible at any given time.
   b. All 88 constellations are visible from any location on Earth at some time during the year.
   c. All constellations are represented in the zodiac.
   d. All constellations are circumpolar.
   e. New constellations are discovered from time to time.

7. Warmer summertime temperatures in the Northern Hemisphere are due partly to
   a. longer days.
   b. a more oblique angle of the Sun's rays.
   c. Earth's being closer to the Sun in summer.
   d. the Sun radiating more energy in summer.
   e. the tilt of the Northern Hemisphere away from the Sun.

8. Which statement about Moon phases is true?
   a. In waxing phases, the lit portion of the Moon faces the eastern horizon.
   b. The new Moon has its whole face illuminated as seen from Earth.
   c. The Moon rises at sunset every day.
   d. The waning gibbous phase follows the full Moon.
   e. In waning phases, the lit portion of the Moon faces the western horizon.

9. You observe the full Moon just rising in the east at 6:00 P.M. At what time will you observe the rising of the third-quarter Moon later that month?
   a. sunrise (about 6:00 A.M.)
   b. noon (about 12:00 P.M.)
   c. midnight (about 12:00 A.M.)
   d. midmorning (about 9:00 A.M.)
   e. midafternoon (about 3:00 P.M.)

10. Synodic and sidereal months differ because of
    a. the Moon's orbit.
    b. Earth's orbit.
    c. the Sun's orbit.
    d. the fact that Earth's year is not exactly 365 days.
    e. the different number of days in each calendar month.

11. The direct (and most typical) seasonal motion of the planets as observed from Earth is
    a. west to east with respect to the background stars.
    b. east to west with respect to the background stars.
    c. east to west at the same rate as the background stars.
    d. north to south with respect to the background stars.
    e. south to north with respect to the background stars.

12. Using only Stonehenge to calibrate astronomical motions, early people would *not* have been able to tell which of the following? Choose all that apply.
    a. when to plant
    b. when the longest day had come
    c. when Mars would appear on the sky
    d. whether Venus would be visible at night
    e. when winter would begin

13. Which Greek philosopher is most closely associated with first rejecting supernatural explanations and arguing that reason alone can explain phenomena?
    a. Thales
    b. Aristotle
    c. Socrates
    d. Hipparchus
    e. Pythagoras

14. For what significant contribution is Eratosthenes most famous?
    a. inventing trigonometry
    b. measuring Earth's radius
    c. constructing the geocentric model of the Universe
    d. creating the first catalog of bright stars
    e. defining the four basic elements

15. Which of the following statements about parallax is true?
    a. Our two eyes enable us to use parallax to determine distances to objects.
    b. Earth's orbit provides astronomers an opportunity to use parallax.
    c. Most stars do not appear to shift position, because they are too far away for parallax to be observed.
    d. If all stars were on the surface of a celestial sphere and located at the same distance, we would observe no stellar parallax.
    e. all of the above

16. Which of the following was/were elements of Ptolemy's geocentric model? Choose all that apply.
    a. an explanation of why we don't feel a constant strong wind on Earth
    b. epicycles
    c. an explanation of retrograde motion
    d. the belief that Mercury and Venus orbit the Sun
    e. the assumption that all planetary orbits are ellipses

17. A lunar eclipse can occur at which Moon phase(s)? Choose all that apply.
    a. new Moon
    b. first quarter
    c. full Moon
    d. third quarter
    e. waxing gibbous

18. Which of the following statements about solar eclipses is/are correct? Choose all that apply.
    a. A total eclipse is possible because the Sun and Moon sometimes appear to be identical in size.
    b. Solar eclipses occur at full Moon only.
    c. Not all solar eclipses achieve totality.
    d. The shadow cast on Earth by the Moon is always the same size.
    e. A solar eclipse is visible to everyone on Earth equally.

19. True/False: Total eclipses can occur only when both the Moon and the Sun simultaneously pass through the line of nodes.

20. Lunar and solar eclipses are possible only when both the Sun and Moon are at specific positions relative to Earth. How many times each month does this alignment occur? (Ignore the inclination of the Moon's orbit.)
    a. one
    b. two
    c. three
    d. four
    e. It varies widely.

To the Point: Qualitative and Discussion Questions

21. What are some factors that led to the advancement of human civilization and culture?

22. What likely uses did the ancients have for megaliths?

23. Name at least one important contribution associated with each of the following Greek thinkers: Thales of Miletus, Pythagoras, Plato, Aristotle, Eudoxus, Aristarchus, Eratosthenes, Hipparchus.

24. If the Moon crosses the meridian at midnight, what phase must the Moon be in?

25. Suppose that a month ago you saw the star Betelgeuse in the constellation Orion just rising at the eastern horizon at 8:00 P.M. Describe its position at the same time today.

26. Define retrograde motion and explain how Ptolemy's model represented it.

27. Define the celestial sphere. How is it a useful (though imaginary) tool?

28. From what location on Earth can you see every part of the celestial sphere over the course of the year?

29. How does the Sun's path across the sky differ in summer versus winter?

30. What is an analemma, and what gives it its characteristic shape?

31. If Earth's axis had no tilt relative to the plane of its orbit, how would the seasons differ from those we experience today?

32. If Earth's axis had a more significant tilt relative to the plane of its orbit, how would the seasons differ from those we experience today?

33. Explain the difference between sidereal and synodic months.

34. Describe and compare the models of the Universe defined by Aristotle and Ptolemy.

35. How did Aristotle use the lack of measurable parallax to argue against the heliocentric model championed by Aristarchus? Comment on the flaw in Aristotle's logic.

Going Further: Quantitative Questions

36. How many arcseconds are there in 4°?

37. How many degrees are there between the horizon and the zenith?

38. From your location, the Sun is at an altitude of 80° as it crosses the meridian on the summer solstice. Describe the Sun's altitude as it crosses the meridian 1 month later.

39. A star is at the zenith for an observer at latitude 44° north. What is that star's declination on the celestial sphere? (Astronomers use a "+" before the number for north declination and a "−" before the number for south declination.)

40. You observe the Moon's position on the sky at the same time on two consecutive days. Across how many degrees of sky has its position moved? (Hint: The Moon moves 360° during its full orbit.)

41. How many days are there between new Moon and full Moon?

42. You observe Mars with an angular diameter of 18″. What is its distance from Earth in kilometers? (Hint: The diameter of Mars is 6,794 km.)

"MOST PEOPLE COULD NOT ENVISION EARTH SPINNING AROUND, BECAUSE IT SEEMS QUITE SOLID TO US"

3

A Universe of Universal Laws

FROM THE COPERNICAN REVOLUTION TO NEWTON'S GRAVITY

Getting Past Ptolemy: The Copernican Revolution

3.1 Fourteen hundred years is a long time—enough for more than 50 generations to come and go. But for more than 1,400 years—from about 200 CE to 1600 CE—Ptolemy's geocentric vision of the Solar System reigned supreme, though it was continually tweaked by Arab astronomers. Until the 16th century, nearly everyone believed that Earth was fixed at the center not only of the Solar System but of the entire Universe. "Most people could not envision Earth spinning around, because it seems quite solid to us," says Owen Gingerich, one of the foremost historians of astronomy and an expert on the Copernican revolution. Today we know Earth rotates on its axis but the Sun is at the center of the Solar System. How did such a radical change in perspective come about?

Setting the Stage

In the fourteen centuries after Ptolemy, the Western world changed dramatically. The might of the Roman Empire, so vital during Ptolemy's day, had splintered. In western Europe, the Dark Ages had slowed education to a crawl. Eastern Europe fared better, but by 700 CE the torch of scholarship and learning had been passed to the vibrant empires of the Middle East.

In places such as Persia (modern Iran), scientific learning continued (**Figure 3.1**). From 500 CE onward, astronomers were particularly energetic. They improved predictions for eclipses, made accurate measurements of Earth's tilt, and produced highly detailed star maps. The legacy of this dynamism is found in the Arabic names we still use for many

← The Moon as seen through an amateur astronomer's telescope. The craters and dark features (called maria) reveal our satellite to be a body scarred by many impacts over time.

Figure 3.1 Early Middle Eastern Astronomers
Middle Eastern scientists, such as those depicted here working in Turkey, kept the flame of astronomical knowledge burning during the Dark Ages.

"People naturally asked, 'If it's spinning, why don't we fly off?'"

of the brightest stars, including Rigel, Aldebaran, and Deneb. But Ptolemy's geocentric model still held sway.

Ptolemy's work earned deep respect because it explained so much of the night sky's behavior. Ptolemy's astronomy made accurate predictions in accord with an increasing body of observations, all made with the naked eye. Eventually, the powerful Islamic empires waned and western Europe awoke from the Dark Ages. Scholars from Spain to Poland rediscovered Ptolemaic astronomy and classical Greek thinkers such as Plato and Aristotle. This rediscovery occurred as Arab scholars translated Greek works, which were later translated into Latin. By 1400, Ptolemy's view was the dominant astronomical model across Europe.

In Europe, Ptolemy's work held sway because of its predictive power and the worldview it supported. The Catholic Church had dominated there for centuries. Catholic scholars fused interpretations of the Bible with the Earth-centered astronomy of Ptolemy and the Earth-centered physics of Aristotle. Aristotle's physics—with its distinction between the lower, imperfect "sublunar" ("below the Moon") realm and the higher, perfect celestial realm—also fit neatly into the church's vision of a perfect God in a perfect Heaven.

Arguing against a geocentric Universe could lead to charges of heresy. Given the absolute authority of the church and its harsh punishments for heretics (burning at the stake was a common sentence), challenging Ptolemaic astronomy was not something to take on lightly.

A Simpler Plan: Copernicus's Heliocentric Universe

"It was so much against what seemed to be common sense," says Gingerich. He is speaking about the famous book *De revolutionibus orbium coelestium* (*On the Revolutions of the Celestial Spheres*), by Nicolaus Copernicus, who published it just before his death in 1543. The "revolutions" Copernicus wrote about were the orbits of planets, but by changing humanity's conception of the Universe, the book was also revolutionary in a political sense.

Copernicus was born in Poland in 1473, during the rise of the *Renaissance*, a remarkable era of learning (**Figure 3.2a**). He disagreed with two features of the Ptolemaic model (**Figure 3.2b**). First, Copernicus was sure that the Sun, not Earth, holds the central position in the Solar System. Second, Copernicus firmly believed in uniform circular motion. Recall from Chapter 2 that to account for the varying speeds of planets across the sky, Ptolemy's model made planetary motion constant relative to an arbitrary point off center from the orbit. This approach was aesthetically offensive to Copernicus, so he set out to develop a full mathematical account of a heliocentric Solar System built on circles and uniform circular motion.

Thus one of the greatest scientific revolutions in history was driven more by aesthetics than by data and predictions. Copernicus attempted to reorder the Solar System on the basis of planets moving on circular paths. As it turned out, Copernicus's new model did not predict the motion of planets much better than Ptolemy's model had—but it told a completely different story about the Solar System's architecture.

Copernicus's model had three critical features. The first and most obvious is that it placed the Sun at the center of the "planetary system." All planets, including Earth, moved around the Sun. (Copernicus left the Moon in orbit about Earth.) Despite his desire for simplicity, Copernicus was forced to shift the center point of the orbits away from the Sun and add epicycles—though, as we will see, he did not need these in order to make the model reproduce the all-important retrograde motions.

The second critical feature of Copernicus's model was the daily rotation of Earth. Explains Gingerich, "People naturally asked, 'If it's spinning, why don't we fly off?'" Since Copernicus and his followers did not yet have a theory of gravity, it took intellectual

Figure 3.2 Nicolaus Copernicus
a. Nicolaus Copernicus (1473–1543) developed a heliocentric model of the Solar System. His controversial work was published just before his death. **b.** Copernicus's heliocentric model contrasted sharply with Ptolemy's geocentric model. Recall from Chapter 2 that Ptolemy's model includes epicycles on each planetary orbit (not shown).

a.

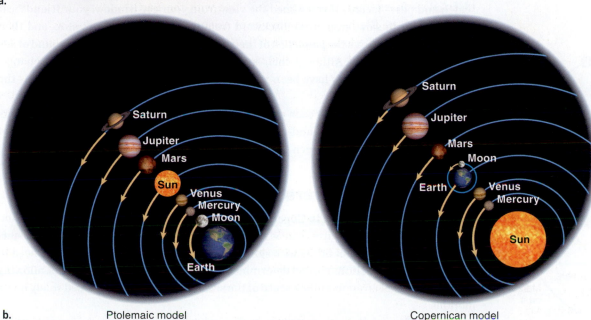

b.   Ptolemaic model   Copernican model

daring to imagine how Earth could spin each day without our feeling rotational effects. (The Copernicans could always argue that we see effects of Earth's rotation every 24 hours, as daylight turns to night and back again.)

The Copernican Model and Retrograde Motion

The third and most important aspect of Copernicus's model was its ability to replicate the retrograde motion of planets that had vexed astronomers for so long. Copernicus came to assume that the periods of planetary revolution—the time planets took to complete one orbit—increased with distance from the Sun. Thus Mercury, which is closest to the Sun, completes an orbit faster than Venus does. Venus completes its orbit faster than Earth. Earth completes its orbit faster than Mars. And so on.

Because of these different speeds, all retrograde motion is a catch-and-pass effect. Faster-moving inner planets "lap" the outer ones. As one planet passes another, its motion—projected against the fixed stars—appears to stop and change direction as it goes into retrograde motion, then again as it returns to prograde motion. This feature of Copernicus's model, along with orbits centered on the Sun, allowed retrograde motion to be explained without resorting to epicycles like those needed for Ptolemy's geocentric model.

To get a feel for how retrograde motion works, suppose you're a passenger in a car driving around a big circular racetrack with well-marked lanes. The track has an outer

INTERACTIVE:
Retrograde Motion

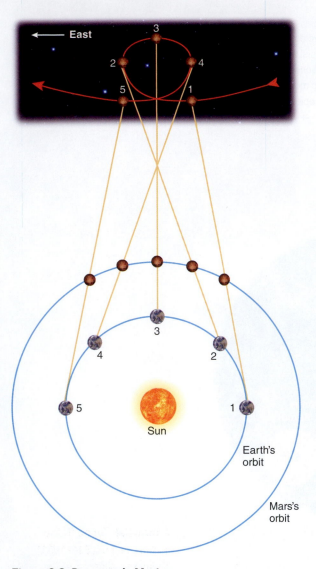

Figure 3.3 Retrograde Motion
The orbit of Mars appears to execute a "loop" against the background of fixed stars when viewed from the perspective of the more rapidly orbiting Earth. In the Copernican heliocentric model, retrograde motion occurs via a catch-and-pass effect because inner planets orbit faster than outer planets.

"Copernicus was reluctant to publish for fear that he would be 'hissed off the stage.'"

wall plastered with ads. Your car stays in its lane at a steady speed. Imagine how the ads on the wall look to you as you speed around the track. Now imagine that your friends are driving in another car, in a lane farther from the center of the track than yours is. They're moving more slowly than you are. Because of the differences in speed and the distance from the center (which determines the difference in each circle's circumference), you will complete one lap faster than they will. Now imagine that you keep your eye on them as you complete a lap. You are still watching their car as it moves against the background of ads on the wall. When you approach their car, you see them moving smoothly against the ads. But then something interesting happens.

As you get close and prepare to lap their car, the line of sight connecting your eye, their car, and a point on the wall swings around. It goes from pointing toward your direction of motion to pointing backward, away from your direction of motion. This is what "catch and pass" means. If you filmed the view from your car window, your friends' car would appear to slow, stop, move backward (relative to the wall), slow, stop, and then move forward again. It's the *projection* of their motion against the fixed background of ads on the wall that matters. Although their speed against the background appears to change, both you and your friends have been moving at constant (though different) speeds the whole time.

Copernicus showed how each planet's motion against the background of fixed stars can be explained by this catch-and-pass effect (**Figure 3.3**). The geometry was complex, but the underlying idea was still much simpler than Ptolemy's.

The Size of the Universe

There was one more benefit to Copernicus's system. In a geocentric model, the size and speed of each planet's orbit was arbitrary. You could assume larger orbits with faster-moving planets or smaller orbits with slower planets. In Copernicus's model, the need to get the retrograde motions correct determined the sizes and speeds of the orbits, allowing Copernicus to get an approximate sense of the size of the Solar System based solely on its motions.

The difference in size between the Ptolemaic and Copernican models was startling. The volume of the Copernican Universe was at least 400,000 times greater than that of the Ptolemaic cosmos. This increase was the first of many times that scientific astronomy would enlarge its estimate of the size of the Universe, making humanity's place ever smaller and less significant.

Copernicus did not publish his book until the end of his life; there has been considerable debate about why he held back. Some scholars assume that Copernicus was afraid of persecution by the Catholic Church. According to Gingerich, it was probably the ridicule of other professional astronomers that Copernicus feared. "Copernicus was reluctant to publish for fear that he would be 'hissed off the stage,'" Gingerich concludes. "He wanted to get things right." (see **Anatomy of a Discovery** on p. 56). But while Copernicus may have been more concerned with his professional reputation than with charges of heresy, politics would soon intrude in astronomy, as Copernicus's idea began to cross Europe.

Section Summary

- Ptolemy's geocentric Solar System was the dominant model of the Universe for 1,400 years.
- Copernicus's heliocentric model correctly depicted the relative positions of the Sun and planets as well as Earth's daily rotation.
- Copernicus's model also solved the mystery of retrograde motion.

CHECKPOINT **a.** What were the significant differences between the Ptolemaic and Copernican Solar System models? **b.** How did Copernicus explain retrograde motion?

Planets, Politics, and the Observations of Tycho Brahe

3.2 Denmark would host the next stage of the Copernican revolution under the guidance of Tycho Brahe (**Figure 3.4**). The son of a Danish nobleman, Brahe was a contentious fighter who, as a youth, lost a considerable chunk of his nose in a sword duel while defending his reputation as a skillful mathematician. For the rest of his life, Brahe wore a bronze nose "bridge" as a memento of the fight.

Brahe developed many powerful methods of observation using only the unaided eye. His astronomical work spanned his entire life; he gained such fame for it that the king of Denmark gave him funds to build a sophisticated observatory on the island of Uraniborg. There, over many decades, Brahe compiled remarkably accurate and extensive naked-eye observations of the sky, including the motion of planets. These observations were the best in the world and would prove crucial in changing our understanding of the Universe. Using these observations, Brahe tried to build a hybrid model of the Solar System in which the planets orbited the Sun but the Sun orbited the Earth.

On November 11, 1572, while looking at the constellation Cassiopeia, Brahe observed the appearance of a tremendously bright "new star," which he called a *nova*. (He actually witnessed what is now known to be a supernova, the titanic explosion of a massive star.) The dramatic brightening of a star was such a rare event that it was deeply disturbing to people of the time, who saw these changing stars as ill omens cast in the otherwise constant sky.

Brahe carried out meticulous measurements of this "new star" before it faded after a few months. With his new methods of observation, Brahe demonstrated that the nova showed no parallax (see Chapter 2). Therefore, the star could not be in the sublunar or atmospheric realm. It was as distant as the other stars—a conclusion that directly challenged Aristotle's ideas about the Universe. Recall that, according to Aristotle, the

> "Brahe always needed mathematically inclined assistants, and he quickly saw that Kepler was a brilliant mathematician."

a.

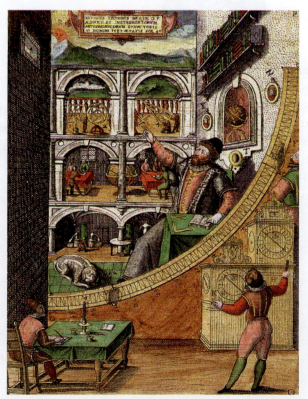

b.

Figure 3.4 Tycho Brahe
a. Tycho Brahe (1546–1601).
b. From his observatory on Uraniborg, Brahe used a variety of instruments to aid his naked-eye observations of planetary motion, recording orbital parameters with great precision.

How influential was Copernicus's *De revolutionibus*?

De revolutionibus orbium coelestium (1543)

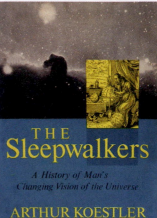

The Sleepwalkers (1959)

hypothesis

The geocentric model was accepted for some 1,400 years. But in 1543, a reluctant **Nicolaus Copernicus** revived an ancient idea that the Sun, not Earth, is the center of the Universe when he published *De revolutionibus orbium coelestium*. This work is often credited with sparking the Copernican revolution.

hypothesis

In 1959, **Arthur Koestler** wrote in *The Sleepwalkers* that *De revolutionibus* was an all-time worst-seller, dubbing it "the book that nobody read."

observation

To assess Koestler's provocative claim, Owen Gingerich sought out every existing copy of the first and second editions of *De revolutionibus*. In the Vatican Library, he reviewed a copy he initially attributed to **Tycho Brahe**. It contained what Gingerich thought were Tycho's earliest drawings of the Tychonic Universe.

failed hypothesis

Owen Gingerich and Robert Westman later found that the drawings were made by Paul Wittich, a mathematician who visited Tycho in 1580 and likely influenced the development of Tycho's model. Gingerich's research demonstrated that some of the revolution's most prominent, and not-so prominent, scientists had read *De revolutionibus* closely. Thus Koestler was wrong. In 2004, Gingerich published *The Book Nobody Read*, recounting his pursuit of the true impact of *De revolutionibus*.

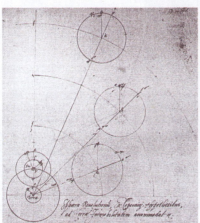

Prototype of Tycho's Universe

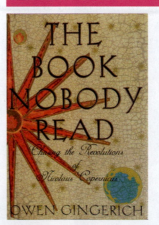

The Book Nobody Read (2004)

heavens are perfect and unchanging. By showing that the new star is situated beyond the Moon in the celestial realm, Brahe put the first nail in the coffin of Aristotle's cosmos.

Section Summary

- Danish astronomer Tycho Brahe designed and used a highly accurate system for capturing naked-eye measurements of planets' motions and meticulously recorded his observations.
- Brahe also observed a supernova, an exploding star, which contradicted ancient Greek ideas about unchanging celestial perfection.

CHECKPOINT   What did Brahe contribute to the changing conceptions of the night sky?

Kepler and the Laws of Planetary Motion

3.3   Religious politics brought together Tycho Brahe and the next figure in our story, Johannes Kepler (**Figure 3.5**). "Johannes Kepler was a brilliant young high school teacher when the Counter-Reformation swept in," says Owen Gingerich. The Counter-Reformation was the often violent Catholic response to the rise of Protestantism. "Suddenly Kepler, along with all the other Protestant teachers, found himself out of a job," says Gingerich. Forced to look for work elsewhere, Kepler accepted an offer to become Brahe's assistant in Prague. "Brahe always needed mathematically inclined assistants, and he quickly saw that Kepler was a brilliant mathematician." In Prague, Kepler took his first steps in formulating a more accurate vision of heliocentric planetary orbits.

Returning to the Copernican Model

The mathematics of Copernicus's model could not be expressed simply in terms of geometry. Kepler was sure that Copernicus's Sun-centered model of the Universe was correct, but he was equally sure that nature's plan should be simple and easy to express.

Kepler was one of the few astronomers to speak openly of the heliocentric model as reality. "Copernicus's work was widely accepted as a recipe book for calculating planetary positions," says Gingerich, "but there was no way astronomers would accept it as a physical reality. It was just so much opposed to common sense." Religion played an important role too. "Remember," says Gingerich, "Psalm 104 says the 'Lord God laid the foundations of the Earth that they not be moved forever.' And Joshua at the Battle of Gibeon commanded the Sun and not Earth to stand still. The reality of a heliocentric Universe was very much seen as a scriptural problem."

Kepler's first mathematical model of the Solar System represented planetary orbits as nested three-dimensional geometric figures called Platonic solids. Kepler quickly wrote his theory into a book called *Cosmic Mysteries*. Although he ultimately found that Platonic solids do not match observations, Kepler's book earned him the attention of influential European astronomers, including Brahe.

During his time in Prague, Kepler gained access to a small portion of Tycho Brahe's data for the motion of the planet Mars. Kepler knew these were the most precise observations in existence, and with that data he was sure he could discern the true nature of planetary orbits—the laws governing their shape and their changes in speed. He begged Brahe to give him the rest of the Mars data. "Tycho was extremely wary about sharing too much data with somebody who was not yet a tried-and-true disciple," explains Gingerich. "It was not until Tycho died and Kepler got access to all the observing books that he could really make great progress."

Most important for the history of science, Kepler's final results could be described in terms of three general laws. The simplicity, mathematical elegance, and compactness of

Figure 3.5 Johannes Kepler
Johannes Kepler (1571–1630) used Brahe's meticulous observations of planet positions to develop his three laws of planetary motion.

Figure 3.6 Kepler's First Law

a. An ellipse is defined by its foci and its semimajor axis. Given these elements, the semiminor axis can be computed. Its eccentricity is defined as the ratio of the distance from the center to one of the foci to the length of the semimajor axis. **b.** All orbiting objects (planets, comets, and so on) have some degree of eccentricity. Comets generally have higher eccentricities than planets.

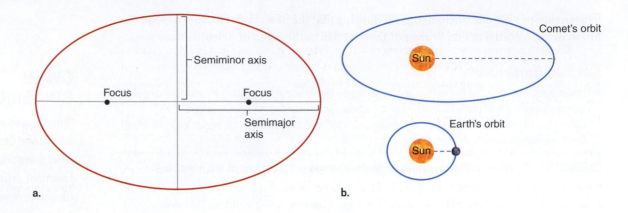

INTERACTIVE:

Ellipses and Kepler's First Law

focus (pl. foci) Either of the two points interior to an ellipse that define its shape. The Sun is always at one focus of a planet's elliptical orbit.

first law of planetary motion The principle, advanced by Johannes Kepler, stating that planets move on elliptical orbits with the Sun at one focus.

semimajor axis Half of the long axis of an ellipse.

semiminor axis Half of the short axis of an ellipse.

eccentricity A measure of the roundness of an ellipse, calculated as a ratio: the distance from the ellipse's center to its foci, divided by the length of the semimajor axis.

second law of planetary motion The principle, advanced by Johannes Kepler, stating that the line connecting a planet to the Sun will sweep out equal areas of its orbit in equal times.

conservation of angular momentum The principle whereby the rotational speed of an object that spins or orbits a central point increases as the distance from the center decreases, and vice versa.

orbital period The time an object takes to complete one revolution of an orbit.

third law of planetary motion The principle, advanced by Johannes Kepler, stating that the square of a planet's orbital period (in years) equals its average orbital radius (in astronomical units) cubed.

Kepler's three laws became a model of scientific descriptions of nature and ushered in a new era of scientific investigation.

Kepler's First Law: Planets Move on Elliptical Orbits

Kepler recognized early on that uniform circular motion would never account for Brahe's detailed observations of planetary motion. Kepler resolved to abandon the 2,000-year-old Greek prejudice for circles and to search for the true form of the orbits. After extensive experimentation, he found that he could model Brahe's precise data for Mars by using an orbit based on the geometric form called an *ellipse*.

The ellipse had been known since the time of Pythagoras. Ellipses look like squashed circles (a circle is actually a form of ellipse). You can draw a circle by pinning a length of string to a sheet of paper, keeping the string taut, and tracing a pencil around the pinned center point. You can construct an ellipse in a similar way, but with two points (called **foci**). Imagine you have a piece of string and two pins. Place two pins at the points that will define the foci of the ellipse and tie each end of the string to the pins. Now pull the string taut around a pencil and trace out a closed line. The shape you trace is an ellipse. Notice that if the two pins are closer to each other, the shape you sketch out becomes more circular. If the pins are farther apart, the shape looks more like a cigar.

Kepler found that Mars traces an elliptical path through space with the Sun at one focus. (The other focus plays no role in the planet's motion.) In time, he determined that all planets move on elliptical orbits. This **first law of planetary motion** was an enormous advance. For the first time, a single shape could account for the path of the planets without need for epicycles or other geometric devices. (Ultimately, the most dramatic example was provided by the highly elliptical orbit of Pluto, though the dwarf planet was not discovered until three centuries after Kepler's death.) Eventually it was recognized that all orbiting bodies move on ellipses.

A few quantities associated with ellipses are important for describing planetary motion. The **semimajor axis** is half the length of the *longest* axis that can be drawn across an ellipse, and the **semiminor axis** is half the length of the *shortest* axis (**Figure 3.6**). Another important quantity is the **eccentricity** (*e*) of the ellipse, which is a ratio: the distance from the ellipse's center to its foci divided by the length of the semimajor axis. An eccentricity of 0 gives a perfectly circular orbit, while an eccentricity approaching 1 gives an orbit that is essentially an oscillation back and forth along a straight line.

Kepler's Second Law: Equal Areas Are Swept Out in Equal Time

Kepler's first law tells us the shape of a planetary orbit, but what about the planet's speed? Kepler still had to account for planets' variable speeds as they trace out their orbits. By closely analyzing Brahe's data, Kepler specified a simple rule of how a planet's speed changes as it completes an elliptical circuit. In doing so, he was able to abandon arbitrary mathematical contraptions such as Ptolemy's equant.

Kepler's **second law of planetary motion** states that the line connecting a planet to the Sun will sweep out equal areas on its elliptical orbit in equal times. To understand what this means, imagine that you map the orbit of Mars on a piece of graph paper with each square equal to one unit of area. Now imagine drawing a line connecting Mars to the Sun, which lies at one of the orbit's foci. Through intense study of Brahe's data, Kepler discovered that, every week, the line connecting Mars and the Sun sweeps out the same area—that is, passes through the same number of squares on the graph paper (**Figure 3.7**). The same relation held true when Kepler changed his unit of time from a week to a day or a month. In fact, it didn't matter what unit of time Kepler used, as long as it was held constant. Measure the area a planet sweeps out in its orbit in any 2 months, 2 weeks, or 2 seconds, and the area swept out in the two (equal) time intervals will be equal.

The importance of this law becomes apparent when you look at how a planet must move in order to fulfill the "equal area, equal time" rule. Kepler's second law tells us how the speed of motion changes as a planet traverses its orbit. Kepler found that planets move fastest at the point closest to the Sun (*perihelion*) and slowest at the point farthest from the Sun (*aphelion*).

Kepler's second law thus tells us more than where the planet's motion is fast and where it is slow. It tells us how to find exactly where the speed changes at every point in the orbit. That is its beauty and elegance.

Kepler's second law is based on an important principle called **conservation of angular momentum**, which relates speed and distance for rotating (or orbiting) objects. (*Momentum*, a basic concept in physics, describes the tendency of a moving object to remain in motion in a given direction.) *Angular momentum* is the tendency of a rotating object to remain in motion around a central point, as in orbits or spin. Conservation of angular momentum tells us that when an object moves around a central point, its speed will increase if its distance from the center decreases. (Likewise, its speed will decrease if its distance increases.) This is the same physical principle that makes a spinning skater speed up when he pulls his arms inward and makes him rotate more slowly if he extends his arms (**Figure 3.8**). Similarly, according to Kepler's second law, a planet orbits the Sun quickly when its distance from the Sun is small (that is, at its perihelion) and orbits the Sun slowly when its distance is large (at its aphelion). The principle of conservation of angular momentum had not been formulated in Kepler's time; in fact, its formulation would take about another 200 years. Kepler, in his genius, invented his own version of it in the "equal area, equal time" law.

Kepler's Third Law: The Period-Radius Rule

From his careful study of Brahe's data, Kepler developed a third powerful law that neatly expresses the relationship between a planet's orbital period around the Sun and its average distance from the Sun. The **orbital period** (*P*) of a planet is the time the planet takes to complete one orbit around the Sun. Earth's period is 1 year. Mars has a period of 1.88 years.

A planet's distance from the Sun would never change if its orbit were perfectly circular (that is, if its eccentricity *e* = 0), because the radius of the planet's orbit would be constant. If, however, the planet moved on an elliptical orbit, its distance from the Sun (located at one focus of the ellipse) would constantly change. In that case, astronomers often consider the *average* radius of the planet's orbit (*R* = the sum of the closest and farthest approaches, divided by 2). Note that the average radius of the orbit is also equal to its semimajor axis.

Kepler's **third law of planetary motion** is a relationship between the orbital period *P* and the average radius of the orbit *R*. When the period is expressed in years

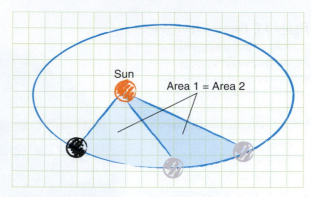

Time interval between planet positions: 3 months

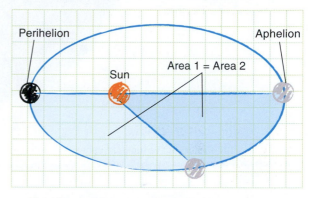

Time interval between planet positions: 6 months

Figure 3.7 Kepler's Second Law

Planets in orbit sweep out equal areas in equal times, regardless of the time interval selected. This is so because the orbital speed of a planet at or approaching perihelion is always greater than when the planet is at or approaching aphelion; the closer the planet is to the Sun, the faster it moves. Here, areas 1 and 2 are the areas bounded by lines connecting the Sun to the planet's position at the start and end of an interval as well as the planet's orbit.

Figure 3.8 Conservation of Angular Momentum

This principle makes a skater spin faster when his arms are pulled in closer to his body.

Figure 3.9 Kepler's Third Law
a. Kepler related the period (in Earth years) and average radius (in astronomical units) of a planet's orbit: $P^2 = R^3$. **b.** A graph of P^2 versus R^3 for the planets Mercury through Mars, and relevant data. All the points lie on a straight line, confirming Kepler's third law.

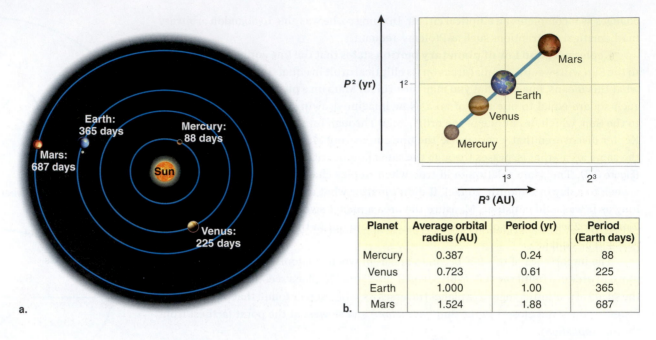

a.

b.

Planet	Average orbital radius (AU)	Period (yr)	Period (Earth days)
Mercury	0.387	0.24	88
Venus	0.723	0.61	225
Earth	1.000	1.00	365
Mars	1.524	1.88	687

and the average radius is expressed in astronomical units (AU), Kepler's third law looks like this:

$$P^2 = R^3$$

(orbital period, in yr)^2 = (average orbital radius, in AU)^3

This is a very powerful relationship because it is universal. Kepler found that every planet orbiting the Sun obeys this rule (**Figure 3.9**). The third law enables astronomers to convert the observation of a planet's period into knowledge of its average distance from the Sun, and vice versa. (**Going Further 3.1** shows how that is done.) Compare the simplicity of Kepler's third law to the description of Ptolemy's model, with its equants and epicycles, to see what a dramatic change had occurred in astronomy in the 100 years separating pre-Copernican astronomy and Kepler's revolutionary discoveries.

With his three simple laws, Kepler was sure he had unveiled the true Copernican Universe. It is worth noting, though, that Kepler was never able to explain why his three laws existed. He imagined they might be the result of magnetic forces holding the planets in orbit around the Sun. The full story of gravity, orbits, and the reasons behind his three laws would have to wait until almost 60 years after Kepler's death for the genius of Isaac Newton. In the meantime, another revolution was initiated by a different genius who also believed in a heliocentric Universe. Far to the south of Kepler, another giant of the age was opening a new window on the night sky.

INTERACTIVE:
Newton's Version of Kepler's Third Law

Section Summary

- Using Brahe's observational data, Kepler formulated three laws that presented a more accurate vision of heliocentric planetary orbits than Tycho Brahe had. Kepler fully supported the Copernican view of the Solar System.
- Kepler's first law states that planets move on elliptical orbits (not on perfectly circular orbits, as had long been assumed).
- Kepler's second law states that a line joining a planet and the Sun sweeps out equal areas in equal time; thus, planets move fastest at the point closest to the Sun (perihelion) and slowest at the point farthest from the Sun (aphelion).
- Kepler's third law describes the period-radius relationship: the square of a planet's orbital period, in years, equals the cube of its average orbital radius, in astronomical units.

CHECKPOINT How did Kepler's three laws advance the understanding of the heliocentric model of the solar system?

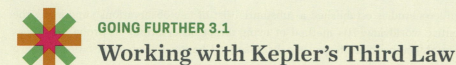

GOING FURTHER 3.1

Working with Kepler's Third Law

Imagine that a new asteroid is discovered orbiting the Sun. When the asteroid is closest to the Sun (at perihelion), the radius of its orbit is $R_{\text{perihelion}} = 3\,\text{AU}$. When the asteroid is farthest from the Sun (at aphelion), its orbital radius is $R_{\text{aphelion}} = 5\,\text{AU}$. How can we use Kepler's laws to find the asteroid's period?

First we need to calculate the average radius of the asteroid's orbit. A simple way to find that average is to add the distance from the planet to the Sun at perihelion ($R_{\text{perihelion}}$) and at aphelion (R_{aphelion}) and divide the result by 2:

$$R_{\text{average}} = \tfrac{1}{2}(R_{\text{perihelion}} + R_{\text{aphelion}})$$

We know that the radii for the asteroid's perihelion and aphelion are 3 AU and 5 AU, respectively, so

$$R_{\text{average}} = \tfrac{1}{2}(R_{\text{perihelion}} + R_{\text{aphelion}}) = \tfrac{1}{2}(3\,\text{AU} + 5\,\text{AU}) = \tfrac{1}{2}(8\,\text{AU}) = 4\,\text{AU}$$

Next we rearrange Kepler's third law

$$P^2 = R^3$$

and solve for the period P. To do so, we take the square root of both sides (which is the same as raising both sides to the power of $\tfrac{1}{2}$):

$$P^{2\left(\tfrac{1}{2}\right)} = R^{3\left(\tfrac{1}{2}\right)}$$

This gives us

$$P = R^{\tfrac{3}{2}}$$

We know that $R_{\text{average}} = 4\,\text{AU}$, so we get

$$P = (4)^{\tfrac{3}{2}} = (2)^3 = 8\ \text{years}$$

Notice that 2 is the square root of 4. In equation form, this relation is written as $(4)^{\tfrac{1}{2}} = 2$. So when you see something like $(4)^{\tfrac{3}{2}}$, you can first take the square root and then raise the result to the power 3: $(4)^{\tfrac{3}{2}} = (2)^3 = 8$.

Galileo Invents New Sciences

3.4 Galileo Galilei was born in 1564 to a cultured family in Pisa, Italy (**Figure 3.10**). His father was a well-known musician and music theorist whose books on the harmonies of musical scales had affected Kepler's thinking about the structure of the Solar System. After briefly considering a life in the church, Galileo began medical studies at the University of Pisa. Mathematics and the sciences captured his interests more than cadavers, however. Soon he was on his way to becoming one of the most famous astronomers in Europe.

Galileo and the Telescope

A key factor in Galileo's success was his quick adoption of the newly developed telescope as a tool for astronomy. Galileo did not invent the telescope. While it is unclear who created its basic form, Hans Lippershey in the Dutch Republic (now the Netherlands) patented the telescope as two lenses placed at either end of a tube that could magnify the image of distant objects. Most people saw these new devices only as military tools. Galileo was the first to recognize the invention's enormous potential for astronomy. In 1609 he built his own telescope. Its magnifying power was dismal by modern standards, increasing image size by a factor of only 8 (called 8× magnification). Later versions that Galileo used had higher magnifying powers (30×).

Galileo systematically observed the night sky with a series of ever-more-powerful instruments. In 1610 he published a short report, *Sidereus nuncius* (*The Starry Messenger*). The book's publication rocketed Galileo to fame and social status, turning him into a kind of Renaissance rock star. Combining his newfound fame, considerable confidence in his own abilities, and firm belief in heliocentric astronomy, Galileo was sure he could win acceptance for Copernicanism.

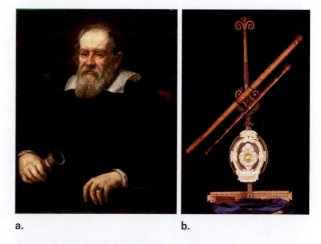

a.   b.

Figure 3.10 Galileo Galilei
a. Galileo (1564–1642), one of the greatest scientists in history, played an essential role in establishing the modern practices of science, such as the use of experiments and direct observations. **b.** One of Galileo's telescopes. They were low-power refractors (much less powerful than those used by amateur astronomers today), but he made discoveries with them that revolutionized astronomy.

Figure 3.11 Lunar Mountains
Galileo used observations of shadows and geometry to compute the heights of mountains on the lunar surface.

Galileo's studies established a substantial list of key observations supporting the heliocentric worldview. His method of using direct observations to probe new ideas helped establish the scientific process discussed in Chapter 1. Such observations led to a number of important discoveries.

MOUNTAINS ON THE MOON. Although even a naked-eye observer can see that the face of the Moon looks mottled, Renaissance astronomers held that this "imperfection" is an optical illusion. Most astronomers accepted Aristotle's claims (and those of the Catholic Church's scholars) that the Moon is a perfect, unblemished orb (like a heavenly Ping-Pong ball). When Galileo trained his telescope at Earth's nearest neighbor, however, to his astonishment he saw the lunar surface pockmarked with craters, ridges, and mountains. When he trained his telescope at the Moon's *terminator*—the line between its sunlit and shadowed sides—he quickly understood that what he saw there were patterns of light and shadow (**Figure 3.11**). The features he observed on the Moon had to be high enough to cast long shadows across the lunar surface. Using his knowledge of geometry, Galileo calculated shadow lengths and mountain heights. His results showed that the mountains rise as high as 5 miles above the lunar surface. Through his observations, Galileo demonstrated that the Moon's surface is much like that of Earth. In so doing, he dealt Aristotle's theory of differing sublunar and celestial realms its heaviest blow.

THE PHASES OF VENUS. Venus always appears relatively near the Sun on the sky. That means Venus often appears as a bright object in the hours after sunset or before sunrise (which is why it is called the "morning star" or "evening star"). With his telescope, Galileo could resolve the disks of Venus and other planets. This was a stunning advance. Previously, the planets had been mysterious, wandering bright points of light, but Galileo was the first to see the face of other planets. By training his telescope on Venus throughout its 225-day orbit, Galileo saw the planet move through phases, as the Moon does (**Figure 3.12**).

Recognition of the phases of Venus presented Galileo with a strong argument against geocentrism. In Ptolemy's astronomy, both Venus and the Sun orbit Earth—and therefore must remain at fixed distances from and orientations to Earth. It was impossible in the Ptolemaic model to create the perspective from which Venus would appear as nearly full. A heliocentric model could naturally predict the correct phases for Venus, however, so Galileo's discovery served as evidence of the Copernican system.

THE MOONS OF JUPITER. Training his telescope on Jupiter, Galileo made his most remarkable discovery. Not only did he make out the face of that planet, but just a few months of observing showed him that Jupiter has its own family of orbiting satellites.

There are four **Galilean moons** (as these largest satellites of Jupiter came to be called): Io, Europa, Ganymede, and Callisto. Their orbital periods range from 16.7 days (for outermost Callisto) to 1.8 days (for innermost Io). These periods are short enough that, even in less than a month of observation, Galileo could see the moons swing from one side of Jupiter to the other. It was clear that the inner moons orbit faster than the outer ones, just as Copernicus predicted for the planets in his Sun-centered Solar System. While the discovery of Jupiter's moons did not prove that Copernicus was right, it was a blow for followers of Ptolemy because it showed conclusively that not every object in the Solar System orbits Earth. In essence, Galileo discovered that Jupiter has its own mini–Copernican system (**Figure 3.13**).

SUNSPOTS. Training his telescope on the Sun, Galileo was able to make out dark spots on the Sun's face more clearly than had anyone before him. Like the mountains on the Moon, the existence of these solar "blemishes" contradicted the church-sanctioned "perfect realm" of the heavens that was the backbone of the Aristotelian cosmos.

Galilean moons The four largest moons of Jupiter: Io, Ganymede, Callisto, and Europa.

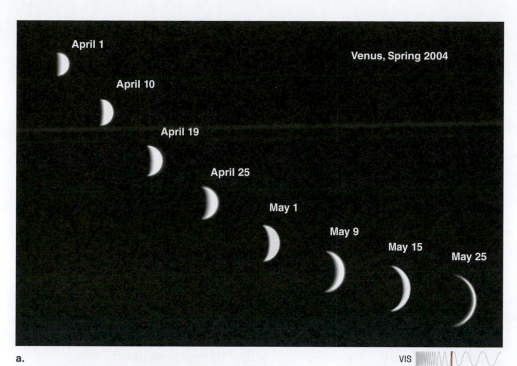

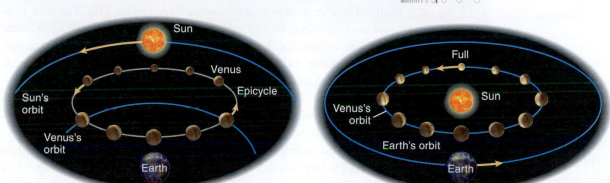

b. **Geocentric model**   c. **Heliocentric model**

Figure 3.12 The Phases of Venus

Galileo observed the phases of Venus with his telescope.
a. The progress of Venus as seen by telescope over 7 weeks.
b. Galileo showed that the phases of Venus required by the geocentric model were not the phases he observed. **c.** His observations thus provided additional evidence of the validity of the Copernican heliocentric model.

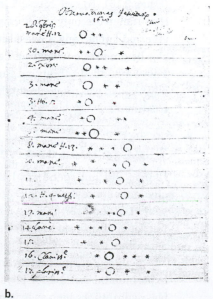

Figure 3.13 Galilean Moons of Jupiter

Observing "stars" executing periodic motions around Jupiter, Galileo reasoned that he had discovered moons orbiting the planet. The fact that Jupiter and its moons represent a mini–Copernican system helped discredit the geocentric model of the Universe.
a. Jupiter and its Galilean moons, photographed by visiting spacecraft. **b.** Galileo's notes (1610) showing the observed movement of the moons.

Figure 3.14 The Milky Way
Galileo was able to resolve some of the individual stars that make up the Milky Way, showing that it is not just nebulosity.

> "He was always trying to convince the hierarchy that the Copernican system was at least a reasonable alternative hypothesis."

nebulosity Clouds in space.

force An interaction between two bodies that, if unbalanced, can change their speed and/or direction of motion.

velocity Rate of change in an object's position and direction per unit of time.

speed An object's change in position per unit of time.

acceleration An object's change of velocity per unit of time.

mass A quantity of matter that is related to an object's resistance to changes in velocity; measured in units of kilograms.

inertia An object's resistance to changes in velocity (speed or direction of motion).

THE MILKY WAY. By telescope, Galileo was able to resolve the structure of the Milky Way (Figure 3.14). Rather than being simply a band of clouds, or **nebulosity** (specifically, clouds in space), as scholars of the day believed, the Milky Way, as Galileo saw, is composed of innumerable stars packed too close together to resolve with the naked eye.

With his newfound fame, Galileo felt sure he could convince the Catholic Church at least to be open to shifting away from its adherence to Ptolemaic astronomy. But while Galileo was a scientific genius, his ability to discern the dark paths of politics appears to have been lacking. After a few famous false steps, he was dragged before the church's most feared judicial body—the Inquisition—on charges of heresy.

Galileo's Trial

For all his abilities, Galileo appears to have been arrogant and vain. As his writings and arguments in favor of Copernicanism accumulated, he acquired many enemies within the church. Galileo took no notice of the storm growing around him. "He was always trying to convince the hierarchy that the Copernican system was at least a reasonable alternative hypothesis," explains Owen Gingerich. In 1616, Galileo was given an order: he could think about the Copernican theory, but he had to stop teaching and writing about it. He agreed to this condition but still maintained that his explanation described the natural world better than any alternative. He was sure that, in time, he would convince the church.

In 1623, Cardinal Maffeo Barberini was named Pope Urban VIII. The new pope favored the geocentric model (as did many in the church) but had been willing to entertain Galileo's Copernican arguments. Galileo believed the time was right to deal his enemies, and the Ptolemaic system, a killing blow. He published a book, the *Dialogue on the Two Chief World Systems*. The *Dialogue* featured a discussion among three gentlemen—Salviati, Sagredo, and Simplicio—on the merits of the geocentric and heliocentric worldviews. In a stunning act of poor judgment, Galileo made Simplicio the proponent of geocentrism. Thus, the pope's favorite arguments for Ptolemaic astronomy were given to a character whose name means "dumb" or "naïve." The publication of the *Dialogue* brought Galileo into the sights of the Inquisition.

"When the Inquisition came to try Galileo, heliocentrism was essentially not discussed," explains Gingerich. "Galileo didn't have a chance to give any kind of arguments in its favor. Instead the trial entirely turned on whether or not he disobeyed papal orders in publicizing the heliocentric system."

In the end, Galileo was condemned for publicly advocating the Copernican model (Figure 3.15). To save himself, he was forced to recant his views on astronomy and claim that they were only intellectual arguments, not the physical truth. In return, Galileo received a relatively modest sentence of permanent house arrest. Galileo's eyesight had also begun to fail. In despair, he wrote to his beloved eldest daughter, "This universe that I have extended a thousand times has now shrunk to the confines of my body." Eventually, Galileo was proven right; in 1992 the church formally acknowledged the error in his treatment.

The Creation of a New Physics

Galileo did not spend his last years mourning his loss of freedom. Just as he had challenged the worldview of Ptolemaic astronomy earlier in his life, in his later years Galileo overturned Aristotelian physics. He accomplished this feat not with a telescope but through a series of ingenious experiments using wooden wedges, balls, and water clocks.

Galileo's greatest achievement during these last years was the development of a true experimental method for physics. For centuries, scholars had relied on the authority of classical authors such as Aristotle in their discussion of scientific topics. They addressed

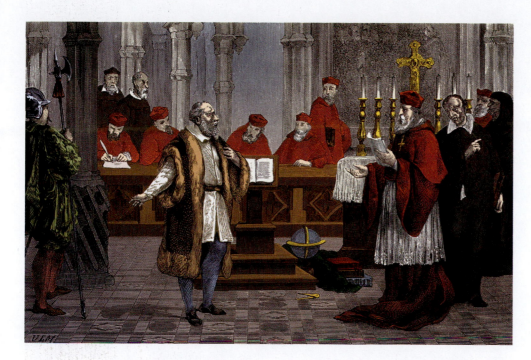

issues such as the relationship between **force** (say, a push or pull) and motion by arguing about what Aristotle had said in his philosophical works. Galileo instead decided to ask nature directly. His experiments tested ideas rather than debating them. In this way he made discoveries that would later prove crucial for the development of Isaac Newton's groundbreaking theories.

The most important outcome of Galileo's investigation was his determination of how velocity, acceleration, and mass are related in falling bodies. **Velocity** is the distance a moving object covers in a certain amount of time (in miles per hour [mi/h or mph] or meters per second [m/s], for example). The words **speed** and *velocity* are sometimes used to mean the same thing. Both measure an object's change in distance with time. Physicists, however, use velocity to mean both the change in distance with time and the direction of that change. Direction is a key aspect in the definition of velocity.

Acceleration is the change in velocity per time (measured in units such as meters per second per second [m/s²]) (**Figure 3.16**). According to Aristotle's physics, a falling object moves downward with an unchanging (constant) velocity. Using experiments, Galileo proved this is not the case. Falling objects accelerate as they drop downward. Galileo measured this *acceleration due to gravity*, and we now know its value to be 9.8 m/s². In other words, for every second a freely falling object drops its velocity will increase by 9.8 m/s.

Galileo also found that the acceleration of falling objects does not depend on their **mass**, a quantity of matter. (Mass is tied to **inertia**, an object's resistance to any change in its velocity.) This observation seems counterintuitive; common sense seems to demand that a hammer should fall faster than a feather. Our common sense is fooled, however, by the force of air resistance, which slows the feather's fall. Galileo's genius included the ability to idealize a physical situation and thus more clearly understand it. Galileo was able to ignore the effect of air resistance and conclude that the acceleration due to gravity occurs at the same rate for all objects. This conclusion contradicted Aristotle and opened the way for Newton to develop his universal law of gravity, discussed in Section 3.5. Galileo's greatness was also revealed by his ingenuity as an experimentalist. He had the remarkable ability to create simple yet powerful experimental setups (including wooden blocks and wedges) that allowed him to test his ideas about mass and motion directly.

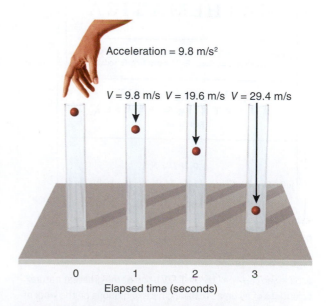

Acceleration = 9.8 m/s²

$V = 9.8$ m/s $V = 19.6$ m/s $V = 29.4$ m/s

| 0 | 1 | 2 | 3 |

Elapsed time (seconds)

Figure 3.16 Acceleration Due to Gravity
At a constant acceleration, velocity is always changing, but in a uniform way. When a ball is dropped at constant acceleration due to gravity into a vacuum cylinder, every second the ball's velocity V increases by 9.8 m/s.

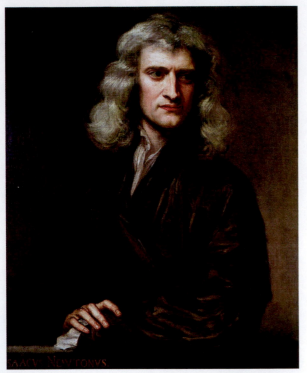

a.

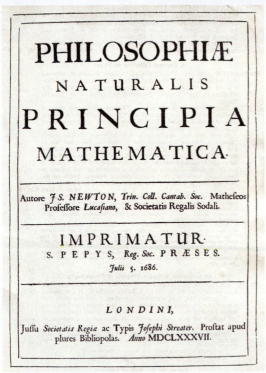

b.

Figure 3.17 Isaac Newton
a. Sir Isaac Newton (1643–1727), one of the greatest natural scientists of all time. **b.** Basing his explorations on the work of Galileo, Kepler, and Tycho, Newton formulated a new physics in his famous book *Principia*.

Galileo wrote his final book, *Two New Sciences*, during the last years of his home imprisonment. It is in many ways his most important and insightful work, establishing many of the key features of what we now call the scientific method. The world owes a great debt to this tireless and courageous Italian. It was Galileo's genius that first shed light on the Universe we now take for granted.

Section Summary

- Using the newly invented telescope, Galileo made remarkable observations that supported heliocentrism, including mountains on the Moon, the phases of Venus, the moons of Jupiter, and sunspots, and he discerned the true nature of the Milky Way as a band of countless stars.
- Galileo could not convince the church to change its views on the heliocentric model of the Solar System. After being tried by the Inquisition, he was forced to recant his beliefs publicly and was placed under house arrest for the remainder of his life.
- Through experimentation, Galileo determined that all objects accelerate due to gravity at the same rate—independent of their mass.

CHECKPOINT a. How did Galileo's observations help argue for a heliocentric Solar System?
b. How did Galileo advance our understanding of motion?

Newton and the Universal Laws of Motion

3.5 "We scientists are always looking for simplicity," says Liliya Williams, reviewing the progression from Copernicus through Kepler to Newton. Williams, a theoretical astrophysicist at the University of Minnesota, has spent a great deal of time thinking about the elegance of nature's laws. "When you find a description of nature that is simple, like an ellipse in the case of Kepler and planetary orbits, that means there is some deeper principle to which you are being led. You just have to walk one step further toward the underlying physics which makes the world behave so elegantly." In the search for the underlying physics of the Copernican model, that one step was taken by Isaac Newton.

It is not an exaggeration to call Isaac Newton (**Figure 3.17**) one of the greatest scientists who ever lived. No one else, except Charles Darwin or Albert Einstein, had so deep a vision of the structure of the world. As the poet Alexander Pope wrote, "Nature and Nature's laws lay hid in night / God said, 'Let Newton be!' and all was light."

Newton's *Principia*

The irony of Newton's great scientific genius is how little time, over the course of his life, he spent doing science. He was born in 1643, just 3 months after the death of his father, and the year that Galileo passed away. Raised in farm country, the young Newton hated farming. Luckily, his gift for mathematics was recognized by a local teacher, who arranged for a scholarship to the University of Cambridge. There, Newton's talent blossomed. Spurning studies of Aristotle, which composed much of the curriculum, the young Newton focused on the "new" astronomical writings of Copernicus, Kepler, and Galileo. In Newton's third year at university, fate and the bubonic plague intervened—setting Newton on the road to immortality.

When the plague broke out in August 1665, the university closed and Newton returned to his rural home. Within a few months, he began developing an entirely original vision of the relationship between force and motion—one that could encompass the physics of Earth and the heavens together. To make his new theory work, Newton had to invent an entirely new branch of mathematics called calculus. He began all this before the age of 26 (though he did not publish much of it until later).

In 1687, Newton published his new physics in a book now known as the *Principia*. Forming the bedrock of the new physics were three universal laws of motion. "The cool thing about Newton's three laws," says Williams, "is that they are applicable to very different situations."

Newton's three laws are described as follows:

1. The **inertial law**. An object in motion at constant velocity along a straight line stays that way unless acted on by a net force. An object at rest stays at rest unless acted on by a net force.

2. The **force law**. The change in an object's velocity due to an applied force is in the same direction as, and directly proportional to, the net force—but inversely proportional to the object's mass.

3. The **reaction law**. For every applied force, there is an equal and opposite force.

Let's unpack each of Newton's laws to understand their meaning and importance. But first it's important to define the "net force" part of the first two laws. A lot of forces can act on an object at the same time. (Some of those forces might oppose each other, such as when you push on a box while a friend next to you pulls on it.) The **net force** is what comes from adding them together.

THE INERTIAL LAW. "The first law was actually discovered by Galileo," says Williams. "It recognizes that objects have this property we call inertia, which means they won't change their state of motion unless literally forced to do so. A wooden block sitting on a table doesn't spontaneously jump up and begin moving. You have to apply force, which sets the block into motion." If you have ever had to push a broken car even a short distance, you know how much force it takes to move a massive object.

Momentum is the mass of an object multiplied by its velocity and is a quantity closely tied to inertia. An object that moves at constant velocity will not accelerate or decelerate unless a net force is applied. Imagine a wooden block moving through space at a constant speed. Unless rockets are attached to the block to make it speed up or slow down, it will continue moving at a fixed rate forever. An object's momentum is a product: the object's mass multiplied by its velocity. We've already encountered angular momentum (the momentum associated with orbits and spin) in our discussion of Kepler's second law.

Newton's first law also includes direction in its definition of velocity. Thus, if an object has uniform velocity, it travels at a constant rate (meaning a constant speed) and in a constant direction. To change an object's direction (even if the speed does not change), a force is needed (**Figure 3.18**). "The first law tells you that if no net force is acting on an object," says Williams, "then the object will either remain stationary or will continue at the same velocity (including direction) forever and ever."

THE FORCE LAW. "The second law," says Williams, "shows you what happens when an object does encounter a force." Whereas the first law states that changes in velocity require a force, the second tells us explicitly what that relationship looks like. Put mathematically, the second law says that the net force F acting on an object is the product of its mass m and acceleration a:

$$F = ma$$

$$\text{force} = \text{mass} \times \text{acceleration}$$

The best way to understand the force law is to rearrange its formula as follows:

$$a = F/m$$

"We scientists are always looking for simplicity."

"The cool thing about Newton's three laws is that they are applicable to very different situations."

"A wooden block sitting on a table doesn't spontaneously jump up and begin moving. You have to apply force, which sets the block into motion."

inertial law The principle, advanced by Isaac Newton, stating that objects in motion at constant velocity along a straight line continue in that way unless acted on by a net force.

force law The principle, advanced by Isaac Newton, stating that the change in an object's velocity due to an applied net force is in the same direction as, and directly proportional to, the force but inversely proportional to the object's mass.

reaction law The principle, advanced by Isaac Newton, stating that for every applied force, there is an equal and opposite force.

net force The sum of all forces acting on an object.

momentum The mass of an object multiplied by its velocity. Only the application of a force can change an object's momentum.

Figure 3.18 Newton's First Law

The inertial law: Objects in motion at constant velocity *V* along a straight line stay that way unless acted on by a net force. **a.** An astronaut floating in deep space (with no massive objects nearby) remains motionless. **b.** An astronaut moving at constant speed keeps moving at that speed unless acted on by a force. **c.** To change velocity (speed up, slow down, or change direction), the astronaut must exert a force, such as by using a rocket backpack.

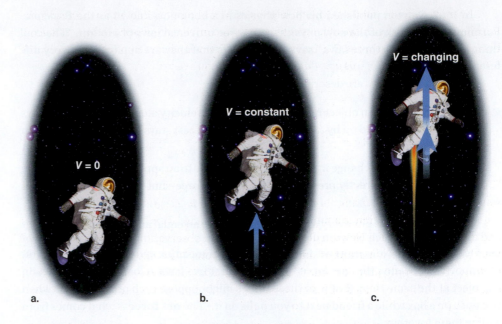

a.   b.   c.

> "The amazing thing about Newton's second law is it never specifies what kind of force we are talking about. [It's] general and universal."

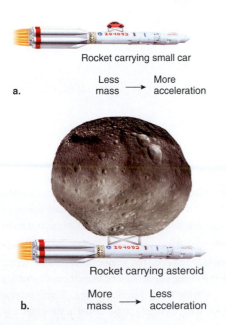

Rocket carrying small car

a.   Less mass → More acceleration

Rocket carrying asteroid

b.   More mass → Less acceleration

Figure 3.19 Newton's Second Law

The force law: The same force when applied to a large mass will yield less acceleration than when applied to a smaller mass. If the same rocket (same force) was attached to **a.** a Volkswagen Beetle (less mass) and **b.** an asteroid (more mass), the Beetle would experience much higher acceleration than the asteroid.

This formula shows that when a force is applied to an object, the resulting amount of acceleration will depend on the object's mass. Notice that mass appears in the denominator (that is, the acceleration depends *inversely* on the mass). This means that a massive object (large *m*) will accelerate less (*a* will be small) than a lower-mass object (small *m*) with the same force acting on it (*a* will be large). If you strap a weak rocket to a mountain-sized asteroid, the asteroid won't accelerate much. Strap the same rocket onto a Volkswagen (in space), and the car will take off (that is, it will experience high acceleration; **Figure 3.19**).

"The amazing thing about Newton's second law," says Williams, "is it never specifies what kind of force we are talking about. The second law holds for any kind of force: gravity, magnetism, a horse pulling a cart. It doesn't matter. The second law is general and universal."

THE REACTION LAW. Newton's third law tells us that for every action, there is an equal and opposite reaction. When a skateboarder kicks off against a wall to get started, the wall pushes back. Perhaps the best example of the third law is a rocket engine operating in space. The fuel ignites in the rocket nozzle, accelerating the gas outward. That's the action. The reaction is the acceleration of the entire rocket in the other direction (**Figure 3.20**).

Together, Newton's three laws were revolutionary. For the first time, people could understand, predict, and even control the behavior of moving objects and the forces that affect them in situations as different as the motion of planets, the design of bridges, and the trajectory of cannonballs.

Newton's Universal Law of Gravity

In the century following Copernicus's publication of *De revolutionibus*, the dominance of Aristotle's concept of distinct sublunar and celestial realms slowly eroded. But it was Newton's conception of a universal law of gravity that united the heavens and Earth under one common physics.

Recall that Newton's second law (*F = ma*) never specifies what force is being discussed. Instead, it is a general law that tells us how an object (of mass *m*) will respond, via an acceleration *a*, to a force *F*.

Newton then provided an explicit relationship for the strength of the particular force he called **gravity** between two objects of masses m_1 and m_2 that are separated by a distance R between their centers of mass. (Basically, the *center of mass* is where the average of an object's total mass is located.) For this *gravitational force*, Newton's second law takes the form

$$F_g = \frac{Gm_1m_2}{R^2}$$

$$\text{force of gravity} = \frac{(\text{gravitational constant})(\text{mass}_{\text{object 1}})(\text{mass}_{\text{object 2}})}{(\text{distance between objects})^2}$$

This equation is Newton's **universal law of gravity**. The subscript g on the force F_g reminds us that we are talking about one specific kind of force: gravity.

The G in the equation is a number that doesn't change (it is the same for all masses and all distances), which Newton derived directly from observations. "G is an example of what physicists call a **constant of nature**," says Williams. "The size of G sets the strength of gravity. It's simply part of nature's architecture." In the course of this book, we will come across other constants of nature associated with other forces and other processes that shape our world.

Newton's law of gravity tells us that force depends on the mass of each of the two objects multiplied together: mass 1 × mass 2. If you increase the mass of one object by a factor of 2, then the force of gravity between them will also increase by a factor of 2 (see **Going Further 3.2** on p. 71).

Newton's law of gravity also tells us that the force between the two masses will decrease as the square of the increase in distance between them. In other words, gravity weakens with the *inverse square of the distance*. (Expressed mathematically, $F_g \propto 1/R^2$, where the symbol $\propto$ means "is proportional to.") Thus Newton's gravitational force law is sometimes called an **inverse square law**: If you double the distance between two

LILIYA WILLIAMS

Liliya Williams knew early on she was headed toward the abstract side of science. "I was born with an interest that steered me towards physics," she says. Telescopes never worked for her. "When I was 9 or 10, somebody gave me a book on planets with colored pictures in it. My dad noticed my interest in astronomy and got me a telescope. I didn't enjoy it at all. But later I got hold of another book that was more mathematically inclined. That one I liked a lot." Now a University of Minnesota professor, Williams is an expert on gravitational lensing—the bending of light as it passes a massive object in space.

"*G* is an example of what physicists call a constant of nature. The size of *G* sets the strength of gravity. It's simply part of nature's architecture."

Figure 3.20 Newton's Third Law

A rocket taking off from a launchpad is an excellent example of Newton's third law, the reaction law: For every applied force, an equal and opposite force arises. As the rocket pushes hot gas backward out of the nozzle of the rocket, the gases push forward on the nozzle of the rocket.

gravity An attractive force between any two massive bodies that depends on the product of the bodies' masses and the inverse square of the distance between them.

universal law of gravity The dependence of the force of gravity between two objects as the product of their masses multiplied by Newton's constant divided by their distance of separation squared.

constant of nature A term that does not change from one situation or time to another. The value of a constant of nature must be measured to be determined.

inverse square law A relationship whereby a quantity (such as gravity) decreases in proportion to the square of a variable (such as distance) as the variable increases.

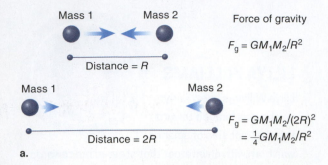

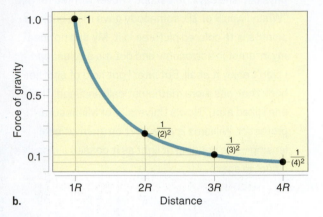

Figure 3.21 Inverse Square Law for Gravity

a. The force of gravity between two objects varies according to the inverse square of the distance between them, thus decreasing as the distance increases. **b.** If the distance between two objects increases by a factor of 2, the force of gravity between them decreases by a factor of 4.

> "Galileo showed that if you dropped a heavy object and a light object from the tower of Pisa, they would hit the ground at the same time, but he never showed the physics behind it. That is where Newton's law of gravity comes in."

AT PLAY IN THE COSMOS THE VIDEOGAME You can use Newton's version of Kepler's third law to find an iron-rich asteroid to repair your ship's hull in Mission 3: Time for Repairs.

objects, the force between them drops by a factor of $2^2 = 4$. If you increase the distance between the objects by a factor of 3, the force drops off by $3^2 = 9$ (**Figure 3.21**). Newton's law of gravity does not explain *why* this is true; Newton simply discovered that the force of gravity depends on mass and distance.

"When you combine the inverse square law for gravity," says Williams, "with [Newton's] second law of motion, $F = ma$, then you can find how an object accelerates in response to gravity." By putting this inverse square law and Newton's second law together, we can see why an object's acceleration due to gravity does not depend on its mass. Recall what Galileo found in his experiments. "Galileo showed that if you dropped a heavy object and a light object from the tower of Pisa, they would hit the ground at the same time," says Williams. "But he never showed the physics behind it. That is where Newton's law of gravity comes in. The acceleration due to gravity is the same for all objects because of the explicit form the law of gravity takes" (see **Going Further 3.3** on p. 72).

Newton's law of gravity shows the important distinction between mass and weight. Mass is an intrinsic property of an object. Your mass would not change if you moved from one planet to another. **Weight** is the downward force on the object produced by gravity. You experience weight when you stand on a planet's surface and your legs resist the downward pull of the planet's gravitational force on your body's mass. If you traveled to a planet whose mass or radius differed from Earth's, your weight would change.

Using his law of gravity, Newton was also able to show that all three of Kepler's laws are simply a consequence of the inverse square law. Speaking of Kepler's first law of elliptical orbits, Newton once said, "Kepler guessed it was an ellipse, but I have proved it." As another example, with a little mathematical manipulation, Kepler's third law, which states that a planet's period squared equals its average radius cubed, just pops out of Newton's laws of motion and his law of gravity. Even more important, Newton's version of Kepler's third law includes the mass of orbiting objects and thus allows astronomers to use their observations to measure the mass of distant planets, stars, and other celestial bodies (see **Going Further 3.4** on p. 73).

Newton's law of gravity holds for objects on Earth and in the heavens. Aristotle's distinction between the two realms was finally proved false.

Newton's Gravity and Orbits

Newton's laws also provided an explanation for orbital motion, which is the foundation of much of astronomy. Newton's first law states that objects in motion will stay in uniform motion unless acted on by a net force. Remember that "uniform motion" means constant velocity, and velocity is speed *and* direction. If you want to change an object's direction, you need to apply a force in that direction. Even a planet that orbits the Sun with a constant speed has a direction that is always changing. For a planet to continue moving in orbit around the Sun, there must be a force that constantly changes its direction of motion. In general, any force that keeps an object moving in a circle is called a *central force* because it points toward the center of the circle. For orbiting bodies, that central force is gravity.

If forces produce accelerations, what kind of acceleration is at work when speed is constant in an orbit and only the direction of motion changes? **Centripetal acceleration** is the change in velocity (change in direction) that occurs as a planet orbits the Sun (**Figure 3.22**).

You have direct experience with centripetal acceleration. When you're in a car that is going around a sharp turn, you feel yourself being flung outward, away from the direction of the turn. The outward "force" you feel is actually an illusion called **centrifugal force**.

Here is what's really happening. Before the car starts turning, you're moving in a straight line. As the car starts its turn, your body tries to keep moving in that straight line

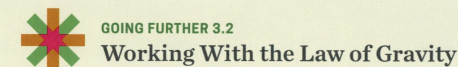

GOING FURTHER 3.2

Working With the Law of Gravity

So far, the mathematical computations in this book have focused on one example at a time. Another way to approach quantitative topics is to make comparisons. With respect to Newton's law of gravity, suppose astronomers discover two planets, named (imaginatively) A and B. They are the same size, meaning they have the same radius. However, planet A is 4 times more massive than planet B.

How much more force would a bowling ball on the surface of planet A experience compared with a bowling ball on planet B? In a comparative approach, the force equations for the two planets would look like this:

$$F_A = \frac{GM_A M_{ball}}{R_A{}^2}$$

$$F_B = \frac{GM_B M_{ball}}{R_B{}^2}$$

We can then divide planet A's equation by planet B's equation:

$$\frac{F_A}{F_B} = \left(\frac{G}{G}\right)\left(\frac{M_A}{M_B}\right)\left(\frac{M_{ball}}{M_{ball}}\right)\left(\frac{R_B}{R_A}\right)^2$$

We can cancel all the quantities that are the same within each set of parentheses:

$$\frac{F_A}{F_B} = \left(\frac{M_A}{M_B}\right)\left(\frac{R_B}{R_A}\right)^2$$

The planets are the same size ($R_A = R_B$), but planet A is 4 times more massive than planet B ($M_A = 4M_B$). Substituting $4M_B$ for M_A and R_B for R_A into the previous equation gives

$$\frac{F_A}{F_B} = \left(\frac{4M_B}{M_B}\right)\left(\frac{R_B}{R_B}\right)^2 = 4$$

The force on the bowling ball on planet A would be 4 times larger than that on planet B.

(consistent with Newton's first law). Since you are connected to the car by only a seat belt, at first you feel your body slide outward along the seat, away from the turn. It feels as if a force, a *centrifugal force*, pushes you away from the direction of the turn. Only when the seat belt grabs hold of you is the real *centripetal force* applied to your body, forcing you to change direction along with the car. It was friction of the car's wheels on the road that applied the centripetal force to the car itself.

Newton's laws also provide two relationships for understanding the properties of simple orbits. An astronomer often needs to ask, "What's the velocity of an object in orbit?" The question is relevant for planets orbiting the Sun, moons orbiting a planet, or a space probe orbiting Mars. Newton's laws provide an equation for the simplest kind of

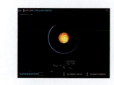

INTERACTIVE:

Circular Orbits

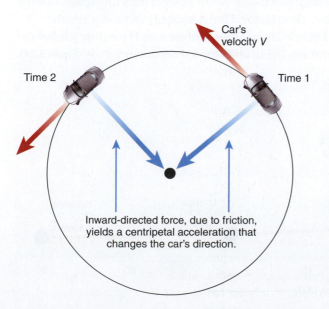

Figure 3.22 Centripetal Acceleration

The acceleration required to change an object's velocity (change in direction) when it moves in a circular path, such as when a car rounds a corner at constant speed or a planet travels in its orbit, is centripetal acceleration. For the car, the force of friction between the tires and the road produces a centripetal acceleration.

Car's velocity *V*

Time 2

Time 1

Inward-directed force, due to friction, yields a centripetal acceleration that changes the car's direction.

weight The downward force that gravity produces on an object.

centripetal acceleration Acceleration involving a force that keeps an object moving on a circular path. The acceleration is directed toward the path's center of curvature.

centrifugal force The illusion of outward force when an object is moving on a curved path.

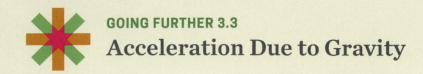

GOING FURTHER 3.3
Acceleration Due to Gravity

Let's say object 1 is Earth (call its mass M_E) and object 2 is an iron ball A (call its mass M_{ballA}), and the distance between them (from Earth's center of mass, which is approximately its geometric center, to the center of the iron ball) is R. To find the acceleration a_{ballA} of an iron ball when it drops, we use Newton's second law as follows:

$$F_g = M_{ballA}a_{ballA}$$

Next we use Newton's law of gravity:

$$F_g = \frac{GM_E M_{ballA}}{R^2}$$

Therefore,

$$M_{ballA}a_{ballA} = \frac{GM_E M_{ballA}}{R^2}$$

Something interesting happens in this equation. The mass of the iron ball (M_{ballA}) appears on both sides of the equation. That means we can divide by it and cancel it out of the equation:

$$a_{ballA} = \frac{GM_E}{R^2}$$

So the acceleration of the iron ball does not depend on its mass. If we used a second, less massive iron ball (call it ball B, where $M_{ballB} << M_{ballA}$), the same cancellation would occur and we would get

$$a_{ballB} = \frac{GM_E}{R^2} = a_{ballA}$$

Thus, the acceleration due to gravity for the massive iron ball and the less massive iron ball is the same. Both balls fall with the same acceleration.

orbit: a perfect circle. Here is the expression for the **orbital velocity** V_c of a satellite in a circular orbit at a distance R from the center of its parent body, with mass M:

$$V_c = \left(\frac{GM}{R}\right)^{\frac{1}{2}}$$

The numerical value of the orbital velocity (the orbital speed) decreases as a satellite moves farther from the parent body. Recall from Section 3.1 that this is exactly what Copernicus hypothesized when he introduced his heliocentric model: the inner planets orbit faster than the outer planets. Newton used his simple and elegant laws to show astronomers why Copernicus had to be right (**Figure 3.23**).

Another basic question for astronomers (and rocket engineers) is how fast something must be moving to be launched from the surface of a parent body into space. Newton thought about this problem quite a bit as he considered what an orbit really entails.

Think about a cannonball launched from a powerful cannon. If too little gunpowder is used, the cannonball will arc through the air and drop a short distance away (**Figure 3.24**).

INTERACTIVE:

Escape Velocity

Figure 3.23 Orbital Radius and Speed

A planet's orbital speed decreases as the radius of the planet's orbit increases, as the velocity formula for circular orbits—derived from Newton's laws—predicts. (Keep in mind that planetary orbits are slightly elliptical.)

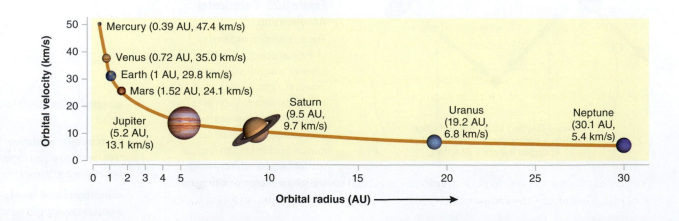

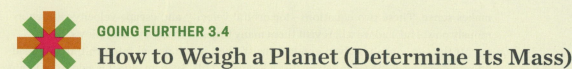

GOING FURTHER 3.4

How to Weigh a Planet (Determine Its Mass)

In providing a mathematical basis for Kepler's third law, Newton's law of gravity was an enormous advance, giving astronomers a way to measure the masses of distant objects (among other things). Here's how it works.

Newton showed that for a planet (of mass M_p) orbiting the Sun (of mass M_{Sun}), the true form of Kepler's third law looks like this:

$$P^2 = \left[\frac{4\pi^2}{G(M_{Sun} + M_p)} \right] R^3$$

where P is the orbital period. According to this equation, the bracketed value should be slightly different for every planet, since a planet's mass (M_p) appears in the denominator. Why didn't Kepler have to adjust his third law a little for each planet?

As it turns out, the Sun's mass, M_{Sun}, is so much greater than the mass of any planet, M_p, that you can ignore the planet's mass in the previous equation. So, to a fair approximation, we can write Newton's form of Kepler's third law like this:

$$P^2 = \left[\frac{4\pi^2}{GM_{Sun}} \right] R^3$$

This equation can be used for every planet orbiting the Sun, confirming Kepler's discoveries. But how do we know the mass of the Sun? You can't just put the Sun on a scale. By rearranging Newton's form of Kepler's third law, however, we can "measure" the masses of astronomical objects. If we know a planet's period and its distance from the Sun, we can solve the previous equation for M_{Sun}, the Sun's mass:

$$M_{Sun} = \frac{4\pi^2 R^3}{GP^2}$$

To this day, astronomers use this relationship to "weigh" an astronomical object by watching something else in orbit around it. In this way, the masses of everything from asteroids to moons, planets, stars, and galaxies have been calculated. (Appendix 1 in the back of this book lists the masses of Solar System objects, including the mass of the Sun.)

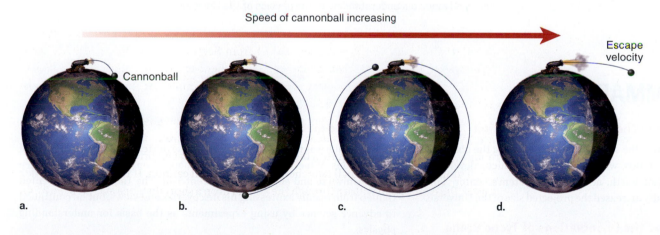

Speed of cannonball increasing

Cannonball

Escape velocity

a.   b.   c.   d.

Figure 3.24 Cannonballs and Orbits

How fast must an object move to be launched into space? Newton thought about a cannonball launched from a powerful cannon. **a, b.** With too little gunpowder, the cannonball will not reach the value of the escape velocity and falls back to Earth. **c.** Just the right amount will launch the cannonball into orbit. **d.** Even more gunpowder will launch the cannonball with much more than the escape velocity, sending it into deep space on an unbound orbit.

With more gunpowder, the cannonball will range farther. Earth is essentially a sphere, so the farther the cannonball travels, the more it will fall toward a surface that is, at the same time, curving away from it.

Extending this idea, Newton was able to find a relationship for the speed that the cannonball would need in order for it to escape the pull of Earth's gravity. This **escape velocity** V_e is

$$V_e = \left(\frac{2GM}{R} \right)^{\frac{1}{2}}$$

Notice that the escape velocity is bigger than the orbital velocity by a factor of the square root of 2. That means the escape velocity is always bigger than the orbital velocity, which

orbital velocity The velocity required to keep a body in orbit around another body.

escape velocity The velocity required to escape the pull of a massive body's gravity.

makes sense. These two equations—for orbital velocity and escape velocity—are enormously powerful, and we will revisit them many times in the story that follows.

If the cannonball is given enough thrust, then it will not go into orbit; it will head off into deep space, never to return (as in Figure 3.24d). Such an object is said to be in an **unbound orbit**. An object in a **bound orbit** always returns, repeating its orbital motion over and over again (as in Figure 3.24c).

Our cannonball example also allows us to understand two important ideas related to orbits and space travel. If you're in a spaceship orbiting a planet, then, like the cannonball, your ship will always be falling toward the planet even though the ship's velocity keeps moving in a circular orbit. That means you're falling toward the planet, too. Since both you and the ship are constantly falling toward the planet, you would not feel yourself pulled to the ship's floor. You, the floor, and the whole ship are in what's called **free-fall**—you freely fall under the force of gravity with no other forces acting on you. Free-fall has no "up" or "down" orientation. In free-fall you experience a state of **weightlessness**—you have no sense of your own weight.

Section Summary

- Newton formulated three laws that revolutionized science: the inertial law, the force law, and the reaction law. They apply to all forces and all objects.
- Newton determined that the force of gravity between two objects is proportional to their masses multiplied together and inversely proportional to the square of the distance between them.
- The acceleration an object experiences because of the gravity of another object depends only on the mass of the other object.
- Newton's law of gravity provided the basis of Kepler's three laws and the means to calculate the mass of a distant object.
- Newton's laws also provided the basis for understanding orbits and the mathematical relationships for orbital velocity and escape velocity.

CHECKPOINT   How did Newton's three laws of motion and his discovery of the universal law of gravity advance our understanding of the physics of the Universe?

unbound orbit An object's motion that carries the object away from a planet permanently.

bound orbit An object's orbital motion around a planet.

free-fall Motion occurring solely due to gravitational force.

weightlessness The absence of the sensation of weight during free-fall.

CHAPTER SUMMARY

3.1 Getting Past Ptolemy: The Copernican Revolution

Ptolemy's geocentric model was the standard for interpreting the observed motions of the sky until the 1500s, when Nicolaus Copernicus developed a model that put the Sun, not Earth, at the Solar System's center. This heliocentric model dramatically increased the projected size of the Universe.

3.2 Planets, Politics, and the Observations of Tycho Brahe

Tycho Brahe made meticulous measurements of planet positions and motions. Although he mistakenly believed the Universe to be geocentric, his data played a critical role in the later understanding of planetary orbits.

3.3 Kepler and the Laws of Planetary Motion

Johannes Kepler believed in the Copernican model and produced three laws of planetary motion: (1) Planets move on elliptical orbits. (2) A line connecting a planet to the Sun sweeps out equal areas in equal time. (3) The square of the period of a planet's orbit, in years, equals the cube of the average radius (which is also the semimajor axis) of the orbit, in astronomical units.

3.4 Galileo Invents New Sciences

Galileo Galilei was the first astronomer to use the telescope for astronomical observations. What he saw drastically changed our view of Earth's place in the Universe and supported heliocentrism. The Inquisition prevented Galileo from openly expressing his discoveries and views, but he continued to advance science by using experiments as the basis for understanding physics.

3.5 Newton and the Universal Laws of Motion

Building on the advances of Copernicus, Kepler, and Galileo, Isaac Newton conceived his three laws of motion: (1) the inertial law, (2) the force law, and (3) the reaction law. Equally important, Newton discovered the universal law of gravity, an inverse square law. He deduced that the acceleration due to gravity on the surface of a celestial body depends only on the body's mass and radius and is thus the same for objects of different masses.

QUESTIONS AND PROBLEMS

Narrow It Down: Multiple-Choice Questions

1. Ptolemy's model of the Universe was accepted for over a thousand years. Which of these is *not* one of the reasons for its acceptance?
   a. It predicted the ever-changing positions of planets fairly well.
   b. An Earth-centered Universe was consistent with Catholic Church doctrine.
   c. Greek philosophy permeated many aspects of life (politics, medicine, physics).
   d. The lack of observable parallax allowed for a stationary Earth.
   e. It correctly put the Sun at the center.

2. On which of the following point(s) was the Copernican model correct, according to our current knowledge? Choose all that apply.
   a. the Sun as the center of the Solar System
   b. use of epicycles
   c. rotation of Earth
   d. shape of planetary orbits
   e. relative orbital speeds of the planets

3. Tycho Brahe is noted for all of the following achievements *except*
   a. observation of a nova.
   b. meticulous recording of observations.
   c. capturing accurate orbital data for Mars.
   d. innovative use of the telescope.
   e. developing a hybrid Sun- and Earth-centered model.

4. Which of the following statements about ellipses is true?
   a. A circle is a form of ellipse.
   b. An ellipse is a form of circle.
   c. All ellipses have four foci.
   d. Orbits are always ellipses with eccentricity equal to 0.
   e. Semimajor and semiminor axes are never equal.

5. Two ellipses have the same semimajor axes but different eccentricities. Which of the following statements is true? Choose all that apply.
   a. The ellipse with greater eccentricity has a longer semiminor axis.
   b. The area inside the ellipse with the smaller eccentricity is smaller.
   c. The semiminor axes of the two ellipses may be equal.
   d. One ellipse may be a circle.
   e. Both ellipses may be circles.

6. Which of the following statements is consistent with Kepler's second law?
   a. All planets move at the same speed when nearest the Sun.
   b. Orbits are actually areas, not paths.
   c. The orbital speed of a planet decreases when the planet nears the Sun.
   d. "Equal time, equal area" implies that planets move at constant speed in orbit.
   e. Planets always move faster when they are close to the Sun than when they are farther away.

7. A spinning figure skater spins faster as he brings his outstretched arms inward. This is an illustration of the physical principle underlying
   a. Kepler's first law.
   b. Kepler's second law.
   c. Kepler's third law.
   d. the inertial law.
   e. none of the above

8. Which of the following statements is true regarding Kepler's third law?
   a. Periods and orbital radii are correlated for planets in our Solar System only.
   b. It says that the period and size of an orbit are correlated.
   c. It implies that orbital speed increases as orbit size increases.
   d. It works for orbits with eccentricity equal to zero only.
   e. The relationship changes to $P^2 = R^4$ for the outer planets.

9. Which of Galileo's observations disputed the perfect nature of the heavens above the sublunar region? Choose all that apply.
   a. mountains on the Moon
   b. phases of Venus
   c. individual stars in the Milky Way
   d. sunspots
   e. Galilean moons

10. Which of the following does *not* describe a contribution that Galileo made to science?
    a. He laid the groundwork for the practices of modern science, including observation and experimentation as the best processes for understanding nature and the Universe.
    b. With his discovery of moons of Jupiter, he demonstrated that the Universe has more than one center of motion.
    c. He argued that objects of unequal mass fall at the same rate.
    d. He theorized that a mutually attractive force proportional to distance explains how and why planets orbit the Sun.
    e. He used his observations of the phases of Venus to argue that Venus clearly orbits the Sun, not Earth.

11. An astronaut is traveling in deep space, far from any star or planet. To change his course, he must fire his rockets. Which of Newton's laws does *not* apply in this situation?
    a. Newton's law of inertia
    b. Newton's law of force
    c. Newton's law of reaction
    d. Newton's universal law of gravity
    e. Newton's third law

12. A gun recoils when it fires. Which principle does this example most clearly illustrate?
    a. Newton's law of inertia
    b. Newton's law of force
    c. Newton's law of reaction
    d. Newton's universal law of gravity
    e. none of the above

13. In the universal law of gravity, *G* is
    a. the gravitational force specific to Earth.
    b. a constant related to the strength of the gravitational force.
    c. the gravitational force specific to the Solar System.
    d. a value that differs depending on the objects that are nearby.
    e. immeasurable.

14. If the mass of one object increases while the distance from a second object increases, how does the magnitude of the gravitational force change?
    a. It always increases.
    b. It always decreases.
    c. It always remains the same.
    d. It depends on the values of the changes in mass and distance.
    e. It depends on the local value of G.

15. If the mass of one object remains unchanged while both the distance to a second object and the second object's mass are doubled, how does the force of gravity between the two objects change?
    a. It always increases.
    b. It always decreases.
    c. It depends on the values of the mass and distance.
    d. It depends on the local value of G.
    e. It cannot be determined.

16. Which value is the same regardless of whether you are standing on Earth or on the Moon? Choose all that apply.
    a. your mass
    b. your weight
    c. your weight-to-mass ratio
    d. the acceleration due to gravity
    e. the value of G

17. Moon α orbits a planet at orbital radius X. Moon β orbits the same planet at orbital radius $4X$. The two moons have equal masses. How does the (much larger) planet's pull of gravity on these moons compare?
    a. The difference is so small that it can be considered zero.
    b. It is 2 times weaker for moon α.
    c. It is 4 times weaker for moon α.
    d. It is 4 times weaker for moon β.
    e. It is 16 times weaker for moon β.

18. A spaceship will either orbit just above the surface of a planet of radius R or attempt to escape the planet. Which of the following statements is true about orbital velocity versus escape velocity? Choose all that apply.
    a. They may be the same, depending on the value of radius R.
    b. Orbital velocity is always greater.
    c. Escape velocity is always greater.
    d. They differ by a factor of 2.
    e. They differ by a factor of $(2)^{\frac{1}{2}}$.

19. Which of the following describes Kepler's approach and contributions to astronomy?
    a. He constrained his models of planetary motion with Tycho Brahe's positional data.
    b. He maintained the use of epicycles in planetary motion.
    c. He correctly showed that planetary orbits are based on geometric shapes called platonic solids.
    d. He ignored the work of important astronomers who preceded him.
    e. He understood that planetary motions result from mutually attractive gravitational force.

20. Which of the following characteristics were included in Ptolemy's model of the Universe? Choose all that apply.
    a. Earth rotates on its own axis.
    b. The celestial sphere is perfect and unchanging.
    c. Planets move in circles on top of circles called epicycles.
    d. All of the planets, as well as the Sun and the Moon, revolve around Earth.
    e. The nested set of crystalline celestial spheres is maintained as proposed by Aristotle.

To the Point: Qualitative and Discussion Questions

21. In what ways did the Greek philosophers contribute to the advancement of science? In what ways did they hinder it?

22. What factors contributed to the decline of the Catholic Church's influence in matters of science during the Renaissance?

23. What does the current use of star names such as Aldebaran and Deneb tell us about some of the early astronomers?

24. Compare the features that Ptolemy's model required to explain retrograde motion with those in Copernicus's model.

25. Describe what an observer sees when a planet displays retrograde motion, and explain its cause.

26. Newton is often described as one of the greatest contributors to our modern understanding of the natural world. Summarize his most important contributions.

27. Kepler's second law implies that a planet increases in orbital speed as it gets closer to the Sun. What underlying principle of physics explains this effect?

28. What strongly held beliefs of Ptolemy and Copernicus were later proved incorrect?

29. Describe the relationship between centripetal acceleration and a central force. Comment on the relationship between the direction of a central force and the direction of the velocity of an object executing circular motion.

30. What do we mean when we say that Newton's universal law of gravity is an inverse square law?

31. How would the value of the acceleration due to gravity for Earth, which is 9.8 m/s², differ if Earth's radius were larger but its mass were the same? Why?

32. The equation for Newton's universal law of gravity gives the magnitude of the gravitational force between two objects and includes the masses of both objects. The equations for both escape velocity and orbital velocity depend on the mass of only one of the objects. Explain why the escape velocity and orbital velocity are independent of the mass of the object whose velocity is being calculated.

33. What factor(s) determine(s) how free-fall for other worlds versus for Earth would differ?

34. What scientists or visionaries in your lifetime would you consider comparable to those described in this chapter, in terms of their long-term impact?

35. The stories of early great astronomers and their contributions were shaped by the intellectual climates that existed during their lives. How does today's intellectual climate encourage or discourage new discoveries in science? How might it differ for future generations?

Going Further: Quantitative Questions

36. What would be the semimajor axis, in astronomical units, of an elliptical orbit for a planet whose perihelion was at 4.5 AU and aphelion was at 6.8 AU?

37. What would be the average distance from the Sun, in astronomical units, of a comet with a period of 47 years?

38. What would be the period, in years, of a planet in our Solar System with an average orbital radius of 26 AU?

39. Asteroid A has a mass of 2×10^{20} kilograms (kg), and asteroid B has a mass of 4×10^{18} kg. Assuming that the same force (a shock wave from a supernova, for example) is applied to both, what would be the ratio of A's acceleration to B's acceleration?

40. How many times greater is the force of gravity on a 3-kg object lying on the surface of a moon than on a 3-kg object orbiting at a distance of three moon radii above the surface?

41. By what factor would the gravitational force of the Earth-Moon system change if the Moon were 3 times as far and 3 times as massive?

42. What is the ratio of the orbital speeds of two satellites, each in circular orbit around Earth, if satellite A orbits 1.7 times as far from Earth's center as satellite B? Give your answer in terms of B's speed to A's speed.

43. NASA is designing a spaceship that will land and launch from two moons of Jupiter. To plan correctly for the amount of fuel the spaceship will carry, NASA must calculate the escape speeds from each of the moons. Moon A has a radius twice that of moon B and a mass 5 times that of moon B. If moon B has an escape speed of 1.2 kilometers per second (km/s), what is the escape speed of moon A?

44. What is the mass, in M_{Sun} units, of a star observed to have a planet with an orbital radius of 4 AU and a period of 15 years?

45. A star is orbited by a planet at an orbital radius of 2.2 AU with a period of 1.6 years. How does the star's mass compare with that of the Sun?

"I KNEW RIGHT THEN THAT THIS WAS GOING TO BE A REALLY BIG DEAL"

4

A Universe of Universal Laws

HOW LIGHT, MATTER, AND HEAT SHAPE THE COSMOS

Light: The Cosmic Envoy

4.1 Five months in Puerto Rico was not what Alyssa Goodman expected when she signed up for graduate studies in astrophysics at Harvard. But that was exactly where her graduate adviser sent her. "We had written this proposal to the Arecibo Observatory," she explains. "That's a giant radio telescope that lives in a collapsed limestone bowl down in Puerto Rico. It's 1,000 feet across, deep in the jungle, and very dramatic" (**Figure 4.1**).

Goodman's proposal was to use the extremely sensitive telescope to detect radio waves from vast clouds of interstellar gas and dust where new stars form. "We hoped to detect magnetic fields in the gas clouds using properties of the radio waves," says Goodman. "Everyone expected the signal to be so faint that we had to ask for a lot of telescope time. I mean a lot of time, like a half a year."

For a grad student living through wet winters in Boston, this project was not necessarily a bad thing. "I got a really good tan, okay, and yes, there was a pool, and yes, I lived in one of those little huts on the cliff." In the middle of the project, however, Goodman had time on her hands and began digging into data she had already accumulated. "So I added up all the data from the previous 20 days of observing and, to my astonishment, there it was! The signal was already there staring back at me. We had already actually detected the magnetic field!" The signals were coming from a star-forming cloud designated B1.

"I knew right then that this was going to be a really big deal," Goodman recalls, smiling. "I called up my adviser and said, 'What if I told you that I detected a magnetic field in B1?'" Goodman's thesis supervisor paused for a minute and then replied, "I would say, 'Would you like to graduate tomorrow?'"

← The Horsehead Nebula is a star-forming cloud about 1,500 light-years away. This combined image shows light from newly formed stars and from the surrounding gas.

Figure 4.1 Arecibo Observatory
This observatory in Puerto Rico is among the world's largest single-dish radio telescopes and largest single-aperture telescopes, 305 meters in diameter.

"Spectrographs turned astronomy into astrophysics. They let astronomers interrogate the light they catch in their telescopes and use it to figure out conditions in the stars."

Goodman, now a professor of astrophysics at Harvard and a world expert on observations of star-forming clouds, looks back fondly on that moment. Her work is a prime example of how light captured in telescopes can be used to unlock the secrets of distant astronomical objects. By linking the absorption and transmission of light to **matter** (anything made of subatomic particles such as quarks or electrons), astronomers can determine the strength of a magnetic field, the mass of an interstellar cloud, and the temperature of a star.

Astronomy Becomes Astrophysics

Newton's law of gravity opened the floodgates to a new understanding of the heavens. Using his inverse square law for gravitational force (see Chapter 3), astronomers turned telescopic observations of celestial motions into detailed descriptions of forces between massive objects, even when those objects were invisible.

But gravity was only part of the story of astronomy from the 1700s to the 1800s. Advances in the design and construction of telescopes during this period enabled astronomers to see deeper and with greater resolution. *Optics*, the study of light and its properties, was the science driving the advance in telescopes (**Figure 4.2**). With these powerful instruments, astronomers mapped the Solar System, determined the distances to the stars, and began a census of the Milky Way.

Telescopes could catch light, but they could not reveal the origin of the light. For that, astronomers needed the help of physicists, who had made strides toward understanding the nature of light and its relationship to matter. Most important, physicists had developed a new instrument called a **spectrograph**. "Spectrographs turned astronomy into astrophysics," says Goodman. "They let astronomers interrogate the light they catch in their telescopes and use it to figure out conditions in the stars."

Waves, Particles, and Light

What is light? Some scientists, such as Newton, argued that light is made of particles, like little bullets. Others, such as the Dutch physicist and astronomer Christiaan Huygens, claimed that light is made of waves, like ripples on a pond. The differences between particles

a.

b.

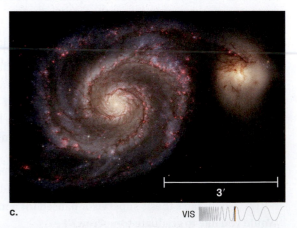

c.

VIS

Figure 4.2 The Leviathan of Parsonstown
a. The 72-inch telescope built by William Parsons, a cofounder of the Royal Astronomical Society, at Birr Castle in Ireland. Used to carry out decades of astronomical observations, the telescope (nicknamed the "Leviathan of Parsonstown") was the world's largest from 1845 until the early 20th century. **b.** Parsons studied and sketched "spiral nebulas" long before the existence of galaxies was generally accepted. **c.** His sketch of M51, the Whirlpool Galaxy, is remarkably accurate, as this photo of M51 shows.

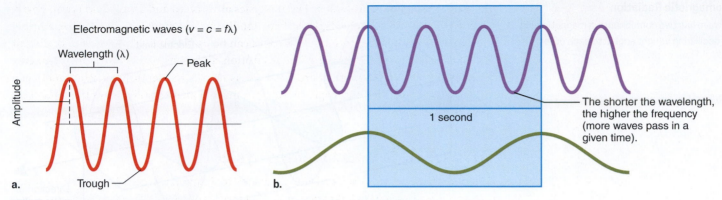

Figure 4.3 Wave Measurements
a. Wavelength and frequency are related properties in waves (like those of electromagnetic radiation—that is, light); amplitude is half of the wave's height. **b.** As wavelength λ increases, frequency f decreases when wave speed v is constant, as it is for light waves in a vacuum (for which v = c).

and waves are significant: a particle can be in only one place at a time, while a single wave can spread over an extended region of space. By the early 1800s, experiments had shown light clearly demonstrating wavelike behavior, and scientists built strong foundations for understanding light in terms of wave properties.

Nature offers many examples of waves: wind-blown ripples on a pond, ocean waves, and sound waves moving through air. But any wave, including a light wave, can be described by three basic characteristics: wavelength, wave frequency, and wave amplitude. Any wave type (whether sound waves, ocean waves, or waves on a taut string) can appear in a range of wavelengths, frequencies, and amplitudes. **Wavelength** is the distance between successive peaks or troughs of a wave (**Figure 4.3**). A wave's **frequency**, measured in hertz, is the number of peaks (or troughs) that pass a point in space in a single second. The **amplitude** of a wave is a measure of its strength.

We must consider an additional property of waves: their speed. *Wave speed* is the distance an individual peak travels in one unit of time. For example, sound waves in air travel at approximately 343 meters per second (m/s). All light waves travel at the same speed when moving through a vacuum. Physicists call this speed *c*, and measurements yield $c = 3 \times 10^8$ m/s (more specifically, 2.998×10^8 m/s). That's about a billion kilometers per hour (km/h). The tremendous speed of light makes "live" television feeds from distant events like the Olympics on the other side of the world possible. The TV signals are beamed around the planet via radio waves, making the trip in milliseconds.

For light waves, the frequency (*f*), wavelength (λ), and speed (*c*) are related in this simple way:

$$c = f\lambda$$

$$\text{speed of light} = \text{frequency} \times \text{wavelength}$$

Because for light the speed *c* is constant, this relationship enables us to convert easily from measurements of wavelength to measurements of frequency.

The Electromagnetic Spectrum

What are light waves made of? Water waves are water molecules being sloshed back and forth. Sound waves are traveling oscillations of air molecules being compressed and then relaxed. What about light waves?

Light is made of a different type of wave than water or sound. In the 1860s, physicist James Clerk Maxwell derived a set of equations predicting that light is composed of

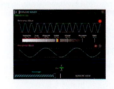

INTERACTIVE:
Waves

matter Any substance made up of particles with mass that occupies space.

spectrograph A device for separating light from a source into its component wavelengths and measuring the energy at each wavelength.

wavelength The distance between successive peaks or troughs of a wave.

frequency The number of peaks (or troughs) of a wave that pass a point in space in a single second.

amplitude A measure of a wave's strength, equal to half the wave's total height.

Figure 4.4 Electromagnetic Radiation

An electromagnetic wave has two components, a magnetic field and an electric field, oscillating at right angles to each other.

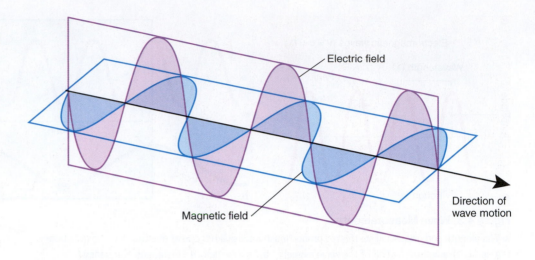

Electric field

Direction of wave motion

Magnetic field

---

electromagnetic radiation Oscillating electric and magnetic fields traveling through space.

visible light The form of light to which the human eye responds, with a wavelength ranging from 400 nm to 700 nm.

electromagnetic spectrum The full range of electromagnetic radiation at different wavelengths, from gamma rays (very short) to radio waves (very long).

radio wave An electromagnetic wave with a wavelength longer than 0.1 meter.

microwave The shortest-wavelength radio wave, with a wavelength of 1 mm (1×10^{-3} meter) to about 0.1 meter.

infrared wave An electromagnetic wave with a wavelength of 700 nm (1×10^{-9} meter) to 1 mm, just longer than that of visible light.

ultraviolet light An electromagnetic wave with a wavelength of 10 nm to 450 nm, just shorter than that of visible light.

X-ray An electromagnetic wave with a wavelength between 10^{-11} and 10^{-8} meter.

gamma ray A high-energy electromagnetic wave with a wavelength shorter than 10^{-11} meter.

refraction The bending of light as it passes from one transparent medium to another, such as from water to air.

spectrum (pl. spectra) The distribution of light intensity versus wavelength (or frequency).

intensity The energy of light that an object emits—either in total or at a specific wavelength.

---

waves of **electromagnetic radiation**: oscillating electric and magnetic fields traveling through space (**Figure 4.4**). In an electromagnetic wave, the electric and magnetic fields are perpendicular to each other, and both are perpendicular to the direction the wave travels. The laws Maxwell derived dictate that the electric-field oscillation induces changes in the magnetic field, which then induce changes in the electric field. Thus, unlike sound waves or water waves, electromagnetic waves do not require a medium and can travel through the virtual vacuum of space.

According to Maxwell's theory, **visible light**—the kind to which your eye responds—is just one form of electromagnetic radiation, with a particular range of wavelengths and frequencies. Maxwell predicted the existence of an entire **electromagnetic spectrum** of waves with a range of wavelengths running from very short to very long (**Figure 4.5**). In 1887, radio waves were first generated in a laboratory, proving Maxwell's theoretical predictions. Since then, we have come to use the electromagnetic spectrum as the basis for everything from cell phones to computer chip manufacturing. The electromagnetic spectrum is made of waves with every possible wavelength, and visible light comprises only one narrow band of possible wavelengths. For that reason, in this book all forms of electromagnetic radiation (not just those at visible wavelengths) are referred to as *light*.

Red light is the longest-wavelength visible light, representing electromagnetic waves with wavelengths of about 700 nanometers (1 nm = 1×10^{-9} meter). Violet light is the shortest-wavelength visible light, at about 480 nm. Our eyes are particularly sensitive to electromagnetic waves in the middle of this visible range, corresponding to yellow-green light at approximately 550 nm.

Radio waves, used in cell phone technology, are at the long-wavelength end of the electromagnetic spectrum, with wavelengths longer than 0.1 meter (in the frequency modulation or FM part of the radio spectrum). **Microwaves**, such as those produced by microwave ovens, fall on the short-wavelength end of radio waves, having wavelengths of 1 millimeter (1 mm = 1×10^{-3} meter) to about 0.1 meter. **Infrared waves**, which we feel as heat, have wavelengths just longer than those of visible light, ranging from 700 nm to 1 mm. Sunburn-producing **ultraviolet light** has wavelengths just shorter than those of visible light, from 10 nm to 450 nm. On the shorter end are **X-rays**, with wavelengths between 0.01 nm and 10 nm and many medical applications. **Gamma rays**, with wavelengths shorter than 0.01 nm, or 1×10^{-11} meter, are the shortest-wavelength electromagnetic radiation known and are used to kill cancer cells. "From radio to X-rays, we experience a good chunk of the electromagnetic spectrum quite a bit in our everyday lives," says Goodman. "We use low-frequency radio waves for communications, we use X-rays for medical imaging, we use ultraviolet for bug zappers."

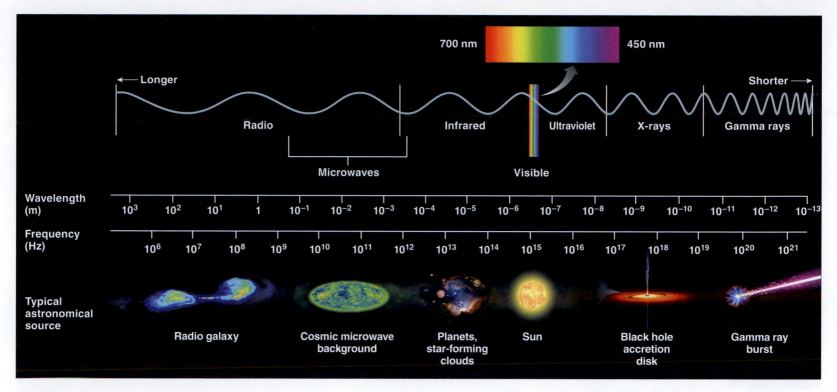

Figure 4.5 The Electromagnetic Spectrum
All parts of the electromagnetic spectrum are "light," but our eyes are attuned to wavelengths in the visible range. The wavelength (and frequency) of light varies across a continuum. While most astronomical objects radiate across a wide span of wavelengths, many emit a significant fraction of their light in a more limited wavelength region, as shown.

Section Summary

- Light is a wave, whose characteristics can be described in terms of wavelength, frequency, and amplitude.
- In a vacuum, light travels at a constant speed (c), which is related to frequency (f) and wavelength (λ) as $c = f\lambda$.
- Light is electromagnetic radiation that exists in a broad range of wavelengths and corresponding frequencies.

CHECKPOINT If the frequency of a wave increases but its speed does not change, does the wavelength increase, decrease, or stay the same?

> "We use low-frequency radio waves for communications, we use X-rays for medical imaging, we use ultraviolet for bug zappers."

Astrophysical Spectra

4.2 Light from the Sun is called *white light* because it is composed of electromagnetic waves of many different wavelengths—which, in the visible part of the spectrum, correspond to different colors. When sunlight passes through a glass prism, the bending of the various wavelengths at different angles breaks up the white light into its constituent colors. This bending of light as it passes from one transparent medium to another is called **refraction**. Separating wavelengths like this is what astronomers call "taking a spectrum" (**Figure 4.6**). A **spectrum** tells us how much energy—also referred to as **intensity**—the energy source has put into each wavelength (or frequency) of light. For example, a blue star emits more energy in the blue portion of the spectrum than a red star does. Notice the distinction between the spectrum of an object

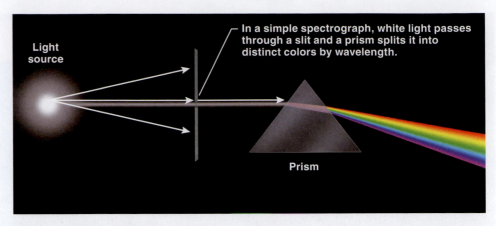

Figure 4.6 Taking a Spectrum
Scientists investigate the interactions of matter and light by separating the light into the wavelengths that compose it. A prism separates a beam of "white" light into its component wavelengths (colors).

INTERACTIVE:
Kirchhoff's Laws

"We all are radiating blackbody radiation right now."

emission line A bright line at a particular wavelength in a spectrum. Specific patterns of emission lines indicate the presence of a specific chemical element in the source.

absorption line A dark line in a continuous spectrum. Specific patterns of absorption lines indicate the presence of a specific chemical element in the source.

energy The ability to perform work. Energy comes in many forms, such as kinetic, thermal, gravitational, and electromagnetic.

kinetic energy Energy that is due to an object's bulk motion.

gravitational potential energy Energy that an object possesses because of its position in a gravitational field.

thermal energy or heat energy Energy that is due to random motions of atoms.

and the electromagnetic spectrum, described in Section 4.1. The electromagnetic spectrum is the entire range of wavelengths (and frequencies) of electromagnetic radiation. The spectrum of any particular object is the pattern of energy and wavelengths that object emits or absorbs.

Kirchhoff's Three Laws

The invention of the spectrograph enabled physicists to take sensitive readings of the amount of energy emitted in each wavelength of light. They immediately began pointing spectrographs at different luminous sources, such as wood fires, a chunk of burning sulfur, and the Sun. Recording spectra from these objects, physicists were astonished to find that each source gives off a distinct mix of colors. The spectrum from some objects contained a smooth, continuous blend of wavelengths called a *continuous spectrum*. For others, only a few wavelengths appeared in the spectrum. For still others, dark lines were superimposed on a continuous spectrum, as if energy had been taken out of the continuous spectrum at just a few wavelengths. Only after considerable effort were physicists able to group the different kinds of spectra and link them to specific physical conditions from various sources.

In unraveling the mystery of spectra, physicists had to take a crucial step that is common to the development of any science. Science always begins with a mass of observations that must be sifted through; the goal is to categorize these observations and identify a pattern.

This is what happened with all the early spectra that scientists acquired. By the 1860s, building on the contributions of many other scientists, the German physicist Gustav Kirchhoff found patterns hiding in the spectral data. Now known as Kirchhoff's laws, these patterns define the kinds of spectra that different sources of electromagnetic radiation produce. In short, Kirchhoff's laws state the following:

1. An opaque body emits a continuous spectrum (**Figure 4.7a**). Example: the hot filament of an illuminated lightbulb.

2. Relatively hot, low-density gas emits a sequence of bright spectral lines, called **emission lines** (**Figure 4.7b**). Example: a neon sign.

3. Relatively cold, low-density gas *in front of a hotter, opaque body* produces a continuous spectrum with dark spectral lines superimposed, called **absorption lines** (**Figure 4.7c**). Example: light from the solar surface passing through cooler gas in the atmosphere.

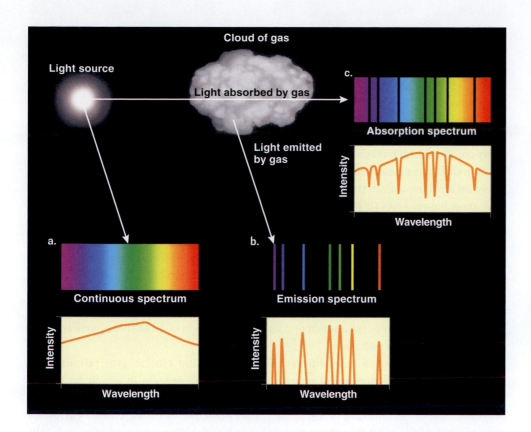

Cloud of gas

Light source

Light absorbed by gas

c.

Absorption spectrum

Intensity

Wavelength

Light emitted
by gas

a.

Continuous spectrum

Intensity

Wavelength

b.

Emission spectrum

Intensity

Wavelength

Figure 4.7 Kirchhoff's Three Laws
Kirchhoff's laws linked the observed differences among the three kinds of spectra—**a.** continuous, **b.** emission, and **c.** absorption—to the physical situation required to generate each one.

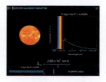

INTERACTIVE:
Blackbody Radiation

Kirchhoff's laws established a key link between light and matter, which astronomers needed in order to understand the distant objects they observed by telescope. The next step was to explain why the patterns of continuous spectra, emission lines, and absorption lines exist.

Blackbody Radiation: Turning Heat into Light

Kirchhoff's first law focuses on continuous spectra. *Continuous* means that the distribution of energy across wavelengths is smooth. Light at all wavelengths is, to some degree, present in the spectrum—but some wavelength bands have more energy than others. Matter can produce a continuous spectrum in a variety of ways. The most important for physics and astrophysics is the link between heat and light, leading to *blackbody radiation*.

ENERGY AND HEAT. Physicists define **energy** as the ability to do work (such as lifting a weight). Energy can take many forms—kinetic, thermal, gravitational, electromagnetic—all of which can be precisely defined and described mathematically. By the early 1800s, scientists had figured out how to express the energy locked up in an object's motion, called **kinetic energy**. A freight train moving at 80 km/h has much more kinetic energy than a bicycle rolling along at 10 km/h because it is both more massive and moving faster. Energy is measured in units called *joules* (J). A baseball pitched at 90 miles per hour (equivalent to 40 m/s) has about 117 J of energy. Scientists had also come to understand **gravitational potential energy**—the energy an object has in a gravitational field. A book on a shelf 3 meters above a floor has more gravitational potential energy than the same book on a shelf just 1 meter above the floor because it has a longer distance to fall.

Heat is the transfer of thermal energy. **Thermal energy** (sometimes called **heat energy**) is the total kinetic energy locked up in the random motions of atoms and mol-

ALYSSA GOODMAN

Not every astronomer has been featured on a game show for kids, but Alyssa Goodman, a professor of astronomy at Harvard University, has had that experience and so is not your everyday astronomer. Her interest in imaging led her to seek an understanding of how other fields represent data. "We need a way of putting our images together into something that looks more like a 3D picture. People in other fields, like health science, figured out how to do what we needed with our radio telescope data. We call the project astronomical medicine; it gives people 3D views of very large regions of space." Goodman's passion for visualization extends to her work as an artist. Her interest in astronomy has taken her to telescopes around the world. "It's amazing to be up at 14,000 feet working all night under the stars," she says.

Figure 4.8 Thermal Energy

The random motions of the atoms in matter generate thermal energy. Thermal radiation is the emission of photons resulting from the rapid motions of atoms in a dense object.

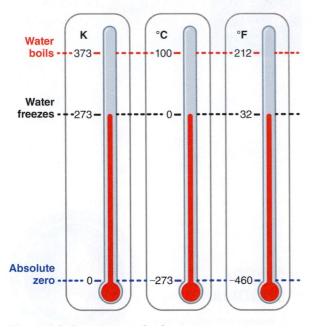

Figure 4.9 Temperature Scales

A comparison of the Kelvin, Celsius, and Fahrenheit temperature scales.

AT PLAY IN THE COSMOS THE VIDEOGAME You can use blackbody radiation and its peak temperature to earn some much-needed credits in carrying out a survey for the Corporation in Mission 9.

ecules in a substance (**Figure 4.8**). Just as the kinetic energy of a freight train is associated with the bulk motion of the whole train moving in one direction, so is the thermal energy of the hot gas in an oven associated with motion. In a hot oven, gas atoms randomly collide with one another and the oven's walls. Faster random atomic motion means more heat.

Temperature, a measure of the average kinetic energy of the randomly moving atoms and molecules in a substance, is expressed as a scale where low is considered relatively colder and high is relatively hotter. Scientists measure temperature in units called **kelvins (K)**. In the Kelvin scale, 0 is called *absolute zero*; at this lowest value, there is absolutely no motion—a state that is not actually possible. For comparison, water freezes at 32 degrees Fahrenheit (32°F), 0 degrees Celsius (0°C), and 273 K (**Figure 4.9**). Note that the term *degrees* is not used in the Kelvin scale.

"Everything has a temperature," says Alyssa Goodman, "and temperature is a measure of how fast atoms are moving inside an object. If the object is very, very cold, the atoms or the molecules aren't moving very much. If something is very hot, they're moving a lot." Even in a solid, where atoms are locked in place, temperature translates into the vibrations of the atoms back and forth. Goodman is quick to point out the crucial relationship for an astronomer between heat and light. "The motion of all of these particles," she says, "generates lots of radiation."

Although energy can have different forms (including kinetic, gravitational, and thermal), the *total energy* cannot change in a system or process in which no energy is added or removed. (Consider a closed thermos, a vacuum-insulated container that prevents heat from entering or escaping.) This principle is called **conservation of energy**—one of the most important principles in physics. Conservation of energy tells us that even though energy can be converted from one form to another, such as gravitational energy being converted into kinetic energy in a falling object, the sum of all of the different forms of energy cannot change.

THERMAL ENERGY AND BLACKBODY RADIATION. The radiation produced by the motion of atoms takes the form of electromagnetic waves, which are generated when electrically charged particles accelerate or decelerate. The random motions in a warm, dense collection of atoms naturally produce accelerations and decelerations as particles jiggle around. The adjective *dense* is important—if enough atoms are packed into a small space, their collective random jiggling motion (acceleration and deceleration) produces a characteristic pattern of light waves. This pattern is a special kind of continuous spectrum called a *blackbody spectrum*. **Blackbody radiation** is the electromagnetic radiation produced by the thermal motions of atoms in a dense, opaque object.

"Max Planck was the one who figured out the relationship between an object's temperature and how much radiation it emitted at every different wavelength," says Goodman. To make his breakthrough in understanding blackbody radiation, Planck first had to realize that light exists in tiny bundles, or *quanta*, of energy (*quantum* is the German word for "package"). He then figured out a relationship between the energy (E) and frequency (f) of these light quanta, which came to be called **photons**:

$$E = hf$$

energy = Planck's constant × frequency

The new constant of nature h was called, appropriately, *Planck's constant*, which is equal to 6.626×10^{-34} m² kg s⁻¹. Planck's constant would come to define the world of the very small, including atoms and their building blocks. Using this relationship between energy and frequency, Planck made a remarkable theoretical leap. He showed how any object that can be considered a blackbody emits a *characteristic spectrum*, and that spectrum depends on temperature only.

"It's really amazing," says Goodman. "Temperature is the only property of the source that sets the shape of its blackbody spectrum. A gold ball at $T = 300$ K (80°F) will produce the same spectrum as a plastic ball at the same temperature." Remember from Kirchhoff's first law that dense objects produce continuous (that is, blackbody) spectra. "If that definition seems really general," says Goodman, "that's because it needs to be. We all are radiating blackbody radiation right now."

Max Planck showed that all blackbody radiation has the same characteristic spectrum: a wide range of emitted wavelengths with a strong peak around a single wavelength (**Figure 4.10**). That peak represents the wavelengths (or, equivalently, frequencies) around which most of the energy is emitted. A simple formula called **Wien's law** relates an object's temperature T to the wavelength λ_{max} where the maximum emission occurs:

$$\lambda_{max} = (0.0029 \text{ m K})/T$$

The unit "meter kelvins" (m K) is needed so that dividing by T, measured in kelvins, gives the wavelength in meters.

"Wien's law explains a lot of things," says Goodman. "You heat up an iron poker and first it turns red, and then brighter red, then bright orange, until it finally glows white. The different colors reflect the change in temperature of the iron and where the wavelength of the spectrum's peak is." In this example, the temperature of the "red hot" iron poker puts the peak wavelength λ_{max} in the infrared portion of the spectrum, but in the visible portion it is dominated by red light. As the temperature rises, the peak shifts to shorter wavelengths and includes enough colors for the poker to appear "white hot." Wien's law also tells us that the iron poker emits radiation even before you stick it in the fire; the peak wavelength is just too long (in the infrared region) for your eyes to detect. This is, therefore, also the spectral region where human beings emit the peak of their blackbody radiation (**Figure 4.11**).

Wien's law is remarkably powerful for astronomy. Stars are big, dense balls of hot gas, so they fulfill the necessary conditions to be blackbodies. Just a glance at a star's color can tell you something about the conditions on its surface, even if the star is billions and billions of kilometers away. "We can't just go stick a thermometer in a star," says Goodman, "but using the properties of blackbody radiation, we can measure its temperature just by taking its spectrum." The Sun's spectrum peaks in the green

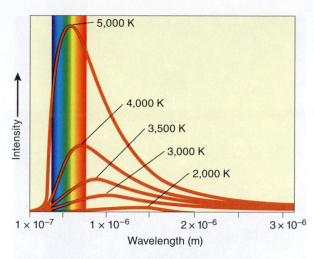

Figure 4.10 Wien's Law and Blackbody Spectra
A blackbody (an object emitting blackbody radiation) has a spectrum in which the wavelength of peak radiation depends only on the object's temperature.

"We can't just go stick a thermometer in a star, but using the properties of blackbody radiation, we can measure its temperature just by taking its spectrum."

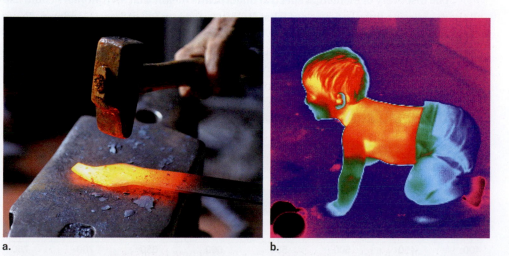

a.   **b.**

Figure 4.11 Blackbodies
Any warm, dense object—including the human body—produces a blackbody spectrum. **a.** An iron rod, heated in a blacksmith's oven, glows in the red region of the spectrum. **b.** When it cools, it will radiate most strongly in the infrared, as does this baby.

temperature A measure of the average kinetic energy of all particles in a substance.

kelvin (K) A unit of temperature in the Kelvin scale, where 0 means absolutely no motion. One kelvin has the same magnitude as 1°C.

conservation of energy The principle whereby total energy remains constant in a system or process in which energy is not added or removed, even though the energy can be converted from one form into another.

blackbody radiation Electromagnetic radiation from a dense opaque body, whose spectrum depends only on the body's temperature.

photon A particle of light.

Wien's law The mathematical relationship between a blackbody's temperature and the wavelength at which it radiates with peak intensity. A hotter object peaks at a lower wavelength.

Figure 4.12 Stars Radiate as Blackbodies
Astronomers can tell the approximate surface temperature of a star just from its color. The Sagittarius Star Cloud presents a wide range of temperatures—from hot, massive blue stars to cooler red ones.

"Astronomers could use all those emission lines to figure out what stuff in space is made of."

luminosity A light source's total energy output per unit of time.

Stefan-Boltzmann law The mathematical relationship between a blackbody's luminosity and its surface area and temperature. Hotter, larger objects glow more brightly.

nebula An interstellar cloud of dust, hydrogen, helium, and other ionized gases.

portion of the visible-light region at $\lambda_{max} = 502$ nm. Using Wien's law, you would find that the Sun's peak wavelength corresponds to a temperature T of 5,777 K (**Figure 4.12**; see **Going Further 4.1** on p. 90).

THE STEFAN-BOLTZMANN LAW AND LUMINOSITY. One last consequence of blackbody radiation deserves mention. Just as Wien's law tells us that a blackbody's peak wavelength depends on its temperature, the **luminosity** (total energy output per unit of time) also depends on temperature as well as on the total surface area of the blackbody. The exact relationship is this:

$$L = \sigma A T^4$$

luminosity = Stefan-Boltzmann constant $\times$ surface area of blackbody $\times$ (temperature)^4

This formula is called the **Stefan-Boltzmann law**, and σ (pronounced "sigma") is a constant of nature called the *Stefan-Boltzmann constant* (equal to 5.7×10^{-8} watts m^{-2} K^{-4}). Notice how strongly luminosity depends on temperature in this law: $L \propto T^4$. Thus, the law tells us that if two blackbodies have the same surface area but temperatures that differ by a factor of only 2, their energy output will differ by a factor of $(2)^4 = 16$.

Emission and Absorption Spectra

By connecting random, thermal motions of atoms in dense collections of matter to the radiation they emitted, Planck explained Kirchhoff's first law. Explaining Kirchhoff's next two laws would prove to be more complex. Linking absorption and emission spectra to the structure of matter would demand redefining the very meanings of matter, light, and physics.

Physicists discovered that a pure sample of a chemical *element* (matter composed of one type of atom) produces its own characteristic emission spectra. Heating a sample of hydrogen gas produces a very different set of bright emission lines than does heating a sample of helium. When mercury vapor is heated until it glows, the light it emits shows a few prominent bands of color in the visible part of the spectrum. Sodium, in contrast, produces two very bright lines in the yellow region of the spectrum (**Figure 4.13**). Careful study showed astronomers that each element has its own emission line "fingerprint."

"The discovery of elemental spectral fingerprints meant that astronomers could use all those emission lines to figure out what stuff in space is made of," says Goodman. "Now

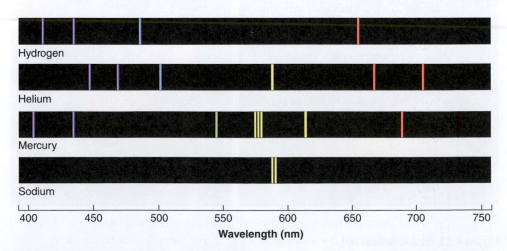

Figure 4.13 Emission Spectra of Elements
The emission spectra of hydrogen, helium, mercury, and sodium reflect the wide variation in energy levels, which are unique to each element.

that is really remarkable." Kirchhoff's second law tells us that relatively hot, tenuous gas is the source of emission lines. (*Tenuous* means the gas is not opaque to the radiation we're interested in; see Figure 4.7b.) Space is full of hot, tenuous gases in the form of diffuse clouds. Using emission lines produced by these astronomical **nebulas** (**Figure 4.14**), astronomers began an inventory of the gas clouds' chemical composition. Says Goodman, "Astronomers started taking a census of the elements in the Universe."

Absorption line spectra proved to be just as useful for astronomers as emission line spectra were. Examining starlight with spectrographs, astronomers found the continuous blackbody spectra expected from a hot, dense gas. Superimposed atop the stellar blackbody emission, however, was a series of dark bands (see Figure 4.7c). These absorption lines turned out to occur at exactly the same wavelengths as the emission lines that scientists had mapped out when studying heated samples of pure elements. The same elemental fingerprints showed up, this time as absorption lines. But where was the gas that was doing the absorbing in those stellar spectra?

Kirchhoff's third law says that absorption lines occur when a relatively cold, tenuous gas is in front of a source of continuous radiation. Blackbody radiation from stars comes from their dense surfaces. Above the surface is a rarefied (low-density) atmosphere. As stellar blackbody radiation passes through this colder, thinner atmospheric gas, light is absorbed, producing the dark lines that astronomers see with spectrographs (**Figure 4.15**). In this way, absorption lines enable astronomers to determine the chemical composition of stellar atmospheres and, by inference, the composition of the stars themselves. Absorption spectra were also critical for astronomers like Cecilia Payne-Gaposchkin, who used them to help create a way of categorizing stars that led to understanding stellar evolution. And for astronomers like Meghnad Saha, absorption spectra were the key to understanding the physical structure of a star's atmosphere (see **Anatomy of a Discovery** on p. 91).

None of the discoveries related to Kirchhoff's laws explained why emission and absorption lines are produced in the first place. "It was very hard for physicists and astronomers to imagine what was going on," explains Goodman. "Why was there lots of light at some particular wavelength, and no emission at some other wavelength? And why did they get absorption or emission depending on the different circumstances in the gas? It was all kind of mystifying." Answers to these questions came only when new insight was developed into the nature of atoms, and scientists found that the "nanoworld" of atoms behaves very differently from the macroworld we know.

Section Summary

- Light is emitted and received in a spectrum—a distribution of wavelengths and intensities. Kirchhoff's laws define the kinds of spectra that different sources produce.
- The thermal motions of a dense object produce blackbody radiation, so an object's blackbody spectrum reveals its temperature.
- Each chemical element can emit or absorb only certain specific wavelengths of light, as shown in emission or absorption spectra.

CHECKPOINT   What are the differences between the three types of spectra and what can scientists learn from each type?

Figure 4.14 Nebular Spectra
Once the links between matter and the emission of light were fully understood, astronomers could examine the spectra of nebulas such as this one (NGC 6326) to determine the elements that compose them.

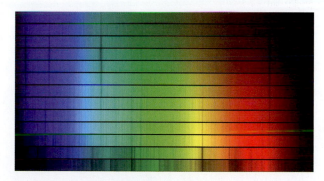

Figure 4.15 Stellar Spectra
Spectra from a number of stars. Superimposed onto the continuous spectra of the stars are the absorption lines created when the star's light (a blackbody spectrum) passes through its own cooler atmosphere.

AT PLAY IN THE COSMOS THE VIDEOGAME   Use your knowledge of atomic spectra to identify much-needed fuel for your ship in the Crab Nebula in Mission 4.

INTERACTIVE:
Atomic Structure: Absorption and Emission

Spectra and the World of the Atom

4.3   The first important step to explaining Kirchhoff's laws came when physicists recognized the dual nature of light. For more than 200 years scientists had believed that light is a wave. At the beginning of the 20th century, they realized light can also act like a particle. Recall from Section 4.2 that Planck recognized that light energy comes in quanta: discrete bundles of light particles now called photons.

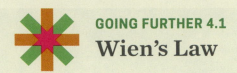

GOING FURTHER 4.1
Wien's Law

Wien's law enables us to find the peak wavelength λ_{max} of emission coming from a blackbody at temperature T. At what peak wavelength does a star with surface temperature $T = 2{,}980$ K radiate? Let's begin with the Wien's law equation and plug in 2,980 K for T:

$$\lambda_{max} = (0.0029 \text{ m K})/T = (0.0029 \text{ m K})/(2{,}980 \text{ K})$$

$$= 9.73 \times 10^{-7} \text{ m}$$

Now we must convert meters to microns (μm), the unit astronomers often use for infrared radiation, where

$$1 \, \mu\text{m} = 1 \times 10^{-6} \text{ m}$$

We can convert the previous result as follows:

$$\lambda_{max} = (9.73 \times 10^{-7} \text{ m}) \times \left(\frac{1 \, \mu\text{m}}{1 \times 10^{-6} \text{ m}}\right) = 0.97 \, \mu\text{m}$$

Thus, most of the energy from the star radiates at a wavelength of 0.97 μm, which is in the infrared portion of the spectrum.

We can do the same for the Sun's spectrum, which peaks in the green portion of the visible-light region, at $\lambda_{max} = 502$ nm. First we convert nanometers to meters (at this wavelength, there's no need to work in microns):

$$502 \text{ nm} \times (1 \times 10^{-9} \text{ m/nm}) = 5.02 \times 10^{-7} \text{ m}$$

We rearrange Wien's law to solve for temperature T and get the Sun's temperature:

$$T = (0.0029 \text{ m K})/\lambda_{max}$$

$$= (0.0029 \text{ m K})/(5.02 \times 10^{-7} \text{ m}) = 5{,}780 \text{ K}$$

How can light be both a particle that exists in only one place at one time and a wave that spreads out through space? Physicists and astronomers needed a way to reconcile this *wave-particle duality* of the photon—its ability to act as either a wave or a particle. Ultimately, the study of atoms and light opened up a fascinating, new field that physicists called **quantum mechanics** or **quantum physics**. Quantum mechanics is the science of the very small—the world of molecules, atoms, and smaller particles.

"What's happening inside atoms to create spectral lines?"

Developing a Model of the Atom

Atomic spectra are Pat Hartigan's bread and butter. Like Alyssa Goodman, Hartigan studies star formation. He also uses spectral lines as a tool for understanding conditions in star-forming clouds hundreds of light-years away. "Every spectral line is associated with a particular kind of atom," says Hartigan. "So the questions scientists had to answer were, What's happening inside atoms to create spectral lines? Why do individual atoms emit light at only certain wavelengths?"

Atoms are composed of three basic types of particles: neutrons, protons, and electrons. At the center of every atom is a **nucleus**, composed of **neutrons** with no net electric charge and **protons** with positive electric charge. Surrounding the nucleus are **electrons**, which carry negative electric charge. By themselves, atoms are electrically *neutral*, meaning that the overall electric charge is zero because the number of electrons orbiting the nucleus matches the number of protons in the nucleus.

"In the early 20th century, physicists thought atoms might be like little solar systems," says Hartigan. "The electrons were like planets, and the nucleus played the role of the Sun." The hope was to use this solar-system model of the atom to explain line spectra as electrons shifting in their orbits around the nucleus—but the model could not explain why light is emitted only at specific wavelengths. That explanation did not come until the young Danish physicist Niels Bohr made an imaginative leap.

quantum mechanics or quantum physics A branch of physics dealing with physical phenomena at the level of molecular, atomic, and subatomic scales.

atom The smallest unit of matter that retains the properties of a given chemical element.

nucleus (pl. nuclei) The dense region at the center of an atom, consisting of protons and neutrons.

neutron A subatomic particle with no net electric charge.

proton A subatomic particle with positive electric charge.

electron A subatomic particle with negative electric charge.

What are the temperatures and composition of stars?

observations

Edward Pickering and Harvard College Observatory "Computers"

Late-19th-century astronomers, mostly men, at the Harvard College Observatory collected thousands of spectral images of stars on glass photographic plates. The massive amount of data the plates held was painstakingly studied and characterized by brilliant women, revolutionizing astrophysics.

hypothesis

Edward Pickering, the observatory's director, hired "computers" including **Williamina Fleming**, Antonia Maury, and Annie Jump Cannon to do this work. At first they classified spectra according to the *strengths* of hydrogen (H) absorption lines.

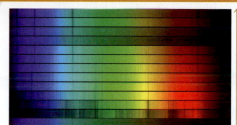

Temperature

Spectra ordered by H line strength

revised hypothesis

Correlating absorption line *breadths* to star colors, **Annie Jump Cannon** reordered the spectra by temperature. She assumed that broader, not stronger, lines indicate hotter stars. At the time (1901), astronomers did not know why the hottest stars have the weakest H absorption lines. Nonetheless, her system worked and is still used.

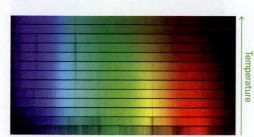

Temperature

Spectra reordered by H line breadth

confirmed by observation

In 1920, **Meghnad Saha** described mathematically how temperature affects the creation of ions of each element in a star. In 1925, **Cecilia Payne-Gaposchkin** applied his work to the formation of absorption lines in stellar atmospheres. She showed how ionization in the atmospheres of the hottest stars produces the weakest H absorption lines, confirming Cannon's system and revealing stars to be composed primarily of hydrogen and helium. Her discoveries, enabled by the Harvard Computers, are fundamental to our understanding of the Universe.

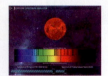

INTERACTIVE:

Spectrum Analyzer

Bohr imagined a very different kind of behavior, *quantum behavior*, for electrons in an atom. His model, the **Bohr atom** (Figure 4.16), still had electrons orbiting a central nucleus, but the young physicist's radical change was to restrict the location of the orbits. In the Bohr atom, electrons can "live" only on a specific set of well-defined orbits at well-defined distances from the nucleus. The innermost orbit has a fixed radius r_0 and the lowest energy; it is referred to as the **ground state**. An electron in the Bohr atom can never get closer to the nucleus than the ground-state orbit. The next orbit, called the first **excited state**, has a larger radius than the ground state: $r_1 > r_0$. The second excited state has an orbit with an even larger radius, and so on, up to a certain fixed number of orbits for each element. In this way, orbits in the Bohr atom form a series of concentric circles of ever-greater radius, each centered on the nucleus.

Spectra and the Bohr Atom

By "quantizing" the orbits (allowing only certain orbits to exist), Bohr offered a simple explanation of atomic spectra. In Bohr's model, both emission and absorption spectra occur when electrons jump from one orbit to another. Jumps from higher to lower orbits lead to the **emission** of a photon of a specific wavelength. Jumps from lower to higher orbits require the **absorption** of a photon of specific wavelength. Physicists later learned that electrons are actually smeared out and that their orbits are best imagined as "probability clouds" (Figure 4.17). Still, the idea of jumps from one orbit to another persisted.

"The secret to the Bohr atom is the energy difference between the orbits," says Hartigan. Each orbit has a specific energy associated with its distance from the nucleus, so physicists call an orbit and its discrete energy value an **energy level**. Bohr's crucial step came in recognizing the relationship between energy levels and the photons' wavelengths. In the Bohr atom, the energy difference between two orbits ($E_{\text{difference}}$) corresponds exactly to the photon energy (E_{photon}) associated with an electron jumping between those orbits ($E_{\text{difference}} = E_{\text{photon}}$). If the electron jumps down from a higher orbit to a lower one, it loses energy, $E_{\text{difference}}$, and gives up a photon that has energy E_{photon}. "The Bohr atom doesn't allow light with some random amount of energy to be emitted," says Hartigan. "Only photons with energy equal to jumps between orbits are going to appear."

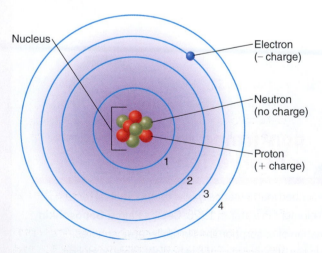

Figure 4.16 The Bohr Atom

In this classic model, neutrons and protons make up the nucleus of the atom, while electrons move around the nucleus on specific orbits. This model served only as a starting point for understanding the quantum mechanical nature of atoms.

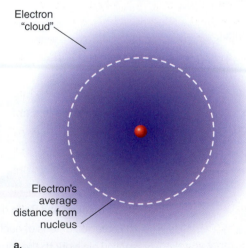
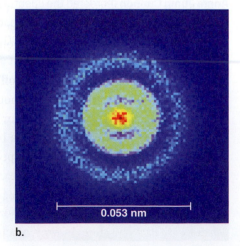

Figure 4.17 Electron Orbits

Quantum mechanics implies that electrons spread out around the nucleus in a probability cloud. As electrons are excited into higher orbits, their cloud changes shape. **a.** Hydrogen's first excited state. **b.** A hydrogen atom in the third excited state. This photo was produced by combining many images of electrons positioned around the nucleus.

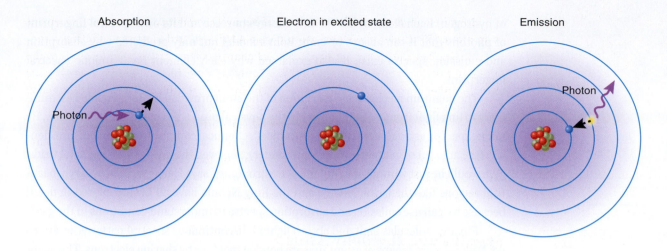

Absorption   Electron in excited state   Emission

Figure 4.18 Energy Levels of an Atom
Niels Bohr imagined—correctly—that electrons can occupy only certain discrete energy levels around the nucleus of the atom. Energy levels, each higher than the preceding one, can be envisioned as successively larger orbits of the electron(s) around the nucleus. Moving between energy levels requires the absorption (jumping to a higher level) or emission (jumping to a lower level) of a specific wavelength of photon.

A photon's energy and frequency are linked (remember the equation $E = hf$ from Section 4.2). That means the frequency f of an emitted photon depends on its energy E_{photon}. So an electron jumping down between orbits can emit only a photon whose energy exactly balances the energy difference between those orbits. And if the photon's energy is specified by the jump, so is its frequency (and wavelength, via the equation $c = f\lambda$ from Section 4.1). The origin of emission spectra lies in the fact that the energy difference between orbits must equal the energy of the emitted photon. Only wavelengths corresponding to quantum jumps between energy levels can be part of the spectrum.

Absorption is just the opposite process. An electron must absorb a photon to go from a lower energy level to a higher one. The energy of the absorbed photon must equal the energy difference between the two orbits. "Bathe an atom in light representing a continuous spectrum, and the atom will see photons with all possible wavelengths," says Hartigan, "but only photons that can kick an electron between specific energy levels can be absorbed." Thus, absorption spectra show a continuous distribution of light with only certain wavelengths missing. Those are the wavelengths of photons that can be absorbed and bump electrons up to higher energy levels (**Figure 4.18**).

Alyssa Goodman uses a common analogy to understand the process of line emission and absorption in the Bohr model. "Think of it like a staircase," says Goodman. "It takes a certain amount of energy to go up a step." In the Bohr atom, however, not all "steps" would have the same height. The old solar-system model for the atom, which allowed any orbit and any energy, was more like a ramp than a staircase. "You can be anywhere you want on the ramp," says Goodman. "But you are only allowed to be on one step or on another. You can't just be floating in between them."

The Bohr model explained Kirchhoff's laws in terms of atomic structure. According to Kirchhoff's second law, a relatively hot, tenuous gas produces an emission spectrum. *Hot* is a critical part of the description for the Bohr atom explanation. In a hot gas, atoms move at high speed, careening off each other in wild collisions. Every time two atoms smash together, an electron can be bumped from the ground state to a higher orbit. That's why a lot of atoms in a hot gas are in excited states with electrons in higher orbits that could emit a photon and jump back down. That is why hot gases produce emission spectra. The same logic in reverse applies to absorption spectra. Electrons in the atoms of a cold gas are in the ground state. In front of a light source generating a continuous spectrum, the electrons of these atoms are ready to absorb emitted photons—but only those photons whose energy matches upward jumps to higher orbits.

In the Bohr model, each element has a different number of protons and electrons, so atoms of different elements have different sets of quantized orbits. In other words, the energy levels of helium (with two protons and two electrons) are unlike the energy levels

> "Think of [the energy levels in an atom] like a staircase. It takes a certain amount of energy to go up a step."

Bohr atom A model of the atom, devised by Niels Bohr, in which electrons in discrete orbits surround a positively charged nucleus. Electron "jumps" between the discrete orbits determine the emission or absorption of light.

ground state The lowest energy state of an atom.

excited state Any energy state that is higher than the ground state of an atom.

emission The release of a photon of a specific wavelength when an electron jumps from a higher to a lower energy level.

absorption The capture of a photon of a specific wavelength, causing an electron to jump from a lower to a higher energy level.

energy level One of the certain discrete values of orbital energy possible for an atom.

of hydrogen. Each element has its own energy staircase and its own spectral fingerprint of photons that it can absorb or emit. Bohr's model not only predicted why absorption and emission spectra exist; it also explained why each element has a unique spectral fingerprint.

The quantum mechanical description of the atom provided by Bohr enabled astronomers to link light from a distant object to the physical properties of that object directly. By taking spectra, astronomers could tell an object's chemical composition, temperature, density (the amount of matter in a cubic centimeter), and even the presence of magnetic fields. "We would be lost without spectral lines in astronomy," says Hartigan. "Those large, beautiful regions like the Orion Nebula and the Ring Nebula have all been explored in detail because we can use emission and absorption spectra to understand conditions in the gas."

Finally, molecules emit and absorb light in discontinuous spectral patterns, as atoms do. *Molecules* are groups of atoms that are bonded together by sharing electrons. The water molecule, for example, is composed of two hydrogen atoms and one oxygen atom (H_2O). Just as a heated sample of atoms of a single element produces a characteristic emission spectrum, a heated sample of a single kind of molecule also produces a unique emission spectrum. We will see that emission by molecules plays an important role in aspects of astronomy.

The Hydrogen Atom

"The hydrogen atom has only the one proton in the nucleus, so its energy levels are the simplest of them all," says Hartigan. Hydrogen's simplicity is good news for astronomers because about 75 percent of all the so-called normal matter (the stuff that appears on the periodic table) in the cosmos is hydrogen. The spectrum of hydrogen is the key to understanding the properties of many astrophysical environments. This spectrum also explains why so many gas clouds appear red.

In a diagram of hydrogen's energy levels, the first two steps are the most important for astronomers, because that's where electrons spend most of their time. Transitions into and out of the first two levels form two series of emission and absorption lines that are often very prominent: the Lyman series and the Balmer series.

The **Lyman series** involves photons produced by transitions that start or end in the ground state (**Figure 4.19**). As discussed in later chapters, Lyman-series photons, which have wavelengths outside the visible range (in the ultraviolet or UV range), are critical to understanding the evolution of matter in the Universe.

Photons involved in the **Balmer series** are produced by transitions that either drop downward to the first excited state or jump upward from it (Figure 4.19). Astronomers often use the letter H (for "hydrogen") to refer to lines in the Balmer series. The jump from the first excited state to the second excited state is called the *H alpha* (or *Balmer alpha*) *transition* and requires a Balmer alpha photon. "Unlike the Lyman series, the brightest line of the Balmer series shows up in the visible part of the spectrum," says Hartigan. In fact, H alpha photons have a wavelength of 656.3 nm, in the red part of the spectrum. "There is lots of hydrogen gas in these clouds," says Hartigan. "That can mean a lot of red H alpha photons for our telescopes to capture."

Spectra and the Doppler Shift

Something else light can tell us is the velocity of an object toward or away from Earth. Astronomers use spectral lines as a kind of cosmic speed gun to help map the motions of the entire Universe. To do so, they take advantage of a phenomenon called the **Doppler shift**, which is the change in wavelength (and therefore frequency) of waves brought about by an object's motion.

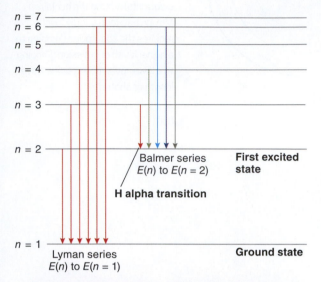

a. Electron transitions for the hydrogen atom

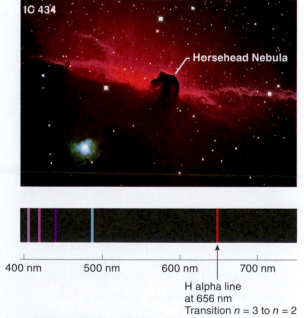

b. Emission nebula IC 434

Figure 4.19 Lyman and Balmer Series
a. Even the simplest atom, hydrogen, has many energy levels—labeled *n*—and therefore many energy transitions—labeled *E(n)*. Of these, the H alpha transition is responsible for the bright-red color of nebulas. **b.** Surrounding the Horsehead Nebula is IC 434, an emission nebula that produces a spectrum with a prominent H alpha line.

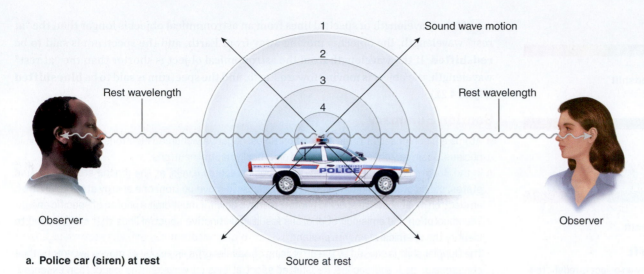

1   Sound wave motion

2

3

4

Rest wavelength

Rest wavelength

Observer

Observer

a. Police car (siren) at rest   Source at rest

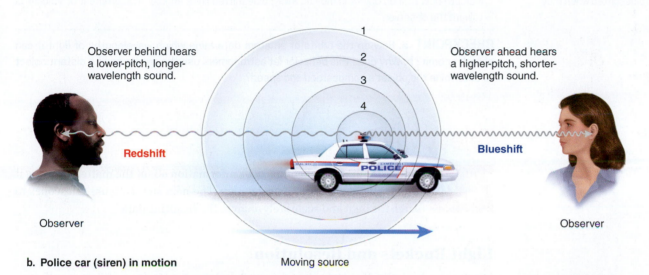

1

2

3

4

Observer behind hears a lower-pitch, longer-wavelength sound.

Observer ahead hears a higher-pitch, shorter-wavelength sound.

Redshift   **Blueshift**

Observer

Observer

b. Police car (siren) in motion   Moving source

Figure 4.20 Doppler Shift
a. The pitch of a police car's siren remains constant when the car is not moving. The numbered circles represent individual sound waves. **b.** The Doppler shift is the reason that the sound of the siren increases in pitch as the car moves toward us but decreases in pitch as the car moves away.

A Doppler shift occurs when either the receiver or the source of the waves is in motion. Consider sound waves, such as those produced by a stationary police car and its siren (**Figure 4.20a**). Waves of equal wavelength spread in all directions. But when the police car is moving toward you, sound waves pile up as the car catches up a little with each wave that has been emitted (**Figure 4.20b**). That means the sound's wavelength decreases, and the frequency increases (the siren has a higher pitch). When the police car is moving away from you, the sound waves are stretched out as the vehicle gets a little farther from each emitted wave. The wavelength increases, and the frequency decreases, giving the siren a lower pitch.

The same basic process occurs for light emitted by an astronomical object. Both the wavelength and the frequency of light change when the source is in motion. To measure the object's motion accurately, however, you have to know the light's original, unshifted wavelength. That's where emission and absorption lines are so powerful. For example, astronomers have measured the precise "at rest" wavelengths for all hydrogen spectral lines in Earth-based laboratories. Astronomers use these known wavelengths as a measuring stick, comparing them with the wavelengths observed from an astronomical object. Any difference between the laboratory measurement of the spectral-line wavelengths and the astronomically observed wavelengths means that the object is moving. "The whole spectrum with all the lines gets shifted," says Hartigan, "so astronomers can use the whole pattern of lines to determine velocity."

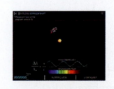

INTERACTIVE:
Doppler Shift

Lyman series Emission or absorption lines of the hydrogen atom as an electron moves between the ground state and excited states.

Balmer series Emission or absorption lines of the hydrogen atom as an electron moves between the first excited state and higher energy states.

Doppler shift A change in wavelength that results because either a wave source or an observer moves relative to the other.

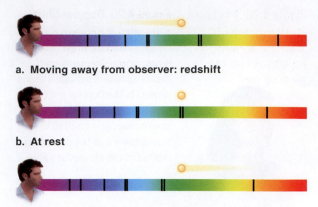

a. Moving away from observer: redshift

b. At rest

c. Moving toward observer: blueshift

Figure 4.21 Redshift and Blueshift
The pattern of spectral lines emitted by an object is redshifted when objects move away from us and blueshifted when they move toward us.

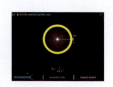

INTERACTIVE:
Inverse Square Law

---

redshift An increase in the wavelength of light that results when the light source moves away from the observer or the observer moves away from the source.

blueshift A decrease in the wavelength of light that results when the light source moves toward the observer or the observer moves toward the source.

aperture The diameter of a telescope's main light-collecting lens or mirror.

angular resolution A measure of a telescope's ability to separate or distinguish features in a distant object.

magnification A measure of a telescope's ability to enlarge appearances in an image.

refractor An optical telescope that uses lenses to collect light and form an image.

objective In a refractor, the large lens that gathers light from the object being observed and focuses the light rays to form an image.

If the wavelength of spectral lines from an astronomical object is longer than the "at rest" wavelength, the object is moving away from Earth, and the spectrum is said to be **redshifted**. If the wavelength from the astronomical object is shorter than the "at rest" wavelength, the object is moving toward Earth, and the spectrum is said to be **blueshifted** (Figure 4.21).

Section Summary

- Atoms are composed of three kinds of particles: negatively charged electrons and, in the atomic nucleus, positively charged protons and electrically neutral neutrons.
- In the Bohr model, electrons occupy only specific energy levels, at the ground state or excited states, associated with orbits around the nucleus. To move up from one energy state to another, an electron must absorb a specific photon; to move down, it must emit a photon of specific energy.
- The absorption and emission of photons result in distinctive spectral lines that can be used to identify the chemical elements present.
- The Doppler shift is caused by the stretching of wavelengths as an object moves away or by their compression as it approaches. Redshifted spectral lines (at wavelengths longer than expected) indicate how fast an object is moving away; blueshifted lines enable measurement of velocities toward the observer.

CHECKPOINT a. How do the orbits of an atom determine which wavelengths of light it can absorb or emit? **b.** Why does this behavior let astronomers use the spectrum of a distant object to determine the object's composition and speed?

Telescopes

4.4   Understanding how light encodes information about the matter emitting it is only the first step for astronomers. The next step is to use telescopes to catch the light and accurately analyze the resulting data.

Light Buckets and Resolution

"A telescope basically does two things for you," explains Pat Hartigan. "First, it collects a lot of light and concentrates it into one spot." Astronomers like Hartigan often think of telescopes as "light buckets." The bigger the bucket, the more light you can collect. For a telescope, the "bucket size" is measured by the size of its principal light collector (such as a lens, mirror, or radio dish), referred to as the telescope's **aperture**. The larger the collecting area, the more light the telescope can gather.

Astronomers measure the aperture of a circular collector in terms of its diameter D (with radius $R = D/2$). The light-collecting power of a telescope is directly proportional to the collector's area. The area A of a circular collector is

$$A = \pi R^2 = \pi(D/2)^2 = \pi D^2/4$$

Thus, the light-collecting power of a telescope increases with the square of its diameter. Build a telescope with an aperture 10 times larger in diameter than the one in your neighbor's telescope, and you will be able to catch 100 times more light—which, in turn, means being able to see objects 100 times fainter.

"The other thing that a telescope allows you to do," says Hartigan, "is to separate two objects that appear very close together on the sky so you can see them independently." Hartigan is describing the ability of telescopes to increase **angular resolution**. The resolution of an optical device (such as your eye or a telescope) tells you the smallest angle on the sky that it can cleanly "see." If a telescope has a resolution of 100 arcseconds (100″), then any details of an astronomical object that extend across an angle less than 100″ will appear as a single blur and be "unresolved" (Figure 4.22). Notice that resolution is distinct

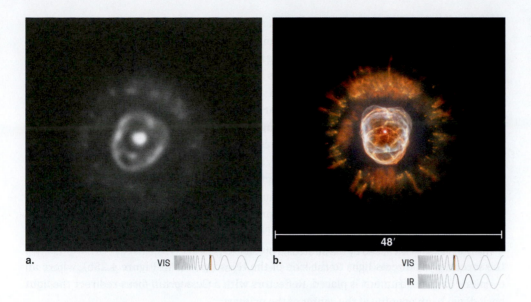

a.   VIS

b.   VIS

48′

IR

Figure 4.22 Telescope Resolution
Planetary nebula NGC 2392 as seen with **a.** low resolution and **b.** much higher resolution. For a given wavelength, the larger the aperture of the telescope astronomers use, the finer the resolution.

from magnification. Although part of a telescope's job is **magnification**—to make something look bigger—it's the resolution that matters most.

Resolution depends not only on a telescope's aperture but also on the wavelength of light the telescope collects. Wavelength matters because as waves enter the telescope, they bend and spread out in a process called *diffraction*, which degrades resolution. The best angular resolution a telescope can achieve can be written as

$$a = (2.5 \times 10^5)\lambda/D$$

smallest angle resolvable $= (2.5 \times 10^5) \times$ wavelength of light/aperture diameter

where angular resolution a is measured in arcseconds and both wavelength λ and aperture diameter D are measured in meters. The human eye has an aperture of about 6 mm, which means that its resolving power in visible light (600 nm) is approximately 30″, or 1/120 of a degree. The Hubble Space Telescope (HST) has a 2.4-meter aperture. Therefore, the HST has a resolution of 0.06″ at the same wavelength—500 times better than the human eye.

The angular-resolution equation shows that if a telescope is used to observe the same object with a shorter wavelength, its resolving power will be even higher. Observations in blue light are better resolved than observations in red light. (Recall that most objects emit at many wavelengths.)

Early on, astronomers looking through a telescope drew whatever phenomena they saw on the sky. After the invention of photography, film was used to record telescopic observations; a good observatory had a collection of "plates" that were photographic maps of the sky. Today observations are usually recorded electronically using *charge-coupled devices*, or CCDs. A CCD is a device that converts light into electric charge. When a telescope captures photons, they are directed onto a CCD, which builds a charge that can then be used to form a digital image.

Telescope Designs: Refractors and Reflectors

The first telescopes, called **refractors** (Figure 4.23a), used a large glass lens—the **objective**—at one end of a long tube as a light collector. Recall from Section 4.2 that refraction occurs when light is bent in passing from one transparent medium to another, such as from air to a glass prism. Light waves pass through the large glass objective lens in a refracting telescope and are gathered at a spot, called the *focus*, at the other end of the

"Why was there lots of light at some particular wavelength, and no emission at some other wavelength?"

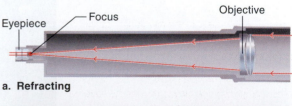

a. **Refracting**

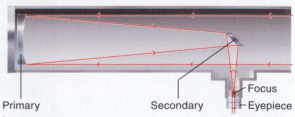

b. **Reflecting**

Figure 4.23 Telescope Types
a. Refractors use lenses (objective and eyepiece) to collect light. **b.** Reflectors use mirrors (primary and secondary) to collect light.

tube. An image of the source emitting the light forms there, and the **eyepiece**, a smaller lens at the back of the tube, magnifies that image.

A problem with making large refracting telescopes is the weight of the objective lens. Attempts to build large-aperture refractors led to glass lenses so heavy that they could not keep their shape. Another problem with glass lenses is that they bend blue light more strongly than red light. This **chromatic aberration**—different colors are refracted at different angles—can be corrected by modifying the shape of the lens, but it's impossible to fix completely. These restrictions have limited the utility of refractor telescopes.

Most telescopes used for research today are **reflectors** (**Figure 4.23b**). Instead of using a lens at the front of the telescope to collect light, reflectors use a large, curved mirror (called the *primary*) at the back end. The primary reflects and focuses the light to a second, smaller mirror (called the *secondary*) positioned above it, which gathers and redirects the light to the focus—wherever that may be. "The secondary redirects the light to different locations depending on the design," says Hartigan. In reflectors with a *Newtonian focus*, the secondary redirects light to the side of the telescope (as in Figure 4.23b), where an eyepiece or an instrument is placed. Reflectors with a *Cassegrain focus* redirect the light back down to an opening at the center of the primary.

Reflectors can be made much shorter than refractors because a mirror redirects light far more effectively than a lens can. Reflectors can also be made much larger because the mirror can be supported from below, helping it retain its shape. The typical aperture size of a modern research telescope is 10 meters across (**Figure 4.24**), more than 5 times larger

a.

b.

Figure 4.24 The Subaru and Yerkes Observatory Telescopes
a. Shown in its housing atop Mauna Kea volcano in Hawaii, this reflector is a modern large research telescope. Its primary mirror is more than 8 meters in diameter and observes in visible and infrared light.
b. The 100 cm Yerkes Observatory telescope, built in 1895, remains the largest aperture refractor ever built. It ceased operation in 2018, as its capabilities were eclipsed by larger and more agile reflectors.

than the height of a typical human adult. Because reflectors provide a larger aperture for much less cost, amateur astronomers usually choose reflectors over refractors. Currently, 20- and 30-meter reflecting telescopes are under construction.

To avoid sagging and distortion from the weight of a mirror, astronomers build large reflectors from mirrored segments (usually hexagonal) that are fitted together. To maintain the mirror's proper shape, small positioning motors shift the segments to focus the light properly as the telescope moves. (As we will soon see, this technique is called *active optics*.)

All telescopes must eventually use lenses and/or mirrors to bring light to the focus, where the light is gathered. The light can then be directed to a set of astronomical instruments, such as a spectrograph or an imager. These instruments take the collected light and manipulate it for astronomical analysis. In designing observational projects, astronomers consider both the telescope and the instruments that are available with it. For example, some telescopes have instruments that are well suited for carrying out "surveys" in which large areas of the sky are observed to search for particular kinds of objects. Special instruments, or computers connected to them, may have to be designed to gather and analyze large amounts of survey data.

Section Summary

- The aperture of a telescope's main light collector determines its light-gathering power. Its angular resolution depends on the size of the aperture and the wavelength of light being gathered.
- Refractors, using lenses at each end of a tube to gather and focus light, are hindered by chromatic aberration and distortion due to the excessive weight of the lenses as aperture diameter is increased.
- Reflectors, using mirrors instead of lenses, can be built with much larger apertures.

CHECKPOINT **a.** How does increasing a telescope's aperture size help in making astronomical observations? **b.** What happens if the wavelength of the observations must be decreased?

"Each transmission window means you can do astronomy from the ground in that wavelength."

Atmospheres and Their Problems

4.5 Earth's atmosphere keeps most incoming electromagnetic radiation at bay. The relatively thin layer of air (containing nitrogen, oxygen, carbon dioxide, and other gases) acts like a blanket, absorbing most wavelengths of light and keeping them from reaching Earth's surface. This is, in general, a good thing. For example, sunburn is caused by UV radiation, which just makes it through the atmospheric blanket. If more-damaging wavelengths such as X-rays penetrated to the ground, our planet could not sustain life.

Where the Light Shines Through

The fact that we can see the Sun and the stars means that our atmosphere allows the **transmission** of light at visible wavelengths; that is, visible light passes through without being absorbed. Astronomers use the term *transmission window* to describe the ability of light in a band of wavelengths to penetrate the atmosphere. "Each transmission window means you can do astronomy from the ground in that wavelength," says Pat Hartigan. Radiation cannot be observed if the atmosphere is opaque at that wavelength. **Opacity** is the degree to which light is absorbed when passing through a material.

"You can see the 'band pass' windows," says Hartigan, "where Earth's atmosphere is transparent. But there are other areas where it's completely opaque, because of carbon dioxide or water vapor or other types of absorbers in the atmosphere." The atmosphere's opacity drops strongly (so, transparency soars) in the visible part of the

eyepiece In a telescope, the small lens through which the observer looks.

chromatic aberration The failure of a lens to focus all colors to the same point.

reflector An optical telescope that uses curved mirrors to collect light and form an image.

transmission The passing of light through matter without being absorbed.

opacity The degree to which light is absorbed when passing through a material.

Figure 4.25 Transmission Windows and Opacity

Not all wavelengths of electromagnetic radiation can be observed from Earth's surface. The atmosphere is transparent to visible and radio light but not to gamma-ray, X-ray, and most UV light.

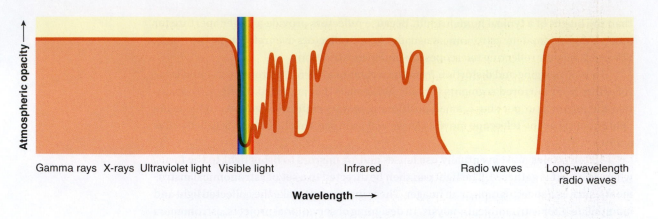

spectrum (**Figure 4.25**). It also drops strongly in the radio part of the spectrum, making ground-based radio astronomy a fruitful endeavor. Notice that the atmosphere also allows some wavelengths of infrared radiation to penetrate. The atmospheric opacity in other parts of the infrared spectrum is low enough that radiation may not reach the ground but can penetrate far enough that telescopes mounted on high-flying airplanes can pick it up.

TWINKLING STARS AND THEIR CHALLENGES. Even if an astronomer wants to observe the sky in a transmission window, the atmosphere is beset by **turbulence**, the random jostling motions of moving air. If you have flown on an airplane, you may have had the uncomfortable experience of being bounced around by turbulence. A related form jostles light on its way down through the atmosphere, making stars appear to twinkle. Light rays from distant sources are bounced around, leading to blurry images. For astronomers, atmospheric turbulence poses a major problem. Even the most powerful telescope with the largest aperture produces blurry, low-resolution images if the light it catches has passed through a turbulent atmosphere.

Astronomers have developed various strategies for dealing with the problems the atmosphere creates. The first is to place the telescope as high above sea level as possible. Modern visible-light and infrared telescopes are located on mountaintops because the density of Earth's atmosphere drops quickly with elevation. Many state-of-the-art telescopes have been built on the extinct volcano Mauna Kea in Hawaii, which stands at 13,803 feet (4,207 meters) above sea level. Its peak lies above 40 percent of the atmosphere and above 90 percent of atmospheric water vapor. Atmospheric opacity and turbulence are thus less of a problem at Mauna Kea's summit than near sea level. A second strategy, used in pursuing some kinds of longer-wavelength studies, such as infrared astronomy, is to put observatories in very dry climates, because water vapor is the primary cause of opacity at these wavelengths. The Atacama Desert in Chile is one of the highest, driest locations on Earth, and a number of telescopes operate there.

QUICK THINKING: ADAPTIVE AND ACTIVE OPTICS. Another strategy for dealing with intermittent atmospheric turbulence is a technique called **adaptive optics**. "Adaptive optics allows you to change the shape of the mirror depending on what the atmosphere above you is doing," explains Hartigan. The computer-controlled motors that keep the segments of the mirror in place can also change the positions of those segments many times per second, adjusting the mirror's shape to sharpen an image that has been blurred by turbulence. "You can either look at a guide star [a reference point] or shine a laser out of your telescope," says Hartigan. "The reflected laser light becomes a guide star. Then you look to see how the image is shimmering around and adjust the mirror rapidly to compensate for the atmosphere's motion." Using adaptive optics, astronomers can achieve very sharp vision in at least some wavelength bands, including infrared.

"Adaptive optics allows you to change the shape of the mirror depending on what the atmosphere above you is doing."

turbulence The random jostling motions of moving air.

adaptive optics A telescope technology in which mirror segments are rapidly adjusted to account for atmospheric turbulence and improve image resolution when needed.

active optics A telescope technology in which mirror segments are constantly adjusted to prevent deformation by external factors.

interferometry The technique of combining the signals from many smaller telescopes to achieve the resolving power of a larger one.

Large reflecting telescopes also continuously employ a technique called **active optics**. The very large mirrors used in some telescopes must be built with lightweight materials to keep the total weight of the telescope manageable. But as temperature and wind change and the telescope moves, the mirrors must keep a very precise shape. In active optics, motors mounted behind the mirror segments push or pull on those segments and constantly maintain their shape to extreme accuracy. Active-optic devices don't need to act as fast as those for adaptive optics.

LIGHT POLLUTION. In the modern world, astronomers must also deal with *light pollution*— artificial light coming from the ground rather than transmitted through the sky. Light pollution is worst near highly populated areas that have street lights. In many parts of the world, the stars can barely be seen because countless sources of artificial light keep the sky from ever truly getting dark.

More with Less: Radio Telescopes and Interferometry

For most of human history, astronomy meant optical astronomy—that is, observation via visible light. Once the entire electromagnetic spectrum was discovered, scientists recognized that studying the sky in other wavelengths would lead to fundamental new discoveries. Beginning in the 1950s, radio astronomy quickly became a major contributor to our understanding of the cosmos.

Most radio telescopes operate like reflectors. They have a curved dish that acts as a collector, bouncing light to a smaller detector at the collector's focus. Because radio waves from distant cosmic sources are so faint and their wavelengths are so long, the telescopes must have large collecting areas.

The long wavelengths of radio waves can be quite useful to astronomers. Electromagnetic radiation tends to be absorbed by solid bodies that have a size similar to the radiation's wavelength. Many dust particles have a radius that is close to or longer than the wavelength of visible light, so absorption causes dusty regions of space to appear dark in optical images. Since the galaxy is full of dusty clouds, studies at visible wavelengths cannot see across galactic distances. Radio waves, with their long wavelengths, can penetrate dusty clouds, allowing radio astronomers to see from one end of the galaxy to the other.

The long wavelengths of radio light do, however, limit the resolution of radio telescopes. Recall from the angular-resolution equation of Section 4.4 that a telescope's minimum angular resolution depends directly on the wavelength of light being used. In other words, even a large radio telescope has lower angular resolution than an optical telescope has. But, in one of the major innovations of modern science, astronomers have found a way around this limitation.

"If you took a large telescope," explains Hartigan, "and blocked out a spot in the middle, you would find the brightness of your image would go down, but your resolving power, meaning the ability to split apart two closely spaced image features, would be just the same." Hartigan is describing the principles behind **interferometry**, a technique that combines the signal from an array of smaller telescopes to create the resolving power of a much larger one. "What matters for resolution," says Hartigan, "is the distance from one side of the aperture to the other. But you don't need a continuous mirror or collector to make an aperture." By carefully combining signals from multiple telescopes that are widely separated, astronomers can achieve images with extremely high resolution.

One of the best-known radio interferometers is the Atacama Large Millimeter Array (ALMA) in Chile. ALMA consists of 66 giant radio telescopes that can be placed in different configurations. Combining the signal from so many individual telescopes, ALMA astronomers can create a single instrument with an aperture diameter of 16 km **(Figure 4.26)**.

Figure 4.26 Atacama Large Millimeter Array (ALMA)
With 66 radio telescopes, ALMA in Chile is the largest astronomical project in the world. Using interferometry, it acts as a single telescope with a 16-km-diameter aperture. ALMA has provided groundbreaking observations, including exoplanets, planet-forming discs around young stars, asteroids, and exploding stars.

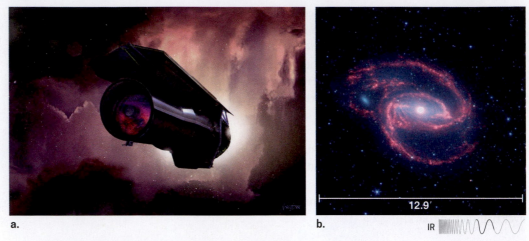

a.   b.   IR

Figure 4.27 Spitzer Space Telescope (SST)
a. Artist's rendering of the SST, which gathers infrared light— particularly useful to astronomers for its ability to penetrate dust. **b.** Spitzer image of galaxy NGC 1097, showing its spiral structure, with a central ring of stars.

Interferometry is being applied in other wavelength bands as well. Infrared telescopes have been combined to increase their resolution, and optical interferometry has been explored, but with limited success. "It's partly an instrumental problem and partly a problem with Earth's atmosphere and turbulence, which need to be taken into account when combining signals from different telescopes," says Hartigan. "It's just more difficult to do when the wavelengths are shorter."

"You don't need a continuous mirror or collector to make an aperture."

Telescopes in Space

Getting a telescope completely above Earth's atmosphere is the best option for eliminating problems of turbulence and absorption. For X-rays and gamma rays, there is no other option because our atmosphere has no transmission windows at these wavelengths. But space telescopes often cannot be reached for maintenance after they have been launched. In addition, designing precision equipment to work in the hostile environment of space requires extensive testing.

INFRARED. Infrared light extends from 700 nm to 350,000 nm (but astronomers use microns for measuring infrared light, where 1,000 nm = 1.0 μm). The atmospheric window for infrared light allows ground-based studies only at the shorter wavelengths, called the near infrared (0.7–5 μm). Observations in the mid infrared (5–40 μm) and far infrared (40–350 μm) can be carried out only by a telescope that is positioned above most or all of the atmosphere. Being above the atmosphere means that the entire infrared spectrum is available to astronomers, so they can use the longer wavelengths of infrared light to see deep into objects such as star-forming clouds and planet-forming disks.

The most famous infrared satellite has been the Spitzer Space Telescope (SST; **Figure 4.27**). Spitzer was the last of NASA's Great Observatories, a series of space missions designed to investigate the sky across a wide range of wavelengths. Spitzer was launched in 2003 with a mirror 85 centimeters (cm) in diameter and a 5-year supply of liquid helium to keep the spacecraft cool (at a temperature of about 5.5 K) and reduce heat radiation from its own "self-emission." Spitzer is in a special orbit that keeps it far from Earth and our planet's infrared heat radiation. (As of late 2019 it was still active, though reduced to

a.   b.   VIS
IR

Figure 4.28 Hubble Space Telescope (HST)
a. One of the most successful scientific instruments in history, the HST works mainly in visible light (with some capacity in UV and infrared). **b.** A Hubble Extreme Deep Field image, covering a speck of the sky only a small fraction of the diameter of the full Moon, shows 5,500 galaxies, some of which are in the early stages of evolution 13 billion years ago.

a smaller range of wavelengths than it could handle when launched.) The James Webb Space Telescope (JWST; **Figure 4.29**), which is NASA's next large-scale space telescope platform, is planned to launch in 2021 and is being designed to work mostly at infrared wavelengths as well.

VISIBLE AND UV. By eliminating atmospheric turbulence, space telescopes that operate in the visible range offer better resolving power than do similar-sized ground-based telescopes. The Hubble Space Telescope (HST) was the first of NASA's Great Observatories, launched in 1990 (**Figure 4.28**). It has been one of the most successful optical space missions to date.

The HST also has the (limited) capacity to carry out UV observations, which cannot be made from Earth because of atmospheric absorption. Ultraviolet photons are produced in stellar atmospheres, at the inner edges of disks surrounding black holes, and in other environments, making UV studies critical to astronomy. An early and important UV telescope was the International Ultraviolet Explorer (IUE). Launched in 1978 with a 45-cm mirror, it carried out observations in wavelengths between 115 and 320 nm. The Extreme Ultraviolet Explorer (EUVE) and the Far Ultraviolet Spectroscopic Explorer (FUSE) pushed farther out in the UV spectrum, observing wavelengths between about 10 and 120 nm.

X-RAYS. The X-ray portion of the spectrum extends from 0.01 to 10 nm. Using Planck's energy-frequency relationship, $E = hf$, you can see that such short wavelengths (high frequencies) mean high energy. Thus, X-rays come from extremely energetic astrophysical regions and events, such as supernovas, neutron stars, and galaxy clusters.

The only way to carry out a robust observational study of the sky in X-rays is to launch telescopes into space. However, because X-rays are so energetic, they tend to pass right through mirrors meant to collect and focus them onto a detector. To get around

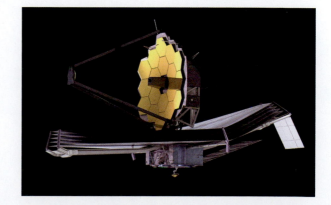

Figure 4.29 James Webb Space Telescope (JWST)
The 18 hexagonal mirrors of the JWST, under construction, will operate at mid-infrared wavelengths.

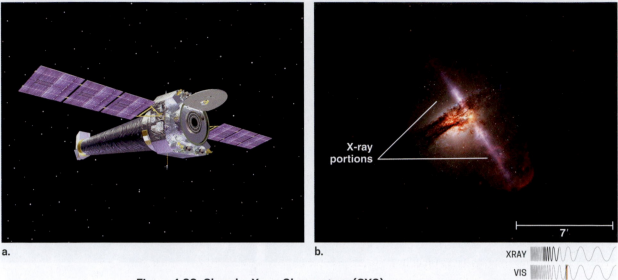

a.   b.

XRAY
VIS

Figure 4.30 Chandra X-ray Observatory (CXO)
a. Artist's rendering of CXO, which presents astronomers with startling new details visible only in X-ray light. **b.** The secrets of galaxy Centaurus A—including jets powered by a supermassive black hole—spring into view.

this problem, engineers build grazing-incidence mirrors, which gently deflect incoming X-rays many times to focus them on detectors in large X-ray telescopes.

The first X-ray telescopes rode on balloons and high-altitude rockets, but X-ray astronomy began in earnest with the launch of the Uhuru satellite in 1970. The current workhorse of X-ray studies is the Chandra X-ray Observatory (CXO), launched in 1999 (**Figure 4.30**). Chandra, another Great Observatories mission, continues to operate today.

GAMMA RAYS. Gamma rays are the light with the shortest wavelength, the highest frequency, and therefore the highest energy of all electromagnetic radiation. Earth's atmosphere is opaque to gamma rays, so gamma-ray astronomy must be done from space. Unlike X-rays, however, gamma rays cannot be focused efficiently. Therefore, gamma-ray telescopes are really gamma-ray detectors that provide very low angular resolution. A gamma-ray telescope provides very little detail about the sources it "sees" compared with even X-ray telescopes. The Compton Gamma Ray Observatory was the second of NASA's Great Observatories to be launched (in 1991) and the first to map the entire sky in gamma rays. More recently, the Fermi Gamma-ray Space Telescope, launched in 2008, observed the "gamma-ray sky" with better resolution capabilities and flew with a suite of powerful new instruments.

Since gamma rays are so energetic, they are produced in only the most energetic celestial objects, such as exploding stars, supernovas, active galaxies with black holes at their centers, and neutron stars. At this extreme high energy, gamma rays are not produced in large numbers; a typical gamma-ray observation may catch only a few photons per hour. However, a special class of objects, called gamma-ray bursts (GRBs), has puzzled astronomers for decades. GRBs appear as brief flashes of gamma rays that do not repeat, and their source remains highly debated. Much of gamma-ray astronomy has focused on the nature of GRBs.

Space-based astronomy will continue to play a critical role as detectors become more sophisticated and astronomers become more ambitious in their designs. The ability to place a detector above the atmosphere and atmospheric complications makes space an attractive place to carry forward astronomical studies. "Once you get above Earth's atmosphere, images are steadier and you can see things a lot more sharply," says Hartigan, who

"Once you get above Earth's atmosphere, images are steadier and you can see things a lot more sharply. That really makes a huge difference for getting to the science."

has spent years using the HST to study regions of star formation. "That really makes a huge difference for getting to the science."

Section Summary

- Earth's atmosphere limits the effectiveness of ground-based observations. Strategies to reduce atmospheric limitations include placing instruments as high as possible in the atmosphere, and using adaptive optics and radio telescopes with interferometry.
- Putting telescopes in space has limitations—mainly cost and size of telescopes—but enables astronomers to observe in many kinds of light (including X-rays and gamma rays) that are not visible from the ground.

CHECKPOINT Describe the strengths and weaknesses of ground-based radio telescopes and space telescopes (infrared, visible and UV, X-ray, and gamma ray).

Telling Light's Stories

From Newton's law of gravity to blackbody radiation, we have covered a lot of ground in these first four chapters. There is still more physics that goes into astrophysics (you will meet Einstein and his theory of relativity in Chapter 14), but from here on you can start using all that you've learned about the basic structure of the physical world to begin gaining a perspective on the cosmos as a whole. In the remaining chapters of this book, we will explore the grand understanding of the Universe that astronomers have painstakingly built using the tools (both conceptual and technological) we've just discussed.

CHAPTER SUMMARY

4.1 Light: The Cosmic Envoy

The ways that light, heat, and matter are linked provide information about distant objects, with the help of advances in optics by means of instruments including telescopes and spectrographs. Light acts as both a wave (with qualities of wavelength, frequency, and amplitude) and a particle (a photon). The electromagnetic spectrum represents all forms of light, with wavelengths running from very small to very large (gamma-ray, X-ray, UV, visible, infrared, microwave, or radio).

4.2 Astrophysical Spectra

Spectra show the distribution of intensity (related to energy emitted) versus wavelengths in a light source. The kind of spectrum observed reflects physical characteristics of the source. Kirchhoff's laws describe the various conditions that produce each basic type of spectrum (continuous, absorption, or emission). Blackbody radiation comes from dense objects whose random thermal motions of atoms create a characteristic spectral shape. The temperature of an object emitting a blackbody spectrum is determined by the spectrum's peak wavelength via Wien's law. A blackbody's luminosity also depends on temperature (and the object's surface area), as described by the Stefan-Boltzmann law. The higher the frequency of light, the higher its energy. Unique spectral lines in emission and absorption spectra show which elements are present.

4.3 Spectra and the World of the Atom

Quantum mechanics describes the behavior of matter at the molecular, atomic, and subatomic levels. Niels Bohr developed a model of the atom in which protons and neutrons in the nucleus are surrounded by electrons on discrete orbits. The orbit closest to the nucleus is the ground state. Successively higher orbits are the first excited state, the second excited state, and so on. Electrons jump directly between levels, emitting energy in the form of photons when moving to a lower level and absorbing energy when moving to a higher level. A photon's wavelength is unique to the energy transitions available in the atom. Thus, every type of atom (each chemical element) has a different absorption or emission spectrum. The Doppler shift—the apparent stretching or compression of light's wavelength based on the source object's motion away from or toward the observer—is used to derive the source's speed along the line of sight.

4.4 Telescopes

Telescopes gather and focus light. Larger apertures gather more light and increase the angular resolution of distant objects. Resolution also depends on the light's wavelength. Larger wavelengths require larger apertures. Refractors are telescopes that use lenses as their light-gathering objectives; reflectors use mirrors.

4.5 Atmospheres and Their Problems

Atmospheres such as Earth's limit the transmission of light, being transparent to some kinds of light through transmission windows while being opaque to others. Optical telescopes are often built at higher altitudes to limit atmospheric effects; space telescopes operate from space to observe some kinds of infrared light, UV light, X-rays, and gamma rays effectively. Radio light can be captured from Earth's surface. Interferometry allows multiple radio telescopes to function together as if they were one very large telescope. The Hubble Space Telescope observes mostly in visible light, and its image quality is greatly enhanced by its location above the atmosphere.

QUESTIONS AND PROBLEMS

Narrow It Down: Multiple-Choice Questions

1. Which of the following is *not* a characteristic of electromagnetic waves?
   a. density
   b. wavelength
   c. frequency
   d. amplitude
   e. speed

2. Which of the following equations correctly gives the speed of light c in a vacuum?
   a. $c = f\lambda$
   b. $c = \lambda/f$
   c. $c = f/\lambda$
   d. $c = \lambda^2/f$
   e. $c = f^2/\lambda$

3. Which of the following are *not* found in the electromagnetic spectrum? Choose all that apply.
   a. gamma rays
   b. cosmic rays
   c. sound waves
   d. UV waves
   e. infrared waves

4. Which of the following correctly lists the regions of the electromagnetic spectrum from lowest to highest frequency?
   a. X-ray, gamma-ray, UV, infrared, radio, visible
   b. gamma-ray, X-ray, UV, visible, infrared, radio
   c. gamma-ray, infrared, visible, radio, UV, X-ray
   d. radio, infrared, visible, UV, X-ray, gamma-ray
   e. visible, radio, infrared, UV, gamma-ray, X-ray

5. A spectrum typically displays which two characteristics of light?
   a. frequency and wavelength
   b. speed and frequency
   c. intensity and wavelength
   d. intensity and amplitude
   e. speed and amplitude

6. Which of the following best explains the presence of dark lines at specific wavelengths in an otherwise continuous spectrum?
   a. The original source does not radiate light at all wavelengths.
   b. The original source radiates light most intensely at the wavelengths corresponding to the missing photons.
   c. A relatively hot gas between observer and source radiates light at the wavelengths corresponding to the missing photons.
   d. A relatively cool gas between observer and source absorbs light at the wavelengths corresponding to the missing photons.
   e. The spectrograph is not sensitive at the wavelengths corresponding to the dark lines.

7. A spectrum of a distant object reveals a sequence of known absorption lines, all of which are shifted to shorter wavelengths. What can you conclude about the object?
   a. It must be very massive.
   b. It must be highly magnetized.
   c. It must emit cosmic rays.
   d. It must be moving away from us.
   e. It must be moving toward us.

8. Which of the following is true about the spectra of blackbodies at different temperatures?
   a. The peak of emission by the hotter object is at a longer wavelength.
   b. The cooler object shows greater emission at infrared wavelengths than the hotter object does.
   c. The peak of emission by the cooler object is at a longer wavelength.
   d. The cooler object emits more total energy per area than that emitted by the hotter object.
   e. The hotter object emits less at UV wavelengths than the cooler object does.

9. The oxygen emission line from a warm gas cloud is expected to appear at 500 nm but is observed at 525 nm instead (5% longer). Which of the following is true?
   a. The object is moving away from the observer at 5% of the speed of light.
   b. The object is moving toward the observer at 5% of the speed of light.
   c. A chemical change has occurred.
   d. The speed of the light is greater than expected.
   e. The observer's equipment is inaccurate, since emission lines do not change.

10. The formation of absorption lines in a spectrum emitted by a blackbody indicates which of the following? Choose all that apply.
    a. the existence of an intervening cloud of material cooler than the emitting source
    b. the specific elements of the atoms present in an intervening cloud
    c. the existence of an intervening cloud of material hotter than the emitting source
    d. a violation of the conservation of energy
    e. the interaction of matter and light

11. Which statement related to the energy levels of electrons is true? Choose all that apply.
    a. An electron in an atom can occupy any of the atom's energy levels.
    b. Electrons within an atom require the absorption or emission of a photon of any wavelength to change energy levels.
    c. The wavelengths of the photons emitted or absorbed are unique to a specific element.
    d. The energy level farthest from the nucleus is the first excited state.
    e. Electrons can sometimes be found between the energy levels.

12. Which of the following statements concerning the Doppler effect is *not* correct?
    a. The pitch of the whistle on an approaching train is higher than that of a stationary train.
    b. The pitch of the whistle on a receding train is lower than that of a stationary train.
    c. The light from a very distant, receding galaxy appears redshifted with respect to a nearby dwarf galaxy whose motion is observed to be negligible.
    d. The light from the approaching Andromeda Galaxy appears redshifted with respect to a nearby dwarf galaxy whose motion is observed to be negligible.
    e. The pitch of a train whistle is unchanged to a passenger on the train.

13. Stars A and B can be considered blackbodies. The peak wavelength λ_{max} of star A is longer than that of star B. What conclusions can you draw? Choose all that apply.
    a. A is cooler than B.
    b. The blackbody curves of the two are identical, since both are stars.
    c. B must be closer to the observer.
    d. The colors of the two stars differ.
    e. Emission lines seen in the spectrum of A will be of greater intensity than those of B.

14. A typical amateur telescope is a 10-inch reflector. How does the light-gathering power of the Hooker 100-inch reflector on Mount Wilson compare with that of the amateur telescope?
    a. It is the same.
    b. It is 10× greater.
    c. It is 100× greater.
    d. It is 1,000× greater.
    e. It is 10,000× greater.

15. Which of the following is a disadvantage of large refractors?
    a. weight
    b. size
    c. chromatic aberration
    d. sagging lens
    e. all of the above

16. True/False: Optical telescopes are now routinely linked together through interferometry to produce the equivalent of a single telescope with a much larger aperture telescope.

17. True/False: Light cannot behave as both a particle and a wave, since particles and waves are mutually exclusive.

18. Which of these NASA space-based observatories was designed primarily to study wavelengths of light to which Earth's atmosphere is transparent?
    a. Hubble
    b. Compton
    c. Chandra
    d. James Webb
    e. Spitzer

19. Why do stars twinkle?
    a. The stellar surface where the light is emitted is very turbulent.
    b. The brightness of stars changes constantly.
    c. The distribution of absorption lines in their atmospheres is changing.
    d. The motion air in Earth's atmosphere creates the appearance of twinkling.
    e. Our eyes are constantly moving.

20. The spectrum of star I has the same pattern of absorption lines as that of star II but peaks at longer wavelengths. Which statements are correct?
    a. Star II cannot have the same composition as star I.
    b. Star II must have the same temperature as star I.
    c. Star II has the same composition and temperature as star I.
    d. Star II has the same composition as star I but higher temperature.
    e. Star II has the same composition as star I but lower temperature.

To the Point: Qualitative and Discussion Questions

21. Name and describe three parameters that we use to describe an electromagnetic wave.

22. Describe, in terms of frequency, wavelength, and energy, where visible light falls on the electromagnetic spectrum with respect to other regions of the spectrum.

23. What are the most important characteristics of objects that emit blackbody spectra?

24. Earth's atmosphere is mostly opaque to which regions of the electromagnetic spectrum? To which regions is it mostly transparent?

25. What characteristics of stars can astronomers learn by observing their starlight with a spectrograph?

26. Using Kirchhoff's laws, describe how two observers can view the same gas cloud but one sees an emission spectrum while the other sees an absorption spectrum.

27. In your own words, describe the relationship established by Wien's law.

28. The interaction of light with matter deals directly with the absorption and emission of photons. What happens to electrons in an atom when a photon is emitted? What happens when a photon is absorbed?

29. How do refractors differ from reflectors?

30. Under what circumstances can an element be observed to emit light with a wavelength that is different than normal for that element?

31. For which wavelengths of light are there transmission windows from Earth's surface?

32. What is the advantage of locating telescopes on mountaintops?

33. Describe how adaptive-optics systems improve observations made from ground-based telescopes.

34. For each of the following regions of the electromagnetic spectrum, name the NASA Great Observatory that was designed to observe in that region: infrared, visible/UV, X-ray, and gamma-ray.

35. How might life on Earth differ if all wavelengths of light could pass through the atmosphere?

Going Further: Quantitative Questions

36. An electromagnetic wave has a frequency 5 times higher than that of another electromagnetic wave. How do their wavelengths compare?

37. How does the frequency of one light wave compare with the frequency of another, if the ratio of their wavelengths is 1:3?

38. The spectrum of a blackbody has a peak wavelength of 7.7×10^{-7} meter. What is its temperature, in kelvins?

39. The blackbody spectrum of a star with a surface temperature of 8,000 K will peak at what wavelength, in meters?

40. Two stars have the same luminosity but the temperature of one is twice that of the other. What is the ratio of the surface area of the cooler star to that of the hotter star? (Hint: Use the Stefan-Boltzmann relationship.)

41. Star A is 2 times hotter than star B. How much shorter is the peak wavelength for star A compared with that of star B? (Use Wien's law.)

42. You have a reflecting telescope with a 6-inch aperture. Your sister has one with a 10-inch aperture. What is the ratio of the light-gathering power of her telescope to that of yours?

43. Telescope A has a diameter of 2 meters, while telescope B's diameter is 5 meters. What is the ratio of the angular resolution of B to A when both observe the same wavelength of light?

44. How would the angular resolution of each of the two telescopes in question 43 change if the observed wavelength was 3 times as long?

45. If the aperture diameter of the HST were doubled, how would the HST's angular resolution compare with its current resolution?

"IN THE TIME BETWEEN JUST TWO PICTURES WE TOOK THAT NIGHT, WE COULD TELL WE HAD A DISCOVERY"

5

Planetary Systems

THEIR BIRTH AND ARCHITECTURE

The Rest of the Solar System

5.1 David Jewitt, a planetary astronomer at UCLA, remembers the moment the rest of the Solar System was discovered. "It was 1992 and my former student Jane Luu and I were sitting in the observing room of the University of Hawaii's 2.2-meter telescope at the top of Mauna Kea," he recalls. Jewitt was searching for the elusive **Kuiper Belt objects (KBOs)**—debris left over from the Solar System's construction almost 5 billion years ago. The Kuiper Belt was *supposed to exist* at the edges of the Solar System beyond Neptune's orbit. Up to that time, the Kuiper Belt was just an idea proposed by Gerard Kuiper in 1951. No one knew if the Kuiper Belt was out there or was a theory gone wrong. Trying to decide that, Jewitt had already spent 5 years staring at fuzzy images. "We had to compare pictures taken just 20 minutes or so apart to see if anything moved." Stars would not move across the sky that fast, but an object orbiting the Sun, even out beyond Neptune's orbit, would.

"In the time between just two pictures we took that night, we could tell we had a discovery," says Jewitt, speaking of the object they called 1992 QB1 (**Figure 5.1**)—a modest name for a discovery that would rewrite textbooks. "We took more and more images of the thing as the night progressed. We were able to follow its motion, which was exactly what we expected for a KBO: the right direction, the right speed, everything. After spending night after night after night after night not getting detections, finally nailing a KBO was pretty exciting."

The 168-kilometer (km) chunk of icy rock that Jewitt and Luu discovered opened a new window on the structure and origin of the Solar System. As Jewitt puts it, "Most of the real estate in the Solar System occupied by significant bodies has only been discovered in the last 20 years." The discovery of 1992 QB1 began a renaissance in Solar System study and was a crucial first step in a chain of events that eventually kicked Pluto out of the club

← The Tarantula Nebula, within the Large Magellanic Cloud, is the most active starburst region in the Local Group of galaxies and one of the largest (diameter 200 parsecs). Astronomers study such regions to learn about the processes by which stars and planets are born.

Kuiper Belt object (KBO) A small body made of ice and rock that orbits in the Kuiper Belt, beyond Neptune's orbit.

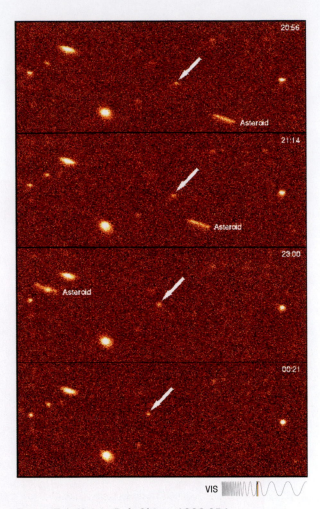

VIS

Figure 5.1 Kuiper Belt Object 1992 QB1
This series of images provided evidence of an orbiting KBO, seen as a bright, moving dot (indicated with an arrow) in an otherwise unchanging field of stars. The identification of this KBO, followed by the discovery of similar objects, validated the idea that the Kuiper Belt exists beyond Neptune's orbit.

Figure 5.2 Giordano Bruno
Bruno (1548–1600) advocated radical astronomical ideas, such as the existence of planets around other stars. For his heretical ideas, he was burned at the stake by the Inquisition.

of planets (as we shall see in Section 5.2). "What we really discovered that night," says Jewitt, "is that the Solar System is not a 'known' place."

Jewitt and Luu's groundbreaking discovery followed decades of revolutionary space exploration. Once the space age began in the late 1950s, humanity began sending robotic probes to other Solar System planets—almost all of which have now been visited at least once. Asteroids and comets have also been targeted by space missions, returning valuable information (and some samples of material). These missions, along with intense study from ground-based telescopes, have provided a new understanding of our Solar System, allowing astronomers to piece together the story of its origin and evolution.

Planets and the 2,000-Year-Old Question

It was a cold day in 1600 as the hooded figure was led out to the plaza. The crowd, there to watch the spectacle of an execution, was hushed as Giordano Bruno—an ex-Dominican monk and philosopher who had been declared a heretic—was tied to the stake. The torches were readied, the straw at his feet was set aflame, and Bruno became a martyr for science (**Figure 5.2**).

The charges of heresy against Bruno focused on details of Catholic Church theology, but undeniably, Bruno's astronomy had led him to this fateful moment. Bruno was a confirmed Copernican who had argued vehemently that the Sun—and not Earth—is the center of the Solar System. But Bruno had gone further. The Sun is not alone in hosting a family of planets, he claimed. Other stars have their own worlds in orbit, and some of these planets are inhabited. It was a radical argument made at a time when radicalism was not tolerated.

Bruno was not the first to ask whether other planets exist beyond our own Solar System. Even in the age of classical Greece, more than 2,000 years ago, philosophers wondered whether other worlds, orbiting other stars like our own, might exist. Only in our own time has this most ancient of astronomical puzzles been answered.

In 1995, two Swiss astronomers announced the discovery of a Jupiter-sized planet orbiting the Sun-like star 51 Pegasi. As of today, thousands of other **exoplanets**, or **extrasolar planets**, have been discovered in orbit around other nearby stars. After millennia of having only one example of a planetary system to study, astronomers can finally ask how our Solar System compares with others. "We have found planetary systems everywhere," says Jewitt, "and most of them do not look like ours." Knowing that fact alone will help astronomers answer many other questions.

Exoplanet system studies, along with the study of our own Solar System, have enabled astronomers to take the first strong steps toward a complete theory of how planets form. With these steps, they have also moved closer to understanding the grandest question of all: Where might life arise elsewhere in the Universe?

This chapter describes the basic architecture of planetary systems. It begins with the eight planets and assorted debris orbiting our own star and then proceeds to examine the diversity of worlds orbiting other stars. With this galaxy-wide view, the chapter then explores how planetary systems form and what can be expected as astronomers probe ever deeper into the Universe in search of life-harboring worlds.

Section Summary

- Planetary systems are now known to be common in the Universe, and recently discovered exoplanetary systems have provided astronomers with the long-awaited answers to questions about the nature of our planet and Solar System.

CHECKPOINT a. How did the discovery of KBOs change astronomers' view of our own solar system? **b.** Why was the discovery of planets around other stars important for astronomy?

Just the Facts: A Solar System Census

5.2 The process of science always begins with the act of collecting data. And before we can ask how planetary systems form, we have to know what they look like. Exploration of the planets orbiting our Sun reveals a few basic facts that will set the stage for an *explanation* of how this system ended up as it did. The first step is to take a census of the Solar System (cataloging what's out there) and describe its architecture (how things are arranged). But those data provide only a partial view. As David Jewitt points out, "We're still pretty ignorant about what's out there, far beyond the known planets."

"We have found planetary systems everywhere, and most of them do not look like ours."

The Inhabitants of Our Solar System

Planets are the most obvious inhabitants of the Solar System—after the Sun. As we saw in Chapter 1, a planet is a body that orbits a star and is massive enough for self-gravity to have pulled it into a spherical shape and to have cleared smaller objects from, and near, its orbit. Our Solar System has eight planets. In order of increasing distance from the Sun, they are Mercury, Venus, Earth, Mars, Jupiter, Saturn, Uranus, and Neptune. Their distances from the Sun range from 0.39 astronomical units (AU) (for Mercury) to 30.10 AU (for Neptune). Recall that $1 \text{ AU} = 1.5 \times 10^8$ km. At least 168 known satellites, or **moons**, orbit the eight planets. Most of these moons orbit the "giant" planets: Jupiter has at least 79 confirmed moons; Saturn has at least 61; Uranus, at least 27; and Neptune, at least 14. The moons of the Solar System range in size from Ganymede and Titan, with radii greater than that of Mercury, to the giant planets' tiny shepherd moons, with radii less than a kilometer (**Table 5.1**).

Although Neptune is the outermost planet in the Solar System, there are many more objects at greater distances from the Sun. Pluto used to be considered a planet, but astronomers now recognize that its small size—and the fact that its gravity is not sufficient to clear its neighborhood of debris—make it a member of a family of objects called **dwarf planets**. One occupies the **asteroid belt**, the region between Mars and Jupiter. Astronomers suspect that most dwarf planets reside in the region beyond Neptune's orbit called the **Kuiper Belt**

exoplanet or **extrasolar planet** A planet orbiting any star other than the Sun.

moon A natural satellite of a planet.

dwarf planet A body, smaller than a planet, that orbits a star and has enough mass to become spherical by self-gravity but that has not cleared its neighboring region of small objects.

asteroid belt The disk-shaped Solar System region 2.1 to 3.5 AU from the Sun where many asteroids orbit.

Kuiper Belt The disk-shaped Solar System region 30 to 50 AU from the Sun, outside Neptune's orbit, containing small bodies of ice and rock.

Table 5.1   Largest Moons of the Solar System

Name	Average Radius (in Moon radii)	Average Radius (km)	Sidereal Period (days)	Average Orbital Distance from Planet (km)	Planet	Planet Type
Ganymede	1.5	2,634	7.2	1.07×10^6	Jupiter	Gas giant
Titan	1.5	2,576	16.0	1.22×10^6	Saturn	Gas giant
Callisto	1.1	2,403	16.7	1.88×10^6	Jupiter	Gas giant
Io	1.0	1,821	1.8	4.22×10^5	Jupiter	Gas giant
Moon	1.0	1,737	27.3	3.84×10^5	Earth	Terrestrial
Europa	0.9	1,565	3.6	6.71×10^5	Jupiter	Gas giant
Triton	0.8	1,353	5.88 (retrograde)	3.55×10^5	Neptune	Ice giant
Titania	0.5	789	8.7	4.36×10^5	Uranus	Ice giant

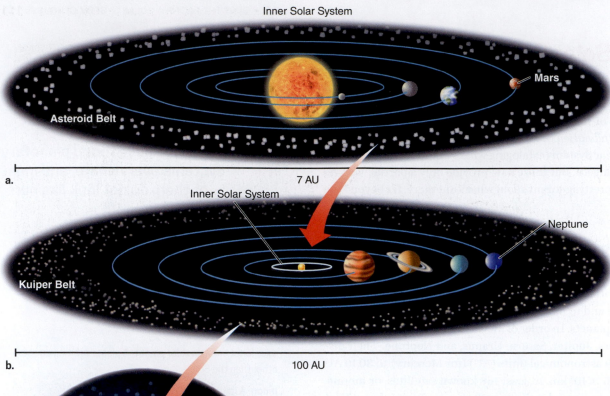

Figure 5.3 The Scale of the Solar System
Different scales are required to display all major components of the Solar System: Sun, planets, asteroid belt, Kuiper Belt, and Oort Cloud. Planets and Sun are not shown to scale because their diameters are vastly different. (The Sun's diameter is 109 times that of Earth, for example.) **a.** The inner Solar System, including the terrestrial planets and the asteroid belt. **b.** The gas giant/ice giant region and the Kuiper Belt. **c.** The scale increases dramatically in the outer Solar System, where the Oort Cloud spans more than 100,000 AU.

(the region David Jewitt was exploring), which may be home to as many as 100,000 objects with diameters of more than 100 km made of ice and rock (**Figure 5.3**).

The discovery of the Kuiper Belt and the recognition of dwarf planets as a separate class of objects occurred in just the past decade or two, and both are important parts of the story of planetary system formation. Beyond the Kuiper Belt is another recently discovered region called the **scattered disk**. The disk, which may be considered an extension of the Kuiper Belt, is estimated to contain tens of thousands of icy, rocky bodies similar to those in the Kuiper Belt. Their highly elliptical orbits may reach perihelion as low as 30 AU and aphelion as high as 100 AU.

Along with planets, moons, and dwarf planets, the Solar System is home to many smaller objects, such as asteroids, comets, and meteoroids. **Asteroids** are mostly rocky objects that range in diameter from hundreds of meters to 1,000 km. The Solar System holds as many as several million asteroids with diameters greater than 1 km. Most reside in the asteroid belt, but some orbit the Sun in the company of Jupiter (preceding and following that giant planet in the same orbit), and a handful of others move through the inner Solar System in orbits that can take them close to Earth.

Comets are made up of rock and ice and, like asteroids, range in size from hundreds of meters to 1,000 km. Most comets move on highly elliptical orbits that take them close to the Sun and then far out into space. About 4,000 comets are known, but astronomers estimate that more than a trillion populate the Solar System. Most comets reside in the **Oort Cloud**, a vast spherical region that has a radius of about 50,000 AU and surrounds the Solar System (see Figure 5.3).

Finally, **meteoroids** are smaller chunks of rock ranging in size from a grain of sand to a boulder. (*Meteorites* are meteoroids that land on Earth.) Because meteoroids are so small—and so numerous—astronomers cannot accurately estimate their number.

Three Flavors of Planets

"There are three different kinds of planets," says Jewitt—terrestrial planets, gas giants, and ice giants—"and each one represents a separate domain of the Solar System" (**Figure 5.4**). **Table 5.2** displays the properties of the eight planets and a few other major Solar System

scattered disk The disk-shaped region beyond the Kuiper Belt that contains many small bodies made of ice and rock scattered by the gravity of the planets to outer regions of the solar system.

asteroid A rocky object that orbits the Sun and has a diameter from hundreds of meters to 1,000 km.

comet A body of rock and ice, ranging in size from hundreds of meters to 1,000 km, that typically orbits the Sun on a highly elliptical orbit.

Oort Cloud A spherical region at the outer limits of the Solar System that has a radius of about 50,000 AU and may contain trillions of comets.

meteoroid A rock that ranges in size from a grain of sand to a boulder and orbits the Sun.

Figure 5.4 The Planets
Each of our eight planets is unique, but they are grouped as terrestrial, gas giant, or ice giant worlds,
shown at proper size scale. (Distance is not to scale because distances between the planets' orbits are
so large and varied. Saturn, for example, orbits 10 times as far from the Sun as Earth does.)

Table 5.2   Properties of Solar System Objects

Object	Class	Mass (in Earth masses)	Average Radius (km)	Average Radius (in Earth radii)	Average Density (kg/m³)	Orbital Radius (AU)	Orbital Period (yr)	Number of Known Satellites (as of 2018)
Mercury	Planet	0.055	2,440	0.38	5,430	0.39	0.24	0
Venus	Planet	0.82	6,052	0.95	5,240	0.72	0.62	0
Earth	Planet	1.00	6,378	1.00	5,520	1.00	1.00	1
Mars	Planet	0.11	3,397	0.53	3,940	1.52	1.88	2
Jupiter	Planet	317.82	71,492	11.21	1,330	5.20	11.86	79
Saturn	Planet	95.16	60,268	9.45	700	9.54	29.42	61
Uranus	Planet	14.54	25,559	4.01	1,300	19.19	83.75	27
Neptune	Planet	17.15	24,764	3.88	1,760	30.10	163.72	14
Pluto	Dwarf planet	2.2×10^{-3}	1,185	0.19	1,879	39.48	248.02	5
Eris	Dwarf planet	2.8×10^{-3}	1,163	0.18	2,520	68.05	561.40	1
Ceres	Dwarf planet	1.5×10^{-4}	476	0.07	2,090	2.77	4.60	0
1992 QB1	Kuiper Belt object	Unknown	80	0.01	Unknown	43.9	291.00	0
4 Vesta	Asteroid	4.47×10^{-5}	251	0.04	3,456	2.36	3.63	0
Earth's Moon	Moon	0.012	1,738	0.27	3,340	2.6×10^{-3} (3.84×10^5 km from Earth)	27.32 days	0
Sun	Star	3.33×10^5 (1.989×10^{30} kg)	6.96×10^5	109.13	1,409	—	—	Trillions

objects. Looking over the table, do you see any patterns that might offer clues to the history of planetary systems? Where, for example, are the massive planets located compared with the smaller ones?

The inner Solar System contains a class of **terrestrial planets** (*terrestrial* comes from the Latin word *terra*, which means "land"). With orbits relatively close to the Sun, Mercury, Venus, Earth, and Mars are fairly small terrestrial planets; they are composed mostly of rocky, solid material. While all of the terrestrial planets have some degree of atmosphere, these gases represent only a tiny fraction of each planet's mass.

Beyond the terrestrial planets are the giant planets, which come in two forms: gas giants and ice giants. Jupiter and Saturn are **gas giants**, massive worlds that are composed mainly of gases that are compressed into liquids or other, more exotic phases and may have relatively small, rocky cores. "They're primarily made of hydrogen and helium," says Jewitt. "They are similar, but not equal, to the Sun in their composition, having an extra component of heavier elements." Jupiter's mass is 318 times the mass of Earth, and Saturn contains about 95 **Earth masses**. Farther out are Uranus and Neptune, considered the **ice giants**, fairly large planets that have rocky cores but are composed mostly of ices. Uranus has about 15 Earth masses, and Neptune has 17. "Most of their mass is not in the form of gas," says Jewitt. "Of the 15 or 17 Earth masses in Uranus and Neptune, only one or two of those are hydrogen and helium. The rest consist of molecules like water, methane, carbon dioxide, and lots of other compounds frozen into ices."

The differences between the terrestrial and giant planets are not confined to location, size, mass, and composition. Other properties distinguish these types of worlds and provide clues for understanding their origin and evolution. The terrestrial planets have relatively slow rotation rates, while the giants tend to rotate quickly (Jupiter's "day" is just 10 hours). Terrestrial planets have few or no satellites; each giant has 14 or more moons. Terrestrial planets have weak magnetic fields, while the giants have strong magnetic fields. No terrestrial planet has rings, but each giant is surrounded by a ring system. The inner terrestrial planets are relatively close to each other, their orbits separated by less than 1 AU. In contrast, each giant planet's orbit lies many astronomical units from the next.

The strong distinctions between the properties of giant and terrestrial planets and between the gas and ice giants must be due to the physics of planetary systems. "The Solar System clearly has these three planetary domains of rock, gas, and ice," says Jewitt. "It's probably divided up this way because of the processes by which the planets were built." Astronomers have been seeking a theory of planetary system formation that naturally explains these domains and their different properties for centuries.

A critical clue for the development of that theory comes from the different densities of the planets. *Density* is defined as the *mass per volume* of a material and is measured in kilograms per cubic meter (kg/m^3). A gaseous atmosphere is less dense (has fewer kilograms of matter per cubic meter) than a chunk of rock. To peel apart the story of the Solar System, however, astronomers have to do more than just look at density. They must also take gravity out of the picture. The **uncompressed density** of a planet is the density the planet would have if there were no pressure due to gravity (if the core weren't squeezed by the weight of material above it), in addition to the usual interatomic forces that hold matter together. Once gravity is factored out, astronomers can see the original density of the material making up each planet. "The uncompressed density is what you would get if you measured a sample of the stuff in the lab," says Jewitt.

Looking at just the terrestrial planets, astronomers found a clear pattern: a decrease in uncompressed density of the planets as the orbital radius increases (**Table 5.3**). The fact that the uncompressed densities of terrestrial planets drop with distance shows astronomers something important: the process of planet formation must somehow be sensitive to the distance from the Sun. We will return to this critical clue in Section 5.5.

> **"There are three different kinds of planets, and each one represents a separate domain of the Solar System."**

terrestrial planet A planet consisting primarily of solid material.

gas giant A planet that consists mostly of hydrogen and helium and has a mass many times that of Earth.

Earth mass The mass of Earth, 5.97×10^{24} kg, used as a relative unit of measure for other objects.

ice giant A planet that consists mostly of ices (water, methane, carbon dioxide, and other compounds) and has a mass many times that of Earth.

uncompressed density The density an object would have if the effect of gravity were excluded.

Table 5.3 Compressed and Uncompressed Densities of the Inner Planets

Planet	Average Compressed Density (kg/m³)	Uncompressed Density (kg/m³)
Mercury	5,430	5,300
Venus	5,240	4,400
Earth	5,510	4,400
Mars	3,930	3,800

Note: The uncompressed densities of the terrestrial worlds clearly decrease as orbital radius increases.

Section Summary

- The primary categories of objects in the Solar System are planets and dwarf planets, moons, comets, asteroids, and meteoroids.
- The Kuiper Belt, scattered disk, and Oort Cloud are disk-shaped regions beyond the region of the planets.
- Our Solar System contains three types of planets: terrestrial planets (small, dense, rocky, and orbiting relatively close to each other nearest the Sun); gas giants (made of hydrogen and helium and on larger orbits); and ice giants (composed of gas and heavy compounds mostly frozen into ices and orbiting farthest from the Sun).

CHECKPOINT Cite an example of each type of planet and describe how it exemplifies that type.

And Pluto Too! Asteroids, Comets, Meteoroids, and Dwarf Planets

5.3 The Sun contains 99.86 percent of the mass of the Solar System. As Table 5.2 shows, the planets, with Jupiter leading the pack, make up the bulk of the remaining mass. The other occupants of the Solar System—smaller orbiting bodies between and beyond the planets, including asteroids, comets, meteoroids, and dwarf planets—account for only a small percentage of the Solar System's mass. But they too are important. Many of these nonplanetary bodies exist in Solar System regions far outside the domain of the planets. Some of the most interesting new discoveries are being made in these outermost regions, such as the Kuiper Belt (see Section 5.1). Many of these bodies can be regarded as debris left over from the construction of the Solar System. By studying them, astronomers have gained essential insights into the distant past, when the worlds of our Solar System were being assembled.

Asteroids

Asteroids are small, irregular objects (compared with planets and moons) composed mostly of rock and metal (**Figure 5.5**). The best-known asteroids reside in the asteroid belt, lying between Mars (at 1.52 AU) and Jupiter (at 5.20 AU). Most of these "main-belt" asteroids have orbits with average radii ranging between 2.1 AU and about 3.5 AU from the Sun.

VIS

Figure 5.5 Asteroids
Images of asteroids Mathilde, Gaspra, and Ida clearly illustrate their rocky, irregular nature. They lack sufficient mass for gravity to pull them into spheres.

IN THE ASTEROID BELT. Science fiction movies tend to portray the asteroid belt as a dense swarm of tumbling, house-sized rocks. The truth is a little less cinematic. The real asteroid belt is mostly empty space, and the average distance between its pieces of Solar System debris is about a million kilometers. This distance is so large that it would take a passenger jet months to cross. But given the force of gravity and enough time, collisions between asteroids do occur, and astronomers have benefited from this process. "We actually know quite a bit about these main-belt asteroids," says David Jewitt, "because we have pieces of them here on Earth in the form of meteorites left over from collisions."

Half the mass of the entire asteroid belt lies in that belt's four largest objects. Ceres is the largest, with a radius of 940 km and a mass of 9.4×10^{20} kg. Ceres is so large, in fact, that its gravity has pulled it into a quasi-spherical shape (and, as we shall see later, astronomers consider it to be a dwarf planet, along with Pluto). The next largest known object in the asteroid belt is asteroid 4 Vesta, with a diameter of 540 km and a mass of 2.6×10^{20} kg. The third and fourth largest main-belt asteroids are 2 Pallas and 10 Hygiea, both greater than 400 km in diameter. Most of the 250,000 known main-belt asteroids are far smaller and less massive than these four objects, and only a few reach a diameter of 100 km.

For many years, astronomers debated the origin of the asteroid belt. Some saw it as the remains of a planet that had been broken apart by either a collision or the force of Jupiter's gravity. However, various facts cast doubt on this initially popular idea. "It's very difficult to smash a planet once you've made it," explains Jewitt. "You'd have to hit a fully formed planet with something just as big and have it moving very fast. There's no evidence that anything like this kind of collision happened." In addition, if the "broken-planet" theory were accurate, all asteroids in the belt should have a similar composition. Instead, asteroids have a range of compositions, suggesting that they did not originate from a single large body.

Today a more prevalent theory is that the belt represents a desert of **planetesimals**, rocky objects that might have been the building blocks of planets. According to this second theory, planetesimals in the region between Mars and Jupiter could not form into a planet because Jupiter's gravity kept the planetesimals from coagulating into a larger body. "It looks a lot more like the asteroids came from a number of objects," explains Jewitt, "with different initial formation histories. Those origins get reflected in their properties, like the different ratios of rock to iron we see in different asteroids."

One of the most important characteristics of the main belt is the presence of **orbital resonances** in the orbits of its asteroids. When two objects orbit the same body, such as an asteroid and a planet orbiting a star, they periodically pass each other. That's when the gravitational force between them is strongest. Usually, these tugs happen at different points in the orbits and the effects average out. But in some instances, the close approaches and strong gravitational interactions happen at the same place (or places) over and over again. These repeated, synchronized gravitational accelerations can either stabilize or destabilize the two objects' orbits (**Figure 5.6**). The orbit of the less massive object is always destabilized, or perturbed, more than that of the more massive object. Thus each close passage provides an extra gravitational tug, and across many orbits these tugs tend to pull the smaller object into a slightly different orbit.

Orbital resonances between Jupiter, with its enormous mass, and main-belt asteroids have shaped the belt in significant ways. For example, there are gaps in the distribution of asteroids in the main belt where orbital resonances with Jupiter occur. Very few asteroids are found at these locations, called **Kirkwood gaps**. Any asteroid that was originally in an orbit within a Kirkwood gap was eventually pulled away into another orbit by periodic tugs from Jupiter.

> "It's very difficult to smash a planet once you've made it."

planetesimal An irregular rocky object, typical of those from which planets are believed to have formed by accretion.

orbital resonance The synchronized, periodic gravitational influences that arise when objects orbiting a body have periods that are whole-number ratios of each other.

Kirkwood gap A band of orbits in the main asteroid belt where very few asteroids are found due to orbital resonances with Jupiter.

Trojan asteroid An asteroid co-orbiting with Jupiter in one of two groups: one group preceding Jupiter and the other group following it, both at the same distance.

Earth-crossing asteroid (ECA) An asteroid whose highly elliptical orbit crosses Earth's orbit and is therefore a potential risk for collision with Earth.

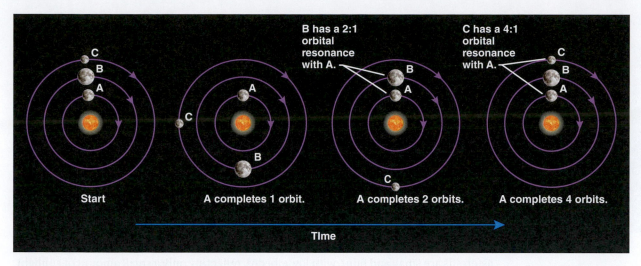

Figure 5.6 Orbital Resonances

Objects orbiting the same central body (as with asteroids orbiting the Sun in the asteroid belt) develop orbital resonances. They exert gravitational influences on one another over the course of many orbits, eventually developing whole-number orbital ratios.

OUTSIDE THE ASTEROID BELT. Beyond the asteroid belt, there are at least two other important collections of asteroids in the Solar System. The **Trojan asteroids** are two groups of objects that move around the Sun along the same orbit as Jupiter, with one group always preceding and the other group always following that gas giant (**Figure 5.7**). The orbits of these bodies are gravitationally stable, meaning they can remain in them for long periods of time. There may be a million or more of these Trojan asteroids with sizes greater than 1 km.

A third collection of asteroids is of enormous importance for human civilization. Most asteroids move along orbits with fairly low eccentricity ($e < 0.3$). Recall from Chapter 3 that low eccentricity (e close to 0) means circular orbits, while high eccentricity (e close to 1) means highly elliptical orbits. The distance from the Sun will not change very much for an asteroid on a low-eccentricity orbit, so the asteroid will not intersect the orbits of any planets. However, the members of one family of asteroids, **Earth-crossing asteroids (ECAs)**, have high eccentricities and periodically cross Earth's orbit. "These Earth crossers are mostly coming from the asteroid belt," explains Jewitt. "They were kicked onto elliptical orbits by Jupiter, and now they cut across the paths of the terrestrial planets." ECAs last only so long. "Either they're going to hit one of the planets," Jewitt says, "or they're going to experience a slingshot effect from a planet's gravity into the Sun. They might even get shot out of the Solar System entirely."

As their name suggests, some Earth-crossing asteroids may become Earth-striking asteroids, and the consequences can be apocalyptic. "We know the surfaces of the terrestrial planets have been hit by asteroids, because they're all covered with craters," says Jewitt. (High-speed impacts produce *craters*, roughly circular surface depressions.) "And we know this is true for Earth as well, even though craters here tend to get weathered away over time." Geologists have catalogued more than 180 impact craters on Earth, some of which must have been caused by asteroid-sized objects. "Some of them are frighteningly big," says Jewitt. "They can stretch across hundreds of kilometers." The most famous impact, forming the Chicxulub crater in Mexico 65 or 66 million years ago, is associated with the extinction of the dinosaurs. "It was probably a 10-km-sized body that hit us back then," says Jewitt. "It was a global extinction event that killed almost every surface-dwelling species." A much smaller body exploded

> "It was probably a 10-km-sized body that hit us back then. It was a global extinction event that killed almost every surface-dwelling species."

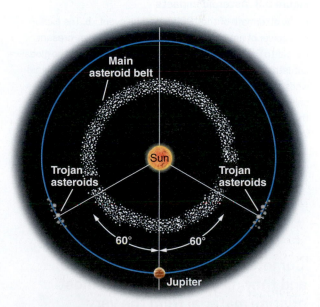

Figure 5.7 Trojan Asteroids

These groups of asteroids move along Jupiter's orbit, at 60° preceding and following the planet, and are not in the asteroid belt. They lie at stable points where the gravity of Jupiter and the Sun balances the motion of Jupiter in its orbit.

a.

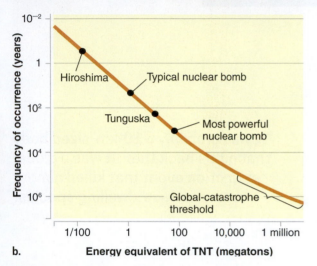

b.

Figure 5.8 Asteroid Impacts

a. The aftermath of the 1908 Tunguska event. **b.** The explosive power of asteroid impacts represents an ever-present danger to Earth. Impacts with energies greater than the global-catastrophe threshold create mass extinctions. The asteroid whose impact resulted in the extinction of the dinosaurs was only about 10 km in diameter, but it had nearly 10,000 times the power of the nuclear bomb that leveled Hiroshima during World War II.

"Comets are made up of a lot of ice, and they may be some of the most pristine material from the very beginning of the Solar System."

over Tunguska, Siberia, in 1908, flattening forests for thousands of square kilometers (**Figure 5.8a**).

In 2002, a 100-meter-wide asteroid called 2002 MN passed a mere 120,000 km from Earth, well inside the Moon's orbit. An object that size has the energy of a 50-megaton nuclear weapon and could do catastrophic damage if it made impact near a populated area. Although the odds of a collision in any given year are quite small, the consequences are so far-reaching that in the late 1990s astronomers began a systematic effort to find all ECAs. Since then, thousands of ECAs have been discovered, hundreds of them listed as "potential hazards." This dubious distinction is given to ECAs with sizes of 150 meters or more and with orbits that bring them within 7,500,000 km of Earth. **Figure 5.8b** shows the frequency of impacts as a function of the impact energy.

Why don't astronomers know where all ECAs are and why can they only estimate the number of asteroids in the Solar System? Finding asteroids is an extremely tricky business. Astronomers use the term **albedo** to describe how much light an object reflects. Most asteroids are small and faint, with low albedos, reflecting only a small amount of sunlight. That means determining both the total number of asteroids and their orbits requires long hours of effort with as-yet-incomplete results.

ASTEROID PROPERTIES. Determining asteroid properties has also been a difficult task. Identifying composition, size, shape, or like properties requires indirect methods, such as interpreting reflected sunlight. From these studies, astronomers have determined that a number of classes of asteroids exist, two of which account for most of the asteroids in the Solar System. Asteroids of the first main class are called *carbonaceous* (rich in carbon), or **C-type asteroids**. These are very dark and reflect little sunlight. Asteroids of the second main class, called **S-type asteroids**, contain more **silicates** (rocks made with silicon and oxygen) and are more reflective. At least 75 percent of the Solar System's asteroids are C-type, 17 percent are S-type, and the rest fall into other classes.

The distinction in mineral contents of these two major classes of asteroids highlights different early evolutionary phases of the Solar System. "From the ages of meteorites, we know that all the asteroids were formed around 4.5 billion years ago," says Jewitt, "but the C-type asteroids were never strongly differentiated." **Differentiation** is the process by which gravity within forming planetesimals forces heavier material (such as iron and nickel) to sink to the center, creating layers of different compositions. "It appears that many C-type asteroids escaped differentiation. They represent more primitive material." The S-type asteroids are probably chunks of planetesimals that were on their way toward forming larger bodies, so they were subjected to heat, compression, and differentiation (all from gravity) when collisions smashed them apart.

STUDYING ASTEROIDS. Although telescopes and meteorites tell us a lot about asteroids, to learn more we have to visit them—as a flotilla of spacecraft has done since 1990. A number of missions from several regions, including the United States, the European Union, and Japan, have journeyed to at least eight asteroids. In 2005, for example, the Japanese *Hayabusa* space probe reached asteroid 25143 Itokawa with the intention of bringing samples of the rocky body back to Earth.

From these missions, astronomers have gained important insights into the nature, structure, and origins of asteroids. Two of the most important asteroid encounters came in 1997 and 2000, when the Near Earth Asteroid Rendezvous (NEAR) mission visited the C-type main-belt asteroid Mathilde and the S-type near-Earth asteroid Eros. Images showed a crater on Mathilde so large that scientists wondered how the object could still be intact. "The answer," says Jewitt, "came when we looked at Mathilde's density and found it to be so low that it wasn't really a solid object." Instead of being a solid rock, Mathilde must be a rubble pile—a loosely porous collection of smaller rock fragments and dust held together by gravity. Images and measurements of Eros showed that, unlike Mathilde, it is solid and heavily cratered from millions of years of collisions with other smaller Solar System objects.

Recently the *Dawn* spacecraft visited 4 Vesta, entering its orbit in 2011. Neither a C-type nor an S-type asteroid, 4 Vesta was found to have a heavily cratered surface and likely has an iron core.

Comets

Like asteroids, comets are a form of debris left over from the formation of the Solar System, but the two are very different in composition. "Comets are much more a primordial object," says Crystal Brogan, an astronomer who studies how solar systems form. "Comets are made up of a lot of ice," she continues, "and they may be some of the most pristine material from the very beginning of the Solar System." While rock and metal are dominant in asteroids, comets are up to 50 percent ice combined with other rocky materials. Astronomers liken comets to "dirty snowballs" (although "icy mudballs" may be more accurate). In their properties, locations, and orbital trajectories, comets offer clues to the Solar System's formation and even the history of our own planet.

The difference between asteroids and comets hints at the processes at work in the early Solar System. Rocky asteroids form within a critical distance from the Sun, known as the **snow line**. "The snow line is the distance from a star at which water has to be frozen," explains Brogan. The temperature of material that orbits a star and is exposed to its radiation decreases the farther an object is from the star. Close to the star, so much stellar radiation is absorbed that temperatures are above the freezing point of water. "But," says Brogan, "there will have to be an orbital radius beyond which it is too cold for water not to be frozen." Inside the snow line, a star's radiation can vaporize water; beyond the snow line, water stays as ice. Thus comets must have formed beyond the snow line because they contain so much frozen water and other icy compounds.

Comets differ from asteroids not just in composition but also in *where* we find them and how they move. "Asteroids generally live in gravitationally stable places," says Brogan. Objects orbiting the Sun are gravitationally stable when their motion is not easily perturbed by the gravitational pull of other bodies, such as the planets (particularly the giants). Over time, the combined gravity of the planets has shepherded asteroids into locations where their orbits stop changing. "That's why you often find asteroids in stable places like the main belt or the location of the Trojans near Jupiter," says Brogan. "But comets originate from much farther out in the Solar System. In some cases we may be seeing them on their first trips inward, and that means they haven't had as much time to interact with the planets."

Comets often make dramatic celestial appearances (**Figure 5.9a**). Taking just a few weeks (or, at most, months) to move across the fixed background of stars, comets first appear as bright, glowing objects. Over time, they grow larger and reveal luminous tails that can extend many degrees of arc before the comet eventually fades and disappears. In premodern times, the appearance of a comet was often interpreted as an omen (usually a bad one) and, as such, was a source of considerable anxiety (**Figure 5.9b**). Astronomers now understand the physics of a comet's fleeting appearance.

A comet's glow comes from changes in its **nucleus** as it approaches the Sun. A comet's nucleus, which ranges in size from 100 meters to 50 km, is a mix of rock, dust, water ice, and icy frozen gases such as carbon monoxide, carbon dioxide, ammonia, and methane. Although these ices make up the bulk of the comet's mass, space missions have shown that comets' nuclei have rocky surface crusts that are remarkably dark. Comet surfaces have very low albedos, just 2–4 percent—about half that of the black tarmac that covers many roads. The comparison with road tar is appropriate: some scientists believe comet crusts are dark because surface ice, burned away by sunlight, leaves behind a sticky goo of rock and dust the color of asphalt. But if the rocky, dusty, tarlike surface of a comet

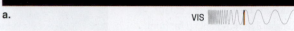

a.

VIS

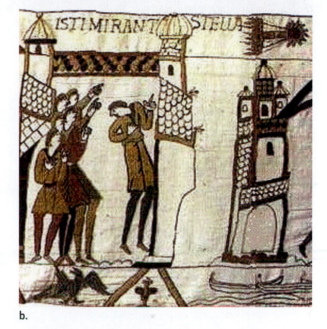

b.

Figure 5.9 Comet Appearances
a. The magnificent Comet Hale-Bopp, discovered in 1995, was visible to the naked eye for 18 months. **b.** Part of the Bayeux Tapestry (which tells the story of the 11th-century Norman Conquest of England) bears a comet (top right), believed to be Halley's Comet. Comets were viewed as ominous events.

albedo The reflecting power of an object, measured as the ratio of light reflected to light received.

C-type asteroid An asteroid that is carbonaceous (rich in carbon) and is typically dark.

S-type asteroid An asteroid that contains mostly silicate rock and is typically more reflective than a C-type asteroid.

silicate A mineral compound containing silicon and oxygen.

differentiation The process in planet formation whereby, during a planet's molten stage, the denser materials sink to the center while less dense materials rise to the surface, forming distinct layers.

snow line The distance from a star beyond which the temperature is cold enough for water to exist as ice.

nucleus (pl. nuclei) The main body of a comet, composed of ice and rock.

Figure 5.10 Comet Structure
a. The coma is wide and diffuse, but the outer surface of a comet's nucleus is composed of crusts left over as ice is burned away. **b.** The crusted nucleus, coma, and jets of Halley's Comet, photographed by the European Space Agency's *Giotto* spacecraft from 600 km. **c.** Nucleus of Comet Tempel 1, photographed far from the Sun's radiation and solar wind; the coma is completely absent.

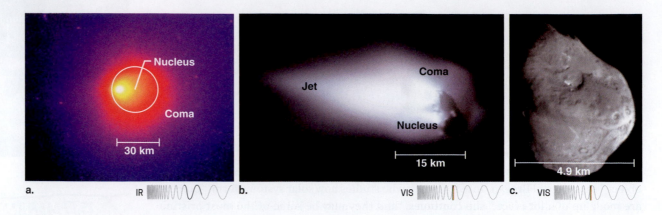

nucleus reflects so little light, what is it that glows brightly when a comet appears on the sky (**Figure 5.10**)?

Most comets move in highly elliptical orbits, taking them from the cold, outer Solar System into the domain of the terrestrial planets and eventually very close to the Sun. (Some end up plunging into the Sun.) As a comet nucleus enters the inner Solar System, sunlight warms the nucleus, vaporizing the ices below its crust. **Jets** of high-pressure water, carbon dioxide, methane, and ammonia vapor break through the surface. The jets of vapor shoot into space and are ionized by the Sun's radiation. In this way the comet's nucleus becomes surrounded by the **coma**, a glowing cloud of gas and dust. Light reflected off dust blown into the coma and direct emission lines from its ionized gases are what make the coma visible from Earth. Recall from Chapter 4 that emission lines form when electrons in atoms or ions are driven into excited orbits; the electrons then emit light when they cascade back down.

The tails of comets, the most spectacular part of their appearance, form as a comet warms up on its journey through the inner Solar System. Two tails—an ion tail and a dust tail—form as the comet moves along its orbit. The **ion tail** is made of ionized gas blown off the coma. The Sun constantly drives a flow of material off its surface, called **solar wind**, which fills interplanetary space. The solar wind sweeps up ions ejected from a comet's coma and carries them outward, away from the Sun. In contrast, the **dust tail** is composed of small grains of solid matter blown off the comet's nucleus. Sunlight gently pushes on the dust particles, forcing them away from the comet. Once outside the coma, they move on their own orbits, so the dust tail tends to be curved. Both tails always point away from the Sun, even when the comet heads back toward the outer Solar System (**Figure 5.11**). While a typical comet nucleus may be only a few kilometers across, the coma of a comet may be 10^6 km in diameter (larger than the Sun), and an ion tail can stretch across an entire astronomical unit (10^8 km).

In addition to ions and dust, larger debris is also blown off the comet. Sand-to-finger-nail-sized bits of rock are driven into space via the jet eruptions. As we will soon see, these larger forms of cometary debris can become the source of meteor showers.

Comets move on orbits with higher eccentricities than any other class of Solar System objects. According to Kepler's second law, this means they spend most of their time at aphelion, their farthest distance from the Sun. Because comets do not become bright until they enter the inner part of the Solar System, astronomers must use observations and detailed calculations to reconstruct comets' orbital paths. These studies show that comets originate from two distinct locations in the Solar System and therefore fall into two distinct classes, short-period and long-period comets, which differ in orbital period and orbital inclination. (*Inclination* is a measure of the tilt between an object's orbit and the ecliptic—the plane of Earth's orbit.)

Short-period comets have orbits that take 200 years or less to complete. These orbits have low inclination, meaning they are usually aligned with the ecliptic plane. In

Figure 5.11 Ion Tail and Dust Tail
Comets have two tails, both of which point away from the Sun. The tails lengthen as the comet approaches the Sun and essentially disappear as the comet approaches aphelion.

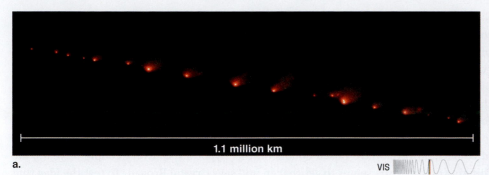

1.1 million km

a.   VIS

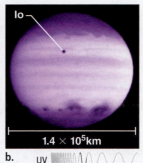

Io

1.4 × 10⁵km

b.   UV

Figure 5.12 Comet Impact with Jupiter
a. In 1994, Jupiter's gravity broke apart comet Shoemaker-Levy 9 into 21 fragments. **b.** The fragments struck Jupiter with spectacular results. Dark regions in the southern hemisphere of the planet in this Hubble Space Telescope image of Jupiter's atmosphere represent explosive energy depositions as comet fragments plunged into that atmosphere.

addition, their direction of orbital motion is the same as that of all the planets in the Solar System. Halley's Comet, with a 76-year orbit, is a short-period comet. Its aphelion lies just beyond Neptune.

Long-period comets, in contrast, can have periods of thousands of years or even a million years. In general, the orbits of long-period comets are not confined to the ecliptic, and their aphelia take them far beyond the domain of the outer planets. These comets tend to have large eccentricities, and they come from all directions relative to the Sun with equal probability.

Of known comets, 80 percent are the long-period type and just 19 percent are short-period. Comets whose orbits cannot be determined or that might leave the Solar System altogether make up the other 1 percent. From orbital analysis, astronomers can tell where the different classes of comets originate.

"The short-period comets come from the Kuiper Belt," says Brogan. "Gravitational tugs from Neptune can yank these comets out of this distant region of the Solar System and onto new orbits." Since the Kuiper Belt is a disk distribution of objects aligned with the ecliptic plane of planetary orbits, the short-period comets have relatively small eccentricities and inclinations. While the relatively circular, relatively uninclined orbits of short-period comets are consistent with an origin in the Kuiper Belt, some may have originated in the scattered disk region beyond the Kuiper Belt (see Section 5.2). The scattered disk is composed of bodies that were kicked by the gravity of the planets (scattered) to outer regions of the Solar System.

The long-period comets, however, come from much deeper in space. In the 1950s, the Dutch physicist Jan Oort recognized that the Solar System must be surrounded by a spherical "cloud" of comet nuclei, 50,000 AU or more from the Sun, named the Oort Cloud. "It's so enormous compared to the size of the inner parts of the Solar System," says Crystal Brogan. "The Oort Cloud is a giant sphere of cometary bodies, and its presence really forces us to expand our ideas about the 'size' of the Solar System." The Oort Cloud is now recognized as the source of the long-period comets. In the same way that the gravitational influence of an outer planet can knock a Kuiper Belt comet nucleus onto a higher-eccentricity orbit, a passing star sometimes comes close enough to disturb comets in the Oort Cloud and knock them onto Sun-bound elliptical orbits. Recently astronomers discovered the first object on an orbit that meant it was just passing through the Solar System. Called Oumuamua (which means "scout" in Hawaiian), it was likely an asteroid or spent comet nucleus from another star system.

Comets on highly elliptical orbits must, by definition, cross the orbits of some or all of the planets. Considerable evidence suggests that encounters with gas giants alter the orbits of comets over time. But gravitational interactions with planets can do more than simply adjust cometary orbits. Collisions between planets and comets had long been hypothesized, but in 1994 Jupiter captured the comet Shoemaker-Levy 9 as it passed the giant planet on its way toward the inner Solar System. Astronomers around the world watched with delight as the comet was torn apart by Jupiter's gravity and the fragments eventually plunged into the planet's atmosphere (**Figure 5.12**).

> "The Oort Cloud is a giant sphere of cometary bodies, and its presence really forces us to expand our ideas about the 'size' of the Solar System."

jet A stream of liquid, gas, or small solid particles shot outward from a comet nucleus or other object in a focused beam.

coma The nebulous envelope around the nucleus of a comet, formed when the comet passes close to the Sun.

ion tail The tail of a comet that is composed of gas from the coma and carried backward by the solar wind and thus always points directly away from the Sun.

solar wind A flow of material from the surface of the Sun that sweeps through interplanetary space.

dust tail The tail of a comet that is composed of dust driven from the nucleus by solar radiation; it points away from the Sun as the dust particles move on their own orbits.

short-period comet A comet that originated in the Kuiper Belt or scattered disk and has an orbital period of 200 years or less.

long-period comet A comet that originated in the Oort Cloud and has a highly eccentric orbit and an orbital period longer than 200 years.

DAVID JEWITT

"I grew up in an industrial part of London, where the city lights were pretty bright," David Jewitt says. "One night I saw a bunch of meteors crossing the sky. I must have seen 10 or 15." The meteor shower set the course for his life. "I saw something that my parents didn't understand. Nobody could tell me what it was."

Jewitt ended up as a professor in the Earth and Space Sciences Department at UCLA. "There are some things we do know," he says, "and there are lots of things we don't know. And then there are things we thought we knew, but we're learning we have to go back and rethink them. It's that last aspect of science that keeps it all interesting."

hydrated Containing water.

Centaur An object that has a mix of asteroid-like and comet-like properties.

meteor A meteoroid that has entered Earth's atmosphere. Heat from friction causes radiation that may be seen as a very brief streak of light or as a fireball.

meteorite A rock that was a meteoroid, then a meteor, and has reached Earth's surface.

meteor shower The appearance of frequent meteors within a few hours or several days. The meteors seem to emanate from a common location on the sky as Earth passes through the debris deposited along a comet's orbit.

According to one hypothesis, collisions between Earth and comets in the early history of the Solar System may have been responsible for bringing our young world the bulk of its water. "Earth formed hot," says David Jewitt. "If you overbake a cake in the oven, it's going to come out as something dry and crisp. The early Earth was probably a lot like that overbaked cake. So Earth's water and other volatile materials had to be added later, after the surface cooled down. Since comets are big, icy things, they are the natural candidates for water delivery systems."

Comparisons between the composition of Earth's oceans and the ice composition of comets, however, have raised doubts on this hypothesis. "Every couple of years, the thinking on this issue seems to get shaken up," says Crystal Brogan. To test the idea that comets delivered a significant fraction of the water to the primordial earth, astronomers compared the water released from comets with what they found in the Earth's oceans. The comparisons didn't always match up. "The problem," explains Brogan, "is that sometimes you'll observe a comet and the results are relatively in good agreement with the hypothesis that Earth's water comes from comets. But sometimes you observe another comet and the results are very different."

These differences led some scientists, David Jewitt among them, to theorize that asteroids—not comets—delivered the early supply of water to Earth. "There is a new class of main-belt 'comets,' some of which are basically water-rich asteroids," says Jewitt. "There are also asteroids which, even though they don't have ice, have minerals in them which include water as part of the molecular structure." If there are significant quantities of these **hydrated** (water-containing) minerals in asteroids, then an impact with Earth would break the minerals apart along with the asteroids. The water molecules would then be freed and might have become part of Earth's oceans. "These kinds of asteroids," says Jewitt, "tell us that a lot of water must have been present in the Solar System in the past in a variety of forms, so there are different ways Earth's oceans may have formed."

A number of probes have intercepted comets on their journeys through the inner Solar System. Humanity's first up-close comet encounter came when the European Space Agency (ESA) flew its spacecraft *Giotto* past Halley's Comet in 1986. In 2004 NASA captured particles from the dust tail of comet Wild 2 and returned them to Earth for study. In 2005 the *Deep Impact* probe drove an almost 400-kg projectile into Comet Tempel 1 and sent back spectral analysis of the resulting vapor blow. One of the most remarkable encounters was in 2014, when ESA's *Rosetta* spacecraft rendezvoused with Comet 67P/Churyumov-Gerasimenko. *Rosetta* found an unusually shaped comet nucleus approximately 4.3 × 4.1 km across with a narrow "neck" connecting two wider regions (**Figure 5.13a**). Although *Rosetta*'s lander *Philae* was unable to take much data due to the comet's hard subsurface ice layer, *Rosetta* continued to orbit the comet as it swung around the Sun. The spacecraft returned remarkable high-resolution images of the comet's surface (**Figure 5.13b**) as well as jets that became more active as solar radiation heated the surface (**Figure 5.13c**).

Consider one more class of Solar System objects: the **Centaurs**. In Greek mythology, Centaurs were half-human, half-horse creatures. In astronomy, they have properties that seem to be a mix of comet and asteroid. They tend to live on unstable elliptical orbits that take them between Jupiter and Neptune, repeatedly crossing those of the outer planets. The largest Centaur, 10199 Chariklo, is 260 km across—the same as a midsized asteroid from the main belt. A few Centaurs have been shown to produce comet-like comas. The Solar System is estimated to hold more than 40,000 Centaurs.

Meteoroids

Meteoroids are the smallest class of debris in the Solar System, ranging in size from a grain of sand to a boulder. When meteoroids are captured by Earth's gravity and plunge into the atmosphere at speeds as high as 20 km per second (km/s), or more than 70,000 km

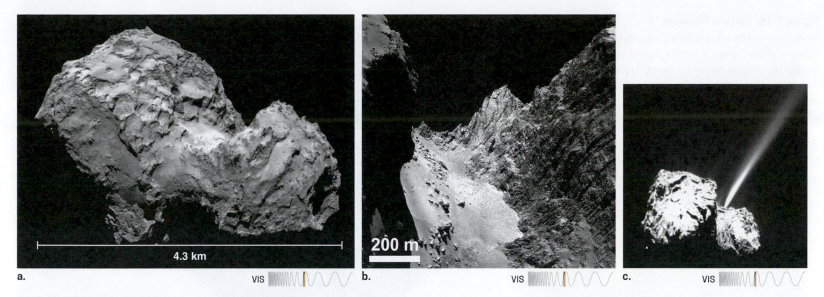

Figure 5.13 Comet 67P/Churyumov-Gerasimenko
a. Image taken by *Rosetta*'s probe in 2014 just 29 km from Comet 67P/Churyumov-Gerasimenko.
b. Close-up view of the comet's neck. **c.** One of the comet's brightest outbursts illustrates the jet typical of an active comet.

per hour (km/h), they are then called **meteors**. They act like supersonic jets as they plow through the air. Extremely high temperatures build up in front of a fast-moving meteor as air molecules pile up. Once inside Earth's atmosphere, meteors leave bright trails of gas— among the most striking events people can see on the sky (**Figure 5.14**). Millions of meteors appear in the atmosphere each day (although they can usually be seen only at night); most of them occur at heights between 50 and 95 km above Earth's surface.

A particularly remarkable example of a meteor occurred on February 15, 2013, over the city of Chelyabinsk, Russia. Thousands of people watched as a fiery ball streaked across the sky and flared brightly (**Figure 5.15**). Soon afterward, a shock wave swept across the region, shattering windows and damaging thousands of buildings. The event was caused by an object 20 meters in diameter with a mass of more than 12 million kg. It was the largest known natural object to enter Earth's atmosphere since the Tunguska event.

Most meteor-causing meteoroids are the size of small rocks and usually burn up (evaporate) in the atmosphere. Larger meteoroids can make it to the ground, at which point they are called **meteorites**. These lumps of space rock are treasured finds for astronomers and planetary geologists, offering a rare glimpse of the raw material of Solar System formation. "It's rare that someone tracks a meteor all the way to the ground and finds a meteorite," says Jewitt. "The majority of meteorites are discovered by accident long after they fell." Many are found near the South Pole. "They fall onto the ice, get buried, and then ice flow under the surface brings them to the surface years later," says Jewitt. "A meteorite is pretty obvious on an ice field, so a lot of scientists travel to the South Pole to hunt for new finds."

Studies of meteorite composition and analysis of meteor tracks on the sky show that most meteorites are chunks of asteroids from the main belt. But not all meteors originate in asteroids. During **meteor showers**, unusually high numbers of meteors occur, all of which appear to radiate from a single point on the sky. Meteor showers occur at fixed times each year. For example, the Leonid meteor shower, in which meteors appear to radiate from a point in the constellation Leo, occurs in November.

The source of these showers has been tracked to material blown off comets. Over time, sand-sized particles and pebbles ejected from a comet become distributed along its orbit. As Earth moves around the Sun, it will pass through this trail of debris if it crosses

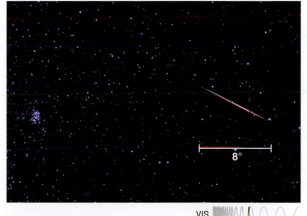

Figure 5.14 Meteors
This streak of light was created by a meteor, a small piece of solid matter being heated by friction with Earth's atmosphere.

Figure 5.15 Chelyabinsk Fireball
The Chelyabinsk meteor created this fireball as it exploded 29.7 km above the ground in 2013. Even at that height, the blast wave from the airburst damaged thousands of buildings.

Figure 5.16 Meteor Showers

When Earth's orbit takes it through clouds of small particles typically left by comets, a meteor shower occurs. Although meteor showers are named for the constellation from which they appear to originate, there is no connection between the meteoroids and the constellation's stars.

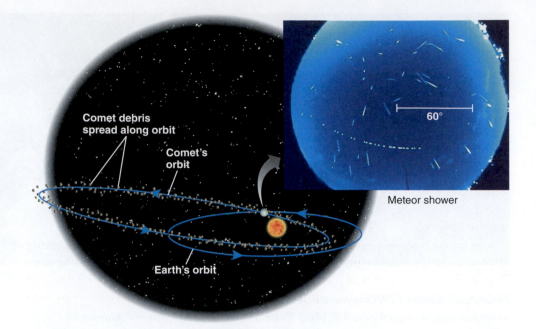

Comet debris spread along orbit

Comet's orbit

Earth's orbit

60°

Meteor shower

"Meteorites give us access to a huge number of different source bodies in the asteroid belt. They put all our theories into a context."

the comet's orbit. All of the resulting meteors will, from the perspective of someone on the ground, appear to radiate from the same location on the sky. That point is the position of the comet's debris trail (**Figure 5.16**).

The detailed study of meteorite structure and composition has provided a unique window into the history of the Solar System. In terms of origin, there are two main meteorite classes: primitive and processed. **Primitive meteorites** are ancient remnants of the original disk from which the planets were born (**Figure 5.17a**). Cutting open a primitive meteorite reveals a mix of rock interspersed with metal flakes. The chemistry of primitive meteorites indicates that they suffered little change since they formed near the time the Solar System assembled. **Processed meteorites**, however, are likely to be fragments of larger asteroids, which means they formed in violent collisions (**Figure 5.17b**). Processed meteorites that come from the differentiated cores of nearly planet-sized objects are mostly metallic. Those that come from the outer parts of such an object are mostly rocky. The *most* primitive meteorites are a special group, called the **carbonaceous meteorites**, which are rich in carbon and volatiles such as water (**Figure 5.17c**). From their composition, scientists can tell they likely formed in cooler regions of the disk of gas and dust that built the Solar System.

Remarkably, some meteorites appear to be chunks of other planets. Meteorites that are chemically similar to the Moon or even Mars have been found. These finds provide strong evidence that asteroid and comet impacts with planets and moons can eject chunks

Figure 5.17 Meteorites

The origin of meteorites helps determine their composition and structure. **a.** Primitive meteorites mix rock with metal flakes. **b.** Processed meteorites are either metallic or rocky. **c.** Carbonaceous meteorites are rich in carbon and volatiles.

a.

b.

c.

of these bodies into space. Eventually, some of these fragments are captured by Earth's gravity and fall to the ground.

Ample evidence shows that Earth is routinely struck by large meteoroids capable of considerable (though not apocalyptic) damage. The Barringer Meteor Crater in Arizona stretches 1.2 km across and is 170 meters deep (**Figure 5.18**). It is the result of an impact 50,000 years ago by a 200,000-ton object that was some 50 meters wide (too small to be considered an asteroid or comet). Atmospheric and geologic forces tend to erase the evidence of meteor craters on Earth, but space-based imaging has helped astronomers find other examples. "A good example of a recent meteor strike is the Tunguska impact in Siberia in 1908," explains Jewitt. "It was a rather small body, maybe only 50 meters across, that exploded in midair, but it delivered a big enough punch to knock over 2,000 square miles of pine forest" (see Figure 5.8). The Tunguska blast had the force of a 5- to 10-megaton nuclear weapon.

If they do not wipe out cities, meteorites can be a great boon to astronomers, giving them a chance to examine pristine material left over from the birth of the Solar System. "Meteorites give us access to a huge number of different source bodies in the asteroid belt," says Jewitt. "They put all our theories into a context." The clues that meteorites offer, along with clues from other sources, provide a strong foundation for building a story of Solar System formation.

Dwarf Planets

You might say that the planet Pluto died in 2006. That year the International Astronomical Union (IAU) held a special meeting to provide a concrete and working scientific definition of the word *planet*. While you might think that it's easy to know what a planet is, for the smaller bodies in the Solar System, the definition has shifted over time. Ceres (**Figure 5.19a**), the largest object in the asteroid belt, was first considered to be a planet when it was discovered in the early 1800s. The *Dawn* spacecraft's visit to Ceres in 2015 was compelling, as initial images showed a large bright spot with an albedo of 40 percent. Later the spot was identified as a collection of highly reflecting regions at the center of a larger crater, named Occator, which were likely due to ice (**Figure 5.19b**). Further observations showed that there is a great deal of ice (and hence water) as well as water vapor on Ceres.

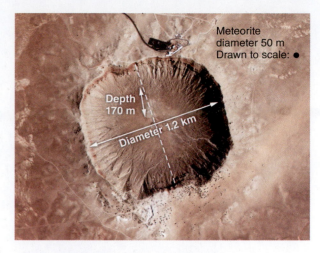

Figure 5.18 Barringer Meteor Crater
Most Solar System debris that reaches Earth's surface is in the form of small meteorites, which provide us with much information. Larger objects have enormous destructive power. The meteorite that struck in Arizona and created this crater 50,000 years ago had the approximate energy of 10 megatons of TNT.

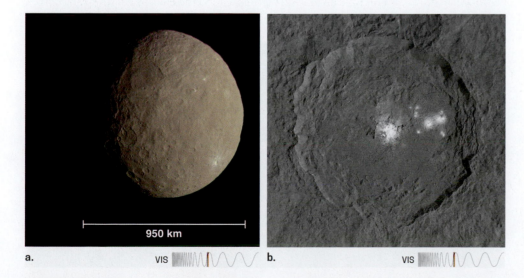

a.   VIS   **b.**   VIS

Figure 5.19 Ceres
Images captured by the *Dawn* spacecraft in 2015. **a.** Ceres—the only dwarf planet in the inner Solar System—is the largest object in the asteroid belt and contains about one-quarter of the belt's mass. **b.** An impact crater that bears intriguing bright spots.

primitive meteorite A type of meteorite that formed, largely unchanged, from the disk that produced the planets of the Solar System.

processed meteorite A type of meteorite that formed when a larger asteroid broke apart during a violent collision.

carbonaceous meteorite The most primitive type of meteorite, which formed from the components that built the Solar System.

When more and more objects in the Kuiper Belt began to be identified, it became clear that Pluto's status might need an update. As we saw earlier, the Kuiper Belt is a disk-like region of space beyond Neptune where swarms of icy bodies reside. In 2003 a team of researchers discovered Eris, a large object in the outer regions of that belt. What made Eris so special is that it is essentially the same size as Pluto. (Eris is slightly more massive than Pluto but has slightly less volume.) Eris even has a moon, named Dysnomia. Soon other bodies comparable to Pluto were found in the Kuiper Belt and beyond (**Figure 5.20**). The two largest after Eris are Makemake and Haumea.

After these discoveries, Pluto began to look less like a planet and more like just a large Kuiper Belt object. That was why the IAU took Pluto off the list of planets and placed it on a new list of objects called dwarf planets. In the formal IAU definition, a dwarf planet orbits the Sun and is massive enough for gravity to draw it into a spherical shape. The definition's important distinction, however, is that dwarf planets are not massive enough to have cleared out material in the neighborhood of their orbits. Currently there are only five known dwarf planets in the Solar System: Ceres, Eris, Makemake, Haumea, and Pluto. Astronomers also use the term **trans-Neptunian object** to describe bodies such as dwarf planets and asteroids that orbit with an average distance beyond Neptune. Comets, with their highly elliptical orbits taking them into the inner solar system, are not considered trans-Neptunian objects.

Pluto did not go quietly into the night, however. Considerable public outcry followed the IAU decision to "demote" it to a dwarf planet and reduce the number of Solar System planets to eight. The controversy highlighted to the general public how dramatically our understanding of the Solar System has changed over the last few decades. In coming to see and understand objects beyond Neptune, astronomers have gained insight into how much more of the Solar System is out there and learned new stories about its formation and evolution.

In 2016 fans of Pluto got the chance to see this dwarf planet up close as the NASA probe *New Horizons* became the first craft to visit a trans-Neptunian object. Even though the probe is one of the fastest human-made objects, it took almost 10 years to reach Pluto. The stunning images it sent back revealed a surface that, despite having a temperature of just 40 K, is surprisingly geologically active in terms of processes reshaping it. The most striking feature that *New Horizons* saw on Pluto is the vast Tombaugh Regio, a smooth heart-shaped region named for Pluto's discoverer, Clyde Tombaugh (**Figure 5.21a**). Its lack

> **"Most of the planets spin in the same direction, and that's the same direction that the Sun spins."**

Figure 5.20 Largest Known Trans-Neptunian Objects
Pluto, Eris, Makemake, and Haumea are dwarf planets that orbit in the Kuiper Belt. Bodies such as asteroids and dwarf planets (but not comets) that orbit the Sun at a greater average distance than Neptune, 30 AU, are trans-Neptunian objects.

trans-Neptunian object A minor body in the Solar System that orbits the Sun at a greater average distance than Neptune, 30 AU.

Figure 5.21 Pluto

a. *New Horizons* image of the dwarf planet includes Tombaugh Regio—the bright, heart-shaped region. **b.** Pluto and its large moon Charon have the largest ratio of moon-to-planet radius in the Solar System. **c.** Pluto's thin atmosphere is primarily nitrogen vaporized on the icy surface.

a. Pluto

2,370 km

of craters indicates that it may be composed of flowing nitrogen ice. Surrounding the Tombaugh Regio are mountains as high as the Rockies that are likely composed of water ice. The Pluto that *New Horizons* revealed is a landscape shaped by the slow flow of ices—a process that has shaped many of the bodies of the outer Solar System, including the moons of the giant planets.

New Horizons also imaged Pluto's large moon Charon (**Figure 5.21b**), one of five natural satellites orbiting the dwarf planet. The probe, as it shot past Pluto, caught Pluto's remarkable landscape and its thin nitrogen atmosphere (**Figure 5.21c**).

Section Summary

- Asteroids, found primarily in stable orbits in the main asteroid belt between the orbits of Mars and Jupiter, are rocky debris left over from the assembly of planets.
- Comets, whether in short-period or long-period orbits around the Sun, are composed of ice and rock. As comets approach the Sun, they form a bright coma (from gases and dust blown off the nucleus by solar wind) and two tails—an ion tail and a dust tail—that point away from the Sun.
- Meteoroids are chunks of rock and metal ranging in size from a grain of sand to a boulder.
- The five known dwarf planets in the Solar System meet some criteria for planets but fail to clear the area around their orbits.

CHECKPOINT Describe the differences and similarities among asteroids, comets, meteoroids, and dwarf planets, and explain how these features contribute to our understanding of the Solar System's history.

In Mission 6, you must save a mining colony from a rogue comet by determining its composition so that you can blow it to smithereens.

Our Solar System and Others

5.4 Now that we have a complete census of the Solar System, we can begin to ask what insights its structure (shape) and dynamics (motion) offer us as we try to build a theory of its origin. Our next step will be to compare the information we have about our own Solar System with what we know about planets orbiting other stars.

Evidence of the Origin of Our Own Solar System

Looking down on the Solar System from "above"—that is, looking down at the Sun's north pole—reveals the first clue about the origin of our Solar System. All planets in the Solar System orbit the Sun in the same counterclockwise direction. The majority of asteroids and most short-period comets also have that orbital direction. But it is not only the orbits that have the same rotational direction: "Most of the planets spin in the same direction, and that's the same direction that the Sun spins," says Crystal Brogan. Almost all of the planets spin counterclockwise.

Venus is the only planet with a different spin direction (clockwise). Most astronomers, however, believe that Venus began with the same direction of spin as the other planets and the Sun. Its current reverse rotation, they suggest, resulted from a powerful collision with another large body early in the Solar System's evolution.

The fact that almost everything is spinning and orbiting in the same direction is, says Brogan, "important for thinking about the origin of the Solar System." It points to a common origin of the material that formed the planets and the Sun. Having spins and orbits aligned any other way would be very unlikely. If planets were randomly captured bodies, for example, formed in other locations in the galaxy, it would be unlikely that every one of them would settle into an orbit with the same rotational sense.

The common direction of spin and rotation tells astronomers that both the Sun and the planets must have formed out of a large object, such as a cloud of **interstellar gas** that was itself rotating. The parent cloud must have imparted its sense of rotation to all of the spinning/orbiting material in the newly formed Solar System.

The next important clue about the Solar System comes not from its motions but from its shape. If we could look down along the Sun's spin axis, we would see the planets orbiting at different distances. But if we looked at the Solar System along the Sun's equator, we would find that all the planets move on orbits that are roughly aligned in the same plane—an ecliptic plane (**Figure 5.22**). "All the planet orbits in our Solar System have pretty small inclinations," explains Crystal Brogan. The largest inclinations are about 6° or 7°, and the smallest is about 3.5°. "The planetary orbits are all aligned pretty close to a plane," Brogan continues. "And that means all the planets probably formed together in a disk."

There are other indications that the primitive material that formed the planets must have had a disk shape. The asteroid belt looks like a disk, or *annulus*, of orbits; and the Kuiper Belt, which extends beyond Neptune, has the shape of a fat disk or doughnut of orbits. The exceptions to this rule, however, are the long-period comets. They originate in the spherical Oort Cloud and dive into the Solar System on highly inclined orbits. But the material in the Oort Cloud probably didn't form out there. The comets of the Oort Cloud are likely to have formed in the primordial disk and then were gravitationally scattered to longer distances by close encounters with the giant planets.

Thus, the shape and motions of material in the Solar System tell us something elemental about how it must have formed. The rotations and spins of the planets tell us that the Solar System must have originated from a single spinning object. The orientation of planetary orbits tells us that they must have formed from some kind of disklike configuration of source material.

In Section 5.5 we will see how these details about our Solar System—including the shape, the motions, and even the composition of the planets—have contributed to the modern story of Solar System formation. For now, the next step in thinking about *a general theory of planetary system formation* is to see what has been learned from other systems. "Until we started finding other planets beyond those orbiting the Sun," explains David Jewitt, "we did not know whether our Solar System was a representative example. Are we kind of average, or are we some kind of freak? Now we know the answer to that question, and it turned out to be pretty surprising."

> "The problem with finding planets in systems beyond our own is very simple to state and very difficult to solve."

interstellar gas Ionic, atomic, or molecular gas found in diffuse clouds in the regions between the stars.

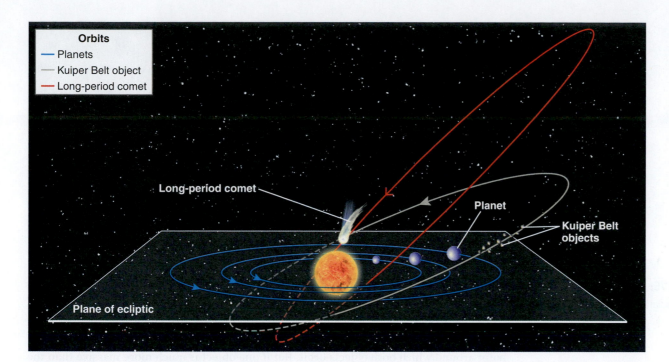

Orbits
— Planets
— Kuiper Belt object
— Long-period comet

Long-period comet

Planet

Kuiper Belt objects

Plane of ecliptic

Figure 5.22 Inclination of Orbits
Planets and asteroids orbit very close to the plane of the Sun's equator, while the orbits of Kuiper Belt objects are typically somewhat inclined. By comparison, long-period comets from the Oort Cloud have orbits with far greater inclination.

Discovering Exoplanets

Do planets exist outside our Solar System? Answering that 2,000-year-old question was bound to be vexing. "The problem with finding planets in systems beyond our own is very simple to state and very difficult to solve," says Heather Knutson. Knutson is an astronomer at Caltech in Pasadena who studies exoplanets. "These planets are very far away," she explains. "They are very faint, and they are parked right smack next to very, very bright objects: their parent stars."

Knutson provides an analogy that underscores the difficulty of exoplanet searches. "Imagine that someone in California sets up a large spotlight, like the kind they have at those Hollywood movie openings," she begins. "Now imagine they point the spotlight at New York City. Next to the spotlight they let some fireflies out of a can. The fireflies flutter around a few feet away from the spotlight. Your job in New York is to take a picture that can separate the fireflies from the spotlight." As Knutson emphasizes, that would be a pretty difficult task—but "It's the same-size problem astronomers have to deal with in finding planets around other stars."

It took a lot of ingenuity and hard work before scientists could be sure that they had discovered conclusive evidence of exoplanets. For many years astronomers employed the same method used to study binary stars (two stars in orbit around each other) to hunt for planets. In the 1880s, for example, a claim was made that the gravitational effect of an unseen planet could be detected on the orbits of the binary stars 70 Ophiuchi. In the 1950s and 1960s, attention turned to Barnard's Star (just 0.14 times the Sun's mass and 6 light-years from Earth) and claims for an entire family of planets residing there. As late as 1990, two radio astronomers claimed to have found a planet orbiting a *neutron star*, the dead cinder of a once massive sun. In each instance, the excitement grew and then faded as detailed examination of the evidence revealed the "planet" to be no more than an illusion. (In 1992, however, planets were confirmed to orbit a type of dead star called a *pulsar*, as we shall see in Chapter 13.) How do we know now that planets orbit other stars? Technological innovations in the past 30 years have brought about two accurate methods of determining the existence and properties of extrasolar planets: the radial velocity method and the transit method.

The first method is based not on finding the planet directly but on finding its gravitational influence on the parent star. Although we say that one object orbits another—such

HEATHER KNUTSON

"I was an undergraduate at Johns Hopkins University," says Heather Knutson. "I wasn't sure what I wanted to study, but right across the street from the physics department was the Space Telescope Science Institute." STScI is the operations center for the Hubble Space Telescope (HST). Technicians and astronomers there oversee the orbiting telescope.

"I was able to work there and got to meet real astronomers. I got my hands on real data from the HST and had a chance to see how science really works. I saw how I could take the physics I was learning in class and apply it to the unsolved astronomical problems." Knutson is now a professor at Caltech. "The time I spent studying extrasolar planets while in grad school was so enjoyable."

Figure 5.23 Center of Mass
A planet does not orbit its star; rather, the star-planet pair orbits its common center of mass. Observations of regular changes in the star's velocity—moving toward and then away from us—provide evidence of an otherwise unseen planet.

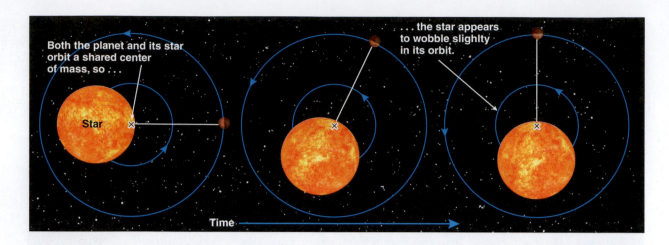

Both the planet and its star orbit a shared center of mass, so . . .

Star

. . . the star appears to wobble slighlty in its orbit.

Time

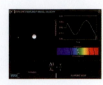

INTERACTIVE:
Exoplanet Radial Velocity

"So if you have a mass and you have a radius, you can calculate a density, which means you are suddenly in the business of categorizing planets."

INTERACTIVE:
Exoplanet: Transits

center of mass The point around which two orbiting bodies move. It lies closer to the more massive body.

reflex motion The movement of a star in response to the gravity of an orbiting planet.

radial velocity method An exoplanet detection method that measures changes in a star's velocity as it orbits a common center of mass with one or more planets.

transit method An exoplanet detection method that measures changes in the brightness of a star as a planet passes in front of it.

as a planet orbiting a star—in actuality both objects revolve around a common **center of mass**, which we can think of as the average location of all total matter in the system. To understand this concept, let's consider two situations. If two equal-mass objects orbit each other, then their center of mass lies at the midpoint of their separation. That is the point around which both objects orbit. But if one object is much more massive than the other (for instance, the Sun is more than 300,000 times as massive as Earth), then their center of mass lies almost at the center of the more massive object.

Almost at the center is not the same as *at the center*. If the Solar System consisted of only Earth and the Sun, the Sun would still complete a small orbit around the Earth-Sun center of mass, which is 449 km from the Sun's center. Earth would also orbit that point, but at a distance of 1 AU. Because the Sun has a radius of 696,000 km, the gravitational influence of Earth, if it were alone, would induce a tiny (449 km) wobble in the Sun's position at a speed of 0.09 meter per second (m/s). What astronomers needed was a way of detecting another star's **reflex motion**—these tiny back-and-forth motions created by the gravity of its orbiting planets (**Figure 5.23**).

In the mid-1990s, high-precision instruments were developed that tracked Doppler shifts from stars moving at velocities as small as a few meters per second toward and away from the observer (**Figure 5.24a**). With this precision, astronomers could finally look for the reflex motion of stars. This approach is called the **radial velocity method**, or the *wobble* method.

Once the new instruments were tested, the search was on. To claim that they had detected a planet, astronomers knew they would have to watch the star (and hence the planet) move back and forth for a full orbit. "As the star moves toward you (and away from you), you see this little Doppler shift back and forth in wavelength," says Knutson. "By measuring the period of that shift, you get the planet's orbital period, and by measuring the amplitude of that shift, you get the mass of the planet. Massive planets mean bigger wobbles." This means that gas giants exert a greater gravitational pull and make their stars wobble back and forth more than lower-mass terrestrial-type planets do. However, the radial velocity method cannot reveal the inclination of the orbit, so the mass it provides is always the minimum mass the orbiting planet might have.

When the radial velocity method was first applied, astronomers' impressions of a planetary system were based on the one example they had—our own Solar System. Because the gas and ice giants in our system have very large orbits that take years to complete, astronomers expected they would have to wait years to find the signature of an orbiting gas giant. To their surprise, the first radial velocity detection showed a planet with an orbit much shorter than those of our gas giants. In 1995, Swiss astronomers Michel Mayor and Didier Queloz announced the discovery of a planet orbiting the star 51 Pegasi and named it 51 Peg b. From the gravitational reflex motion, they derived its minimum mass to be about half that of Jupiter. But the new planet's orbit was the real surprise: it lasts just 4.2 days (**Figure 5.24b**). An orbit of such short period suggests that this giant planet circles its star at a distance of 0.05 AU, well within the orbit of Mercury in our system. The very first exoplanet detected is a gas giant hugging its own star. Clearly, there are other planetary systems that look nothing like our own.

Since the discovery of 51 Peg b, astronomers have found more than 700 extrasolar planets by the radial velocity method. Up to now, it has been one of the most successful means of detecting the presence of planets beyond our Sun. By careful examination of a star's radial velocity graphs, astronomers can even see evidence of multiple planets orbiting other stars. To date, astronomers have found evidence of hundreds of planetary systems—stars with more than one planet orbiting them. The more distant a planet is from its star, the longer its orbital period is—and the longer astronomers must wait to detect "outer planets" in that system.

Although the radial velocity method has been extremely successful, it does not tell astronomers what kind of planet is being discovered. For example, how large is the planet? For this kind of information, astronomers use the **transit method**, in which they look for small decreases in starlight as the planet passes in front, or *transits*, the distant star (**Figure 5.25**). When combined with the radial velocity method, the transit method is enormously powerful. "If you know the planet passes in front of the star, then you know the orbital inclination," says Knutson, "and you know the true mass of the planet from your Doppler measurements." But the transit method can provide even more information.

"By measuring the amount of light the planet blocks when it goes in front of its star, you learn that planet's radius," explains Knutson. "So if you have a mass and you have a radius, you can calculate a density, which means you are suddenly in the business of categorizing planets. You can say, 'Is your planet a gas giant like Jupiter? Is it a small, rocky planet more like Earth?' "

Even more exciting for astronomers is the fact that, for a brief window of time, starlight passes through the exoplanet's atmosphere (if it has one). As stellar photons glance along the edge of the alien world, some are absorbed by the atoms and molecules in the planetary atmosphere. Absorption lines from the atmosphere become superimposed on the spectrum of the star. Astronomers can derive the constitution of the alien atmosphere once they separate the stellar spectrum from the planetary spectrum.

The transit method has been the most successful for small exoplanets. The *Kepler* mission, which used a space telescope designed to find transits, revolutionized the field

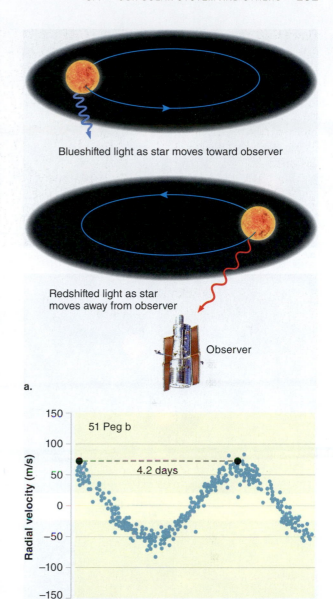

Blueshifted light as star moves toward observer

Redshifted light as star moves away from observer

Observer

a.

Figure 5.24 Radial Velocity Method
a. Doppler shift of light allows astronomers to determine the motion of a star toward and away from Earth. **b.** This method identified the presence of planet 51 Peg b, whose very short orbital period indicates the planet's close proximity to its star. (Negative velocity signifies motion away from the observer.)

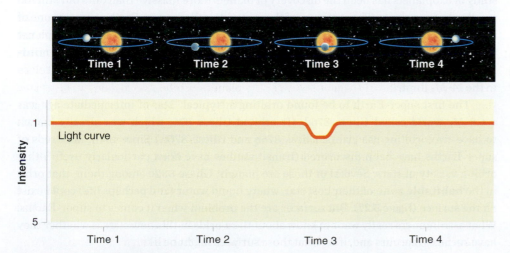

Figure 5.25 Transit Method
This method detects the periodic decrease in starlight received when an orbiting exoplanet passes in front of its star. By accurately tracking the amount of starlight blocked, astronomers can infer the planet's presence and many of its properties.

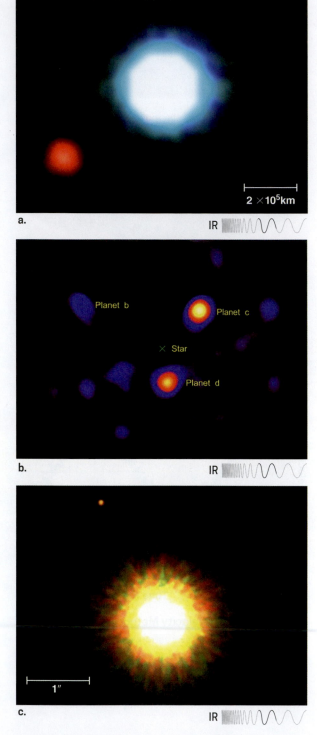

Figure 5.26 Direct Imaging of Exoplanets
a. The brown dwarf 2M1207 and a giant planet candidate, in perhaps the first direct image captured of an exoplanet (2004). **b.** Hale Telescope image (2010) of three gas giants orbiting the star HR8799. **c.** Gemini Observatory image (2008) of a gas giant orbiting a Sun-like star at a distance of more than 300 AU.

by discovering thousands of new exoplanet candidates. "To detect something Earth-sized moving at 1 AU from its star, we would have to be able to see Doppler shifts of a few centimeters per second," says Knutson. This remains too low for current Doppler-based instruments.

What astronomers want most is a direct image of an extrasolar world (**Figure 5.26**). "You can just go straight for the really hard thing," says Knutson, "which is taking an image of the planet where you can see the star in one part of the image and the planet separate from the star in the other part. We're starting to be able to do that. We have images where we see planets separate from stars, but as you would imagine, they are mostly very large planets, very far from their stars—many times the distance of Jupiter." A number of astronomical teams have directly imaged an exoplanet. In a few cases, these astronomers have relied on the planet's own infrared glow to capture the image. In general, however, direct detections remain rare.

Astronomers are also hard at work developing methods to screen out the light of the star. A number of future space-based missions are planned whose sole objective is the direct detection of extrasolar worlds.

Perhaps the most remarkable aspect of exoplanet searches is that astronomers are now able to make educated predictions about how many planets orbiting other stars are likely to be found. "There is still a lot of debate, but most astronomers would accept that somewhere between 30 and 50 percent of all stars like the Sun have low-mass planets orbiting around them," says Knutson. "When you consider that there are hundreds of billions of stars like the Sun, you see that there are a lot of planets. When we talk about the possibility for life elsewhere in the galaxy and the Universe, I think it starts to sound a little bit more plausible when you consider that planets must be so easy to make."

The Planets We Don't Have and the Ones We Do: Super-Earths

If all we knew was the Solar System, we would think there are just two major classes of planets: rocky terrestrial worlds and giants (the massive gaseous kind and the less massive icy kind). In our Solar System, Earth is the most massive rocky world. The next-most-massive planet after Earth is the ice giant Uranus, with about 14.5 times the mass of Earth ($14.5\ M_E$). Thus, no Solar System planets exist in the range between $1\ M_E$ and $14\ M_E$.

This absence, however, is not the rule. One of the most exciting developments in the study of exoplanets has been the discovery of planets more massive than ours but still too small to be true giants. The structure of these worlds, called super-Earths, remains one of the most intriguing mysteries in modern astronomy. Formally, a **super-Earth** is a planet with a mass larger than Earth's but substantially smaller than Uranus's. The term **mini-Neptune** is sometimes applied to worlds with masses below that of the ice giants but close to the 14-M_E limit.

The first super-Earth to be found orbiting a "typical" star of intermediate age was a 6.8-M_E world called Gliese 876d. (It orbited Gliese 876, which was already known to have two orbiting gas giants: Gliese 876b and Gliese 876c.) Since then, hundreds of super-Earths have been discovered (transit studies have been particularly useful) that orbit a variety of stars. Several of those are planets, Gliese 832c among them, that orbit in the **habitable zone** of their host star, where liquid water (and perhaps life) could exist on the surface (**Figure 5.27**). But surfaces are the problem when it comes to super-Earths. What astronomers really want to know about the worlds in this new class is whether they have surfaces like ours and, if so, what those surfaces might be like.

In many cases, it is not yet possible to determine what a super-Earth should look like, given its mass and radius (which then yields its density). From these data, a range of configurations is possible for some worlds with mass greater than Earth's. The super-Earth might have a rocky core with a thick gaseous atmosphere, or it might be a "water world"

Figure 5.27 Super-Earth Gliese 832c
An artist's depiction of Gliese 832c (compared with Earth). This 5.2-M_E "super-Earth" orbits within its host star's habitable zone, where liquid water (and hence potentially life) can exist on the surface.

composed mostly of H_2O molecules in liquid and gaseous forms. It is not yet possible for observations to distinguish among the various models predicted for these different types of planets (**Figure 5.28**).

Astronomers have also been finding Earth-sized planets via transit methods. In 2014 the first Earth-sized exoplanet in the habitable zone of its star was found. The planet, called Kepler-186f, is one of at least five planets orbiting a star half the mass of our Sun. Such *M-dwarf stars* are likely to play an important role in the search for life in the Universe because they are the most numerous kind of star in our galaxy.

Learning from Exoplanets

With 20 years of exploration and thousands of exoplanets discovered, what have astronomers learned about planetary systems that they could not have learned from the study of our Solar System? More important, what new insights has the discovery of exoplanets provided into how planetary systems are assembled? "I think the biggest thing that we've learned," says Knutson, "is that we have a lot to learn. We see that most planetary systems don't look anything like ours."

The first big surprise for astronomers was the location of exoplanets. In our system, there is a clean division between terrestrial planets, which define an inner Solar System, and giant planets (of the gas and ice varieties), which define an outer planetary domain. But

"I think the biggest thing that we've learned is that we have a lot to learn. We see that most planetary systems don't look anything like ours."

Figure 5.28 Predicted Sizes of Equal-Mass Planets with Different Compositions
Six versions of a planet, all of which have the same mass but different elemental and chemical abundances. A planet's size is determined by its composition as well as its mass. Astronomers must use detailed models to determine the structure, and hence radius, of a planet based on composition (among other things).

super-Earth A planet with a mass greater than Earth's but substantially less than 15 Earth masses.

mini-Neptune A planet with a mass less than 15 Earth masses but substantially greater than Earth's mass.

habitable zone The region around a star where liquid water can exist on a planet's surface.

Figure 5.29 Hot Jupiter

An artist's rendering of a hot Jupiter, a gas giant exoplanet orbiting much nearer to its star than our Jupiter orbits the Sun. A hot Jupiter probably did not form so close to the star; its orbital radius likely decreased over time.

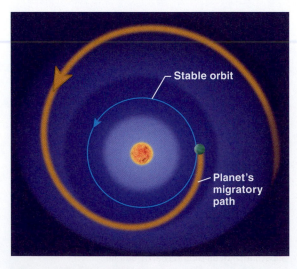

Figure 5.30 Planet Migration

A planet undergoing migration slowly spirals inward and ultimately reaches a stable orbit closer to its star.

the first exoplanets discovered showed that this tidy distinction is not universal. "When we started looking for planets around other stars," says Knutson, "the very first thing that we found was a population of Jupiter-like planets (in terms of mass) that orbited right next to their host stars."

All of the initial discoveries were **hot Jupiters**—a new class of planet with a massive Jupiter-sized body that moves on an orbit of extremely small radius (**Figure 5.29**). Astronomers use the term *hot* to refer to any planet orbiting 0.1 AU or less from its star. One particularly important hot Jupiter is HD 209458 b, an exoplanet orbiting a Sun-like star (in terms of mass, radius, and temperature) in the constellation Pegasus, 150 light-years from Earth. HD 209458 b has a semimajor axis of just 0.045 AU and a period of just 3.5 days. What makes HD 209458 b special is that it was the first hot Jupiter that astronomers detected by the transit method, so they could determine its mass with fairly high precision at 0.69 times that of Jupiter. With such a large mass, the exoplanet must be a gas giant. Transit observations confirmed that it is, showing the atmosphere of HD 209458 b to be rich in hydrogen but also containing carbon, oxygen, and water vapor.

Detailed calculations of hot Jupiters such as HD 209458 b show that exposure to intense stellar radiation heats and puffs up their atmospheres, extending their outer boundaries and lowering the planet's overall density. Transit observations have confirmed this result, adding yet another dimension to the depth with which this new class of planets has been explored.

"The hot Jupiters were a huge surprise," explains Knutson, "because we know such a massive planet can't form so close to its star." The reason is that intense stellar winds and radiation drive away the gases and ices that make up the bulk of a gas giant from regions close to the star while its planets are forming. Thus, the massive planets found on "hot" orbits must have formed at greater distances from the star and then migrated inward. "You must form these planets out beyond the snow line," says Knutson, "and then move them in closer to the star." The recognition that **planet migration**—a change in a planet's orbit after it forms—is a common occurrence in planetary systems is one major conclusion of exoplanet studies (**Figure 5.30**). "Somehow you must take a planet from the outer part of your system and move it in close," says Knutson. "But how do you get it close without shoving it all the way into the star?" This question is one of the new challenges facing astronomers in their quest to build a complete theory of planetary system formation.

Even as astronomers debate the mechanisms driving planet migration, they recognize that its consequences are dramatic. With migration, the architecture of a planetary system when it forms may not be what it ends up with. Many fall into the "hot" category, and a sizable fraction of these are hot Jupiters. Astronomers have also found planets with a range of masses from Neptune-sized bodies to super-Earths in *hot orbits* (**Going Further 5.1** on p. 136), so migration is an almost universal phenomenon.

Most planets discovered to date, however, have orbits with radii that keep them farther from their parent stars. Hot Jupiters, mini-Neptunes, and super-Earths were dramatic discoveries, but the bulk of massive planets are not on hot orbits, although their orbits still are smaller than what we see in our own system. Many exo-Jupiters and exo-Neptunes, for example, have orbital radii of a few astronomical units or less, which in our Solar System is the domain of the terrestrial planets (and asteroid belt). Extremely large orbits also occur in exoplanetary systems. In some cases, there is evidence of gas giant–sized or ice giant–sized worlds with orbits of 100 AU or more, which is much larger than the orbit of any planet in our Solar System.

Location of exoplanet orbits was just one surprise awaiting astronomers. The *shapes* of many exoplanet orbits, and the histories they imply, have forced astronomers to revise their thinking about planetary systems and their formation. "When we started looking at extrasolar planets," explains Knutson, "not only did we see very big planets very close

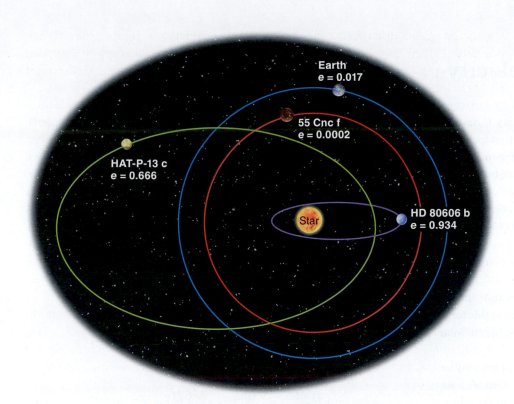

Figure 5.31 Eccentricities of Exoplanet Orbits
Earth's orbit is compared with three exoplanet orbits. Many exoplanets do not exhibit the nearly circular orbits typical of the planets in our Solar System. Highly eccentric orbits are evidence of scattering—gravitational interactions that change planetary orbits.

to their stars; we also saw lots and lots of planets with large orbital eccentricities. That means these planets move on very elliptical orbits instead of the more circular orbits we see in our Solar System" (**Figure 5.31**).

The planet with the highest-eccentricity orbit in our Solar System is Mercury, with $e = 0.2$, and six of the eight planets have eccentricities of 0.05 or less. In contrast, more than 80 percent of all exoplanets discovered to date have eccentricities above $e = 0.1$, and nearly one-fifth have eccentricities of 0.5 or greater. How do planets end up with such highly eccentric orbits? Like migration, gain in eccentricity may be an essential part of early planetary system evolution. "If two young planets have a close encounter, perhaps because one is migrating, then a strong gravitational interaction occurs that can really shake things up," explains Knutson.

The process is called **scattering**—the rearrangement of orbits due to gravitational interaction. Scattering can take a planet that was on a circular orbit and fling it onto a highly eccentric orbit. "If you have a bunch of planets that begin at different distances from the star," explains Knutson, "and close encounters lead to scattering, then in time all the orbits can be moved around, and they can all attain high orbital eccentricities."

When planets end up on highly eccentric orbits, a planetary system can become far less stable than our own. Venus will never feel Jupiter's gravitational influence up close because both planets are on fairly circular orbits with very different radii. But a planet on a highly eccentric orbit will have many chances to cross paths and interact with other orbiting bodies.

"It seems like we might have been lucky in our Solar System," says Knutson. "All our planets are on fairly circular orbits, so everything is pretty stable. There probably haven't been any catastrophic encounters between planets except for a long time in the past, and what we are left with has become very safe."

Evidence of scattering in exoplanetary systems also comes in the form of the orbital inclinations of planets relative to the host star's spin. While almost all Solar System planets have a low inclination angle (that is, they rotate in a plane that is roughly perpendicular to the Sun's spin axis), exoplanets show a wide range of inclinations. Again, scattering interactions between exoplanets are the likely culprit, tossing orbiting bodies in different

> "It seems like we might have been lucky in our Solar System."

hot Jupiter A Jupiter-sized planet orbiting very close to its star, usually at a fraction of an astronomical unit.

planet migration A change in a planet's orbit after the planet has formed.

scattering Rearrangement of the orbits of planets due to mutual gravitational interactions.

Orbit, Speed, and Velocity

In Chapter 3 you learned an expression for orbital velocity V_c, the velocity of an object in a circular orbit. Now we can use that formula to get an idea of how orbital properties change, depending on the architecture of a gaseous disk orbiting a young star where new planets may be forming.

Let's start with the formula for the object's orbital velocity at a distance R from its star of mass M [see Chapter 3 (Section 3.5)]:

$$V_c = \sqrt{\frac{GM}{R}}$$

With it we could, for example, calculate that Earth orbits the Sun with a speed of about 30 km/s. (You can try it yourself.) Let's use this formula to examine a kind of orbit very different from those in our Solar System.

Imagine we're exploring the space around a star that is Sun-like but has 50 percent more mass: $M = 1.50$ solar masses $= 1.50\,M_{Sun}$. As we've seen, astronomers have found planets orbiting other stars in so-called hot orbits very close to the star. Let's consider a planet in a hot orbit around this star at a distance of $R = 0.05$ AU and calculate the planet's velocity.

The orbital-velocity formula tells us the orbital speed:

$$V_c = \sqrt{\frac{GM}{R}} = \sqrt{\frac{(6.67 \times 10^{-11}\,\text{m}^3\,\text{kg}^{-1}\,\text{s}^{-2}) \times [1.50\,M_{Sun} \times (1.99 \times 10^{30}\,\text{kg})/M_{Sun}]}{0.05\,\text{AU} \times (1.50 \times 10^{11}\,\text{m/AU})}}$$

$$= 1.63 \times 10^5\,\text{m/s} = 163\,\text{km/s}$$

where we use $M_{Sun} = 1.99 \times 10^{30}$ kg and 1 AU $= 1.50 \times 10^{11}$ m to convert the units.

Now let's find this planet's orbital period. We cannot use Kepler's formula $P^2 = R^3$ because, as noted in Chapter 3, it does not account for the different mass of the star. Even if we do not recall "Newton's version" of Kepler's third law (which does include the star's mass), we can reason out what the period should be by remembering how speed, distance, and time are related.

The period is the time it takes a planet to orbit its star once. The distance a planet covers in one orbit of average radius R is the circumference of a circle of that radius: $C = 2\pi R$. This means the planet's period in its orbit around a star is

$$P = \frac{2\pi R}{V_c}$$

Substituting the formula for V_c, we get

$$P = \frac{2\pi R}{\sqrt{\dfrac{GM}{R}}} = \frac{2\pi R\sqrt{R}}{\sqrt{GM}} = \frac{2\pi\sqrt{R^3}}{\sqrt{GM}}$$

(If you square this formula, you will get back Newton's version of Kepler's third law.) When we use this formula for our hot orbit, we find

$$P = \frac{2\pi\sqrt{[0.05\,\text{AU} \times (1.50 \times 10^{11}\,\text{m/AU})]^3}}{\sqrt{(6.67 \times 10^{-11}\,\text{m}^3\,\text{kg}^{-1}\,\text{s}^{-2}) \times [1.50\,M_{Sun} \times (1.99 \times 10^{30}\,\text{kg})/M_{sun}]}}$$

$$= 289{,}226\,\text{seconds} = 3.35\,\text{days}$$

Thus, a planet in a hot orbit has a "year" that lasts just a few days.

directions, including out of the plane of the original disk from which the system formed. "It's clear from exoplanet studies that planetary systems may be far more dynamic than we thought," says Knutson. "You form things with one configuration, but the final configuration that you see may be very different."

Section Summary

- All planets in our Solar System orbit the Sun in the same direction the Sun rotates (and most spin in that direction). The alignment of planetary orbits in a disk strengthens the evidence that they formed in a disk.
- Exoplanets are difficult to find, but three primary methods have been used to identify them: (1) the radial velocity method, (2) the transit method, and (3) direct imaging.
- Some exoplanets, such as super-Earths and hot Jupiters, have characteristics quite different from those of Solar System planets. Some of their differences provide evidence of planet migration and scattering between sibling planets.

CHECKPOINT Name two things about other solar systems that make them similar to ours and two things that make them different.

Developing a Theory of Planetary System Formation

5.5 After touring planetary systems near and far, we are now in a good position to take all the observed facts and distill them into a set of critical points that must be explained by a theory of planetary formation.

Let's begin with four key points we know about our own Solar System (as summarized in **Table 5.4**):

1. All planets in a system tend to orbit in the same direction, which is the same direction as the parent star's spin.

2. Planets tend to orbit with low inclination; that is, all planetary orbits tend to be closely aligned, and they are arranged in a disklike configuration.

3. Planets in our system are highly differentiated in terms of composition and location. The inner planets are rocky and dense; the outer planets have lower density and contain significant amounts of gas or ice.

4. Our Solar System contains debris from the era of planet construction, in the form of asteroids, comets, and Kuiper Belt objects.

Now let's add two further points from studies of exoplanets:

5. Planets can migrate. Their orbital position may change from the location of their formation. Close encounters between planets or planetesimals are likely to occur during the migration process and can lead to catastrophic gravitational interactions in the form of scattering.

6. Planets can exist on highly eccentric orbits. These eccentric orbits may be the result of scattering events and can, in turn, lead to further scattering. The planets in some planetary systems have highly inclined orbits.

Table 5.4 Key Features of Planet Formation Models

Feature	Effect	Cause
Orbit and rotation	Planets generally orbit and spin in the same direction as their star's spin.	Conservation of angular momentum. Planets formed from a single rotating cloud.
Inclination	Planets (in Solar System) orbit with low inclination and are arranged in a disklike configuration.	Conservation of angular momentum.
Composition and location	Planets (in Solar System) are highly differentiated in composition and location: inner planets are rocky and dense, while outer planets are less dense and contain much gas or ice.	Planets formed from disk with a temperature that decreased with distance from the star.
Debris	Debris from planet construction remains, represented in Solar System by asteroids, comets, and Kuiper Belt objects.	Planets formed from collisions between smaller objects.
Migration (extrasolar planets)	Planets can change orbital position from the location of their formation.	Gravitational interactions between planets and disk or other planets are common.
Eccentric orbits (extrasolar planets)	Planets can exist on highly eccentric orbits.	Gravitational interactions between planets are common.

1 ly

VIS

Figure 5.32 Interstellar Dust Clouds
Hubble Space Telescope image of dense clouds of interstellar dust, from which stars (and hence planets) form. (The black rectangle at top left is an artifact.)

INTERACTIVE:
Conservation of Angular Momentum

Condensation Theory and Conservation of Angular Momentum

The goal of any successful theory of planetary system formation is to explain all six facts in one simple, all-encompassing model. Scientists believe they have a candidate in the **condensation theory** (also sometimes called *nebular theory* or *nebular condensation theory*). Remarkably, elements of the condensation theory have been around for almost 400 years, owing to philosophers such as René Descartes (1596–1650) and astronomer Pierre-Simon Laplace (1749–1827).

This theory begins with a cold, slowly rotating cloud of interstellar gas. "The cloud started with a balance of forces that kept it supported against its own gravity," says Crystal Brogan. "But all it took was a little nudge to knock it out of balance. That's when the cloud started to collapse inward on itself." The collapsing, spinning cloud is condensation theory's starting point. From there, the theory goes on to tell a detailed story of gravity, gas flows, radiation, and chemistry to explain how a system of worlds is constructed.

Dense interstellar dust clouds (**Figure 5.32**) are well known to astronomers and will be explored in Chapter 12 in association with star formation. For now, imagine one of these clouds floating in space. Astronomers assume that the entire cloud slowly rotates with some initial spin that is a remnant of the cloud's formation from a larger region of turbulent interstellar gas. The inward pull of gravity in the cloud is balanced by the outward push of pressure from internal thermal energy in the cloud. "The balance of forces in the cloud is inherently unstable," says Brogan, which is why any nudge (such as a collision with another cloud) lets gravity overwhelm the pressure and triggers the cloud's collapse.

At this point, a fundamental principle of physics called *conservation of angular momentum* takes over, shaping the cloud's collapse and setting the stage for planet formation. Recall from Chapter 3 the example of a spinning figure skater. He begins the spin with his arms extended. As he pulls his arms inward, his rate of rotation increases until he is spinning so fast as to be just a blur. "For skaters, the principle called conservation of angular momentum is really essential," says Brogan.

Conservation of angular momentum *relates changes in an object's rate of rotation to changes in its size*. In particular, this principle states that when a spinning object decreases in size, its rate of rotation must increase. (The opposite holds true when the size of a spinning object increases.) When applied to a collapsing cloud, conservation of angular momentum tells us that, like the skater, the spin rate of the cloud around its axis of rotation must increase as the cloud gets smaller (see **Going Further 5.2** on p. 144). But while the skater decreases in size by only a factor of 2 (pulling his arms in), the cloud decreases in size by a factor of 1,000.

Any gas that begins on the cloud's axis of rotation is not spinning. That material will drop straight to the center and start forming a star. But parcels of gas that do not lie on the axis of rotation have a different fate. As this gas falls inward, its rotation forces it to trace out a spiral path. The closer it gets to the center, the more its rate of rotation increases because of conservation of angular momentum. Eventually, rotation can balance gravity and stop the gas from getting any closer to the center. But there is more to the story. If we imagine that the cloud starts as a sphere, then some of the gas begins its spiral above or below the sphere's equator. These gas parcels eventually collide at the midplane as they spiral inward. The swirling disk of gas that now surrounds the newly formed star is referred to as a **protoplanetary disk**. In this way, what began as a spherical rotating cloud ends up as a rotating disk surrounding a new star (**Figure 5.33**).

"What conservation of angular momentum does is take the initial spinning cloud and cause it to form a flattened disk," says Brogan, "and that takes you a long way to

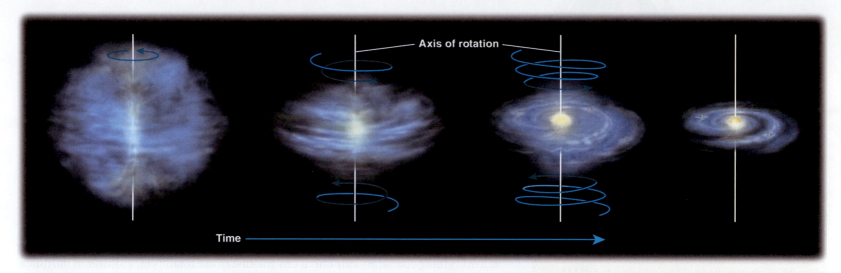

Axis of rotation

Time

Figure 5.33 Formation of the Protoplanetary Disk
A rotating cloud of interstellar material collapses, forming a spinning disk of gas and the newborn star it surrounds. The young star forms when material along the cloud's axis of rotation falls straight inward. The protoplanetary disk forms along the cloud's midplane when material from other regions of the cloud spirals inward.

explaining the properties of planetary systems." By beginning with a collapsing, rotating gas cloud, we can already explain why all planets in our Solar System orbit in a plane and in the same direction as the Sun's spin. Conservation of angular momentum also explains why all but one of the planets spin in the same direction. When material in the disk eventually creates planets, conservation of angular momentum tells us that the direction of a planet's spin will be the same as the original cloud's overall direction of rotation. So, using just the force of gravity and the principle of conservation of angular momentum, scientists can understand many of the Solar System's properties.

Condensation Theory, Dust Grains, and Planet Formation

The rest of the story, including the formation of different kinds of planets, requires more than just conservation of angular momentum (and gravity). This is where condensation comes into play in condensation theory. To understand the formation of planets, astronomers must focus on the physics occurring within the protoplanetary disk. In particular, they focus on the disk's density, its temperature, and the kinds of dust grains that can form at different locations.

The cloud that forms the star and disk is full of **interstellar dust grains**, tiny bits (1 micron [μm] or 0.001 millimeter [mm]) of solid matter that form in the winds of dying stars and are ejected into space. What matters in planet formation is the fate of these grains at different locations in the disk. The temperature in the dusty disk close to the star (where it absorbs the most light) is very high, and it drops at greater distances. Calculations show that the inner edge of the disk (at about 0.1 AU) can have temperatures of 1,500 K, while the temperature at 10 AU (where Saturn is today) would be only 100 K. "Interstellar dust grains can't survive at the high temperatures close to the star," says Heather Knutson. "They get vaporized." But the grains are less affected in the outer parts of the disk. As the entire disk evolves, it slowly cools by radiating energy out into space. Except for regions very close to the star, the cooling of the disk allows new dust grains to condense out of the gas.

"Dust condensation is a bit like the way raindrops form in a cloud," explains David Jewitt, "where the drops condense out of water vapor." But the new grains that form in the disk are not like the original grains. While the original gas cloud contained a fairly uniform mix of dust grain sizes and compositions, the new grains that condense out of the cooling disk are very sensitive to the location of their birth.

"Interstellar dust grains can't survive at the high temperatures close to the star. They get vaporized."

condensation theory A theory of planetary formation in which planets are created from a disk of gas and dust around a newly formed star.

protoplanetary disk The disk that surrounds a young star, from which planets may form.

interstellar dust grain Tiny bit of solid matter found in diffuse clouds in the regions between the stars.

"The spatial sequence of condensation temperature and radius—first metals, then rocky material, then simple ices, then more complex ices—does a wonderful job of explaining the sequence of planets."

As we have seen, temperature in the disk falls with distance from the star, and different materials condense into grains at different distances from the star (Figure 5.34). In higher-temperature regions, comparable to the location of the planet Mercury, the only kinds of material that can condense from the vapor to the solid state (grains) are metals. Farther out, about where Venus, Earth, and Mars are now, the gas temperatures are lower. At these distances and temperatures, silicates (mixtures of silicon and oxygen) and other rocky materials can also begin to form dust grains. Beyond the snow line, the temperature gets low enough for water ice to form, and at even larger radii, ices of compounds such as ammonia and methane can condense.

"The spatial sequence of condensation temperature and radius—first metals, then rocky material, then simple ices, then more complex ices—does a wonderful job of explaining the sequence of planets," says Knutson. The farther out in the Solar System you peer, the more the composition of the planets begins to look like that of the rest of the Universe. Recall that the most abundant elements in the cosmos are hydrogen and helium, and everything else represents only a tiny fraction of the total. Temperatures in the inner disk keep hydrogen and helium from condensing into grains at those distances, so these most abundant elements do not become part of the inner planets. Beyond the snow line, when large enough rocky cores form, they can rapidly accrete surrounding gas, building up their mass into the range of gas giants and ice giants.

Accretion, Fragmentation, and Planet Building

The existence of asteroids and comets in our Solar System also fits the general picture of condensation theory and adds more detail to the story. These smaller bodies have provided crucial information about the next steps after condensation. For example, how do young planetary systems go from making dust grains to making rocky planets or the rocky cores of giant planets?

"The answer to that question," explains Jewitt, "is a process called **binary accretion**, in which collisions between pairs of objects let larger and larger structures get put

binary accretion The process by which the collision of two objects results in the formation of one larger object.

fragmentation The shattering of solid bodies, such as planetesimals, caused by collisions.

core accretion model A model of gas giant formation in which an icy terrestrial planet grows by binary accretion up to a critical point and then rapidly pulls in gas from the surrounding disk.

hydrodynamic instability model A model of gas giant formation in which small regions collapse, forming planets within a gravitationally unstable disk.

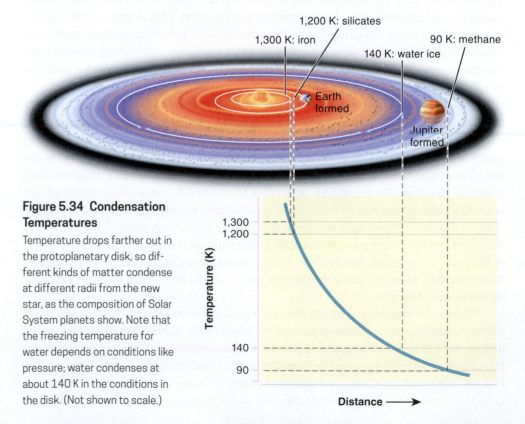

Figure 5.34 Condensation Temperatures

Temperature drops farther out in the protoplanetary disk, so different kinds of matter condense at different radii from the new star, as the composition of Solar System planets show. Note that the freezing temperature for water depends on conditions like pressure; water condenses at about 140 K in the conditions in the disk. (Not shown to scale.)

together." Dust grains are small and sticky; when they collide, they stick together, forming larger grains. This process repeats itself until pebbles have been built. Likewise, when pebbles collide and stick together, they form larger objects that we would consider to be rocks. In the same way, collisions between rocks lead to boulders, and collisions between boulders lead to planetesimals, rocky bodies the size of asteroids. We can imagine how grains can stick together if they are covered in ice—but how this "sticking together" process works for rocks and larger objects is still hotly debated.

If this accretion process happens far enough from the star (beyond the snow line), the planetesimals will include significant quantities of ices. This is the likely origin of comets. Eventually, the objects are large enough for gravity to begin compressing and heating the planetesimal interiors. As planetesimals grow even larger, gravity pulls them into a spherical shape, and the heaviest elements differentiate, sinking to the center of the body. In this way, iron and nickel form the dense metallic cores of the young planets. Eventually, full-sized terrestrial planets, the rocky cores of gas giants, and moons form by accretion.

On the journey from microscopic grains to full-sized planets (or moons), the competing process of **fragmentation** can occur along with accretion. Shattering collisions break asteroid-sized planetesimals apart into smaller bodies. Fragmentation is still at work in the asteroid belt and is the source of many meteorites. In particular, iron-rich meteorites come from fragmented cores of differentiated planetesimals.

The timescale for binary accretion depends on the density of gas and dust in the protoplanetary disk: higher densities mean more frequent collisions. In the inner part of our pre–Solar System disk, the density was high, and it took only 100 million years to build the terrestrial planets. The formation of the outer planets is a more complicated story; scientists are still not sure how the gas and ice giants formed.

"There are basically two models for the formation of giant planets like Jupiter," says Jewitt. In the first model, an icy terrestrial planet grows by binary accretion up to a critical mass of about 5–10 Earth masses. "When that small core planet reaches the critical mass, it has enough gravity to start pulling in gas from the surrounding disk. You get a very rapid flow of gas onto this core, taking the planet all the way up to Jupiter's or Saturn's mass." This idea is the **core accretion model** (Figure 5.35). According to Jewitt, it is the model that most Solar System astronomers prefer for gas giant formation.

The problem with core accretion is that it may take too long to build up the rocky center. If the disk loses its gas before the core has a chance to reach its critical mass, the model cannot work. "That is why there is this other model," says Jewitt. "It starts with an enormous disk of gas that would be unstable to its own gravity." Just as star and disk began as a gravitationally unstable cloud, some astronomers claim that planets may form from the gravitational collapse of gas within the disk itself (see **Anatomy of a Discovery** on p. 143). "Parts of the disk would just contract under their own gravity. The planets would form directly without needing a core." This second idea is the **hydrodynamic instability model** (Figure 5.36). "It tends to be more popular for people who study extrasolar planetary systems," says Jewitt. "The fact that we don't know which model is appropriate, and we still talk about this very basic issue, shows our state of ignorance on the subject. That is exciting because it shows we have lots to learn."

Astronomers are certain, however, that far enough out in our Solar System, low disk densities meant that the process of turning planetesimals into planets never finished. This is because where the density of dust grains, pebbles, and rocks is very low, collisions among them are rare. At the outer edges of the early Solar System, collisions didn't happen often—and it takes so long to go from dust grains to planetesimals that planets never got fully built out there. This explains the presence of the Kuiper Belt and objects such as Pluto. The Kuiper Belt represents the outer regions of the Solar System where planet formation began but could never be completed.

While everything about the condensation theory seems to fit our Solar System, what have we learned from exoplanets? Do the profound differences in planetary systems other

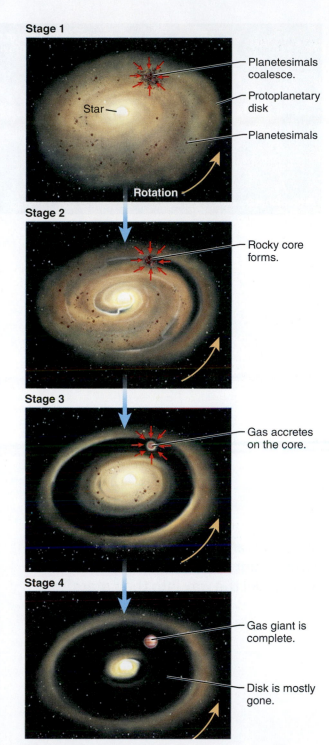

Stage 1

Star

Rotation

Planetesimals coalesce.

Protoplanetary disk

Planetesimals

Stage 2

Rocky core forms.

Stage 3

Gas accretes on the core.

Stage 4

Gas giant is complete.

Disk is mostly gone.

Figure 5.35 Core Accretion Model

According to this model of giant planet formation, collisions between planetesimals result in a rocky core. The gravity of the core then pulls in gas—more and more over time, until the core accretes gas and eventually forms a gas giant.

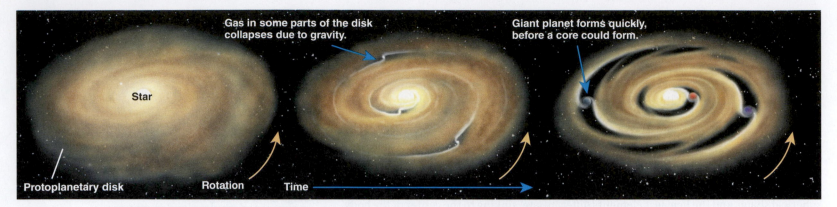

Gas in some parts of the disk collapses due to gravity.

Giant planet forms quickly, before a core could form.

Star

Protoplanetary disk

Rotation

Time

Figure 5.36 Hydrodynamic Instability Model
According to this model, small regions of denser gas in the disk begin attracting material around them in a runaway process that quickly forms a giant planet.

than our own contradict the theory? The answer is that the basic ideas of condensation theory still hold, but studies of exoplanets show us the overwhelming importance of gravitational interactions among planets, planetesimals, and disk material.

"As early as 1980, well before exoplanet discoveries, we already had evidence that Jupiter must have formed slightly farther out in the Solar System than its current location," says Jewitt. By gravitationally interacting with planetesimals early in the history of the Solar System, Jupiter must have scattered smaller objects. "All those scattering events flung the planetesimals out of the Solar System," says Jewitt, "and they let Jupiter move slowly inward."

In many exoplanet systems, this process may be much more dramatic than it appears to have been in ours. Early in the history of an exoplanetary system, young Jupiter-sized planets can have direct gravitational interaction with nearby gas in the disk, thus causing planet migration. A giant planet can form in the outer system, scatter material in the disk, and migrate all the way to the inner system, where it can become a hot Jupiter. "We still aren't entirely sure what halts inward migration," says Heather Knutson. "Some planets may migrate all the way into their stars."

Once new planets have settled into relatively stable orbits, the remaining gas in the disk is disturbed by winds and radiation from the star. In time, all that is left of the former gaseous disk is whatever stable planets managed to form, accompanied by a swarm of rocky and icy planetesimals farther out. There are enough of these asteroid-sized planetesimals early on that their interactions with planets can still cause planet migration. But migration does not always take a planet inward to an orbit closer to the star. "If conditions are right, the planet can migrate outward," says David Jewitt. This process most likely moved Saturn, Uranus, and Neptune out to longer radii than those where they formed and may be responsible for the very large orbits that some exoplanets exhibit.

The gravitational interactions between planets and planetesimals produce not only migration but also scattering, kicking small bodies into highly eccentric and highly inclined orbits. In our own Solar System, the spherical Oort Cloud of comets at large distances from the Sun was probably the result of scattering between icy planetesimals and giant planets. In exoplanetary systems, the predominance of observed high-eccentricity orbits also may be a result of scattering between planets.

Thus, the condensation theory provides a useful framework for understanding the formation of both our own Solar System and the exoplanetary systems we continue to discover. But it is a framework only. Many questions remain unanswered, such as how giant planets form. Many more surprises are likely as astronomers continue their observations and discoveries.

"We still aren't entirely sure what halts inward migration. Some planets may migrate all the way into their stars."

How do planetary systems form?

hypothesis

In the 18th century, scientists and philosophers such as **Pierre-Simon Laplace** and **Immanuel Kant** hypothesized how a planetary system like ours could have formed from a rotating, contracting cloud of gas and dust.

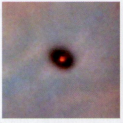

Young planetary system in Orion Nebula

Orion Nebula

confirmed by observation

In 1992, **C. R. O'Dell** pointed the Hubble Space Telescope at the Orion Nebula. He observed hundreds of young stars surrounded by thick, dusty disks in silhouette against the background cloud, in perfect agreement with Laplace and Kant's hypothesis.

hypothesis

By the mid-2000s, astronomers such as **Sebastian Wolf** made simulations of what images of circumstellar disks harboring young planets would look like with the right telescope. These simulations predicted that, while forming, planets sweep up gas and dust along their orbits about the central star, clearing gaps in the disks.

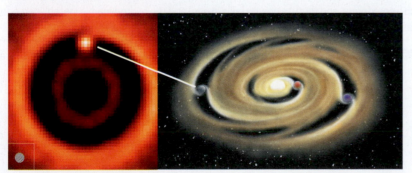

Computer simulation   **Artist's conception**

confirmed by observation

In 2014, the Atacama Large Millimeter Array's first high-angular-resolution photo of a nearby million-year-old Sun-like star became instantly iconic, revealing, as predicted, gaps within the disk of HL Tau. The image captures planet formation in action, again confirming astronomers' hypotheses. **Crystal Brogan** was thrilled to be the first person to see the calibrated image.

HL Tau

GOING FURTHER 5.2

Conservation of Angular Momentum

Physicists are always looking for "conserved" quantities. In physics, *conservation* means "no change." When examining the evolution of complicated systems such as a gravitationally collapsing cloud of interstellar gas, knowing what does not change (what is conserved) and what does change is often the key to calculation and understanding.

In its simplest terms, conservation of angular momentum relates changes in an object's size to changes in its speed. Imagine we're following a blob of gas of mass M in a collapsing, rotating interstellar cloud. The blob begins rotating at speed V_1 around the cloud's center at a distance R_1 from the center. Conservation of angular momentum relates those initial values of radius and speed to the blob's radius R_2 and speed V_2 sometime later in the collapse:

$$MR_1V_1 = MR_2V_2$$

mass × radius at time 1 × speed at time 1 = mass × radius at time 2 × speed at time 2

It is assumed that the blob's mass M does not change, so we can cancel it and solve for the later speed V_2:

$$V_2 = V_1(R_1/R_2)$$

A typical interstellar cloud has a size of $R_1 = 1$ parsec (pc) $= 3 \times 10^{16}$ meters. The typical size of a protoplanetary disk is only $R_2 = 100$ AU $= 1.5 \times 10^{13}$ meters. Thus, conservation of angular momentum tells us that the rotational speed of the gas blob will increase 2,000-fold as it spirals inward. Once the speed of rotation reaches the orbital speed (see Chapter 3), the blob will stop collapsing and go into orbit around the new star.

Section Summary

- Condensation theory states that planetary systems form from a spinning, collapsing interstellar cloud of gas and dust. It accounts for many of the observed characteristics of our Solar System and others.
- Condensation produces different kinds of matter at different distances from the central star, because melting points of interstellar dust grains determine where grains condense.
- As pairs of objects collide and stick together by binary accretion, grains become pebbles, then boulders, then planetesimals, then planets.
- Two theories have been posited about formation of gas and ice giants: the core accretion model (by which a core forms, and then its gravity pulls in gas) and the hydrodynamic instability model (by which gases within the disk collapse).

CHECKPOINT What is the sequence of events that led to the formation of the solar system's planets?

CHAPTER SUMMARY

5.1 The Rest of the Solar System

The Solar System is known to include several regions far beyond the realm of the eight planets, as well as numerous smaller objects. Recent discoveries show that ours is not the only planetary system.

5.2 Just the Facts: A Solar System Census

The primary regions of the Solar System include the region of the planets, the Kuiper Belt, the scattered disk, and the Oort Cloud. The eight planets fall into three classes differentiated primarily by location, composition, and size. The four terrestrial planets are small, rocky worlds with solid surfaces that orbit nearest the Sun. Gas giants Jupiter and Saturn, composed mostly of hydrogen and helium, orbit beyond the terrestrial planets and are much more massive. Farthest from the Sun, the midsized masses of ice giants Uranus and Neptune contain mostly various types of ices. Other objects in the Solar System include dwarf planets, Kuiper Belt objects, asteroids, comets, and meteoroids.

5.3 And Pluto Too! Asteroids, Comets, Meteoroids, and Dwarf Planets

Some of the material in the Solar System is found in remnants of the construction of larger bodies. Asteroids are irregular rocky or metallic objects, typically carbonaceous or silicate, the largest of which are less than 1,000 km in diameter. Collections of asteroids are found in the asteroid belt between the orbits of Mars and Jupiter, within Jupiter's orbit (the Trojan asteroids), and on Earth-crossing orbits. Comets, composed of ice and rock, originate in the Kuiper Belt or Oort Cloud and generally follow long, elliptical, highly inclined orbits. As a comet approaches the Sun, material is blown off its nucleus, forming jets and a coma (a cloud of gas); an ion tail and a dust tail then form, driven by solar emissions and gravitational effects. Rocky meteoroids are the smallest kind of Solar System debris and may have originated from asteroids, comets, or planets. When they encounter Earth's atmosphere, they are called meteors, seen as glowing trails of gas. Meteorites are meteoroids large enough to survive their plunge through the atmosphere. The five known dwarf planets, which meet all criteria for planets but are not massive enough to have cleared the area around their orbits, are Pluto, Ceres, Eris, Makemake, and Haumea. The discovery of dwarf planets that orbit in the Kuiper Belt increased our understanding of the scope of the Solar System.

5.4 Our Solar System and Others

In our Solar System, all planets, most asteroids, and many comets orbit in the same direction, and most planets also rotate that way. The planets' orbits are aligned in a disk, as are the asteroid belt, Kuiper Belt, and scattered disk. Exoplanets have been found by the radial velocity method (looking for changes in a star's velocity as it orbits a common center of mass with one or more planets), the transit method (looking for dips in a star's brightness as a planet passes in front of it), and direct imaging. A few thousand exoplanets have been confirmed, and astronomers believe that as many as half of all stars have planets. Super-Earths have been found, including some in the habitable zone of their stars, but their suitability for life is unknown. Differences between exoplanets and Solar System planets provide insights into how planetary systems originate. Hot Jupiters—gas giants very close to their stars—reflect planet migration. The highly eccentric, highly inclined, and exceptionally large orbits of some exoplanets suggest that scattering by gravitational interaction has occurred.

5.5 Developing a Theory of Planetary System Formation

The prevailing model of planetary system formation is the condensation theory, which states that planetary systems form from a spinning cloud of interstellar gas and dust that collapses into a disk, following the principle of conservation of angular momentum. Dust grains condense in the disk at different distances from the central star depending on their melting points: first metals, then silicates, then water ice, then ices of heavier compounds. Grains then grow by binary accretion into larger and larger objects orbiting in the disk, the composition of each determined by its distance from the star. Differentiation, collision, and fragmentation change the sizes, shapes, and structures of the objects. The largest bodies, planets, become spheres by self-gravity, and gravitational interactions among them may cause their orbits to move and change.

QUESTIONS AND PROBLEMS

Narrow It Down: Multiple-Choice Questions

1. What is the defining characteristic of the Trojan asteroids?
   a. They are the largest known asteroids.
   b. They are rocky.
   c. They orbit ahead of and behind Jupiter.
   d. They orbit within Saturn's rings.
   e. They are the remnants of an inner terrestrial planet.

2. What determines the path of a comet's ion tail?
   a. the Sun's gravity
   b. the direction of the solar wind
   c. Earth's gravity
   d. conservation of angular momentum
   e. the solar magnetic cycle

3. What makes a typical Earth-crossing asteroid difficult to detect and track?
   a. its density
   b. its albedo
   c. its lack of an atmosphere
   d. the low eccentricity of its orbit
   e. its low speed relative to background stars

4. A rocky object 10 cm in diameter is found atop otherwise untouched Antarctic ice. It is determined to have originated from space. Therefore, the rock is
   a. a meteorite.
   b. a meteoroid.
   c. a meteor.
   d. an asteroid.
   e. a planetesimal.

5. Detailed observations of a long-period comet near aphelion, far from the Sun, would most likely *not* include which of the following? Choose all that apply.
   a. jets
   b. nucleus
   c. coma
   d. dust tail
   e. ion tail

6. Which of the following characteristics of our Solar System does *not* directly support the condensation theory of planetary system formation?
   a. the planets' orbital direction
   b. Saturn's rotational direction
   c. the positions of the ice planets
   d. Venus's rotational direction
   e. the positions of the gas giants

7. Using technology similar to that available on Earth today, a distant alien astronomer searching for evidence of planets around the Sun is likely to be most successful with which of the following?
   a. direct imaging of the terrestrial planets in visible light
   b. detection of Mercury via the transit method
   c. detection of Pluto via the transit method
   d. detection of Mercury via the radial velocity method
   e. detection of Jupiter via the radial velocity method

8. The earliest observations of exoplanetary systems showed exoplanets to be similar to planets in our Solar System in which way(s)? Choose all that apply.
   a. They showed planets with high orbital eccentricity.
   b. They showed planets orbiting in the same direction around the host star.
   c. They showed gas giants orbiting very close to the host star.
   d. They showed planets with masses between 1 and 15 M_E.
   e. They showed planets with atmospheres.

9. You notice one meteor during 1 hour of observing the night sky. This meteor was most likely
   a. from a comet.
   b. part of a meteor shower.
   c. composed of ice.
   d. from an asteroid.
   e. a Kuiper Belt object.

10. Which of the following is an official criterion for being called a planet that is *not* met by Pluto?
    a. It has enough mass that its self-gravity has caused it to assume a nearly round shape.
    b. It orbits the Sun.
    c. It has at least one moon.
    d. It has gravitationally cleared the neighborhood around its orbit.
    e. Its name has a Greek origin.

11. Which of the following lists items in sequence from smallest to largest?
    a. Trojan asteroids (radius), 10 AU, Kuiper Belt (radius), Oort Cloud (radius), 1 light-year (ly)
    b. 10 AU, Trojans, Oort Cloud, Kuiper Belt, 1 ly
    c. Kuiper Belt, Trojans, 10 AU, 1 ly, Oort Cloud
    d. 1 ly, 10 AU, Trojans, Oort Cloud, Kuiper Belt
    e. Trojans, Kuiper Belt, 10 AU, Oort Cloud, 1 ly

12. Use your knowledge of orbital velocity and the orbital radii of the following objects to list the sequence from highest to lowest *average orbital velocity.*
    a. Ceres, Earth, 1992 QB1, Halley's Comet, Jupiter
    b. Earth, Ceres, Jupiter, Halley's Comet, 1992 QB1
    c. 1992 QB1, Halley's Comet, Jupiter, Ceres, Earth
    d. Halley's Comet, Earth, 1992 QB1, Ceres, Jupiter
    e. Earth, 1992 QB1, Ceres, Jupiter, Halley's Comet

13. Which of the following processes is *not* evident in the current structure and distribution of planets in our Solar System?
    a. ongoing planet scattering
    b. previous eras of planet differentiation
    c. previous eras of planet formation inside or outside a snow line
    d. previous eras of planet formation within a disk
    e. ongoing comet collisions with planets

14. The search for exoplanets is challenging because of a number of constraints. Which of the following is *not* a constraint?
    a. the orientation of many planetary orbits in relation to their star
    b. the proximity of planets to their star
    c. the distance to exoplanetary systems
    d. the existence of appropriate Sun-like stars
    e. the orbital periods of many exoplanets

15. Which of the following would you expect to differ significantly between a long-period comet and an asteroid? Choose all that apply.
    a. radius
    b. distance between location of formation and the Sun
    c. inclination of orbit
    d. albedo at perihelion
    e. distance at aphelion

16. In which of the following ways are the terrestrial and giant planets in our Solar System similar?
    a. average number of moons
    b. approximately spherical shape
    c. elemental composition
    d. average rotation speed
    e. orbital velocity

17. Which of the following is/are potential results of planet scattering? Choose all that apply.
    a. change in orbital inclination of a planet
    b. change in the proximity of a planet's orbit to its star
    c. presence of a gas giant very close to its star
    d. presence of a terrestrial planet very far from its star
    e. highly eccentric orbit of a planet

18. In terms of orbital inclination, what conclusion can we draw by comparing Pluto with Solar System planets?
    a. It is higher, implying that Pluto likely underwent gravitational interactions with the planets.
    b. It is higher, implying that Pluto likely was captured from deep space.
    c. It is lower, implying that Pluto likely formed by a different mechanism than the planets.
    d. It is lower, implying that Pluto likely has never had gravitational interactions with other bodies.
    e. none of the above

19. Observations yield radius and mass measurements for an exoplanet. Which of the following properties can be determined from these two quantities? Choose all that apply.
    a. the existence of a core in the planet
    b. the planet's status as a water world
    c. whether the planet is a gas giant
    d. the eccentricity of the planet's orbit
    e. the planet's density

20. True/False: An exoplanet with an uncompressed density of 980 kg/m³ is likely to have migrated to its current position, 0.5 AU from its Sun-like star, after formation.

To the Point: Qualitative and Discussion Questions

21. How is the discovery of the Kuiper Belt connected to Pluto's reclassification as a dwarf planet?

22. How can Solar System objects be identified among background stars?

23. How do we know the Kuiper Belt and Oort Cloud exist?

24. How do the orbits of the Trojan asteroids, Ceres, and 2002 MN differ?

25. How would the spectrum of a comet's nucleus compare with that of its coma?

26. What category of comets has the greatest potential to be hazardous to Earth?

27. Why does a meteor shower occur at roughly the same time each year?

28. How does the wavelength of a star's light change as a result of the Doppler shift when the star is being pulled by an orbiting planet *away from* us on our line of sight? When it is being pulled *toward* us on our line of sight?

29. The melting and freezing temperatures of material that makes up planets correlate strongly with each planet's distance from the Sun. What does this correlation suggest about the formation of planetary systems?

30. How does the condensation theory explain the characteristics of our Solar System?

31. List several everyday occurrences that demonstrate conservation of angular momentum.

32. What is your reaction to the evidence that exoplanets exist in abundance?

33. What is your definition of a day? How does a day on Earth compare with a day on Jupiter?

34. What actions should Earth's inhabitants take with regard to Earth-crossing asteroids? How might the danger these asteroids represent affect Earth, both positively and negatively?

35. What characteristics would a super-Earth need for life to be possible on it?

Going Further: Quantitative Questions

36. What is the ratio of the orbital speed of a terrestrial planet orbiting at 3 AU from its star to that of a giant planet orbiting at 12 AU?

37. How many times greater is the escape velocity from a giant planet with a mass of 298.5 Earth masses (M_E) and a radius of 10 Earth radii (R_E) than that of a terrestrial planet with mass 2.4 M_E and radius 2 R_E? (See Chapter 3.)

38. What is the ratio of the orbital speed of Halley's Comet at perihelion (0.6 AU) to that at aphelion (35.1 AU)?

39. Calculate the orbital speed, in kilometers per second, of a planet orbiting 7.8 AU from a star of mass 4 M_{Sun}.

40. What is the ratio of the period of an asteroid orbiting at 1.8 AU to that of another asteroid orbiting at 3.3 AU?

41. How many times greater is the period of a planet orbiting its star compared with the period of a sibling planet orbiting at half the distance?

42. A spinning molecular cloud has decreased in radius by a factor of 23. By what factor has its speed increased?

43. A portion of dust in a protoplanetary disk is rotating at 450 m/s. If the disk decreases in size by a factor of 400, what will the rotational speed of the dust be?

44. How much greater is the speed of a blob of gas rotating in an interstellar cloud at a radius of 0.74 pc than that of another blob of gas at a radius of 1.7 pc?

45. What would be the period, in years, of a Kuiper Belt object with an average orbital radius of 52 AU? (See Going Further 3.1 in Chapter 3.)

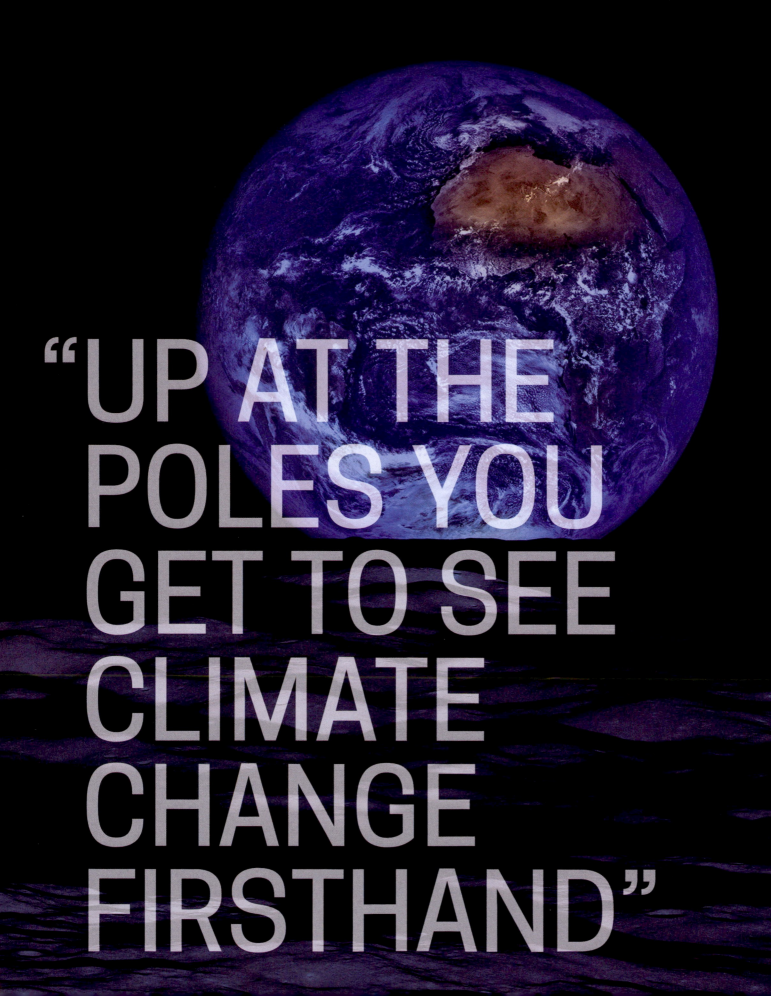

"UP AT THE POLES YOU GET TO SEE CLIMATE CHANGE FIRSTHAND"

6

Home Base

EARTH AND THE MOON

Discovering Change: Arctic Crocodiles and the Earth-Moon System

6.1 There are no hotels or convenience stores deep in the Arctic. When research takes John Tarduno, a geophysicist at the University of Rochester, to desolate ice-bound islands, it's all tents, cookstoves, and very warm sleeping bags. Tarduno studies *paleomagnetism*—the history of Earth's magnetic field. "We are looking for rocks with a lot of iron in them," he says. "When these rocks formed millions of years ago, the iron particles in them got oriented in the direction Earth's magnetic field had back then. Since Earth's magnetism is constantly changing, we go up there looking for what are, essentially, fossil compasses."

On one of his expeditions to Ellesmere Island, just 500 miles from the North Pole, Tarduno and his students found unexpected fossils. "We found a champsosaur," explains Tarduno, "which is a kind of prehistoric crocodile-like creature" (**Figure 6.1**). Because it was suited to life in warm waters, its discovery meant that millions of years ago, the Arctic that is now a frozen tundra was once a warm swamp. "That was a big 'Wow!' moment for us," says Tarduno, "one of the most exciting I have ever had."

Tarduno's trips to the Arctic have shown him not only how much Earth has changed but also how much the planet remains in flux. "Up at the poles you get to see climate change firsthand," explains Tarduno. "The first time I was in the Arctic was in 1990. Back then you never thought about walking around without a parka, even in the summertime. That's how cold it was all the time. Now I take students up there and they walk around in T-shirts. We have seen a lot of dramatic changes over just that short a timescale."

The history of a planet such as Earth is a history of change driven by powerful forces and enormous energies. The research of atmospheric physicists, ocean scientists, and geophysicists shows us how Earth and the Moon have evolved, creating a unique home for life in the Solar System. Through the intense study of our planet, scientists have laid the groundwork for understanding all the worlds in the Solar System. Thus we begin our

Figure 6.1 Champsosaur Skeleton
John Tarduno's Arctic expedition found the unexpected: a champsosaur, or prehistoric crocodile, providing evidence that swamp conditions once prevailed where extreme cold and very low humidity exist today.

← Earth and the Moon (its surface in the foreground) as seen by the *Galileo* spacecraft on its way to Jupiter.

Figure 6.2 Earth and Its Moon

Earthrise, an iconic image captured by *Apollo 8* as it orbited the Moon. For many people this image offered a new vision of Earth's unity and fragility.

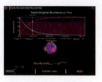

INTERACTIVE:
Radiometric Dating

study of **comparative planetology**—comparing the different Solar System planets—with Earth and its Moon (**Figure 6.2**).

Earth and the Moon: A Unique System

Earth and the Moon do not resemble any other planet-satellite system in the Solar System. We know from the technique of **radiometric dating** that Earth, the largest terrestrial planet, is about 4.5 billion years old (see **Going Further 6.1**). Unlike the other planets, Earth's surface is covered in liquid water. It has supported a complex biosphere for billions of years. Some of the differences between Earth and other planets are less immediately obvious, however. For example, the Moon is so large (27 percent of Earth's radius) that an alien visiting our system might consider Earth and the Moon to be a binary planet system rather than a planet and its satellite. (In astronomy, a **satellite** is a small body that orbits a planet.) "The Moon is about 1 percent the mass of Earth," says Jonathan Lunine, a planetary geologist at Cornell University who has studied many Solar System bodies. "Although that sounds small," Lunine continues, "it's actually a far larger ratio compared to the moons of Jupiter, Saturn, Mars, Uranus, and Neptune" (**Figure 6.3**).

Earth's role as a life-bearing world and the fact that it has such a large moon are probably related. As we shall see, because of its large mass, the Moon has had an important effect on Earth's evolution by creating tides and lengthening the day. Just as important, the Moon's gravitational force has kept the spinning Earth from wobbling too much, keeping Earth's climate relatively steady. No other moon in the Solar System has played such a key role in its parent planet's life.

Yet for all their uniqueness, Earth and the Moon reveal much about the evolution of the other worlds in our Solar System. We can apply what we learn from a detailed study of Earth and the Moon to other planets, and we can apply what we learn

Figure 6.3 Comparison of Planet and Moon Sizes

In our Solar System, only the dwarf planet Pluto and its moon Charon are closer to each other in relative size than Earth and the Moon are. Jupiter's largest moon, Ganymede, whose radius is 27 times smaller than the gas giant's, is more typical of planet-moon systems in our Solar System.

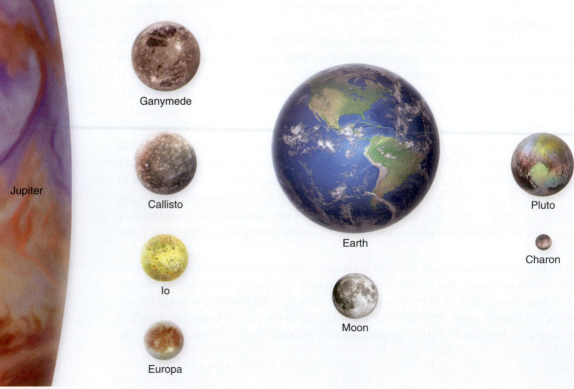

GOING FURTHER 6.1
Measuring Eternity: Radiometric Dating

Although humans were not around for the formation of ancient meteorites or the early eras of Earth's geologic activity, we have found a kind of "stopwatch" that reveals the age of rocks and meteorites. Using radiometric dating, scientists can measure the relative amounts of certain types of atoms to measure remarkably precise ages and recount the story of Earth's evolution.

As discussed in Chapter 4, an atom of a given element can have only a fixed number of protons. Carbon, for example, always has six protons. The number of neutrons in an element is not fixed, however. **Isotopes** are versions of a given element that have different numbers of neutrons. Some isotopes are stable, but others are *radioactive*—they are unstable and will transform over time into another isotope or even another element, emitting particles and energy. This process is called **radioactive decay**.

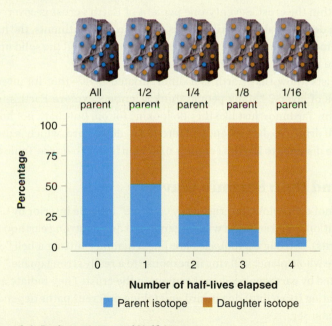

Figure 6.4 **Radioactivity and Half-Lives**
The predictable average decay times of radioactive parent elements with known half-lives into their daughter elements provides an accurate means of determining the ages of Earth or Moon rocks and meteorites.

Radioactive decay is the clock hiding in atomic nuclei. Scientists cannot predict when a single unstable nucleus will decay (it's an inherently random quantum process), but they can predict with great certainty how large collections of unstable nuclei will behave. Each unstable isotope has a distinctive timescale called a **half-life**. The total number of nuclei of that isotope drops by one-half during each half-life (**Figure 6.4**).

To see how radioactivity works, let's look at the formula for how the fraction F of an unstable *parent* isotope with half-life T changes over a specific time interval t to a *daughter* isotope:

$$F = \left(\frac{1}{2}\right)^{t/T}$$

$$\frac{\text{fraction of unstable parent left}}{\text{relative to initial amount of parent}} = \left(\frac{1}{2}\right)^{\frac{\text{time interval}}{\text{half-life of unstable parent}}}$$

Thus after one half-life, or $t = T$, the fraction of the original unstable isotope left is 1/2. The other 1/2 has decayed to the daughter isotope. What is the fraction left after 5 half-lives, or $t = 5T$?

$$F = \left(\frac{1}{2}\right)^{\frac{5T}{T}} = \left(\frac{1}{2}\right)^{5} = \frac{1}{32} = 0.031$$

$$0.031 \times 100 = 3.1\%$$

After 5 half-lives, only 3.1 percent of the parent isotope remains.

Many unstable isotopes occur in nature, and all have different half-lives. Uranium-238 (92 protons + 146 neutrons), for example, has a half-life of 4.5 billion years. Thus if we began with 1,000 uranium-238 nuclei, only half (500) would be left 4.5 billion years later. Since uranium-238 decays into "daughter" nuclei of lead-206 (82 protons + 124 neutrons), you would expect that if we began with no lead-206, then after 4.5 billion years, there would be 500 lead-206 nuclei. If scientists know the ratio of the isotopes present when a meteorite (or other object) formed, they can then use the ratios measured today to determine the age of the meteorite.

from other planets to understanding our own world. What we always find is change: planets forming and *differentiating* (that is, separating into layers); surfaces changing from forces within and without; and in many cases, atmospheres growing to blanket their worlds, profoundly changing the surface histories of those planets. And for Earth, at least, we must add the all-important evolution of life, which changed land, sea, and sky.

comparative planetology The study of planets' characteristics based on comparison with other known planets.

radiometric dating A method used to date a sample of material by comparing relative amounts of radioactive atoms with the "daughter" atoms into which they decay.

satellite A small body orbiting a planet.

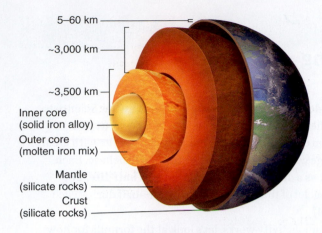

5–60 km

~3,000 km

~3,500 km

Inner core
(solid iron alloy)

Outer core
(molten iron mix)

Mantle
(silicate rocks)

Crust
(silicate rocks)

Figure 6.5 Earth's Interior Structure
Earth has four major layers: the thin, rocky crust; the mantle, where convection occurs; and a two-part core. The outer core remains molten, while the inner core is solid because of the extreme pressure.

isotope One of several forms of an element with different numbers of neutrons in the nucleus.

radioactive decay The process by which some isotopes of an element transform over time into other isotopes or elements, emitting subatomic particles and energy in the process.

half-life The period of time over which the total number of nuclei of a radioactive isotope will drop by one-half.

core Earth's innermost layer, composed primarily of iron. The inner core is solid; the outer core is in a fluid state.

mantle The silicate-rock layer between Earth's core and crust. High pressures cause material there to deform and flow, except in the uppermost mantle.

crust Earth's solid outer layer.

basalt A type of volcanic silicate rock that forms at the ocean floor.

lithosphere Earth's solid outer layers, consisting of the crust and the uppermost regions of the mantle.

seismic wave A propagating wave in Earth's interior caused by an earthquake (or other energetic event).

P-wave Pressure wave or primary wave: a seismic wave in which atoms and molecules oscillate back and forth in the direction of the wave's propagation.

S-wave Shear wave or secondary wave: a seismic wave in which atoms and molecules oscillate perpendicular to the direction of the wave's propagation.

Section Summary

- Earth is unique among planets of the Solar System in having liquid water covering much of its surface.
- The Moon's large size relative to Earth has played a key role in making Earth suitable for life.

CHECKPOINT How has the relatively large size of the Earth's moon affected the Earth's history?

Earth Inside and Out

6.2   Earth's internal structure is dominated by a metallic **core**, made mostly of iron, with a radius that is about one-half the radius of the entire planet (**Figure 6.5**). The inner part of the core is solid; the outer core appears to be a fluid molten iron mixture. Surrounding the core is an extended **mantle**, a layer composed of silicate rocks (made of silicon, oxygen, and other elements) that range from soft and deformable to solid and brittle. Lying atop the mantle is a thin layer of **crust**, also composed mainly of silicate rocks, that includes both the ocean floor and the continents (**Table 6.1**). However, the ocean floor is composed of **basalt**, a volcanic silicate rock (so, it forms at the surface), and the continents are composed of *granite*, a silicate rock that forms at great depth within the crust. Seventy-one percent of Earth's crust is covered in oceans, which can reach depths of about 11 kilometers (km). On the continents, the highest mountains extend almost 9 km above sea level. Together, the crust and the solid upper portion of the mantle are referred to as Earth's **lithosphere**.

How do geophysicists know that Earth is a differentiated body—that its internal structure has this series of layers? "Many people think scientists explore Earth's interior by drilling," says John Tarduno, "just like they do looking for oil. But that's not how it works." The deepest well drilled to date extends a mere 12 km downward, which is only a fraction of the crust. The distance to Earth's center is 6,000 km below that (see Table 6.1).

Shake, Rattle, and Roll: Seismic Waves

Although scientists cannot drill below the crust, they can "see" into the interior by using a kind of planetary vibration called **seismic waves**, generated during strong earthquakes (or other energetic events). "Strong earthquakes are like a hammer hitting a bell," says Tarduno. "In a sense, the whole planet will ring in response to a really strong quake." The seismic vibrations created by earthquakes do not remain in the crust. They radiate away from the quake, propagating (traveling) downward and taking different paths depending on the material they encounter.

Since Earth is a sphere, most seismic waves eventually make it back to the surface hundreds or thousands of miles away. "A strong earthquake sends out so much energy," explains Tarduno, "that the waves it generates can make it all the way through the planet." Scientists pick up the vibrations from large earthquakes by using sensitive instruments called *seismographs*, which can detect even tiny surface vibrations. By mapping the global distribution of vibrations from different earthquakes at different locations on the planet, scientists have built up a detailed picture of Earth's interior.

Two kinds of seismic waves are used in mapping Earth's internal structure: P-waves and S-waves. The *P* in **P-waves** stands for *pressure* (or *primary*). "In a seismic P-wave," says Tarduno, "you are actually seeing compression and expansion of different materials at different conditions deep within Earth." P-waves travel through Earth's interior much the way sound waves travel through air—in a back-and-forth motion. The atoms and molecules of the medium through which a P-wave passes oscillate back and forth in the direction of the wave's propagation. The *S* in **S-waves** stands for *shear* (or *secondary*); they travel more slowly than P-waves do. In an S-wave, matter is distorted in a direction

Table 6.1   Earth's Layers

Layer	Depth from Surface	Primary Material	Temperature	Major Processes
Continental crust	30–60 km Greatest height above sea level: 9 km	Silicate rocks	Average: 287 K Range: 184–344 K at surface, 450–650 K at boundary of mantle	• Earthquakes • Movement of tectonic plates • Volcano formation
Oceanic crust	5–10 km	Silicate rocks	450–650 K at boundary of mantle	• Earthquakes • Movement of tectonic plates • Volcano formation
Oceans	Greatest depth below sea level: 11 km	Water	309 K in Persian Gulf, 271 K near North Pole	
Mantle	From crust to 2,890 km	Silicate rocks	750–1,150 K at boundary of crust About 4,000 K at boundary of core	Convection that drives movement of tectonic plates
Outer core	2,890–5,150 km	Molten iron-nickel-sulfur-oxygen mix	4,000–5,000 K (estimated)	Dynamo effect creating geo-magnetic field
Inner core	5,150–6,370 km	Solid iron-nickel alloy	5,000–6,000 K (estimated)	

perpendicular to the wave's motion. "If you take a rope and yank it sideways," says Tarduno, "you can make an S-wave."

What makes these two kinds of waves so important for studies of Earth's interior is a key difference in the way they propagate. P-waves can travel through solids and liquids, but S-waves can pass through solids only. Thus, S-waves traveling into a molten layer are stopped entirely (**Figure 6.6**). In addition, both types of waves change direction—they bend—as they enter a material of different density. This bending, called *refraction*, says Tarduno, is "the same thing that makes a ruler half-submerged in a glass of water appear bent." (We saw this process in Chapter 4 for light.) As a result of refraction at the core-mantle boundary, P-waves cannot reach the surface at certain locations relative to the earthquake's origin. By tracking the time it takes P-waves and S-waves from a strong quake to reach various points on the surface, scientists can calculate the range of densities the waves must encounter as they move through Earth.

Earth's Interior

Seismic waves have provided information about Earth's core, which lies below the mantle. Scientists have seen that S-waves from earthquakes never reach seismic stations located directly on the other side of Earth. That means S-waves do not pass directly through, or close to, the center of the planet; instead, they are absorbed (Figure 6.6). In contrast, most P-waves make it from one side of Earth to the other. This difference in the motions of S-waves and P-waves tells scientists that Earth must have at least a partially liquid core. "The core is actually made of two components," says Tarduno. "We have the solid inner core, and we have the molten outer core. Both the inner and outer cores are mostly iron."

From studying P-waves and S-waves, scientists understand too that the part of Earth we experience, the crust, is just a thin layer of rock that literally *floats* on top of denser rock below. Under the continents, the crust reaches depths of 60 km; under the oceans, it may extend downward for only 5–10 km.

Seismic studies also show that below the crust is a remarkable layer of flowing rock—the mantle—whose density increases with depth from Earth's surface. "The material in

"Strong earthquakes are like a hammer hitting a bell. In a sense, the whole planet will ring in response to a really strong quake."

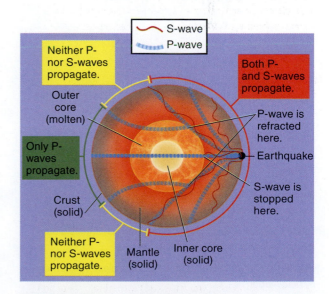

Figure 6.6 P-waves and S-waves
Seismic waves caused by earthquakes propagate through Earth (shown in cross-section). Two kinds of seismic waves, P-waves and S-waves, pass through liquid and solid matter in different ways. Scientists track these waves to probe the properties of Earth's deep interior. Waves change direction (refract) as they travel through material of different densities.

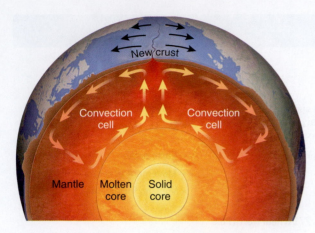

Figure 6.7 Convection in the Mantle
A schematic of the convection patterns occurring in Earth's mantle. Convection is the engine that drives the motion of plate tectonics.

"When radioactive elements in the bottom layer of the mantle decay, they release heat. That is what sets the mantle convection in motion."

plume A rising column of molten rock from a planet's mantle that breaks through the crust.

shield volcano A volcano that forms from a mantle plume and tends to have shallow, sloping sides.

convection The transfer of heat from one place to another by fluid motion.

erosion or **weathering** The process by which surface features are worn away by the action of wind and water (including ice).

plate tectonics The movement of Earth's lithospheric plates due to underlying mantle flow.

plate A large rigid block—crust plus solid upper-mantle material—that floats on the deformable part of the mantle.

the mantle is not melted," says Tarduno, "but the high pressure in there allows it to move slowly." How slow is the mantle rock's motion? "The mantle moves fastest at the upper layers," says Tarduno. "There you get top speeds of around 10 centimeters a year." The idea of rocks that move may seem unusual, but the tarmac that makes up most roads behaves the same way. Under a hammer blow, tarmac is solid, either shattering or bouncing the hammer back. Under high pressures, however, such as the pressure of a million-kilogram weight, the tarmac will ooze and deform like warm taffy.

Although the mantle is deep below the surface, scientists can get a good look at its makeup during volcanic eruptions. A volcano is a region where *magma*—melted mantle material (molten rock)—manages to break through the crust. At that point, the molten rock is called *lava*, which may spew out in a violent eruption or slowly ooze out. Locally, where temperature and pressure conditions are high enough, magma has completely melted, so it flows like a liquid and may well up to the surface. These rising columns of magma, known as mantle **plumes**, are the source of one kind of slow, relatively gentle volcanic activity that forms **shield volcanoes**. The Hawaiian Islands are shield volcanoes that formed by this mechanism. Plumes also provide evidence that the material in the mantle is in a state of constant movement—a slow, steady series of rising and falling motions called **convection**.

Convection occurs when material deep in a planet (or a star) is heated. The heat makes the material expand, becoming less dense than its surroundings. Like a balloon held underwater, the lower-density rock becomes buoyant and begins to rise. As the material rises, however, it cools (unless it is part of a plume) and eventually begins to sink back downward, where it heats up again and the whole process restarts, forming a *convection cell* (**Figure 6.7**). The presence of mantle convection also shows that Earth is hotter in the center than it is in the outer layers. "When radioactive elements in the bottom layer of the mantle decay, they release heat," says Tarduno. "That is what sets the mantle convection in motion."

The presence of the iron-rich core and the rocky but convecting mantle means that early in Earth's history the planet must have differentiated. The heavy elements that formed the core sank to the planet's center.

Earth's Changing Plates

The planet's surface is in a state of constant change driven by a variety of forces, including volcanic eruptions and **erosion** (also called **weathering**) via the flow of wind and water. The most dramatic surface changes, however, occur so slowly that they are virtually invisible in human lifetimes. Over long periods, they have reshaped the planet's surface again and again (**Figure 6.8**).

The idea that Earth's continents move over time—a theory known as *continental drift*—was proposed by Alfred Wegener in 1912 (**Figure 6.9**). He noticed, for example, that the east coast of South America and west coast of Africa appear to fit together like puzzle pieces, and he studied fossils on one side of the Atlantic that were identical to those on the other side. Wegener hypothesized that the continents were once part of a single landmass and had slowly drifted apart over time. Wegener was laughed out of scientific meetings when he proposed his theory. Not only did contemporary scientists discount his evidence, but he could not explain why the continents would move. Not until other lines of evidence—including detection of changes in Earth's magnetic field, the discovery of undersea mountain ranges, and seismographic studies revealing the properties of Earth's interior—became available was the force driving continental drift, called **plate tectonics**, recognized. "Plate tectonics was a complete paradigm shift," says Tarduno. "It was not accepted until the 1960s." Wegener died in 1930 and never got to see his radical idea become mainstream.

Scientists now understand that it is not just continents that move; instead, large rigid blocks, or **plates**, float on the nonrigid portion of Earth's mantle. Tectonic plates make up the *lithosphere*, which is crustal material plus solid upper-mantle material. There are seven major plates and many minor ones, and, says Tarduno, "they are always moving

a.

c.

b.

Figure 6.8 Earth's Surface

Throughout its evolution, Earth has been resurfaced by processes originating within its interior, such as **a.** volcanic eruptions; **b.** the shifting of rigid pieces of the crust and upper mantle, which created this escarpment; and **c.** earthquake-induced tsunamis.

a.

b.

Figure 6.9 Continental Drift

a. Alfred Wegener (1880–1930) during an expedition to Greenland that cost him his life. In 1912 he developed the theory of continental drift, but it was not generally accepted until the 1960s. He based his theory partly on how the continents seem to fit together. **b.** An artist's rendering of today's continents imagined as a single landmass by joining their boundaries.

Figure 6.10 Tectonic Plates
Color-coded map of Earth's tectonic plates. The darkest shades within each plate represent continents. Earthquakes occur with greatest frequency and force at boundaries between plates where pent-up energy is released in sudden plate movement (red dots). The red arrows indicate the direction of movement at some plate boundaries.

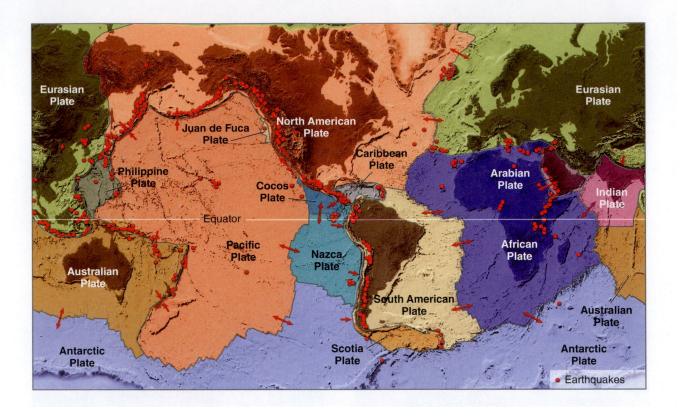

relative to one another." As the buoyant mantle material reaches the top of its circulating motion in a convection cell, it pushes the surface plates along. Boundaries between plates are locations of powerful and often violent geologic activity. **Figure 6.10** shows where the majority of volcanoes and earthquakes take place on Earth. Overlaid on this map is the location of various plates. Wherever two plates meet, there is a strong likelihood of violent geologic forces being unleashed. Volcanic eruptions and earthquakes occur because the plates slide past each other, collide, or spread apart. In all three cases, tremendous energies locked up in moving trillions of tons of rock from one part of the planet to another are released at these plate boundaries, often with devastating consequences.

In regions where plates slide past each other, the movement is not continuous. Pressure builds up at the plate edges until it is released suddenly as an earthquake, and the plates move a few fractions of a centimeter past each other. The San Andreas Fault in California, responsible for many of the region's large and small tremors, comes from shearing (sideways motions) between the Pacific Plate and the North American Plate.

Plates may collide rather than slide past each other. When two plates containing continental crust collide, the plate material may be thrust upward, forming mountain ranges. This process of "mountain building" is occurring now in the Himalayas, where the Indian Plate is colliding with the Eurasian Plate. When at least one of two colliding plates includes (thinner) oceanic crust, the collision between them forces one plate to dive downward in a process called **subduction**. In subduction zones, the sinking plate starts to melt as water in the crustal portion of the plate lowers the melting temperature of the mantle. Eventually the plate is reabsorbed into the mantle.

Collisions between plates can also cause volcanic activity. Where one tectonic plate slips under another, melting plate material rises to the surface, creating a *composite volcano*. This form of volcanism is usually violent as gases build up, only to vent in a tremendous explosion. Composite volcanoes tend to rise sharply and have steep slopes because they form rapidly. In 1980, the powerful eruption of Mount St. Helens, a composite volcano, in Washington State killed 57 people and flattened trees for 230 square miles. Other well-known composite volcanoes are Mount Shasta in California and Mount Fuji in Japan.

subduction A geologic process in which one edge of a tectonic plate is forced below the edge of another.

seafloor spreading The creation of new lithospheric material due to the upwelling of magma as tectonic plates spread apart.

Youngest
island

Oldest
island

Movement of tectonic plate

Mantle
plume

**Figure 6.11 Origin of
Shield Volcanoes**

Volcanic island chains form when
a tectonic plate moves over a
(stationary) mantle plume, forming island after island over millions
of years.

Plate tectonics can also affect shield volcanoes, the type that forms above mantle plumes. The Hawaiian Islands formed in that way, but because the Pacific Plate (the seafloor beneath Hawaii) is moving, the plume below that plate created one island after another. The northwesternmost island, Kure Atoll, is older than the southeasternmost island, Hawaii, by about 28 million years because the plate continued to move while the plume stayed in the same location (**Figure 6.11**).

At the center of the Atlantic Ocean is an underwater mountain range called the Mid-Atlantic Ridge, where two plates are moving away from each other. Magma, molten material from the mantle, wells up into the gap formed as the two plates move apart, emerging as lava at the ridge's crest. As the lava cools, it becomes new ocean floor and is carried along with the separating plates (**Figure 6.12**). This process of creating new lithosphere at a mid-ocean ridge is called **seafloor spreading**—an essential part of plate tectonics. "You can see that the oceanic crust has been constantly re-forming," says Tarduno. "The oldest oceanic crust is 200 million years old. The continents, on the other hand, are billions of years old."

Tectonic plates move at a rate of about 10 centimeters (cm) per year, the same speed at which the mantle flows below them. (This is just a bit faster than your fingernails grow.) Although it takes hundreds of millions of years to alter the shape of the continents appreciably, scientists have been able to use highly accurate global positioning system (GPS) measurements to track the shifting plates.

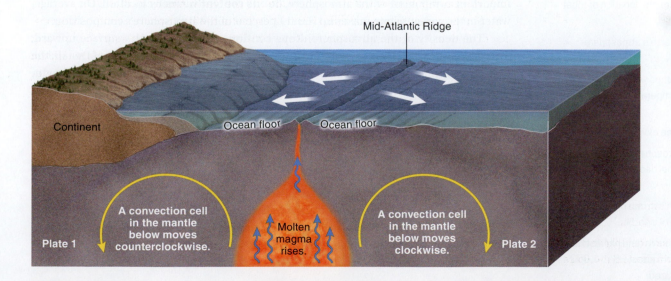

Mid-Atlantic Ridge

Continent

Ocean floor

Ocean floor

A convection cell
in the mantle
below moves
counterclockwise.

Molten
magma
rises.

A convection cell
in the mantle
below moves
clockwise.

Plate 1

Plate 2

**Figure 6.12 Formation
of the Ocean Floor**

Tectonic plates move apart at a
mid-ocean ridge, allowing magma
to well up. The lava that emerges
from the ridge crest then cools,
forming new lithosphere at the
bottom of the ocean.

Combining current data with fossil records and geologic studies, scientists have reconstructed plate tectonic movement over the entire history of our planet. With this long view they can see, for example, that 250 million years ago all of Earth's continents formed a single giant landmass called Pangaea. Before the era of Pangaea, the continents were separate, and afterward they separated again. "If you visited Earth 250 million years in the future," says Tarduno, "it would be a very different-looking place. The maps we have now wouldn't make any sense."

Section Summary

- Earth's interior has four major layers: the solid inner core; the molten outer core; the partly deformable mantle; and the thin, rocky crust, which floats atop the mantle.
- Earthquakes generate two kinds of seismic waves, which provide insight into Earth's interior. P-waves pass through solids and liquids; S-waves pass through solids only.
- Plate tectonics is the slow movement of plates of lithosphere (crust plus solid upper mantle), driven by convection in the mantle. Plate movements can generate earthquakes and are associated with the locations of some volcanoes.

CHECKPOINT How is the interior structure of the Earth linked to the activity you can see on the surface?

Earth's Near-Space Environment

6.3 When astronauts orbiting Earth look out over the planet's curved horizon, they can see a thin layer of blue-white haze blanketing the oceans and continents below. Earth's atmosphere constitutes a small fraction of the planet's mass (1/1,200,000), but it has been a principal driver of Earth's history as a life-bearing world. "Sure, the atmosphere has affected the evolution of life," says John Tarduno, "but it has worked the other way around too. Life has actually had a significant effect on the evolution of the atmosphere."

A Thin Life-Giving Veil: Earth's Atmosphere

The primary gases that compose the atmosphere (in order of volume) are nitrogen (78.09 percent), oxygen (20.95 percent), argon (0.93 percent), and carbon dioxide (CO_2; 0.039 percent). Small amounts of other gases, including methane (CH_4), helium, and ozone (O_3), can play an important role in regulating Earth's climate or the kinds of solar radiation that penetrate the atmosphere and reach the ground. Water vapor is also an important component of the atmosphere, but its content varies by location. On average, water in the gaseous state makes up about 1 percent of the atmosphere's composition.

The density of the atmosphere drops continuously from Earth's surface upward. This decrease in density is so swift that by the time you reached the summit of Denali, the tallest mountain in the United States and only about one-tenth the atmosphere's height, less than one-half of the atmosphere's mass would lie above you. The temperature of the atmosphere does not, however, decrease evenly with altitude; this variation creates fairly well-defined atmospheric layers (**Figure 6.13**).

The lowest layer of the atmosphere, the **troposphere**, contains 80 percent of its mass. The troposphere extends 17 km above ground level at Earth's equator and just 7 km above at the poles. Sunlight warms the ground and is then reradiated upward; this is the principal mechanism warming the troposphere. Because this heating starts at the ground and the atmosphere's density decreases with altitude, the tropospheric temperature also decreases with altitude. The troposphere is where the majority of Earth's weather is generated; most forms of clouds occur within this layer. Convection of air, driven by the absorbed sunlight, also occurs in the troposphere. The circulation of warm air upward and cold air downward

> "If you visited Earth 250 million years in the future, it would be a very different-looking place. The maps we have now wouldn't make any sense."

troposphere The lowest layer of Earth's atmosphere, reaching a maximum altitude of 17 km; the location of most of Earth's weather and clouds.

stratosphere The second layer of Earth's atmosphere, extending from about 17 to 50 km.

thermal inversion An increase in temperature that occurs with increasing altitude (whereas temperature usually decreases with increasing altitude).

ozone A molecule composed of three oxygen atoms: O_3.

ozone hole A region within the stratosphere above Earth's poles that shows a substantial decrease in ozone abundance.

mesosphere The third layer of Earth's atmosphere, extending from about 50 to 85 km.

ionosphere or **thermosphere** The fourth (and highest) layer of Earth's atmosphere, extending from about 85 to 1,000 km, where most of the atoms are ionized.

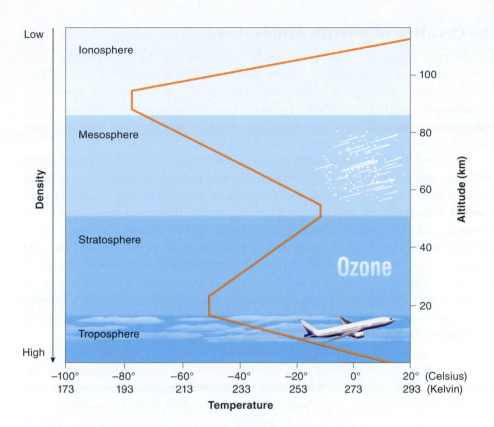

Figure 6.13 Earth's Atmospheric Layers

Our atmosphere decreases in density as altitude increases. Temperature (zigzag line) *increases* in the stratosphere and again in the ionosphere, forming a thermal inversion. Clouds (weather) and airplane travel are confined to the troposphere. Meteors typically burn up within the mesosphere. The region we normally think of as "space," where the International Space Station orbits (although well above 100 km), is the ionosphere.

makes the layer relatively turbulent, meaning the air does not flow smoothly in sheets but circulates chaotically. Commercial jetliners cannot always avoid turbulence, because they never reach altitudes of much more than 12 km, which in most areas is well within the troposphere.

The **stratosphere** is the next layer, extending up to altitudes of 50 km. Temperature in the stratosphere, unlike that in the troposphere, *increases* with height. This reversal is called a **thermal inversion**. It occurs because the stratosphere contains a layer relatively rich in **ozone**, a molecule composed of three oxygen atoms (O_3). Ozone is extremely effective at absorbing ultraviolet light from the Sun, which leads to an increase in atmospheric temperature at these altitudes. Ultraviolet light is harmful to biological compounds. "There is a thin region of the stratosphere where ozone can form and remain stable," says Tarduno. "It's really critical for life on Earth, as we've learned over the past few decades."

The emission of certain chemicals, called *chlorofluorocarbons* (CFCs), into the atmosphere has been shown to destroy stratospheric ozone. In response, every year giant **ozone holes**—substantial, localized decreases in ozone abundance—now appear above the Antarctic and the Arctic, because the large seasonal changes in sunlight there affect ozone chemistry. Worldwide bans of CFCs have helped stabilize the decrease in stratospheric ozone, but many scientists remain concerned about its long-term health (and hence our own).

Above the stratosphere lie the **mesosphere** and **ionosphere** (also called **thermosphere**). Stretching from about 50 to 85 km, the mesosphere is the layer in which most meteors burn up. Its temperature drops with altitude. The ionosphere, extending up to 1,000 km from the surface, gets its name because most atoms in this layer are *ionized*—that is, they've lost one or more electrons. During solar storms, the Sun emits high-energy particles, which collide with atmospheric ions and drive them into excited states. These interactions with solar radiation heat the low-density outer regions of the ionosphere. As in the stratosphere, temperatures in the ionosphere increase with altitude; they plateau at about 300 km (**Figure 6.14**).

Although scientists consider the ionosphere part of the atmosphere, its upper reaches are what most people consider "space" because space shuttles and the International Space Station have orbited in this layer.

> "There is a thin region of the stratosphere where ozone can form. . . . It's really critical for life on Earth, as we've learned over the past few decades."

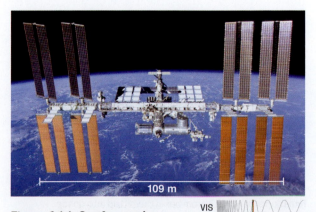

Figure 6.14 Our Atmosphere

In 2010, astronauts aboard space shuttle *Atlantis* docked with the International Space Station for 7 days. While undocking, *Atlantis* took this image that included Earth and the denser portions of the planet's atmosphere, while orbiting within the ionosphere (about 300–500 km above Earth).

INTERACTIVE:
Retaining an Atmosphere

JOHN TARDUNO

John Tarduno's research as a geophysicist has taken him to remote corners of Earth, from frozen Arctic wastelands to the mountains of New Zealand. Tarduno's field research for paleomagnetic samples has gotten him into dangerous situations. "One time working in the Arctic, there was this narrow, steep-walled ravine that I wanted to explore," Tarduno recalls. "So I had the chopper put me down at one end. I was lucky because I found a natural terrace that I could move along, but it was 500 feet down and if things hadn't worked out I would have been in trouble." The samples he retrieved that day were key to recovering the record of a particularly turbulent part of Earth's magnetic history. "I guess it was worth it," he says, smiling.

thermal velocity The average velocity of particles in a gas, which depends on the temperature of the gas.

volatile Easily vaporized (transformed into a gaseous state) at relatively low temperatures; or, a volatile substance.

greenhouse gas An atmospheric gas that absorbs infrared radiation and keeps it from being reemitted into space.

greenhouse effect The phenomenon by which the fraction of incoming solar energy that would have been radiated back into space is trapped by greenhouse gases, leading to an increase in the average temperature of the planet.

greenhouse forcing An increase in greenhouse gas concentrations that leads to a change in a planet's climate.

The Creation of Earth's Atmosphere

The history of life on our planet is intimately connected with the history of the atmosphere, and life has played a strong role in determining the atmosphere's form. In fact, the layer of gas surrounding the planet today is not its original atmosphere. Earth's first atmosphere was probably very similar to the composition of the solar nebula and the gas giant planets—containing mostly hydrogen, helium, and other light gases. Early in Earth's history, however, these light elements escaped into space.

The ability of a planet to retain different gases directly affects what kind of atmosphere it will have. The composition of a planet's atmosphere depends on the balance between its escape velocity (see Chapter 3) and atmospheric particles (atoms and molecules). **Thermal velocity** is the velocity that a particle obtains from random collisions as it careens around. "Like the name implies, the thermal velocity depends on temperature," says Tarduno. "In a hot atmosphere, atoms and molecules move at high velocity. In a cold atmosphere, they move more slowly." If an atmospheric particle's thermal velocity is higher than the planetary escape velocity, the particle will eventually escape into space. But the thermal velocity also depends inversely on the particle's mass (that is, thermal velocity increases as mass decreases). Therefore, hot atmospheres are prone to lose lightweight particles before they lose heavy ones (see **Going Further 6.2** on p. 164).

Earth's escape velocity is low enough and the temperature of its first atmosphere was high enough that all of the lightest components eventually escaped into space. "The formation of the Moon, which was likely a pretty violent event, may also have contributed to losing the first atmosphere," says Tarduno. What replaced the primal atmosphere were *outgassed* compounds—gases released by geologic processes such as volcanoes. Much of the second atmosphere may also have been delivered to Earth via comets and other planetesimals that were rich in heavier **volatile** (easy-to-vaporize) materials, such as methane and ammonia.

While there remains intense debate about the early stages of the atmosphere's evolution, scientists agree that the rapid increase in oxygen was one of its most important transitions. As Tarduno puts it, "Clearly one of the most fundamental events in Earth's history is something called the *Great Oxygenation Event*." Roughly 2.3 billion years ago, the oxygen content of the atmosphere increased dramatically. The cause was photosynthesis by single-celled organisms called cyanobacteria, which produce oxygen as a byproduct. (Like plants, they "breathe in" CO_2 and "breathe out" O_2.) Minerals absorbed most of the oxygen initially. "All those red rocks you see out West are basically from this period," says Tarduno. "Those rocks had iron in them, which absorbed the oxygen, making the minerals in the rocks rust colored." Eventually all the oxygen levels became so high that the minerals could not absorb it all. That is when oxygen began to accumulate in the atmosphere.

Taking the Heat: The Greenhouse Effect

Each second, the Sun-facing side of Earth encounters 1.366 kilowatts per square meter (kW/m^2) of solar radiation. Thus, more energy falls on the hemisphere facing the Sun in 1 second than all power plants on Earth generate in a year. What happens to this energy once it reaches the planet?

For the most part, the Sun can be considered a blackbody. Using Wien's law (Chapter 4) and the Sun's surface temperature, 5,778 K, we see that the Sun's radiation peaks in the visible part of the electromagnetic spectrum. Earth's atmosphere is transparent to those wavelengths, so most of the solar energy falling on the upper layers of the atmosphere reaches Earth's surface. The surface, including the oceans, is quite effective at absorbing incoming solar energy, and in doing so, the surface heats up. If there were no atmosphere, we could predict what would happen next: Earth's surface temperature would increase until the energy it radiated back into space (as each planet also acts as a blackbody) would balance the energy it received from the Sun. The temperature at which that balance would be achieved is calculated to be about 250 K ($-23°C$, or $-10°F$), well below the freezing point of water. (At this temperature, Earth's outgoing radiation is mostly in the infrared.) But our planet is not a giant frozen snowball, and since the incoming energy rate from the Sun is fairly constant over short times, the extra warming must be due to the atmosphere. Somehow the atmosphere must trap some of the infrared energy that Earth would otherwise radiate back into space, holding it in and warming the planet.

What raised Earth's temperature above freezing? The Sun-warmed surface of Earth reemits most of its energy in the infrared. Certain molecules in the atmosphere, called **greenhouse gases**, such as CO_2, water vapor, and methane, are extremely effective at absorbing the infrared radiation emitted by Earth's surface and keeping it from escaping into space. The relative effects of greenhouse gases differ, and some are more effective than others. The most effective on Earth, based on concentration, is water vapor, followed by carbon dioxide. By trapping a fraction of the energy Earth radiates, these gases create an atmospheric **greenhouse effect** (Figure 6.15). "Without the greenhouse effect, the planet would be a terribly cold place," says Tarduno. "Just as the windowpanes in a greenhouse allow solar energy in but do not allow Sun-warmed air (or infrared radiation) out, greenhouse gases raise the temperature of a planet." In essence, the greenhouse effect increases the fraction of incoming solar energy that stays within Earth's surface and atmospheric system. The greenhouse effect is an essential part of atmospheric dynamics for many of the planets in the Solar System, including Venus and Mars.

"We have tremendous evidence that Earth has been warmer and much cooler over the geological record," explains Tarduno. "All the evidence points to these changes coming from natural variations in the concentration of greenhouse gases from something we call **greenhouse forcing**." *Forcing* in this case means changing the physical and chemical conditions, which can lead to climate change. Climate variations have been driven not only by CO_2 but also by natural variations in methane. "Methane gets stored in many places in Earth, including permafrost," says Tarduno. (*Permafrost* is a soil layer that remains frozen year-round, largely limited to polar regions.) "Melting the permafrost can release methane, a more potent greenhouse gas than CO_2." Drops in methane may also have strongly reduced the greenhouse effect in the past and led to snowball-Earth phases, super ice ages when much of the planet was covered in snow and ice.

Thus, the greenhouse effect is crucial to Earth's ability to maintain a comfortable environment for life. Unfortunately, defining "comfortable" may become problematic for human civilization, as the greenhouse effect has taken on new importance for us.

INTERACTIVE:
Greenhouse Effect

"Without the greenhouse effect, the planet would be a terribly cold place."

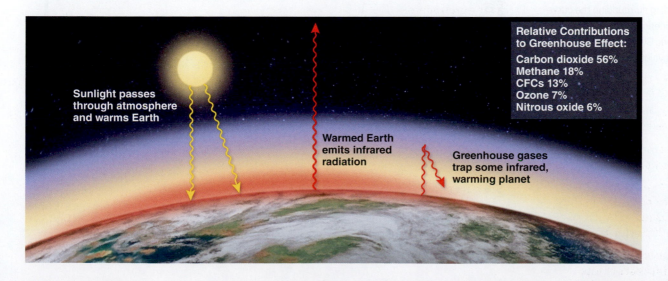

Relative Contributions to Greenhouse Effect:

Carbon dioxide 56%
Methane 18%
CFCs 13%
Ozone 7%
Nitrous oxide 6%

Sunlight passes through atmosphere and warms Earth

Warmed Earth emits infrared radiation

Greenhouse gases trap some infrared, warming planet

Figure 6.15 The Greenhouse Effect
Sunlight in primarily visible wavelengths warms Earth's surface, which then reradiates energy back into space in the infrared. Greenhouse gases absorb the infrared radiation, thus increasing the planet's overall temperature. Relative contributions of human-emitted greenhouse gases to current global warming are also shown.

Carbon dioxide is a by-product of combustion—that is, burning things. The industrial revolution, which began in the late 1700s and led directly to the society we have today, was based on the combustion of *fossil fuels*, such as coal, oil, and natural gas. Before the industrial revolution, the combustion of these materials was negligible. Now, however, fossil fuel combustion occurs at a rate that produces about 21.3 billion tons of CO_2 per year. Over the last century, the atmospheric CO_2 levels have increased by 30 percent, and they are expected to continue rising by 4 percent each decade. After exhaustive study, scientists are confident that all of this CO_2 added to the atmosphere is causing an increased greenhouse effect, warming Earth's surface (**Figure 6.16**). As more of the incoming solar radiation is trapped, more energy is added to Earth's atmospheric and ocean systems. The result will be an intensification of climate change. *Climate* is long-term (decades-long) behavior of weather systems. While Earth as a whole is getting warmer, humanity can expect long-term changes in regional rainfall, snowfall, ice coverage, the number and intensity of storms, and so on.

It is currently impossible to predict with certainty how severe the change in climate will be. Because climate affects everything from agricultural production to transportation, human civilization is sensitive to climate change. Efforts to halt continued human-driven climate change, as well as to deal with effects we cannot change, are likely to play an important part in the next century.

"Earth is *not just* a bar magnet. Something else is going on."

Our Magnetic Blanket: The Magnetosphere

"Earth has a pretty strong magnetic field," says Tarduno, "just like a number of other planets in the Solar System." The field extends from deep in Earth's interior far out into space. The region of space where a planet's magnetic field exerts a strong influence is called the **magnetosphere**. Only recently have scientists come to understand how Earth's magnetic field is generated and the important role it plays in the life of the planet.

Overall, Earth's magnetic field has the same shape as a bar magnet—what scientists call a **dipole field**. It has a **north magnetic pole** and a **south magnetic pole**, connected by magnetic-field lines that bow outward (**Figure 6.17a**). (Magnetic-field lines are imaginary lines depicting the direction of a magnetic field.) Earth's magnetic field shows more variation than that of a simple bar magnet and its dipole field, however. At places on the

Figure 6.16 Temperatures at Earth's Surface
Reconstruction of Northern Hemisphere temperatures over 1,000 years. Since about 1900, the average temperature of the planet has risen. (The light blue shading surrounding the curve shows the range of uncertainty.) Evidence has tied rising temperatures to increases in atmospheric CO_2 concentrations that have occurred since the industrial revolution.

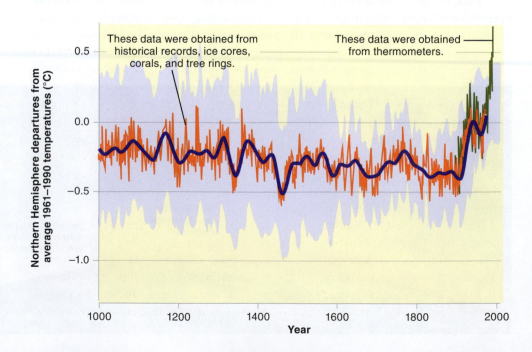

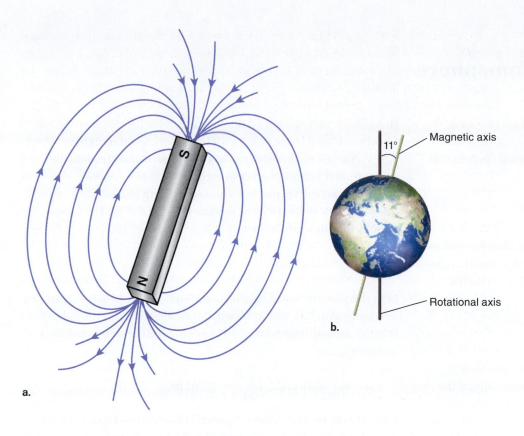

a.

 b.

Magnetic axis

11°

Rotational axis

Figure 6.17 Earth's Magnetic Axis

a. Earth's magnetic field is similar in structure to that of a simple bar magnet, including two poles, although it is much bigger and more complex. Earth's geographic north pole is actually a magnetic south pole. **b.** Earth's rotational and magnetic axes are not perfectly aligned with each other; they differ by 11°.

planet's surface called *magnetic anomalies*, the field is stronger or weaker than average by as much as a few percent.

In addition to magnetic anomalies, Earth's north and south magnetic poles are not aligned with the axis of Earth's rotation. The spin axis and magnetic axis are misaligned by 11° (**Figure 6.17b**). Even more important, the magnetic poles move. During the 20th century the north magnetic pole wandered 1,100 km. The origin of the anomalies and the movement of the magnetic poles are all critical clues to the processes that generate Earth's magnetic field. "Earth is *not just* a bar magnet," says Tarduno. "Something else is going on."

Earth's core is composed partly of liquid metal (mostly iron). Metal is an excellent conductor of electricity. When liquid metal moves, its flow becomes an electric current—the movement of charged particles—and produces a magnetic field. (Each current-carrying wire will generate a magnetic field. In your home, a magnetic field surrounds every wire with current running through it.) Thus, Earth's fluid metal core helps explain its magnetic field. Many astrophysical objects (galaxies, stars, planets, and moons) possess magnetic fields, and the mechanisms that create these fields are probably similar to what happens inside Earth. The moving, electrically charged fluid that generates a large-scale magnetic field in a planet, a star, or any other astronomical object is called a **magnetic dynamo**. "Magnetic anomalies tell us that Earth's dynamo is not steady," says Tarduno. "The flows at the core generate the fields by changing their patterns over time."

The most dramatic evidence for change in the dynamo and Earth's magnetic field consists of **field reversals**. On timescales of many hundreds of thousands of years, Earth's field loses its large-scale dipole pattern. The north and south magnetic poles disappear, only to reappear with their orientation reversed. A record of the field-flipping is preserved in the rocks composing the spreading seafloor at mid-ocean ridges. "You can see the field direction changing back and forth in the orientation of iron filaments in the seafloor rocks," explains Tarduno.

magnetosphere A region of space around a planet where the planet's magnetic field exerts a strong influence.

dipole field The magnetic-field configuration consisting of north and south magnetic poles connected by magnetic-field lines.

north magnetic pole The point on Earth's Northern Hemisphere at which the planet's magnetic field points vertically downward.

south magnetic pole The point on Earth's Southern Hemisphere at which the planet's magnetic field points vertically upward.

magnetic dynamo The flow of electrically charged fluid that can generate a large-scale magnetic field in a planet, a star, or any other astronomical object.

field reversal A reorientation of north and south poles within a magnetic field such that the two poles swap positions.

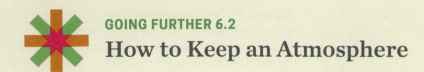

GOING FURTHER 6.2

How to Keep an Atmosphere

A particle's thermal velocity depends on its temperature and mass. Suppose that we measure the temperature T in kelvins (K) and the particle mass m in multiples of the hydrogen atom's mass (so m has no units). The thermal velocity V_t is then

$$V_t = 0.16 \text{ km/s} \sqrt{\frac{T}{m}}$$

For instance, nitrogen molecules are 28 times heavier than the hydrogen atom, so $m = 28$ in multiples of the hydrogen atom's mass. If a gas made of nitrogen molecules has a temperature of 350 K, its thermal velocity is

$$V_t = 0.16 \text{ km/s} \sqrt{\frac{350}{28}} = 0.57 \text{ km/s}$$

This equation helps explain why planets retain or lose their atmospheres (or at least some atmospheric constituents). Recall the concept of escape velocity from Chapter 3. If a planet's mass M and radius R are expressed in terms of multiples of Earth's values and G is the gravitational constant, the escape velocity V_e is

$$V_e = 11.2 \text{ km/s} \sqrt{\frac{M}{R}}$$

An atmospheric particle whose thermal velocity V_t is much higher than the planet's escape velocity V_e will eventually experience a collision that knocks it upward, and it will escape into space. Since different types of atmospheric particles have different masses, the lightest atmospheric constituents could have $V_t > V_e$, while heavier constituents have $V_t < V_e$. Therefore, the lighter atoms and molecules will eventually escape from the atmosphere, leaving only the heavier ones.

Imagine that a planet has a mass one-tenth that of Earth ($M = 0.1$) but a radius twice as large as Earth's ($R = 2$). The escape velocity for this world is then

$$V_e = 11.2 \text{ km/s} \sqrt{\frac{0.1}{2}} = 2.5 \text{ km/s}$$

Let's also assume that the planet has a temperature of 350 K. Nitrogen molecules with $V_t = 0.57$ km/s will not escape from its atmosphere. What about hydrogen with $m = 1$? Let's use the thermal-velocity equation again:

$$V_t = 0.16 \text{ km/s} \sqrt{\frac{350}{1}} = 3.00 \text{ km/s}$$

Since $V_t > V_e$ here, we would expect this world to end up eventually with a nitrogen-rich, hydrogen-depleted atmosphere.

An important point: Thermal velocity is just an average of all the velocities that an atom or molecule has at the atmosphere's temperature. Some particles will move more slowly than V_t, and some will move faster. Any particles moving faster than the escape velocity will eventually be lost, regardless of the average thermal velocity—so the atmosphere may slowly lose that type of particle.

Although the motion of charged particles is what generates magnetic fields, the influence works both ways: magnetic fields exert forces on charged particles. For example, the trajectory of an electron moving through a region with a magnetic field will be deflected by the field. Strong-enough fields deflect the electron's motion so completely that the particle ends up trapped by the magnetic field. If you imagine magnetic-field lines moving in one direction, a charged particle entering the field will move in a helix, rotating *around* the (imaginary) field lines while also traveling *along* the field lines (**Figure 6.18**). Understanding these basic features of magnetic phenomena provides the building blocks for understanding all planetary magnetic fields, including Earth's.

Although the terrestrial portion of Earth's magnetic field was discovered centuries ago, scientists did not know that the field extends far out into space until the dawn of the space age, when the satellite Explorer 1, launched in 1958, discovered strong currents and magnetic fields as it orbited the planet. The magnetosphere is the result of Earth's magnetic field interacting with the solar wind—the stream of energetic particles flowing off the Sun. Because of this interplay between solar wind and terrestrial field, the magnetosphere is not symmetric. The solar wind exerts pressure on the magnetic field, driving

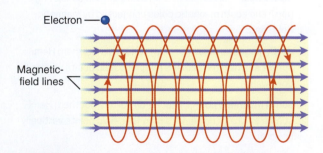

Figure 6.18 Electron Moving in a Magnetic Field
An electron that enters a magnetic field becomes trapped by the field and takes a helical path around the field lines while continuing its forward motion in the direction of the field lines.

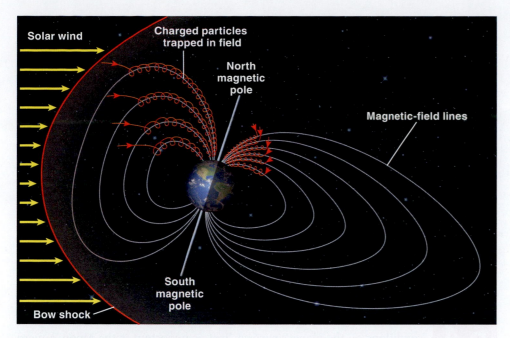

Figure 6.19 Earth's Magnetosphere
Earth's magnetic field extends far into space and interacts with the solar wind at a region called the magnetosphere. The magnetosphere shields Earth by deflecting charged, high-energy particles in the solar wind away from the planet. Particles that are not deflected are trapped in Earth's field lines and spiral toward the poles.

Figure 6.20 Auroras VIS
An aurora is the result of charged particles from the Sun becoming trapped in Earth's magnetic field.

it into a teardrop shape. The magnetosphere extends only about 70,000 km on the side facing the Sun, but it can extend 20 times farther on the side facing away from the Sun **(Figure 6.19)**.

The distortion of Earth's magnetic field by solar wind shows us that the field acts as a planetary buffer against the powerful and deadly stream of solar particles. As charged particles from the Sun flow past Earth, its magnetic field deflects some of them via a **bow shock** (Figure 6.19). "The bow shock is a lot like the bow wave that forms in front of a fast-moving boat," says Tarduno. Particles that are not deflected away from Earth at the bow shock are trapped in Earth's field. They spiral around the field lines and simultaneously move toward the poles. Collisions with these charged particles drive the atoms and molecules in the atmosphere at the poles into excited states. As they drop back to lower levels, they emit photons, which produce the glowing **auroras** visible in the polar regions (Figure 6.20).

When powerful solar storms occur, blasts of matter and magnetic fields from the Sun erupt into space. If these eruptions hit Earth, they can trigger powerful "space storms" in the magnetosphere, leading not only to bright auroras but also to dangerous torrents of energetic charged particles. These high-energy particles can destroy satellite electronics, pose a serious radiation threat to orbiting astronauts, and, on rare occasions, lead to so much current flowing to Earth that power grids are quickly overloaded. In 1989 a space storm sent enough current surging into the atmosphere that it took only 90 seconds to bring down Quebec's electric grid, leaving more than a million people without electricity.

Earth's magnetosphere may have played an important role in the early history of life on our planet. If energetic solar wind particles routinely reached Earth's surface, the damage they would have caused to biological molecules could have increased the rate of genetic mutation, wreaking havoc on living systems. Thus, we may all have the magnetosphere to thank for our existence.

bow shock The boundary where a strong flow runs into an obstacle.

aurora The emission of light in the upper atmosphere, visible at higher latitudes, caused by charged particles from the solar wind flowing along Earth's magnetosphere.

Figure 6.21 High and Low Tides
The Bay of Fundy, on the Atlantic Coast border of the United States and Canada, exhibits an extreme difference between **a.** low tide and **b.** high tide.

tide The effect of differential gravity on an extended object, leading to distortion of spherical mass distributions, among other effects.

differential force of gravity The different strengths of the gravitational force from a massive body that different parts of an extended object "feel" because the force decreases as the inverse square of the distance between them.

tidal bulge A distortion from spherical symmetry that is caused by tidal force.

spring tide A tide that is higher than average because it is reinforced by the linear alignment of Sun, Earth, and Moon at the new and full Moon phases.

neap tide A tide that is lower than average because the Sun's tidal force partially cancels out that of the Moon at the Moon's first and third quarter phases.

tidal force Force acting on an object that is due to differential gravity from a second object.

Section Summary

- The layers of atmosphere from lowest to highest altitude are the troposphere, stratosphere, mesosphere, and ionosphere. Their density continuously decreases with altitude, but temperature varies with altitude because of other factors.
- Earth's original atmosphere, composed of light gases, was probably lost to space. The current atmosphere was likely created from a combination of incoming materials, outgassing, and the activity of life.
- Increased combustion of fossil fuels has increased atmospheric greenhouse gases (including CO_2), contributing to climate change as Earth heats up.
- Earth has a dipole magnetic field—with magnetic-field lines connecting the north and south magnetic poles—generated by a magnetic dynamo.

CHECKPOINT **a.** List the layers of Earth's atmosphere and describe their features. **b.** Describe the relationship between the Earth's core and its magnetic field.

The Closest Desolation: Earth's Moon

6.4 The magnetosphere is not the only entity in space close to Earth that affects conditions on the planet's surface. Just as important is the Moon, a mere 384,400 km away, which exerts an enormous influence on Earth and its daily life.

Tides and the Earth-Moon Connection

Anyone spending a day at the beach can notice the periodic rise and fall of Earth's oceans. Walk along the shoreline in the morning, and the waves will come only so far up the beach. Take the same walk a few hours later, and the waves will roll in much farther. Carry out the same experiment a few weeks later, and the timing of the risings and fallings will have shifted. This is the phenomenon of **tides**—a twice-daily change in the local height of sea level—with regular modulations over periods of months to a year.

In the middle of the ocean, the typical variation in sea level is about a meter. The impact of this change on a particular coastline—how far it moves any waterline inland—depends on the slope of the coast and other factors. If the topology of the land is such that water is funneled through channels, the tidal variation can be extreme. Tides in the Bay of Fundy in Nova Scotia, Canada, where the seawater must flow through a narrow causeway, reach daily variations of 17 meters—large enough for ships to be grounded during low tide (**Figure 6.21**).

The tides have been known since prehistory. (The first known sea voyages were braved in canoes some 20,000 years ago.) Understanding the cause of the tides, however, had to wait for Isaac Newton and his theory of gravity. Recall from Chapter 3 that the strength of the gravitational force is inversely proportional to the distance between two objects. If the objects are *extended*, meaning they are not simply tiny points of mass, then astronomers must also consider the differences in force from one side of the object to the other. The origin of Earth's tides lies in this concept of the **differential force of gravity** due to the Moon. Jonathan Lunine explains how tides work. "A small particle that's located on the surface of Earth, facing the Moon, is going to feel a stronger gravitational force from the Moon than a particle that's at the center of Earth or a particle that's all the way on the other side of Earth. It's the differential force of gravity that matters in making tides."

Ocean water on the side of Earth closest to the Moon feels a stronger gravitational pull than that on the side of Earth farthest from the Moon. Thus the Moon creates a **tidal bulge** of water on Earth. There are two bulges at once: one pointing toward the Moon and one pointing away. The second bulge may seem counterintuitive, given the preceding arguments about differential force. But just as Earth's seawater on the Moonward side is pulled toward the Moon, Earth itself is pulled a little toward the Moon, relative to the ocean on the opposite side of the planet. The differential force of gravity means that the oceans on the side of Earth facing away from the Moon are "left behind," creating a second tidal bulge. As Earth rotates once each day, the tidal bulges maintain their relative positions, pointing toward *and* away from the Moon. Thus, Earth rotates under the tidal bulges, which, from the perspective of the land, become giant waves flowing around the planet twice a day (Figure 6.22).

The Moon is not the only astronomical body controlling the tides. Although the Sun is about 400 times farther from Earth than the Moon is, the Sun is 27 million times more massive than the Moon. Since the force of gravity depends directly on an object's mass and inversely on the square of its distance, the Sun still exerts a strong differential force on Earth's oceans. "Because the Sun is so massive," Lunine explains, "we now have a very complex situation where we have actually two bodies raising the tides." Thus, variations in the height of the daily tides depend on the positions of Earth, Sun, and Moon. When the Moon is either new or full, the Sun and Moon line up relative to Earth, and their combined gravitational effects produce large tidal variations called **spring tides** (Figure 6.23a). When the Sun and Moon are at right angles to each other (when the Moon is in its first or third quarter phase), the tidal effects of the Sun and Moon tend to cancel each other partially, creating weak tidal variations called **neap tides** (Figure 6.23b).

"**Tidal forces** shape a lot of the behavior we see in the Solar System," says Lunine. Consider a local example: the direct, long-term consequences that the Moon and its tidal forces have had on the evolution of Earth's day.

Earth rotates on its axis once every 24 hours—but this was not the length of the day 20 million years ago. Nor will it be the length of the day 20 million years in the future. Earth's rotation, and hence the length of its day, have gradually slowed down since Earth formed. Fossil records of animals such as coral, whose daily growth cycles create patterns that can be examined today, show that rather than 365 daily cycles in a year, there used to be as many as 400 day-night cycles per year. The culprit slowing Earth's day down is again the tidal effect of the Moon (and to a lesser degree the Sun).

Although the reasons for this deceleration of Earth's spin are complicated, one way to understand them is to think about tidal bulges within Earth itself. Differential gravity not only raises a tidal bulge in the oceans but also deforms the crust, the mantle, and the entire Earth. More important, these tidal bulges do not line up directly with the Moon. Instead, Earth's rotation keeps the tidal bulge slightly "ahead" of the imaginary line connecting Earth's center with the Moon's center. That means there is always a small but important component of gravitational force between the Moon and Earth's tidal bulge.

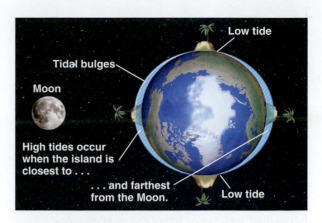

Figure 6.22 Tidal Bulges
The Moon causes two tidal bulges in Earth's oceans. The first occurs on the side facing the Moon as the oceans pull toward the Moon; the second occurs on the far side as Earth itself pulls toward the Moon. Earth's rotation leads to two high tides each day (shown schematically for an island, viewed from above the North Pole).

"Because the Sun is so massive, we now have a very complex situation where we have actually two bodies raising the tides."

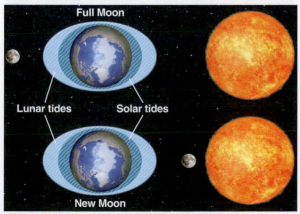

a. Spring tides

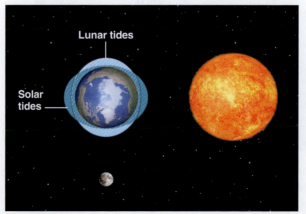

b. Neap tides

Figure 6.23 Spring and Neap Tides
a. Tidal effects are greater when the Sun and Moon are aligned.
b. Tidal effects are smaller when the Sun and Moon form a right angle relative to Earth, as the tidal effect of the two bodies on Earth tend to cancel each other out.

JONATHAN LUNINE

Jonathan Lunine grew up a few blocks from New York's Hayden Planetarium. "My parents took me there all the time." Tragedy, however, almost steered him away from astronomy. "I was 14 or 15 when my father died, and I began to get this guilt complex that I needed to help save the world. I decided I was not going to become an astronomer but instead would become a medical doctor."

Lunine asked his doctor for medical school catalogs. "He talked me out of it," Lunine recalls. The doctor offered him a ticket to a talk by astronomer Frank Drake. "After Frank Drake was done lecturing, I remember coming back down in the elevator and thinking, 'I really want to be an astronomer, not a doctor.'"

This force pulls back, *against* the direction of Earth's spin, decreasing its rate of rotation and slowing the length of the day (**Figure 6.24**). The effect is small, only 0.0017 second every century, but over time it adds up. In 500 million years, the expression "there are only 24 hours in the day" will have to be changed—each day will be about two and a half hours longer.

The same tidal interaction between Earth and the Moon that increases Earth's day length also leads to a slow increase in the size of the Moon's orbit. Using reflectors placed on the Moon by *Apollo* astronauts, scientists have measured precise changes in the Moon's orbit. They found that the Moon actually gets farther away from the Earth by 3.8 cm every year.

Another example of how one body tidally affects the rotation of another is the dramatic effect Earth has had on the Moon's rotation. According to Newton's third law (Chapter 3), for every force, there is an equal and opposite force. Thus, while the Moon raises tides on Earth, Earth is also raising tides on the Moon. Since Earth is far more massive than the Moon, it has a greater effect in distorting the Moon's shape and creating tidal drag. The effect of Earth's differential gravity on the Moon's spin is so strong that our satellite's rotation has been permanently altered. In an effect known as **tidal locking**, the Moon now rotates only once for every orbit that it completes around Earth (**Figure 6.25**). "The perfect match between the Moon's rotation and its orbit is entirely a tidal effect," says Lunine. "It's Earth's gravity acting on bulges in the Moon, braking the Moon's rotation." Tidal locking, a process by which larger bodies exert tidal effects on smaller orbiting ones, occurs throughout our Solar System.

From the creation of ocean tides to changing day length, the gravitational effect of having a relatively large moon has helped shape the environment in which Earth's life evolved. As we will see, the Moon had other effects on life as well, including keeping the orientation of Earth's spin axis (and therefore its climate) from changing in ways that might have been dangerous to life. In many ways, Earth's Moon has benefited Earth's life.

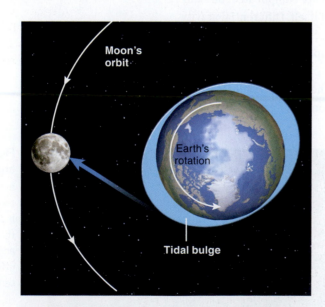

Figure 6.24 The Moon's Effect on Earth's Rotation
The Moon's tidal forces have changed Earth's rotation speed slightly over long periods of time. The Moon pulls on the tidal bulge, slowing Earth's rotation and lengthening the day.

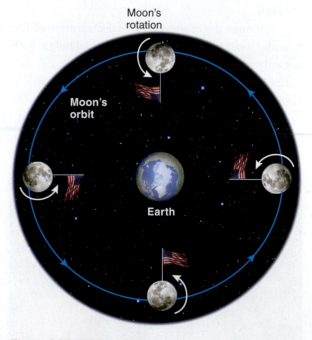

Figure 6.25 Tidal Locking
The Moon's rotation period and orbital period are equal. Because of this tidal locking, the same side of the Moon always faces Earth.

a.

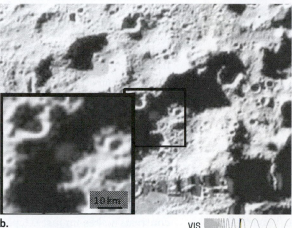

b.

VIS

Figure 6.29 *LCROSS* Mission
a. An artist's rendering of the *LCROSS* spacecraft and Centaur rocket, which NASA purposely crashed into a shadowed crater to search for water on the Moon. **b.** The collision site, appearing as a cloudy spot seen 20 seconds after impact. Observers on Earth who analyzed the debris plume did indeed find proof of water on the Moon.

Inside the Moon

"Of course, we know less about the Moon's interior than we do about Earth's," says Lunine. "Much of what we know comes from the *Apollo* missions. The astronauts left seismic detectors on the lunar surface [**Figure 6.30**]. Then, NASA generated artificial earthquakes by sending spent rocket stages to crash into the Moon's surface." Using the S-wave and P-wave data returned from the lunar seismic stations, scientists have been able to build a fairly complete picture of the Moon's interior.

The structure of the Moon is, in many ways, similar to that of Earth. Both bodies have a metal core that is mostly iron, with smaller amounts of elements such as oxygen and sulfur. The Moon may have a two-layer core like Earth's, with fluid iron in the outer zone and solid iron in the inner zone, but the Moon has no large-scale organized magnetic field like Earth's. What does exist is weak, indicating that no dynamo process is ongoing in the core (**Figure 6.31**).

VIS

Figure 6.30 Moonquakes
Apollo 11 astronaut Buzz Aldrin, beside a seismic detector placed on the Moon to record potential moonquake activity or impacts from asteroids on the Moon's surface.

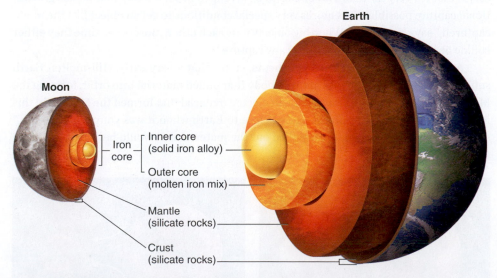

Moon

Earth

Iron core

- Inner core (solid iron alloy)
- Outer core (molten iron mix)
- Mantle (silicate rocks)
- Crust (silicate rocks)

Figure 6.31 Comparison of Earth and Moon Interiors
The interiors are strikingly similar (although the Moon's iron core is smaller proportionally), suggesting that Earth and the Moon originated from the same material and providing clues to the Moon's formation. It is not definitely known if the Moon's outer core is molten.

lunar highlands Old, heavily cratered portions of the Moon's surface lighter in color than the maria.
Late Heavy Bombardment A hypothetical epoch, 4.1–3.8 billion years ago, during which a large number of asteroids collided with bodies in the inner Solar System.

MIKI NAKAJIMA

When she was a little girl, Miki Nakajima didn't want to study the Moon; she wanted to draw comics. "I really like manga, and I wanted to be a comic artist," she says. "But I thought it would be really competitive."

Though her love for *Dragon Ball* was firm, once Nakajima reached college her real decision became a career in physics versus one in Earth science. "I kind of chose both," she explains. Her research eventually took her to geophysics, which uses the methods of physics to study how planets evolve. "I was really happy with my choice," says Nakajima. "It's not just the science, but also the people that I work with that really makes it fun."

"There's been a long debate about the origin of Earth's Moon."

"The big difference between Earth and the Moon," says Lunine, "is the size of the core." The Moon's core constitutes only 20 percent of its radius, compared with a full 45 percent for Earth. "That fact immediately tells us something important," says Lunine. "The Moon is greatly impoverished in iron relative to Earth. It's a clue we can use to understand the Moon's history."

The Moon's mantle layer, like Earth's, is composed of rocks under high enough pressure for the mantle to flow slowly, but there are no longer large-scale circulation patterns in the Moon's mantle. Regular, though weak, moonquakes have been detected and are believed to originate deep within the mantle. They occur every month or so, indicating that the tidal stresses associated with Earth continue to affect the Moon's interior. The mantles of both the Moon and Earth are low in iron relative to what is found in primitive meteorites, suggesting that both Earth and the Moon differentiated as they formed.

Above the lunar mantle lies a rigid layer of crust. Unlike Earth's crust, where the ocean floors and continents are composed of different materials, the Moon's crust shows no such difference. "On the Moon," says Lunine, "there's just a single type of crust, which probably has not changed much since its original formation." The Moon's crust is quite thick relative to its size, making up 3 percent of the Moon's total radius. Earth's crust, in contrast, makes up just 0.6 percent of the planet's radius. In addition, Earth's crust has been shaped by plate tectonics and, at the surface, by weathering, while the Moon's crust has been shaped primarily by volcanism and impacts.

Making Earth's Moon

"There's been a long debate about the origin of Earth's Moon," says Miki Nakajima, a scientist who uses sophisticated computer simulations to study how the Moon formed. "The *Apollo* missions brought back a few hundred kilograms of Moon rocks, and that helped a lot in getting the right story." Rock samples returned by both American and Russian missions were chemically analyzed, giving scientists the hard data they needed to track the Moon's history all the way back to its formation.

One of the earliest models of the Moon's formation was the **capture hypothesis**, which argued that the Moon formed somewhere else in the Solar System but was captured as it neared Earth. "The idea was that the Moon was flying past Earth and then got caught by Earth's gravity," says Nakajima (**Figure 6.32a**). One persistent problem with this hypothesis is that it's very hard for two bodies to be arranged in such a way that makes gravitational capture possible. "It requires very special conditions to get an object like the Moon captured," says Nakajima. "When big objects pass each other, most of the time they either collide or just pass each other without any capture."

A second idea, the **fission hypothesis**, states that a very early, still-molten Earth suffered a close encounter with another body that pulled material into orbit, forming the Moon or a rapidly spinning molten Earth ejected material that formed the Moon. "In this idea, another planet-sized object passed close to Earth when it was young and molten," says Nakajima. "Then tidal forces pulled out the material that would become the Moon"

Figure 6.32 Hypotheses of Moon Formation
Two early explanations of how Earth acquired the Moon: **a.** According to the capture hypothesis, Earth's gravity snared a passing Moon-sized object. **b.** According to the fission hypothesis, rotation pulled material away from a molten Earth and into orbit.

a. Capture hypothesis

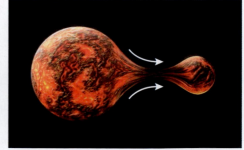

b. Fission hypothesis

(**Figure 6.32b**). This hypothesis, however, is inconsistent with scientists' understanding of Earth's history, as our planet was probably never fully molten. Although a magma ocean may have covered Earth's surface early in its history, the planet became mostly solid shortly after its formation.

Other ideas, such as the Moon condensing from material orbiting around the young Earth, have also been proposed. "People have also proposed that the Moon material was ejected from a rapidly spinning young Earth," says Nakajima. This proposal too runs into difficulties with facts now known about the planet's early history. "Earth was probably never spinning fast enough to throw off a Moon-sized amount of mass," explains Nakajima.

Despite their difficulties, each hypothesis on its own provides a piece of the puzzle that the modern theory of lunar origin has embraced.

The model that is generally accepted today arose from two striking clues provided by the lunar missions. The first clue was that rocks on the Moon contain very little iron compared with rocks on Earth. The second clue came from lunar seismic studies showing that the Moon's core accounts for a much smaller fraction of total volume than does Earth's core. Together these data pointed to a Moon that is seriously depleted in iron. Lunar rocks are also low in elements, such as potassium, that are often found in Earth minerals. The lack of lunar potassium was an important discovery because potassium has a low melting temperature—so it can be considered a volatile material that is easily vaporized. The lunar samples did, however, show some chemical similarities to Earth rocks. The relative abundance of oxygen isotopes—forms of oxygen with more or fewer neutrons in the nucleus—was the same in Moon rocks as in Earth rocks.

After several years of analyzing the *Apollo* samples, scientists found a coherent story that fit these data. "In the classic model people work with now," explains Nakajima, "Earth gets hit with a very large body but the collision is not direct." The idea is that Earth and a Mars-sized body, both of which had already differentiated into a mantle and core, underwent a glancing collision. As a result, a piece of Earth's crust and mantle was torn out but the core was missed. "Material from the collision would have the right ratio of elements," explains Nakajima. Because it came from the mantle, the matter torn off in the collision would have been iron-poor. Mantle material blown off Earth by the impact would also have been very hot, so the volatile elements (including potassium) would have evaporated away quickly. The ratios of oxygen isotopes, however, would have been unaltered from what is in Earth's mantle. "Most of the material from the collision goes into orbit around the Earth and forms a disk," explains Nakajima. "Eventually the disk coalesces into a young Moon."

This basic idea, called the **impact hypothesis** (**Figure 6.33**), accounts for much of what is known about lunar structure and chemistry (see **Anatomy of a Discovery** on p. 174).

> "Impacts may have played a role in a lot of what happened in shaping the solar system."

capture hypothesis A hypothesis of the Moon's origin stating that the Moon formed elsewhere and was captured intact by Earth's gravity.

fission hypothesis A hypothesis of the Moon's origin stating that the Moon formed from material pulled into orbit from the still-molten Earth by a close encounter with another body.

impact hypothesis The currently accepted hypothesis of the Moon's origin, stating that a large, Mars-sized body struck Earth and forced mantle material into orbit, which gravitationally coalesced into the Moon.

Figure 6.33 The Impact Hypothesis
The generally accepted explanation of the Moon's formation: Earth collided with a Mars-sized body, causing a segment of Earth (part of the crust and mantle) to be torn away. This material then formed a ring that eventually coalesced into the orbiting Moon.

Where did the Moon come from?

Moon
Density: 3,344 kg/m³

Earth
Density: 5,514 kg/m³

observation

The origin of Earth's relatively large Moon is tied to the processes that formed the Solar System but remains unsettled. Early on, Earth and its Moon were assumed to have formed in close proximity when the Solar System was young. However, the two differ significantly in density and composition (the Moon has relatively little iron). Why would objects that formed together be so different?

competing hypotheses

In the late 1800s, **George Darwin,** son of Charles, formulated the *fission hypothesis*: a molten Earth rotated so rapidly that a blob of matter spun off, forming the Moon.

In the late 1940s, **Harold Urey** devised the *capture hypothesis*: the Moon formed earlier and in a different location than Earth, which later captured it.

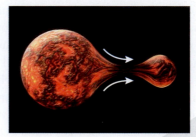

Fission hypothesis

Capture hypothesis

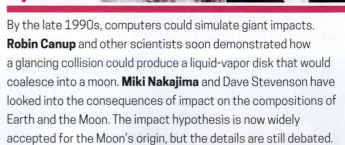

hypothesis

Neither the fission hypothesis nor the capture hypothesis accounts for known differences in composition, and it was proved unlikely that Earth could have captured a Moon-sized object. So, in the 1970s, **William Hartmann** and **Donald Davis** suggested the *impact hypothesis*: an impact between two large planetesimals can result in two bodies with very different densities.

Impact hypothesis

supported by model

By the late 1990s, computers could simulate giant impacts. **Robin Canup** and other scientists soon demonstrated how a glancing collision could produce a liquid-vapor disk that would coalesce into a moon. **Miki Nakajima** and Dave Stevenson have looked into the consequences of impact on the compositions of Earth and the Moon. The impact hypothesis is now widely accepted for the Moon's origin, but the details are still debated.

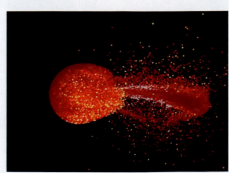

Nakajima-Stevenson simulation

One critical trial of the impact hypothesis is to figure out what kinds of impacts could make the idea work. Almost all impacts would either cause the material to be ejected, leaving Earth's orbit, or cause most of the material to fall back to Earth. Nakajima has spent years exploring the outcome of the impact hypothesis, using computer models, and she has seen both its successes and challenges. "With impacts you can get a lot about the Earth-Moon system right," says Nakajima. "That includes the angular momentum in the system and the composition of elements. But there are still many details that seem to require starting with just the right kind of initial setup."

The acceptance of the impact hypothesis was an important turning point, not only in lunar studies but in all planetary science. "Once people accepted that the Moon formed from a big impact," says Nakajima, "it opened the door to seeing that impacts may have played a role in a lot of what happened in shaping the Solar System."

Section Summary

- The Moon causes two tidal bulges on Earth; they manifest as ocean tides. Earth's gravity has tidally locked the Moon, making the Moon's periods of rotation and orbit equal.
- Asteroids and meteoroids that impacted the Moon billions of years ago created craters as well as maria. There is strong evidence of water delivered to craters via impacts.
- The Moon may have a two-layer metal core, a mantle, and a crust but no atmosphere. Its crust is much thicker relative to its size than Earth's, and the Moon has no tectonic plates.
- The Moon's small iron core and lack of iron in its crust support the impact hypothesis of the formation of the Moon.

CHECKPOINT Compare the different theories for the Moon's origin. What is the evidence for and against each one?

CHAPTER SUMMARY

6.1 Discovering Change: Arctic Crocodiles and the Earth-Moon System

The relative sizes of Earth and the Moon are unusual for our Solar System; most planets have much smaller satellites. The Moon's influence on Earth has played a large role in the evolution of life.

6.2 Earth Inside and Out

Earth's layers include the two-part core, the mantle, and the crust. The layers below the crust (which is 60 km at its deepest) have been identified through the study of two kinds of seismic waves that propagate through Earth: P-waves and S-waves. Most of the mantle's rock is so hot and under such pressure that it flows at a rate of centimeters per year. Earth's lithosphere (crust and solid upper mantle) floats on the deformable part of the mantle, and its movements, known as plate tectonics, are driven by convection in the mantle. Earth's layered structure is evidence that our planet differentiated early in its history.

6.3 Earth's Near-Space Environment

Earth's atmosphere is primarily nitrogen and oxygen. Its layers, from the planet's surface upward, are the troposphere, stratosphere, mesosphere, and ionosphere. Their density decreases with altitude, but temperature variations are more complex. Much of the Sun's energy directed toward Earth passes through the atmosphere. Some is reemitted outward, but some is absorbed by greenhouse gases. This greenhouse effect causes Earth's temperature to be higher than it would be without an atmosphere. The increase of greenhouse gases since the industrial revolution is raising Earth's average temperature. Earth has a magnetic field created by the combination of its rotation and its liquid core, as well as a magnetosphere that is shaped by solar wind and helps protect the planet from harmful radiation and high-energy particles.

6.4 The Closest Desolation: Earth's Moon

On Earth, tidal forces from the Moon create a tidal bulge that is observed primarily as daily cycles of ocean level changes. The Sun also exerts tidal effects on Earth, magnifying or reducing tide levels based on the relative positions of Moon and Sun. Tidal effects have lengthened Earth's rotational period and have locked the Moon's rotation period equal to its orbital period. The Moon's surface lacks Earth's weathering and plate tectonic effects, so impact craters remain highly visible. The near side of the Moon has smooth maria; the far side is more uniformly cratered. Evidence shows that the Moon contains water delivered via comet and asteroid impacts. The Moon has the same general internal structure as Earth, and their relative sizes and compositions provide evidence of the Moon's formation. In the currently accepted hypothesis, a large planetesimal collided with Earth in a glancing blow, ejecting crust and some mantle; the orbiting debris then coalesced into the Moon.

QUESTIONS AND PROBLEMS

Narrow It Down: Multiple-Choice Questions

1. Which of the following does *not* describe an effect that the Moon has had on Earth?
   a. limiting the wobble in Earth's spin
   b. lengthening Earth's day
   c. creating the tides
   d. stabilizing Earth's climate
   e. decreasing Earth's orbital speed

2. Which of the following items describe(s) features of Earth believed to be unique among the eight planets in our Solar System? Choose all that apply.
   a. presence of life
   b. liquid water on the surface
   c. solid surface
   d. presence of one or more moons
   e. presence of an atmosphere

3. Where is solid material found in the layers of Earth's structure?
   a. crust only
   b. continents only
   c. crust, upper mantle, and inner core
   d. crust and outer core
   e. mantle and core

4. In which of the following ways are Earth and the Moon similar? Choose all that apply.
   a. length of orbital period with regard to the Sun
   b. presence of similar structural layers
   c. relative sizes of structural layers
   d. length of day
   e. geologic activity

5. The distance from New York City to San Diego, CA, is 3,940 km. If you traveled downward through Earth that distance, where would you find yourself?
   a. in the crust
   b. in the mantle
   c. in the liquid core
   d. in the solid core
   e. outside Earth

6. The motion of cars in stop-and-go traffic is similar to which kind of wave?
   a. S-wave
   b. P-wave
   c. both S-wave and P-wave
   d. neither S-wave nor P-wave
   e. It is not a wave.

7. When volcanoes erupt, the spewed material comes from which layer(s) of Earth? Choose all that apply.
   a. crust
   b. mantle
   c. molten core
   d. solid core
   e. ocean floor

8. Which of the following is/are considered evidence of tectonic plates? Choose all that apply.
   a. matching fossils in separated landmasses
   b. the shapes of the continents
   c. convection in the mantle
   d. observed motion of the continents
   e. distribution of hurricanes

9. Arrange the layers of Earth's atmosphere in the correct order of increasing altitude.
   a. troposphere, stratosphere, mesosphere, ionosphere
   b. ionosphere, mesosphere, stratosphere, troposphere
   c. stratosphere, troposphere, mesosphere, ionosphere
   d. troposphere, ionosphere, mesosphere, stratosphere
   e. stratosphere, mesosphere, troposphere, ionosphere

10. Earth's first atmosphere escaped because of
    a. the planet's low density.
    b. the planet's high temperature.
    c. lack of O_2.
    d. the high velocity of Earth's spin.
    e. violent winds.

11. The temperature at Earth's surface directly depends on which of the following? Choose all that apply.
    a. the amount of energy received from the Sun
    b. the presence of radio waves in the Sun's spectrum
    c. the concentration of CO_2
    d. the concentration of CH_4
    e. the presence of cosmic rays at the surface

12. The Moon's surface shows more cratering than Earth's because
    a. it was hit more than Earth because of its location.
    b. it formed earlier than Earth, when more debris was present in the Solar System.
    c. its surface is softer and thus less resistant to cratering.
    d. the forces that resurface Earth do not exist on the Moon.
    e. Earth deflected Solar System debris, which then bombarded the Moon.

13. Which of the following features may occur when an object impacts another body, causing a crater? Choose all that apply.
    a. rays
    b. a peak
    c. terracing
    d. melted crust
    e. dust in the atmosphere

14. Which of the following statements regarding magnetic fields is/are true? Choose all that apply.
    a. Earth's North Pole has always been the magnetic north pole.
    b. The presence of a magnetic field implies the presence of a solid iron core.
    c. Any large, rotating body has a significant magnetic field.
    d. Earth's magnetic field protects it from solar wind.
    e. Earth's magnetosphere is symmetric.

15. Which of the following statements about the Moon is true?
    a. The side of the Moon not visible from Earth is always dark.
    b. The two hemispheres of the Moon (near and far) are symmetric in structure and similar in appearance.
    c. Maria appear only on the near side of the Moon.
    d. Most of the Moon's craters have disappeared because of weathering effects.
    e. An impact crater on the Moon always takes the exact shape of the object that created it.

16. The strongest tides occur at which Moon phase(s)? Choose all that apply.
    a. first quarter
    b. waxing crescent
    c. full Moon
    d. third quarter
    e. new Moon

17. If the rate of change in the length of Earth's day remains constant at its current level, what will the day length likely be in 100 million years?
    a. 24 hours
    b. a little longer than 24 hours
    c. a little shorter than 24 hours
    d. many times longer than 24 hours
    e. The value cannot be determined from the available data.

18. If Earth were tidally locked with the Sun but the Moon's orbital period remained as it is now, what would be the frequency of high lunar tides experienced on Earth?
    a. once per year
    b. twice per year
    c. once per lunar orbit
    d. twice per lunar orbit
    e. The frequency cannot be predicted from the information given.

19. True/False: If the Sun were cooler but Earth's average temperature was the same as it is today, a weaker greenhouse effect could be the cause.

20. True/False: The greenhouse effect has been detrimental to life on Earth throughout its evolution.

To the Point: Qualitative and Discussion Questions

21. Describe the two kinds of seismic waves.

22. How do scientists use seismic waves to "see" inside planets and moons?

23. What evidence indicates that tectonic plates continue to move today?

24. If there were no convection in the mantle, how might Earth's geologic history have differed?

25. What are the possible outcomes when two tectonic plates interact?

26. How does Earth's magnetic field protect life on the planet?

27. How does Earth's Moon differ from the moons around other planets?

28. In which layers of Earth's atmosphere does temperature increase while density decreases? Why does that occur?

29. How long is a day on the Moon? In other words, how long does one cycle of day and night last?

30. Compare the layers of Earth and the Moon. How are they similar, and how do they differ?

31. What is required for a planet or moon to have a magnetosphere? Describe the shape of Earth's magnetosphere.

32. How are plate tectonics tied to the structure of Earth's interior?

33. What does the degree of cratering on the Moon's surface tell astronomers about the timing of events that shaped that surface?

34. What are the major three hypotheses regarding how Earth acquired the Moon? Describe the model that is generally accepted today.

35. How has Earth's atmosphere changed over Earth's history, and why?

Going Further: Quantitative Questions

36. Gravity decreases with the square of the distance between objects. How much greater is the force of the Moon's gravity on the side of Earth closer to it versus the opposite side of Earth? Use 384,000 km as the average distance from the Moon's center to Earth's center and 12,756 km as Earth's diameter. Give your answer as a percentage.

37. Calculate the escape velocity, in kilometers per second, of a moon with radius 2,173 km and mass 1.1×10^{23} kg.

38. The escape velocity of Earth's Moon is 2.38 km/s. The moon of a distant planet has a mass 2 times that of Earth's Moon and a radius 3 times that of Earth's Moon. What is its escape velocity, in kilometers per second?

39. By what factor would the gravitational force between Earth and the Moon be greater if the mass of each body were twice as great and the distance were half as great as they are today?

40. If a rock sample initially contained 5,000 uranium-238 nuclei 6.75 billion years ago, how many lead-206 nuclei would it contain today?

41. A rock sample is determined to have a concentration of uranium-238 that is consistent with 0.5 half-life having passed. What percentage of uranium-238 remains, and how old is the rock?

42. An analysis of a rock sample indicates that 12.5% of the expected concentration of uranium-238 remains. How many half-lives have passed, and how old is the rock?

43. A planet has a mass twice that of Earth and a radius 1.5 times Earth's. What is the planet's escape velocity? Is this planet likely to have retained hydrogen in its atmosphere if the temperature is 350 K?

44. Would oxygen be retained in the atmosphere of a planet whose surface temperature was 250 K and whose escape velocity was 4.6 km/s? (Molecular oxygen is 32 times heavier than a hydrogen atom.)

45. What is the thermal velocity of nitrogen in the atmosphere of a planet whose surface temperature is 380 K? Would it retain nitrogen if the planet had 3 times the mass of Earth and 4 times Earth's radius?

"ALL THAT WORK AND IT COMES DOWN TO FIVE MINUTES OF HELL"

7

Sibling Worlds

MERCURY, VENUS, AND MARS

Planet Stories

7.1 "All that work and it comes down to five minutes of hell," says Jim Bell of Arizona State University. Bell helped develop high-resolution stereo cameras for NASA's wildly successful *Spirit* and *Opportunity* Mars rovers. "The rover project started 10 years before the landings, and it was an enormous task. We knew whatever we built would have to survive the shock of launch, the shock of landing, and the harsh Martian environment." The camera development was just one part of building the complicated rovers.

On June 10, 2003, the rocket carrying *Spirit* lifted off the pad at Cape Canaveral, Florida. Seven months later, the truck-sized capsule that housed the rover dove into the Martian atmosphere. "We were all there in mission control as *Spirit* started its descent," recalls Bell. "It was crazy and exhilarating and terrifying all at the same time. You have a decade of your life wrapped up in this project. At that moment, you think, 'Oh my God, did I waste 10 years and it's all about to come to nothing?' And then a second later you're thinking, 'Oh, my God, I'm about to get a view of this world nobody's ever seen before!'"

Bell did not face his fear and excitement alone in the control room. "You are going through all this and there's CNN, MSNBC, and all the other news crews with the cameras trained on you. The whole blasted world was watching over our shoulders live to see if we screwed up."

They didn't screw up. Both rovers survived their landings (*Opportunity* arrived in 2004) and almost immediately began sending images from Mars. The rover mission, which was planned to last a mere 3 months, turned into a years-long exploration that finally ended in 2019. In 2012, *Opportunity* was joined on the Martian surface by the rover *Curiosity*, which had its own harrowing high-speed 7-minute ride through the atmosphere to nail a perfect landing (**Figure 7.1**).

The rovers that Bell helped build are another step in humanity's ongoing exploration of the Solar System over the last 50 years. After centuries of questions about the planets, we finally have enough answers to begin a true science of comparative planetology.

← Earth or Mars? Rovers exploring the surface of the red planet have found landscapes that look remarkably like the deserts of our world. This image, taken by the *Opportunity* rover, shows Cape St. Vincent, one of the cliffs of the Victoria Crater.

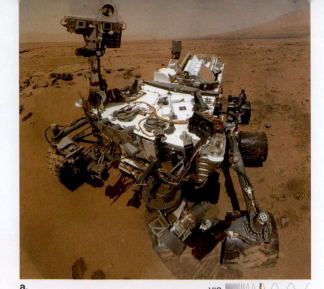

a.   VIS

b.   VIS

Figure 7.1 Mars Rover
Approximately 6 years after the rovers *Spirit* and *Opportunity* landed on Mars, **a.** a larger rover named *Curiosity* safely landed and began exploring the planet. **b.** A landscape image of the Martian surface captured by *Spirit*.

"The first billion or so years of a planet's life set the stage for much of what is to come."

thermal history The evolutionary pattern of heating and cooling of a planet.

planetary differentiation The separation by gravity of a planet into different layers, from densest at the core to least dense near the surface.

planetary cooling The process by which a planet loses the heat acquired during its formation.

cratering The change in a planet's surface features due to impacts with comets and asteroids.

magma flooding The creation of smooth lowlands via lava flows that fill in craters.

Comparative Planetology: Guiding Principles

Life as we know it requires liquid water. Hence the overarching goal of the Mars rover program was simple: Follow the water. Earlier missions revealed tantalizing hints that Mars—a barren, frozen desert world today—might once have been a water world with streams, rivers, and perhaps oceans on the surface. NASA has used the hunt for H_2O as the *guiding principle* in developing new Martian orbital satellites and the rovers.

To begin our study of comparative planetology, we too must establish guiding principles. Chapter 6 explored Earth and the Moon. Knowledge of our world and its satellite provides a foundation to ask about the other planets in our Solar System: How are they like Earth? How are they different? If we're not careful, though, we will quickly become lost in a sea of facts. Which facts are important, and how do they relate to one another? What questions and guiding principles will help us make sense of the data?

In light of the recent discovery of exoplanets (planets orbiting other stars), the first question and guiding principle for organizing our exploration of the Solar System might be, What do the planets in our Solar System tell us about the origins of planetary systems? A second question and guiding principle might be, What do those planets tell us about the evolution of planetary systems? And a third question and guiding principle might be, How well would each planet in our Solar System fare as a home for life? Extending this inquiry into the past and into the future, we ask whether, at any time in the past, conditions on a certain planet could have supported life. Similarly, we look toward the next few centuries and imagine which planets might be best suited for building permanent bases or colonies for human habitation. The presence of water in any accessible form is a related issue.

This chapter focuses on the inner "rocky" worlds of the Solar System: Mercury, Venus, and Mars. The discussion of each planet proceeds in four sections, starting at the center and moving outward:

1. *Interior structure.* What does the internal structure of the planet tell us about its origins and evolution? Does it have a core? Does it have a mantle?

2. *Surface evolution.* What surface features exist on the planet today, and what do they tell us about the planet's origin, evolution, and suitability for life? Is there volcanic activity? Is there plate tectonic activity? Is there evidence of water on or below the surface?

3. *Atmosphere.* Does the planet have an atmosphere? If so, how has it evolved over time, and how have the changes affected the surface? Was the atmosphere ever conducive to life?

4. *Near-space environment.* Does the planet have a magnetic field? Does the planet have satellites? Is there anything about the planet, such as its location or the shape of its orbit, that has affected its evolution?

With these questions and our guiding principles in hand, we are ready to begin. But before we turn to any one terrestrial planet, we can use what we learned from our study of Earth and the Moon to examine the evolution of rocky planets in general.

The Five Stages of Terrestrial Planetary Evolution

Since the 1970s, increasingly sophisticated data-gathering techniques have helped scientists understand that all terrestrial planets have shared certain key phases in their evolution. "The inner planets all condensed out of the hotter inner parts of the *protosolar disk*," says Jim Zimbelman, a planetary scientist in the Center for Earth and Planetary Studies at the Smithsonian National Air and Space Museum. "That means they ended up with silicates and metals rather than ices and gas. They were dense rocky worlds, and that shaped their future development in ways that led to very different evolution from that of planets further out."

The early years of a planet's life (like those of a human being's life) have a huge impact on its subsequent history. "The first billion or so years of a planet's life set the stage for much of what is to come," says Zimbelman. "One thing we know from basic principles is that small planets lose their heat faster than big planets, and a planet's **thermal history** drives its whole geophysical and geochemical evolution" (see **Going Further 7.1** on p. 183). Some scientists divide the evolution of the terrestrial worlds into five stages: differentiation, cooling, cratering, magma flooding, and weathering.

Planetary system formation begins when rocky planets are created by collisions between planetesimals. As these mountain-sized bodies collide, they build up a planet's mass. In the first key stage, **planetary differentiation** (Figure 7.2a), material is melted first by the violence of collisions and then by the force of gravity squeezing in on material in the newborn protoplanet. Heavy elements, such as iron, sink to the center of the newly formed world, and lighter materials, such as silicates, rise toward the surface. Says Zimbelman, "This process of differentiation is why all the terrestrial planets have a core-mantle-crust structure" (**Figure 7.3**).

The second evolutionary stage is **planetary cooling** (Figure 7.2b). "A planet's internal heat flows up to the surface and escapes into space," explains Zimbelman. The process is similar to what happens when freshly baked bread cools once it is out of an oven. "Some of the heat a planet has will be left over from its original formation. Some of it may be from the decay of radioactive materials. The process of differentiation itself—making a core—actually releases enormous heat. It takes time for all of that heat to work its way out as the planet cools." Since the cooling time depends on the planet's size (longer for large planets, shorter for small ones), the cooling stage may overlap with subsequent evolutionary phases.

The battering of a planet's surface by comet and asteroid strikes marks the third stage, **cratering** (Figure 7.2c). "The cratering story happens throughout the evolution of the planets," says Zimbelman, "but early on it is really intense." In our Solar System, most of the planetesimals that had not yet become part of newly formed planets were cleared away during the Late Heavy Bombardment phase of Solar System evolution (see Chapter 6). During this epoch, about 4 billion years ago, the gravity of the Solar System's young planets swept up many planetesimals, leading to intense cratering. Very early in the planets' lives, this bombardment may have kept the planetary surface partially molten.

The Late Heavy Bombardment phase is the period of most dramatic cratering, but even later large impacts remain a force that can reshape planetary surfaces. The effects of these later impacts are most dramatic if the planet is still hot enough to allow for volcanism and the existence of partially melted rock close to the surface. "The later impact phases weren't catastrophic like the Heavy Bombardment," says Zimbelman of our Solar System's history, "but there was a period when collisions could form big basins that were later flooded by lava. On the Moon this led to the maria we see today, but it happened on other worlds too." During this **magma flooding** stage (Figure 7.2d), smooth lowlands are created after asteroid collisions or through volcanism. If significant amounts of water (or other liquids) exist on the surface, a second kind of flooding creates oceans.

The last evolutionary stage that occurs on planets with atmospheres (and perhaps oceans) is the slow process of *weathering* (Figure 7.2e; see also Chapter 6). Over time, the flow of atmospheric gases, as well as their condensation and evaporation into ice or water, alters a planet's surface features. Weathering can both erase older features, including craters, and create new ones, such as extended regions of sand dunes. Even on planets without atmospheres, solar wind particles and the impact of smaller bodies can also change the surface over time.

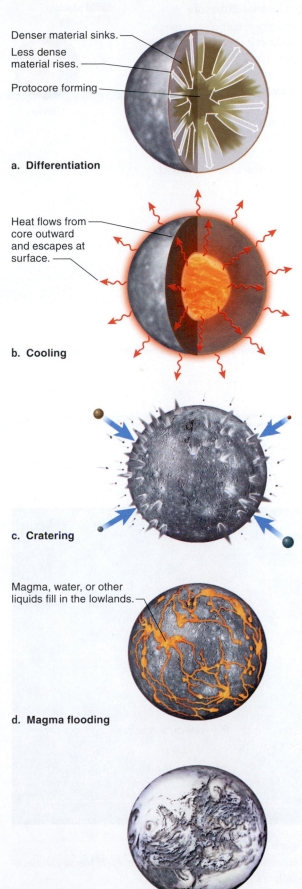

Denser material sinks.
Less dense material rises.
Protocore forming

a. Differentiation

Heat flows from core outward and escapes at surface.

b. Cooling

c. Cratering

Magma, water, or other liquids fill in the lowlands.

d. Magma flooding

e. Weathering

Figure 7.2 Evolutionary Stages of a Terrestrial Planet
a. During planetary differentiation, denser material sinks to the planet's core. **b.** During planetary cooling, heat from the interior escapes to space. **c.** During cratering, impacts with smaller bodies shape the surface. **d.** During magma flooding, magma from below and substances such as water fill in lowlands. **e.** During weathering, surface features are eroded by winds and liquid flows.

Figure 7.3 Differentiation and Interior Structure

After differentiating, each terrestrial planet is composed of similar layers of increasing density from surface to center—from crust to mantle to core (which may have both molten and solid portions). The layers' relative thicknesses differ for each planet, depending on the composition of the material from which each planet formed.

Mercury

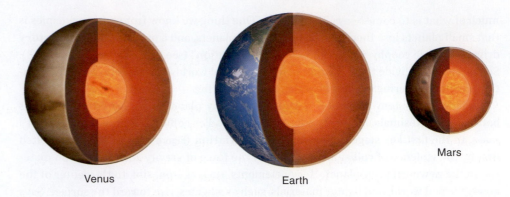

Crust
Mantle
Core

Venus   Earth   Mars

Section Summary

- We can consider three guiding principles of comparative planetology: the need to understand a planet's origin, its evolution, and its suitability for life.
- Comparative planetology focuses on four areas of comparison: planetary internal structure, surface evolution, atmosphere, and near-space environment.
- Scientists believe that all terrestrial planets experienced five basic stages in their history: planetary differentiation, planetary cooling, cratering, magma flooding, and weathering.

CHECKPOINT   What might a terrestrial planet's surface look like during each of the five stages of terrestrial planet evolution?

Mercury: Swift, Small, and Hot

7.2   Mercury (**Figure 7.4**) is the Solar System's innermost planet. Its history has been shaped by two important facts. First, its 88-day, 0.39-astronomical-unit (0.39-AU) orbit means that the Sun has exerted a dominant influence on its evolution. Second, with a radius of 2,440 kilometers (km), Mercury is a small world. It is just 38 percent the size of Earth and 1.4 times the size of the Moon. In many ways, Mercury is like the Moon's larger cousin. Because of their relatively small sizes, the two bodies have experienced similar histories, and they are equally inhospitable to life.

Spacecraft have visited Mercury just twice. In 1973, *Mariner 10* made three sweeps past Mercury, photographing its surface and using other instruments to probe conditions on the planet. And in 2011, the *Messenger* spacecraft entered orbit around the innermost terrestrial world, giving astronomers another chance to explore the planet.

Mercury's Interior

At the radius of Mercury's orbit, the original disk of gas and dust that formed the planets would have been very hot. The torrent of solar radiation tearing through the gas so close to the young Sun would have kept most elements in a vapor state. Only material with the highest condensation temperatures, such as iron and other metals, could have "frozen out" and formed solid grains. Therefore, according to the condensation theory of planet formation (see Chapter 5), Mercury should have a high percentage of metals relative to planets farther from the Sun. In fact, Mercury's density is measured as 5,430 kilograms per cubic meter (kg/m^3)—a value so high that the planet must have a large iron core. Mercury's iron content is so high that even the condensation theory cannot account for it. Scientists have theorized that early in its history Mercury must have collided with a metal-rich body. "It is certainly possible that Mercury was subject to a large impact," says Jim Zimbelman. "We've been finding evidence of really big whacks in other places, and we've realized that

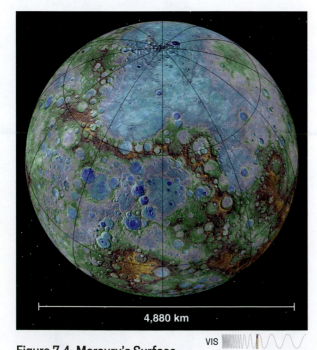

4,880 km

Figure 7.4 Mercury's Surface

A *Messenger* spacecraft false-color image of Mercury shows a heavily cratered surface.

VIS

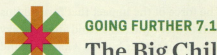

The Big Chill: The Rate of Planetary Cooling

A major factor governing a planet's evolution is its size, which determines how rapidly the planet cools. Recall from Chapter 4 that all dense objects radiate heat into space as blackbodies. Thus, planets continuously lose their internal heat. Since internal heat ultimately powers a planet's volcanism, plate tectonics, and dynamo-generated magnetic field, any significant cooling of a planetary interior puts an end to the planet's evolution.

How can we determine how fast one planet will cool relative to another? A planet's store of internal heat depends on, among other things, its volume. The greater a planet's volume, the more total heat energy it holds. A planet's volume V is the volume of a sphere of radius R:

$$V = \frac{4}{3}\pi R^3$$

However, the rate at which heat flows *out* of the planet depends not on its volume but on its surface area, since the surface is the only place the heat can escape. The surface area A of a sphere is

$$A = 4\pi R^2$$

We now have a way to consider the planet's *cooling time*—the characteristic time it takes the planet to lose most of its interior heat. The total heat (which depends on the planet's volume) divided by the rate of heat escaping (which depends on its surface area) tells us the cooling time t_{cool}:

$$t_{cool} \propto \frac{\frac{4}{3}\pi R^3}{4\pi R^2} \propto R$$

(The $\propto$ symbol means that t_{cool} is *proportional* to volume divided by area.) So we are left with cooling time proportional to R. To write out the full expression for the cooling time, we would have to include factors related to the planet's composition and temperature. But all we need to see is how cooling time relates to a planet's volume and surface area, because that relationship alone shows us that cooling time is proportional to the planet's radius, which makes sense. Smaller planets cool faster than larger planets do.

To see this explicitly, take the ratio of the cooling times for a large planet A and a smaller planet B, with radii $R_A > R_B$:

$$\frac{t_{cool(A)}}{t_{cool(B)}} \propto \frac{R_A}{R_B} > 1$$

Thus bigger planets (like planet A) will always have longer cooling times than smaller planets (like planet B). We can expect small planets to cool down more quickly than big ones.

---

these large impacts are more prevalent and more important to the overall history of the terrestrial objects than was previously believed."

The delivery of the extra metals would have left Mercury, after differentiation, with an iron core that takes up a far larger fraction of its volume (42 percent) than the core of any other terrestrial world. Earth's iron core, for example, accounts for only 15 percent of its volume. Mercury is essentially a cannonball with a rocky crust.

After its formation and differentiation, Mercury's internal history was set by its size. Being a small object, Mercury radiated the internal heat from its formation into space relatively quickly (see Going Further 7.1). Once Mercury lost most of its internal heat, most of its geologic activity such as volcanism ended. The loss of heat led the planet to contract, however. As the inner zones shrank, the crustal material was compressed, squeezing shut fissures and other paths for molten material to make it to the surface. Thus, while the Moon and Mercury bear many surface similarities, the extensive lava flows that occurred on the Moon did not occur on Mercury. It's worth noting that the *Messenger* mission to Mercury found evidence that the planet is still cooling and contracting (though slowly) and, for this reason, is considered tectonically active.

Mercury's Surface

Having lost most of its heat and volcanism early, Mercury retains a fairly pristine record on its surface of encounters with Solar System debris. The surface shows intense cratering; in many ways the entire planet looks much like the Moon's jagged highlands without broad maria. One of Mercury's most prominent features is the Caloris Basin, a giant

INTERACTIVE:
Anatomy of Terrestrial Planets

"We've been finding evidence of really big whacks, and these large impacts are more prevalent and more important to the overall history of the terrestrial objects than was previously believed."

JIM ZIMBELMAN

Jim Zimbelman's interest in planets began early. "When I was in middle school, I was interested in all sorts of 'sciency' kinds of things," he says. "But the turning point came when my dad got a small telescope. That's what really got me hooked specifically in astronomical things. But what I really loved best of all were the planets."

Zimbelman, a former chair of the Smithsonian Center for Earth and Planetary Studies, has played a significant role in the study of volcanism on the terrestrial planets. He attributes his broad perspective on celestial science to his family. "One thing I appreciated about my father . . . , he was open to listening to discussions of things." This open-mindedness has always encouraged Zimbelman in his studies of the sky.

impact crater stretching 1,500 km (**Figure 7.5a**). The force of the impact that formed this basin was so great that the planet's surface rippled like the surface of a pond after a heavy stone is thrown in. Rings of mountains 1,500 km in radius and as high as 3 km surround the impact crater (**Figure 7.5b**).

One surface feature is unique to Mercury: a system of long lines of scarps (cliffs) that run for hundreds of kilometers. Discovery Rupes, for example, is an escarpment on Mercury approximately 650 km long (**Figure 7.6**). The steep cliffs run straight across craters, so they must have formed *after* the Late Heavy Bombardment period. Since Mercury has no plate tectonics, the best explanation for what formed the scarps is the cooling and shrinking of the planet's interior. When Mercury's interior shrank, its skin (the crust) had to wrinkle, fold, and collapse to adjust, like the skin of a dried fruit. "The existence of the cliffs gets us back to the size of the core and the speed of Mercury's cooling," says Zimbelman. In going from a liquid to a solid, the volume of the core shrinks a lot when it freezes. The shrinkage forces so much "crustal shortening," or wrinkling of the crust, that scarps form.

The *Messenger* spacecraft took data in orbit around Mercury between 2011 and 2015. *Messenger*'s observations of the scarps led to a remarkable discovery. Based on the heights of some scarps, scientists were able to see that Mercury is not entirely geologically dead; the planet is still slowly shrinking and creating new scarps.

Mercury's Atmosphere

Mercury has no atmosphere. It does manage to hold on to a thin layer of hydrogen and helium, but this layer is simply solar wind particles that Mercury's gravity can only temporarily trap. Eventually, even this thin blanket of gas (including tiny amounts of other gases released through surface interactions with the solar wind) escapes back into space. The reason for Mercury's inability to retain an atmosphere can be traced back to its small size and proximity to the Sun. Going Further 6.2 showed how the ability to retain atmospheric particles depends on a planet's gravity and temperature, as well as on the mass of the particles. Mercury's low mass means that its escape velocity is also low (you can calculate it yourself now). Its proximity to the Sun means that its surface temperature (on its daytime side) rises as high as 700 K. Such enormous tem-

Figure 7.5 Caloris Basin

a. A *Messenger* image of one of the Solar System's largest impact basins. The false colors enhance the chemical, mineralogical, and physical differences among the rocks on Mercury's surface. **b.** The mountains surrounding the Caloris Basin probably formed from the ejecta of the powerful impact that created the crater.

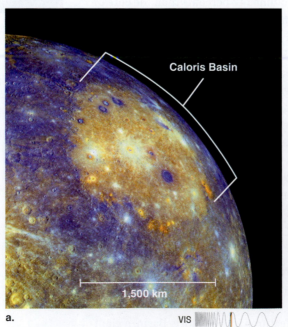

Caloris Basin

1,500 km

a. VIS

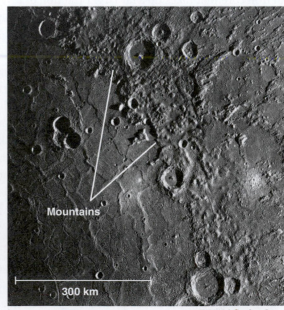

Mountains

300 km

b. VIS

peratures drive even the heaviest atmospheric atoms to speeds high enough to escape into space.

The temperature at Mercury's poles, however, where the sunlight glances off the planet's surface, can be 93 K, far below freezing (273 K). Therefore, craters near Mercury's poles could contain frozen water, just as deeply shadowed craters near the Moon's poles are known to retain large quantities of water ice.

Mercury's Near-Space Environment

Mercury supports a weak magnetic field. The strength of its magnetism is only one ten-thousandth that of Earth. (In other words, the ratio of Mercury's field strength to Earth's is 10^{-4}.) With such a large iron core, Mercury might be expected to host a strong magnetic field. Mercury's core, however, may have solidified 4 billion years ago. If so, no dynamo like Earth's would be possible now. Recall from Chapter 6 that a *magnetic dynamo*, the motion of electrically charged fluids in a planet's interior, produces a magnetic field. In terrestrial planets, dynamos occur when a molten metal such as iron exists in the core and is set in motion by planetary rotation. Mercury's fields are most likely remnants of **fossil magnetism** (magnetic fields imprinted in the planet's rocks and minerals), left over from the period when its iron core was still melted and generated its own field. Currently, Mercury shows a dipole magnetic field whose center is located not at the planet's center but north of that, along the rotational axis. It's worth noting that data from the *Messenger* spacecraft suggest to some scientists that a liquid core may still exist on Mercury, making its magnetism an area of ongoing research.

Mercury's proximity to the Sun has affected its evolution in ways even more profound than driving its temperatures to such extremes. Mercury is close enough to the Sun that its rotation period has been strongly modified by the Sun's gravity. If Mercury were in a circular orbit, the Sun's tidal effects would have locked its rotation to its 88-day orbital period; that is, Mercury would always present the same face to the Sun, as the Moon does to Earth. The real situation is a bit more complicated. Mercury's orbit, with an eccentricity of $e = 0.206$, is fairly elliptical by our Solar System's standards. That means Mercury has a larger variation in orbital velocity and distance as it swings in close to the Sun (perihelion, at 0.31 AU) and swings farther from the Sun (aphelion, at 0.47 AU). These swings make simple tidal locking impossible. Instead, Mercury ends up in a state of **tidal synchronization**, with its spin modified to present the same face to the Sun every *other* orbit. Thus, Mercury's solar day (the time between two successive noons) lasts two full orbits, or 176 days. The time it takes to spin once around on its axis, however, is just two-thirds of an orbit, or 59 days (**Figure 7.7**).

VIS

Figure 7.6 Discovery Rupes

This dramatic line of cliffs was produced by the shrinkage that occurred as Mercury cooled. Because these cliffs run across craters, the cliffs must have formed after the impacts.

fossil magnetism Magnetic fields that are imprinted in a planet but were created by an earlier active dynamo phase

tidal synchronization Coordination of the rotation and orbital periods of a satellite (such as a planet or moon) by the gravity of the body it orbits. (For example, the satellite completes two rotations for every three of its orbits.)

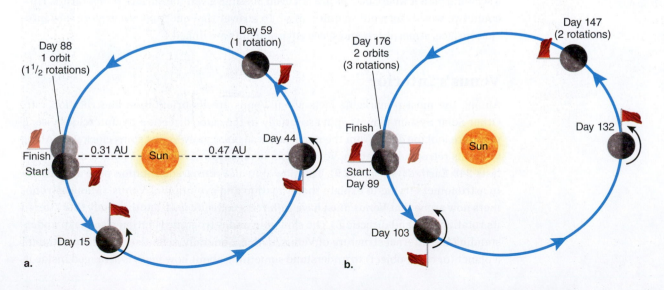

Figure 7.7 Mercury's Orbit and Rotation

Because Mercury's orbit is fairly eccentric, tidal forces from the Sun have not produced tidal locking. **a.** Thus, Mercury spins on its axis approximately every 59 days, whereas its orbit is 88 days. **b.** This means there are three rotations for every two orbits—an example of tidal synchronization. As a consequence, it takes two orbits for the Sun to be overhead at the same spot on Mercury.

a. Day 88 1 orbit (1½ rotations) · Day 59 (1 rotation) · Finish Start · 0.31 AU · Sun · 0.47 AU · Day 44 · Day 15

b. Day 176 2 orbits (3 rotations) · Day 147 (2 rotations) · Finish Start: Day 89 · Sun · Day 132 · Day 103

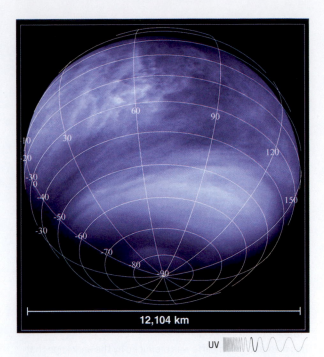

UV

Figure 7.8 The Clouds of Venus

An image of Venus, taken by the European Space Agency's *Venus Express* probe, revealing the structure of the planet's clouds. Latitude and longitude lines show the locations of poles and equator.

"Venus is the poster child for why we don't want the greenhouse effect to go too far."

With strong tidal influence from the Sun; an airless, heavily cratered surface; and long-extinct volcanism, Mercury does indeed appear to be a larger cousin to Earth's Moon.

Section Summary

- Mercury's large iron core is consistent with the condensation theory but may involve other factors, such as a collision with another iron-rich body.
- Because of Mercury's small size, it lost heat early and is believed to be relatively geologically inactive though the planet continues to shrink, leading to some slow changes in the crust.
- Mercury's surface features include dense cratering, signs of very large impacts, and long lines of cliffs caused by rapid cooling and shrinking.
- Because of Mercury's extreme heat and low gravity, no atmosphere was retained, but the poles could contain frozen water.
- Mercury's elliptical orbit has caused its rotation period to be tidally synchronized, but not tidally locked, with the Sun.

CHECKPOINT If Mercury were bigger, would its surface have more or fewer scarps, and why?

Venus: Hothouse of the Planets

7.3 As we take one step outward from the Sun and a step deeper into comparative planetology, we encounter Venus, which should be a slightly less massive cousin of Earth. Exploration of Venus, however, has shown how two planets that began from very similar places have led very different lives.

In terms of the most basic planetary characteristics, Venus and Earth are near-twins. Venus has a radius of about 6,050 km, 95 percent the size of Earth. Its mass is 4.87×10^{24} kg, about 82 percent that of Earth. Together these numbers give Venus an escape velocity of 10.4 kilometers per second (km/s), 92 percent of Earth's. Venus's density is 5,240 kg/m³, and its uncompressed density is 4,400 kg/m³. Both values are also very close to those for Earth. With an average orbital radius of 0.72 AU, Venus is closer to the Sun than Earth is (at 1 AU). That 28 percent difference in distance means that Venus receives almost twice as much solar energy as Earth gets, but this fact alone cannot account for the most stunning item on Venus's fact sheet. While Earth has an average temperature range above freezing (which is 273 K), the temperature on Venus is about 737 K. It is a world covered in a thick blanket of perpetual cloud (**Figure 7.8**) and extreme heat, all a result of trapped solar energy that has run amok. "Venus," says Jim Zimbelman, "is the poster child for why we don't want the greenhouse effect to go too far."

Earth today is a temperate world rich with liquid water and life, while Venus now is a scalding planet where oceans are not even possible given the surface temperature. How could two worlds turn out so differently? To answer that question, we explore how interior, surface, atmosphere, and space environments are linked.

Venus's Interior

Among the most remarkable facts about Venus are its orientation and rotation rate. Other Solar System planets spin essentially in the same direction (counterclockwise, if you were looking down from the north celestial pole); only Venus spins clockwise. Along with this retrograde rotation, Venus's spin is very slow. A "day" on Venus (one rotation) lasts 243 Earth days (**Figure 7.9**). The slow spin and retrograde rotation pose a challenge to astronomers trying to explain the formation and evolution of Venus. (Many astronomers now argue that Venus must have suffered a collision with another body that altered its rotation rate and direction.) The slow spin and retrograde rotation also make understanding the internal structure of Venus harder. Generally, scientists use the rotation of a planet (or other object) to understand something about how mass is arranged inside it.

But because Venus spins so slowly, this method does not work well. Astronomers are left with Venus's density as a principal clue to its internal structure.

There is, however, another way to "see" the internal arrangement of a planet's mass. In the 1990s, the *Magellan* probe mapped Venus by penetrating the clouds with radar and returning topographic information. "By accurately mapping the shape of the surface with radar and simultaneously following the precise motion of the spacecraft in orbit, we could figure out a lot about how Venus's mass is distributed," explains Zimbelman. "When we combined this kind of data with models of the planet's interior, it really helped us understand what is going on below the surface."

Because Venus's density is so similar to Earth's, scientists could use the models they had constructed for our planet's structure and adjust them for a Venus-sized planet. They concluded that its structure must look a lot like that of Earth. Both planets have an iron core of roughly the same size, overlain by a mantle of silicate rocks, also of roughly the same size. We might thus expect both planets to have cooled at about the same rate, so Venus should still retain a significant amount of heat from its formation, as well as the heat generated by radioactive decay.

Venus's Surface

The similarities in core and mantle might lead us to expect Venus's crust to look similar to Earth's, but Venus's high atmospheric temperatures have changed the crustal properties. "Raising the surface temperature hundreds of degrees changes how heat escapes from the crust," explains Zimbelman. "The surface is already hot, so you don't have to go down very far before the crust melts. That means the crust is thinner, and Venus may have thousands of plates. The thinner crust means that the kinds of plate tectonics we're used to here won't happen on Venus. There the thin crust gets scrunched up easier, leading to entirely different kinds of behavior that manifest in a very different kind of surface." Some scientists still hold that Venus's crust is thicker than Earth's, but most evidence at this time appears to support a thinner crust. As is often the case, competing views can coexist until the scientific community develops new techniques to obtain new data and answer questions.

The first ground-view images of Venus were transmitted in 1975, when the Soviet space probe *Venera 9* (**Figure 7.10**) parachuted through the planet's dense atmosphere and dropped heavily onto the surface. Where almost all other Venus probes had failed, *Venera 9* functioned in that scorching, toxic environment for a full 53 minutes, beaming back images. The pictures were not encouraging. The surface looked flat, dry, and broken—a world smoldering under a sulfuric haze. None of the subsequent (all Soviet)

Length of day in Earth days

Figure 7.9 Relative Day Lengths
Each Sun symbol represents one Earth day (24 hours) to show the relative length of one complete day-night cycle on the terrestrial planets and the Moon.

a. **b.** VIS

Figure 7.10 *Venera 9*
In 1975, the Soviet Union produced the first-ever images from the surface of another planet: Venus, where atmospheric and surface conditions limited the spacecraft's functioning life span to minutes or hours. **a.** *Venera 9*. **b.** A surface image from one of the landers.

Figure 7.11 Venus's Surface

a. A computer-enhanced view of the surface, obtained by radar data from the *Magellan* mission. The color coding represents elevation. Computer-enhanced three-dimensional perspectives show surface features such as **b.** the dome-like hills of the Alpha Regio region and **c.** the volcano Maat Mons.

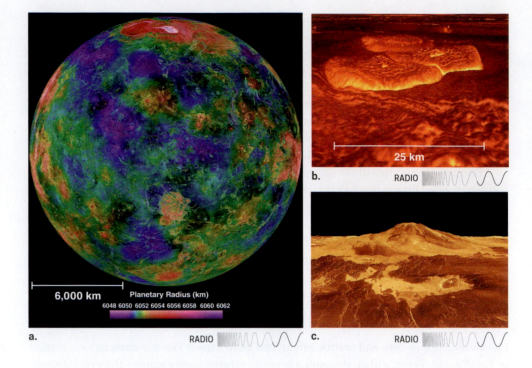

spacecraft that landed on Venus lasted very long, so we have few images at optical wavelengths of the planet's surface. That limitation has not, however, prevented the study of Venus's geology.

While the unyielding clouds covering Venus keep its surface invisible to us at optical wavelengths, radio waves penetrate the clouds. With *Magellan*'s radar mapping of Venus, scientists developed detailed surface "images" down to scales of 100 meters. By bouncing radio waves from orbiting spacecraft down to the surface and timing the delay in return signals, scientists build topological maps of the entire planet's surface (**Figure 7.11**).

"Everybody assumed we would find lots of impact craters on Venus," says Zimbelman, "just like on the Moon, Mercury, and the older parts of Mars. That didn't happen." Only 900 or so craters were discovered. "That's not many for an object as big as Venus," he explains. "The implication is that the entire surface somehow resurfaced itself within the last billion years."

The relative lack of craters is consistent with volcanism. Meteor craters on Venus tend to be quite large. In many cases, irregularly shaped **ejecta blankets** (layers of ejected material) surround the craters, indicating that the meteors fragmented into irregular shapes in the dense atmosphere before they hit the ground in many pieces (**Figure 7.12**).

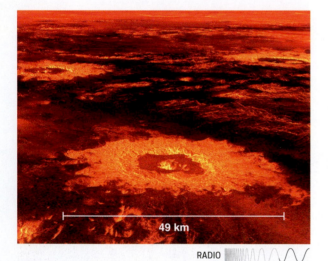

Figure 7.12 Venus's Danilova Crater

A *Magellan* spacecraft view shows this impact crater's broad, irregular ejecta blanket.

Volcanism has strongly shaped the rest of Venus's surface features. Ishtar Terra is a region of raised highlands that extends across 5,600 km—larger than Australia (**Figure 7.13**). In its center is a plateau, most likely a great lava flow, called the Lakshmi Planum. Maxwell Montes, an 11-km-high mountain, lies at one edge of Ishtar Terra. Jumbled mountainous terrain defines much of Ishtar Terra's boundaries.

Whereas Earth's continents are composed of low-density granites and thus "float" on the mantle below, the material composing Ishtar Terra and most of Venus's terrain appears to be the same basaltic material that makes up Earth's seafloors. Basalts well up from lava flows and form a variety of structures. "Venus's thin crust," explains Zimbelman, "means that molten materials like those that generate lava flows on Earth or Mars or anywhere else had good access to the surface."

ejecta blanket   A layer of pulverized material that is blown out of a crater on impact.

caldera   The depression at the top of a volcano.

corona   A large circular or oval structure surrounding volcanoes on some Solar System bodies, formed by upwelling.

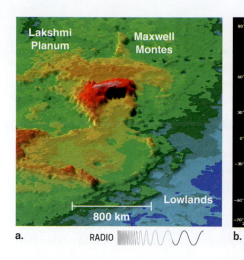

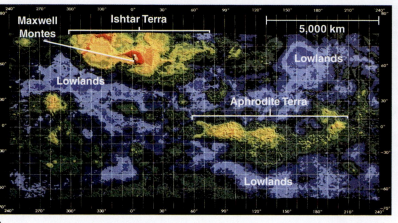

a.

b.

Figure 7.13 Venus's Terrain
a. A perspective view of Ishtar Terra from *Pioneer* Venus radar imaging. Larger than Australia, Ishtar is a steep-sloped plateau rising 3.3 km above the surrounding lowlands. Maxwell Montes, nearly 11 km high, is the highest elevation on Venus and thus the coolest, only 380°C. Color-coded altimetry shows elevations in 0.5- and 1-km intervals.
b. A color-coded map of elevations on Venus. Aphrodite Terra is a plateau about the size of South America.

All volcanoes on Venus seem to be shield volcanoes (**Figure 7.14**), like the Hawaiian Islands (see Chapter 6). They form when lava plumes rise from deep within the mantle, swelling up under crustal material. This type of volcano tends to release its energy gradually and have sloping sides. "The lava can flow fairly easily, and that's what makes the shield's gentle slopes," explains Zimbelman.

As discussed in Chapter 6, composite volcanoes, caused by collisions between tectonic plates, are found in various places on Earth and are characterized by violent eruptions and steep sides. "We don't see steep-sloped volcanoes like Mount Fuji on Venus, and that's consistent with the planet not having Earth-like plate tectonics," says Zimbelman.

The fact that only shield volcanoes are seen on the surface of Venus tells scientists about both the thinness of the crust and the flow of materials below the crust. Circular domes of lava are seen in many places on Venus's surface, resulting from the upwelling of magma that distorts crustal material. Larger, broad, circular volcanoes are also seen with craterlike **calderas** at or near their centers where lava has welled up to the surface and then withdrawn. (A caldera is the well at the top of a volcano.) The largest shield volcanoes on Venus are surrounded by **coronas**, which are circular or oval uplifted regions. Coronas show evidence of extensive surface distortions and huge plumes of lava pushed upward. The lava plumes on Venus do not appear to be part of general convection systems in the mantle, but there is evidence that limited mantle flows may have forced crustal material to buckle into the mountainous regions of the highlands.

Any theory of the surface evolution of Venus must be grounded in one fact: scientists believe that its surface features are less than 1 billion years old. Something catastrophic appears to have happened in Venus's relatively recent past that led to a complete resurfacing. "It may have been an era of supervolcanism," says Zimbelman, "or it may have been the crust breaking apart and overturning like ice floes in the Arctic (you might think of Venus's crust as made of 'rockbergs'). The jury's still out, though."

Whatever its nature, this dramatic and powerful event limits our ability to understand the deep history of Venus. For instance, the planet might not always have had extreme temperatures and a runaway greenhouse effect (to be discussed soon). Perhaps very early on, more than 4 billion years ago, Venus was a water world like Earth. At present there is no way to know, and the mystery of Venus's surface only underscores the fact that planetary surfaces and atmospheres can change dramatically over time.

"Everybody assumed we would find lots of impact craters on Venus."

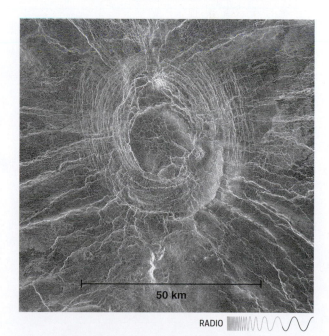

Figure 7.14 Shield Volcanoes on Venus
Coronas surround the largest shield volcanoes on Venus—evidence of extensive surface distortions and huge plumes of lava pushed upward.

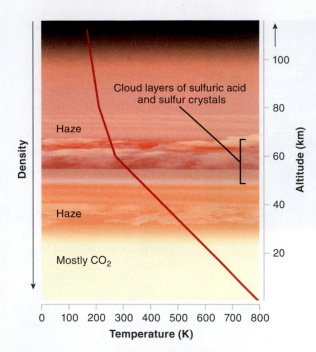

Figure 7.15 Venus's Atmosphere
The layers differ in composition, density, and temperature. Both temperature and density decrease steadily with altitude, so the atmosphere is hottest and densest near the surface. Although the atmosphere is more than 96.5 percent carbon dioxide by mass, several distinct sulfur-based cloud layers exist.

> "A runaway greenhouse effect means a situation where once you pass a certain point in temperature, you can't go back."

runaway greenhouse effect An unstable situation in which a planet's temperature rises, leading to an increased greenhouse effect, which raises the temperature higher still; this feedback drives planetary temperatures to very high levels.

Venus's Atmosphere

You run out of superlatives when describing the atmosphere of Venus. Unlike the other terrestrial worlds, Venus hosts an extremely dense blanket of gases that make it inhospitable to life (**Figure 7.15**). With a surface temperature around 730 K and sulfuric acid rain falling within the planet-enveloping layer of clouds that never parts (the rain never reaches the ground), it is a planet of extremes.

On Earth, carbon dioxide (CO_2) constitutes just 0.041 percent of the atmosphere. Because of volcanism, however, Venus's atmosphere is 96.5 percent CO_2, by mass. Carbon dioxide is a powerful greenhouse gas. The bonds between the carbon atom and two oxygen atoms are extremely effective at absorbing and reradiating heat energy (light in the infrared). Earth and Venus have the same total amount of planetary CO_2. However, much of the CO_2 on Earth is bound up in rocks and trapped under the oceans. Venus may have had an atmosphere similar to Earth's long ago—but when it lost its oceans, CO_2 was liberated into its atmosphere, raising Venus's surface temperatures.

While its small proportion of atmospheric CO_2 has allowed Earth to maintain mild temperatures (see **Going Further 7.2**), Venus's overwhelming abundance of this compound has driven the planet into its extreme state via a **runaway greenhouse effect**. The term *runaway* refers to evidence that Venus may not always have been as it is today. "A runaway greenhouse effect means a situation where once you pass a certain point in temperature, you can't go back," says Zimbelman. "The hotter it gets, the more the greenhouse effect increases, which makes it even hotter still." Scientists are not sure what triggered Venus's runaway greenhouse effect, but the culprit may have been the loss of water early in the planet's history. The weathering of rocks by water allows new rocks to form with CO_2 bound up in them. In this way, CO_2 is taken out of the atmosphere. If a planet's water is lost, atmospheric CO_2 levels continue to grow—magnifying the greenhouse effect and raising the surface temperature.

Venus's atmosphere is incredibly dense as well—about 90 times as dense as the air we breathe on Earth. The density and heat drive the formation of at least three permanent layers of clouds surrounding the planet. These clouds contain both sulfuric acid droplets and tiny sulfur grains.

Venus's Near-Space Environment

Venus is a world without a substantial magnetic field, which, given its size, came as a surprise. It's not clear yet why Venus has no field, but it may have to do with the planet's slow rotation rate. "Without a decent rotation rate, you can't produce the circulating flows of molten iron in the core that create a magnetic field through the dynamo process," says Zimbelman.

The lack of a magnetic field has important implications for Venus's atmosphere. While the sheer mass of Venus's gaseous blanket protects it from disruption by the solar wind, the interaction with that wind may have changed the atmosphere's chemistry over time. "With no magnetic field acting as a buffer, the solar wind can strip away particles at the top of the atmosphere, and this is what may have happened to water on Venus," says Zimbelman. Ultraviolet radiation also played an important role in removing the water from Venus. Ultraviolet light dissociates the H_2O molecule into its component elements. The liberated hydrogen atoms easily escaped Venus's gravitational hold. Once a planet loses its hydrogen atoms, it has lost its water for good. Thus, dissociating water may have led to Venus's runaway greenhouse effect, and the planet is one of the driest environments in the Solar System.

It is notable that only Mercury and Venus, the two innermost worlds, are free of satellites. Every other planet from Earth to Neptune has at least one moon. The lack of moons may be due to the nature of debris in the vicinity of Mercury and Venus during their formation or to the kinds of collisions they endured. As we've seen, Venus's unusual rotation

Planets without a Blanket: Temperature before Greenhouse Warming

What would be the average surface temperature of a planet with no greenhouse effect? Some calculation shows what would happen on a planet containing no greenhouse gases in the atmosphere. Specifically, we want to know what surface temperature a planet with no atmosphere would attain just through the surface's absorption and reemission of solar energy.

Let's begin with the energy radiated into space by the Sun—its luminosity L_{Sun}. To calculate this energy, we use the Stefan-Boltzmann law (see Chapter 4), which relates L_{Sun} to the Sun's radius R_{Sun} and temperature T_{Sun} for a spherical Sun with surface area $A_{Sun} = 4\pi R_{Sun}^2$:

$$L_{Sun} = A_{Sun}\sigma T_{Sun}^4 = 4\pi R_{Sun}^2 \sigma T_{Sun}^4$$

Now, how much of that solar radiation is captured by a planet of radius R_p orbiting at a distance D (that is, orbital radius D) from the Sun? We call this $L_{captured}$:

$$L_{captured} = L_{Sun}\left(\frac{\pi R_p^2}{4\pi D^2}\right)$$

Here the factor $4\pi D^2$ is the surface area of the sphere of radius D across which all of the radiation emitted by the Sun must pass. The term πR_p^2 is the planet's cross-sectional area. The ratio of the two tells us how much of the original solar radiation is captured by the planet.

Next we have to account for how much of the captured solar energy is reflected back into space. The albedo a of the planet is the fraction of incoming energy that is reflected (see Chapter 5). Then the amount of energy that the planet absorbs is $L_{absorbed}$, where

$$L_{absorbed} = L_{captured}(1 - a)$$

Finally, we expect that the planet, bathed in sunlight, will eventually come to a stable temperature T_p, at which it will radiate as a blackbody with energy $L_{radiated}$:

$$L_{radiated} = 4\pi R_p^2 \sigma T_p^4$$

We find the actual value of T_p by equating the blackbody energy the planet radiates into space with the solar energy it absorbs. That is, we set $L_{radiated} = L_{absorbed}$ and substitute:

$$4\pi R_p^2 \sigma T_p^4 = (1 - a)\left(\frac{\pi R_p^2}{4\pi D^2}\right)(4\pi R_{Sun}^2 \sigma T_{Sun}^4)$$

To solve for T_p, we first divide both sides by the constants 4π and σ and cancel them out:

$$R_p^2 T_p^4 = (1 - a)\left(\frac{R_p^2}{4D^2}\right)(R_{Sun}^2 T_{Sun}^4)$$

Now we divide both sides by the planet's radius R_p to cancel it out, and we take the fourth root of both sides (just like taking the square root twice). This gives us the final answer,

$$T_p = (1 - a)^{\frac{1}{4}}\left(\frac{R_{Sun}}{2D}\right)^{\frac{1}{2}} T_{Sun}$$

Notice that the planet's radius is not part of the final answer (it canceled out). Thus, the planet's temperature depends only on its albedo, its distance from the Sun, and the properties of the Sun (radius R_{Sun} and temperature T_{Sun}).

Let's use this formula to find the temperature T_E that an airless Earth would have. Using orbital radius $D = 1.5 \times 10^{11}$ meters, $T_{Sun} = 5,778$ K, $R_{Sun} = 6.96 \times 10^8$ meters, and albedo $a = 0.306$ for Earth, we find

$$T_p = T_E = 254 \text{ K}$$

This is well below freezing (273 K). Thus, without the atmosphere and its greenhouse gases, our planet would be a dead, frozen rock.

is thought to be the result of a dramatic collision with another large body. The energy of this early impact may have gone into reorienting Venus's rotation rather than generating an orbiting moon (as scientists believe happened to an early Earth).

At one time, Venus might have been very Earth-like in its ability to harbor an environment conducive to life. But planetary evolution was very different on Venus than on Earth. At some point 4 billion or so years ago, the runaway greenhouse effect began on Venus, which ended up remarkably hot and dry. Venus shows us something very important in our exploration of planets: things change. This potential for transformation will be even more apparent as we move on to the final stop in our tour of terrestrial planets: the fabled world of Mars.

Section Summary

- Venus is Earth's twin in radius and mass, but it had a very different history that left it wildly unsuitable for life. Its density, core size, mantle, crust, and composition are similar to Earth's, but its slow retrograde rotation is a Solar System anomaly.
- Lava flows, large shield volcanoes, and jumbled mountainous terrain dominate the topography, but it lacks plate tectonics like that on Earth. Venus likely experienced a major resurfacing a billion years ago.
- The extremely dense, CO_2-dominated atmosphere and the loss of water to space have led to a runaway greenhouse effect on Venus.
- Venus lacks a core dynamo and has no moons.

CHECKPOINT   Venus's atmosphere is far more massive than Earth's. How does that affect conditions on the planet's surface?

Mars: The Red Planet of Change

7.4   Late in the 19th century, astronomer Percival Lowell looked at Mars through a 24-inch refracting telescope and convinced himself he saw the handiwork of a dying civilization (**Figure 7.16**). In 1877, astronomer Giovanni Schiaparelli observed what he thought was a network of straight lines across the planet; he called them "canali," which means *channels* in Italian. When the term was mistranslated into English as *canals*, Lowell picked up on Schiaparelli's observations and hypothesized that Martians had built these canals to transport water across a planet that had become an arid wasteland, desperately trying to save their civilization.

While it soon became clear that Lowell was mistaken, he was not alone in his fascination with life on Mars. H. G. Wells had already published his book *War of the Worlds* (1898), in which Martians undertake a nearly successful invasion of Earth. That story was reimagined as a fake radio news program in 1938, inspiring widespread panic, and in films.

The irony of this fascination with the red planet and Martians is that Mars may hold the best chance as another home for life in the Solar System. But current conditions there suggest that we're in no danger of invasion anytime soon. Today Mars is a cold, dry, and empty world (**Figure 7.17**). Billions of years ago, however, it might have sported a far bluer and perhaps even greener hue than the dust-swept planet we see now.

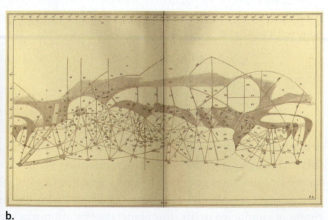

a.   b.

Figure 7.16 The "Canals" of Mars
a. Percival Lowell (1855–1916) erroneously interpreted the markings he observed on Mars. **b.** A drawing of features that Lowell believed to be canals transporting water away from the polar regions. He concluded that intelligent life exists on Mars.

"Mars is a relatively small world by Solar System standards," says Jim Bell. "Its mass is just a tenth of Earth's, and it's only about half the size of our world." The lesser mass gives Mars a surface gravity that is just a bit more than one-third as strong as Earth's. Orbiting at 1.52 AU, Mars receives 56 percent less sunlight at its surface than Earth does. Its larger orbit also gives it a longer year, lasting almost 687 days. However, its day is quite similar to ours: Mars rotates once on its axis every 24 hours and 37 minutes.

Mars's Interior

In 1976 the first two landers to touch down on the Martian surface—*Viking 1* and *2*—carried seismographs. "Unfortunately, they both failed," says Bell. "The first seismograph never turned on, and the second was unable to filter out the effects of strong winds flowing past the lander." In the absence of seismographic data, planetary scientists used the same methods to infer the internal structure of Mars that they used for Mercury and Venus: density, rotation, volcanic activity, and the presence or absence of magnetic fields. It's worth nothing that a seismograph was one of the instruments carried in 2018 by NASA's successfully delivered *InSight* lander, so new data about the Martian interior can be expected.

"Mars is pretty widely thought to have a core," says Bell, "but we don't know much about its size." The uncompressed density of Mars is 3,800 kg/m³, the lowest of all terrestrial planets, so Mars must have the lowest metal content. The presence of a magnetic field can also tell scientists whether Mars has a planetary dynamo created by a molten iron core. None of the probes that have carried a magnetometer to the red planet have detected a strong global magnetic field. In 1997 the orbiting *Mars Global Surveyor* found evidence of fields as high as 0.1 percent of Earth's, but these appeared to be local anomalies related to differences in crustal composition. "Mars probably used to have a magnetic field," says Bell. "That implies that, like Earth, Mars once had a partially molten iron core." The present size of Mars's iron core remains a contentious issue. "Altogether there's some uncertainty, but the data are consistent with the presence of a smaller core on Mars than we might have expected."

Even though the seismographs on the first landers failed, the presence of so many orbiters responding to Mars's gravitational field and landers giving precise measurements of the planet's spin rate enabled extremely accurate determination of its rotation. That, in turn, gave scientists a measure of how the mass inside Mars is distributed. "Based on these kinds of measurements, it's likely there is a basic mantle and crust structure very similar to that on Earth and the other terrestrial planets," says Bell. In addition, the ability to probe the soil and rock on Mars directly has provided a unique view of Mars's composition. "The volcanic rocks on the surface of Mars are very similar to garden-variety volcanic rocks on Earth. It's a lot like the stuff coming out of the volcanoes on Hawaii, for example." These rocks are basalts, similar to those on Earth's seafloor and on Venus. "Mars, Earth, Mercury, Venus are all basically made of the same stuff," says Bell, "which is not surprising since it's completely consistent with the standard condensation model of planet formation."

Mars's Surface

No terrestrial planet other than Earth has such a rich variety of surface features. Along with the volcanoes and meteor craters seen on other terrestrial worlds, Mars hosts continent-spanning canyons, polar caps of water ice and frozen CO_2, and a host of features that seem to point to a time when liquid water ran free on the surface. And while the atmosphere of Mars is thin, its winds are strong enough to have shaped the surface terrain over time.

Most of the craters visible on Mars are large, indicating that small meteoroids tend to burn up in the atmosphere and do not make it to the surface. In addition, wind erosion

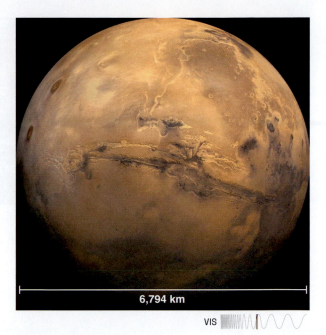

6,794 km

VIS

Figure 7.17 Martian Terrain
The surface of Mars has dramatic features such as huge canyons and towering volcanoes. Many forces, from volcanism to weathering, have shaped Mars over time.

"Mars, Earth, Mercury, Venus are all basically made of the same stuff."

Figure 7.18 Tharsis Bulge

a. This broad, elevated region of Mars is home to active shield volcanism. **b.** The volcano Olympus Mons, nearly 25 km in elevation, is the highest mountain in the Solar System and covers an area the size of Arizona.

c. A computer-generated view of Olympus Mons created from laser altimeter measurements made on the *Mars Global Surveyor* space-craft. Despite this volcano's great height, its slopes are only about 6 degrees, gentler than Mauna Loa's in Hawaii.

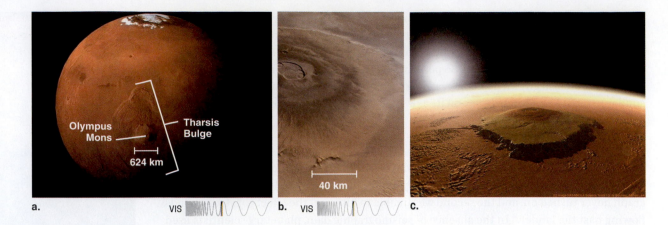

"Mars seems to be a one-plate planet."

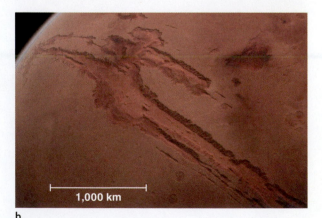

Figure 7.19 Valles Marineris

a. This system of canyons, resembling an enormous gash in the side of Mars, extends more than 4,000 km in length, 600 km across, and 7 km deep at its extremes. **b.** A computer-generated three-dimensional representation of Valles Marineris created from data from Mars's orbiters.

has smoothed away many craters, so some of the record of the Late Heavy Bombardment phase has been lost.

Some of the most dramatic features on Mars are the result of volcanism. "Most of the volcanoes on Mars appear to be of the shield type," says Bell. "The slopes are long and broad." Bell notes that there is some evidence of explosive volcanism on Mars. The planet also hosts the Solar System's largest volcano, Olympus Mons (**Figure 7.18**), a towering shield volcano that rises 24 km above the average surface level. The volcano is so high that the caldera at its peak is above the bulk of the planet's atmosphere. The giant Olympus Mons is part of an extended region called the Tharsis Bulge, which is home to a number of volcanoes. On average, the Tharsis region rises 10 km above the planet's surface.

In a sense, the entire Tharsis Bulge is a shield-volcanic region. Continual upwelling of magma from deep inside the Martian mantle pushed Tharsis and its volcanoes to such enormous heights. The fact that these volcanoes have grown so large offers compelling evidence that moving plates cannot exist. "If there were plate tectonics on Mars," says Bell, "you would have a chain of volcanoes like what you see in Hawaii instead of the super-volcano that is Olympus Mons. Mars seems to be a one-plate planet." Although little is known about why any planet has, or does not have, plate tectonics, many scientists believe that the difference between Mars and Earth with respect to plate movements may be the presence of water. "Mars may have had more water in the past," says Bell, "but it did not have a gigantic ocean interacting with the crust like Earth does. All that water may be what's lubricating the whole process on our world."

Mars is home to one of the largest canyon systems in the Solar System. Valles Marineris is a giant network of canyons that appears like a gash across the planet's face (**Figure 7.19**). Stretching more than 4,000 km, it's longer than the North American continent. At its widest, the canyon network stretches 600 km across. In some places, Valles Marineris drops 7 km to the canyon floor. (The Grand Canyon is 2 km deep.) This system of canyons appears to be a direct result of crustal "faulting" associated with the nearby Tharsis Bulge. As the bulge rose, the surrounding surface was forced to respond, breaking into sections; some sank as others rose. The Valles region is relatively old, but along with the Tharsis region, it demonstrates that Mars was once a far more dynamic planet than it is today. "As Mars cooled, its surface activity slowed," says Bell. "But at one time it must have been really dramatic."

Mars, like Earth, has polar ice caps, though neither Venus nor Mercury does. Even a small Earth-based telescope can see Mars's northern and southern caps and their seasonal changes. The presence of the ice caps and their yearly growth and retreat were reasons that some 19th- and early-20th-century astronomers thought Mars is much like Earth and possibly inhabited.

Seasonal ice caps, which grow and retreat, are distinct from **residual ice caps**, which are present all year long (**Figure 7.20**). Earth's polar ice caps are currently residual, but global warming may lead to a seasonal melting of the Arctic (northern) ice sheets.

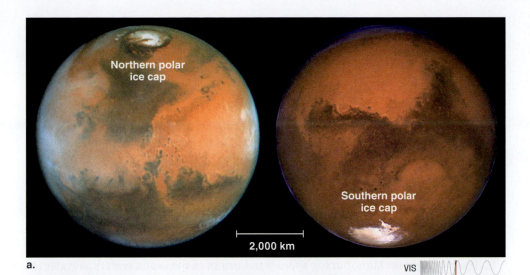

Figure 7.20 Polar Ice Caps on Mars
a. Northern and southern polar ice caps. **b.** The ice caps change seasonally (as the northern polar ice cap shows).

Since the orbital tilt of Mars is similar to Earth's, Mars experiences seasons in which the strength of sunlight varies as the planet orbits the Sun. Thus, in the northern winter, temperatures drop at the poles while the south experiences summer and rising temperatures. The northern and southern poles are not symmetric, however, because of Mars's elliptical orbit. The difference in size between the residual and seasonal caps in both the north and south is noteworthy. The southern seasonal cap covers 4,000 km; the northern seasonal cap covers 3,000 km. The northern residual cap is 1,000 km across, while the southern residual cap barely crosses 300 km. In terms of composition, the seasonal caps are composed mainly of frozen CO_2.

One important difference between the northern and southern caps is the role of water. Spectral measurements made from orbit reveal more atmospheric water vapor above the northern residual cap than above the southern cap, leading scientists to conclude that prodigious quantities of water may exist underground in the north. In 2008 the *Phoenix* lander touched down in the northern polar regions and immediately found supporting evidence when its mechanical scoop revealed water ice a few centimeters below the ground.

Mars's Atmosphere

"As important as Mars's atmosphere is to the planet," says Bell, "there is actually not much of it to go around." Pressure at ground level, the base of the Martian atmosphere, is $\frac{1}{150}$ of that on Earth. Another way of looking at the sparseness of Mars's atmospheric blanket is

seasonal ice cap A polar region of frozen material such as CO_2 or water that increases and decreases in area and thickness seasonally.

residual ice cap A polar region of frozen material such as CO_2 or water that stays constant in size with seasonal changes.

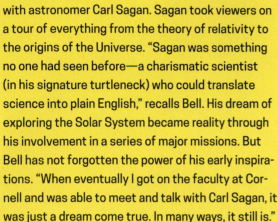

JIM BELL

One astronomical landmark in Jim Bell's early life came from TV: the program *Cosmos* with astronomer Carl Sagan. Sagan took viewers on a tour of everything from the theory of relativity to the origins of the Universe. "Sagan was something no one had seen before—a charismatic scientist (in his signature turtleneck) who could translate science into plain English," recalls Bell. His dream of exploring the Solar System became reality through his involvement in a series of major missions. But Bell has not forgotten the power of his early inspirations. "When eventually I got on the faculty at Cornell and was able to meet and talk with Carl Sagan, it was just a dream come true. In many ways, it still is."

> "From these humble beginnings you can end up with vast planet-engulfing dust storms."

to note that the mass of gas surrounding Mars is a thousandth of that surrounding Earth and a ten-thousandth of that around Venus. Most of Mars's atmosphere is CO_2, which accounts for 95 percent of the gas by mass. Nitrogen makes up 1.6 percent, and oxygen makes up only 0.13 percent. Clearly, anyone who visits Mars will have to bring oxygen tanks.

While Mars's atmosphere is thin, there is enough of it to give the planet weather. As Mars landers such as *Viking, Pathfinder,* and *Phoenix* dropped to the ground, they measured the atmospheric conditions. The data show that the Martian atmosphere is layered like Earth's (**Figure 7.21**). In the Martian troposphere, CO_2-ice clouds form at high altitudes, and water-ice clouds form below them. The cycle of surface heating by sunlight and cooling by thermal radiation drives Martian weather via convection. The thin atmosphere retains little heat, leading to almost 100-K changes between daytime and nighttime temperatures. While daytime temperatures on Mars can, at best, be almost comfortable for humans, nighttime temperatures would be deadly cold.

It never rains on Mars, but fog appears to form in canyons and craters as water ice evaporates as the Sun rises. Days are usually marked by, at most, light breezes. However, periodic strong dust storms, appearing to begin in the southern highlands, pick up light surface dust. Convection then carries the dust high into the troposphere. "From these humble beginnings you can end up with vast planet-engulfing dust storms," says Bell. "The storms can sweep across the planet, taking weeks before they subside." Wind speeds are extreme, sometimes reaching 400 km per hour (km/h). Because Mars's atmosphere is so thin, these storms are unlikely to pose much of a challenge to future astronauts. Bell says that the strong winds in Martian storms "will feel only like being pummeled by feathers to a properly suited astronaut."

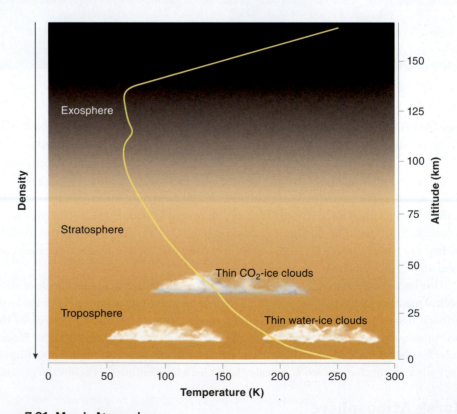

Figure 7.21 Mars's Atmosphere
The thin atmosphere exhibits layers of different densities and temperatures. In general, both temperature and density decrease steadily until an altitude above 100 km, well into the exosphere, where the temperature rises again as solar ultraviolet radiation is absorbed.

The difference between the atmospheres of Earth and Mars resides in the absence of life-forms on Mars and the disparity in mass. Earth's oxygen comes from biological activity. Since Mars has no biological activity (that we know of), the oxygen in its atmosphere has become bound in rocks. The lower atmospheric pressure (low atmospheric mass) on Mars, however, is directly related to the planet's lower total mass. The escape velocity of Mars is 5.0 km/s, compared with Earth's 11.2 km/s. Both water vapor and lighter elements such as nitrogen and argon, important parts of Earth's atmosphere, cannot be held by Mars's low gravity; over time they have escaped into space. Thus, Mars's atmosphere may have been far heavier and denser early in its evolution. The lesser mass of Mars also makes the atmosphere more fragile with respect to large-scale impacts. Some scientists have suggested that early asteroid impacts blasted significant fractions of the planet's gases into space.

Mars's Near-Space Environment

As mentioned, the Martian magnetic field is many times weaker than Earth's and does not envelop the entire planet. Without such a protective magnetic blanket, Mars's atmosphere has suffered erosion at high altitudes through direct interaction with the solar wind. This interaction is yet another reason for the current depleted state of the atmosphere.

While Mars lacks a magnetic field, it is the only terrestrial planet other than Earth to host any natural satellites. Mars has two moons: Phobos and Deimos (**Figure 7.22**). Phobos, with a diameter of just 22.4 km, and the smaller Deimos, with a diameter of 12.4 km, are tiny compared with Earth's Moon. Both moons are tidally locked and have orbits aligned with Mars's equator. Phobos orbits just 9,377 km above the Martian surface. Being so close to the planet, its orbital period is only 7.7 hours. This is less than the Martian rotation period, meaning that Phobos would appear to move backward (relative to the Sun) across the Martian sky twice in each day-night cycle. Deimos orbits farther out, at a distance of 23,460 km, giving it a 30.4-hour orbit.

Both Martian moons are irregularly shaped objects with dark, low-albedo surfaces. The dominant surface feature on Phobos is a giant, 9-km-diameter crater called Stickney. Phobos is also covered in unusual "grooves" as much as 100–200 meters wide and 20 km long. While many of them appear to radiate away from Stickney, analysis by the European Space Agency's *Mars Express* orbiter showed they are centered at the apex of Phobos's distorted oblong geometry, making their origin unclear. Deimos also has craters, although the largest is only 2.3 km across. Deimos's surface has a unique appearance that holds other surprises. "It's very smooth," explains Bell. "Deimos looks like some powdery material is covering much of the surface. We haven't seen any asteroids that look like that." Scientists still do not understand what gives Deimos its unique appearance.

Given the moons' small sizes and proximity to the asteroid belt, it's tempting to think of Phobos and Deimos as captured asteroids. "That's the standard story," says Bell. "Both moons are very small, they're both irregular in shape, and they have spectra that look somewhat asteroidal." This theory, however, doesn't quite work. Both the orbital dynamics of these moons and some of their surface features point to a more complex history, and calculations suggest that both moons would have been disrupted by tidal forces and torn apart long before they could have been captured into orbit. "From theoretical studies we find it's impossible to capture an object the size of Phobos so close to Mars," says Bell. "People are starting to consider whether some unique situation enabled these objects to be captured, while others are exploring if a really giant impact might have launched material into orbit to form these moons."

"Phobos is extremely close to Mars," says Bell, "and it's slowly spiraling downward. In 10 million years or so (the blink of an eye in Solar System timescales), it will crash into Mars." To scientists, the short life of Phobos is an oddity. "Has Phobos been there for billions of years and it's just a coincidence that we are seeing it near the end of its existence?" asks Bell. "That's an argument that doesn't carry a lot of weight."

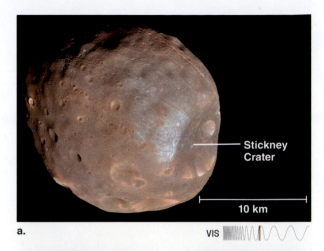

Stickney Crater

10 km

VIS

a.

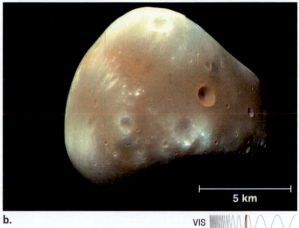

5 km

VIS

b.

Figure 7.22 The Moons of Mars
The origin of Mars's two small moons, **a.** Phobos and **b.** Deimos, remains a subject of debate.

Whatever the origin and fate of tiny Phobos and Deimos, the lack of a large moon like Earth's has had a dramatic impact on Mars and its evolution in a way that may have affected its ability to maintain life. The difference has to do with Mars's obliquity, the angle between its orbital plane and its spin axis. While Earth's spin axis does slowly rotate like a wobbling top (a process called precession), its obliquity is remarkably stable, varying only slightly from its 23.5° tilt (see Chapter 2). Mars has not been so lucky. "The obliquity of Mars has changed by as much as 60° over timescales of just tens of millions of years," says Bell. "It must have been traumatic for the planet's climate."

The cause of Mars's wild swings is the gravitational influence of Jupiter and the other giant planets. On Earth these forces are mediated by the gravitational influence of our large Moon. "Our Moon keeps Earth's spin stable," says Bell. "Mars has no such luck, and it has suffered for it."

> "The pictures don't tell you whether it's water or beer that was flowing on Mars."

Evidence of Water on Mars

NASA's Mars Exploration Program has been successful in its attempts to find evidence of water on the planet. Remember, liquid water cannot exist on Mars's surface today; the atmospheric pressure is so low that any water poured onto the surface would quickly turn to water vapor. But through exploration with orbiters, landers, and rovers, the evidence that Mars's surface *once* supported liquid water has become almost incontrovertible. The evidence of *substantial* quantities of water existing today on Mars, in the form of subsurface ice, has also grown. Recent studies have even shown evidence of a subsurface lake below the southern ice cap.

High-resolution images of the Martian surface show ample evidence that liquid water once ran freely, sculpting features similar to those on Earth. Long, meandering features called **runoff channels** stretch for hundreds of kilometers on Mars. These features—found most often in the southern highlands—take the same form as river systems on Earth. They provide evidence that rivers flowed on Mars long ago, transporting water from high elevations down to the lowlands. With water often comes flooding, and the Martian surface also shows evidence of dramatic, perhaps catastrophic, flows in the form of **outflow channels**, which created teardrop-shaped "islands" around craters and broad systems of plateaus and cliffs. Vast amounts of water flowing around or clearing away obstacles appear to have scoured out these channels. Images taken by *Mars Global Surveyor* even indicate the presence of vast river delta systems, where soils transported from upstream in the highlands were dumped as a river met a larger body of water.

Can astronomers conclude that Mars was once a "blue" world of water lakes and water oceans? For Bell and some other scientists, that's pushing it too far. "We just can't make a claim yet with great confidence," he says. "The pictures don't tell you whether it's water or beer that was flowing on Mars, but they do show that there was liquid, and even that is a big difference from what we see on Mars now."

The best evidence of water on Mars has come from studying minerals on the surface. "The story of the rovers," says Bell, "is one of direct geologic investigation. Only by getting close can you really find the smoking gun." The rovers were equipped with instruments that collected and analyzed samples using techniques such as spectrometry. These instruments have enabled scientists to determine the composition and mineralogy of the Martian surface. "Just like the periodic table of elements is universal," says Bell, "lots of minerals are universal too. What we discovered was a variety of hydrated minerals. That means minerals requiring water for their formation."

An early water-related discovery made by the rover *Opportunity* was the detection of tiny spherical, pebble-like structures in the soil of the Meridiani Planum region just south of the Martian equator. NASA scientists nicknamed these **Martian spherules** "blueberries" (**Figure 7.23**) because of their appearance in false-color images. Analysis showed they are composed of hematite, a mineral associated with the presence of water. Their

VIS

Figure 7.23 "Blueberries"
A false-color image of tiny spherules scattered on the Martian surface. These "blueberries," discovered via the *Opportunity* Mars rover, consist of hematite, a mineral associated with the presence of water.

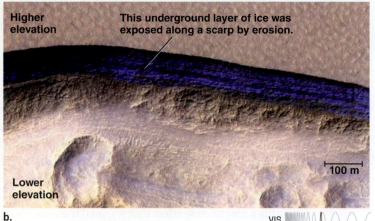

Figure 7.24 Water Ice on Mars
a. The *Phoenix* Mars lander scooped up Martian soil, finding evidence of water ice just below the surface (seen as white patches). **b.** A false-color image, captured by NASA's HiRISE camera, of a thick, steep cross-section of clean subterranean ice exposed by erosion on Mars.

spherical shape led some scientists to believe that the blueberries crystallized out of a water-rich solution. (Mineral deposition in water tends to form somewhat spherical crystals.) Other processes, such as volcanism, can also create spherical shapes. The discovery of blueberries both on the surface and mixed uniformly in deeper layers of Martian soil, however, has led *Opportunity* mission scientists to add them to their list of evidence of water. "As far as I am concerned," says Bell, "this is really unambiguous evidence that liquid water must have been present on Mars when these rocks formed."

The *Phoenix* lander confirmed the presence of subsurface water ice today at the Martian north pole (**Figure 7.24a**). In addition, ice mounds have been seen in craters and in sheets along exposed cliffs (**Figure 7.24b**).

Perhaps the most striking new evidence of water existing on Mars today came in 2018 via the *Mars Express* orbiter. Using a ground-penetrating radar device, researchers found reflections from more than 1 km below the southern polar cap that are best explained by a subsurface "lake" of water approximately 20 km across (**Figure 7.25**). Scientists hypothesized that the water could remain liquid if it is extremely salty. (Salt lowers the freezing temperature of water.) If this result stands up to further study, it may

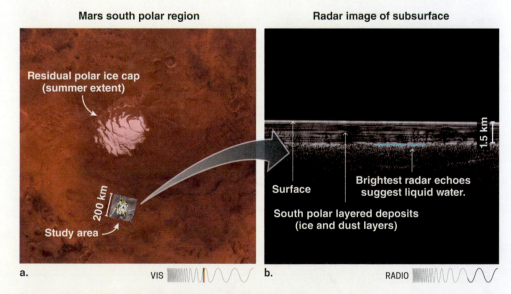

Figure 7.25 A Subsurface Lake of Liquid Water on Mars
The *Mars Express* found ground-penetrating radar evidence of a lake of water buried below Mars's southern polar cap. **a.** The 200-km-square study area (beneath the winter extent of the polar ice cap). **b.** Radar profile from the Martian surface, icy in this region, down to the subsurface. Analysis of the details of the reflected signals yields properties that correspond to liquid water.

runoff channel A long, meandering feature on Mars that resembles a river system on Earth.

outflow channel An extended region of scoured ground on Mars that includes features indicating high rates of fluid flow.

Martian spherule A tiny pebble-like structure ("blueberry") that was found in Martian soil and may have crystallized out of a water-rich solution.

mean that Mars holds some significant liquid water deposits underground to this day (see **Anatomy of a Discovery**).

The History of Mars

Mars has evolved and changed over time, just as Earth has. After intense study, scientists can even identify the key periods in Mars's history.

Scientists call the period from about 4 to 3.5 billion years ago in Mars's history its **Noachian period** (**Figure 7.26**). "Mars was much more active early in its history," explains Bell. "It had a magnetic field. It had a partially molten core. It had an active interior, active volcanoes on the surface. It's completely reasonable to assume that CO_2, sulfur dioxide, and water were coming out of those volcanoes, so the planet had a thicker atmosphere, providing a greenhouse effect that warmed the planet's surface above the melting point of ice. Put it all together, and Mars may have been a much more Earth-like place than what we see today." Bell cautions that *Earth-like* does not mean just like Earth. "There may have been ponds, lakes—and they may have been intermittent—but that doesn't mean Mars was an ocean world. The evidence so far doesn't point in that direction."

Still, conditions in the Noachian period (such as a thicker atmosphere with more greenhouse warming) might have made it conducive for life to begin. "Early Mars was probably really interesting," says Bell. "By our usual definition, it would have been a habitable place."

Some 3.5 to 3 billion years ago, in the **Hesperian period**, everything changed. "The core may have cooled and solidified," says Bell. "We are not sure exactly why, but the planet stopped being volcanically active." The magnetic field declined as the core solidified. With the field's shielding effect reduced, the solar wind was then able to erode the top of the atmosphere, and particles were lost to space because of Mars's low escape velocity. Any

> "Put it all together, and Mars may have been a much more Earth-like place than what we see today."

Figure 7.26 The History of Mars
The periods in the history of Mars are well defined, although uncertainty remains about the dates of transitions. Mars's evolution is characterized in general by decreases in temperature, atmospheric density, and the probability of the existence of microbial life over time.

Noachian period The evolutionary period of Mars from about 4 to 3.5 billion years ago, marked by a thick atmosphere and warmer climate.

Hesperian period The evolutionary period of Mars from about 3.5 to 3 billion years ago, marked by the end of volcanic activity and the slow loss of much of Mars's atmosphere.

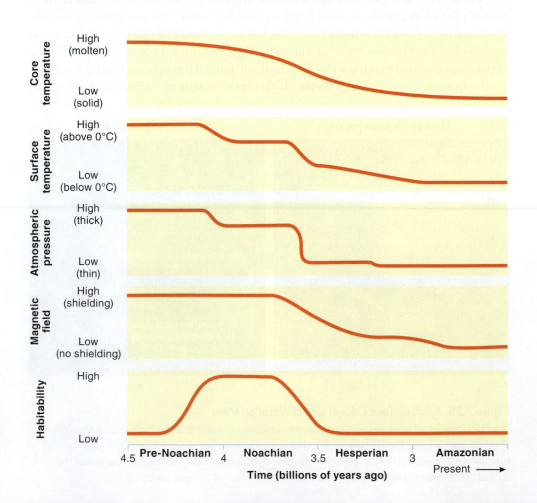

Does water flow on the surface of Mars?

observations & hypothesis

In the late 19th century, **Giovanni Schiaparelli** depicted the Martian surface as covered by "canali" (channels). His drawing fueled speculation that the red planet is covered with water and structures that an intelligent civilization may have built. **Percival Lowell** sketched a similar map of what he believed were canals built to transport water. He built an observatory to study Mars's surface.

Schiaparelli's 1890 map of Martian "canali"

Warrego Valles (*Mariner 9*)

Dendritic channels (*Viking 1* lander)

1 km

40 km

observations

In the 1970s, scientists began sending robotic spacecraft to Mars to search for evidence of water and life up close. Some images returned from the *Mariner 9* and *Viking 1* missions revealed several kinds of Martian surface features typically associated with flowing water on Earth.

confirmed by observation

The orbiting *Mars Global Surveyor* found evidence of hematite, a mineral that tends to form in the presence of water, on Meridiani Planum. The rover **Opportunity**, carrying a camera built by **Jim Bell**'s team, landed there in 2004 and discovered little spherules of hematite. Dubbed "blueberries," they confirmed that water once existed on the surface.

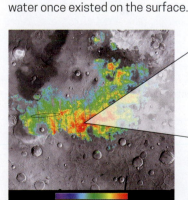

Abundance of hematite on Meridiani Planum

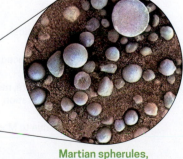

Martian spherules, aka "blueberries"

confirmed by observation?

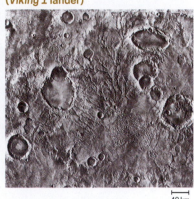

High-resolution imagery captured by NASA's **Mars Reconnaissance Orbiter** in 2011 discovered the seasonal appearance of new gully channels suggestive of flowing, salty, liquid water.

Seasonal gullies in Newton Crater

greenhouse effect diminished over time. "Give these processes a billion years to work . . . ," says Bell, "and you end up with a pretty icy planet."

Once the surface froze, Mars no longer provided what scientists think of as habitable conditions for life. "It might be," speculates Bell, "that microbes, if they did exist, took refuge underground where there was a flux of heat coming from deeper regions."

Eventually the planet entered its deep freeze in the current **Amazonian period**. "Since the beginning of the Amazonian about 3 billion years ago," says Bell, "there has just been basically nothing going on except the wind moving across the surface, the occasional impact crater, or maybe an occasional burst of very, very late volcanism."

Mars shows that planets that begin in the range of habitability can change with time. That is a conclusion of fundamental importance for us as inhabitants of the only planet in our Solar System still harboring conditions favorable to life.

The key features of each terrestrial planet are summarized in **Table 7.1**.

Amazonian period The evolutionary period of Mars that began about 3 billion years ago, marked by low atmospheric mass and dry, cold surface conditions.

Table 7.1   Terrestrial Planet Fact Sheet

	Mercury	Venus	Earth	Mars
Radius (km)	2,440	6,052	6,378	3,397
Radius (Earth radii)	0.38	0.95	1	0.53
Mass (kg)	3.30×10^{23}	4.87×10^{24}	5.97×10^{24}	6.42×10^{23}
Uncompressed Density (kg/m³)	5,300	4,400	4,400	3,800
Magnetic Field	Fossil magnetism only	No	Significant	Very weak or none
Atmosphere	None	Primarily carbon dioxide	Nitrogen and oxygen	Primarily carbon dioxide
Atmospheric Density Compared to Earth	N/A	90 times as dense	1	1/150th as dense
Surface Temperature (K)	93–703	730	260–293	183–268
Polar Caps	No	No	Water ice	Carbon dioxide and water ice
Orbital Radius (AU)	0.39	0.72	1	1.52
Orbital Period (Earth days)	88	224	365	687
Rotation Period (Earth days)	59	243	1	1.03 (24 h 37 min)
Liquid Water	No	No	Yes	No?
Tectonic Activity	Yes	No	Yes (plate tectonics)	No
Age of Surface (years)	4 billion	1 billion	Continually resurfaced	3–4 billion
Surface Features	Many craters; cliffs, high mountains	Few craters; shield volcanoes, high mountains	Oceans, continents, mountains, volcanoes, vegetation	Canyon network, shield volcanoes, craters
Tidal Effects	Tidal synchronization of rotation and orbit	No	Ocean tides due to Moon's orbit	No
Moons	No	No	One	Two: Phobos and Deimos
Other	Orbital eccentricity 0.206	Sulfuric acid rain; retrograde rotation	Abundant life	Evidence of surface water in past

Section Summary

- From direct measurements and calculations, scientists believe that Mars has a mantle and crust like Earth's, surrounding a small, dense core.
- Surface features on Mars are dramatic: vast canyons, high mountains, and broad shield volcanoes. Mars has seasons, with CO_2 ice caps that regularly grow and shrink on both poles; the northern polar cap includes water ice.
- Mars's low gravity allowed heavier gases to escape its surface. CO_2 is the most plentiful atmospheric gas; oxygen is only a trace element.
- Mars lacks a global magnetic field. It has two tiny moons; without the stabilizing factor of a larger moon, Mars's obliquity has varied greatly over time.
- Evidence strongly suggests that Mars once had liquid water on its surface and harbors some subsurface water today.

CHECKPOINT **a.** Why can't liquid water exist on the surface of Mars today? **b.** Why might it have existed in previous eras of Mars's history?

Outward Ho!

With our tour of the terrestrial planets complete, we are ready to turn our gaze outward to the outer planetary domains of the Solar System. While our concern with the inner planets focused on surfaces and their evolution, our concern with the outer planets will have to shift. The outer planets are gas and ice giants; they do not even have surfaces. What they do have, however, are environments unlike anything we experience on Earth. From a cornucopia of water-rich moons to gossamer ring systems, the outer Solar System is a frontier that has only recently been opened to human exploration. It is strange and awesome territory, and it has its own story to tell us about planetary evolution and the possibilities of life.

CHAPTER SUMMARY

7.1 Planet Stories

Comparative planetology is the study of similarities and differences between planets in order to understand their natures and the underlying processes that shaped them. All of the terrestrial planets in our Solar System have a core-mantle-crust configuration, but they differ in interior structure, surface evolution, atmosphere, and near-space environment. The thermal history of these planets has driven their evolution through five distinct stages: differentiation, cooling, cratering, magma flooding, and weathering.

7.2 Mercury: Swift, Small, and Hot

Mercury, closest to the Sun, is hot and lifeless and orbits faster than any other Solar System planet. Its high density suggests a large iron core. It has been mostly geologically dead for 4 billion years and bears many ancient craters. Because of its small size, it cooled quickly, leaving distinctive surface features such as scarps. Any initial atmosphere evaporated, and surface temperatures range widely, from 700 K in daylight to 93 K at the poles. Mercury has a weak dipole magnetic field offset from its center. Mercury's rotation period has been influenced by tidal synchronization with its orbital period, creating a solar day lasting 176 Earth days.

7.3 Venus: Hothouse of the Planets

Venus, swathed in a dense atmosphere (mostly CO_2), is uniformly hot, with a surface temperature of 730 K because of a runaway greenhouse effect. The nearly Earth-sized planet has retrograde rotation that takes 224 Earth days, most likely the result of a collision early in its history. Radar studies reveal a flat, dry, broken surface with few craters and many shield volcanoes. These findings suggest that Venus's thin crustal plates have allowed volcanism to reshape the surface relatively recently in the planet's history. Venus has no magnetic field.

7.4 Mars: The Red Planet of Change

Mars is likely to have harbored water on its surface earlier in its history. Mars and Earth are similar in rotation period, seasonal changes, and the presence of polar ice caps. Mars's surface has been studied by orbiters, landers, and rovers, which have produced evidence of subsurface water but not current or past life. Volcanism on Mars has produced vast shield volcanoes (such as the enormous Olympus Mons). A thin CO_2 atmosphere creates strong winds and planet-enveloping dust storms. Two small, irregular moons orbit Mars. Astronomers posit three distinct evolutionary periods for Mars: the Noachian, the Hesperian, and the Amazonian.

QUESTIONS AND PROBLEMS

Narrow It Down: Multiple-Choice Questions

1. Which of the following characteristics do the four terrestrial planets have in common? Choose all that apply.
   a. a significant magnetic field
   b. plate tectonics
   c. an atmosphere
   d. the presence of one or more moons
   e. a rocky crust

2. Which of the following statements accurately describe(s) all terrestrial planets? Choose all that apply.
   a. They condensed from the part of the solar disk that had no solid hydrogen compounds (ices).
   b. They differentiated into core-mantle-crust.
   c. They maintain a presence of liquid water.
   d. They are warmed by means of a greenhouse effect.
   e. They display some evidence of cratering.

3. Which of the following is *not* a source of heat in the interiors of the terrestrial planets?
   a. friction between atmosphere and surface
   b. collisions from other bodies
   c. force of gravity
   d. radioactive decay
   e. differentiation

4. Four planets that formed together have equal density but different radii. Ignoring any other factors, which planet would cool fastest?
   a. planet W, with radius 2,000 km
   b. planet X, with radius 2,500 km
   c. planet Y, with radius 2,700 km
   d. planet Z, with radius 5,000 km
   e. There is no way to know.

5. A caldera is which of the following?
   a. the depression at the top of a volcano
   b. a long ridge formed by planetary shrinkage
   c. a depression that filled with lava
   d. a crest formed by the meeting of two tectonic plates
   e. a channel dug by the glancing blow of an asteroid

6. Which of the following bodies exhibit(s) rotation patterns clearly influenced by tidal effects? Choose all that apply.
   a. Earth's Moon
   b. Mercury
   c. Venus
   d. Deimos
   e. Mars

7. Which body has the longest "day" (a complete day-night cycle)?
   a. Mercury
   b. Venus
   c. Earth
   d. Mars
   e. Earth's Moon

8. Which of the following statements about Venus is/are *not* true? Choose all that apply.
   a. It rotates retrograde.
   b. All of its volcanoes are shield volcanoes.
   c. Its surface was mapped by Russian landers.
   d. It may have crustal plates.
   e. On average, it has as many craters per square meter as Mercury does.

9. Which of the following geologic features is/are found on Mars? Choose all that apply.
   a. Ishtar Terra
   b. Valles Marineris
   c. Discovery Rupes
   d. Olympus Mons
   e. Stickney Crater

10. Rank the terrestrial planets by the ages of their surfaces, oldest to newest.
    a. Mercury, Venus, Earth, Mars
    b. Mercury, Mars, Venus, Earth
    c. Mars, Earth, Venus, Mercury
    d. Venus, Mercury, Mars, Earth
    e. Mars, Mercury, Earth, Venus

11. Which of the following statements about the atmosphere of Venus is *not* true?
    a. Its composition has changed significantly over time.
    b. It is 90 times less dense than Earth's atmosphere.
    c. It is mostly CO_2.
    d. It contains sulfuric acid clouds.
    e. Its heat and acidity quickly disable spacecraft that land on the surface.

12. Imagine Minutia, a terrestrial planet with mass m and radius r. A second planet, Garganzo, has mass $1.4m$ and radius $1.4r$. How does the surface gravity on Garganzo compare with that on Minutia? (See Going Further 3.2.)
    a. It is about the same.
    b. It is greater.
    c. It is less.
    d. It depends on the density of the two planets.
    e. It depends on whether either or both have an atmosphere.

13. Which two planets are closest in rotation period?
    a. Mars and Venus
    b. Mars and Mercury
    c. Mercury and Earth
    d. Venus and Earth
    e. Earth and Mars

14. Which of the following describe(s) commonalities in the ice caps of Earth and Mars? Choose all that apply.
    a. Both planets have ice caps on both poles.
    b. Both planets' ice caps have residual components.
    c. The composition of the ice is the same on both planets.
    d. Seasonal changes occur in the ice caps of both planets.
    e. Each planet has roughly identical ice caps in its northern and southern hemispheres.

15. If Deimos were orbiting at its current orbital radius around Earth instead of around Mars, how would its orbital velocity differ?
    a. It would be the same.
    b. It would be slower because of Earth's greater radius.
    c. It would be slower because of Earth's greater mass.
    d. It would be faster because of Earth's greater radius.
    e. It would be faster because of Earth's greater mass.

16. Which of the following affect(s) the climate of a planet? Choose all that apply.
    a. greenhouse gases
    b. obliquity
    c. orbital radius
    d. rotation period
    e. presence of a large moon

17. During which of the following periods of Martian evolution would life have been most likely to form?
    a. Noachian
    b. Hesperian
    c. Pre-Noachian
    d. Amazonian
    e. Paleozoic

18. The radius of planet APC-11 is r, and the radius of planet APC-23 is $2.2r$. How do their cooling times compare?
    a. That depends on the volumes of both planets.
    b. Both planets will take about the same time to cool.
    c. APC-11 will take about twice as long to cool as APC-23.
    d. APC-11 will take about half as long to cool as APC-23.
    e. APC-11 will cool about 4 times as fast.

19. A planet orbits at 2 AU from a distant 5-billion-year-old star. The planet has minimal cratering, a strong magnetic field, tall composite volcanoes, and a larger obliquity angle than Earth's. From observations of our Solar System, which of the following is/are likely to be true of the planet? Choose all that apply.
    a. It has a large molten core.
    b. It has multiple moons.
    c. It has tectonic plates.
    d. It has been geologically dead for billions of years.
    e. It experiences no seasons.

20. Which of the following is the proper order of the stages of planetary evolution?
    a. weathering, magma flooding, differentiation, cratering, cooling
    b. cratering, magma flooding, differentiation, cooling, weathering
    c. differentiation, magma flooding, cratering, cooling, weathering
    d. differentiation, cooling, cratering, magma flooding, weathering
    e. cooling, differentiation, magma flooding, cratering, weathering

To the Point: Qualitative and Discussion Questions

21. Describe at least three surface features or characteristics of Mercury's surface, and indicate how each provides evidence of the planet's history.

22. Describe at least three surface features or characteristics of Venus's surface, and indicate how each provides evidence of the planet's history.

23. Describe at least three surface features or characteristics of Mars's surface, and indicate how each provides evidence of the planet's history.

24. What is planetary differentiation, and what is its end result?

25. Rank the four terrestrial planets by their current degree of visible cratering, from least cratering to most cratering.

26. Where in our Solar System is plate tectonics known to exist, and why does it not exist on the other planets?

27. Describe the different kinds of volcanoes found on each of the terrestrial planets.

28. What are the requirements for the formation of a planetary magnetic field?

29. What factors have changed on Mars that have made it less suitable (even hostile) for the evolution of advanced life-forms?

30. Describe the differences in the number, size, and shape of moons for the terrestrial planets.

31. Views change frequently about the evidence of life on Mars. Explore recent scientific articles on this topic. How convincing do you find the evidence?

32. Impacts by planetesimals, asteroids, and comets on young planets are a normal part of planetary evolution. For each of the terrestrial planets and moons, name at least one effect caused by such collisions.

33. The Martian day is slightly longer than Earth's day. Engineers and scientists working with Mars orbiters and rovers must sometimes adjust to Martian time for maximum mission efficiency. How might this adjustment affect rover mission personnel?

34. An enhanced greenhouse effect on Earth due to climate change could create a more Venus-like surface and atmosphere on Earth. Describe how such changes might affect life on Earth.

35. What would be some of the greatest challenges to the physical and psychological well-being of the astronauts involved in a manned spaceflight to Mars?

Going Further: Quantitative Questions

36. If Venus had its current radius but a mass equal to Earth's, how many times greater would Venus's escape velocity be? (See Going Further 6.2.)

37. Calculate the escape velocity, in kilometers per second, of a planet with mass equal to that of Mars and radius equal to that of Mercury.

38. What is the orbital speed, in kilometers per second, of a planet with 2 times the orbital radius and 4 times the mass of Mars? (See Going Further 5.1.)

39. The orbital speed of Venus is 35.0 km/s. What would this value be, in kilometers per second, if Venus's orbital radius were half as great?

40. The radius of an exoplanet is 3 times that of Earth. What is the ratio of Earth's cooling time to the exoplanet's cooling time?

41. A terrestrial planet with radius r cooled to 287 K over 1.4 billion years. How long would it take another terrestrial planet that formed with the same initial temperature and with radius $2.7r$ to cool to 287 K?

42. Use Kepler's third law to calculate the orbital periods of two hypothetical planets that are 1.2 and 2.5 AU, respectively, from the Sun.

43. Calculate the stable temperature that Mars would have without an atmosphere. (Mars's albedo is 0.15.)

44. Find the stable temperature of an airless world that has an albedo of 0.220 and orbits 7.8×10^{11} meters from a star with radius 1.7×10^9 meters and surface temperature 6,400 K.

45. A planet that has no atmosphere, an albedo of 0.550, and an orbital radius of 9.1×10^{12} meters orbits a star with temperature 8,700 K and radius 4.5×10^{10} meters. What is the planet's stable temperature?

"IT FELT LIKE I WAS WALKING ACROSS SATURN'S RINGS AND LOOKING THROUGH THEM TO THE STAR"

8

Gas, Ice, and Stone

THE OUTER PLANETS

Giant Planets on a Roll

8.1 Linda Spilker dreamed of seeing Saturn's rings up close. But she never imagined that the first detailed view would come on computer paper rolled out across a hallway in NASA's Jet Propulsion Laboratory (JPL).

Spilker was fresh out of college when she joined JPL and the *Voyager* team. The two *Voyager* probes, launched in 1977, were designed to make a grand tour of the giant planets (**Figure 8.1a**) before heading into interstellar space. Spilker, one of few women on the *Voyager* science team then, had a ringside seat for humanity's early explorations of Saturn and its beautiful, mysterious rings.

"We were studying the rings by watching how a background star dimmed and brightened as *Voyager 2* passed behind ring material," Spilker recalls. *Voyager 1* images had shown a wealth of substructure in the rings that looked like grooves on a phonograph record. "Those images were just incredible," says Spilker, "but with our **stellar occultation** method, where we used the blocking of starlight by ring material, we were able to detect structures much smaller: less than 100 meters."

This was 1979, however. "In those days you couldn't easily put a plot up on your screen," she recalls. "We ran the long sheet of paper through a printer and had a computer plot the points on it as the star went behind the rings. I remember unrolling it in the hall and walking down its length. It felt like I was walking across Saturn's rings and looking through them to the star, just as *Voyager* was doing so far out there in the outer Solar System. It was an amazing feeling."

Beyond the Snow Line

The outer realms of the Solar System bear little resemblance to the terrestrial worlds. "It's a different kind of Solar System out there," says Amanda Hendrix, the deputy project scientist of the *Cassini* mission to Saturn from 2010 to 2012 (**Figure 8.1b**). Only through the most advanced technologies, including the robot deep-space probes that Spilker and

IN THIS CHAPTER

you will learn about the giant planets—the large, ringed, moon-rich siblings of the terrestrial planets. After reading through each section, you should be able to do the following:

8.1
Giant Planets on a Roll Explain how the vast distances between the Sun and the giant planets create conditions quite different from those on terrestrial planets.

8.2
The Giant Planets: Structures and Processes Summarize the similarities and differences in the processes shaping the giant planets and the structures that have evolved as a result of those processes.

8.3
Jupiter: King of Planets Describe the characteristics that make Jupiter and its satellites unique and important to the structure and evolution of the Solar System.

8.4
Saturn: Lord of the Rings Describe the characteristics of Saturn and its satellites.

8.5
Uranus and Neptune: Ice Giants Discovered in Twilight Distinguish the characteristics and properties of the ice giants from those of the gas giants.

← Jupiter, the largest planet in our Solar System, captured by NASA's *Juno* spacecraft during a close flyby of the gas giant in 2016. The banded structure in its atmosphere results from the planet's rapid rotation and heat flowing from its interior. The white spot is a storm—a frequent occurrence.

stellar occultation The blocking of a star's light by an object between the star and an observer's view.

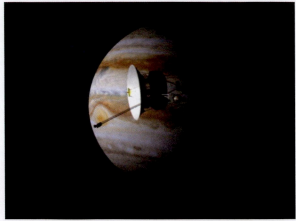

a.

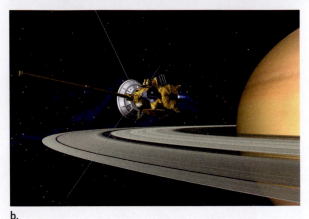

b.

Figure 8.1 *Voyager* and *Cassini*
a. An artist's representation of one of the twin *Voyager* spacecraft, passing Jupiter. *Voyager 1* and *2* were launched in 1977. **b.** An artist's conception of NASA's *Cassini* spacecraft, framed by Saturn's rings. The *Cassini* mission, launched in 2001, explored Saturn and its rings and moons from 2004 to 2017.

"There is so little light that we have to adjust the cameras on probes like *Cassini* to get good images."

INTERACTIVE:
Equilibrium Temperature

Hendrix oversaw, have we been able to see any part of that dark domain up close and explore the outer Solar System.

To get a sense of the difference between the Solar System's inner and outer domains, let's consider two fundamentals: light and warmth. "In the outer Solar System, the Sun is faint, it's small, and it does not provide as much heat," says Hendrix. The farther you travel from a light source, the fainter it appears (**Figure 8.2**). Jupiter, the closest of the outer planets, orbits at 5 astronomical units (AU) from the Sun. "At this distance," Hendrix explains, "sunlight has dropped in its intensity by 25 times compared with what we get on Earth." Not only does the Sun appear dimmer; it appears smaller too. Rather than the 32-arcminute (32′) diameter of the solar disk that we see on our sky, for an observer on Jupiter the Sun would span just 6′—looking only one-fifth as large as it appears from Earth. Thus Jupiter, Saturn, Uranus, and Neptune exist in half-light. "There is so little light," she explains, "that we have to adjust the cameras on probes like *Cassini* to get good images. The exposures have to be a lot longer than what you need around Earth or Mars, because of the Sun's distance."

In addition to illumination, sunlight brings heat. Without the full effect of the Sun's warmth that we enjoy, the outer realms of the Solar System are colder as well as darker. Going Further 7.2 showed how to calculate the surface temperature of a planet with no atmosphere—its *equilibrium temperature*—if we know the planet's distance from the Sun. At the location of Earth, a planet without an atmosphere would be chilly but bearable, with a temperature of about 254 K (or −2°F). For Jupiter, however, that temperature drops to 124 K (−236°F), and the equilibrium temperature at the distance of Neptune's orbit is just 59 K (−353°F).

The decreasing temperatures far from the Sun also greatly influenced the formation of the giant planets. Recall from Chapter 5 that the giants lie outside the *snow line*. "The snow line is an imaginary boundary at somewhere around 3 AU from the Sun (depending on how it's calculated)," says Hendrix. "Inside the snow line, a newly forming planet accumulates lots of silicates and metals. Outside the snow line, water is stable as a solid. That means once you get beyond the snow line, you should expect water ice and other frozen compounds to play a big role in both planets and moons."

Even beyond the snow line, sunlight still plays a role. "We see seasonal variations on the giant planets in response to how much sunlight they get, and we see some kinds of chemistry happening on both the giant planets and their moons that depend on energy from the Sun," says Hendrix. "So the Sun is still affecting these outer worlds, but it just doesn't play as important a role as it does for the inner planets." In addition, the solar wind

Views of the Sun and its relative effect on each planet

	Earth	Jupiter	Saturn	Uranus	Neptune
Sun's angular size (arcminutes, or ′)	32′	6′	3′	2′	1′
Sun's brightness relative to Earth	1	0.037	0.01	0.003	0.001
Planet's equilibrium temperature (K) (no atmosphere)	254	124	95	59	31

Figure 8.2 The Sun and the Giant Planets
The Sun appears much smaller and less bright on the giant planets' skies than from Earth and correspondingly provides less light, leading to much lower equilibrium temperatures.

and its magnetic field stream through these regions as they do through the inner Solar System.

As we begin to explore the outer planets, we must again develop questions and guiding principles, as we did for the inner planets. This chapter, like Chapter 7, focuses on planetary evolution and the possibilities for life. The outer Solar System giants, however, raise profound new questions about both issues. What kinds of processes made these planets so different from the inner, rocky worlds? Must we think more broadly about where and how life might exist?

Section Summary

- Because of their great distance from the Sun, the outer planets receive less light and warmth than the terrestrial planets do. Solar radiation still affects the outer planets and produces seasons.
- The giant planets lie beyond the snow line, where water and other compounds freeze.

CHECKPOINT How would the location of the snow line change if the Sun were much brighter?

The Giant Planets: Structures and Processes

8.2 Just as all terrestrial planets share certain structural similarities, so do the giant planets, and these similarities anchor our understanding of their evolution. "A lot of what we see in the outer Solar System reflects the conditions out there beyond the snow line when the planets formed from the protosolar nebula," says Amanda Hendrix. "In the end, the questions come down to what kinds of physics and chemistry the temperatures in that initial disk of gas and dust allowed."

Raw Facts: Size, Mass, Density, and Composition

The giant planets are huge in both size and mass, compared with their terrestrial cousins. Jupiter, the largest giant, has a radius 11 times larger than Earth's radius (**Figure 8.3**), and its mass is a whopping 318 times that of Earth. Jupiter's volume is so great that 1,321 copies of Earth could hide beneath Jupiter's clouds. Saturn, the second-largest planet, has a radius that's slightly smaller than Jupiter's, though its mass is only about one-third that of its larger sibling. Uranus and Neptune are smaller and are almost twins in size and mass: both planets' radii are roughly 4 times that of Earth, and their masses are roughly 15 and 17 times Earth's, respectively. "The sheer volume and mass of these big planets determine a lot of the amazing behavior we see around them," says Hendrix.

> "The sheer volume and mass of these big planets determine a lot of the amazing behavior we see around them."

Figure 8.3 Giant Planet Radii
The relative sizes of gas and ice giants, compared with Earth (the largest terrestrial planet). All giant planets have ring systems, though not all are visible in these images.

Jupiter   50,000 km   Saturn   Uranus   Neptune   Earth

VIS

As **Table 8.1** shows, the giant planets have much lower densities than the terrestrial planets. These densities are found by using the motion of a planet's satellites to determine its mass (via Newton's version of Kepler's third law; see Chapter 3) and then dividing that mass by the planet's volume, which can be found via measurements of its radius. (The volume of a sphere is $\frac{4}{3}\pi R^3$.) "The low densities mean that the giants must be made mostly of material that is lighter than the silicate rocks and metallic cores composing the terrestrial worlds," says Hendrix. Jupiter and Saturn are composed mainly of the lightest elements, hydrogen and helium, with the outer layers in gaseous form; they are therefore called *gas giants*. Uranus and Neptune, in contrast, contain larger amounts of heavier elements such as carbon, oxygen, and nitrogen. These heavier elements are locked into chemical compounds with the abundant hydrogen, creating substances such as methane (CH_4), ammonia (NH_3), and water (H_2O). At the distances of Uranus's and Neptune's orbits, temperatures are low enough for these compounds to have frozen into something like a slush. Thus, the outermost planets likely have a rocky core surrounded by icy material and are called *ice giants*.

Internal Structure

The giant planets lack surfaces, so there are no distinct transitions between their atmospheres and their interiors. If you could dive downward into a giant planet's atmosphere, you would find the density and pressure increasing continuously. Rather than eventually crashing into a hard surface, you would pass smoothly from the gaseous state of the atmosphere to the liquid state of the interior as the pressure around you increased. At their centers, all giant planets contain a core. "Certainly there's some kind of metal/rock core in each of these worlds," says Hendrix, "but the conditions will be extreme and there is a lot we have to learn about how materials behave under these kinds of conditions" (**Figure 8.4**).

These extreme conditions are, in part, the result of enormous pressure. The masses of the giant planets are so great that gravity squeezes down hard on their interiors, creating pressures high enough for unusual kinds of physics to kick in. The surface pressure on Earth is measured in a unit called a *bar*. (The pressure at sea level is about 1 bar.) At Earth's center, underneath all that rock and metal, pressures can be 4 million times surface pressure, or 4 **megabars**. Pressures in the central regions of Jupiter may reach 30–45 megabars. Greater mass thus creates greater pressure, which determines what kinds of structures exist between the core and atmosphere.

In addition, the giant planets produce more heat than they receive from the Sun (although for Uranus, the excess is tiny). The source of this heat is thought to be gravitational contraction. As the same amount of mass is squeezed into a smaller space, atoms collide more frequently and more energetically with one another, generating heat. This heat eventually makes it to the surface, where it escapes as radiation. In other words, gravitational energy is converted into thermal energy and then radiated away as radiant energy, in accordance with the law of conservation of energy.

Atmospheres and Rotation

The gas giants Jupiter and Saturn are among the most beautiful objects in the Solar System, in large part because of the banded cloud structures in their atmospheres. The atmospheric structures of Uranus and Neptune are similar but less dramatic. "The remarkable atmospheric patterns we observe in the gas giants, and to a lesser degree in the ice giants, originate in the development of clouds forming at different layers," explains Hendrix. Depending on the pressures and temperatures in each layer, different compounds condense into clouds at different altitudes, giving each planet a distinctive appearance.

The atmospheres of the inner and outer worlds differ in significant ways. "Lots of basic processes are the same from one planet to another," says Hendrix. "But then when you talk about Jupiter and Saturn, you're talking about much more gas and much more

INTERACTIVE:
Anatomy of Gas and Ice Giant Planets

megabar A unit of pressure equal to a million times Earth's surface pressure.

Table 8.1 Giant Planet Fact Sheet

	Jupiter	Saturn	Uranus	Neptune
Type	Gas giant	Gas giant	Ice giant	Ice giant
Mass (Earth masses)	318	95	15	17
Radius (km)	69,911	58,232	25,362	24,622
Radius (Earth radii)	11.0	9.1	4.0	3.9
Volume (Earth volumes)	1,321	764	63	58
Density (kg/m³)	1,330	690	1,300	1,760
Magnetic Field	Strong	Strong	Strong	Strong
Atmosphere	Primarily H, He	Primarily H, He	Primarily H, He	Primarily H, He
Surface Temperature (K)	~165	~124	~73	~73
Orbital Radius (AU)	5.2	9.5	19.2	30.1
Orbital Period (Earth years)	12	29	84	165
Rotation Period	9 h 50 min	10 h 14 min	17 h 14 min	16 h 6 min
Liquid Water	No	No	No	No
Plate Tectonics	No	No	No	No
Age of Surface	Gas: constantly changing	Gas: constantly changing	Ice: young compared to terrestrial planets	Ice: young compared to terrestrial planets
Upper Atmosphere Features	Storms including Great Red Spot; banded clouds	Storms including north polar hexagon; banded clouds	Storms; banded clouds	Storms; banded clouds
Tidal Effects	Rings; impacts on moons	Rings; impacts on moons	Rings; impacts on moons	Rings; impacts on moons
Number of Known Satellites (in 2018)	67	62	27	14
Primary Elements and Compounds	H, He	H, He	H, He, CH_4, NH_3, H_2O	H, He, CH_4, NH_3, H_2O
Other	Large magnetosphere	Large, complex ring structure	Rotates nearly on its side (98° inclined to ecliptic)	Largest moon (Triton) orbits retrograde

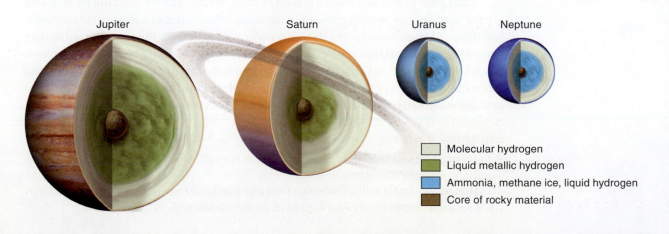

Jupiter   Saturn   Uranus   Neptune

- Molecular hydrogen
- Liquid metallic hydrogen
- Ammonia, methane ice, liquid hydrogen
- Core of rocky material

Figure 8.4 Giant Planet Interiors

The relative densities of the giants indicate that gas giants consist largely of hydrogen in increasingly dense forms, including a planet-covering "ocean" of molecular hydrogen atop a liquid metallic hydrogen layer surrounding a relatively small "rocky" core. Ice giants have a larger rocky core relative to the overall mass of the planet with overlying layers of water, ammonia, and methane, mainly in ice form, and hydrogen.

AMANDA HENDRIX

It can take decades for a project such as *Cassini* to go from proposal to design to assembly to launch and, finally, arrival. Amanda Hendrix, now a senior scientist at the Planetary Science Institute, was with the project from its launch in 1997. Seven years later, her most cherished moment in an impressive career came when she viewed the *Cassini* data.

"We were having a little image-viewing celebration as the data were coming back from *Cassini*," says Hendrix. "We were looking at images of Saturn's outer moon Rhea. There was so much detail I was stunned. I am a moon person, so this was really exciting to me. Finally I was seeing one of those moons up close!"

"There is a lot of water on these moons, and that makes us wonder if they might be places where life evolved."

pressure variation. That makes for a lot more drama." Another key difference between terrestrial and giant planet atmospheres is the giant planets' remarkably rapid rotation. Saturn, in particular, rotates so fast that the shape of the planet is noticeably distorted. "The rotation of the giant worlds makes the behavior in their atmospheres—including storms—a lot more dramatic and more long-lived." Recall from Chapter 6 that rotation shapes Earth's large-scale patterns of circulation, as convection drives hotter gas to rise and cooler gas to fall. "The rapid rotation of the giant worlds sends this process into overdrive," says Hendrix.

The gas giants in particular spin so fast as a result of the way they formed and the conservation of angular momentum. These planets gravitationally pulled in a lot of gaseous material from the protostellar disk. This gas was drawn from a wide region (a band) around the planetary orbit, and all of it had angular momentum as it spun around the new Sun. Because the gas was gathered into the relatively small space of a planet (even a giant one), conservation of angular momentum requires that the planet be a fast rotator. (Think of the spinning skater in Figure 3.8.)

Magnetic Fields

Another structural similarity across the giant planets is the presence of strong, large-scale magnetic fields. "Everything's bigger in the giant planets, so it's not terribly surprising that they have these giant magnetic fields around them," says Hendrix. The giant planets' rapid rotation and the presence of circulating gas flows from lower to higher depths drive the fields. "What is surprising is how much diversity there is in the structure of the fields among the giants." Differences in the location of conducting layers, where the flows creating a dynamo are active, lead to remarkable differences in the orientation and alignment of the giant planets' magnetic fields. In all cases, the fields extend well beyond the planets' atmospheres, forming protective regions around their local systems of moons (**Figure 8.5**).

Moons and Rings

Although moons are rare in the inner Solar System, every gas and ice giant hosts a family of satellites. Jupiter and Saturn each have more than 60 natural moons, 27 satellites have been found to date orbiting Uranus, and 14 have been found around Neptune (**Table 8.2**). "There are a lot more moons orbiting the giant planets," says Hendrix, "because they have so much mass and, therefore, stronger gravity." What is even more remarkable than the sheer numbers of these moons is the fact that some show evidence of thick layers of ice, under which may lie kilometers-deep subsurface oceans. "There is a lot of water on these moons," says Hendrix, "and that makes us wonder if they might be places where life evolved."

Each giant planet also hosts a system of rings. While Saturn's beautiful banded arcs are the most famous, robotic probes and telescopic observations have revealed narrower, less dramatic rings surrounding the other outer worlds as well. Rings are not solid; they appear to be made of billions of relatively small orbiting particles of ice and rock. In most cases, the particles don't stay in orbit for long, which suggests to astronomers that the moons are the origin of ring particles. Small bits of rock and ice that were blasted off the orbiting moons may continuously supply the rings with new material.

Section Summary

- The giant planets are larger than the terrestrial planets in radius and mass, have no solid surfaces, and form pressure-related interior layers.
- The gas giants (Jupiter and Saturn) consist mainly of the lightest elements. The ice giants (Uranus and Neptune) formed in the colder reaches of the Solar System.

Table 8.2   Selected Moons of the Giant Planets

Unique to the giant planets are the large number of moons orbiting them and the variety of sizes and orbital patterns of those moons.
Regular moons have prograde orbits that are aligned with the planet's equatorial plane; irregular moons may have retrograde orbits and/or tilted orbits.

Planet	Moon	Radius (km)	Mass (kg)	Orbital Period (days)	Unique Features
Jupiter	Io	1,821	8.93×10^{22}	1.8	Active volcanism; lava and sulfur plumes
	Europa	1,565	4.80×10^{22}	3.6	Iceberg surface; deep subsurface ocean
	Ganymede	2,634	1.48×10^{23}	7.2	Grooved surface; subsurface water
	Callisto	2,403	1.08×10^{23}	16.7	Subsurface water ocean but geologically dead
Saturn	Pan	14	4.95×10^{15}	0.6	Orbits in Encke Division
	Prometheus	43	1.60×10^{17}	0.7	Shepherd moon for F ring
	Pandora	40	1.37×10^{17}	0.7	Impacts F ring
	Mimas	198	3.74×10^{19}	1.0	Gravitational effects on Cassini Division
	Enceladus	252	1.08×10^{20}	1.4	Liquid water geysers
	Titan	2,575	1.35×10^{23}	16.0	Only Solar System moon with substantial atmosphere; liquid methane lakes
Uranus	Miranda	236	6.59×10^{19}	1.5	Faults, ridges, valleys; innermost moon
	Ariel	579	1.35×10^{21}	2.6	Long surface cracks
	Umbriel	584	1.17×10^{21}	4.2	Geologically dead
	Titania	788	3.53×10^{21}	8.8	Ancient craters
	Oberon	761	3.01×10^{21}	13.5	Ancient craters
Neptune	Proteus	210	5.04×10^{19}	1.2	Active geysers; orbits in rings
	Triton	1,352	2.14×10^{22}	8.9	Retrograde orbit; thin atmosphere
	Nereid	170	2.70×10^{19}	360.2	Very eccentric orbit

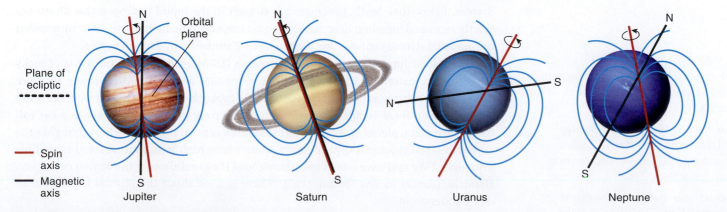

Figure 8.5 Giant Planet Magnetic Fields
The four giant planets, shown with their spin axes positioned relative to their planes of orbit around the Sun, the ecliptic, their magnetic axes, and the relative inclinations and orientations of their magnetic fields. Each giant planet has a significant magnetic field, but their orientation varies greatly, and some fields are off-center.

- The atmospheres of the giant planets are much deeper and more layered than those of terrestrial planets, with strong pressure variations that lead to banded cloud layers.
- Each giant planet exhibits rapid rotation (which contributes to conditions such as storms); a large, strong magnetic field; many diverse moons; and a ring system.

CHECKPOINT How does the position of the outer planets relative to the snow line affect conditions on those planets, compared with conditions on the inner planets?

Jupiter: King of Planets

8.3 Jupiter, named after the ruling Roman god, is without a doubt the king of the planets. Excluding the Sun, this gas giant accounts for 63 percent of the Solar System's mass, and its enormous gravity profoundly affects much of the other 37 percent. In 1994, astronomers watched in amazement as the comet Shoemaker-Levy 9 was pulled apart by Jupiter's gravity and then plunged in fragments into the planet's atmosphere. It was the first time a collision between a planet and a comet was seen, though the Solar System's impact craters give ample evidence that such collisions occurred in the past. "It just so happens that we were lucky to catch this one," says Amanda Hendrix.

Astronomers gained a treasure trove of new insights watching Shoemaker-Levy 9 hit Jupiter. "We learned about Jupiter's atmospheric chemistry," says Hendrix. "We learned about its atmospheric structure. We even learned [more] about Jupiter's rings because the passing comet fragments created a tilt or warp in the rings." Most important, however, astronomers saw direct evidence that giant planets may act as a kind of planetary vacuum cleaner, sweeping the Solar System of wandering debris that could pose a threat to Earth.

Jupiter's Interior

The composition of Jupiter is much like that of the Sun: by mass it contains about 89 percent hydrogen and 10 percent helium, with the rest made up of heavier elements. Jupiter's balance of elements reflects the composition of the interstellar cloud out of which the Sun and its planets originated.

Using Jupiter's known mass, size, and elemental abundances, scientists can build models of the planet from atmosphere to interior. Beginning where the atmosphere merges smoothly into a liquid is a molecular hydrogen ocean that extends about 16,000 kilometers (km), making it the largest internal planetary structure in the Solar System. Below that depth, pressures are so high in the liquid hydrogen that atoms are tightly squeezed together. Electrons in those atoms become mobile, which in turn makes the material strongly conducting. Here a layer of **metallic hydrogen** begins.

Models of Jupiter predict a 15-Earth-mass ($15\text{-}M_E$) core made from a mix of rocky and metallic materials. The weight of the other 303 M_E of H and He above the core drives temperatures at the center to 30,000 K—more than 5 times hotter than the Sun's surface. Given these temperatures, it would be wrong to think of Jupiter as a gas ball with a terrestrial planet at its center. Instead, the central regions are best thought of as an Earth-sized spherical core of heavy elements at high temperatures and incredible pressures. "We still have lots of questions about the conditions at the Jovian core," says Hendrix. (*Jovian* means "of Jupiter.") "There is a lot about the physics there that we don't understand."

Because it is still contracting, Jupiter gives off approximately 1.7 times more energy than it receives from the Sun. The contraction is slow, its radius shrinking about 2 centimeters (cm) a year. The energy released by the planet as it contracts keeps the interior temperatures high and drives the planet's remarkable atmospheric convection.

metallic hydrogen Hydrogen at a pressure and temperature high enough for electrons in closely spaced atoms to be able to move freely, as in a metal.

droplet condensation The transformation of molecules from gaseous to liquid state in the form of small droplets in an atmosphere. It generally occurs where temperature is decreasing.

zone A white, rising cloud region that forms a horizontal band in Jupiter's atmosphere because of convection and the planet's rapid rotation.

belt A dark, sinking cloud region that forms a horizontal band in Jupiter's atmosphere because of convection and the planet's rapid rotation.

Jupiter's Atmosphere

In 1995 the *Galileo* spacecraft released a small probe that descended into the Jovian atmosphere. Falling first through a clear sky composed of nearly pure hydrogen, the probe then dropped into the upper cloud layer, made up of ammonia droplets. A middle layer of ammonium hydrosulfide lies below the first, and deeper still is a layer of water clouds. "Each layer exists because temperature and pressure vary, and with them, the composition of droplets that can condense under those conditions," says Hendrix (**Figure 8.6**). Such **droplet condensation** also occurs in Earth's sky, when water vapor condenses from the gaseous state into the liquid state, forming clouds.

Clouds in Earth's sky are white, but Jupiter's clouds show a remarkable range of colors. This is surprising because all of Jupiter's cloud constituents (ammonia, ammonium hydrosulfide, and water) should form pure-white crystals. Astronomers hypothesize that small amounts of complex compounds form when sunlight strikes the upper atmosphere, "polluting" Jovian clouds and yielding the hallucinogenic mix of colors. "This is one of the places on Jupiter where sunlight really does matter," says Hendrix.

The most striking feature of Jupiter's clouds is their horizontal banded structure. Astronomers call these bands *belts* and *zones*. While older definitions focused on the brightness of the bands, over time better observations have changed our understanding of them. The best interpretations today suggest that the brighter bands, the **zones**, represent material rising from deeper in the atmosphere and that the darker bands, the **belts**, consist of material sinking back down into the atmosphere (**Figure 8.7**). Convection initiated by heat released from the interior is believed to be the engine driving the belts and zones.

In 2016 the *Juno* probe began a new series of investigations designed to see deeper into the upper layers of Jupiter's atmosphere. *Juno* images have revealed a remarkable array of flow patterns inside and between the bands, including many vortices of swirling gas (**Figure 8.8**).

On Earth, weather is driven by the movement of winds that circulate in cyclonic patterns because of both convection and Earth's rotation. Jupiter's rotation is so much more rapid than Earth's that its high- and low-pressure systems have been permanently stretched out in the counterrotating bands of rising zones and sinking belts. Winds between the belts and zones can reach speeds of hundreds of kilometers per hour and produce circulating hurricane-like patterns. Some of these features can grow into vast storms, such as Jupiter's famous long-lived Great Red Spot, which is larger than Earth. Winds in the Great Red Spot have been clocked at 250 km per hour (km/h), making it, by Earth's standards, a category 5 hurricane that has persisted for at least 300 years, though it changes in size and shape over time (**Figure 8.9**).

Jupiter's Magnetic Field

The 50,000-km-deep layer of metallic liquid hydrogen in Jupiter's interior is the source of the planet's powerful magnetic field. Discovered in the 1950s via radio emission from Jupiter, this magnetic field was first detected directly in the 1970s by the *Pioneer* and *Voyager* flybys. The *Galileo* mission in the 1990s then gave scientists a detailed laboratory for exploring Jupiter's magnetism and its effect on the nearby environment.

Jupiter's magnetic field, like Earth's, extends into space, creating an elongated magnetosphere around the planet. Other than the Sun's field, Jupiter's is the largest in the Solar System. Its magnetosphere extends more than 600 million km backward, reaching all the way to Saturn's orbit.

Jupiter's magnetosphere acts as a trap for charged particles. Some of the particles come from the solar wind; some are ions ejected from Jupiter's highly volcanic moon Io. Streams of charged particles are channeled by the magnetic lines of force to the northern

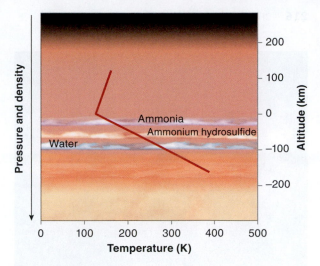

Figure 8.6 Jupiter's Atmosphere
Each of the three distinct cloud layers of Jupiter's atmosphere—consisting of ammonia, ammonium hydrosulfide, and water—condenses at altitudes where the pressure and temperature allow droplets of that compound to form. The 0 point of altitude marks the reference point where temperature stops decreasing with distance from the planet's center and the hydrogen-dominated gas atmosphere layer begins.

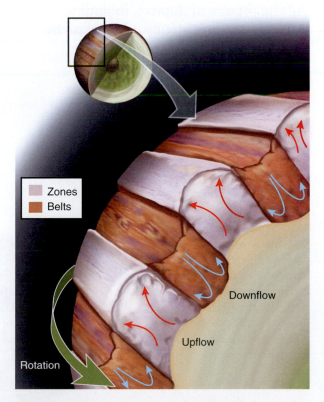

Figure 8.7 Zones and Belts: A Cross-Section
Zones are cloud regions that are rising. Belts are cloud regions that are sinking lower in the atmosphere.

Figure 8.8 *Juno* **and Jupiter's Atmospheric Flows**
Four color-enhanced images taken over an 8-minute interval by the *Juno* spacecraft in 2017. *Juno* flew close enough to the top of Jupiter's atmosphere—just 12,143 to 22,908 km above the cloud tops—to reveal new details of its flow patterns.

Figure 8.9 Jupiter's Great Red Spot
This dynamic, evolving high-pressure system has been observed for over 300 years. Striking new detail is visible in these images from the Juno mission.

and southern Jovian poles, where they collide with atmospheric atoms and create vast auroras thousands of times larger and more energetic than those on Earth (**Figure 8.10**).

Jupiter's powerful magnetic field is also linked to intense radiation belts. The field traps charged particles from Jupiter's atmosphere, its moon Io, and the solar wind, forming a vast **ion torus**. In this doughnut-shaped region of emission, ions spiral around the magnetic field and emit radio photons. The radiation belts are so strong that they can damage electronics in visiting spacecraft. NASA's *Juno* probe had a special orbit that dove close to the planet and then swung far outward to minimize the radiation's harmful effects.

Jupiter's Moons

Jupiter is known to host more than 79 moons; the exact number is bound to change, since many of these satellites are just captured asteroids. Four of Jupiter's moons stand out: Callisto, Ganymede, Europa, and Io are large enough to be seen with a small telescope or even binoculars (**Figure 8.11**). In fact, Galileo discovered these four rapidly orbiting moons with his first telescopes. These "Galilean" moons hold clues to the origin of Jupiter (and the Solar System) and may be critical in understanding where and when life might form outside Earth.

> "All the moons of Jupiter, including Callisto, are powerfully influenced by tidal forces from the giant planet."

Figure 8.10 Jupiter's Magnetic Field

a. Jupiter's enormous magnetic field includes a torus of material, some of which is blown off the moon Io. **b.** An aurora around Jupiter's north pole, photographed by the Hubble Space Telescope (HST), provides further evidence of magnetic forces focused at the poles.

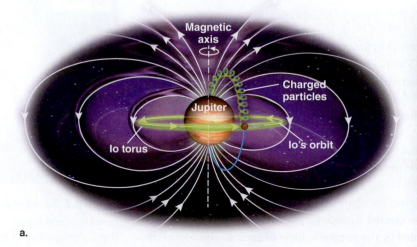

VIS

Figure 8.11 The Galilean Moons

The four Galilean moons of Jupiter, so named because they were first observed by Galileo, have been the subject of intense study because of their varied surfaces and internal structures.

CALLISTO, DEAD WORLD FULL OF WATER. The farthest Galilean moon from Jupiter, Callisto orbits at 1.9 million km from the planet's center with an orbital period of 16.7 days. "All the moons of Jupiter, including Callisto, are powerfully influenced by tidal forces from the giant planet," says Hendrix. "That means all their rotation rates have been locked to their orbit periods." The tidal forces arise because one side of the moon feels Jupiter's gravity more strongly than the other. (Recall that the force of gravity decreases as the inverse of the distance squared.) The difference in force leads to both a squeezing and a stretching of the moon's interior, which increases internal temperature. The internal heating is minimal, however, as Callisto is far enough from Jupiter that Jupiter's tidal forces are weaker than on most Jovian moons. Thus, Callisto is mostly geologically inactive. The surface features visible today have not changed much since they formed billions of years ago.

Similar to the way the planets formed from a disk orbiting a young Sun, the Galilean moons formed from a disk swirling around young Jupiter. Thus, the range of densities in the moons is similar to what we see in the planets (**Table 8.3**). Callisto is farthest out and has the lowest density: 1,830 kilograms per cubic meter (kg/m^3). "At that distance," Hendrix explains, "temperatures in the disk would have been lowest." Thus, Callisto contains significant amounts of water ice mixed with heavier material, such as silicate rocks. It does not appear to have differentiated to produce a core.

Callisto's surface is composed of dark ice shattered by craters from meteorite impacts (**Figure 8.12**). Most of the giant planets' moons with icy surfaces have darkened with age as small meteorite impacts have vaporized the ice, leaving larger proportions of rock behind. The presence of significant amounts of water also changes the shape of larger impacts.

The *Galileo* space probe carried sensitive magnetometers as it orbited the gas giant and explored its moons. When Jupiter's powerful tilted magnetic field swept across

Table 8.3 Densities of the Galilean Moons	
Moon	**Density (kg/m^3)**
Io	3,530
Europa	3,010
Ganymede	1,940
Callisto	1,830

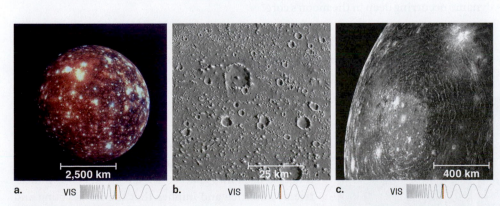

a. VIS **b.** VIS **c.** VIS

Figure 8.12 Callisto

Galileo captured **a.** Callisto's dark surface, which is marked by bright patches from impacts, and **b.** craters **c.** A multiringed basin is surrounded by ejecta and rings of mountains from the impact that created it.

ion torus (pl. tori) A doughnut-shaped region of emission where ions spiral around the magnetic field.

Callisto, *Galileo* saw variations in magnetic energy, indicating the presence of liquid water beneath the moon's crust. "We think a layer of liquid water 10–100 km deep may be hiding beneath Callisto's ice-rock crust," says Hendrix. "That was a really exciting discovery, and Callisto is likely not alone in harboring such an underground ocean."

GANYMEDE, AN ALMOST ACTIVE MOON. Unlike Callisto, Ganymede appears to have remained geologically active for some time after forming. With a higher density, 1,936 kg/m³, it may have differentiated into a rocky core below an ice-rock mantle. Ganymede's higher density is an indication that its location in the early moon-forming disk had a higher temperature, making the moon less able to hold on to volatiles. Its proximity to Jupiter, however, has made its interior more susceptible to the big planet's tidal forces. "The extra internal heating left its mark on Ganymede's surface," says Hendrix.

Numerous craters appear on Ganymede's icy surface, as on Callisto's; however, Ganymede has vast "grooved" regions where resurfacing must have occurred. These grooves contain fewer craters, indicating that they are newer, and cut across older craters, erasing half of their structures. The shapes of these grooves, along with their spectra, indicate that water-rich flows of muddy material streamed across the surface before turning into the hardened ice-rock structures seen today (**Figure 8.13**). Although the grooves are clearly newer than the cratered regions around them, we do not know whether Ganymede is still active. "If activity really has stopped," says Hendrix, "then the question is, why? What made this moon so active earlier in its history and then shut that activity down?"

One answer may lie in Ganymede's orbit. A perfectly circular moon on a perfectly circular orbit around a perfectly circular planet would feel tidal forces, but those forces would not change over time. Jupiter's moons, however, are not on perfectly circular orbits. The gravitational interactions between these moons, as well as with Jupiter, have nudged the moons into somewhat elliptical orbits. Thus, a moon feels more tidal stretching in its interior as it nears Jupiter than when it is more distant. These changes in tidal stretching that occur throughout its orbit release energy. Ganymede's orbit has a relatively low eccentricity now but in the past may have been more elliptical, which might explain the energy that drove the grooved flows and other younger surface features. If Jupiter's massive gravitational field eventually pulled Ganymede into a more circular orbit, then the heating would have lessened and the moon's surface activity would have diminished.

The *Galileo* probe's magnetometer found evidence of subsurface water on Ganymede. "The evidence points to a 5-km-thick layer of liquid water lying more than 150 km below Ganymede's surface," says Hendrix. In addition, Ganymede is the only Solar System moon with its own relatively strong, well-organized magnetic field. The field likely comes from a dynamo occurring deep in the moon's core.

EUROPA, WATER WORLD. Europa, the second-closest Galilean moon to Jupiter, has become the focus of intense interest and debate over the last three decades. Europa has a radius of 1,565 km, making it slightly smaller than Earth's Moon. With a density of 3,010 kg/m³, Europa is clearly rockier than Callisto or Ganymede, and scientists believe it has a differentiated metal core.

Europa's surface has an albedo of almost 0.7—almost 70 percent of the sunlight striking the surface is reflected back into space. Such a high albedo is indicative of water ice and implies that the ice is young and relatively pure. Unlike the ice on Jupiter's two outer moons, ice on Europa's surface is young enough that meteor impacts have not vaporized the water away, so

> "Something remarkable is happening on Europa."

Figure 8.13 Ganymede

a. This moon retains many craters from its early existence but also shows **b.** many long grooves that formed because of resurfacing driven by internal heating from Jupiter's tidal forces. **c.** Enki Catena, a 13-crater feature, is consistent with the idea that Ganymede once collided with a disintegrating comet that Jupiter's gravity had torn apart.

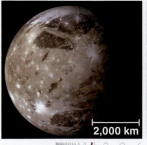

a. VIS

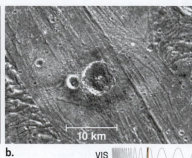

b. VIS

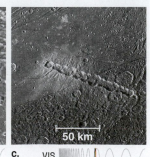

c. VIS

there is less dark, "dirty" material (**Figure 8.14**). "Something remarkable is happening on Europa," says Hendrix.

A relative lack of craters confirms the apparent youth of Europa's surface. The brightest feature, called Pwyll (named for Pwyll Pen Annwn, a character from Welsh mythology), appears as a circular bull's-eye (Figure 8.14a). "Pwyll looks more like concentric cracks on the surface of an icy pond than an impact crater on a rocky crust," says Hendrix. That pond analogy can be taken much further. A close inspection of Europa's surface indicates the absence of any rocky surface. Instead, Europa appears to be covered entirely by ice in constant motion. Jumbled, blocky features, including spikes of ice, are seen along with long systems of cracks, some of which extend more than 1,600 km. Taken together, these features show that Europa's surface is composed of moving "icebergs." When two icy plates move apart, water from below wells upward and freezes in the open space.

Magnetic-field measurements provide strong evidence that below Europa's jumbled icy surface lies a liquid ocean potentially 100 km deep. Thus, despite its dense rocky interior, Europa is a water world: it is entirely covered by an ocean that, in turn, lies below kilometers of thick ice. The presence of so much water, especially in liquid form, makes Europa a prime candidate for life forming elsewhere in the Solar System.

The secret to Europa's oceans may rest, once again, with tidal forces. "Being closer to Jupiter than either Ganymede or Callisto is, Europa is constantly stretched and compressed," says Hendrix. "As that heat works its way toward the surface, it keeps the water ocean liquid and drives all the resurfacing of the icy crust."

IO, VOLCANIC CHILD. Tidal forces are also the key to understanding Io, Jupiter's innermost Galilean moon. Io's density, 3,530 kg/m³, makes it the densest of the Galilean moons. We might expect Io to be covered with impact craters, since all of the debris captured by Jupiter's gravity would pass by it. Instead, on Io, even more than on Europa, craters disappear almost as fast as they form. "Io's surface is not just young," says Hendrix, "it's newborn."

When it passed Io in 1979, *Voyager* saw a moon bathed in sulfurous reds and yellows. Instead of craters, it saw active lava flows stretching across hundreds of kilometers. The

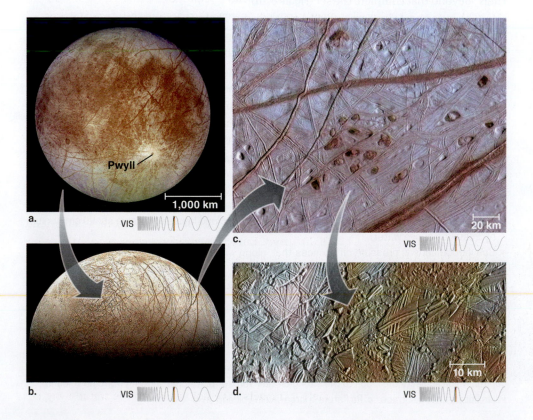

a. Pwyll

1,000 km

VIS

b.

VIS

c.

20 km

VIS

d.

10 km

VIS

Figure 8.14 Europa

a. The bright, young surface of Europa is composed of ice and lies atop a water ocean 100 km deep. **b.** The planet is constantly resurfacing itself as "plates" move or break apart, allowing water from underneath to rise up and **c, d.** form new surface ice, shown in false color.

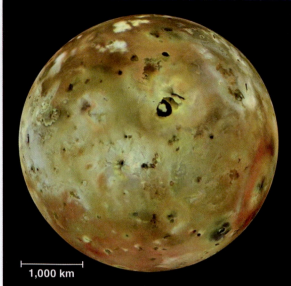

a.

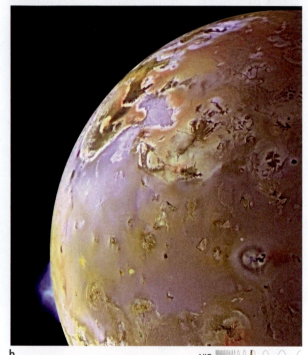

b. VIS

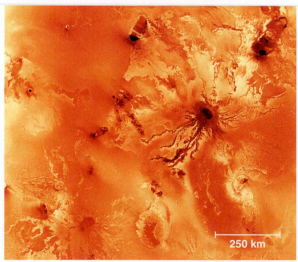

c. VIS

space probe's cameras caught a stunning plume of gas rising from one of Io's 150 visible active volcanoes. Sixteen years later, when *Galileo* pulled into orbit, a few of the older volcanoes were still visible, but new features had appeared, including curtains of bright lava erupting from faults in the surface (**Figure 8.15**). "We could watch as the surface of Io was continuously being formed and re-formed," says Hendrix. "It was simply amazing."

Io's continuous volcanism is a direct result of intense tidal forces working within the moon. Io moves on a fairly elliptical orbit because of gravitational interactions with the other Galilean moons. The changes in tidal stretching caused by the elliptical orbit continuously pump energy into Io's interior on every orbit, resulting in an ongoing state of volcanism. That activity has shaped Io's evolution and prevents it from having a subsurface ocean, as all the other Galilean moons are believed to harbor in some form.

Plumes of sulfuric gas released by Io's volcanoes are blown hundreds of kilometers into space. While much of this material escapes the moon, some of it returns to the surface. Brighter patches on Io may be sulfur dioxide that is released as a gas in volcanic eruptions, condenses, and drifts back to the surface as snow. Other elements, mostly sulfur, remain gaseous, forming a thin atmosphere.

Ionized gases escaping Io's meager gravity are swept up by Jupiter's powerful magnetic field. As we saw, the ions blown off Io in volcanic eruptions spiral around the planet's magnetic field and emit radio photons, as part of Jupiter's ion torus.

Jupiter's Rings

Before *Voyager* passed Jupiter, scientists did not know that the giant planet is a ringed world. As the probe passed, it turned its camera back toward Jupiter's dark side and revealed the faint outlines of a ring. These images made it clear that Jupiter, like Saturn, hosts a thin assembly of orbiting particles.

Detailed study by the *Galileo* probe demonstrates the presence of four separate rings around Jupiter. The innermost ring is a relatively thick torus of dust called the "halo ring." Beyond it is the exceptionally thin "main ring": two relatively thick but faint "gossamer rings" beyond that complete the set (**Figure 8.16**).

The outermost rings are named Amalthea and Thebe in honor of the two tiny moons that appear to be the principal source of their dust particles. The other two rings are also the result of dust ejected off small moons (Metis and Adrastea) as they are bombarded by meteoroids. Astronomers believe that, as with most Solar System rings, their constituent particles are not long-lived. "The dust won't be in the rings forever," says Hendrix. "Eventually the particles will spiral down and fall into Jupiter's atmosphere." While uncertainties abound, the lifetime of a dust particle in Jupiter's rings appears to be on the order of 100–1,000 years. "That means there has to be a continual resupply of ring material," says Hendrix. "Without the moons, the rings would not last."

Section Summary

- Jupiter's core of heavy elements is surrounded by a layer of liquid metallic hydrogen, overlain by a molecular hydrogen ocean. In its atmosphere are three cloud layers (ammonia above ammonium hydrosulfide above water).
- Convection causes pressure differences that, along with rapid rotation, create counterrotating belts and zones and giant, long-lived storms.
- Jupiter's magnetic field creates a large magnetosphere and belts of high-energy particles.

Figure 8.15 Io

a. Io, the most actively volcanic globe in the Solar System, owes its activity to its position closest to Jupiter, which subjects the moon to the strongest tidal forces. **b.** A plume spewing from the volcano Pillan Patera on Io's surface. **c.** Ra Patera, a large shield volcano on Io.

■ Jupiter has more than 79 moons. Four of these moons stand out and offer clues to the planet's origin: Callisto, Ganymede, Europa, and Io. Jupiter has a system of four rings, fed by several of its smaller moons.

CHECKPOINT Imagine you are in a space probe descending into Jupiter. How would conditions change as you dive from the outer atmosphere to the core?

Saturn: Lord of the Rings

8.4 Galileo's first telescopic observations of Saturn in 1610 demonstrated that something is clearly different about the planet, which was then thought to be the Solar System's outermost world. Through the low resolution of a refracting telescope with a diameter of less than an inch, all Galileo could see were earlike features on either side of Saturn. But Saturn's spin axis is tilted 27° to the plane of its orbit, and its rings possess the same tilt. Therefore, as Saturn moves through its orbit, its rings change their orientation relative to the Sun (**Figure 8.17**).

When the ring plane was aligned with Earth's line of sight, Galileo was astonished to see Saturn's "ears" disappear. "Has Saturn eaten his children?" Galileo wrote. (Saturn, the Roman god of agriculture, was said to have devoured his own children to keep them from taking his place.) Not until 1659, when Dutch astronomer Christiaan Huygens trained a larger telescope on Saturn, was the planet recognized to be surrounded by rings, not ears. Today we rely on powerful cameras aboard space probes such as *Voyager* and the 1997 Saturn probe *Cassini* to enhance our understanding of Saturn's rings (**Figure 8.18**).

Saturn's Interior

With a radius of 58,232 km, Saturn is the second-largest planet in our Solar System. "It would take about 764 Earths to fill up the volume of Saturn," says Linda Spilker.

Saturn's elemental composition is similar to Jupiter's. "You're looking at 80–90 percent hydrogen in Saturn's atmosphere, with helium coming next in abundance," says Spilker. "Then there are also trace amounts of various other gases." Despite the compositional similarities, Saturn's interior is not identical to that of its larger cousin. Like Jupiter, Saturn has a layer of molecular hydrogen over a layer of metallic hydrogen, all surrounding a rocky/metallic core of heavy elements, but Saturn has a uniquely low average density: 690 kg/m³. It is the only world with an average density less than that of water.

The cause of Saturn's low density is likely tied to its lesser mass combined with its rapid rotation (see **Going Further 8.1** on p. 224). With less mass than Jupiter, Saturn has less gravity to pull the gas inward. Saturn also spins at a high rate, rotating once every 10.6 hours. That rapid rotation counteracts Saturn's lower gravity (relative to Jupiter's), reducing both the density of gas in the equatorial regions and the average density of the whole planet. The rapid rotation also affects the planet's shape; it is noticeably flattened

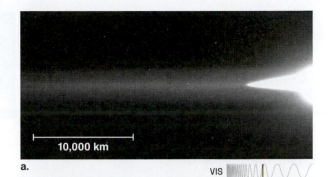
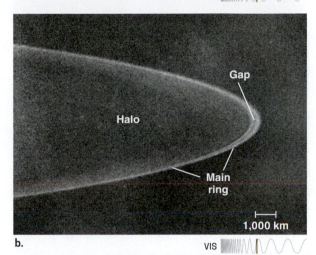

a.

VIS

Gap

Halo

Main ring

1,000 km

b. VIS

Figure 8.16 Jupiter's Rings
a. This *Galileo* image shows Jupiter's main ring (right) and its faint, outer gossamer ring stretching out three Jovian radii from the planet's center. **b.** *New Horizons* captured another view of the main ring, with a fainter, broad area of fine dust inside it.

"Without the moons, the rings would not last."

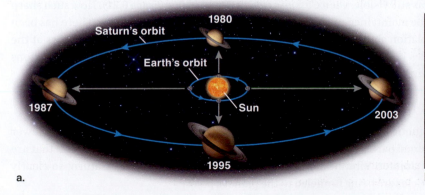

a.

1980

Saturn's orbit

Earth's orbit

1987

Sun

1995

2003

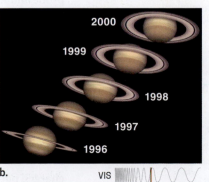

b. VIS

2000

1999

1998

1997

1996

Figure 8.17 Saturn's Tilt
a. As Saturn moves through its orbit, its rings change their orientation relative to the Sun. **b.** Hubble Space Telescope images show how the view of the rings from Earth changes from practically invisible to broad in a continuous cycle.

Figure 8.18 Saturn's Rings

a, b. The earliest images of Saturn are those drawn by prominent 17th-century astronomers, who saw the planet's rings without understanding their true nature or origin **c–e.** NASA space missions in the 20th and 21st centuries have provided stunning details about the ring's nature and composition.

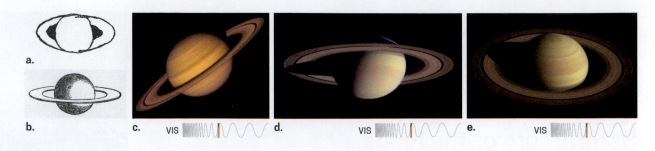

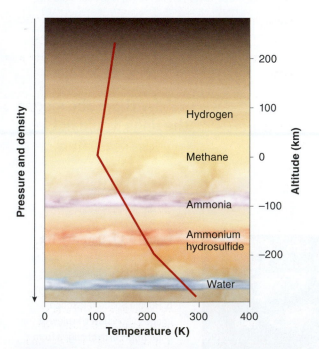

Figure 8.19 Saturn's Oblateness

The gas giants, especially Saturn, have a noticeable oblate shape because of their rapid rotation. The red circle shows what the planet would look like if it were perfectly spherical.

Figure 8.20 Saturn's Atmosphere

Saturn's colder atmosphere, compared with Jupiter's (see Figure 8.6), produces lower cloud layers and a haze that hides deeper layers.

at the poles—that is, **oblate**. "Saturn is the most oblate of all the planets," says Spilker. "Its radius at the equator is 10 percent higher than the radius at the poles" (**Figure 8.19**).

Like Jupiter, Saturn emits more energy than it receives from the Sun—about two and a half times more energy. This extra energy is more than astronomers would expect from their models and is something of a mystery. "While some of the heat could be left over from Saturn's formation," says Spilker, "the numbers don't really add up."

For some scientists, the solution to this mystery might lie in a helium "rain" going on inside Saturn. Astronomers suspect that, deep in Saturn's ocean of hydrogen, the temperature and pressure may be just right for the initially well-mixed helium atoms to condense into individual droplets. Just as water droplets condensing in Earth's atmosphere are denser than the air, the condensed helium is denser than hydrogen, so the droplets sink deeper into Saturn's interior, releasing gravitational energy and heating the planet's interior.

Saturn's Atmosphere

Saturn's atmosphere does not offer the same visual spectacle as Jupiter's. "Saturn is smoother and has this overall golden haze," says Spilker. "We don't tend to see huge colorful storms on Saturn like the Great Red Spot of Jupiter." Astronomers believe the difference between Saturn and Jupiter may be simply that Saturn is farther from the Sun, so its atmosphere is colder. "The colder temperatures may allow a layer of haze to form at the top of the atmosphere," says Spilker, "and that masks cloud structure deeper down" (**Figure 8.20**).

Using the *Cassini* space probe, scientists such as Spilker have been able to view these cloud structures with imagers tuned to longer-than-visual wavelengths. "When we go deeper, we do start to see an atmosphere that's more dynamic and looks more like Jupiter's." But while Saturn shows the same belt and zone structure as Jupiter, there are important differences. Wind patterns on Saturn show fewer alternating bands than do those on Jupiter, and the wind speeds are enormous, reaching 1,800 km/h (almost 5 times as high as those on Jupiter). "The high wind speeds are partly due to Saturn's fast rotation," explains Spilker.

One Saturnian mystery (not seen on Jupiter) is the presence of strong, strangely shaped polar storms. "There's a hurricane-like structure at the south pole that we were able to image," says Spilker. "We actually saw the shadow of the huge wall of clouds that makes the hurricane's eye. That was amazing. The storm at the north pole is even more peculiar. It has a six-sided hexagon-shaped pattern." That storm was first seen with the *Voyager* probe and was still visible when *Cassini* arrived 25 years later (**Figure 8.21**). How such sharp corners can be maintained on a storm vortex was a puzzle for some time. "There has been lots of speculation about how you end up with two very different kinds of vortices at the different poles on Saturn," says Spilker. Recent experiments and computer simulations have, however, revealed that they can be a natural result of complex rotating flows.

Saturn's Magnetic Field

Saturn's magnetic field is 20 times weaker than Jupiter's. The weaker field leads to fewer trapped, charged particles, so Saturn does not show the extended radiation belts that are seen around Jupiter. Vibrant auroras are, however, observed around Saturn, so clearly charged particles are being funneled to the planet's poles.

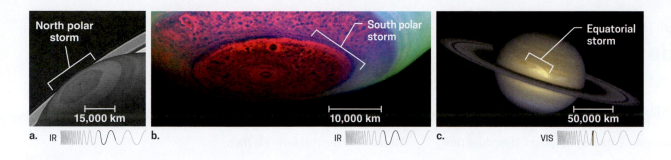

Figure 8.21 Storms at Saturn's Poles
a. A 2016 *Cassini* image of Saturn's north pole, which is encircled by a hexagonal feature believed to be a storm at least decades old. **b.** A 2007 *Cassini* image of a massive cyclone-like storm at the south pole. **c.** A 1994 HST view of an enormous storm at Saturn's equator.

The near-perfect alignment of Saturn's magnetic field with its rotation axis is a huge puzzle for scientists. "Is it really so perfectly aligned, or are processes in the atmosphere just making it appear symmetric?" asks Spilker.

This magnetic field must be formed by swirling convective motions in the metallic hydrogen layer deep within Saturn. But Saturn's lesser mass and lower gravity (compared with Jupiter) suggest that the region where compression can squeeze hydrogen into metallic form is smaller than on Jupiter, accounting for Saturn's weaker field. Whether this difference in size of the metallic hydrogen layer causes such strong alignment between the magnetic and rotational poles is still unknown.

Saturn's Moons

Scientists classify moons as regular or irregular. Of Saturn's 62 moons, 24 **regular satellites** tend to move on prograde orbits (in the same direction as Saturn's rotation) and stay close to the plane defined by Saturn's equator. "These moons probably formed out of the same disk of material which gave birth to Saturn," says Spilker. The 38 **irregular satellites** tend to be farther from Saturn and move on highly inclined orbits. "The irregular satellites can be retrograde or prograde," says Spilker, "and are most likely material Saturn captured at some point later in its history." Two of the regular moons that astronomers have found most fascinating are Titan and Enceladus.

TITAN. Saturn's largest moon, Titan, is larger than the planet Mercury. "If Titan had formed on its own," says Spilker, "it would be a planet in its own right." Titan's uncompressed density is just 1,500 kg/m³, meaning that it must contain significant amounts of ice. But what really sets Titan apart is that it is the only moon in our Solar System with a substantial atmosphere. In 2005, *Cassini* released the *Huygens* probe, which parachuted into Titan's atmosphere and returned remarkable views of the icy moon (**Figure 8.22**).

Titan's atmosphere is so dense that the pressure on that moon's surface is 50 percent higher than what we experience on Earth's surface. The composition is also remarkable.

> "While some of the heat could be left over from Saturn's formation, the numbers don't really add up."

oblate Roughly spherical with some bulging at the equator.

regular satellite A satellite orbiting in the same direction as its planet and with a low inclination to the planet's orbital plane.

irregular satellite A satellite orbiting in the opposite direction from its planet and/or on a highly inclined orbit in relation to the planet's equator.

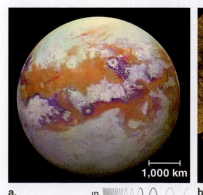
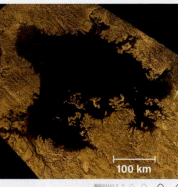
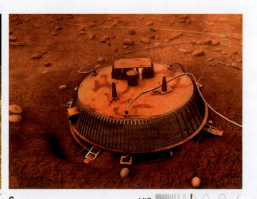

Figure 8.22 Titan
a. Although visible light shows only the cloud tops of Titan's atmosphere, infrared light and false colors show details of Titan's surface. **b.** A mosaic of Titan's north polar region, revealing large lakes and tributary networks of liquid ethane and methane. **c.** The *Huygens* probe landed on Titan's surface and beamed back images from the surface for 90 minutes before ending operation.

GOING FURTHER 8.1
Rotation and Planetary Oblateness

Why does rotation change a planet's shape? Both Jupiter and Saturn spin fast enough to turn them from perfect spheres into fairly oblate objects that are wider at the equator than at the poles. How does this happen?

In general, all planets rotate at speeds much below the speed a satellite needs to stay in orbit just above a planet's surface (the orbital speed; see Chapter 3). Even a mostly gaseous planet spins mostly as a single body because internal forces (such as the force between atoms in rocks or the pressure on gases) provide support against gravity and keep everything moving at the same rotational speed. But for very rapidly spinning planets, we must consider the increased role of the *centripetal force*. Recall from Chapter 3 that this is the inward-directed force needed to keep a freely orbiting object moving in a circle. Whenever an object of mass m moves at speed V in a circle with radius R, we can express the centripetal force F_c as

$$F_c = \frac{mV^2}{R}$$

We can use this formula to think about the shape of the gas giants.

Think about a blob of gas in the atmosphere of a rapidly spinning spherical world. Material at the equator lies at the largest distance (R) from the spin axis, so if we approximate the planet as spinning as a solid body, it would have the highest rotation velocity. So that material requires the highest centripetal acceleration to keep it moving with the rest of the rapidly rotating planet. Material at the poles is not spinning at all.

The ratio of the centripetal force to the force of gravity can give us an idea of how much the demands of rotation are beginning to affect the planet's dynamics. Recall from Chapter 3 that the force of gravity F_g between a blob of mass m_{blob} and a planet of mass m_p is

$$F_g = \frac{Gm_p m_{blob}}{R^2}$$

where G is Newton's gravitational constant. (We're using R as both the planetary radius and the radius of the blob's circular path. That's because we imagine that our gas blob sits in the outer parts of the planet and we're considering gas at the equator, the widest part, only.) The ratio of the two forces is then

$$\frac{F_c}{F_g} = \frac{m_{blob}V^2/R}{Gm_p m_{blob}/R^2}$$

When we can cancel all of the identical terms, we find that

$$\frac{F_c}{F_g} = \frac{V^2 R}{Gm_p}$$

Now let's compare this ratio for Earth, Jupiter, and Saturn:

Earth: $\dfrac{F_c}{F_g} = 0.0001$

Jupiter: $\dfrac{F_c}{F_g} = 0.0021$

Saturn: $\dfrac{F_c}{F_g} = 0.040$

These numbers tell the whole story. For Earth, the centripetal force at the equator is a tiny fraction of the gravitational force. That means internal forces are what keep Earth rotating as a solid body, and they have no problem keeping Earth an (almost) perfect sphere. For Jupiter, that ratio is still small, but it is 20 times larger than Earth's. For Saturn, the ratio of forces is 20 times as high as on Jupiter. Note that the difference between Jupiter and Saturn is because of difference in the planets' mass and thus their gravity.

Thus, we should expect that internal forces alone cannot keep the gas giants rotating as purely spherical bodies and that they are distorted the most at the equator, where the spin matters most. Equatorial gas moves slightly outward, reducing the centripetal force (with its $1/R$ dependence). Gas at the equator therefore bulges outward relative to the poles, making both planets fairly oblate. Saturn's high ratio of centripetal force to gravitational force makes it the most oblate of the planets, and this high oblateness contributes to its low density.

"If Titan had formed on its own, it would be a planet in its own right."

hydrocarbon An organic compound consisting entirely of hydrogen and carbon.

Roche limit The closest a satellite can get to its central object before that object's tidal forces pull the satellite apart, when the satellite's own gravity can no longer hold it together.

"Titan's atmosphere is composed mostly of nitrogen, very similar to Earth's atmosphere," says Spilker, "but there's pretty much no oxygen." Titan's atmosphere does, however, have a wealth of complex **hydrocarbons**: methane, ethane, propane, acetylene, and others. Together these compounds make for strange weather on Titan.

"It's very cold on Titan," says Spilker. "Daytime temperatures are only about 94 K." With so little heat available, hydrocarbons in the atmosphere can condense into rain. Astronomers have observed lakes of liquid methane. "We can even see river channels where the methane rain collects and flows downhill to fill the lakes," says Spilker. The composition and low temperatures on Titan mean that the remarkable mountains seen there are likely composed of water ice rather than rock.

ENCELADUS. A tiny moon only about 480 km in diameter, Enceladus has become the other celebrity in Saturn's family of satellites. Given its tiny size, Enceladus should be frozen solid and inactive. "That is not the case," explains Spilker. "We were all absolutely amazed and delighted to see active geysers of water vapor and ice crystals blasting out at Enceladus's south pole" (**Figure 8.23**).

The plumes of material erupting from Enceladus are important for a number of reasons. First, they appear to be the source of one of Saturn's outer rings (called the E ring). Second, the presence of liquid water on Enceladus again raises the possibility that life can form on a moon of the outer Solar System. "When we flew near these plumes to investigate their composition," says Spilker, "we saw compounds with carbon, nitrogen, and oxygen in them, as well as salts—table salt, potassium salt, and so on." These are the kinds of minerals scientists would expect if an underground ocean was providing the water for Enceladus's plumes.

Some scientists have argued that tidal forces from one of Saturn's other moons, Dione, might be the plumes' energy source, but other researchers have questioned whether the amount of heat generated by another moon is sufficient to keep water in liquid form on Enceladus. For now, the plumes remain a wonderful mystery.

Saturn's Rings

Although Saturn's main rings extend 7,000–80,000 km from the planet's edge, they are a mere 10 meters thick vertically. From Earth, astronomers can see that Saturn's spectacular rings are actually a system of separate smaller ring systems. Earth-based telescopes resolve three main sections, called the A, B, and C rings. With space probes such as *Voyager* and *Cassini*, scientists saw new rings, as well as fascinating new features within the larger systems (**Figure 8.24a, b**).

What is the composition of Saturn's rings? That is the most pressing question. "They are made primarily of water ice," explains Spilker. "The best data show the rings to be 99 percent water ice with a tiny amount of non-icy material contaminant." Contaminant mass likely comes from micrometeorite bombardment adding to the icy material over time. The ring particles range from dusty ice grains all the way up to objects meters in diameter. "Even though you have this wide range," explains Spilker, "the typical size of material seen in the large rings is about a centimeter" (**Figure 8.24c**).

When and how did the icy material get there? "When we got to Saturn with *Voyager* and saw these bright, pristine, icy rings," says Spilker, "we immediately concluded they must be made of material much younger than the age of the Solar System." The rings had been thought to be fairly young—about 100 million years old—because radiation would destroy the ice crystals over time. "But based on what we learned from *Cassini*," Spilker continues, "our studies show it may be more complicated." According to scientists, the massive B ring might have formed with Saturn many billions of years ago. The other rings, however, might have formed in other ways. "Maybe a comet got a little too close and was broken apart by tidal forces from Saturn," says Spilker. This can occur when a comet (or other body—even a moon) comes within a planet's **Roche limit**, the closest distance an object can approach and not be torn apart by tidal forces. Once inside the Roche limit, the comet would be pulled apart and its material would become part of the rings (**Figure 8.25**). In addition, the smaller rings may be linked to material stripped off from moons, such as the E ring, which is fed by Enceladus's plumes.

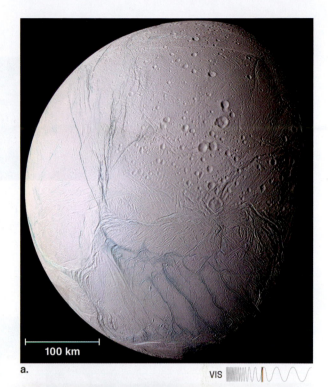

a.   VIS

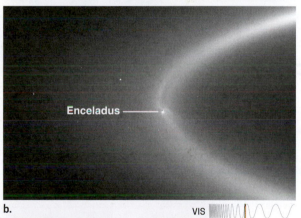

Enceladus

b.   VIS

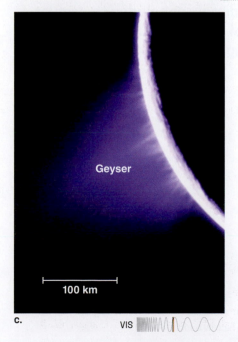

Geyser

c.   VIS

Figure 8.23 Enceladus
a. The highly reflective surface of Enceladus, imaged by the *Cassini* mission, is believed to be composed of water ice. **b.** Enceladus, which orbits in Saturn's E ring, acts as a shepherd moon. **c.** Geysers vent water mixed with salt and hydrocarbons from cracks in the moon's surface.

100 km

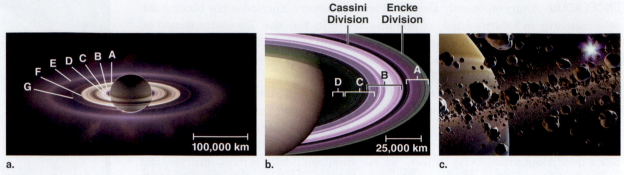

Figure 8.24 Saturn's Ring Structure and Particles
a. The individual rings and **b.** gaps, or divisions, of Saturn's rings have been observed for centuries, although the dimmer rings were discovered when visiting spacecraft beamed back images. **c.** An artist's conception of a ring's many particles, which range from grain-sized ice granules to boulder-sized objects. The average size is about 1 cm (0.01 meter).

Figure 8.25 Inside the Roche Limit
The Roche limit is the closest distance an object can approach to a large body and not be torn apart by tidal forces. Once inside the Roche limit, the smaller body is pulled apart and its material may form a ring.

Moons play another important role in Saturn's rings: a moon's gravitational force can keep ring particles in place. **Shepherd moons**, for example, maintain narrow rings. The motion of ring particles within the thin, braided F ring is confined by gravitational forces of the moon Prometheus, orbiting just inside it (**Figure 8.26**). Without the gravity of the shepherd moon, the material in the F ring would slowly wander from its current orbit and the ring would eventually disappear (see **Anatomy of a Discovery**).

> **"We immediately concluded they must be made of material much younger than the age of the Solar System."**

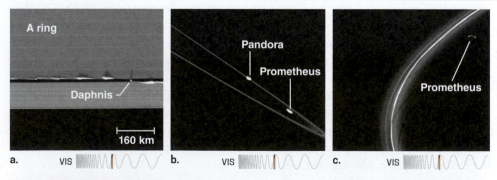

Figure 8.26 Shepherd Moons
Shepherd moons of Saturn seen in action by the *Cassini* orbiter. Saturn's rings are remarkably stable, thanks in part to the moons that sweep out gaps and provide gravitational "shepherding." **a.** Daphnis creates ripples in the A ring. **b.** Prometheus guides the F ring, despite the gravitational influence of nearby Pandora. **c.** Prometheus near some of its sculpting of the F ring, seen as places where the ring particles have been scattered by the moon's gravity.

shepherd moon A planetary satellite whose gravity confines a ring to a narrow band.

How does Saturn get its rings?

Galileo's sketch

Huygens's sketch

observations

In 1610 Galileo Galilei was the first to notice something unique about Saturn's appearance: through his telescope, it appeared to have "ears."

Christiaan Huygens concluded in 1655 that what looks like ears is instead a giant ring.

In 1675 **Giovanni Cassini** discovered large gaps, indicating that the giant ring is actually a system of rings.

hypothesis

Voyager *1* and *2* visited Saturn in the 1980s, discovering that the ring particles are icy and pristine. Scientists hypothesized that the rings must be young (that is, they did not form when Saturn did) and continuously replenished.

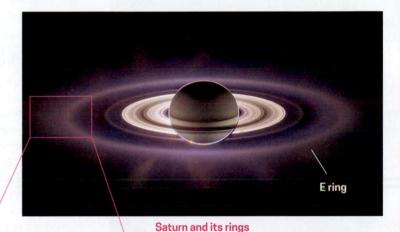

E ring

Saturn and its rings

observation

Cassini, reaching Saturn in 2004, confirmed a related *Voyager* discovery: some Saturnian moons, such as Prometheus, act as "shepherds." The gravity of these shepherd moons holds ring material on stable orbits for long periods of time, challenging the premise that *all* ring particles must be young.

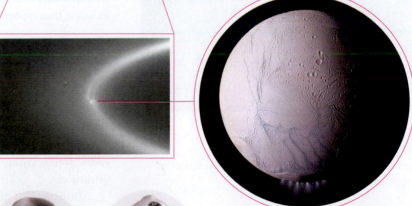

Artist's rendering of Enceladus

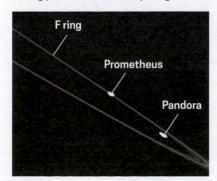

F ring

Prometheus

Pandora

Two of Saturn's moons and F ring

original hypothesis refuted

In 2005 the *Cassini* team, including **Amanda Hendrix** and **Linda Spilker**, found direct evidence that Saturn's moon Enceladus forms the E ring. Active geysers near this moon's south pole spew icy material into orbit about Saturn, replenishing the E ring with new ring particles. This discovery established moons as sources of ring material (and showed that icy moons can have subsurface oceans of liquid water).

a.

Great Dark Spot

b.

Figure 8.27 Uranus and Neptune

Ice giants **a.** Uranus and **b.** Neptune are similar in size, structure, appearance, and composition.

AT PLAY IN THE **COSMOS** THE VIDEOGAME   Fly through and learn about the rings of Saturn while battling enemy ships in Mission 10.

Cassini Division A 4,800-km-wide region between Saturn's A and B rings that has been shaped by Saturn's moon Mimas.

Encke Division A 325-km-wide gap within Saturn's A ring that has been shaped by Saturn's moon Pan.

The larger rings exhibit many substructures that also probably originated in small satellites embedded within the rings. Saturn's ring system has two big gaps: the **Cassini Division** and the **Encke Division** (Figure 8.24b). Moons help govern the ring particles' orbits: material in the Cassini Division responds to gravitational tugs from the moon Mimas, which lies farther from Saturn, while the Encke Division is maintained by the moon Pan, which orbits inside it. The gravitational force from these moons clears the space around them, creating gaps devoid of ring particles.

The phenomenon of shepherd moons in Saturn's rings highlights the role of *orbital resonances*, which are responsible for the Kirkwood gaps in the main asteroid belt (see Chapter 5). Recall that orbital resonances are synchronized, periodic gravitational influences that arise when objects orbiting a body have periods that are whole-number ratios of each other.

If many bodies are involved, such as a moon and ring particles, then an orbital resonance that produces many close encounters will destabilize the orbits of the ring particles and clear material away, leading to a gap. For example, the Cassini Division is produced by an orbital resonance with Mimas. In contrast, an orbital resonance that avoids close encounters allows ring particles to stay in their orbits.

Section Summary

- Saturn has an oblate shape (because it rotates so rapidly), is composed mainly of hydrogen and helium, and is the least dense Solar System planet.
- Haze partly obscures Saturn's cloud layers, but high winds and polar storms have been observed. Saturn's magnetic field is much weaker than Jupiter's.
- Saturn has more than 60 moons. Titan, the only Solar System moon with a substantial, hydrocarbon-rich atmosphere, has methane lakes and surface rain. The tiny moon Enceladus spouts geysers of water that contain salts (suggesting an underground ocean).
- Saturn's rings, much more complex than Jupiter's, differ in sources and histories. Individual ring particles range widely in size and consist of water ice with contaminants. The gravity of Saturn's moons forms and regulates the ring divisions and other features.

CHECKPOINT Identify and explain two important differences between Saturn and Jupiter, including their moons.

Uranus and Neptune: Ice Giants Discovered in Twilight

8.5 The outermost planets in our Solar System, Uranus and Neptune (**Figure 8.27**), are almost twins in terms of their size and composition. Like any human twins, however, they have acquired distinct properties through time and evolution.

During most of humanity's history, only five planets were visible from Earth: Mercury, Venus, Mars, Jupiter, and Saturn. In 1781, William Herschel spotted a fuzzy object that appeared too large to be a star. Watching carefully over many nights, Herschel found his mysterious object moving against the background stars. He soon recognized it to be a seventh planet. Using Newtonian mechanics, he calculated its orbit and found the new world moving around the Sun at a distance of 19 AU and a period of 84 years. While he was inclined to name the new planet "George" in honor of the British king George III. After some debate, astronomers decided to retain the mythological tradition of planet naming and called it Uranus for the Greek god of the sky, the father of Cronus (Saturn) and grandfather of Zeus (Jupiter).

While the discovery of Uranus came through serendipity, the discovery of Neptune was anticipated—and a triumph of Newtonian science. After Uranus was discovered, astronomers began charting its orbit in great detail. But a discrepancy was soon found in the new planet's motion that became more pronounced with time. All solutions to Newton's equations that described Uranus moving solely under the gravitational influence of the Sun and other known planets failed to predict the motions astronomers saw. Only when astronomers added an as-yet-unknown planet tugging on Uranus could they predict the observed motions.

Using Newtonian mechanics, astronomers predicted where this eighth planet should be and estimated its mass. In 1846, the planet was found almost exactly where predictions had put it, on an orbit with average radius of 30 AU from the Sun and a period of 165 years. The confirmation of a prediction as specific as the location of Neptune is one of the hallmarks that distinguishes science from other human endeavors.

Ice Giant Interiors

Given how similar in size, mass, and density Uranus and Neptune are, astronomers do not expect the internal structures of the two planets to differ significantly. The lower densities of Uranus and Neptune indicate large reservoirs of light elements. The planet's densities are, however, high enough to also indicate the presence of a rocky core at the center of each planet. Mathematical models of internal structure suggest that the core takes up about 20 percent of the planetary radius of Uranus, and somewhat more for Neptune. These cores are then about the size of Earth but probably contain 10 times the mass of our planet. Above each core lies a slush of light-element ices such as methane and ammonia mixed in a bath of liquid hydrogen. Above the slush is a large ocean of mostly liquid molecular hydrogen. But we have a lot to learn about the interior structure of the ice giants, and not all scientists agree with this description of their internal layers. It is clear, however, that unlike the gas giants, neither Uranus nor Neptune possesses enough mass (and hence gravity) to squeeze the liquid hydrogen into metallic form.

Ice Giant Atmospheres

The only way astronomers can get direct information on the composition of ice giant atmospheres is by spectroscopy. From these measurements, both planets appear to have chemical makeups similar to those of the gas giants: about 80 percent hydrogen and 18 percent helium. Methane is the next-most-abundant component, at less than 2 percent on each world (**Figure 8.28**). One key difference between the gas and ice giants is that Uranus and Neptune show much less ammonia in their atmospheres. The lower temperatures on these worlds allow ammonia to condense into crystals, which fall to the interior.

When *Voyager 2* visited the ice giants (the only time a probe has approached them), it saw Uranus as a hazy greenish orb and Neptune as far more blue. Methane is very effective at absorbing longer-wavelength (red) visible light. Neptune has more methane than Uranus, so more long-wavelength light is absorbed out of the sunlight striking Neptune, leading to a deeper blue color there.

The atmospheres of Uranus and Neptune appear far less structured than those of the gas giants in terms of bands of different colors (**Figure 8.29**). Because these outermost planets are colder, clouds form lower in the atmosphere (at higher temperatures) below the overlying haze. Storms and some banded cloud structures are seen on both planets, however. Infrared observations that can see deep into the planets' atmospheres have shown individual bands of clouds rotating around Uranus at 500 km/h. On Neptune, Earth-sized storms, such as the Great Dark Spot seen during *Voyager 2*'s visit (Figure 8.27b), have been observed to appear and vanish within a few years.

One significant difference between Uranus and Neptune is the inclination of their spins. Neptune rotates around an axis that is inclined 28° to the plane of the ecliptic (within range of most other planets), but Uranus's axis of rotation is inclined a remarkable 98° from the ecliptic. Its spin axis is almost in the plane of its orbit (**Figure 8.30**). Uranus

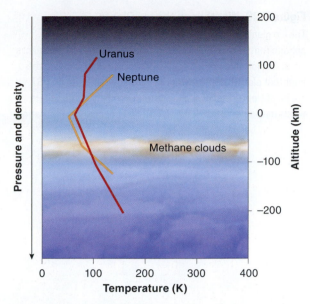

Figure 8.28 Ice Giant Atmospheres
With only a single cloud layer each, the atmospheres of the ice giants display far fewer features than those of the gas giants.

LINDA SPILKER

When Linda Spilker was 9, her parents got her a telescope. "I remember looking at Jupiter and seeing the little moons as pinpoints of light. From that point, I was hooked," she says. Her guidance counselor told her that girls don't go into science. "I remember watching all the astronaut missions from *Mercury* to *Gemini* and then *Apollo*. I think that really fixed my attention on space."

Despite the guidance counselor's discouraging words, Spilker's parents knew their daughter could do anything she set her mind to. After getting a degree in physics at California State University, Fullerton, Spilker worked for JPL and later received a PhD from UCLA in geophysics and space physics. As the *Cassini* project scientist, she was responsible for the science operations of the entire billion-dollar mission.

Figure 8.29 Weather on the Ice Giants
The ice giants exhibit high winds and other complex atmospheric features, though to a lesser degree than the gas giants do. **a.** A color-enhanced, Keck II telescope image showing the brightest cloud feature ever observed on Uranus. **b.** An HST image of Neptune showing the strong band structure, where winds may exceed 300 km/h (670 miles per hour) in alternating directional bands.

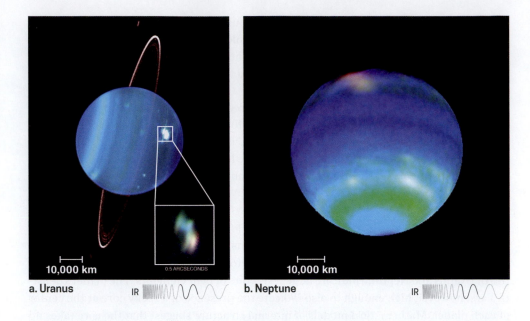

a. Uranus   IR
b. Neptune   IR

experiences the most dramatic seasons possible as it orbits the Sun. During its northern "summer," its north pole points almost directly at the Sun, which therefore never sets over the entire northern hemisphere. Half a Uranian year later, the south pole points toward the Sun and the northern hemisphere is enshrouded in darkness. Astronomers are still at a loss to explain how Uranus ended up with such an extreme axial tilt. A collision with another large object is one possibility, but that far out in the Solar System, where the density of planetesimals would be lower, it is unclear how probable such an event would be.

Ice Giant Magnetic Fields

Without the oceans of metallic hydrogen that the gas giants have, the ice giants might be expected to lack strong magnetic fields. However, both ice giants possess fields that are 100 times stronger than Earth's.

The most peculiar aspect of these fields is not their existence but their inclination. Both Uranus's and Neptune's magnetic axes are significantly offset from their planets' spin axes—by 59° and 47°, respectively. The magnetic field on each ice giant is not centered in the planet but is shifted to a position some distance from the planet's core (see

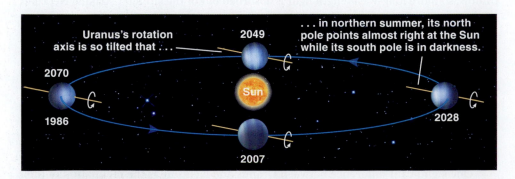

Figure 8.30 Uranus's Axial Tilt
Because its spin axis is nearly aligned with its orbital plane, the north and south portions of Uranus experience long periods of darkness during its 84-year orbit around the Sun.

Figure 8.5). Scientists have not definitively identified the cause of these off-center fields, but potential explanations rest on currents of ammonia ions flowing in the slush layers deep within Uranus and Neptune.

Ice Giant Moons

THE MOONS OF URANUS. At least 27 moons orbit Uranus. Most are less than 400 km in diameter; even Titania, the largest, is smaller than Earth's Moon. Remarkably, the orbits of all of Uranus's moons align with the planet's wildly tilted equator. Thus the moons experience the same seasonal extremes as Uranus. All of these moons have low albedos, which may result from solar and cosmic radiation breaking down surface molecules. In addition, the less geologically active a moon is, the darker its surface should be, since any geologic events would bring up new, lighter material to the surface. Uranus's five largest moons have densities less than 1,700 kg/m³, indicating a composition that includes significant amounts of water ice. This is to be expected for objects forming so far beyond the snow line.

The outermost major moons, Titania and Oberon, have surfaces pocked with ancient craters. Both moons appear geologically dead, experiencing no resurfacing; perhaps that means they were born with Uranus. Neither Umbriel nor Ariel, next in order of decreasing orbital distance, show evidence of ongoing geologic activity. However, Ariel has long cracks running across its surface, perhaps the result of tidal stresses experienced during the moon's evolution.

Miranda, the innermost major Uranian moon, is the most intriguing. Miranda shows some cratering, but its face is gashed by enormous faults, ridges, and valleys (**Figure 8.31**). Instead of a geologically dead world, as some astronomers expected, Miranda shows a buffet of planetary features. Some researchers have proposed that it may have been torn apart by some unknown disaster and gravitationally reassembled a number of times.

THE MOONS OF NEPTUNE. Neptune has seven regular moons and seven irregular ones—satellites that orbit either retrograde or nonequatorially (**Figure 8.32**). The largest Neptunian moon, Triton, is the only satellite of its size in the Solar System with a retrograde orbit. Some scientists speculate this retrograde orbit means that Triton did not form along with Neptune but was captured later. The orbits of Neptune's regular moons lie inside Triton's; the orbits of the other irregular moons lie outside Triton's.

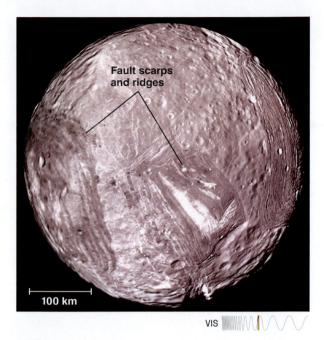

VIS

Figure 8.31 Miranda

This innermost major Uranian moon presents a variety of dramatic surface features. Uranus has a broad array of moons—some related to its rings in properties and origins, some with highly irregular orbits.

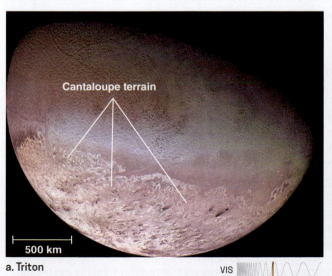

a. Triton   VIS

b. Proteus

c. Nereid

Figure 8.32 Major Moons of Neptune

a. Neptune's moons are dominated by Triton, the only spheroidal one, which accounts for 99.5 percent of the mass of Neptune's satellites. Its other major moons, **b.** Proteus and **c.** Nereid, are tiny by comparison and have irregular surfaces.

Geologically, Triton is bright and appears to reshape its surface continuously. When *Voyager 2* passed Triton, scientists were amazed to see geysers of nitrogen blown from the surface into space. These geysers appear to be the source of Triton's thin nitrogen atmosphere (it is $\frac{1}{100,000}$ as dense as Earth's). The remarkable low temperatures on Triton (just 37 K) allow some of the nitrogen to turn to frost and drift back down to the planet.

Along with the nitrogen geysers, the lack of craters indicates that Triton has been active fairly recently. Circular features resembling lakes suggest that liquid from below the surface may periodically well up, flooding landforms before it freezes. Strangest of all is the "cantaloupe" terrain in the moon's southern regions, composed of numerous dimples, ridges, and valleys (Figure 8.32a). The jumbled features in these regions indicate a surface that has experienced significant activity, probably the result of Neptune's tidal forces.

Ice Giant Rings

Both Uranus and Neptune host rocky rings (**Figure 8.33**). In fact, Uranus's rings were discovered before Jupiter's or Neptune's were. In 1977, astronomers were using the stellar occultation method (see Section 8.1) to study the planet's atmosphere. As the planet's path took it close to a background star, the astronomers kept careful watch for unseen Uranus-orbiting objects that might block out the starlight. The star winked on and off 40 minutes before Uranus blocked the starlight and again 40 minutes after. The symmetry of the occultations suggested that they were caused by rings rather than moons.

Uranus hosts at least 11 distinct rings with radii ranging from 37,000 to 51,000 km. Each ring is relatively narrow. Called 1986U2R, the innermost ring is the widest, at 3,500 km, but it is diffuse and difficult to see. The other rings have widths of 100 km or less. The rings are separated by gaps as large as 1,000 km wide.

All of Uranus's rings are within the planet's Roche limit. The rings are presumably composed of comet or asteroid material torn apart as it wandered too close to Uranus. These rings are quite dark, which means they are probably not composed of relatively pure ices, as are Saturn's large rings. Instead, the rocky particles have likely been coated with material blown off Uranus's moons. As in other ring systems, Uranus's moons play a shepherding role, their gravity keeping the rings sharp and preventing the rings' particles from drifting away.

All of Neptune's five rings lie within Neptune's Roche limit as well, so those rings are likely to be similar in origin to Uranus's. Neptune's rocky rings are a mix of both broad and narrow particle systems, however. Of particular interest is the outer Adams ring, which is only 50 km wide and shows bright knots or clumps. In shorter exposures (when the camera collects little light), only arcs of material are seen rather than a true ring. The origin of the clumps is not yet clear, though an unseen shepherd moon may be the culprit.

Section Summary

- Both Uranus and Neptune consist of a rocky core beneath a layer of ices and liquid hydrogen and a top layer of liquid molecular hydrogen. These planets are composed mostly of hydrogen and helium, with a small amount of methane.
- Neptune's spin axis is similar to that of most other planets, but Uranus's spin axis lies almost in its orbital plane, giving Uranus extreme seasonal variations. The ice giants' strong magnetic fields are off-center and greatly inclined.
- Uranus has more than 25 moons. All of its large moons have orbits aligned with its tilted equator and likely contain significant quantities of water ice. Uranian rings are dark with dust, widely spaced, and narrow.
- Neptune has seven regular moons and seven irregular ones, many with highly eccentric orbits. Its moon Triton, on a retrograde orbit, has a bright, young surface; a thin atmosphere; nitrogen geysers; and rough terrain. Neptunian rings vary in width.

CHECKPOINT Identify the important ways in which the ice giants differ from the gas giants.

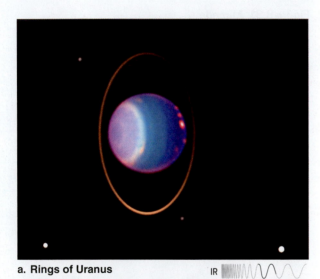

a. Rings of Uranus IR

b. Rings of Neptune VIS

Figure 8.33 Ice Giant Rings

Images from HST of the thin, rocky rings of **a.** Uranus and **b.** Neptune.

CHAPTER SUMMARY

8.1 Giant Planets on a Roll

The gas and ice giants formed outside the snow line from molecules and compounds that became solid (froze) at lower temperatures. Located at successively greater distances from the Sun, they receive much less solar radiation than the terrestrial planets do.

8.2 The Giant Planets: Structures and Processes

The giant planets are greater in mass and radius than the terrestrial planets, but they are much less dense. The giants lack hard surfaces and increase in density and pressure from the outer edges of their atmospheres to their rocky cores. The enormous internal pressure due to the gravity of overlying material creates extreme internal conditions, such as liquid hydrogen oceans. Via slow contraction, each giant produces more energy than it receives from the Sun. Rapid rotation drives strong weather systems in their atmospheres, such as banded winds and giant storms, and strong magnetic fields. Each giant planet has a ring system and many diverse moons.

8.3 Jupiter: King of Planets

With two-thirds of the Solar System's mass (excluding the Sun), gas giant Jupiter has played a stabilizing, protective role by gravitationally sweeping up comets and asteroids. Its interior layers consist of a molecular hydrogen ocean, metallic hydrogen, and a rocky and/or metallic core at high temperature. The energy from Jupiter's continued contraction helps drive convection in the atmosphere's counterrotating belts and zones. Winds reach hundreds of kilometers per hour, and the hurricane-like Great Red Spot has lasted at least 300 years. A vast, elongated magnetic field draws ions from the tidally locked moon Io. Of the more than 65 known moons of Jupiter, the four largest are the tidally locked Galilean moons. Callisto and Ganymede are geologically dead moons whose surfaces hide water oceans, though Ganymede may once have been extremely active. Many plates of ice, overlying a water ocean, are in constant motion on Europa's surface.

Io is covered with active volcanoes that derive their heat from Jupiter's strong tidal forces. Jupiter has dark, segmented rings that are not visible from Earth.

8.4 Saturn: Lord of the Rings

Although it has the same structure as Jupiter, gas giant Saturn has a lower density because of its lesser mass and rapid rotation. (It is the only planet with a density less than that of water.) The rapid rotation has also given Saturn an oblate shape, drives high wind speeds within the belt/zone atmospheric structures, and is responsible for large polar storm systems. The more than 60 known moons are either regular (orbiting prograde in the plane of Saturn's equator) or irregular (with highly inclined and some retrograde orbits). The largest moon, Titan, has a dense atmosphere with complex hydrocarbons. Geysers of water emanate from the surface of the moon Enceladus. Saturn's rings have a complex structure, with distinct ring systems and gaps that are shaped and maintained by shepherd moons.

8.5 Uranus and Neptune: Ice Giants Discovered in Twilight

Ice giants Neptune and Uranus are similar to each other in size, mass, density, and composition. Their structures include rocky cores; slushes of methane, ammonia, and hydrogen; and oceans of liquid molecular hydrogen. Their atmospheres contain hydrogen, helium, and methane and display storms, belts, and zones. A higher methane concentration in Neptune's atmosphere gives it a bluer color, compared with greenish Uranus. While Neptune shows a typical axial tilt, Uranus lies nearly on its side, thus experiencing dramatic seasons. The magnetic fields of both planets are offset significantly from their spin axes. Among Uranus's more than 25 moons, most noteworthy is Miranda, with dramatic faults, ridges, and valleys. Neptune's 14 moons are split evenly between regular and irregular types. The largest, Triton, is the only large satellite in the Solar System to orbit retrograde; it has nitrogen geysers on its jumbled surface.

QUESTIONS AND PROBLEMS

Narrow It Down: Multiple-Choice Questions

1. Which of the following planets formed within the snow line? Choose all that apply.
    a. Earth
    b. Mars
    c. Jupiter
    d. Saturn
    e. Uranus

2. In which of the following environments are ice giants and gas giants most likely to harbor life?
    a. atmosphere of the planet
    b. surface of the planet
    c. subsurface ocean on one of the planet's moons
    d. atmosphere of one of the planet's moons
    e. none, since these planets orbit where it is too cold for life to exist

3. If the spectrum of an exoplanet reveals significant amounts of methane with lesser amounts of ammonia, the exoplanet might be similar to which planet in our Solar System?
    a. Earth
    b. Mars
    c. Jupiter
    d. Saturn
    e. Uranus

4. Metallic hydrogen layers in a giant planet are characterized by which of the following? Choose all that apply.
    a. presence of mobile electrons
    b. molecules of hydrogen
    c. shiny surface
    d. high pressure
    e. liquid state

5. Which of the following is the most likely source of the heat radiated by the giant planets?
    a. solar radiation
    b. magnetic fields
    c. falling rain
    d. planet contraction
    e. nuclear fusion

6. Arrange the following planets by rotation period, longest to shortest:
    a. Mercury, Venus, Mars, Saturn, Uranus
    b. Venus, Mercury, Mars, Uranus, Saturn
    c. Saturn, Uranus, Mars, Mercury, Venus
    d. Mars, Mercury, Venus, Uranus, Saturn
    e. Uranus, Saturn, Mars, Venus, Mercury

7. If a distant exoplanet has rings, based on examples in our Solar System, what is the likelihood that it will also have moons?
   a. quite high
   b. quite low
   c. There is no way to know; rings and moons are not related.
   d. It is a certainty, since rings and moons always occur together.
   e. zero

8. Which of the following does *not* describe characteristics or effects of Jupiter?
   a. It has benefited life on Earth by sweeping up Solar System debris.
   b. It has a ring system.
   c. It has the greatest mass of any object in the Solar System.
   d. It has a strong tidal influence on its moons.
   e. It has distinct cloud layers varying in color.

9. In what way is Earth similar to Jupiter?
   a. The heaviest elements sank to the center of the planet during its formation.
   b. Its atmosphere has three distinct cloud layers.
   c. Each acquired a large moon by the same process.
   d. The planet is still shrinking.
   e. The mass ratio of planet to moon(s) is the same.

10. What is the main reason that Earth does not show the patterns of atmospheric belts and zones seen on Jupiter?
    a. Jupiter is farther from the Sun than Earth is.
    b. Jupiter has more mass and hence stronger gravity than Earth has.
    c. Jupiter's rotation rate is higher than Earth's.
    d. Jupiter has more moons than Earth has.
    e. Jupiter's magnetic field is stronger than Earth's.

11. What evidence tells us that the Galilean moons evolved together with Jupiter?
    a. the fact that all are tidally locked with Jupiter
    b. their relative densities
    c. the orientation of their orbits
    d. their orbital periods
    e. the quantity of hydrogen in their composition

12. In which of the following ways is Io unique among the Galilean moons?
    a. It is strongly influenced by tidal forces from Jupiter.
    b. It has a very high concentration of volcanoes.
    c. Its orbital period is measured in (Earth) days rather than (Earth) months or years.
    d. It is bigger than Earth's Moon.
    e. It was first observed by Galileo.

13. Which of the following characteristics is unique to Saturn, among the giant planets?
    a. ring system
    b. density less than water
    c. moon larger than Mercury
    d. storms observed to last decades or longer
    e. atmospheric structure of belts and zones

14. Why weren't the rings of Jupiter, Uranus, and Neptune discovered until the 20th century—centuries after Saturn's rings were discovered?
    a. They are made of dust rather than ice and thus reflect less sunlight than Saturn's rings do.
    b. They have large magnetic fields.
    c. They have no shepherd moons.
    d. They are composed of highly energetic particles that don't emit light.
    e. The planets are farther from Earth than Saturn is.

15. If Saturn's axis were tilted 8° instead of 27° to the plane of its orbit, which of the following would be true about the appearance of the rings?
    a. They would be easier to observe with a small telescope.
    b. From Earth we would be able to see the rings as if looking down from Saturn's north pole.
    c. The maximum apparent extension of rings above and below the equator of Saturn would be less.
    d. They would never disappear.
    e. They would never be visible.

16. If the following five Solar System planets were placed on the surface of a vast, deep cosmic water ocean, which would sink fastest?
    a. Saturn
    b. Mercury
    c. Mars
    d. Jupiter
    e. Uranus

17. Which of the ice giants have all of these characteristics: less distinct bands and zones than the gas giants; spin axis inclination within a few degrees; rings; shorter-lived storms than gas giants; orbital inclination within a few degrees of Earth?
    a. only Uranus
    b. only Neptune
    c. both Uranus and Neptune
    d. neither Uranus nor Neptune
    e. It cannot be determined from the information given.

18. Which characteristic is *not* likely to be present in a moon that formed from the protoplanetary disk with its giant planet? Choose all that apply.
    a. prograde orbit
    b. prograde rotation
    c. higher density than a similar moon of the planet with greater orbital radius
    d. highly inclined orbit
    e. composition identical to planet's

19. Which of the following features differs significantly between Uranus and Neptune? Choose all that apply.
    a. inclination of the rotational axis
    b. presence of cloud bands
    c. strong inclination of the magnetic field to the spin axis
    d. degree of seasonal impacts
    e. density

20. True/False: Two planets spin at the same rate, but planet A is more massive than planet B; planet A is therefore less oblate than planet B.

To the Point: Qualitative and Discussion Questions

21. How is the temperature within versus beyond the snow line related to the kinds of materials found in each region?

22. How are seasons related to the tilt of a planet's axis?

23. Would the stellar occultation method work for discovering rings around exoplanets? Explain.

24. Describe the key differences between the gas giants and the ice giants.

25. How do the densities of the terrestrial worlds compare with those of the giant planets, and what does this difference tell us?

26. How do the rotation periods of the giant planets compare with those of the terrestrial planets? What conditions on the giant planets are determined by their rotation rates? (See Table 7.1)

27. Moons orbit within and near the rings of the giants. In what ways do these moons and rings interact?

28. In what ways has the presence of Jupiter been favorable to conditions for life on Earth?

29. Name two important characteristics of each of the four Galilean moons of Jupiter. Which are candidates for microbial life?

30. How do tidal forces influence conditions on the moons of the giant planets? Name two dramatic examples.

31. Describe five key features of Saturn's complex rings.

32. Describe the ways in which Uranus and Neptune are close enough to be considered "twins." In what significant ways do the two planets differ?

33. Saturn is 10 times as far from the Sun as Earth is and rotates in about 10 hours. Describe how day and night, as experienced from an observer near Saturn's surface, would differ from our experience on Earth.

34. Astronomers had a good opportunity to learn about Jupiter's atmosphere by studying the effects of the Shoemaker-Levy 9 comet collision with the giant planet. Discuss two other events or situations that have offered similar opportunities in astronomy.

35. From what you've learned about the eight planets of the Solar System, which seem the most compelling as places to explore? Why?

Going Further: Quantitative Questions

36. Calculate the equilibrium temperature, in kelvins, of an airless planet that orbits the Sun and has an albedo of 0.45 and an orbital radius of 7 AU (see Going Further 7.2).

37. Use Kepler's third law to calculate the orbital radius, in astronomical units, of an imaginary planet orbiting the Sun with an orbital period of 46 years (see Going Further 3.1).

38. Jupiter's mass is 318 times that of Earth. Given that the gravitational pull of any object extends forever and depends on the object's mass and distance, how many times greater would Jupiter's gravitational force be, compared with Earth's, on an object twice as far from Jupiter as it is from Earth? (See Going Further 3.2.)

39. Calculate what Jupiter's mass, in kilograms, would be if Io were to orbit at twice its distance from Jupiter with a period 2.5 times as long, compared with Io's current orbital parameters (see Going Further 3.3).

40. Calculate the mass, in kilograms, of a planet whose moon orbits with a period of 5 days and an orbital radius of 6.0×10^6 km (see Going Further 3.4).

41. Planet Z's ratio of centripetal force to gravitational force is, in decimal form, 0.0013. If Z's rotational speed were to triple but its radius and mass remained the same, by what factor would this ratio change?

42. What would the ratio of centripetal force to gravitational force be for Saturn if its mass were to double but its rotational speed and radius remained the same?

43. Planet X's ratio of centripetal force to gravitational force is, in decimal form, 0.022. What would the ratio be if X's rotational speed were 80 percent as fast, its mass increased by 50 percent, and its radius increased by 15 percent?

44. Viewed from the surface of Jupiter at its average orbital distance from the Sun, what angular size in arcseconds would the Sun have? (See Going Further 2.1.)

45. At perihelion, Uranus is 2.74×10^9 km from the Sun. If you were able to view Uranus from that distance, what angular size in arcseconds would it have? (See Going Further 2.1.)

"WE MIGHT LEARN SOMETHING ABOUT HOW LIFE ON THIS PLANET GOT STARTED"

9

Life and Planets

THE SEARCH FOR HABITABLE WORLDS

The Origin of Life and a Trip to Antarctica

9.1 Chris McKay takes his science very seriously. In pursuit of his research, he has hiked through arid deserts, scrambled up dangerous cliffs, and chainsawed through 10 feet of solid Antarctic ice. Daunting as these undertakings are, the kind of determination they require may be the only way to understand questions as big as the origin of life on Earth and elsewhere in the Universe.

A recent stint at the bottom of the world gives a glimpse of McKay's workdays. "We went to the Antarctic to study these special microbial 'masses' living at the bottom of Lake Untersee," McKay explains. "But the lake is permanently frozen over, so we had to fight our way through the ice." Once they broke through, McKay and his team squeezed into wet suits for a dive into the lake's frigid waters. "Diving is the only way we can get down to the lake bottom and study the microbes." McKay and other scientists believe that the microbes in Lake Untersee may be cousins to some of Earth's first life-forms. "If we can understand the biology of these critters, we might learn something about how life on this planet got started."

Camping out in the Antarctic to scuba dive at the bottom of a frozen lake may seem like a strange way to reach back 4 billion years to the time when life began on Earth. It is, however, just one example of the way research in **astrobiology** is carried out. The goal of astrobiology, a relatively new science, is to place life into its planetary (and cosmic) context. The study of astrobiology has recently become a diverse, worldwide endeavor, incorporating telescopic studies of newly discovered exoplanets, biochemical experiments on how molecules "learn" to self-replicate, and searches for signals from extraterrestrial civilizations. More important, it represents a turning point in the way humanity asks the most profound of questions: What is life, and where does it exist?

Life, the Universe, and the Greatest of All Questions

For thousands of years, human beings have asked whether planets exist outside our Solar System (see Chapter 5). Just as ancient is the question of whether life exists on planets besides our own. As yet, no definitive evidence has been found that life—intelligent or otherwise—exists anywhere else in the Universe. But what would it mean if we never

astrobiology The interdisciplinary study of life in its astronomical context.

← An artist's conception of what one of the thousands of recently discovered exoplanets (planets orbiting other stars) might look like. Here, we see a relatively Earth-like world with oceans and continents.

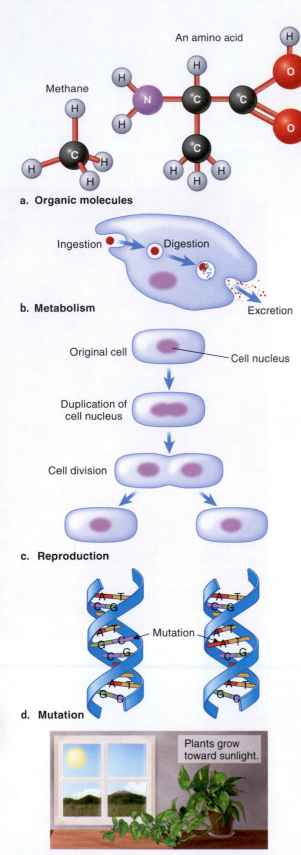

An amino acid

Methane

a. **Organic molecules**

Ingestion   Digestion

b. **Metabolism**

Excretion

Original cell — Cell nucleus

Duplication of cell nucleus

Cell division

c. **Reproduction**

Mutation

d. **Mutation**

Plants grow toward sunlight.

e. **Sensitivity**

Figure 9.1 Five Basic Characteristics of Living Things
While there is no perfect definition of life, the five characteristics represented here provide a good way for scientists to think about what life is and how it behaves.

discovered life anywhere other than Earth? If we were to learn that our planet is the only world in the entire galaxy harboring life, how would our perspective be different? Would we see all life as sacred or feel a profound sense of meaninglessness?

On the contrary, it is not hard to imagine how finding (and perhaps meeting) other intelligent species could lead to dramatic consequences for our culture. Just being able to communicate with an intelligence that does not share our human perspectives about questions of life and death and existence would doubtless lead to a monumental shift in human culture.

But why explore the question of life in the cosmos just after finishing a study of planets? While some scientists have hypothesized that life might form in exotic places such as interstellar clouds, everything we *currently* understand suggests that planets, and perhaps the moons surrounding them, are essential for life to begin and develop through evolution. This chapter therefore focuses on what we can infer from the life we can study on Earth.

Section Summary

- Scientists generally believe that the environments provided by planets or moons will be required for life.

CHECKPOINT   How can we answer questions about astrobiology by studying life on Earth?

What Is Life, and Where Can It Exist?

9.2   To study life in an astronomical context, we must first define what constitutes life. This is particularly important if we are interested in the origins of life—the transition from nonliving (*abiotic*) material into a living system.

From Chris McKay's perspective, life is a *process*, not a thing. He makes an analogy between life and fire. "It's very hard to come up with a closed, concise definition for fire that captures everything," he says. "But we do have a very good scientific (thermodynamic) understanding of it as a process." Thus, asking for a definition of what life *is* may be less useful than asking what life *does*.

The Machinery of Life

Let's start by summarizing five basic features of living things (**Figure 9.1**):

1. *Organic molecules*. In terms of chemical composition, living organisms are made up of organic molecules—that is, molecules that contain the element carbon. Many of these molecules have structural characteristics that allow them to form long chains or rings. Carbon is a unique element in that the structure of its six electron orbits enables carbon atoms to form a diverse array of bonds with other atoms or ions. Some other elements, such as silicon, have similar properties, but carbon is remarkably well suited to building life's chemistry.

2. *Metabolism*. The consumption of energy (often supplied in the form of chemicals) and, in the process, the production of waste is an essential feature of living organisms.

3. *Reproduction*. Through reproduction, living organisms create new versions of themselves by various means. These new versions are sometimes identical copies, but often they are not.

4. *Mutation*. The ability to adapt and evolve, an essential aspect of living organisms, depends on mutation. Mutations are random changes in an organism that occur during reproduction. They may be beneficial, detrimental, or immaterial to survival.

5. *Sensitivity*. Living organisms show sensitivity to their environment—they notice and respond to changes in ways that enhance survival.

All five features are deeply connected with the processes of evolution. For McKay, this is the most important aspect of understanding what is and what is not life. "When you get down to it," McKay says, "the best definition of life is 'a material system that undergoes Darwinian evolution.'"

It is worth mentioning a few exceptions. For example, some nonliving things show one or more of these characteristics. Viruses contain organic molecules and in some ways adapt to their environment, but they don't carry out metabolism and lack the biosynthetic machinery needed for reproduction—they require a host cell. Many biologists don't consider viruses to be alive. Yet some clearly living parasites cannot reproduce outside a host cell either. In addition, crystals, for example, show growth that could be likened to some aspects of living systems. Despite these possible exceptions, the five criteria listed provide a place to start a discussion of where, when, and how life might form on other planets.

Planets and Life: The Habitable Zone

Most scientists agree that for a planet to harbor life, it must have liquid water. The reasoning is that water is a highly effective solvent for biochemical reactions. That is, water readily acts as a medium in which compounds dissolve into components that are ready for further reactions. Scientists therefore tend to define a planetary system's habitable zone as the range of distances from a star where planets can retain liquid water on their surfaces. (Recall from Chapter 5 that the *habitable zone* is the region around a star where life could form.) For example, if a planet is too close to its star, surface water will boil away. If it is too far from the star, surface water will remain eternally frozen. Thus, the most basic definition of a habitable zone relies on a calculation of the range of distances from a star at which liquid water can exist on a planet's surface. The surface temperature (if life forms and evolves on the surface) must lie in the range between water's freezing and melting temperatures. In other words, the planetary temperature T_p must range between 273 K and 373 K (it's worth noting that these temperatures will also depend on atmospheric pressure).

You have already learned how to calculate the surface temperature of a planet that lacks an atmosphere. The formula given in Going Further 7.2 used the Sun's temperature and radius to find a planet's surface temperature, given its distance from our star. We can manipulate this formula to give us the *range of distances* from any star that defines the habitable zone of its solar system (see **Going Further 9.1** on p. 242). Using this formula for our Solar System, we find that between 0.47 and 0.88 astronomical unit (AU) constitutes a first approximation of the Sun's habitable zone. The surprising result that Earth (at 1 AU) is not in the Sun's habitable zone appears only because we used the simplest form of the habitable-zone equation, omitting the all-important effect of atmospheres in trapping solar radiation. Nevertheless, these calculations are an important first step in capturing the basic physics involved as sunlight warms the planet's surface.

When scientists carry out a more detailed calculation that includes the role of atmospheres, they find that the habitable zone of our Solar System today extends from 0.99 to 1.68 AU from the Sun. Notice that, in these detailed calculations, Earth is just at the inner edge of the Sun's habitable zone (**Figure 9.2**). Depending on the way the habitable zone is defined, Venus may also lie within it, but the runaway greenhouse atmosphere of Venus makes that planet too hot for liquid water.

Not all stars are like the Sun, however: some are much hotter and brighter, while most are much cooler and dimmer. (Recall from Chapter 4 that a star's luminosity varies with its temperature and radius.) Therefore, the habitable zones around other stars may look very different from our own. Some of the coolest, lowest-luminosity stars have a narrow habitable zone at small orbital distances. In the stellar classification system (discussed in

"When you get down to it, the best definition of life is 'a material system that undergoes Darwinian evolution.'"

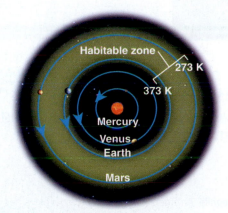

Figure 9.2 Our Habitable Zone
The Solar System's habitable zone includes bodies at a distance of about 1 astronomical unit (AU), making water-based life on Earth possible.

Figure 9.3 Habitable Zones of M-dwarf and O Stars
Each star's habitable zone depends on the star's luminosity, which in turn depends on temperature and radius, so **a.** the habitable zone of a large, hot O star is broader and farther from its star than **b.** the habitable zone of a smaller, cooler M-dwarf star.

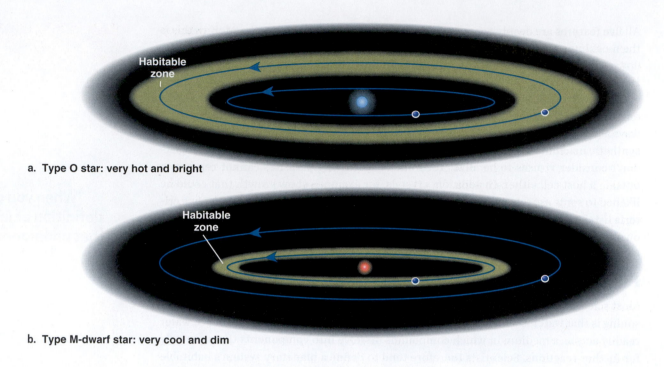

a. **Type O star: very hot and bright**

b. **Type M-dwarf star: very cool and dim**

INTERACTIVE:

Habitable Zone

"We have found ecosystems completely independent of anything that goes on at the surface."

extremophile An organism that thrives under extremes of temperature, pressure, and other conditions.

polymer or **macromolecule** A large molecule containing many atoms in repeating subunits (monomers), often with a chain structure.

monomer A small molecule that, when repeated, forms a polymer.

deoxyribonucleic acid (DNA) A double-stranded polymer that encodes genetic information and is the biological basis of life on Earth. Its two strands form a double-helix shape.

nucleotide A monomer built of a sugar molecule, a phosphate group, and a nucleobase. Sugars and phosphates link together, forming the backbones of DNA and RNA.

nucleobase A nitrogen-containing base; a component of a nucleotide in DNA and RNA.

ribonucleic acid (RNA) A single-stranded polymer that transcribes the code of DNA, placing amino acids in the correct sequence to create proteins.

Chapter 11), such stars are referred to as type M. The habitable zone of a very bright, very hot star, called a type O star, is at large orbital distances from the star (**Figure 9.3**).

We must consider other factors as well. Hot, bright stars tend to have very short lifetimes—some as brief as a few million years, perhaps too short for life to form. Magnetic activity on a star in the form of giant flares might also affect the development of life. Some aspects of the planets' own composition and history may also matter. Is the recycling of a planet's surface material that occurs in plate tectonics needed for life to form and evolve? Planetary magnetic fields also may be important because they shield a planet from harmful space weather caused by the host star's magnetic activity. Such issues add complexity to determinations of habitability and a planetary system's habitable zone.

One of the major discoveries in the march toward understanding planetary life came in 2014 when Kepler-186f, the first Earth-sized exoplanet in the habitable zone of a star (called Kepler-186), was found (see **Anatomy of a Discovery** on p. 244). Astronomer Elisa Quintana led the team that detected it. This was an important step for astronomers, as planets like Kepler-186f will become targets of future telescopic searches for life.

Before we leave this subject, let's consider the possibility of life that does not require water as its foundation. Scientists have explored other compounds that might serve as the solvent supporting biochemistry in the same way water does. Both ammonia and formamide have been suggested as alternatives to water. Studies have been done to understand how life based on them might work—but the consensus so far is that water offers a far richer set of biochemical pathways.

Expanding the Habitable Zone

As scientists have learned more about the limits of life on our own world, new discoveries have expanded the definition of habitability. Since the 1970s, scientists have found a bewildering array of life-forms thriving in conditions so harsh that a biologist in the early 20th century would not have dreamed they were possible. These newly revealed organisms, called **extremophiles**, live in such inhospitable environments as within volcanic springs, in excessively saline lakes, or deep below Earth's surface. "No one really expected it," says Chris McKay of the recent Earth-based discoveries shaking astrobiology. "We have found ecosystems completely independent of anything that goes on at the surface."

Among these news-making extremophiles are giant tube worms, crabs, and shrimp that were discovered in 1977 a mile below the ocean surface around deep-sea hydrothermal vents. The vents, called "black smokers" for the plumes of hydrogen sulfide they belch into the ocean, represent a previously unknown ecosystem that does not need direct sunlight to survive. The biochemistry of these organisms relies on the heat and raw chemical material the black smokers provide (**Figure 9.4**). Unlike anything on the surface, this deep-ocean ecosystem gave scientists a hint that life can form and thrive far from surface sunlight and surface water.

Far more compelling for McKay's thinking about habitable zones, however, is another class of extremophiles: bacteria that thrive deep underground. Scientists discovered colonies of microbes living 5 miles down in the subterranean reaches of South African gold mines. "These creatures get their energy from sources we never imagined," says McKay. "The South African extremophile bacteria are powered by the radioactive decay of unstable molecules in the rocks. Radioactive decay! Sunlight and surface water play no role. Radioactivity sets the biochemistry in motion. It's amazing!"

Extremophiles that feed on nonsolar energy sources represent the first step in a process of discovery that has radically altered the concept of planetary habitable zones. The second step is the direct exploration of alien worlds. "When you take the discovery of liquid water below the surface of ice-covered moons like Europa and Enceladus," says McKay, "and put it together with our understanding of terrestrial extremophiles, you can see why the definition of 'habitable zone' had to change." Direct exploration has shown that liquid water exists in many places in our Solar System where scientists had not expected it. With the discovery of watery moons around ice giants and gas giants and the possibility of extremophile biology, scientists are rethinking the habitats for life.

Molecular Engines of Life: DNA, RNA, and Proteins

For all the diversity of life on our planet—in terms of both the forms organisms take and their ideal environmental conditions—the molecular basis of that diversity is remarkably simple. McKay explains, "If we ask the question 'What is life made of?' the answer is really rather surprising. Life on Earth is not made up of a large number of different types of molecules. Instead, it's made up of a small set of molecules used over and over and over again."

Terrestrial life relies on a few classes of **polymers** (also called **macromolecules**). These are giant molecules that look like long chains, often folded over on themselves. Polymers are made up of repeating subunits of shorter molecules called **monomers**. "The diverse organisms we see on Earth come from the combinatorial complexity of those molecular units," says McKay, referring to the immense number of ways these building blocks can be put together and taken apart. "In a sense, life is like Legos. A Lego kit is a small number of standardized bricks, and you can make all sorts of diverse objects by the way you put these bricks together."

The most important classes of polymers are the biologically active molecules DNA, RNA, and the proteins. **Deoxyribonucleic acid (DNA)** consists of two long polymer strands that wind around each other in a characteristic double-helix shape (**Figure 9.5a**). The DNA molecule consists of monomers called **nucleotides**. Each nucleotide contains a sugar molecule, a phosphate group (a molecule made of phosphorus and oxygen atoms), and a **nucleobase** (nitrogen-containing base). The sugars and phosphate groups form the two strands that make up the "backbone" of a DNA molecule; nucleobases link the strands together. DNA has four types of nucleobases: adenine (A), thymine (T), guanine (G), and cytosine (C). In binding the two backbone strands together, these bases pair up in specific ways—adenine can bond with thymine only (an AT pair), and guanine bonds to cytosine only (a GC pair).

Ribonucleic acid (RNA) is a similar polymer made of nucleotide building blocks (**Figure 9.5b**). In RNA, however, thymine is replaced by another nucleobase, uracil (U), which bonds to adenine only (a UA pair). Unlike DNA, RNA has just a single strand.

a.

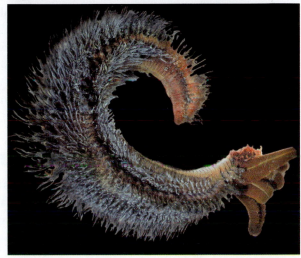
b.

Figure 9.4 Extremophiles in a Black Smoker
a. Giant tube worms flourish far from light, in the hot, sulfurous environment around deep-ocean vents called black smokers.
b. The bacterium *Nautilia profundicola* is a microbe that survives near such vents.

"In a sense, life is like Legos."

GOING FURTHER 9.1
Calculating Habitable Zones

To calculate the habitable zone around a star, recall from Going Further 7.2 the formula for a planet with temperature T_p (and no greenhouse effect), albedo a, and distance D from the star it orbits:

$$T_p = (1 - a)^{\frac{1}{4}}\left(\frac{R_*}{2D}\right)^{\frac{1}{2}} T_*$$

Here R_* is the star's radius and T_* is the star's temperature because we are not referring to the Sun in particular.

We assume that liquid water is necessary for life. So, to calculate the star's habitable zone, we must solve this equation for D and then use the new expression to find the inner boundary of the habitable zone, where the temperature of the planet T_p will just equal the boiling point of water (100°C, or 373 K). We can then find the zone's outer boundary by setting T_p to the temperature at which water freezes (0°C, or 273 K).

We want to consider both water under conditions in which the atmosphere has no greenhouse effect and stars that are more realistic than mere blackbody radiators. Thus we use a modified (and rearranged) version of the previous equation:

$$D = \left(\frac{T_o}{T_p}\right)^2 [(1 - a)L_*]^{\frac{1}{2}}$$

where D is measured in astronomical units and L_* is the star's luminosity, measured in solar units. (That is, our Sun has a luminosity of $L_* = 1$.) The term T_o expresses the role of planetary rotation. We must account for rotation rates because rapidly spinning planets absorb solar radiation more evenly than slowly spinning ones do. For fast-rotating planets, $T_o = 279$ K; for slow-rotating planets, $T_o = 394$ K.

Let's use this second equation to find the inner boundary of the habitable zone of a slow-rotating planet that orbits a star that's 4 times as luminous as the Sun ($L_* = 4.0$). We can use $a = 0.3$ for an Earth-like planet:

$$D = \left(\frac{394\text{ K}}{373\text{ K}}\right)^2 [(1 - 0.3)4]^{\frac{1}{2}} = 1.87\text{ AU}$$

This distance is beyond Earth's orbit. The outer boundary of this star's habitable zone lies at

$$D = \left(\frac{394\text{ K}}{273\text{ K}}\right)^2 [(1 - 0.3)4]^{\frac{1}{2}} = 3.49\text{ AU}$$

This distance is well beyond Mars's orbit. Thus, increasing the luminosity of a star changes the planet's habitable zone considerably.

Recall that none of these calculations include the effect of greenhouse gases, which will warm a planet's surface by trapping heat from the star. If we used this formula alone, we would come to the surprising conclusion that Earth is not in our Sun's habitable zone! Without Earth's atmosphere, the development of life on our planet would have been impossible.

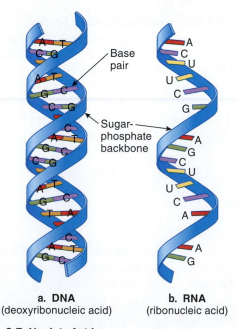

a. DNA
(deoxyribonucleic acid)

b. RNA
(ribonucleic acid)

Figure 9.5 Nucleic Acids
The related complex molecules **a.** DNA (double-stranded) and **b.** RNA (single-stranded) are necessary for life as it exists on Earth.

Proteins, the workhorses of cell biology, are polymers that play many roles within a cell. Some proteins form the cell's architecture—structures that maintain cell shape and allow subcellular components to move around. Other proteins regulate cell activities; still others are enzymes, which help chemical processes proceed. All proteins are made up of monomers called *amino acids*, which are strung together in the polymer chains. The sequence of amino acids in a long chain of a protein is a key factor in determining the protein's shape, which determines its function (the role it plays in the cell). "There are just about 20 amino acids that terrestrial life uses," explains McKay. "By stacking them in different ways, you can make a bewildering array of proteins that do an enormous range of structural and functional things for life."

The complex ways in which all of these polymers' building blocks combine (nucleotides for RNA and DNA, amino acids for proteins) are crucial—and the relationships *among* DNA, RNA, and proteins make life possible. Because of the specific way that nucleobases link up in the double helix, DNA acts as an information storage molecule. The sequence of bases on each DNA strand is a kind of molecular language, or "code," that stores information about how to build the working parts of a cell: proteins. To initiate protein building, the two strands of DNA are unwound, and RNA then reads the DNA code. Other forms of RNA then translate the code into sequences of amino acids, which form the proteins that cells need to do their work.

DNA can *self-replicate*, or produce copies of itself. When a cell grows and splits into two new cells, these daughter cells take copies of the parent's DNA (and its genetic instructions) with them. DNA molecules contain **genes**—bits of genetic code that are passed down from one generation of life to the next.

Section Summary

- Scientists define living things as being composed of organic molecules and exhibiting metabolism, reproduction, mutation, and sensitivity.
- Assuming that life on Earth is typical, scientists consider water to be essential for life. The habitable zone—the region around a star where surface liquid water can exist—is a key consideration in the search for other habitable worlds.
- The discovery of extremophiles on Earth (and subsurface water found on some giant planets' moons) suggests that life can flourish in a broader range of conditions than was previously believed.
- DNA, a self-replicating polymer, consists of nucleotide monomers that form two strands linked by four possible nucleobases. The sequence of bases forms a genetic code. RNA is a single strand of nucleobases that transcribes DNA's code, placing amino acids in the correct sequence to create proteins, which perform many functions.

CHECKPOINT a. How does the habitable zone depend on the location of a planet's orbit? b. What are possible reasons to expand the definition of a planet's habitable zone?

The Origins of Life

9.3 The remarkable interactions among DNA, RNA, and proteins represent the fundamental biochemical machinery of life. But to study the origin of life, scientists must first explain an even more remarkable transition. How could nonliving material have combined in just the right way to be able to transform into living material? The task for scientists who study **abiotic synthesis**—the creation of life from nonlife—is to understand, using only the laws of physics and chemistry, how molecules of DNA, RNA, and proteins could form.

Getting to Biology

The study of abiotic synthesis began in earnest in the 1920s with the work of Russian scientist Alexander Oparin and British researcher J. B. S. Haldane. Both men proposed that **prebiotic molecules**, the precursors to biological molecules, that act as the building blocks of life could form through natural processes in the early Earth. In the modern understanding of biochemistry, amino acids are prebiotic molecules. They are necessary for life but are not living organisms themselves.

In 1953, Stanley Miller and Harold Urey, chemists at the University of Chicago, decided to test one version of this theory (**Figure 9.6**). They attempted to create a simulated version of the early Earth in a test tube. "The essence of the Miller-Urey experiment (which has been repeated a thousand times)," explains Chris McKay, "is to take simple molecules like methane, hydrogen, and water; stick them in a jar; add energy; and see what happens. The results turn out to be astounding." Specifically, Miller and Urey created an "atmosphere" of hydrogen, ammonia, and methane in one flask and connected it to an "ocean" of liquid water in a separate flask. To provide energy for chemical reactions, they set up a high-voltage discharge in the atmospheric flask to act as lightning and a source of ultraviolet light.

After running the experiment for a week, Miller and Urey found that brown goo had formed in their mock ocean. Chemical analysis showed this goo to be a rich soup of prebiotic molecules, such as glycine, lactic acid, and urea. "It turns out," says McKay, "that the Lego blocks life uses for building its macromolecules could be reproduced quite easily from nonbiological processes." What's noteworthy is that 10–15 percent of the carbon atoms in the original flask had been converted into molecules such as amino acids, sugars, lipids (the basis of fat), and a variety of other organic acids.

a.　　　　b.

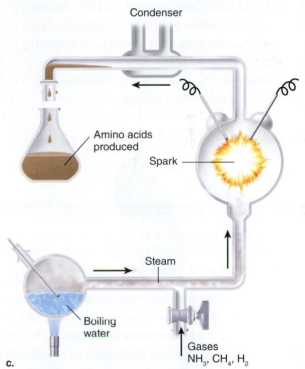

c.

Figure 9.6 The Miller-Urey Experiment

In 1953, **a.** Stanley Miller and **b.** Harold Urey conducted **c.** an experiment to try to replicate the conditions and materials they believed to be present in Earth's early environment. They were able to produce the molecules required for life within 1 week.

protein A polymer, made of amino acid monomers, that performs and assists in cellular functions.

gene A section of a DNA or an RNA molecule that is the code for a particular trait or function.

abiotic synthesis The creation of life from nonlife.

prebiotic molecules Molecules that act as the building blocks of life—necessary for the origin of life.

Do other habitable Earth-sized planets exist?

hypothesis

A statue in Rome, Italy, stands where **Giordano Bruno**, a defrocked Dominican monk, was burned to death in 1600. Among his "blasphemies" was the belief that the Universe is infinite and has an infinite number of inhabited worlds. In his time, there was no evidence that planets orbit other stars; most scholars and theologians believed that Earth is motionless at the center of the Universe.

Radial velocity graph

51 Peg b

4.2 days

Radial velocity (m/s)

150

0

−150

Time

Artist's rendering of exoplanet 51 Peg b

confirmed by observation

Using the radial velocity method in 1995, nearly 400 years after Bruno's execution, **Michel Mayor** and **Didier Queloz** discovered 51 Peg b, the first exoplanet known to orbit a Sun-like star. To the surprise of astronomers, the Jupiter-like planet orbits its host star at a distance one-tenth that of Mercury's distance from the Sun. A new era of exoplanet research began, indicating that planetary systems are abundant and diverse.

observation

In 2009, with a goal of detecting the transits of Earth-sized planets orbiting Sun-like stars, NASA launched the **Kepler Space Telescope**. Kepler monitored the brightness of 150,000 stars over a four-year span. Shortly after Kepler became operational, discoveries of exoplanets of all types began in earnest.

confirmed by observation

In 2014, **Elisa Quintana** led the team that detected the first Earth-sized planet in a star's habitable zone: Kepler-186f. It may have the conditions needed for life, and it orbits a star with half the Sun's mass that hosts at least four other planets. Much work is needed to determine the habitability of such planets, but Kepler-186f signifies that the term f_h in the Drake equation (introduced later in the chapter) is nonzero.

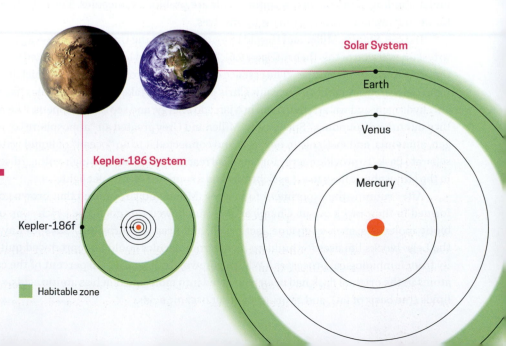

Solar System

Earth

Venus

Mercury

Kepler-186 System

Kepler-186f

Habitable zone

The Miller-Urey experiment was hailed as an instant classic of biochemistry and proof that, in principle, nonliving material, along with the natural processes of physics and chemistry, could create the molecular basis of life. "It was a major breakthrough and a major surprise," says McKay. "People thought that we were just moments away from solving the origin-of-life problem."

Although the Miller-Urey experiment showed that life's building blocks *could* be synthesized under conditions like those of the early Earth, with more research, scientists also began to see how the story is more subtle and more complicated. Studies focusing on the atmosphere of the early Earth, for example, made it clear that there would not have been much hydrogen. Volcanism would have produced much more atmospheric CO_2 and N_2, yielding very different chemical reactions than what Miller and Urey saw. Modern studies have therefore focused on environments such as volcanic vents in the seafloor as molecular factories where prebiotic molecules might be assembled. More important, scientists' perspectives changed as the relationships and functions of proteins, DNA, and RNA came to be understood.

"The Miller-Urey experiment happened back when the origin of life was viewed as a chemistry problem," explains McKay. "Now we know it's also an information-processing problem. It's a software problem as much as a hardware problem." In other words, scientists need to understand how self-replicating molecules, which store information and process it via the construction of new molecules, were created. The famous Miller-Urey experiment was a first step, showing one pathway for assembling life's building blocks, but researchers now look to other processes and perspectives to understand the creation of the first "information-rich" self-replicating molecules on the young Earth.

From Nonlife to Life, and the Heroism of Time

At the molecular level, the story of life's origin focuses on one aspect of our definition of life: self-replication. "That is where the software aspect of the problem comes in," says McKay. "If you make an amino acid in a prebiotic experiment, there's information in that molecule, but it's purely in the structure of the molecule. There is no algorithm for producing more amino acids coded within it. So going from a molecule where the information is the molecule itself to a molecule where information about making new versions of itself is all coded within the molecule, that's quite amazing." McKay uses the analogy of a computer memory chip. "The shape of the chip does not depend on what is written on it. In the same way, you have to figure out how to build a stable molecule (in terms of its structure) in which information—like instructions for making copies of itself—can be coded into the molecule." In other words, the ability to store and pass on information is the most important part of DNA that scientists must understand. The molecule's shape is just a means to that end.

Thus, beginning with prebiotic molecules—such as amino acids or nucleic acids and nucleobases—scientists must imagine how ongoing random chemical reactions led to more complex molecules that could store information and enable self-replication. Scientists consider that the early Earth was acting like a giant chemistry experiment. Chemically active sites such as pools of gooey, warm water or deep-ocean vents allowed for endless trial and error as new molecular combinations came into being and then broke apart. If some of these new molecules could self-replicate even partially, the wheels of self-replication may have been set in motion. Once a self-replicating molecule appeared in a primordial soup of prebiotic chemicals, its numbers would increase relative to other molecules over time. "You can take this kind of logic a step further," says McKay. "If two kinds of self-replicating molecules appear, then whichever one reproduces itself faster will come to dominate in terms of sheer numbers." Imagine that gold beetles in a population breed faster than green beetles; as time goes on, more and more beetles will be gold. The same is true of chemical reactions. Molecules that could self-replicate would eventually have dominated.

How do you build a self-replicating molecule? If DNA contains the information needed to produce copies of itself, could a strand of DNA have appeared in the soup of

CHRIS MCKAY

Chris McKay spends months at a time living in a tent in the remotest parts of the world. Nevertheless, he loves his science. "To do the kind of research that interests me, you have to get pretty extreme. If you are going to go out and find extremophiles, it means you're not going to be sitting in an office somewhere wearing a white lab coat."

McKay went to Florida Atlantic University as an undergraduate and the University of Colorado for graduate school. "It was 1976," he says. "That was when the two *Viking* probes landed on Mars." Watching those missions propelled him into astrobiology. "It was fate in a lot of ways that got me into this," says McKay. "But I am pretty happy with the way it worked out."

Figure 9.7 Evolutionary Adaptations of Animals

a. The bald eagle's keen eyesight, aerodynamics, and sharp beak and talons make it a formidable predator. **b.** At ocean depths, bioluminescence helps organisms find food, hunt, and attract a mate. **c.** The anteater uses its elongated snout and extendable, thin tongue to extract termites from their nests. **d.** The Arctic hare's snow-colored fur helps the animal avoid becoming prey.

Figure 9.8 Charles Darwin

Darwin (1809–1882), an English naturalist, conducted meticulous studies of related species of plants and animals. He established that all species descend from common ancestry, with variations occurring over time, through natural selection. In 1859 he published his findings in *On the Origin of Species*.

genetic drift A change in a gene's frequency due to random events other than gene mutation.

natural selection The process by which an organism with traits that make it well adapted to its environment survives and reproduces, passing on those traits more successfully than organisms without those traits.

random chemical reactions early in Earth's history? Most researchers think not. Studies of chemical reactions and their speeds suggest that the time it would take to assemble a working DNA molecule from random reactions is longer than the age of the Universe (10^{10} years). Researchers now believe that smaller segments (monomers) of RNA formed first. These then combined into longer polymers of RNA, a few of which had the ability to self-replicate (see **Going Further 9.2** on p. 250). Once self-replication began, it would have led to huge populations of the self-replicating molecule, and further modifications ensued.

The key ingredient in this story is *time*. Human scientists can't wait more than a few decades to see whether their experiments pan out. Nature is not so impatient. The prebiotic soup could percolate for hundreds of thousands of years, with countless molecular combinations forming, until self-replication randomly occurred. When people question whether life could have formed from nonlife, they often fail to take into account the vast stretches of time that were available for the process to work.

Evolution 101

Life on Earth exhibits a stunning array of forms and functions, from tiny single-celled diatoms floating in the ocean to massive multicellular elephants roaming the land. In between is everything from mosses to mushrooms to mollusks to muskrats. We see living beings that are astonishingly well tuned to their environments. Walking sticks, a type of insect, can mimic the trees on which they move, and eagles have feathers with special fringes that allow them to dive from high altitudes at tremendous speeds. Such a behavior or trait that helps an organism succeed in its environment is an *adaptation* (**Figure 9.7**). Where did this diversity and adaptation come from?

Scientists believe that once self-replicating molecules emerged, the stage was set for the processes of *evolution* to create more complex and more diverse forms of life over time. Evolution is central to an understanding of the biological world. As McKay puts it, "Without doubt, evolution is the most powerful force in biology. It works because all life carries genetic information, which can be altered."

The mechanisms of evolution were first articulated by Charles Darwin in the 1800s (**Figure 9.8**). Darwin recognized that traits of living organisms—such as the color of a beetle's shell or the shape of a bird's beak—are passed down from parents to offspring. In other words, traits are *hereditary*. Darwin also recognized that individuals differ from others of their species and from their parents. Descent with modification was fundamental to Darwin's concept of evolution.

The basis of evolution, as we know it today, is the idea that living organisms contain genes and that genes control which traits an organism displays. There are genes for shell color in beetles and genes for beak length in birds. These genes are passed from parent to offspring. (Recall that genes are part of the DNA inside every cell.)

If some genes become more prevalent from one generation to the next, or if entirely new genes are created, then new traits may emerge. Four basic processes in living systems lead to such changes in genes and their distribution within populations of organisms:

1. *Mutation.* A mutation randomly changes a gene in an individual. Mutations occur naturally (for instance, when an error is made in replication) but can be caused by certain kinds of radiation, such as X-rays or cosmic rays (high-energy particles). The altered DNA may then produce a structural or functional change, depending on the trait for which the gene codes. For example, a mutation in green beetles might lead to gold-colored offspring. If the offspring pass on the mutated gene to their own offspring, gold beetles become more common in the population (**Figure 9.9a**).

2. *Migration.* In migration, individuals from one population with a certain trait move into a new territory. If they breed with individuals that have different traits, the gene distribution changes. For example, if gold beetles move into a region occupied by green beetles and the two populations interbreed, in time the gene for gold coloring may become more common in the green-beetle population (**Figure 9.9b**).

3. *Genetic drift.* In **genetic drift**, certain traits (and genes) become less common because of random events other than gene mutations. Imagine that several green beetles in a mixed population of gold and green beetles are accidentally killed before leaving offspring. The next generation of beetles could have a higher ratio of gold to green beetles than the previous generation—just by chance (**Figure 9.9c**).

4. *Natural selection.* **Natural selection** is the process by which some genetic traits make one individual more likely than others of its species to survive in its environment and pass on those traits to offspring. If birds can spot (and hence eat) green beetles more easily than gold ones, for example, then gold beetles are more likely to survive, produce offspring, and pass on their genes for color. In the next generation, gold beetles will be more common than in the previous generation.

Natural selection is the most important of these four processes for tuning an organism to its environment. By natural selection, an organism with a trait that's favorable for survival in a particular environment has a better chance to survive and produce more young with the gene for that trait. Environments, however, can change. Imagine that green beetles are well hidden in a forest. If a drought kills the forest, these beetles will suddenly be exposed and vulnerable. But beetles of colors that more closely match this new environment may lead to individuals that are better able to survive and pass on their genes.

Natural selection, migration, mutation, and genetic drift contribute to a "flow of genes" from one generation to the next. Given enough time, these changes in genetic inheritance may become so pronounced that a new species with significant new traits emerges. "It's these processes," says McKay, "that allowed evolution to create organisms that can be as complex as those we see today, and as adapted to their environment."

Time ⟶

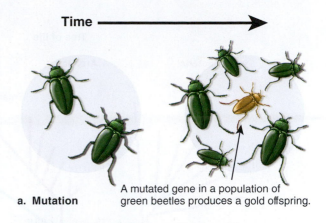

a. Mutation   A mutated gene in a population of green beetles produces a gold offspring.

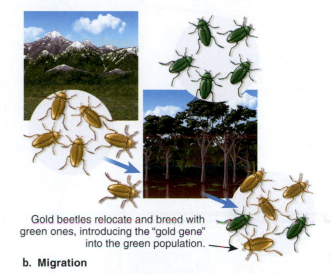

Gold beetles relocate and breed with green ones, introducing the "gold gene" into the green population.

b. Migration

Random acts lower the frequency of green beetles relative to gold ones.

c. Genetic drift

Figure 9.9 Three of the Processes of Evolution
In addition to natural selection, species evolve by **a.** mutation (random change in a gene that may be passed on to offspring), **b.** migration (inflow of genes from one population to another), and **c.** genetic drift (random effects other than mutation that lead to changes in a population's gene frequency).

"Without doubt, evolution is the most powerful force in biology."

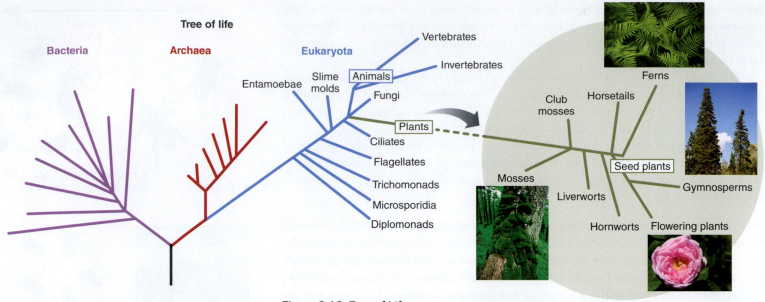

Figure 9.10 Tree of Life

A phylogenetic "tree" represents our current understanding of evolutionary relationships among organisms. Organisms that share more common ancestry are more closely related; for example, roses and pine trees are more closely related than either is to a moss or fern.

The raw material of evolutionary studies consists of the relationships among species—relationships that can be represented as a "family tree" called a **phylogeny** (**Figure 9.10**). Species have evolved over time from other species; an earlier species that is genetically related to later ones is their *common ancestor*. For example, mosses, ferns, pine trees, and roses are very different, but they share enough traits (and genes) that all four plants must have a common ancestor. Pine trees and roses, however, are more similar to each other than either is to a moss or a fern, so pine trees and roses must have had a common ancestor that evolved later than the common ancestor of all four plants.

Lateral Gene Transfer

Although the notion of genetic inheritance from parents to offspring is appropriate for large animals and plants, microorganisms are different. "To think that a microorganism primarily gets its genes from parents is a mistake, because they are constantly swapping genes," says McKay. This process, called **lateral gene transfer**, occurs when, for instance, one microorganism injects its genes into another. "It's a genetic free-for-all for these kinds of organisms compared to what we think of when we talk about plants and animals," McKay adds.

Lateral gene transfer explains one of the most potent examples of evolution: the development of antibiotic resistance among bacteria. McKay explains, "We begin with a population of microbes. Then we introduce a new element into the environment: antibiotics. The microorganisms respond in that most of them die. But some of them don't die; they develop mutations that allow them to survive to deal with the antibiotics. That mutation spreads through the population. We now have antibiotic-resistant strains, a variety of types."

Antibiotic resistance demonstrates the reality of evolution even on timescales that are short compared with a human life. On much longer timescales, we can see evolution's grand drama in the story of the entire planet.

"It's a genetic free-for-all for these [microorganisms] compared to what we think of when we talk about plants and animals."

phylogeny Relationships among species over time.

lateral gene transfer The movement of genetic material that does not proceed from parent to offspring.

anaerobic Non-oxygen-breathing.

Great Oxygenation Event (GOE) The period in Earth's history when oxygen built up relatively rapidly in the atmosphere; about 2.3 billion years ago.

Mileposts in the History of Life

Try this thought experiment: Stand up and hold up both arms on either side of your body. Imagine that your outstretched arms represent the entire history of life on Earth. The tips of your left fingers represent the origin of life some 3.7–4 billion years ago. Where do human beings appear on this time line? By your heart? Near your right armpit? Around your right wrist? Not even close.

Conclusive evidence indicates that humans appeared on Earth at the point represented by the tip of the fingernail of the longest finger on your outstretched right hand. Humanity, with its remarkable intelligence and self-consciousness, is a very new addition to the tree of life (**Figure 9.11**). But if human beings and intelligence are new, what was happening in terms of evolution for those first few billion years? Remarkably, not much.

Most scientists believe that single-celled microorganisms appeared soon after the first self-replicating molecules were synthesized. In these single cells, genetic material floated freely within the cell boundaries. The cell nucleus had yet to evolve. Fossil evidence indicates that these microorganisms lived in giant colonies forming dense "mats" (thick, floating layers) in lakes or shallow-water coastal regions. "For 3 billion years, the story of life on Earth was pretty much nothing but single-celled organisms," says Woody Sullivan, an astrobiologist and astronomer at the University of Washington who has pursued extraterrestrial life for 40 years. "Then something remarkable happened."

One important fact about the first billion or so years of life on Earth is that there was little or no oxygen in the atmosphere. The "respiration" of early life-forms relied on "breathing" in CO_2, breaking it down in biochemical reactions, and then expelling oxygen. (It was really photosynthesis in the cells that liberated the oxygen.) For billions of years, these **anaerobic** (non-oxygen-breathing), single-celled organisms were the only kind of life on the planet (**Figure 9.12**), and the oxygen they created was absorbed by the sea and ground through chemical reactions. But by approximately 2.3 billion years ago, these *sinks* (mechanisms or locations of absorption) of oxygen became saturated. The oceans could no longer soak up the oxygen expelled by a planet's worth of microorganisms. Oxygen built up rapidly in the atmosphere during a period of a few million years called the **Great Oxygenation Event (GOE)**, which led to a dramatic change in the

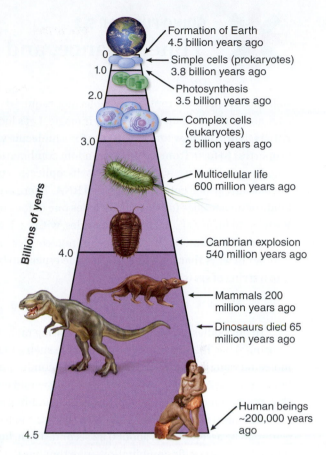

Figure 9.11 Time Line of Life on Earth

This time line spans most of the 4.5 billion years since Earth's formation. Simple cells developed very early, but it took over a billion years more for complex cells to appear. On this scale, human beings appeared very recently.

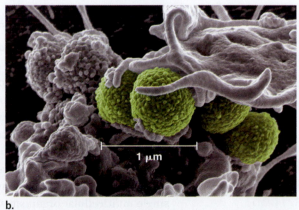

Figure 9.12 Anaerobic Organisms

The earliest life-forms, at a time when the atmosphere lacked oxygen, were anaerobic. **a.** A 3.5-billion-year-old geologic formation in Glacier National Park, Montana, consisting of fossilized stromatolites—layered mats of anaerobic microbes—possibly the earliest known form of life on Earth. **b.** This common anaerobic bacterium is found on human skin and mucous membranes and has become resistant to many modern antibiotics. It carries out aerobic respiration when oxygen is present but can switch to anaerobic respiration in the absence of oxygen.

Time, Chance, and Life

Building the first self-replicating molecules required time for chemistry and random chance to work their magic. Let's look at how scientists figure out how long it will take for a molecule with certain properties to form from a series of random combinations.

Imagine that we want to build a self-replicating molecule out of nucleotides, which comprise DNA and RNA. There are four distinct kinds of nucleotides (since each includes one of four nucleobases). We wish to form a chain of six nucleotides, with each of the six chosen randomly from among the four types of nucleotides available. The number of ways that these four nucleotide types can be combined into a string of six is

$$N = 4 \times 4 \times 4 \times 4 \times 4 \times 4 = 4^6 = 4{,}096$$

In our bodies, one type of RNA-replicating molecule (called a ribosome) has 192 nucleotides. To assemble such a 192-nucleotide molecule randomly, the number of possible combinations would be a huge number: $N = 4^{192} = 4 \times 10^{115}$. How long would it take to create that exact molecule from so many possible combinations?

Chemical studies show that any single 192-nucleotide molecule can be assembled in about 0.32 second. Then, how long would it take to get a specific combination—one that has the property of self-replication? The answer comes from multiplying the number of combinations by the time each one takes to be assembled: it takes $4^{192} \times 0.32$ second $= 1.26 \times 10^{115}$ seconds to run through all possible combinations.

Next we calculate the total age of the Universe, 13.7 billion years, in seconds:

$$(1.37 \times 10^{10} \text{ years})(3.15 \times 10^7 \text{ seconds/year}) = 4.32 \times 10^{17} \text{ seconds}$$

That is much less than the time we calculated for every possibility that leads to a specific 192-nucleotide, self-replicating molecule to be tried, even at a rate of one combination every 0.32 second. The Universe is not old enough for such a molecule to have formed from scratch by chance.

Scientists now believe that building a self-replicating molecule like a ribosome was a multistep process. Research indicates that the smallest molecule that can self-replicate is about 24 nucleotides long. A molecule that size would not provide the information storage even the simplest living organism needs, but it could have served as a subunit of a longer, more information-rich self-replicating molecule.

Studies show that a single 24-nucleotide molecule can be assembled in just 0.04 second. How long would it take to build the *right* 24-nucleotide subunit? Using calculations similar to those above, we find

$$4^{24} \times 0.04 \text{ second} = 1.13 \times 10^{13} \text{ seconds} = 3.6 \times 10^5 \text{ years}$$

Therefore, it takes 360,000 years to form a specific 24-nucleotide, self-replicating subunit from all possibilities. Further calculations show that it takes only another 5,000 years or so to assemble the 192-nucleotide molecule from multiple smaller subunits. Thus, the chance that a large information-rich self-replicating polymer could form by the construction and merging of smaller monomers is reasonable.

"For 3 billion years, the story of life on Earth was pretty much nothing but single-celled organisms. Then something remarkable happened."

nature of life on the planet (**Figure 9.13**). As evolution proceeded, the oxygen content rose even higher.

Much of the existing microbial life—mostly *prokaryotic cells*, which lack cell nuclei—could not survive in the new oxygenated atmosphere. But other kinds of microbes—in particular, **eukaryotic cells**, which store their DNA in the central repository of a cell nucleus—were better able to adapt. In fact, because oxygen allows for energy to be chemically stored and transferred better than CO_2 does, the organisms that could breathe oxygen suddenly found themselves with fuel to burn. Eukaryotic cells that could take advantage of increased cellular activity began to join forces and cooperate, forming a single multicellular organism. Over time, life became more complex and varied. But the real turning point came even later, when oxygen levels approached what we experience today and evolution went through a remarkable period of creativity.

"That is when this tremendous explosion in evolution occurred," says Sullivan. "It was just 540 million years ago, during the Cambrian period. It was like evolution went crazy." During this geologically brief period, called the **Cambrian explosion**, almost all of the basic branches of the tree of life that we know today formed. The Cambrian explosion led to plants such as mosses, ferns, and trees, as well as animals such as sharks, dinosaurs, and mammals (**Figure 9.14**).

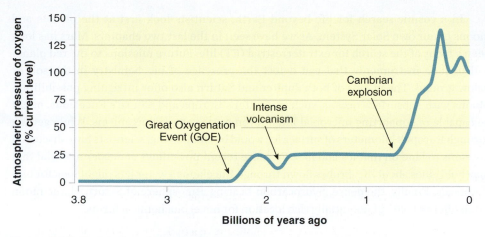

Figure 9.13 Oxygen Levels on Earth over Time

A graph showing atmospheric pressure of oxygen—a proxy for oxygen content—as a percentage of the current value (100 percent) through time. Atmospheric oxygen increased dramatically during the Great Oxygenation Event (GOE) once land and sea had become saturated with oxygen. The atmosphere could then support the growth in number and diversity of oxygen-breathing organisms that would occur later. Subsequent events also produced changes in the atmospheric oxygen content until it reached its present level.

Ultimately, a species of mammal evolved that could walk upright on two feet and use simple tools. Over time (though very little time, compared with the billions of years of life that had already existed), these creatures grew in complexity, progressing from burning twigs for cooking to burning rocket fuel for launching telescopes into space. And they became intelligent enough to wonder whether other intelligent life-forms exist elsewhere in the Universe.

Did the trajectory of life that evolved on Earth happen anywhere else in the galaxy or in the Universe as a whole? Now that we've learned about life on Earth in detail, it is time to ask more detailed questions about life elsewhere.

Section Summary

- Abiotic synthesis, the transformation of nonliving molecules into life, is not fully understood. But by simulating the expected state of Earth's early atmosphere and ocean, the Miller-Urey experiment produced many of the molecules required for life.
- Given sufficient time, self-replicating molecules can be produced randomly by natural chemical processes on a young, habitable planet. Scientists believe that the first self-replicating molecules on Earth were small segments of RNA.
- Darwin defined evolution in terms of descent (heredity) with modification. The modification occurs by changes in gene frequency via mutation, migration, genetic drift, and natural selection.
- After oxygen built up in Earth's atmosphere, eukaryotic cells thrived; their efficient energy storage and transfer enabled multicellular life to evolve. The number and variety of species increased dramatically during the Cambrian explosion.

CHECKPOINT Describe the mechanisms that allow life to change (evolve) on Earth, and cite an example of each that was not given in the chapter.

Searching for Other Life in the Universe

9.4 Scientists now have detailed theories about the evolution of single-celled organisms, the Cambrian explosion, and the evolution of human intelligence. But how common are any of these steps in a cosmic perspective? Could the trajectory of life that occurred on Earth have happened anywhere else in the galaxy or in the whole Universe?

a. Trilobite

b. Dinosaurs

Figure 9.14 Life during and after the Cambrian Explosion

Some of the highly diverse kinds of animal life that emerged **a.** during and **b.** after the Cambrian explosion.

eukaryotic cell A cell that stores its DNA in a cell nucleus.

Cambrian explosion The period about 540–505 million years ago when the diversity of life expanded at an accelerated rate.

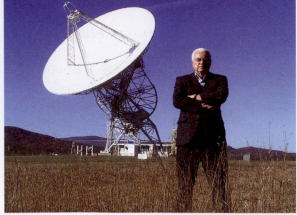

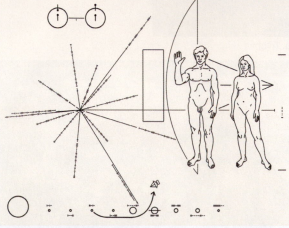

b.

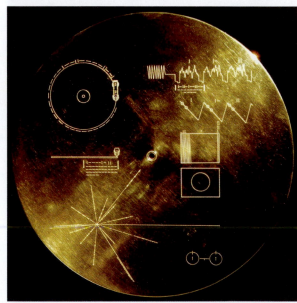

c.

Figure 9.15 Frank Drake and the Search for Extraterrestrial Intelligence

a. Frank Drake, a pioneer in the search for extraterrestrial intelligence, shown with a radio telescope—one of his primary search tools. He developed the Drake equation and helped create two physical messages sent into space: **b.** the *Pioneer* plaque and **c.** the *Voyager* Golden Record.

In a cosmic search for life beyond Earth, scientists look first at the planets and moons of our own Solar System. As we have seen in the last two chapters, Mars has long been a focus of the search for extraterrestrial (ET) life. Future missions to the red planet will include instruments that can detect the presence of biochemistry in the soil or below ground. The moons of both Jupiter and Saturn also offer intriguing possibilities. Moons with subsurface oceans, such as Jupiter's Europa and Saturn's Enceladus, might be capable of supporting microbial life or even more complex organisms. While getting through Europa's kilometers of surface ice would be challenging, scientists have discussed missions to land on that ice and look for more recent upwellings of deeper material that may hold clues about life. Probes flown through Enceladus's water jets might look for biochemicals. Finally, the dense, hydrocarbon-bearing atmosphere of Saturn's large moon Titan offers a cold but potentially rich location for novel biochemical forms.

The Drake Equation: Guesstimating Life in the Galaxy

Frank Drake is one of the first modern astronomers to take the question of life elsewhere in the galaxy seriously. His name is attached to the equation most often associated with that question. "The Drake equation," explains Woody Sullivan, "has been the standard way of understanding what we do and do not know about extraterrestrial life for five decades."

Drake is a radio astronomer by trade (**Figure 9.15a**). His early work focused on using astronomical observing technologies to search for evidence of intelligent life in other parts of our galaxy. He carried out the first astronomical search for signals from extraterrestrial intelligence in 1959. Later Drake helped design "messages" in the form of illustrated plaques and phonograph records that were carried on the first space probes to leave the Solar System (**Figure 9.15b, c**).

In 1961, NASA asked Drake to host a workshop on the problem of interstellar communication. "It was during the preparation for the meeting," says Sullivan, "that Drake came up with the idea for breaking down the problem of intelligent life on other planets into pieces that could be expressed in a single equation." The ultimate goal of the workshop was to determine how likely it was that scientists could find evidence of extraterrestrial intelligence by searching a sample of stars. But the answer to that question depends on how many intelligent civilizations exist in a galaxy. "What Drake was after," says Sullivan, "was a way to formulate that second question."

Drake decided to organize much of the workshop around a discussion of that single issue: how many intelligent, technologically advanced civilizations exist in the Milky Way Galaxy (Drake called the number N_c)? "He set up discussions of all the factors which would determine N_c," explains Sullivan. "There was a half-day discussion of how many planets might be out there and a half-day discussion of how many of these planets might have life on them and another half day of talking about which planets with life might have evolved intelligence and so on."

The most significant outcome of the workshop was the equation that summarized these discussions. Over the years there have been many versions of the Drake equation, but they all seek to express the same idea: how to estimate the number of technological civilizations in the galaxy by using our best knowledge (or best guesses) of the factors that affect the probability of their existence. One often-used form of the Drake equation is this:

$$N_c = N_* \times N_p \times f_h \times f_l \times f_i \times f_c$$

Expressed in words, the number of technological civilizations in the galaxy (N_c) is the product of the number of stars in the galaxy (N_*), the average number of planets orbiting each star (N_p), the fraction of those planets in each star's habitable zone (f_h), the fraction of those planets on which life evolves (f_l), the fraction of those planets with life that develop intelligence (f_i), and the lifetime of an average intelligent civilization expressed as a fraction of the star's life (f_c) (**Figure 9.16**).

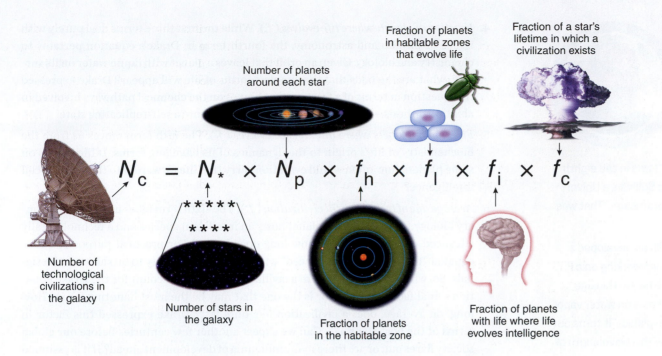

Number of planets around each star

Fraction of planets in habitable zones that evolve life

Fraction of a star's lifetime in which a civilization exists

$$N_c = N_* \times N_p \times f_h \times f_l \times f_i \times f_c$$

Number of technological civilizations in the galaxy

Number of stars in the galaxy

Fraction of planets in the habitable zone

Fraction of planets with life where life evolves intelligence

Figure 9.16 The Drake Equation Illustrated

Several versions of the Drake equation have been used since it was first proposed, but all of them attempt to include the various factors involved in estimating the number of technological civilizations in the galaxy.

"The Drake equation does not allow you to actually predict how many technical civilizations there are in the galaxy," says Sullivan. "It's not an equation like $F = ma$; instead, it's a probabilistic statement. It lets us sum up our ignorance." The Drake equation has been useful in astrobiology because it helps scientists organize their thinking about life and intelligence in the galaxy. "People use it," says Sullivan, "because it forces them to think about the factors that affect life and intelligence. It lets us see how many of those factors we know well and how many we don't know well." Once the Drake equation was formulated, the next step was to figure out the value of each factor. As Sullivan puts it, "Moving from left to right in the equation is an exercise in going from knowledge to guesswork."

Let's examine all six terms on the right-hand side of the Drake equation to get a better understanding of how each factor helps us formulate questions about other civilizations:

1. *Number of stars in the galaxy* (N_*). It is perfectly reasonable to ask whether life can bypass planets and form in something like an interstellar cloud. But given our experience and what we understand about the mechanisms of life, it's far more likely that a solid planetary surface with lots of liquid water and other chemicals is a requirement to get biology going. A focus on planets brings us straight to a focus on stars. To estimate how many planets host exocivilizations, we must have an idea how many planets exist—and that means we first need to know how many stars exist in a given galaxy.

2. *Average number of planets orbiting each star* (N_p). Once we know the number of stars, we can ask how many planets are created *around* these stars. As recently as the first half of the 20th century, many scientists believed that planet formation was very rare because it required two stars passing near each other with their mutual gravity extracting material that then formed a planet. So is planet formation a rare occurrence or a common one? The next term in Drake's equation estimates the average number of planets that orbit each star.

3. *Fraction of habitable-zone planets* (f_h). It is not enough to know whether a star hosts a planet. As we have seen, a planet's orbit around its star is a key factor, because the planets in the star's habitable zone of orbits may provide the best chance of forming life. Thus, we have to estimate the fraction of planets (of those stars that have planets) that are in the habitable zone.

"Drake came up with the idea for breaking down the problem of intelligent life on other planets into pieces that could be expressed in a single equation."

INTERACTIVE:

The Drake Equation

WOODY SULLIVAN

A 9-pound Russian satellite was all it took to ignite Woody Sullivan's scientific ambitions. "I was in the eighth grade on October 4, 1957," says Sullivan, a University of Washington astronomy professor. "That was the day Sputnik was launched."

As an undergraduate at MIT, Sullivan developed a passion for radio astronomy. While working on a PhD at the University of Maryland, he began thinking about SETI. "I did my graduate thesis on water vapor in the interstellar medium," he explains. "It turns out water vapor has a spectral line with a wavelength of 1.35 cm (which is in the radio band). This was a new discovery at the time, and I ended up doing much of the early research on the subject. That's what led me to think about life in terms of water and radio signals."

"It's possible that intelligent technological civilizations might be popping up all over the place."

4. *Fraction of planets where life evolves* (f_l). While the first three terms deal purely with issues of physics and astronomy, the fourth term in Drake's equation pertains to chemistry and biology. Given an orbit that leaves a planet with liquid water on its surface, what are the odds that the simplest forms of life will appear? Drake expressed this question in terms of a fraction that hinges on the chemical pathways involved in abiotic synthesis—conversion of nonliving matter into a self-replicating state.

5. *Fraction of planets where intelligence evolves* (f_i). The fifth term moves us from the biochemistry of life's origin to the dynamics of its changing forms. If life begins on some planets, how often would evolution carry that life forward into the state we call intelligence?

6. *Average age of technological civilization* (f_c). The sixth term takes us from evolutionary biology to sociology. If a planet hosts an intelligent species and a technologically advanced civilization arises, how long does it last? For practical purposes, Drake equated "technologically advanced" with having the capacity to broadcast radio signals. So, while the Romans were a civilization, they don't count for the Drake equation's final factor. This factor is the one that may be the most haunting for us. How long, on average, does a civilization like ours last? Drake expressed this factor in terms of the host star's age. Can we expect another few centuries before our global society flares out, or are there many millennia of development ahead? If it is assumed that technological civilizations have occurred often enough for an average to be well defined, what is their average lifetime? With this last term, Drake accounts for alien sociology. Are most civilizations as aggressive and warlike as our own? Do they become more peaceful as they evolve? How long, on average, can they last without destroying themselves?

The number of stars in the galaxy is reasonably well known: about 400 billion. Until the 1990s, that was the only credibly known factor in the Drake equation. The recent revolution in exoplanet studies, however, has changed that. "The discovery of exoplanets has given astrobiology a tremendous boost," says Sullivan. For example, the Kepler satellite mission, which relied on observations of planet transits, has provided thousands of new exoplanets and exoplanet candidates. "Basically, we now have some idea—based on real data—about how many ecologically suitable planets per star there are out there." Through the database from the Kepler mission and other observation programs, it is now possible to make reasonable estimates for both the N_p and f_h terms in the Drake equation, where N_p is at least a few and f_h is approximately 0.3. The rest of the factors remain in the realm of conjecture.

"The other three terms are also going to require empirical survey—going out and looking," says Sullivan. Scientists don't know enough from just looking at Earth to say how easy it is to get life going on a planet. "And if life does get going," says Sullivan, "we can't say with any certainty if it will develop intelligence." Life on Earth evolved within the planet's first billion years, but the evolution of intelligent life is, relatively, a very recent phenomenon. The terms f_l and f_i are issues of biology and evolution. Since we've seen these processes work on only one planet (ours), it's hard to know how they might work on others. "I call this the $N = 1$ problem," says Sullivan. "If you only have seen biology and evolution on a single world, how can you extrapolate to all possible planets?"

The $N = 1$ problem gets even tougher for the last factor of the Drake equation. If intelligence develops and leads to a technological civilization that can build radio telescopes, how long will that civilization last? "It's possible that intelligent technological civilizations might be popping up all over the place," says Sullivan. "But if each one only lasts a short time, then the odds of two civilizations getting in touch go way down." This final factor has implications beyond the search for intelligent extraterrestrial life. Modern humans have a vested interest in knowing the lifetime of technological civilizations, since that is exactly what we are.

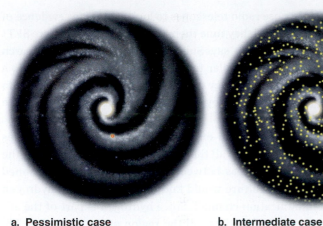

a. Pessimistic case   b. Intermediate case   c. Optimistic case

Figure 9.17 Three Sets of Results for the Drake Equation
At this time, three of the six terms on the right-hand side of the Drake equation are undetermined. Thus, we must make assumptions about their values to calculate the number of technological civilizations in the galaxy (N_c). Results range from **a.** the pessimistic case, that there is only one intelligent civilization in the galaxy and we are it, to **b.** the intermediate case to **c.** the optimistic case, that there are many thousands or more.

Because the last three factors of the Drake equation cannot be directly measured, scientists do their best to make logical arguments based on what we do know of chemistry and biology (f_l), evolutionary theory (f_i), and sociology (f_c). Even when scientists make their best estimates, the numbers for N_c span a huge range (**Figure 9.17**). "The optimistic estimates can put N_c in the tens of billions and make the galaxy appear to be teeming with intelligent life," says Sullivan. "Pessimistic estimates end up with N_c of just a few hundred. And really pessimistic ones can even lead to N_c less than one, which means that you would have to look at many galaxies before you found another technological culture. Those seem pretty depressing."

The Search for Other Minds

The Drake equation has proved enormously powerful for guiding our thinking about life in the galaxy (and the Universe), but ultimately, scientists have to collect more data. "There is no substitute for actually getting out to the telescope and looking," says Sullivan. That is exactly what researchers involved in the Search for Extraterrestrial Intelligence (SETI) are doing.

SETI has been called everything from a wild-goose chase to the most important project in history. Sullivan is well aware of the hint of science fiction that comes with any consideration of extraterrestrial (ET) civilizations. "There have been many stories and movies based on the premise of aliens visiting Earth for one reason or another," he says. "But those of us involved with SETI feel that we shouldn't wait for them to visit, but rather actively search for possible distant extraterrestrial civilizations." Figuring out just what to do constitutes the science of SETI.

The fundamental premise of SETI research is that ET civilizations may be emitting signals we can detect on Earth. Beginning from that premise, SETI researchers ask detailed questions about their nature and the best strategy for finding them.

"One important question," says Sullivan, "is about what kind of radiation the ETs might be emitting." Because radio-band electromagnetic radiation has very long wavelengths, it can propagate over extremely long distances and not be absorbed. That makes radio waves a favorite of SETI scientists. "When we think about what kind of light ETs might use to beam a signal into space," explains Sullivan, "radio almost always comes out as the best candidate" (**Figure 9.18**). Recent advances in technology, however, have made researchers realize that powerful lasers could also be used to transmit information across interstellar distances. A number of research teams have set up searches in the optical or near-infrared ranges.

Anyone who has flipped around the stations on a car radio knows that there are many frequencies (and hence wavelengths) to choose from in any of the electromagnetic

AT PLAY IN THE **COSMOS** THE VIDEOGAME   Find a planet in the habitable zone and discover the aliens' home world in Mission 8.

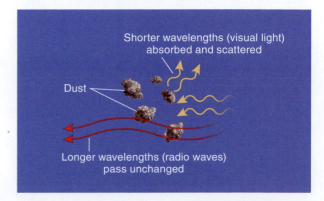

Shorter wavelengths (visual light) absorbed and scattered

Dust

Longer wavelengths (radio waves) pass unchanged

Figure 9.18 Radio Waves Can Travel Long Distances
Unlike light of shorter wavelengths, such as visible light, radio waves are not absorbed by interstellar dust and therefore propagate over long distances. This makes them a favored candidate for SETI programs.

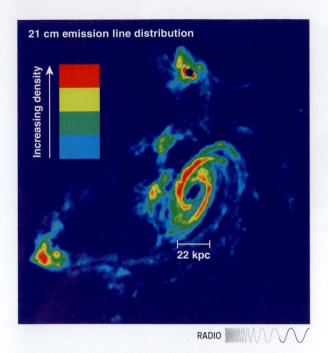

21 cm emission line distribution

Increasing density

22 kpc

RADIO

Figure 9.19 Cosmic Watering Hole
A 21-cm image of galaxy M81, with its size shown in kilo-parsecs (kpc). Emission in this wavelength is ubiquitous. SETI scientists refer to the 21-cm emission line of hydrogen as the "cosmic watering hole," because hydrogen is part of the water molecule and because they believe advanced alien life would naturally communicate at this frequency—like a gathering place on Earth.

"People could make fun of it and say that searching for little green men was a waste of money."

21-cm emission line A wavelength of radio light emitted by hydrogen; considered by some to be a natural communication channel between Earth and other civilizations.

wavelength bands. If SETI scientists use radio telescopes to look at stars for evidence of intelligent life, to what frequencies should they tune their receivers? "This is where SETI becomes an exercise in outguessing the aliens," says Sullivan. In Frank Drake's first search for ET signals, he focused on nearby stars similar to the Sun, tuning his telescope to a wavelength of 21 centimeters (cm), or a frequency of 1,420 megahertz (MHz), similar to what a microwave oven uses. Hydrogen, which is by far the most abundant element in the Universe, naturally gives off radiation at 21 cm.

Since hydrogen is so abundant, the **21-cm emission line** occurs in many places throughout the galaxy. "We use the 21-cm line all the time," says Sullivan, "to study the temperature, density, and motions of the gas between stars." Frank Drake reasoned that any other technologically capable culture would know about the 21-cm hydrogen line and use it as a natural communication channel. Since hydrogen is part of the all-important water molecule, many SETI scientists call the region of the 21-cm line the "cosmic watering hole" (**Figure 9.19**). "The idea is that different civilizations would naturally come to this frequency to communicate," says Sullivan, "just like a watering hole on Earth where people from different villages gather." (Microwave ovens are tuned to about the 21-cm wavelength because it makes water molecules in food vibrate rapidly enough to heat food.)

Drake based his search strategy on the 21-cm line, but some researchers have designed searches that scan as many frequencies as possible. "Some folks felt like too many eggs were being put in the 21-cm basket," says Sullivan, "but the problem with scanning lots of frequencies is the huge amount of data you generate. A search like that can easily overwhelm your processing and analysis resources."

Wavelength isn't the only domain in which SETI researchers have to make informed guesses about ET behavior. The nature of the signal itself is a matter of debate. "The signal is going to have to be different enough from natural astrophysical radio emission that you can tell that it's due to an intelligent source," says Sullivan. "But what form will that take? Should we be listening for some kind of pulsing beacon like a 'beep beep beep,' or should we look for steady signals with information coded into them?" Human beings have already sent purposeful signals to stars encoded with information. In 1974, Frank Drake and colleagues sent a signal toward the star cluster M13 that used binary bits (ones and zeros) to encode information about both our Solar System and humans (**Figure 9.20**).

One way or another, SETI scientists must sift through vast quantities of data. They will have to look at many stars and carry out their observations at many frequencies. To sort through the mountains of data, scientists at the SETI Institute have enlisted the public's help, through the SETI@home project. Participants download the SETI@home software. When their computers are idle, the software uses the central processor to analyze radio observations for possible ET signals (**Figure 9.21**).

Despite the widespread fascination of the subject, SETI work has always been controversial. Most studies so far have piggybacked on more traditional radio astronomy projects. In the 1980s and 1990s, NASA had plans to harness existing radio telescopes with powerful electronics to carry out ambitious searches for ET signals. But pressure from a few vocal members of Congress killed the project before it was completed. "The project was axed because it was an easy target," says Sullivan. "People could make fun of it and say that searching for little green men was a waste of money. It's a pity. The project needed a relatively small amount of money to ask what is really the most important question we have in science: Are we alone?" (**Figure 9.22**).

SETI activity continues, but to date, only a small percentage of the sky has been explored, even in the 21-cm line. We may have to wait a long time before we can make definitive statements about intelligent civilizations in space.

A few times, scientists thought they might have found a signal of ET origin, but each was found to be a false alarm. When Frank Drake began the first SETI project in 1960, his

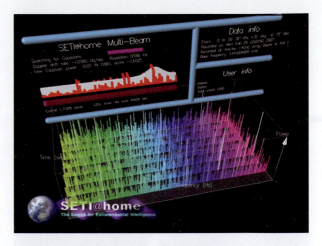

Figure 9.21 SETI@home Screen Saver

In analyzing the voluminous data it captures, SETI has sought assistance from the public. Individuals can use their home computers to download and search data for meaningful patterns via the SETI@home program.

Figure 9.20 Frank Drake's Message to Extraterrestrials

This 1974 message, beamed toward star cluster M13 from the Arecibo radio telescope, used binary bits (ones and zeros) to form a symbolic picture in a 23 × 73 array. Its content, developed by Frank Drake with collaborators such as Carl Sagan, included representations of the fundamental chemicals of life, the formula of DNA, a crude Solar System diagram, and simple pictures of a human and a telescope.

Figure 9.22 Making a Joke of SETI

The concept of contact by aliens is often the subject of satire, as evidenced by this whimsical response to stories of aliens abducting cattle for experiments.

radio telescope recorded a powerful signal. At first Drake and his collaborators thought they might have found evidence of "little green men" before recognizing the radio source was actually being emitted locally and had a quite human origin.

Life and Its Implications

In the 1950s, the Nobel Prize–winning physicist Enrico Fermi was having lunch with friends when the subject of life elsewhere in the galaxy came up. Fermi was known for his brilliance and the quickness with which he could dissect a problem. After thinking for a minute, he asked a penetrating question: If ETs exist, why aren't they here already?

What Fermi realized in that instant was that if even one civilization were to develop interstellar space travel and begin exploring other worlds, then they should have quickly spread across the entire galaxy. Even though it might take hundreds of years to travel

Figure 9.23 Alien Civilization
If other intelligent beings exist, what would they and their civilization be like? Theoretical physicist and cosmologist Stephen Hawking warned that aliens from an advanced civilization might not be as benevolent as **a.** the lovable creature from the movie *E.T. the Extra-Terrestrial* but more like **b.** the predatory, violent creature from the movie *Alien*. Until contact occurs, only the imaginations of science fiction authors and artists can suggest what we might expect.

Fermi paradox The conclusion that no intelligent life exists in our galaxy beyond Earth, since the billions of years following the Milky Way's origin would have allowed advanced civilizations to have spread across interstellar space and reached Earth, or that we have not yet received signals indicating the existence of other technological civilizations.

between nearby stars, the galaxy is billions of years old. Assuming that a civilization was born relatively soon after the galaxy's formation, there should have been more than enough time for it to have swept across our part of space already.

One answer to this **Fermi paradox** is simple: there are no other intelligent species in the galaxy. The implications of this cosmic isolation would be profound. What would it mean for us as a species if we were the only life-forms with self-consciousness in the galaxy—or the entire Universe? Would humanity be destined to spend its entire history knowing only its own perspective of life and death, meaning and existence?

There is another, more ominous answer to Fermi's paradox. Perhaps we have not found other examples of intelligence because they are smart enough *not* to announce themselves. Many wild animals are careful to cover their tracks to avoid predators, so perhaps we are being unwise in allowing our TV, radio, and radar signals to leak so freely into space. While SETI researchers often assume that extraterrestrials will be benign, some scientists, including the cosmologist and science communicator Stephen Hawking, have argued that caution is a better strategy. Hawking's view was that we should not announce our presence by beaming signals into space until we have a better understanding of what's out there (**Figure 9.23**).

Whatever the answer to Fermi's paradox, if and when any evidence of extraterrestrial life is found, it will be one of the greatest discoveries in history. Even finding nothing more than microbes on another world would tell us that life was not a single accident on a single world. Knowing that the process of life's formation and evolution happened on more than one planet would force us to alter dramatically our understanding of the Universe.

What would happen, however, if we discover not just life-forms but *intelligent* life-forms other than our own in the Universe? It is difficult to speculate, but many scientists and philosophers believe such knowledge could change human culture. How would we understand our own history, for example, if we knew there were other histories out there? Would knowing that at least one other civilization exists allow us to assume that technologically adept cultures can survive past our own point in evolution? While much would depend on the nature of what we learned about and from the new civilization, these questions demonstrate how significant and powerful such a discovery would be.

As astrobiology matures through discoveries of exoplanets, explorations of our own Solar System, and new studies of life on Earth, we appear to be moving closer to an answer to these questions. Within your lifetime, astronomers may announce evidence, direct or indirect, of life on another planet. And on that day our view of our place in the Universe will change forever.

Section Summary

- The Drake equation is an attempt to evaluate the probability of extraterrestrial life and technological civilizations existing elsewhere in the Milky Way.
- The Search for Extraterrestrial Intelligence (SETI) is based on the premise that extraterrestrial civilizations are emitting signals, such as radio waves, that we could detect.
- Some searches focus on the 21-cm emission line for hydrogen because hydrogen is so abundant and is a component of water; others scan many frequencies to increase their chances.
- The Fermi paradox asks, "If extraterrestrials exist, why aren't they here already?"

CHECKPOINT How did the Drake equation shape the way scientists think about searching for intelligent life in the Universe?

CHAPTER SUMMARY

9.1 The Origin of Life and a Trip to Antarctica

Planets (and moons) may be essential for life to begin and evolve.

9.2 What Is Life, and Where Can It Exist?

A living organism can be defined as something that is composed of organic molecules and exhibits metabolism, reproduction, mutation, and sensitivity to its environment. Surrounding every star is a habitable zone, where liquid water could exist on a planet's surface. In other planetary systems, the boundaries of such a zone depend primarily on the star's size and temperature. Moons of exoplanets are also potential environments for life. The discovery of extremophiles in deep-sea vents and rocks at great depths has broadened the known range of viable environments for life and helps scientists understand how the earliest life may have evolved on Earth. Polymers (macromolecules) including DNA, RNA, and proteins are the basis for life and its diversity. DNA, a double helix capable of self-replication, is composed of nucleotides and has a sugar-phosphate backbone linked by nucleobases; RNA has a similar composition but is single-stranded. Both carry genetic information. Proteins, made of amino acids, perform various functions within living cells.

9.3 The Origins of Life

The Miller-Urey experiment—an early study of abiotic synthesis (how life formed from nonlife)—showed that prebiotic molecules could readily form from simple compounds that are abundant in the Solar System. Complicated polymers can be built from smaller segments that are then linked. Producing the exact polymer molecules for self-replication is a matter of chance and would have required great stretches of time. Once a self-replicating molecule is built, however, it can proliferate. Genes and their distribution can be altered by mutation, migration of populations with interbreeding, genetic drift, and natural selection. Natural selection is the process by which the genes that code for traits that enhance survival and reproduction increase in frequency. In this way, organisms adapt to various environments. Microbes that undergo lateral gene transfer, in contrast, can swap genes. The path of evolution started with one-celled organisms without nuclei, which evolved into organisms with cell nuclei. The early Earth's atmosphere had little oxygen; the first microbes did not require it for their metabolism but produced it as a by-product. This biological activity led to the buildup of atmospheric oxygen during the Great Oxygenation Event, after which oxygen-breathing organisms came to dominate. The diversity of complex organisms multiplied exponentially during the Cambrian explosion.

9.4 Searching for Other Life in the Universe

The evolution of intelligent life is a very recent phenomenon. The Drake equation specifies the factors believed to be important in assessing the likelihood of such life elsewhere in the Milky Way. SETI (Search for Extraterrestrial Intelligence) is a scientific search for intelligent life on other worlds. Radio signals, particularly at the 21-cm hydrogen line, are considered by some to be the most likely means of detecting intelligent life beyond Earth. Recent discoveries about the prevalence of exoplanets have raised the probability of extraterrestrial life considerably, but the question remains open.

QUESTIONS AND PROBLEMS

Narrow It Down: Multiple-Choice Questions

1. Which of the following is *not* a basic feature of all living things on Earth?
   a. consumption of energy
   b. requirement for oxygen
   c. mutation
   d. sensitivity to environment
   e. organic molecules

2. The original definition of "habitable zone" included which of the following?
   a. average surface temperature of a planet in the range of 273–373 K
   b. presence of surface water
   c. known existence of life-forms
   d. Sun-like central star
   e. availability of carbon

3. Why has the original definition of a habitable zone been reconsidered? Choose all that apply.
   a. Not all life on Earth is carbon based.
   b. Extremophiles have shown that life can exist in a broader range of temperature and pressure conditions than was previously considered.
   c. Not all terrestrial life exists on Earth's surface.
   d. Water has been detected on giant planet moons.
   e. Life has been discovered in the dry regions of Mars.

4. You find a polymer that contains the nucleobases A, G, and C. What kind of molecule can it be?
   a. only DNA
   b. only RNA
   c. either DNA or RNA
   d. an amino acid
   e. DNA, RNA, or an amino acid

5. Which of the following is/are true of nucleotides? Choose all that apply.
   a. Each includes a sugar molecule and a phosphate.
   b. They are components of DNA.
   c. They are components of RNA.
   d. They link via their nucleobases to form long chains.
   e. They are proteins.

6. The pattern of nucleobases in a strand of DNA is TGCAACG. During replication, which nucleobases will attach to it, in what sequence?
   a. TGCAACG
   b. CATGGTA
   c. CGTTGCA
   d. GCAACGT
   e. ACGTTGC

7. The sequence of bases in DNA directly determines which function?
   a. regulating oxygen synthesis
   b. building proteins
   c. facilitating mutation
   d. regulating reproduction
   e. regulating metabolism

8. The results of the Miller-Urey experiment supported which of the following conclusions about the early Earth?
   a. Life on Earth was inevitable.
   b. All available carbon would have been converted into organic molecules.
   c. The molecules produced in the experiment represented the specific molecules that formed life on Earth.
   d. Life was easy to produce.
   e. Amino acids and organic compounds would have formed readily.

9. An exoplanet is determined to have liquid water at or just below its surface and orbits in the habitable zone of a 1-billion-year-old Sun-like star; its spin axis is oriented nearly perpendicular to its orbital plane. Using Earth's evolutionary history as a model, which of these factors makes it unlikely that intelligent life exists there?
   a. the presence of water
   b. the planet's location
   c. the type of star
   d. the age of the star
   e. the orientation of the planet's spin axis

10. Which of the following statements about natural selection is *not* true?
    a. Different traits are more favorable in different environments.
    b. It can lead to better adaptation of an organism to the environment.
    c. Traits that enhance the probability of reproduction tend to proliferate.
    d. Less adaptive traits die out in one generation.
    e. It can begin with a random genetic change.

11. Which of the following terms does *not* describe a process that can affect the frequency of a gene in a population?
    a. migration
    b. mutation
    c. genetic drift
    d. natural selection
    e. replication

12. Which of the following is/are examples of traits that can be tied to heredity? Choose all that apply.
    a. Men who shave their heads as a cultural norm also shave their sons' heads.
    b. The tallest sibling in a family produces taller children than his siblings do.
    c. Pancreatic cancer runs in a family.
    d. A child whose parents lack a specific antigen in their blood also lacks that antigen.
    e. A chameleon can control and change its skin color.

13. Which of the following statements about lateral gene transfer is/are *not* true? Choose all that apply.
    a. It occurs only in reptiles.
    b. It occurs only in some microorganisms.
    c. It facilitates antibiotic resistance.
    d. It involves a two-way swapping of genes.
    e. It occurs within one generation.

14. Which of the following characteristics accurately describe Earth's first life-forms? Choose all that apply.
    a. self-replicating
    b. single-celled
    c. multicellular
    d. eukaryotic
    e. anaerobic

15. The Great Oxygenation Event is believed to have been caused by what? Choose all that apply.
    a. saturation of the ocean and land with CO_2
    b. reduction in CO_2 due to a reverse greenhouse effect
    c. saturation of the ocean and land with O_2
    d. proliferation of anaerobic organisms
    e. proliferation of oxygen-breathing organisms

16. Which of the following accurately describe(s) the Drake equation? Choose all that apply.
    a. It tells us exactly which kinds of stars can harbor planets with life.
    b. It calculates the actual probability of extraterrestrial life.
    c. It names the factors that contribute to the total probability of life elsewhere.
    d. It provides a way to gauge how knowledge of the various factors is progressing.
    e. Its numeric value has not changed since it was first conceived.

17. Recently acquired knowledge based on exoplanet discoveries has addressed which terms of the Drake equation?
    a. N_* (number of stars in the galaxy)
    b. N_p (average number of planets orbiting each star)
    c. f_l (fraction of planets on which life evolves)
    d. f_i (fraction of planets with life that develops intelligence)
    e. f_c (lifetime of an average intelligent civilization expressed as a fraction of the star's life)

18. Why is the 21-cm line believed to be a prime candidate for the wavelength an extraterrestrial would send or recognize if sent? Choose all that apply.
    a. It is radiated by hydrogen, the most abundant element in the Universe.
    b. It falls in the radio part of the spectrum, whose long wavelengths are most likely to travel unimpeded over large distances.
    c. An advanced civilization would be aware of its relevance.
    d. It is definitive proof of water and therefore life.
    e. It is easier to detect than all other frequencies.

19. True/False: In calculating the Sun's habitable zone, astronomers assume that life on Solar System planets requires water. If a planet orbiting 1 AU from a Sun-like star had life based on a molecule with much higher melting and boiling points, that star's habitable zone would be significantly farther from the star than the Sun's is.

20. True/False: All known living things on Earth contain carbon.

To the Point: Qualitative and Discussion Questions

21. What are the distinguishing characteristics of organic molecules?

22. Why are some moons of planets outside the habitable zone in our Solar System considered possible habitats for extremophiles?

23. Why are very hot stars not good candidates for highly evolved life?

24. What sources of energy other than sunlight have been discovered for extremophiles on Earth?

25. Describe how DNA differs from RNA in structure and function.

26. What additions to our understanding of the early Earth changed the interpretation of the Miller-Urey experiment?

27. Discuss the importance of time in abiotic synthesis.

28. As the environment changes, how does a particular species adapt by natural selection?

29. Describe the Cambrian explosion.

30. Why do astronomers believe that radio wavelengths are the most likely means of interstellar communication?

31. What is the Fermi paradox?

32. To date, what evidence is there for the existence of life beyond Earth?

33. Should the level of resources expended on the search for extraterrestrial life be expanded, reduced, or eliminated? Give at least two reasons for your choice.

34. Some people believe that evolution is "only" a theory and that there are other equally valid theories for the creation of different species on Earth. How would you counter that claim?

35. How would it change your religious or philosophical views if the existence of intelligent extraterrestrial life were confirmed? What if it were conclusively refuted?

Going Further: Quantitative Questions

36. Use the Drake equation to calculate N_c if N_* is 100 billion, N_p is 8, f_h is 0.3, f_l is 0.5, f_i is 1, and f_c is 2×10^{-5}. What does your answer mean for the likelihood of finding intelligent life elsewhere in the galaxy?

37. What is N_c if the values of all factors in the Drake equation are the same as in Question 36, except that f_l is 0.005?

38. What would be the minimum length of time for sending a message and receiving a reply from an extraterrestrial on a planet 23 light-years away?

39. Suppose that an RNA molecule, composed of four types of nucleotides, has 144 nucleotides. How many attempts would it take to produce such a molecule from individual nucleotides?

40. Imagine that life that evolved on an exoplanet is based on six unique nucleotides (instead of four) and that an RNA molecule on that planet requires 138 nucleotides built from them. How many attempts would it take to produce the necessary molecule from individual nucleotides?

41. Suppose that an RNA molecule needed for life has 144 nucleotides, based on four unique types, but it can be constructed out of a specific combination of 16-nucleotide subunits. How many different combinations of the 16-nucleotide molecules would there be?

42. A slowly rotating planet with an albedo of 0.5 orbits a star that has a luminosity 2 times that of the Sun at 5 AU. What are the inner and outer boundaries of the star's habitable zone? Is the planet in that zone?

43. A slowly rotating planet with an albedo of 0.3 orbits a star that has a luminosity 5 times that of the Sun at 2.5 AU. What are the inner and outer boundaries of the star's habitable zone? Is the planet in that zone?

44. A fast-rotating planet with an albedo of 0.4 orbits a star that has a luminosity 0.8 times that of the Sun at 0.5 AU. What are the inner and outer boundaries of the star's habitable zone? Is the planet in that zone?

45. A fast-rotating planet with an albedo of 0.7 orbits a star that has a luminosity 3 times that of the Sun at 0.3 AU. What are the inner and outer boundaries of the star's habitable zone? Is the planet in that zone?

"THE SUN SHOWS US THAT UP CLOSE, STARS ARE ASTONISHINGLY COMPLEX"

10

The Sun as a Star

OUR OWN FUSION ENGINE

Living with a Star

10.1 At just 149 million kilometers (km) away, the Sun is the closest star to Earth. Recall from Chapter 1 that stars are balls of gas that produce energy by means of nuclear fusion in their cores. The next nearest star, Proxima Centauri, is nearly 300,000 times more distant. This proximity presents both benefits and problems. "We gather so much data about the Sun that putting it all together is a real challenge," says Lika Guhathakurta. "The Sun is such a remarkable object, and there are so many different questions to ask: Where does it get its power? What is its internal structure? What forces drive the incredible events happening on its surface?" As program manager for key NASA solar physics programs, Guhathakurta has brought together researchers from different fields—such as those who study the solar atmosphere and those who study internal solar rotation—and helped them integrate their findings.

In the 1990s, when space instruments began providing 24-hour views of the Sun, Guhathakurta and other scientists initiated a push to bring together all of the different perspectives on the Sun. From the magnetic storms of its outer atmosphere to the nuclear fires of its core, the combined efforts taught scientists just how remarkable a star can be. "There is an old joke that stars look pretty simple because they are so far away," says Guhathakurta, "but the Sun shows us that up close, stars are astonishingly complex."

From a God to a Star

Many of our ancestors worshipped the Sun as a god—and for good reason. Even at a distance of tens of millions of kilometers, the Sun is extremely powerful. On a clear summer day, looking anywhere near the Sun is enough to hurt your eyes, and staring directly into the Sun can cause blindness. The warmth that reaches Earth from the Sun can be either comfortable or oppressive, depending on the season. Up close, the Sun's blinding light and searing heat reach scales unlike anything you've imagined or experienced. Using existing technology, humans could not reach the "surface" of the Sun. (As we will see, the Sun has no actual surface.) Our machines would be vaporized to nothing more than stray atoms.

IN THIS CHAPTER
you will learn about our Sun, its internal structure, and its outer environment. After reading through each section, you should be able to do the following:

10.1
Living with a Star Describe the structure and composition of our star.

10.2
The Sun's Fusion Furnace Explain the source of the Sun's energy and how we know about it.

10.3
Moving Energy Explain the processes by which energy moves from the Sun's core outward to the surface.

10.4
The Active Sun: Photosphere to Corona and Beyond Discuss the origin and dynamic nature of the Sun's magnetic field and its effect on solar behavior.

← The Sun as seen through the visible light of an emission line of once-ionized calcium and taken with an amateur telescope fitted with a special filter. The blue light shows the chromosphere region of the Sun, approximately 2,000 km above the Sun's photosphere (its "surface").

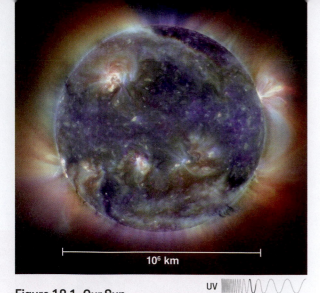

Figure 10.1 Our Sun
The Sun is a star—a massive sphere of superheated, electrically charged gas. This image, taken by the *SOHO* (Solar and Heliospheric Observatory) spacecraft, shows how active its surface is.

UV

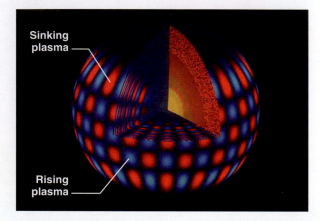

Sinking plasma

Rising plasma

Figure 10.2 Helioseismology
Astronomers investigate conditions in the Sun's internal structure by studying solar oscillations and their propagation. This image from a theoretical model is a snapshot of rising and sinking plasma in the Sun. (The rising and sinking regions switch as the oscillation progresses.) The actual oscillations are a combination of many such patterns.

plasma A gas composed of charged particles such as electrons, protons, and ions.

thermonuclear fusion The transformation of lighter atomic particles into heavier ones, requiring high temperatures.

helioseismology The study of oscillations in the Sun and their propagation, used to develop precise models of solar structure.

hydrostatic equilibrium A balance between gravity and pressure.

differential rotation The rotation of different regions of an object (different latitudes and/or depths) at different rates.

core A star's central region, where high temperatures and densities allow thermonuclear fusion reactions to occur, releasing energy.

This description of the Sun as a source of vast energy is equivalent to its definition as a star. The Sun, like other stars, is a massive sphere of superheated **plasma**—a gas composed of charged particles such as electrons, protons, and ions (**Figure 10.1**). (An *ion* is an atom with some electrons removed or added.) And like other stars, the Sun is a forge in which light elements such as hydrogen undergo **thermonuclear fusion**, a high-temperature process in which light elements fuse together, creating heavier elements such as helium.

This chapter begins our exploration of stars and their life stories. Stars do not live forever; they cycle through processes of birth, middle age, and decay, just we do. The life story of a star involves a remarkable array of physical processes that shape the Universe and life. However, we start at home with the most familiar star, the Sun.

How We Know What We Know about the Sun

If we can't yet send a probe into the Sun that would stand up to the extreme pressures and temperatures, how do scientists know anything about the Sun's deep interior? Much of the current knowledge comes from the remarkable field of **helioseismology**. Scientists use oscillations of the Sun's surface to infer properties of the solar interior (as in seismological studies of Earth after an earthquake). To see those oscillations, astronomers must use spectroscopy, monitoring absorption lines in the Sun's spectrum for wavelength shifts. These changes are then tied to the motion of the Sun's surface. "We can see the Doppler shift of material on the solar surface," says Guhathakurta. "It's like watching the whole Sun ring like a bell. By tracking these solar oscillations, we have been able to work backward and build models of the different layers of the solar interior" (**Figure 10.2**).

Using helioseismology, astronomers in projects such as the Global Oscillation Network Group (GONG) have tracked the Sun's vibrations in exquisite detail. With these data, they then constructed testable predictions about the Sun's structure and developed precise theories describing what makes stars tick. Let's begin with some basic facts and a quick tour of the Sun from its fusion-powered core to the vaporous violence of its outermost layer, the *corona*.

Anatomy of the Sun

"The Sun is far and away the most massive object in the Solar System," says Guhathakurta. It accounts for 99.86 percent of material in the Solar System. The Sun's mass is about 2×10^{30} kilograms (kg). You would need 333,000 Earth-sized planets packed together to equal the Sun's mass. The Sun is enormous as well. Its radius is 696,000 km, over 100 times that of Earth. You could put more than 1.3 million Earths in the volume occupied by the Sun. Says Guhathakurta, "A typical passenger jet would take almost a year to circle the 4.4-million-km equatorial circumference of the Sun."

Remarkably, all that mass is pretty stable; the Sun's radius does not change significantly, except over very long timescales. That is because the Sun is in **hydrostatic equilibrium**: the inward crush of gravity is balanced by the outward push of pressure (**Figure 10.3**). Every star has a layered structure. For a star to be in hydrostatic equilibrium, something must support it against its own gravity; that something is *pressure*—an outward push exerted from one layer to the next. Gravity and pressure must balance at each layer. The pressure at the Sun's center is very high but drops steadily in the layers closer to the surface. Pressure in the core comes from energy released there by nuclear fusion reactions. As long as fusion continues, new energy is released, maintaining the pressure and keeping the Sun from collapsing on itself. If fusion reactions in the core were to stop, no new energy would be generated, pressure in the interior would drop, and the Sun would no longer maintain its hydrostatic equilibrium balancing act. As we will see, the conflict between gravity and fusion-produced pressure determines the fate of all stars.

Like everything else in the Solar System, the Sun is spinning. It takes approximately 26 days for the Sun's equator to complete one rotation, but material near the poles takes almost 8 days more. This difference in spin rates at different locations is called **differential rotation**. "It's not just a curiosity," Guhathakurta explains. "Differential rotation is essential for generating the kind of powerful magnetic field we find on the Sun, which controls so much of what happens on its surface." While not all magnetic fields require differential rotation, it's crucial for the Sun's dynamo (**Figure 10.4**).

At present, the Sun's elemental composition (by mass) is 71.5 percent hydrogen and 27.1 percent helium; other elements, including oxygen, carbon, iron, neon, nitrogen, silicon, and magnesium, each make up a fraction of a percent of the Sun's mass (**Table 10.1**). These values have been determined by spectroscopy. We specified the *present* composition of the Sun, because that composition changes. Fusion reactions in the core convert hydrogen into helium, though the rate at which solar composition changes significantly is measured in billions of years. "If we could wait long enough," explains Guhathakurta, "we would see the helium fraction of the Sun's mass slowly increase and the hydrogen fraction decrease."

As we have seen, the Sun's composition is layered rather than uniform. "Each of these layers has different properties, and each is controlled by different physical processes," says Guhathakurta. From the center outward, these layers are the core, the radiative zone, the convective zone, the photosphere, the chromosphere, and the corona (**Figure 10.5**). The way energy is transported from the core outward is what defines the Sun's structure.

CORE. The **core** is composed primarily of ionized hydrogen along with helium "ash" left over from fusion. "The core is the region where energy is produced," explains Guhathakurta. Extending from the center to about 150,000 km (20–25 percent of the Sun's full radius), it is the Sun's densest and hottest region. Core densities are as high as 160,000 kg per cubic meter (kg/m³), 160 times denser than water and 14 times denser than lead. Temperatures reach 15 million kelvins (K)—more than high enough for the energy-generating fusion reactions to occur.

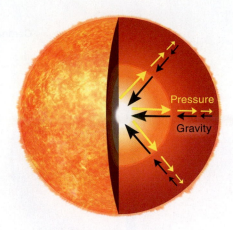

Figure 10.3 Hydrostatic Equilibrium
The Sun is in hydrostatic equilibrium at each layer. The force of its own gravity is balanced by pressure—the radiation pressure created by nuclear fusion in the Sun's core.

INTERACTIVE:
Anatomy of the Sun

"A typical passenger jet would take almost a year to circle the 4.4-million-km equatorial circumference of the Sun."

Table 10.1	Elemental Composition of the Sun
Element	**Percentage of Sun's Mass**
Hydrogen	71.50
Helium	27.10
Oxygen	0.60
Carbon	0.25
Iron	0.14
Neon	0.13
Nitrogen	0.07
Silicon	0.07
Magnesium	0.07
All others	0.07

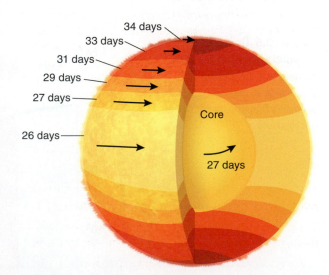

Figure 10.4 Differential Rotation
The Sun's core rotates as a solid body with a period of 27 days, but the outer layers experience differential rotation. That is, different latitudes and depths have different rotation periods, ranging from 26 days at the equator to 34 days at the poles.

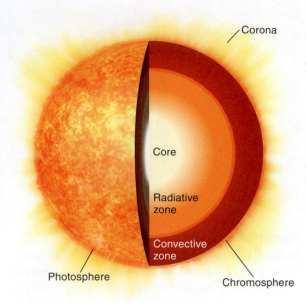

Figure 10.5 Structure of the Sun

Using helioseismology and other data, scientists have constructed detailed models of the Sun's principal layers. Energy released by fusion reactions in the core works its way through the radiative zone, the convective zone, the photosphere (the solar "surface"), the chromosphere, and the corona.

RADIATIVE ZONE. The **radiative zone** is composed primarily of ionized hydrogen and helium. "After energy is generated in the core, it has to work its way to the surface," says Guhathakurta. "The radiative zone marks the first step in that journey." It extends from just above the core to 500,000 km (70 percent of the Sun's radius). Energy is transported outward by means of electromagnetic radiation (the scattering of photons off particles of mass). The density drops by more than a factor of 100—from 20,000 kg/m^3 to only 200 kg/m^3—across this layer. The temperature, however, drops by less than a factor of 4, from 7 million K to 2 million K.

CONVECTIVE ZONE. Although radiation is the means of energy transfer in the radiative zone, in the **convective zone** the motion of gas (convection) is what moves energy around. Motion in the convective zone originates when plasma at the bottom of the zone absorbs heat and then rises to the surface. There it cools by emitting radiation into space and then sinks again, and the process begins anew. This zone stretches from 500,000 km above the Sun's center all the way to its "surface" (the boundary with the photosphere). The density drops from 200 kg/m^3 to a fraction of a kilogram per cubic meter just below the photosphere. The temperature falls dramatically, from nearly 2 million K near the border of the radiative zone to 5,800 K at the Sun's surface.

PHOTOSPHERE. The **photosphere**, the closest thing the Sun has to a surface, is where light from the solar plasma escapes into space. That means it's where the Sun's blackbody spectrum is emitted. "This is the part of the Sun we actually see," says Guhathakurta. Temperatures at the photosphere are 5,800 K, and particle densities are 10^{-6} kg/m^3. The temperature falls with height through the photosphere, and it is in these colder regions that absorption lines form. The progression of temperatures and densities from core to photosphere is shown in **Figure 10.6**.

CHROMOSPHERE. Extending 2,000 km above the photosphere, the **chromosphere** is considered the lower layer of the Sun's atmosphere. In the bottom parts, the temperature drops to 4,100 K. "Once you get to the upper layers of the chromosphere," says Guhathakurta, "the temperature actually begins to rise again." These regions, being

Figure 10.6 Decreasing Parameters within the Sun

The layers of the Sun exhibit decreasing **a.** temperature and **b.** density from the center to the photosphere.

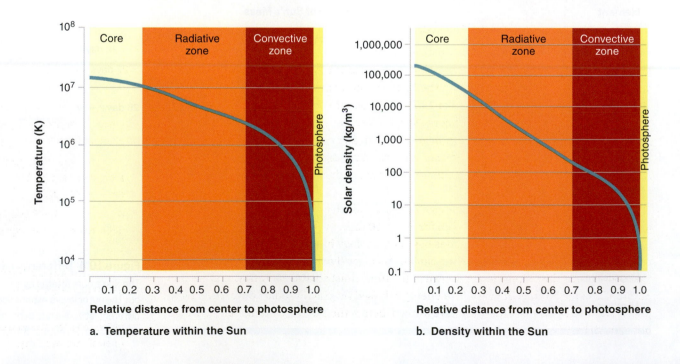

a. Temperature within the Sun

b. Density within the Sun

hotter than the underlying photosphere, produce an emission line spectrum. The density, however, continues to drop with distance from the photosphere, until the chromosphere merges with the very low density corona above.

CORONA. With temperatures rising as high as 1 million K, the **corona** is a region of extremely hot, low-density gas surrounding the Sun. The corona typically extends far out into space, up to a few times the Sun's radius, though gas densities there can be as low as 10^{-14} kg/m³, a hundred-trillionth the density of air at Earth's surface. Unlike the layers below the corona, it is not in hydrostatic equilibrium and its outer regions expand outward into space. The corona is heated to such high temperatures by waves of magnetic energy propagating upward from near the solar surface.

Section Summary

- The Sun comprises most of the mass in the Solar System. From helioseismology, combined with modeling, astronomers know that the Sun comprises the majority of mass in the Solar System and is composed primarily of hydrogen (71.5 percent) and helium (27.1 percent).
- The Sun is in hydrostatic equilibrium, balancing the inward pull of gravity with pressure generated by nuclear fusion.
- The Sun rotates differentially, more slowly at the poles than at the equator.
- Six distinct layers constitute the Sun and its atmosphere. Density and temperature decrease in each successive layer except the corona, where temperature increases dramatically.

CHECKPOINT   What do astronomers mean by the "surface" of the Sun?

The Sun's Fusion Furnace

10.2

For most of human history, the Sun inspired mythologies of gods and their immense powers. It was only natural that the Sun's heat and light were associated with fire, though no one could say what was burning. Even by the mid-1800s, with the industrial revolution in full swing, the solar link with fire was still in place, as some scientists hypothesized that the Sun is a large sphere of burning coal. The true story about the source of the Sun's power came only when scientists learned to unpack the secrets of atomic nuclei.

The Birth of Nuclear Physics

By the early 1930s, spectroscopy had enabled astronomers to deduce that the Sun is composed mostly of hydrogen gas. The question then arose whether the Sun's energy might come from gravity squeezing down on the large gaseous sphere. To the British scientist William Thomson (Lord Kelvin; **Figure 10.7**), such **gravitational contraction** was the answer to the mystery of the Sun's energy source. Kelvin believed that the inward pressure of gravity forces the solar gas to heat up. When some of that heat energy radiates into space at the solar surface, gravity can squeeze a little tighter, shrinking the Sun a bit more—liberating a bit more energy. The process will continue until the Sun is just a cold cinder.

Lord Kelvin calculated the timescale of this slow release of energy via contraction and showed that it could power the Sun for about 20 million years. Although this seemed like a long time, independent estimates based on geology had already put Earth's age at more than a billion years. Since the Sun can't be younger than Earth, it soon became clear that some other kind of physics, unknown at the time, must also be operating to allow the Sun to shine for billions of years. The energy released by gravitational contraction *does* power stars, but only just after they have formed, before their cores are dense enough and hot enough to drive nuclear fusion.

Figure 10.7 Lord Kelvin
William Thomson (the first Baron Kelvin; 1824–1907) assumed that the Sun is powered by the conversion of gravitational potential energy into thermal energy, which then radiates into space. Given the Sun's present luminosity, he (incorrectly) calculated that the Sun is no more than 20 million years old.

> "After energy is generated in the core, it has to work its way to the surface. The radiative zone marks the first step in that journey."

radiative zone The layer of the Sun above the core, where energy is transported outward by electromagnetic radiation (photons scattering off particles of mass).

convective zone The layer of the Sun above the radiative zone, where energy is transported outward by the motion of gas (convection).

photosphere The Sun's thin, "surface" layer, from which radiation escapes into space.

chromosphere The lower layer of the Sun's atmosphere.

corona The extended region of extremely hot, low-density gas surrounding the Sun.

gravitational contraction The shrinking of an object's radius and accompanying increase in its density due to the force of its own gravity.

The solution to this dilemma was first conceived by astrophysicist Arthur Eddington in 1920 when he proposed that energy from nuclear reactions could power stars. That proposal began to be worked out in more detail as the science of nuclear physics matured in the late 1920s and early 1930s. "Nuclear physics studies the cores, or nuclei, of atoms," says Falk Herwig, an astrophysicist at the University of Victoria who specializes in the nuclear physics driving stellar evolution. "Early in the 1900s," he explains, "physicists discovered that all atoms were composed of negatively charged electrons orbiting a central positively charged nucleus."

Hydrogen has a single electron surrounding a nucleus composed of a single proton (see Chapter 4). Go one step up in the periodic table, however, and things get more complicated. Helium has two electrons orbiting a nucleus with two protons. But the nucleus of helium weighs approximately *four* times that of the hydrogen atom. "Since electrons have a very small mass compared with protons," says Herwig, "physicists concluded there had to be another kind of nuclear particle other than a proton." Thus, helium must have two *neutral* particles inside its nucleus. In 1932, this uncharged particle, called the neutron, was discovered, and nuclear physics began in earnest. Physicists found that the nuclei of most elements contain a mix of protons and neutrons.

Although the number of protons is always fixed for a given element, the number of neutrons varies, producing different isotopes of that element (see Going Further 6.1). A nucleus with six protons and six neutrons (carbon-12, or ^{12}C) is called carbon, but so is a nucleus with six protons and seven neutrons (carbon-13, or ^{13}C).

As physicists probed the nucleus, they began mapping out a wide range of nuclear reactions between two or more nuclei. "Nuclear reactions are interactions that transform one nucleus into another," says Herwig. Just a few years after the advent of nuclear physics, scientists such as Hans Bethe and Enrico Fermi began to see how reactions called thermonuclear fusion might be the source of the Sun's energy. During the next two decades, the pieces of a full theory of stellar nuclear power were put into place. By the early 1960s, one of the oldest of human questions had been answered: We finally knew how stars shine.

Thermonuclear Reactions: The Curve of Binding Energy

In transforming one type of nucleus into another, nuclear reactions release energy. The concept of *binding energy* is crucial to understanding why and when nuclear reactions can give up more energy than is needed to initiate them and so act as a power source for stars. **Binding energy** is the energy needed to break up an object being held together by a given force. The same amount of energy is released when the object forms. The binding can be done by any attractive force, such as gravity or electromagnetism (which is attraction between opposite charges). In an atomic nucleus, protons and neutrons are held together by the **strong nuclear force**. (It must be strong to overcome the electromagnetic repulsion associated with the positively charged protons in the nucleus.) The strong force is one of the four fundamental forces in nature, along with gravity, electromagnetism, and the weak nuclear force. The strong nuclear force is the strongest of the four.

To see how nuclear reactions work, we turn to a graph called the *curve of binding energy* (**Figure 10.8**). "The curve of binding energy tells us how effective nature is at keeping protons and neutrons packed into the nucleus," says Herwig. The curve rises as the number of protons and neutrons in an atom's nucleus increases, but this continual increase with the size of a nucleus goes up to the element iron only. The increase means that the energy associated with holding the nucleus together is increasing as the nuclei get larger. Therefore, iron nuclei, having the greatest binding energy, are the most stable atomic nuclei in nature. For nuclei with more protons and neutrons than iron, the curve begins to fall, because the strong force (which acts only over short distances) has a harder time binding the protons and neutrons in a larger, heavier nucleus.

"Since [uranium nuclei] are unstable, you just need to tickle them a little and they split up into two less heavy nuclei, giving up energy in the process."

binding energy The energy needed to break up an object being held together by a given force. The same amount of energy is released when the object forms.

strong nuclear force One of the four fundamental forces of nature, it binds particles together in the nucleus of an atom.

nuclear fission The splitting apart of a nucleus into two or more smaller nuclei.

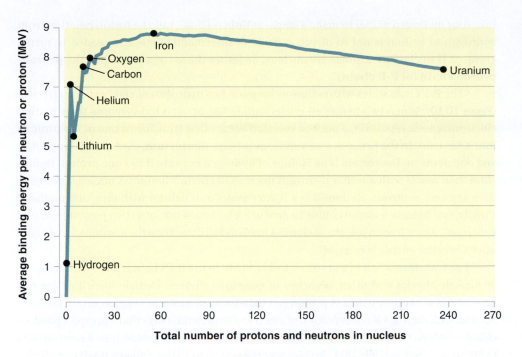

Figure 10.8 The Curve of Binding Energy
This classic graph shows how tightly each element's nucleus is held together. A nucleus's total binding energy rises as the number of protons and neutrons in the nucleus increases. Up to the element iron, energy can be released through fusion (bringing two smaller nuclei together, making a larger nucleus). For elements larger than iron, more energy is needed to create a larger nucleus than is released. Beyond the peak in the curve of binding energy, represented by iron, very large nuclei can release energy when broken apart through nuclear fission. Binding energy in nuclei is usually measured in mega-electron volts (1 MeV = 10^6 electron-volts).

The curve of binding energy helps us understand the two most important kinds of nuclear reactions: fusion and fission. Fusion reactions, which occur when two nuclei are slammed together and produce a larger nucleus, generate energy by fusing nuclei that are lighter than iron. In contrast, **nuclear fission** is a splitting apart of large (heavy) nuclei, which fall beyond iron on the curve of binding energy. A heavy nucleus made of many protons and neutrons can be driven to fission if it is bombarded with neutrons. "With fission," explains Herwig, "you begin with heavy nuclei, such as isotopes of uranium. Since they are unstable, you just need to tickle them a little and they split up into two less heavy nuclei, giving up energy in the process."

All nuclear power stations utilize fission reactions. Most often, uranium-235 (92 protons and 143 neutrons) is bombarded with neutrons and splits into smaller nuclei (**Figure 10.9**). In breaking the heavy nucleus apart, the fission reaction gives back more energy than was supplied to it via the bombarding neutrons.

Thermonuclear Fusion in the Sun: The Proton-Proton Chain

Nuclear fusion reactions are the source of the Sun's power. Every second, the Sun is consuming 6.1×10^{11} kg of hydrogen gas through fusion in the core. But what exactly is fusion? Thermonuclear fusion is the formation of heavier nuclei by combining lighter nuclei. When light nuclei (with mass lower than iron) combine, the mass of a nucleus coming out of a fusion reaction is always a little less than the combined masses of the lighter particles going in. Thus, by binding the protons and neutrons together in the heavier nucleus, the fusion reaction converts some matter into energy. (Einstein's equation $E = mc^2$, where c is the speed of light, shows that mass m is a form of energy E.) "As we add nuclear particles to a light nucleus through the process of fusion," says Herwig, "we're releasing energy, allowing [the particles] to bind together."

Throughout a star's evolution, the fusion of lighter elements into heavier ones is the fundamental energy generation mechanism that enables the star to support itself against its own gravity. Specifically, the Sun (and stars like it) converts hydrogen gas into helium gas. Hydrogen has one particle (a proton) in its nucleus. Helium's nucleus has four particles (two protons and two neutrons). It might seem as though it should

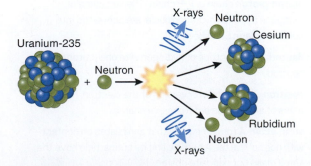

Figure 10.9 Nuclear Fission
Fission, the process used in nuclear reactors, is the opposite of nuclear fusion, the source of energy in stars. In fission, a neutron bombards the large nucleus of a heavy element, which splits into lighter nuclei and other neutrons, releasing energy. The instability of uranium-235 makes it an ideal fuel for fission.

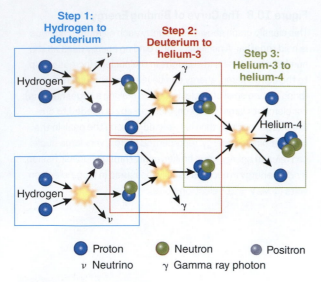

Step 1: Hydrogen to deuterium

Step 2: Deuterium to helium-3

Step 3: Helium-3 to helium-4

Hydrogen

Hydrogen

Helium-4

● Proton   ● Neutron   ● Positron
ν Neutrino   γ Gamma ray photon

Figure 10.10 The P-P Chain

Sun-like stars convert hydrogen to helium via the proton-proton (P-P) chain. The reaction begins when two protons combine, forming a nucleus of deuterium (one proton and one neutron), which is an isotope of hydrogen. Deuterium nuclei then collide with protons, producing helium-3 (two protons and one neutron), a lighter isotope of helium. Helium-3 collisions then yield a stable helium-4 nucleus (two protons and two neutrons). The mass of the final products is less than that of the original protons; the difference represents energy released in the fusion process.

"You need really high temperatures to get fusion going, and the only place you get those is in a star's core."

proton-proton chain or **P-P chain** The sequence of reactions by which hydrogen nuclei are converted into helium nuclei in Sun-like stars.

deuterium An isotope of hydrogen with one proton and one neutron.

positron The antimatter version of the electron, having the mass of an electron but a positive charge.

antimatter A form of matter that annihilates on contact with normal matter. A particle and its antiparticle have the same mass but opposite charge.

neutrino An electrically neutral, weakly interacting elementary subatomic particle often created in nuclear reactions.

solar neutrino problem The apparent discrepancy between the expected number of neutrinos being emitted from the Sun and the number measured experimentally; resolved by the discovery that neutrinos oscillate among three types.

take four hydrogen nuclei to make a single helium nucleus, but the fusion reaction from hydrogen to helium is not so direct. A chain of intermediate steps is needed to turn a group of H nuclei into a single He nucleus. Together these steps are called the **proton-proton chain** (or **P-P chain**).

"The P-P chain starts with collisions between two hydrogen nuclei," explains Herwig (**Figure 10.10**). When two hydrogen nuclei collide fast enough to overcome their mutual electromagnetic repulsion, a nuclear reaction occurs that transforms one of the protons into a neutron. In the process, a new nucleus is born: **deuterium**, containing one proton and one neutron. Deuterium is an isotope of hydrogen because it has one proton. Deuterium then reacts with another hydrogen nucleus to create a helium-3 nucleus (two protons and one neutron). Helium-3 is a lighter isotope of helium with only one neutron. Finally, two helium-3 nuclei collide to produce a helium-4 nucleus (two protons and two neutrons). As we'll soon see, that helium-4 nucleus is lighter than the combined hydrogen nuclei because energy is released.

The transformation of a proton into a neutron in the P-P chain is an essential aspect of nuclear physics and other branches of quantum physics. Particle identities are not fixed in the subatomic world. A proton can turn into a neutron as long as certain laws are obeyed. One such law is conservation of electric charge, which says that a proton's positive charge can't just disappear. The nuclear reaction that turns a proton into a neutron has to create another particle that carries away the positive charge, called a **positron**. "Positrons are positively charged particles with the mass of an electron," says Herwig. They're a type of **antimatter**—a form of matter that annihilates (is destroyed) on contact with normal matter. "If a positron and an electron meet, they annihilate each other, leaving only energy (in the form of photons) behind." The proton does not split into a neutron and positron but instead is instantaneously and discontinuously transformed into these other particles. The neutron and positron do not exist inside the proton before the proton transforms. Such transformations are part of the behavior of the quantum world, on which everything we experience is built.

Another particle, called a **neutrino**, and some energy in the form of gamma-ray photons are also by-products of the P-P chain. Neutrinos are not antimatter but are strange in their own way. "Neutrinos are like ghost particles," says Herwig. "They have very little mass, have no charge, and travel at close to light speed. Most important, they barely interact with the kinds of matter our bodies are made of (protons, neutrons, and electrons)." It would take a slab of lead (the densest metal) a light-year thick to halt a bunch of neutrinos. "And that lead would only stop half of them," adds Herwig.

Fusion reactions, such as the P-P chain that keeps the Sun shining, can occur only under very extreme conditions. "Normally, protons repel each other via the electromagnetic force because they have like charges," says Herwig. "That means the only way to get them to fuse (via the strong nuclear force) is by having them collide at very high speed to overcome the repulsion." Since temperature is a measure of gas particles' random motion, high speed translates into high temperature. "You need really high temperatures to get fusion going, and the only place you get those is in a star's core." Temperatures at the center of the Sun reach 15 million K, high enough for the P-P chain to operate. "As you go farther out from the core," says Herwig, "the temperatures decrease. Fewer particles have high enough energy to overcome the electromagnetic repulsion, and fusion stops."

Other conditions in the core, such as density, also contribute to energy generation. High densities aren't needed for individual fusion reactions, but the *frequency* of fusion reactions depends on how often particles collide. The more particles in a given area, the higher the collision rate—the number of collisions per second. In the Sun's core, matter is packed so tightly that more than 10^{38} protons can be converted into helium every second. How is that big number related to energy production? Recall that in fusion reactions, a fraction of the mass in the constituent particles is converted to energy. "The majority of the hydrogen mass is converted into helium, not energy," says Herwig. "In fact, only 0.7 percent of every kilogram of hydrogen is converted into energy in the fusion process." Thus, for the specific case of H burning in stars like the Sun, we can modify Einstein's formula $E = mc^2$ to $E = 0.007mc^2$. The decimal number is tied to the concept of binding energy. While 0.7 percent may not sound like much, the Sun represents a very large reservoir of mass. Every second, more than 4.3 million tons of hydrogen is converted into energy, which equals 3.8×10^{26} joules (J)—the same amount of energy as in 9.1×10^{10} megatons of TNT (see **Going Further 10.1** on p. 272).

Energy can take many forms, ranging from magnetism to motion. What form of energy does fusion produce? As **Figure 10.11** illustrates, the P-P chain not only yields a nucleus of helium but also produces photons and neutrinos. In terms of a total energy budget, the photons (in the form of gamma rays) carry away most of the energy. But these fusion-generated light particles can't just escape immediately. The solar interior is far too dense for a photon to travel very far before it is absorbed again by a particle of matter. The neutrinos, however, suffer no such restriction.

While the neutrinos represent a small share of the fusion energy budget, their "ghostly" nature does provide physicists with direct evidence of fusion in the solar core. Because neutrinos rarely interact with other matter particles, they immediately escape the solar core. Traveling through the entire Sun at just below the speed of light, the neutrinos quickly emerge into interplanetary space. Some of these solar neutrinos cross Earth's path, where physicists can use huge detectors to capture a few of them. These elusive particles have been used to test theories of solar fusion and the P-P chain directly. Such experiments take enormous effort; most detectors require thousands of tons of target mass (in the form of water or other substances) to catch even a few neutrinos (**Figure 10.12**).

For decades, every effort to match solar fusion theory with solar neutrino experiments failed. There were always 65 percent fewer neutrinos than expected. This **solar neutrino problem** remained unsolved until some physicists imagined that the problem was not with the Sun or with their understanding of nuclear fusion physics but with the

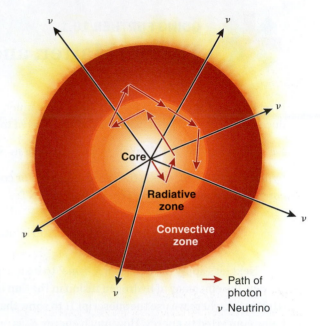

Path of photon
v Neutrino

Figure 10.11 Solar Neutrinos
Neutrinos (v) readily escape from the Sun's interior, since they interact very weakly with matter. Photons, however, are scattered many times as they encounter particles of matter before eventually escaping. It takes roughly 100,000 years for a photon to make it out of the Sun.

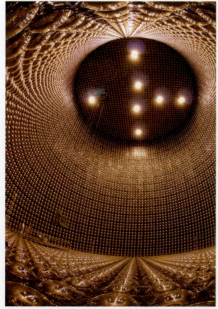

a.

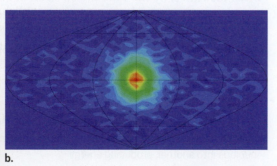

b.

Figure 10.12 Super-Kamiokande (SK) Neutrino Detection Experiment
a. Located 1,000 meters below Mount Ikeno in Japan, this facility is a neutrino observatory consisting of a tank holding 50,000 tons of ultrapure water. The tank's inner surface is covered by detectors sensitive to the flash of light that occurs when a neutrino interacts with an electron or a nucleus in a water atom. Because neutrinos interact very weakly with matter, scientists must gather enormous quantities of them to capture even one interaction. **b.** Using the SK detector, scientists used neutrinos to image the Sun (which would appear as a small dot at the center of the image in visible light).

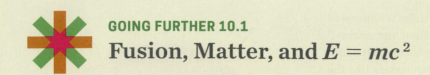

GOING FURTHER 10.1
Fusion, Matter, and $E = mc^2$

First let's use the equation $E = mc^2$ to see how much energy E is produced when $m = 1.0$ kg of matter is *fully* converted to energy:

$$E = (1.0 \text{ kg})(3.0 \times 10^8 \text{ m/s})^2 = 9.0 \times 10^{16} \text{ kg m}^2 \text{ s}^{-2} = 9.0 \times 10^{16} \text{ J}$$

(where $c = 3.0 \times 10^8$ m/s is the speed of light). For comparison, we convert the result in joules (J) to kilotons of TNT:

$$\frac{9.0 \times 10^{16} \text{ J}}{4.2 \times 10^{12} \text{ J/kiloton of TNT}} \approx 21{,}500 \text{ kilotons of TNT}$$

Next let's apply the equation $E = mc^2$ to fusion in the Sun. As we saw, the efficiency of hydrogen fusion in the Sun is such that $E_H = 0.007 m_H c^2$, where we use the subscript H to show that hydrogen (H) is being converted to energy. How much energy E_H is produced when $m_H = 1.0$ kg of hydrogen is converted into helium in fusion?

$$E_H = 0.007(1 \text{ kg})(3.0 \times 10^8 \text{ m/s})^2 = 6.3 \times 10^{14} \text{ J}$$
$$\approx 150 \text{ kilotons of TNT}$$

Only a small amount of hydrogen's mass is released as energy in a fusion reaction. (Remember that most of the hydrogen mass is converted not into energy but into helium mass.)

Given that we know how much energy the Sun produces every second (by measuring its light output), we can figure out how much mass is converted into energy every second in the Sun. The energy output (luminosity) of the Sun is 3.8×10^{26} J/s. That means the Sun produces 3.8×10^{26} J every second. Let's put this into our formula and solve for mass:

$$m_H = \frac{E_H}{0.007 c^2} = \frac{3.8 \times 10^{26} \text{ J}}{0.007(3.0 \times 10^8 \text{ m/s})^2} = 6.1 \times 10^{11} \text{ kg}$$

All of the energy we see coming from the Sun originates in fusion reactions. Since we know the efficiency of the fusion reactions in liberating energy (0.007), we can convert what we see (the Sun's energy output) into what we can't see (the Sun's mass consumption).

neutrinos themselves. In 2002 it was finally discovered that the ghostly neutrinos can transform from one type into another as they travel through space.

There are actually three types of neutrinos, each associated with how it interacts with different kinds of subatomic particles. The original solar neutrino detectors were designed to look for *electron neutrinos*—the kind emitted in solar nuclear reactions. In addition to those, there are *muon neutrinos* and *tau neutrinos* (cousins of the electron type). But as physicists learned in experiments carried out in 2002, all neutrinos can switch identities. By the time the electron neutrinos reached Earth, they had already transformed (the technical term is *oscillated*) into either tau or muon neutrino versions. In this way, advances in basic particle physics, including the discovery of neutrino oscillations, solved a vexing and long-standing problem in basic nuclear astrophysics (see **Anatomy of a Discovery**).

Section Summary

- For each element, the nucleus of every atom has a unique number of protons, but isotopes of that element can have different numbers of neutrons. Thermonuclear fusion in the Sun transforms one nucleus into another, producing energy.
- The curve of binding energy shows that, up to the element iron, fusion gives up more energy than is input. Nuclear fission splits apart nuclei heavier than iron, releasing energy.
- Einstein's equation $E = mc^2$ shows that mass (m) is a form of energy (E) and can be used to calculate the energy locked up in matter.
- By nuclear fusion, stars combine light nuclei into heavier ones and release mass that is converted to energy. These reactions require high temperatures and speeds to overcome the repellent electromagnetic forces of protons.
- In the proton-proton (P-P) chain, a fusion process in the Sun's core, hydrogen nuclei are converted to helium nuclei; 0.7 percent of the hydrogen's mass is converted to energy.

CHECKPOINT   Where does the energy come from when the Sun converts hydrogen into helium?

What powers the Sun?

Sun

hypothesis

By the mid-19th century, scientists realized that chemical burning could not sustain the Sun for thousands, much less millions, of years. A much more efficient energy source is needed. In 1939, **Hans Bethe** proposed the proton-proton chain, a series of nuclear reactions that convert H to He in the Sun's core and could power the Sun for billions of years. The neutrino (ν) is a by-product of these reactions.

observation

Civilizations have always recognized the Sun as a major source of heat and light. If it behaves as heat sources on Earth, then it just might be "on fire." Before discovering that stars are mostly H, hydrogen, and He, helium, scientists believed the Sun might be made of the same stuff as Earth and might generate energy by burning a source as familiar as a carbon-rich fossil fuel.

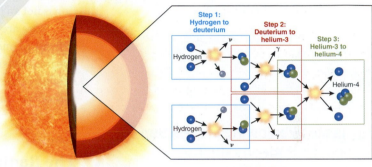

Structure of the Sun

P–P chain

experiment

John Bahcall calculated how many solar neutrinos fusion reactions should produce each second and predicted the number of neutrinos per second that reach Earth. In the 1960s, **Ray Davis** and Bahcall built a neutrino detector deep in a South Dakota gold mine. They detected only one-third of the neutrinos predicted. The "missing" neutrinos launched the solar neutrino problem.

confirmed by experiment

This problem indicated that either the fusion reactions powering the Sun or the nature of the neutrino was not understood. In 2002, with new detectors, such as Super-Kamiokande, scientists including **Takaaki Kajita** and **Arthur McDonald** independently discovered that neutrinos oscillate among three types. Nuclear reactions in the Sun proceed just as Bahcall (and Bethe) predicted.

Ray Davis in neutrino tank

Super-Kamiokande

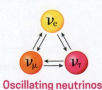

Oscillating neutrinos

Moving Energy

10.3 The majority of the energy released by fusion reactions at the Sun's core makes it all the way to the surface and escapes into space (minus the fraction that immediately escapes as neutrinos). How do we know? "Well, we can see it shining," says Falk Herwig. "Energy is always flowing *off* the Sun into space. That means the Sun is always losing energy from its surface."

Energy transport is any process by which energy moves from one place to another. There are three basic ways in which energy is transported in nature: radiation, convection, and conduction. In radiative transport, energy is moved via light (see Chapter 4). In convection, energy is transported by large-scale fluid motions (see Chapter 6). In *conduction*, energy is transported from a hotter region to a cooler one by collisions of atoms inside matter—such as when the handle of a pot over a lit oven burner warms up. Conduction is not an important mechanism inside stars, but it is inside some planets.

Stars are always transporting heat from the inside out. "If that transport becomes difficult in one region of the star," says Falk Herwig, "then the star will find an alternative mechanism of getting that energy moving again." As discussed in Section 10.1, detailed studies of the Sun's interior (via methods such as helioseismology) have revealed that two of the three mechanisms of energy transport operate within a star: radiation and convection.

The Radiative Zone

The fusion reactions in the core release the bulk of their energy as light, especially gamma-ray photons. When these photons emerge from a fusion reaction, they begin traveling at the speed of light in random directions. "But those gamma rays don't make it very far," says Herwig. The layers above the core constitute a soup of electrons and (mostly) ionized hydrogen and helium. Each encounter between a gamma ray and one of these charged particles **scatters** the light particle, sending it in a different random direction. Scattering occurs when light interacts with matter, changing its direction of propagation. "The average distance a photon can travel before it interacts with a charged particle is just a few centimeters," says Herwig. Only after trillions and trillions of these random scatterings does the photon (and its energy) slowly wander outward from the core.

The movement of energy through this repeated scattering of photons is called **radiative energy transport**; it gives the Sun's radiative zone its name. In radiative energy transport, a given quantity of energy is transported entirely by photons (electromagnetic radiation) as opposed to some other means. "On average," says Herwig, "it will take the energy bundled into a single photon at least tens of thousands of years to make its way out of the radiative zone."

The direction of a photon is not all that changes in scattering events. Recall that gamma rays are at the high-energy end of the electromagnetic spectrum (high energy = high frequency = short wavelength; see Chapter 4). As the photon moves outward, it encounters solar plasma at ever-lower temperatures. The cooler ions and electrons in these layers change the photon's energy as they scatter it, shifting the light to longer wavelengths. "One gamma ray gets converted into two lower-energy X-rays," says Herwig. "Each of those X-ray photons will then be converted into multiple UV photons." Those UV photons are further "downshifted" in wavelength when they scatter off lower-temperature gas farther from the core. In this way, energy originally locked up in a single gamma ray that was released in a fusion reaction in the core is divided into almost 2,000 photons of lower energy that diffuse outward through the Sun's outer layers. In other words, photons that emerge from the photosphere are not the same as those produced by fusion reactions in the core.

> "Energy is always flowing *off* the Sun into space. That means the Sun is always losing energy from its surface."

scatter Change direction; for example, change a photon's direction of propagation.

radiative energy transport The transfer of thermal energy via electromagnetic radiation (photons).

solar granulation The pattern of bright interiors and dark edges on the photosphere; generated by convection cells.

The Convective Zone

"The solar plasma can transport energy in different ways, depending on the local conditions," says Herwig. "Moving energy by radiation is one option, but it's very inefficient because it takes so long. If the conditions are right, then the solar plasma can switch to a second option: moving the heat by moving the gas." In the Sun's next layer, the convective zone, large-scale flows of matter accomplish the energy transport. Hot gas rises, taking its thermal energy up to the top of the convective zone. This mode of energy transport via mass motions is convection, and it occurs in the Sun's convective zone.

To understand how convection works in the Sun, recall that the temperature drops continuously from the Sun's core to its surface. As photons squirm outward through multiple scatterings, they eventually encounter plasma that has cooled enough so that much of the hydrogen gas has become neutral rather than ionized. In this state, instead of simply scattering the light particles, the hydrogen and helium atoms now absorb them and trap their energy. But the energy flow from the core can't be blocked forever, so an alternative method of transport comes into play: motion.

"A pot of water on a stove continually absorbs heat from the burner," says Herwig. "At some point, however, the water becomes so hot that the only way to move the energy absorbed at the bottom is to set the water in motion through boiling. That's sort of what happens in the Sun." The energy absorbed at the top of the radiative zone allows blobs of plasma in the convective zone to expand slightly, making them buoyant (lighter than surrounding gas). The buoyant blobs begin to rise, just as a helium balloon rises into the sky. When the hot, rising plasma blobs reach the surface, blackbody radiation from the hot gas escapes into space. The gas cools and sinks back down toward the convective zone's base, starting the process over again. Thus, energy generated by fusion reactions in the core finally is released as starlight at the Sun's photosphere.

Solar Granulation

The convective zone extends to the photosphere, and, says Lika Guhathakurta, "all that churning leaves its mark there." High-resolution photographs show the Sun's surface to be almost uniformly tiled with a network of bright patches surrounded by thin, dark outlines. In this pattern, called **solar granulation**, the bright spots are plumes of hot gas that have risen from deep within the convective zone to the surface. Each solar granule is a cell of convective motion (**Figure 10.13**). "Once that hot gas hits the photosphere, it spreads out, cools, and then sinks down again," says Guhathakurta. The dark outlines of a solar granule show where the cool gas has begun its downward journey.

The solar surface is extremely dynamic; it's always changing. Using Doppler shift measurements, astronomers can see rising and descending plumes of plasma moving at approximately 0.4 km/s. A typical solar granule extends across 1,000 km (about the size of

FALK HERWIG

Falk Herwig wanted a field of study that was as far as possible from anything practical. "When I was younger, I think I must have had a big chip on my shoulder or something," he says. "I just couldn't imagine using my time to do anything applied to real life."

When he started his PhD studies in Germany, Herwig explored a number of disciplines. "Astrophysics, however, seemed to have just the right mix of distance from day-to-day reality and a reality that included job prospects," he says, laughing. Herwig now models the evolution of Sun-like stars. His work places him at the frontiers of understanding what happens deep within stars as they reach the end of life.

"The solar plasma can switch to a second option: moving the heat by moving the gas."

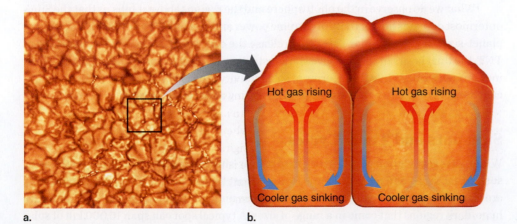

Figure 10.13 Solar Granulation
a. Granules, which live typically 20 minutes or less, are the primary solar surface feature. **b.** Each granule is a convection cell in which hot gas rises toward the lighter center and cooler gas sinks toward the darker edges.

Texas). But despite this enormous size, a granule lasts only 10–20 minutes before it fades and the surface pattern changes.

The reason for granules' characteristic bright centers and dark edges lies in the Stefan-Boltzmann law for blackbody radiation (see Chapter 4). Recall that the energy output per time from a section of a blackbody's surface is the surface's luminosity L, which is proportional to surface temperature T as $L \propto T^4$. This formula tells us that even a small change in temperature results in a large change in brightness (that is, light-energy output). Imagine, for example, that the temperature in the center of a granule is just twice as high as the temperature of the cooler gas at its edge. According to the Stefan-Boltzmann law, the granule's center will be $(2)^4 = 16$ times brighter than its edge.

"That difference might not seem like much," says Guhathakurta, "but try staring into a 10-watt lightbulb versus a 160-watt bulb and see what happens. The 160-watt bulb would be painful to your eyes. Small differences in temperature really mean a lot for how bright surface features appear on the Sun." Guhathakurta notes that, in reality, the temperature at the granule edges is only a few hundred degrees lower than at the granule center, but this difference is enough to create the light-and-dark pattern.

Section Summary

- In the radiative zone, high-energy gamma rays are scattered numerous times and eventually are converted into lower-energy photons as the energy is transported outward.
- In the convective zone, gas is no longer fully ionized, and photons become trapped by neutral atoms. There, energy is transported by convection of gas, which cyclically heats, rises, cools, sinks, and is heated again.
- Blackbody radiation from the hot gas is released at the Sun's surface, the photosphere.
- Solar granulation on the photosphere results from convection cells of hot gas rising from below, cooling, and sinking at the cells' edges.

CHECKPOINT   Give an example in your life of energy being transported in different ways.

The Active Sun: Photosphere to Corona and Beyond

10.4   While we have to rely on helioseismology to "see" what happens inside the Sun, what happens *outside* is on display every day. "The Sun's surface, called the photosphere, is not really a surface in the sense of either a solid or a liquid," Lika Guhathakurta explains. It is merely the region from which light emitted by solar matter escapes to space. Gas in the photosphere has a temperature of approximately 5,800 K. Gas layers above it are transparent to the (blackbody) radiation the photosphere creates. "That's why we 'see' the Sun as a 5,800-K ball of plasma," says Guhathakurta. "If radiation could escape freely from deeper inside the Sun, then we would see a hotter photosphere."

What we do observe in the photosphere and the regions above it tells us that the Sun's outermost domains host activity of extreme power and violence. From sunspots that cover planet-sized regions to solar flares that release the equivalent of 100 billion megatons of TNT, the final layers of the Sun give stark testimony to the vast reservoirs of energy flowing through a star. While these forms of solar activity are fascinating, they present significant dangers to us as we take our steps in becoming a space-faring race.

Sunspots and Their Cycles

While solar granulation constitutes a kind of ever-shifting carpet of structure on the solar surface, far larger and more dramatic features exist there as well. **Sunspots**, in particular, are giant structures that play a critical role in our understanding of the Sun. Sunspots are huge, dark regions that come in a range of sizes. A typical spot can span 10,000 km of solar

> "All that energy locked up in the Sun's magnetic field drives breathtaking solar activity, including sunspots."

sunspot   A large, strongly magnetized region of the Sun's surface that appears darker because it is lower in temperature than its surroundings.

surface (almost the diameter of Earth). Each sunspot consists of a dark center, the *umbra*, surrounded by a brighter area, the *penumbra* (**Figure 10.14**).

Like solar granules, sunspot umbras appear dark because gas there is cooler than surrounding material and therefore emits less blackbody radiation (see **Going Further 10.2** on p. 278). "Sunspots are only dark in contrast to the bright face of the Sun," says Guhathakurta. "If you could cut an average sunspot out of the Sun and place it in the night sky, it would be about as bright as a full Moon."

Unlike solar granules, sunspots are not convective plumes of rising gas. Instead, powerful magnetic fields generate sunspots. "Giant columns of the Sun's magnetic field well up to the photosphere from deep in the convective zone to form spots," says Guhathakurta. Just as magnetic fields are generated in the interiors of many of the Solar System's planets, the Sun also creates its own magnetic field. However, surface fields on the Sun are many times more powerful than the strongest regions of the strongest planetary fields. "All that energy locked up in the Sun's magnetic field," says Guhathakurta, "drives the breathtaking solar activity occurring on and above its surface, including sunspots."

Powerful columns of magnetic fields rise up in a sunspot from below the solar surface. The fields exert force on the solar plasma and thus create their own form of pressure, which pushes plasma out from the sunspot's center. (Recall that plasma is an ionized gas, so plasmas will feel forces exerted by magnetic fields.) The lower gas pressure in the sunspot translates into lower gas temperatures and a darker appearance (again, as explained by the Stefan-Boltzmann law). "A large sunspot might have a temperature of about 4,000 K," explains Guhathakurta, "much less than the 5,800 K of the bright photosphere surrounding it."

Like granules, sunspots are transient; they last from days to as much as months. But unlike the granules, sunspots appear and disappear following a regular pattern. "The average number of spots on the face of the Sun is not always the same; it goes up and down in a cycle," says Guhathakurta. Historical records of sunspot counts show that the sunspot cycle has an average period of about 11 years. Large numbers of spots appear on the Sun at the beginning of each cycle, and over the next 11 years their numbers slowly decrease (**Figure 10.15**). The location of the spots also varies with time. Groups of sunspots first appear at middle latitudes on the solar globe (about 30° from the equator). As the cycle progresses, the spots make their appearance closer and closer to the solar equator.

This progression from many spots to a few and from midlatitudes to the equator is cleanly captured in so-called Maunder butterfly diagrams, named after the British

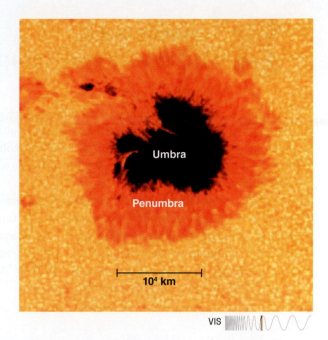

VIS

Figure 10.14 Sunspots
Each sunspot has a darker, cooler umbra encircled by a brighter, hotter penumbra. The umbra is approximately 2,000 K cooler than the surrounding solar surface, making the sunspot appear dark by comparison. The small-scale details reflect the structure of the magnetic fields within the sunspot.

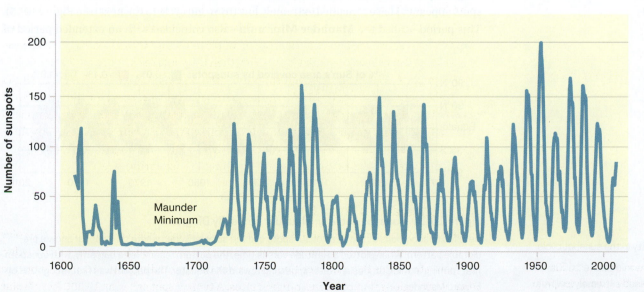

Figure 10.15 Sunspot Cycle
Historical records show that the sunspot cycle has an average period of about 11 years. Notice that the number of sunspots at the cycle's peak varies by a factor of 4 or more. During the "Maunder Minimum" of the mid- to late-1600s, the number of sunspots was greatly reduced.

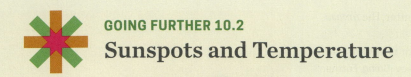

GOING FURTHER 10.2
Sunspots and Temperature

The Sun's surface appears to have dark features, such as sunspots and the edges of solar granulation cells. But are these features really dark, or are they just a result of relative brightness? The solar surface radiates as a blackbody, so we can use all of the formulas we learned for blackbodies to answer this question.

According to the Stefan-Boltzmann law, the luminosity of a blackbody's surface depends very sensitively on its temperature. If we had two same-sized square plates of metal and heated each to a different temperature, we could use this law to understand how much energy one plate would radiate relative to the other. To see this, let's write out the full Stefan-Boltzmann equation:

$$L = A\sigma T^4$$

where L is the luminosity (energy radiated per time), A is the area of the plate surface, σ is the Stefan-Boltzmann constant, and T is the blackbody temperature in kelvins. If we take the equations for plates 1 and 2 and divide them by each other, the area A and the Stefan-Boltzmann constant σ drop out:

$$L_1/L_2 = (T_1/T_2)^4$$

So, if $T_1 = 400$ K and $T_2 = 200$ K, then the ratio of luminosity between the plates is

$$L_1/L_2 = (T_1/T_2)^4 = (400 \text{ K}/200 \text{ K})^4 = (2)^4 = 16$$

A 16-fold difference in energy radiated per second is quite a lot for such a small difference in temperature.

A typical temperature in a sunspot is about 4,000 K, while the rest of the solar surface has a temperature of 5,800 K. Using these numbers, we can calculate the difference in brightness between a sunspot and the gas surrounding it:

$$L_{\text{spot}}/L_{\text{surface}} = (4,000 \text{ K}/5,800 \text{ K})^4 = 0.2$$

Thus, the sunspot produces only about 20 percent as much light as the hotter gas around it does. This difference is enough for optical images (or your eye) to see the sunspot as dark relative to its surroundings. In reality, however, the spot still produces a lot of light.

---

astronomer Edward Maunder, who first "visualized" sunspot data in this way (**Figure 10.16**). These diagrams plot the location and number of spots as a function of time. The butterfly pattern that emerges constitutes the raw data of solar science. Any theoretical explanation of sunspots must explain how this pattern is created.

In addition to this 11-year cycle, the magnetic behavior tied to sunspots shows even longer-term changes. The number of sunspots occurring at the **solar max**—the maximum of the cycle—can vary from more than a hundred to just a few. The record of observations spanning 350 years shows a remarkable period in the mid to late 1600s when the number of sunspots appears to have remained extremely low from one cycle to the next (see Figure 10.15). This period—called the **Maunder Minimum**—also coincided with an extended period of

INTERACTIVE:

Sunspot Cycle

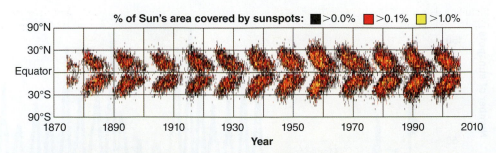

Figure 10.16 Sunspot Activity: Maunder Butterfly Diagram
Detailed observations of the sizes of sunspots and the solar latitudes where they appear show a predictable pattern. When plotted as latitude versus time, the pattern resembles a butterfly, with sunspots first forming at midlatitudes and later appearing toward the equator. The pattern was first recognized by Edward Maunder.

solar max The period of peak activity in the sunspot cycle.

Maunder Minimum The period in the mid to late 1600s when the number of sunspots remained extremely low from one sunspot cycle to another.

cold climate in northern Europe. This kind of evidence suggests a link between Earth's climate and solar activity. Spacecraft measurements have confirmed that the strength of solar activity correlates with the amount of solar energy that Earth receives. But while such evidence does show ties between climate and sunspots, climate scientists have been able to rule out the role of solar activity in present-day climate change (discussed in Chapter 6).

It's worth noting that spots may be a common astronomical phenomenon. "Our Sun isn't the only star with spots," says Guhathakurta. "Just recently, astronomers have been able to detect 'star spots' appearing on other stars." This discovery means that the stellar physics of spots, including the role of magnetic fields, is likely a universal process.

Solar Magnetic Cycle

The 11-year sunspot cycle is only one part of an even longer cycle in which the Sun's magnetic field rebuilds itself over and over again. To study the solar cycle, astronomers need a means of measuring the magnetic field on the Sun. The *Zeeman effect* is one direct way to detect magnetic fields in distant gas. In the Zeeman effect, magnetic fields change the structure of electron orbits in atoms. The fields create multiple emission lines where only one had existed. By measuring these split emission lines from gas in the solar photosphere, astronomers can obtain a direct measure of the Sun's field strength in different places and at different times.

Using such techniques, astronomers have found that nearby groups of sunspots are linked by magnetic-field lines that rise from one set of spots and descend into the other. Thus, one sunspot constitutes the northern pole of the magnetic field, and another constitutes the southern pole.

Sunspot groups in both the northern and southern hemispheres of the Sun show these kinds of magnetic links. "But there are differences between the behavior of sunspot groups in the two hemispheres," explains Guhathakurta. "And the differences shift from one cycle to the next." In any one cycle, all sunspot groups in the northern hemisphere show the same field direction. For example, trailing spots are the source of a rising field (the north polarity), and leading spots have the south-field polarity, where the field dives back down. Throughout a single sunspot cycle, this pattern holds true in the northern hemisphere, and the southern hemisphere shows the opposite pattern. Says Guhathakurta, "There the polarity between trailing and leading spots is reversed. The trailing spot acts as the field's 'south pole' and the leading spot is the 'north pole'" (**Figure 10.17a**).

"Then," Guhathakurta continues, "things get really weird." As a cycle ends, the number of sunspots gradually goes almost to zero; as the new cycle rises, new groups of spots appear—with flipped polarity compared with that of the previous cycle (**Figure 10.17b**).

LIKA GUHATHAKURTA

Born in Kolkata, India, Lika Guhathakurta moved with her parents to Mumbai when she was 8. From an early age she was introduced to astronomy through books. "I remember these wonderful picture books about stars, comets, and planets," she says with a smile. Eventually, she applied to the University of Colorado to study astrophysics. There, in the Laboratory for Atmospheric and Solar Physics, she began her study of the solar atmosphere.

But Guhathakurta was always interested in seeing the big picture. "I really like bringing people together from the different subdisciplines because it's getting the overview that makes science really fun! People should always try to remember that gaining a grasp of the whole is why we do what we do as researchers."

"Then things get really weird. . . . The Sun's entire field has flipped!"

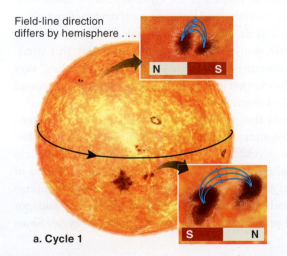
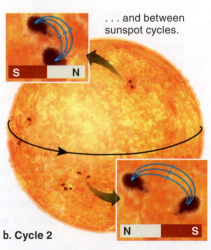

Field-line direction differs by hemisphere . . .

. . . and between sunspot cycles.

a. Cycle 1

b. Cycle 2

Figure 10.17 Alternating Polarity of Sunspots

As the polarity of the entire Sun reverses with each sunspot cycle, the polarity of sunspot pairs by hemisphere reverses as well.

In the new cycle, the orientation of magnetic north and south for the spots in the northern hemisphere is the same as what was seen in the southern hemisphere in the previous cycle, and vice versa. "But it's not just the sunspots that change their magnetic orientation," says Guhathakurta. As the Sun moves from one sunspot cycle to the next, the polarity of the entire Sun reverses. The solar north magnetic pole becomes the south pole, and the south magnetic pole becomes the north pole. "The Sun's entire field has flipped!" exclaims Guhathakurta.

This large-scale reorientation of the Sun's magnetic field is called the **solar magnetic cycle**. It takes two full sunspot cycles (22 years) for the Sun to go through a single magnetic cycle, returning the "north" magnetic pole to its starting orientation.

Solar Dynamo

What explains the Sun's strange magnetic behavior? We've seen how planets generate their own magnetic fields. "Magnetic fields are produced by electric currents," says Guhathakurta. "These currents are generated within the Sun by the flow of the Sun's hot, ionized gases. There are strong flows of plasma on the Sun's surface and within its interior. Nearly all of them contribute in some way to the production of the solar magnetic field."

Any time a charged fluid creates a magnetic field, astronomers say that a dynamo is at work. The solar dynamo works in a similar way to planetary dynamos but has a much shorter timescale for dramatic changes, such as field reversals. "The Sun's field reverses every 11 years," says Guhathakurta, "but the same process can take 100,000 years or more for Earth."

The origin of the solar dynamo rests in the Sun's differential rotation. Recall from Section 10.1 that different regions of the Sun rotate at different speeds. Material near the solar poles completes one full rotation in about 34 days, but material at the equator makes a complete circuit in just under 26 days. In addition, material at different distances from the Sun's center also rotates at different speeds. In particular, plasma at the outer edge of the radiative zone rotates faster than material just above it in the convective zone.

"Magnetic fields within the Sun get stretched out and wound around the Sun by this differential rotation," says Guhathakurta. "It can be useful to think of magnetic-field lines as rubber bands. They're continuous lines of force which, like rubber bands, can be strengthened by stretching and twisting." The magnetic fields increase in strength as they are stretched because energy in the motion of the gas gets converted into energy of the magnetic field.

Imagine magnetic-field lines that run from one pole to the other along the solar surface. Because of differential rotation, the plasma at the equator rotates faster than the plasma near the poles, so the initially straight field lines are soon stretched out along the Sun's waist (**Figure 10.18**). The lines that used to run pole to pole now have a section that's parallel to the equator. As the Sun continues its differential rotation, this winding of the field lines becomes more pronounced. "The Sun's differential rotation," says Guhathakurta, "can take a north-south-oriented magnetic-field line and wrap it around the whole circumference of the Sun in just about 8 months."

This solar dynamo model predicts that the winding of field lines by differential rotation causes the reversal of the sunspot magnetic-field orientation from the northern to the southern hemisphere. Eventually, the winding makes the solar field so complex that it can no longer maintain itself. The large-scale magnetic field of the Sun becomes unstable because of the complex overlaps that the field windings form. Each disruption of the large-scale field marks the end of the 11-year sunspot cycle and either the midpoint or the end of the 22-year solar magnetic cycle. Since

solar magnetic cycle A large-scale reorientation of the Sun's magnetic field occurring every 22 years.

prominence A giant arc of magnetism anchored in the photosphere.

flare An eruption that launches material from the solar surface.

reconnection The breaking and re-joining of magnetic-field lines.

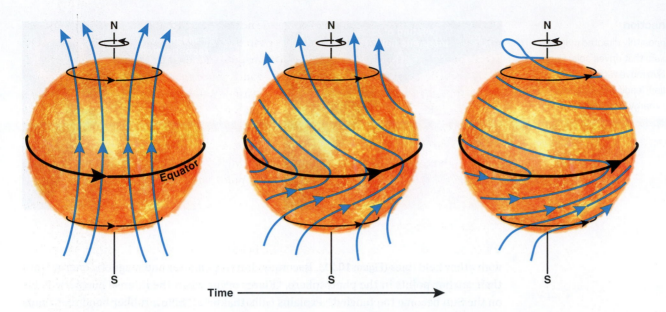

Figure 10.18 Winding of the Sun's Magnetic Field
Because of the Sun's differential rotation, the magnetic field is twisted during the course of a cycle. The field lines become buoyant and rise to the surface as loops, which then define the Sun's large-scale field. Eventually, the field structure's complexity collapses and another dynamo cycle begins.

the solar dynamo model so neatly predicts what is observed, astronomers feel that, despite many unanswered questions, the model gives them a good basis to explain solar magnetism.

Prominences, Flares, and Coronal Mass Ejections

Sunspots come and go in a matter of weeks, and granules appear and fade in 10–20 minutes. "From just these two types of phenomena," says Guhathakurta, "you can see that the photosphere is *active*." But there are forms of solar activity with far more power and violence than either granulation or sunspots have. Prominences, flares, and coronal mass ejections (CMEs) are three related classes of activity. While they differ in behavior, they are all driven by powerful solar magnetic fields.

Prominences (**Figure 10.19**) are giant arcs of magnetism whose legs are anchored in the photosphere. Smaller prominences can be tens of kilometers long and extend into the chromosphere; the largest ones stretch across much of the Sun and rise far into the corona. Two classes of prominences are observed. *Eruptive prominences* explode from the photosphere, only to collapse back down in just a few hours. *Quiescent prominences*, in contrast, maintain their arched magnetic fields for many days.

"Most of the solar atmosphere is so hot that the gas is primarily in the plasma state," says Guhathakurta. "That means electrons are no longer bound to atomic nuclei. The gas is now made up of separated charged particles, mostly protons and electrons." Recall from Chapter 6 that magnetic forces trap charged particles by redirecting the particle motion to spiral around the magnetic-field lines. "When arcs of field erupt from the surface, the plasma is carried along with them. Charged particles making up the plasma reach the top of the magnetic arcs and then drain back down, emitting radiation as they get heated by collisions." The collisions typically drive temperatures in a prominence to 70,000 K. Emission lines produced by this hot gas are one means for astronomers to observe and analyze prominences.

Unlike prominences, **flares** launch material away from the solar surface into space (**Figure 10.20**). "Solar flares are essentially huge explosions on the Sun," says Guhathakurta. In a prominence, the magnetic fields remain tied to the surface, but in a flare, a process called **reconnection** occurs in which the old field lines simultaneously break and re-join

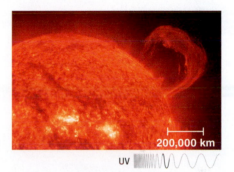

Figure 10.19 Solar Prominence
A solar prominence, as captured by the *SOHO* spacecraft in 2008. An arc from the Sun's magnetic field erupted from the surface, briefly suspending ionized gas along with it.

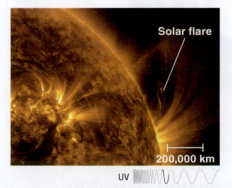

Figure 10.20 Solar Flare
Powerful magnetic forces above an active region on the Sun spew plasma into space. The plasma in this 2014 solar flare comprised a volume larger than several Earths.

Figure 10.21 Magnetic Reconnection

The breaking and reconnecting of oppositely directed magnetic-field lines in a plasma is the mechanism that drives solar flares. As magnetic fields anchored in the Sun move with the solar plasma, field lines can become tangled, leading to reconnection events that release matter and energy, which escape into space.

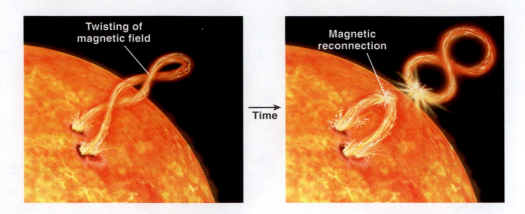

"Solar flares are essentially huge explosions on the Sun."

with other field lines (**Figure 10.21**). Reconnection frees matter and magnetic energy from their anchor points in the photosphere. "Flares occur when the intense magnetic fields on the Sun become too tangled," explains Guhathakurta. "Like a rubber band that snaps when it is stretched and twisted too far, the tangled magnetic fields break and reconnect, releasing huge energies in the process."

In a reconnection event, magnetic-field lines running in opposite directions (north and south) are squeezed close enough together that they cancel each other out. The energy that was locked up in the magnetic field is rapidly converted into light, heat, and motion. Flares are remarkably violent events. "Just for perspective," says Guhathakurta, "the energy emitted by even a small solar flare is more than a million times greater than the energy from a volcanic explosion on Earth." Much of the energy in a flare is converted to radiation across a wide range of wavelengths. "Although solar flares can be visible in white light," says Guhathakurta, "they are often studied via their bright X-ray and ultraviolet emissions."

While much of the energy in a flare travels upward (some of which will escape into space), the flare also affects the photosphere. Astronomers have observed powerful "sunquakes" propagating away from flares, in the same way an earthquake travels away from a slipped fault line.

Coronal mass ejections, or **CMEs** (**Figure 10.22**), are particularly potent phenomena associated with solar flares. "Coronal mass ejections are explosions in the Sun's corona that spew out huge amounts of matter and energy," says Guhathakurta. "And they can be particularly dangerous when they hit Earth." This danger is ever present, even though only a small fraction of CMEs do hit Earth.

CMEs appear as a kind of superflare, but their exact connection with the lower-energy phenomena is still unclear. "Coronal mass ejections often accompany solar flares, but not always," explains Guhathakurta. Astronomers are sure of the connection between CMEs and solar magnetic fields, however. "CMEs and solar flares explode outward from the vicinity of magnetically active regions on the Sun. They are most common during times of peak solar activity, the so-called solar max years of the sunspot cycle." A typical CME will drive 10 billion kg of plasma into space, along with energy equivalent to that of a flotilla of 220 aircraft carriers moving at 500 km/s. "That," says Guhathakurta, "is a whole lot of energy."

Explosions on the Sun that prove devastating to Earth may sound like the stuff of cheesy disaster movies, but they can pose a real threat. **Space weather**, a new term in the lexicon of science, refers to the CMEs blown off the Sun that periodically sweep across interplanetary space (**Figure 10.23**). A particularly potent example occurred on March 13, 1989, when a severe space storm caused a system-wide power failure in Quebec. The time from the onset of the storm as it hit the Earth to total blackout was 90 seconds, and 6 million people were left without power.

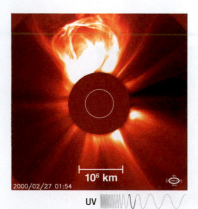

Figure 10.22 Coronal Mass Ejections (CMEs)

A CME captured by the *SOHO* satellite in February 2000. The Sun is blocked so that the light from the corona is visible. (The white circle marks the size of the Sun.) Powerful CMEs can spew a billion tons of matter into space at speeds of millions of miles an hour.

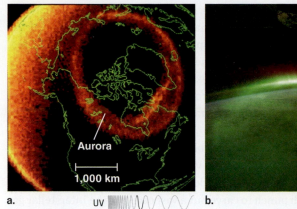

Aurora

1,000 km

a.   UV

b.   VIS

1,000 km

Figure 10.23 Space Weather
Storms caused by CMEs traveling through interplanetary space comprise space weather. **a.** Viewed from high orbit, charged particles from a CME strike Earth and work their way into the atmosphere along magnetic-field lines, creating a large oval aurora (shown in false color). **b.** The International Space Station's view of an auroral glow from charged particles flowing in the atmosphere.

Before we became a high-tech culture, the collision of a CME with Earth was no cause for alarm. The result would, at best, produce beautiful auroral displays. (Recall from Chapter 6 that auroras are emissions of light caused by the flow of charged particles along Earth's magnetosphere.) Today, in a worst-case scenario, the ionizing radiation could knock out orbiting satellites and space stations, possibly proving fatal to anyone working on them.

Space weather represents a potent example of how science and technology raise new issues for politicians and policy makers. The growth of our high-tech society—and, in particular, our reliance on satellites for everything from communications to weather prediction—means that as a society we must consider the dangers that space weather poses. Policy makers who are not scientists themselves must decide whether and how to dedicate scarce resources to prepare for such risks. In such situations, scientists and politicians must find ways to work together to develop optimal solutions to new problems.

The hardest part, however, is to make accurate predictions. "Those supercomputer calculations of a CME's path can take longer to run than the time it takes the storm to reach Earth," says Guhathakurta, who notes that approximately 3 days are required for a CME to get from the Sun to the Earth. "That means predicting space weather is going to remain an inexact science for some time to come." Still, it's a measure of how far we have come as a species that even the weather in space has become our daily concern.

Solar Wind

Flares and CMEs are distinct events in which matter from the Sun is launched into space. Though they are dramatic, they are not the only way the Sun loses mass. Every moment of every day, the solar wind, a continuous stream of particles, is blown off the Sun (see Chapter 5). But don't let the word *particles* fool you. All of those solar protons and electrons add up. "The Sun is flinging more than a million tons of matter into space every second," says Guhathakurta. "And with typical speeds in the solar wind ranging from 200 to 500 km/s (that's about a million miles per hour), it all adds up so you get a lot of matter and a lot of energy plowing into interplanetary space every day."

Like so much else in the Sun, the solar wind is related to magnetic fields and their ability to trap charged particles. "Near the Sun's equator, magnetic-field lines are more likely to loop back on themselves and reenter the Sun's surface," says Guhathakurta. "These 'closed' field lines trap hot coronal gases that emit X-rays. The particles are trapped within the magnetic field and can't escape into space." In X-ray observations, the closed field lines show up as bright regions of intense emission clustered near the Sun's equator.

The situation is reversed near the Sun's poles. There, the large-scale magnetic field does not arc back to the Sun but extends out into interplanetary space through regions called **coronal holes**. As in a deflating balloon, thermal pressure is high enough in these regions to launch hot gas out along the open magnetic-field lines and produce the solar

"The Sun is flinging more than a million tons of matter into space every second."

coronal mass ejection (CME) A large-scale ejection of mass and magnetic energy from the Sun's corona.

space weather Coronal mass ejections sweeping through interplanetary space, which produce auroras when they hit Earth and may pose threats to astronauts and technological systems.

coronal hole A region of low density where the solar wind is generated.

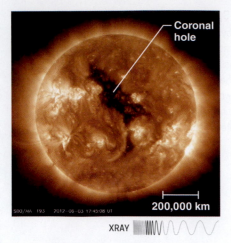

Figure 10.24 Coronal Holes
A coronal hole imaged in June 2012. Coronal holes appear dark in X-ray light because their densities are lower than the surrounding bright regions, which indicate hotter areas of the solar corona, mainly above the sunspot regions.

wind. The term *holes* comes from X-ray observations, in which coronal holes appear dark (**Figure 10.24**). That means the X-ray coronal holes have low emission because their gas densities are low. Note that coronal holes and sunspots are not, in general, connected, since the scale of the magnetic-field variations and their locations are much different.

The formation of solar wind is also part of the expansion of the solar corona. Recall that the corona is the only layer of the Sun not in hydrostatic equilibrium. A component of the solar wind comes from the expansion of the corona's outer regions into interplanetary space.

The particles of matter composing the solar wind blow through the entire Solar System. By our terrestrial standards, however, the wind is not very dense. "The solar wind is pretty thin," says Guhathakurta. "It only has about six particles per cubic centimeter near Earth." Even so, the solar wind packs enough punch to carve out a vast cavity in the interstellar gas that surrounds the Solar System. During the Sun's lifetime, the solar wind has pushed all interstellar gas out to a radius of 15 billion km. Astronomers call this solar wind–dominated cavity the **heliosphere** (**Figure 10.25**). Its boundary is called the **heliopause**. Only recently (August 2012) did the *Voyager 1* spacecraft, launched in the late 1970s, reach the heliopause, transmitting back valuable data about this last stand of the Sun at the edges of interstellar space and providing proof of its location, 18 billion km from the Sun.

Even though the solar wind becomes less dense as it flows outward, it is still an important player in the outer Solar System. For example, solar wind interacts with the magnetic fields of all the outer planets, shaping them in much the same way it does for Earth. And over long timescales, solar wind particles hitting distant Kuiper Belt objects drive changes in their surface properties.

Section Summary

- Sunspots are regions of the photosphere with a powerful solar magnetic field that follow a repeating 11-year cycle. At the end of each cycle, sunspot groups reverse polarity as part of the solar magnetic cycle, a longer-term periodic reorientation of the Sun's magnetic field.
- The Sun generates magnetic fields through a dynamo process driven by differential rotation with both latitude and depth. As magnetic-field lines wind ever tighter around the Sun, areas within the convective zone become buoyant, rise up, and create sunspots on the surface.

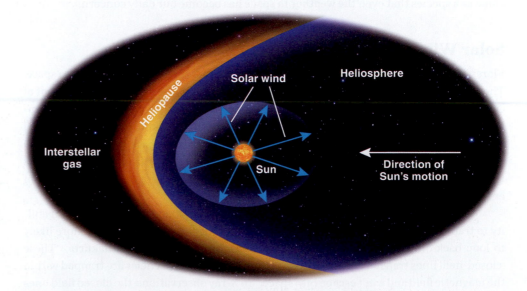

Figure 10.25 The Heliosphere
The outward-flowing solar wind creates this cavity in the interstellar gas surrounding the Solar System. The heliopause marks the boundary between solar wind and the swept-up interstellar gas. The inner bubble around the Sun is a region where the solar wind begins decelerating as it pushes against the interstellar gas farther out.

heliosphere A cavity, created by the solar wind, in the interstellar gas surrounding the Solar System.

heliopause The outer boundary of the heliosphere.

- Prominences are giant arcs erupting from the solar surface. Flares are violent events that launch material into space when magnetic-field lines are broken via reconnection. Coronal mass ejections (CMEs) are the most powerful solar eruptions; when they sweep through interplanetary space, these storms, known as space weather, can damage satellites.
- The Sun continuously blows particles from its surface at high speed. This solar wind creates the heliosphere, a cavity in space that surrounds the Solar System.

CHECKPOINT   What do sunspots, the solar magnetic cycle, and solar activity in terms of prominences, flares, and CMEs have in common?

CHAPTER SUMMARY

10.1 Living with a Star

The Sun, our closest star, contains more than 99 percent of the Solar System's mass. It rotates differentially—faster at the equator than at the poles. The Sun and its atmosphere comprise six layers: core, radiative zone, convective zone, photosphere, chromosphere, and corona. Density continually decreases from the core to the outer layers; temperature decreases as well, except in the hot corona. The Sun is in hydrostatic equilibrium, meaning that the inward crush of gravity is balanced by the outward push of pressure, generated by nuclear fusion. Helioseismology is used to study the Sun's inner workings.

10.2 The Sun's Fusion Furnace

Advances in nuclear physics helped us understand the source of the Sun's energy. Stars generate energy by means of nuclear fusion in their cores. Fusion requires high densities and temperatures above 10 million degrees kelvin. In the proton-proton (P-P) chain characteristic of Sun-sized stars, hydrogen is converted to helium, releasing 0.7 percent of the hydrogen's mass as energy. Neutrinos are also released, and studying them has provided insight into the Sun's nuclear engine.

10.3 Moving Energy

Two processes transfer energy from core to photosphere. In the Sun's radiative zone, gamma-ray photons created in the core are scattered many times; each scattering eventually leads to the creation of multiple lower-energy photons. In the convective zone, gas heated from below (the top of the radiative zone) rises to the solar photosphere (the Sun's visible surface), where it cools by emitting photons and then sinks back downward. Solar granulation on the photosphere is caused by smaller convection cells.

10.4 The Active Sun: Photosphere to Corona and Beyond

Sunspots, caused by the Sun's magnetic-field lines, follow a predictable pattern in an 11-year cycle. The Sun's polarity reverses on a 22-year cycle. The Sun's differential rotation generates a solar dynamo. Magnetic activity on the Sun's surface causes eruptions of varying size and energy, characterized as prominences, flares, and coronal mass ejections. Space weather—surges of matter and magnetic energy from the Sun—can cause power failures and danger for satellites and astronauts in space. The constant solar wind, a stream of particles blown off the Sun's surface, escapes through coronal holes near the Sun's poles. The solar wind clears out a large cavity in the interstellar gas around the Solar System: the heliosphere.

QUESTIONS AND PROBLEMS

Narrow It Down: Multiple-Choice Questions

1. Which of the following statements about differential rotation is/are correct? Choose all that apply.
   a. The average speed varies with the sunspot cycle.
   b. The spin rate is slower at the poles.
   c. The Sun rotates at different speeds at different times of the year.
   d. Different layers of the Sun rotate at different speeds.
   e. The Sun's outer layers rotate at different speeds at different latitudes.

2. Choose the correct order of the Sun's layers from the center outward.
   a. corona, chromosphere, photosphere, convective zone, radiative zone, core
   b. core, magnetosphere, heliosphere, atmosphere
   c. atmosphere, heliosphere, magnetosphere, core, solar wind
   d. corona, chromosphere, convective zone, photosphere, radiative zone, core
   e. core, radiative zone, convective zone, photosphere, chromosphere, corona

3. Which layer(s) of the Sun do/does *not* decrease in temperature as distance from the core increases? Choose all that apply.
   a. radiative zone
   b. convective zone
   c. photosphere
   d. chromosphere
   e. corona

4. How will the composition of the Sun change over the next billion years?
   a. It will not change appreciably.
   b. There will be more hydrogen, more helium, and less of the heavier elements.
   c. The proportions of carbon and iron will increase.
   d. There will be less hydrogen and more helium.
   e. There will be less hydrogen and more of all the heavier elements.

5. Which of these statements about temperature, pressure, and density in all the layers of the Sun and its atmosphere is correct?
    a. The higher the temperature, the higher the pressure.
    b. Temperature decreases from the innermost layer to the outermost.
    c. The deeper below the Sun's surface the layer is, the higher the pressure.
    d. The higher the temperature, the higher the density.
    e. The values of these parameters are unknown for some layers of the Sun.

6. If a Sun-like star could not create energy by nuclear fusion,
    a. thermal energy from gravitational contraction would have sustained it until now, but its life expectancy would be much shorter.
    b. it would have exhausted its energy long ago.
    c. it would last longer because it would not be radiating away its stored energy.
    d. its antimatter and matter would have recombined, making it disappear.
    e. it could have sustained itself by nuclear fission, using its hydrogen as fuel.

7. What particle is the antimatter counterpart of an electron?
    a. positron
    b. boson
    c. neutrino
    d. proton
    e. neutron

8. Nuclear fission is readily achieved on Earth with current technology, but large-scale nuclear fusion is not, because
    a. temperatures required for fusion cannot be achieved and sustained.
    b. there is not enough hydrogen on Earth to fuel fusion adequately.
    c. unlike fission reactions, fusion reactions release deadly amounts of energy.
    d. nuclear fusion requires the reunification of matter and antimatter, but no antimatter is available on Earth.
    e. heavier elements required for fusion are not abundant on Earth.

9. Which statement about energy from nuclear fusion is correct?
    a. Nuclear fusion produces energy by annihilating matter, completely converting it to energy.
    b. The proportion of hydrogen converted to energy in nuclear fusion is 7 percent.
    c. The proportion of hydrogen converted to energy in nuclear fusion is 0.7 percent.
    d. Nuclear fusion requires more energy put in than it gives back.
    e. Fusion converts radio photons into gamma-ray photons.

10. The solar neutrino problem is accurately described in which of the following statements?
    a. The Sun lacks enough neutrinos to sustain fusion.
    b. Too many neutrinos are produced in solar fusion, and they are damaging to Earth.
    c. No neutrinos were found to come from the Sun.
    d. Experimental results found more solar neutrinos than were expected from models of nuclear fusion.
    e. Experimental results found fewer solar neutrinos than were expected from models of nuclear fusion.

11. A gamma ray is produced in the Sun's core. What happens next?
    a. It emerges intact from the photosphere.
    b. It is scattered many times as a gamma ray before emerging from the photosphere.
    c. It is scattered and converted into lower-energy photons many times before emerging from the photosphere.
    d. It remains intact in the Sun's interior.
    e. It undergoes nuclear fusion.

12. Which of the following statements about sunspots is true?
    a. They occur in predictable cycles.
    b. They are permanent features on the Sun's surface.
    c. They are a primary cause of climate change.
    d. They exist in the convective zone.
    e. Each is unrelated to nearby sunspots.

13. The Zeeman effect, in which some individual lines in a stellar spectrum are split into multiple lines, is caused by
    a. gravity.
    b. nuclear fission.
    c. nuclear fusion.
    d. gravitational contraction.
    e. magnetism.

14. Once the Sun's magnetic north pole is located at its geographic north pole, on average how many years will pass before it reverses polarity once?
    a. 11 years
    b. 22 years
    c. 33 years
    d. 44 years
    e. The answer is unknown.

15. Which statement(s) about nuclear fusion and nuclear fission is/are true? Choose all that apply.
    a. Both involve molecular reactions.
    b. Both release energy that had been in the form of mass.
    c. Both convert 100 percent of their fuel to energy.
    d. Both were unknown phenomena until about 100 years ago.
    e. Both require large magnetic fields.

16. What property of neutrinos allows them to pass through most matter?
    a. their very small size
    b. their zero mass
    c. their low probability of interaction with other particles
    d. their high temperature
    e. their immutability

17. A solar feature that lasts about 10 minutes is most likely to be a
    a. granule.   d. prominence.
    b. sunspot cycle.   e. CME.
    c. sunspot.

18. Which statement about the inputs to and outputs from nuclear fusion in the Sun is true?
    a. Outputs have lighter nuclei than their corresponding inputs.
    b. Only elements up to iron can be inputs.
    c. The total mass of outputs is less than the corresponding inputs.
    d. Outputs are always radioactive.
    e. Any element can be an output of nuclear fusion.

19. True/False: Nuclear fusion requires extremely high temperatures, so collisions of protons and neutrons occur at speeds high enough to overcome the strong nuclear force.

20. True/False: If the variance between rotation speeds of different regions of the Sun were less, the sunspot cycle would likely be longer than 11 years.

To the Point: Qualitative and Discussion Questions

21. What is the P-P chain?

22. What are the characteristics of a plasma?

23. The Sun was once considered to be perfect and changeless. Explain how that belief has been disproved.

24. Compare and contrast energy transport in the radiative zone and in the convective zone.

25. How is the Sun's elemental composition changing over time?

26. What layer of the Sun is visible to the naked eye? Explain why that is so.

27. What conditions of the hydrogen in the Sun's core are necessary to create the enormous amount of energy produced there?

28. What phenomenon within Earth is similar to the method of energy transport in the convective zone, and in what layer of Earth's structure does it occur?

29. Describe the cause of the sunspot cycle. In your explanation, discuss how it can be considered both an 11-year cycle and a 22-year cycle.

30. How does a solar flare compare with a CME?

31. Describe the relative quantity and movement of sunspots over the solar cycle.

32. What is the underlying cause of the solar dynamo?

33. What is helioseismology, and what useful information does it yield?

34. Would you expect that nuclear fusion of *any* element to another would convert 0.7 percent of the mass to energy? Explain.

35. What other phenomena are like the Sun's energy in that they are life-giving yet potentially harmful and destructive?

Going Further: Quantitative Questions

36. What is the Sun's mass in units of Mars masses? In units of Saturn masses? In units of (Earth's) Moon masses?

37. How many Mercury diameters are there in the Sun's equatorial diameter?

38. How long, in years, would it take a passenger jet from Earth to arrive at the Sun's photosphere, assuming the jet travels at 800 km/h?

39. How much energy, in joules, is released when 50 kg of hydrogen is converted into helium by nuclear fusion?

40. If 210 kg of hydrogen could be entirely converted to energy, how many joules would be produced?

41. The penumbra surrounding a particular star spot (the equivalent of a sunspot on a star other than our Sun) has a temperature of 5,000 K. The star spot itself has a temperature of 3,500 K. Comparing areas of equal size, how much brighter is the penumbra than the star spot?

42. You are viewing two light sources of the same size at the same distance. One is at 1,000 K and the other is at 2,400 K. How many times brighter is the hotter light source?

43. A solar granule is 25 times brighter at its center than at its edge, for the same size area. By what factor is the temperature of the center greater than that of the edge?

44. The temperature of a sunspot is 0.66 as high as the surrounding photosphere. What is the ratio of the sunspot's brightness to that of an equal-sized area around it?

45. Four kilograms of hydrogen is converted to helium by nuclear fusion. How much of it, in kilograms, remains as matter (and is thus not converted to energy)?

"THAT WAS HOW I WATCHED A STAR EXPLODE ONE NIGHT. I JUST HAPPENED TO BE THERE TO SEE IT"

11

Measuring the Stars

THE MAIN SEQUENCE AND ITS MEANING

The Life of the Stars

11.1 Jim Kaler remembers what astronomy was like before computers controlled telescopes and Google put all of the Universe's information at our fingertips. In those days, if you wanted to find an object on the night sky, you couldn't just plug in its coordinates; you had to find it yourself.

"That was how I watched a star explode one night," says Kaler. Now retired from the University of Illinois, Kaler was a professor of astronomy studying stars and their life cycles. He remains one of the world's experts on extracting stellar properties from stellar spectra.

"I had been following this one object for years," recalls Kaler. "It was a pair of stars orbiting so close to each other that they exchanged mass. We call it a *symbiotic binary system*." With approximately half of all stars in binary systems (**Figure 11.1**), the study of these objects can provide unique information about stellar evolution. The brightness of symbiotic systems changes in regular ways over time, as mass from one of the stars flows onto the other. By tracking the response of the star that was gaining mass, Kaler hoped to learn something general about the physics governing all stars.

"One night," Kaler continues, "I'm at the telescope and something just doesn't look right. The system was much too bright. I was sure something was wrong, so I closed the dome and climbed down into the pit where the telescope's instruments were mounted to make sure everything was working." He reopened the observatory dome and let the stars' light flow once more through the telescope. To his astonishment, the new measurements were the same as before. The instruments were fine. It was the binary system that had gone crazy.

"Usually stars change over very, very long timescales," says Kaler. "To understand how they evolve, we have to measure lots and lots of them at different points in their lives and then deduce their individual stellar life stories. But on that night, stellar evolution was happening at high speed. The atmosphere of one of my binary stars had exploded, and I just happened to be there to see it." Kaler had caught a star in the midst of an intense eruption driven by the mass it gained from its companion. It was the kind of direct view into the life of stars that most astronomers can only dream of.

← Cascading dust surrounds this pair of stars in a group called DI Chamaeleontis, actually a quadruple system that includes a second, unseen binary pair.

IN THIS CHAPTER

you will learn about the discoveries and tools leading to our understanding of stars and their evolution. After reading through each section, you should be able to do the following:

11.1
The Life of the Stars Explain how the discovery of stellar nuclear fusion allowed astronomers to conclude that stars have life cycles from birth to death.

11.2
Measuring Stars List the measured properties needed to understand stellar evolution, and describe the methods astronomers use to determine each of them.

11.3
From Observations to Explanations Explain the importance of the main sequence on the Hertzsprung-Russell diagram, and explain what it tells us about stellar evolution.

11.4
The March of Time: Stellar Evolutionary Tracks Describe how astronomers employ stellar models, tested using globular clusters, to further their understanding of stellar structure and evolution.

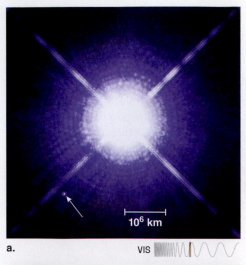

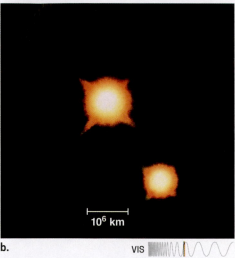

Figure 11.1 Binary Star Systems
a. Bright Sirius A and its white dwarf companion, Sirius B.
b. Sunlike Alpha Centauri A and B, whose smaller third companion, Proxima Centauri, is the closest star to the Sun.

Stellar Stories

For most of human history, people assumed that the Sun and other stars are eternal. The idea that stars have life stories—that they are born, live, and eventually die—seemed absurd. But the human perspective on the stars slowly changed, and with that change came the recognition that even stars evolve.

This idea caught on in the mid-1800s with the science of *thermodynamics*, the study of energy and its transformations. Steam engines—driving everything from trains to factories—had taught scientists the value of thinking about energy and the evolution of a "system." It didn't matter what the system was; any collection of interacting parts fit the bill. Thermodynamics provided laws that describe how transformations of energy from one form to another force systems to evolve. In a steam engine, for example, the chemical energy in coal is transformed into heat energy in steam and then transformed yet again into the train's motion.

By the beginning of the 1900s, astronomers using spectroscopy had discovered that stars are nothing more than giant systems of atoms. As such, stars must also be subject to the laws of thermodynamics. Forty years later the energy source powering the stars was discovered in nuclear fusion. Once it was recognized that stars stay alive by "burning" hydrogen fuel, it was also understood that sooner or later the fuel would run out and the stars would have to die. Stars *must*, therefore, have life stories.

But learning those stories proved a long and difficult journey for astronomers. As is always the case in science, the road began with data. To understand how stars evolve, astronomers had to find links between the properties they could measure and the unknown story of stellar evolution. First, however, they had to acquire accurate measurements of the important stellar properties and figure out which measurements mattered.

Our own journey through the life and death of stars will begin with that most basic of questions: Where, in terms of measurements and data, do we start?

Section Summary
- Stars were once thought to be unchanging, but astronomers now know that they evolve and have life cycles, including birth and death.
- Measurements of the properties of stars form the basis for understanding their life cycles.

CHECKPOINT How can a star be thought of as a system that transforms energy from one form to another?

Measuring Stars

11.2 "There are a handful of properties we need to build up the life story of stars," says Jim Kaler. "Some of them you get directly from measurements. Others get *derived* from measurements. That means you calculate them using other measurements as inputs." Specifically, in their attempts to understand stellar evolution, astronomers focus on six critical properties: distance, luminosity, temperature, size, mass, and composition (**Table 11.1**). One additional property—the velocity of a star through space—is also useful as astronomers attempt to build a picture of stellar evolution in the context of the Milky Way Galaxy as a whole.

How do astronomers get these measurements? You can't stick a thermometer in a star to get its temperature, and you can't run a tape measure around its equator to find its size. Astronomers must rely on their ingenuity and their knowledge of the laws of physics to make these measurements. And, in every case, they must use only the stellar light collected in their telescopes.

> "There are a handful of properties we need to build up the life story of stars. Some you get directly from measurements. Others get *derived* from measurements."

standard candle An object with known luminosity that can be used to measure distance.

Table 11.1	Stellar Evolution: Key Properties	
Stellar Property	**How Determined**	**Relationship Applied**
Distance	(a) Parallax (b) Standard candles: Cepheid variables (and others)	(a) Parallax relationship (b) Period-luminosity relationship
Luminosity	Apparent brightness and distance	Inverse square law for brightness
Temperature	Blackbody spectra	Wien's law
Size (radius)	(a) Observed size; distance (b) Luminosity and temperature	(a) Small-angle formula (b) Stefan-Boltzmann law
Mass	(a) Spectral lines (b) Orbital parameters of binaries	(a) Relative width (b) Newton's law of gravity
Composition	Spectral lines	Comparison to known elements, molecules, and compounds
Velocity	(a) Proper motion and distance (b) Spectral lines	(a) Small-angle formula (b) Doppler shift

The Inverse Square Law for Brightness

Before we begin our exploration of stellar properties, we need one of the most important tools astronomers have for learning about the Universe. Knowing the distance to an object can be critical to understanding its most basic properties. Think about the Moon and Sun. Both have the same angular size in the sky. Are they the same size? No. The Moon is just closer than the Sun, but you couldn't know that if you did not know their distances (**Figure 11.2**). To determine the distance to stars and other astronomical objects, astronomers have to be very clever.

One property of light—brightness—can be used to determine distance. This makes sense intuitively; as light radiates away from its source and travels farther and farther, its energy spreads across a greater area. The apparent brightness of a light source decreases with the square of distance from the source. This inverse square law for brightness looks just like the inverse square law for the force of gravity (see Chapter 3). A detector 2 meters from a lightbulb will record 4 times less energy than a detector 1 meter away (**Figure 11.3**). The brightness B of a light source, as it appears to you at a distance D from the source, depends on its luminosity L (its total energy output per time; see Chapter 4), as follows:

$$B = \frac{L}{4\pi D^2}$$

$$\text{brightness of light source} = \frac{\text{luminosity (that is, intrinsic energy output) of source}}{4\pi(\text{distance to source})^2}$$

The "100 watts" of a 100-watt lightbulb is a measure of luminosity.

This equation holds the key for astronomers trying to determine astronomical distances. For a class of objects whose luminosity L is already known, astronomers can compare that luminosity to the brightness B measured in telescopes and solve the equation for distance (see **Going Further 11.1** on p. 292). A **standard candle** is an astronomical light source whose luminosity L is known. If astronomers are able to designate an object as a

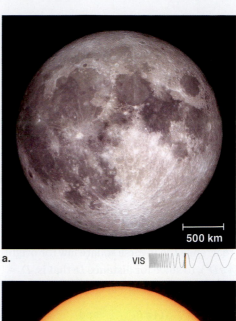

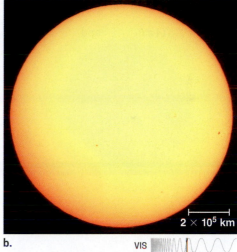

500 km

2 × 10⁵ km

Figure 11.2 Apparent Sizes of the Sun and Moon
a. The Moon and **b.** the Sun have approximately the same apparent size in our sky because of a coincidence: the Moon is 400 times closer while the Sun is 400 times larger.

Figure 11.3 The Inverse Square Law for Brightness
The geometry of a sphere helps illustrate why brightness decreases with the square of distance from the source. With each doubling of radius, the light must spread out over 4 times as much area A.

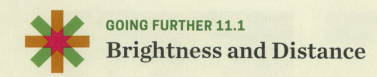

GOING FURTHER 11.1
Brightness and Distance

Imagine that astronomers discover two stars, X and Y, that have the same luminosity L. In the language of math, this means

$$L_X = L_Y$$

Now imagine that star Y *appears* 25 times brighter than star X when viewed from Earth.

$$B_Y = 25B_X$$

We can rearrange the inverse square law for brightness, $B = L/(4\pi D^2)$, to solve for distance D: that is, $D = \sqrt{L/(4\pi B)}$. We can apply that equation to both stars:

$$D_X = \sqrt{L_X/(4\pi B_X)} \quad \text{and} \quad D_Y = \sqrt{L_Y/(4\pi B_Y)}$$

When we divide the first equation by the second, the factors of 4π cancel out:

$$D_X/D_Y = \sqrt{(L_X/L_Y)(B_Y/B_X)}$$

We can now use the fact that $L_X = L_Y$ to cancel luminosity from the result, and then we can substitute $B_Y = 25B_X$ into this equation:

$$D_X/D_Y = \sqrt{(25B_X/B_X)} = \sqrt{25} = 5$$

Solving for D_X, we get

$$D_X = 5D_Y$$

That means star X is 5 times farther than star Y.

---

standard candle, then they can compare how bright it appears with how much light energy it emits (in terms of its intrinsic luminosity). That measurement of brightness can then be converted into a computation of distance by means of the inverse square law for brightness. It's extremely useful.

More than 100 years ago, astronomers discovered that certain kinds of stars behave as standard candles—they have properties that enable astronomers to determine the stellar luminosity. Now astronomers have also learned to use certain types of galaxies, supernovas, and other objects as standard candles. We will meet these objects in later chapters. For now, remember two important points: First, the behavior of brightness as indicated by the inverse square law is what makes standard candles possible. Second, standard candles are rare; astronomers are always on the hunt for new ones in order to map the true structure of the cosmos.

INTERACTIVE:

Parallax

Measuring Distance

The inverse square law for brightness provides a way to determine distance when a star's luminosity is known (see Going Further 11.1) or a way to determine a star's luminosity when the distance to it is known. But how do astronomers determine either of those properties on its own? Let's start with distance. "The distance to a star is often the most important measurement we make," says Jim Kaler. "Many of the properties we need for understanding stars are derived from combinations of other measurements. Distance appears in many of those formulas. If you get that wrong, then lots of other things will be wrong too."

"Distance is not really a property of stars in the same way as, say, temperature," continues Kaler. Some properties of stars are *relational*: they make sense only relative to other stars. Other properties, however, are *intrinsic*: they relate to the star itself. Distance is a relational property; temperature is intrinsic. Luminosity is an intrinsic property of a star, but its apparent brightness—how bright it appears from Earth—is relational.

A standard unit of distance on interstellar scales is the *light-year* (ly), which is the distance light travels in one year (see Chapter 1). Since light moves at $c = 3.00 \times 10^8$ meters per second (m/s) and there are about 3.16×10^7 seconds in a year, by multiplication we see that a light-year equals 9.46×10^{15} meters. However, astronomers tend to use another unit, the *parsec* (pc), where 1 pc = 3.26 light-years = 3.09×10^{16} meters. The parsec is based on parallax, the most direct method used to find distances to the stars.

> **"Distance is not really a property of stars in the same way as, say, temperature."**

a. **Right eye closed**   b. **Left eye closed**

Figure 11.4 Parallax
The apparent angular position of a foreground object changes relative to a more distant background as the observing position changes. Here an observer views her index finger in the foreground, relative to the background of the more distant building while closing first **a.** the right eye and then **b.** the left eye. Parallax is the angular change in position of the finger relative to features on the building.

USING PARALLAX TO DETERMINE DISTANCE. *Parallax* is the apparent change in position of a foreground object relative to a more distant background object when the point of view changes (see Chapter 2). If that sounds confusing, try this experiment: Hold your index finger about 6 inches from your nose. Now close your left eye and notice how your finger lines up relative to objects in the background. Then open your left eye and close your right. What happened? The relative position of the background has shifted to the left. The angular change in position of the background relative to your finger is the parallax that occurs as your point of view shifts from the right eye to the left eye (**Figure 11.4**). The parallax is directly proportional to the distance between your finger and your nose and the distance between your eyes. For example, if you move your finger to an arm's length from your nose, the parallax shift will decrease.

Astronomers use this principle to determine the distance to stars by making measurements 6 months apart—say, in July and January. Making these measurements at the maximum separation across the year is like shifting the point of view from one eye to the other, but now the "baseline" over which the two measurements are made spans the diameter of Earth's orbit (2 astronomical units, or 2 AU) rather than just a few inches (**Figure 11.5**).

If the object is much farther from Earth than the baseline distance of 2 AU, a simple formula for parallax relates the distance D to the object and its observed angular shift p relative to more distant background stars. If p is measured in arcseconds and D is measured in parsecs, then

$$p = 1/D$$

"The parallaxes of stars are small because they are so far away," Kaler explains. "That is why it took so long to be able to measure them. You need a pretty good telescope to resolve their tiny shifts on the sky" (see **Going Further 11.2** on p. 294). In 2013 the Gaia satellite was sent into orbit to measure the parallax of more than a billion stars with extreme precision, giving accurate distances (less than 10 percent error) out to thousands of parsecs. Gaia has been remarkably successful in filling in the three-dimensional map of vast regions of stars.

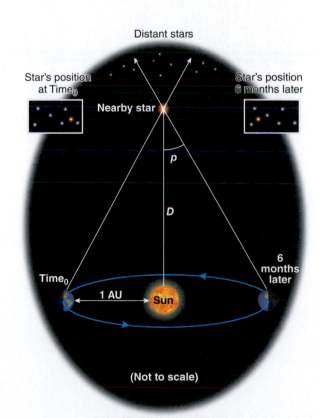

Figure 11.5 Stellar Parallax
In making stellar parallax measurements, the observing position shifts from one side of Earth's orbit to the opposite side 6 months later. Basic trigonometry then allows an accurate calculation of the distance D to a star, based on its angular change in position relative to the more distant background stars (its parallax p).

GOING FURTHER 11.2
Parallax

The parallax formula $p = 1/D$ comes from the basic geometry of triangles when one of the triangle's sides is much shorter than the other two. That is the situation when we view a star at two points in Earth's orbit separated by 6 months, because the diameter of Earth's orbit is much smaller than the distance between Earth and any star other than the Sun. (Figure 11.5 is not drawn to scale; in reality, distance D is much greater than Earth's orbital diameter of 2 AU.)

The formula for parallax shows that there is an inverse relationship between a star's angular shift seen on the sky (p, measured in arcseconds, or ") and the distance to the star (D, measured in parsecs):

$$p = 1/D$$

For a star that is 1 pc (3.26 light-years) away,

$$1'' = 1/(1\ \text{pc})$$

Recall that the Moon is about 1,800" across, as viewed from Earth. Therefore, to see a star 1 pc away requires a telescope capable of resolving shifts that are 1/1,800 the diameter of the Moon.

Let's put these calculations to practical use. Gliese 581 is a star with a planet known to be in its habitable zone. The parallax for Gliese 581 is 0.16". How far away is this star (and its planets) from us?

To calculate the distance D, we rearrange the parallax formula and substitute in $p = 0.16''$:

$$D = 1/p = 1/0.16'' = 6.25\ \text{pc}$$

Now we convert parsecs to light-years to find

$$6.25\ \text{pc} \times 3.26\ \text{ly/pc} = 20.40\ \text{ly}$$

Thus, Gliese 581 lies about 20 light-years away. A spaceship traveling at 10 percent the speed of light (something we're not even close to building) would take 200 years to reach its planets.

USING STANDARD CANDLES TO DETERMINE DISTANCE. To obtain precise distances to objects even farther than Gaia can measure, astronomers need other methods. One particularly powerful method relies on finding standard candles, which, as we saw, are objects whose luminosity is known. Luminosity is related to distance by means of the inverse square law for brightness. Standard candles are worth their weight in gold because they make distance measurements simple and can yield very accurate results. The best example of a standard candle comes from the study of variable stars.

While most stars evolve and their properties change over millions or billions of years, stars called **variable stars** change over much shorter intervals. These stars come in many forms, but the surface properties (temperature and luminosity) of all of them change within years or less (**Figure 11.6**). In most cases, these changes are quite regular and repeat themselves with well-defined periods. A variable star's period is the time it takes for the variation to cycle from low to high and back. "Variable stars are usually in some stage where their surface layers have become unstable," explains Kaler. "Their surfaces are ringing like a gong. The outer layers swell and shrink in cycles. The luminosity and surface temperature of the stars change in step with each repeat of the cycle."

Figure 11.6 Cepheid Variables

a. V1, a Cepheid variable star in the Andromeda Galaxy, is shown at four levels of brightness during one of its cycles. These variations occur as a result of instability in the star's internal structure. The variation in brightness is directly related to the Cepheid's energy output or luminosity. **b.** In graph form, the brightness of V1 cycles between its high and low points at regular time intervals.

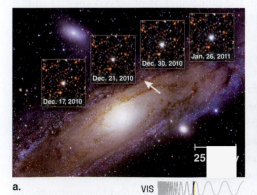

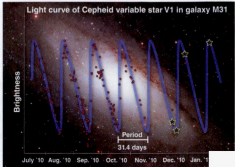

Cepheid variables constitute a class of variable stars that is of vital importance in astronomy because of their role in distance determination. Cepheids are bright stars that usually have 4–20 times the mass of the Sun. Over very well-defined periods of days to months, the brightness of an individual Cepheid changes by a factor of about 10.

The special properties of Cepheid variables were discovered in 1908, when Henrietta Swan Leavitt (**Figure 11.7**) was studying stars in the Magellanic Clouds, two small satellite galaxies that orbit the Milky Way. Because an object's apparent brightness decreases with distance and all stars in the Magellanic Clouds are at roughly the same distance from us, Leavitt could directly compare their luminosities, leading her to a huge discovery.

Leavitt was not officially an astronomer at the time but a "computer" hired by the Harvard College Observatory to process the enormous amounts of data being collected at the time. Not content simply to repeat calculations time and time again, Leavitt began asking her own questions. After trawling through thousands of images, she found a relationship between the average luminosity of a Cepheid and the time it took the star to cycle through its pulses of brightness. After exhaustively confirming her results, she published a paper announcing the **period-luminosity relationship** for Cepheid variables (**Figure 11.8**). Her studies gave astronomers a way of directly determining a Cepheid's intrinsic energy output (its luminosity) purely from an observation of its pulsation period.

Since Cepheids are fairly common, they soon became the quintessential standard candle. A simple measure of a Cepheid's period (P) yields its luminosity (L). Comparing the star's luminosity with how bright it appears on the sky yields an accurate measure of its distance. In addition, because Cepheids are so bright, they can be seen from very far away. If a Cepheid is found in another galaxy, astronomers can determine the galaxy's distance from us. Distance determinations with Cepheids became essential in the history of astronomy, providing the first true measure of the scale of galaxies and the Universe.

Other objects can act as standard candles as well. We will explore the topic of distance determination and its vital role in cosmology—the study of the Universe as a whole—in Chapters 16, 17, and 18.

Measuring Luminosity and the Magnitude Scale

Among a star's most important properties is its luminosity—its total energy output per time. Luminosity is the star's *intrinsic* energy output. Knowing a star's luminosity, we can determine how fast it uses its nuclear fuel. But astronomers can't measure luminosity just by looking at how bright a star appears on the sky. Remember from the inverse square law for brightness, $B = L/(4\pi D^2)$, that apparent brightness B decreases with increasing distance D. Brightness is measured as the energy per second per unit area falling on a detector. At a distance D from the star, brightness is the fraction of the star's energy that reaches Earth—the object's brightness that we see on the night sky. The inverse square law tells us that once we know the distance to a star, we can convert a brightness measurement into a luminosity measurement and hence stellar energy generation.

To make actual measurements of brightness, astronomers use a scale that dates back to the ancient Greeks. When Hipparchus made his first star maps, he categorized their brightness according to what's now called the **magnitude scale**. Hipparchus assigned each star to one of six magnitude classes, with the 20 brightest stars ranked at a magnitude of 1. The stars in the next-brightest group were assigned a magnitude of 2. He gave stars at the limit of detection by the naked eye a magnitude of 6.

Remarkably, astronomers still use this system, but they extended the scale in both directions, since Hipparchus did not include objects such as the Sun, which look much brighter to us than distant stars, or objects that can be seen only through telescopes, which are much fainter. The Sun has a magnitude of about −27 (the scale had to go negative for

Figure 11.7 Henrietta Leavitt's Contribution to Astronomy
Henrietta Swan Leavitt (1868–1921) worked as a "computer" at Harvard College Observatory, earning $10.50 per week. She performed meticulous studies of images of Cepheid variable stars on thousands of photographic plates.

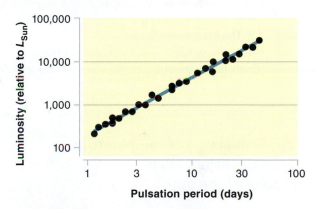

Figure 11.8 The Period-Luminosity Relationship
When Henrietta Leavitt plotted her observations of Cepheid variable stars in the Small Magellanic Cloud (SMC), she found an unmistakable relationship between the period of each star's variation and its luminosity at the brightest point in the cycle: the longer the period, the greater the luminosity.

variable star A star with regular periodic changes in its temperature and luminosity. Periods tend to be on the order of days to years.

period-luminosity relationship The correlation between the luminosity and period of a Cepheid variable star, allowing a star to act as a standard candle for determining distance.

magnitude scale A categorization of stellar brightness in which faint objects are given higher numbers than brighter objects.

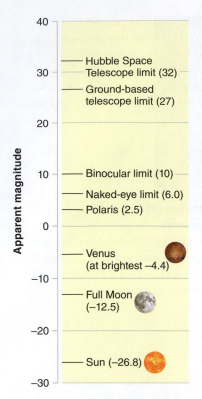

Figure 11.9 The Magnitude Scale
Astronomers use this system to measure brightness. Higher numbers indicate fainter objects.

Table 11.2 Magnitude and Brightness

Star	Apparent Magnitude	Absolute Magnitude	Distance (pc)
Sirius	−1.44	1.45	2.6
Procyon	0.4	2.68	3.4
Vega	0.03	0.58	7.7
Arcturus	−0.05	−0.31	11.1
Capella	0.08	−0.48	12.5
Achernar	0.45	−2.77	50
Canopus	−0.62	−5.53	100
Betelgeuse	0.45	−5.14	125
Rigel	0.18	−6.69	250
Sun	−26.72	4.8	5×10^{-6}

bright objects), and the faintest objects detectable by the Hubble Space Telescope have a magnitude of about 32 (**Figure 11.9**).

Astronomers call the brightness we observe from Earth *apparent magnitude* because it's the brightness that the object appears to have, given its distance. To determine *absolute magnitude*, which is a more direct measure of an object's luminosity, astronomers have to know the object's distance. The absolute magnitude tells us how bright the star would appear (in the magnitude scale) if it were seen from 10 pc away. **Table 11.2** gives the apparent magnitude, the absolute magnitude, and the distance for different objects in the night sky, including the Sun. The Sun's apparent magnitude is a whopping −26.72 because it's so close. Placed at 10 pc away, the Sun would appear on the sky with a brightness of just 4.8 in the magnitude scale—not very bright.

> "A crude measure of temperature can be taken just from the color of the star."

Measuring Temperature

After distance and luminosity, the next property astronomers need is a star's surface temperature, T_*. "A crude measure of temperature can be taken just from the color of the star," says Jim Kaler (**Table 11.3**). The trick is that stars are blackbodies. Recall from Chapter 4 that the spectrum of any blackbody is determined by its temperature, and Wien's law can be used to calculate the temperature from the peak wavelength (λ_{max}). Thus, by observing a star's spectrum, astronomers can measure its surface temperature (**Figure 11.10**). In practice, only the amount of light arriving in a few wavelength bands or "colors" needs to be measured. Comparing the energy emitted in these different wavelengths is often enough to estimate the shape of the blackbody curve and thus the surface temperature.

Table 11.3 Temperature and Star Color

Temperature (K)	Color
Greater than 30,000	Blue
10,000–30,000	Blue-white
7,500–10,000	White
6,000–7,500	Yellow/white
5,200–6,000	Yellow
3,700–5,200	Orange
Less than 3,700	Red

Measuring Size

The size of a star (its radius) is a key property needed for piecing together the story of stellar evolution because radii change as stars evolve. But how do astronomers determine the radius of a star that may be hundreds or thousands of parsecs away?

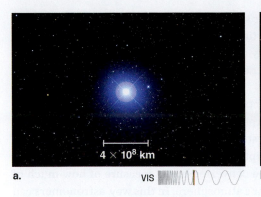

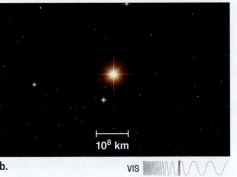

a.   VIS

b.   VIS

Figure 11.10 Star Color and Wien's Law
Because stars radiate as blackbodies, there is a clear relationship between the wavelength where the emission peaks and the surface temperature (Wien's law). Thus we can "take a star's temperature" by noticing its color. Contrast **a.** Rigel, which is hot and blue, with **b.** Arcturus, which is cooler and red.

"In some cases," says Kaler, "you get a direct measure of size by observing a star's angular diameter and combining that with a measure of its distance. If you have both measurements, you can then use the small-angle formula to convert the star's angular size into a physical size" (see Going Further 2.1). Very few stars, however, are close enough and large enough to appear as anything more than a point of light in even the most powerful telescopes (**Figure 11.11**).

Direct measurements of size are rare, but astronomers have a powerful, indirect means of determining a star's radius from its surface temperature and luminosity. To see how, let's start with the Stefan-Boltzmann law for the luminosity L of a blackbody at temperature T: $L = A\sigma T^4$, where A is the blackbody's surface area and σ is the Stefan-Boltzmann constant (see Chapter 4). Since stars are spheres, with radius R and area $A = 4\pi R^2$, we can rewrite this equation as

$$L = 4\pi R^2 \sigma T^4$$

When we rearrange terms and take the square root, this equation yields R:

$$R = \sqrt{L/(4\pi\sigma T^4)}$$

If two stars have the same temperature T, the one with higher luminosity is larger (has a larger radius) than the one with low luminosity. Conversely, if the luminosity L of the two stars is the same, the star with the lower temperature has the larger radius. Thus, the Stefan-Boltzmann law for blackbodies gives astronomers an indirect means of establishing a star's radius.

Reflect on what you've just learned. It is nothing short of astonishing that Earth-based scientists can use just the laws of physics to measure the size of a star that lies across trillions of kilometers of deep space.

Measuring Mass and Composition

At least two other significant properties of a star are needed to understand its evolution: its mass and chemical composition. "Taken together, those two properties pretty much set the entire life cycle of the star," says Kaler. "Mass and composition will tell you the path a star takes from birth to death."

SPECTRAL LINES. The width of spectral lines can, in some cases, give a measure of a star's mass. The width of stellar absorption lines can be attributed to the thermal motions of atoms in the star's atmosphere. The higher the pressure that atoms experience, the faster they

Figure 11.11 Measuring a Star's Size by Direct Observation
Most stars are so distant that they appear only as points of light, making direct size measurements impossible. A rare exception is **a.** the red giant Betelgeuse, in **b.** the constellation Orion. It is about 600 light-years away but large enough to appear as an extended object, allowing its radius to be directly measured.

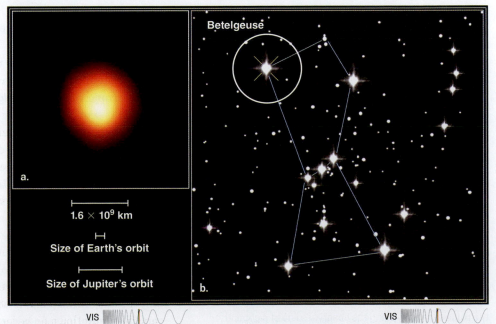

a.

1.6 × 10⁹ km

Size of Earth's orbit

Size of Jupiter's orbit

Betelgeuse

b.

VIS   VIS

"Mass and composition will tell you the path a star takes from birth to death."

JIM KALER

When Jim Kaler was 8 years old, he asked, "Grandma, why do stars all have points?" His grandmother told the young boy that stars don't have points. "Just look up and see for yourself," she told him (like a good scientist). "I did look up," says Kaler. "But this time I really saw! I was looking into the sky at what turned out to be the giant star Arcturus, and sure enough, stars didn't have points at all."

This experience drove Kaler to become an expert in the late stages of evolution of stars like the Sun. "I always viewed astronomy as a very aesthetic science," he says. "It's as much an art form to me as it is a science."

binary star One of two stars in orbit around a common center of mass. Binary stars are born together.

astrometric binary or **visual binary** Co-orbiting stars that have been detected by the proper motions of the components.

spectroscopic binary Co-orbiting stars that have been detected by the Doppler shift of their absorption lines.

eclipsing binary Co-orbiting stars in which one star is observed passing in front of the other.

proper motion An object's motion that is on the plane of the sky relative to other stars and can be detected by observations taken over time.

radial velocity An object's motion that is perpendicular to the plane of the sky along the observer's line of sight and can be detected via Doppler shift measurements.

bounce around. But the pressure on a star's surface is directly related to the gravitational pull of the star itself, which is proportional to the star's mass and radius. Since the radius can be determined if the luminosity and temperature are known, astronomers have learned to exploit the widths of spectral lines and in some cases use them to estimate a star's mass.

Composition is also measured by spectral lines. Recall from Chapter 4 that every element produces a "fingerprint" of absorption or emission lines. Astronomers must first identify all of the absorption lines in a stellar spectrum, connecting the lines to different elements in the atmosphere. The depth of the absorption lines (which indicates how much light has been absorbed) can be converted into a measure of how much of each element or molecule resides in the star's atmosphere. In this way, astronomers can build a complete census of the elements present in a star. "Most stars in our neighborhood of the galaxy have a chemical composition similar to the Sun's," says Kaler. "That means they are mostly hydrogen, a little bit of helium, and a tiny smattering of everything else."

Measurements of composition not only help astronomers build models of stellar evolution but also allow them to explore the galactic makeup of stars. "Different parts of the galaxy show different chemical abundances," says Kaler. "Since stars move quite far from where they were born over time, observations of a star's chemical composition can tell where in the galaxy the star was born. That helps us understand the history of star formation in the galaxy as a whole."

BINARY STARS. A more direct way to measure stellar mass comes from two stars orbiting each other, called **binary stars**. As Kaler's story at the beginning of this chapter suggests, not all stars are "loners." By taking a careful census of stars, astronomers have learned that at least half of all stars in the galaxy may be born as multiples, meaning two or more stars formed together that orbit their common center of mass. Binary star systems, in which two stars spend their lives locked in a gravitational dance, are the most common form of multiple system, accounting for half of the stars in our galaxy.

To determine the mass of a pair of binary stars, astronomers track the motion of the stars to obtain their orbital periods and orbital radii. They then use Newton's equations to calculate the total mass of the pair (Newton's version of Kepler's third law; see Chapter 3). From Newton's laws it can be shown that the ratio of the two stars' masses is the inverse of the ratio of their orbital radii. Therefore, if we observe star A and star B orbiting about their center of mass, then $R_A/R_B = M_B/M_A$. If the two stars' orbital radii are roughly equal, their masses are also roughly equal. And if one star's orbital radius is larger than the other star's, then its mass is lower.

How astronomers track the stars' motion in the first place depends on the type of binary system. From an observational standpoint, binaries are classified as astrometric, spectroscopic, or eclipsing (**Figure 11.12**). In **astrometric binaries** (or **visual binaries**), one star is seen moving back and forth around the other, and astronomers use images taken over many years to determine the orbital periods and radii. The motions of stars in **spectroscopic binaries** are detected via Doppler shifts (see Chapters 4 and 5) because absorption lines are blueshifted when either star is headed toward us in its orbit and redshifted when it moves away. In **eclipsing binaries**, one star passes in front of the other during an orbit, blocking out some of the light of the combined system. Astronomers use a combination of light curves and Doppler shifts to determine the motions of stars in this type of binary system. Notice that these three categories relate to how we can observe binary stars but say little about the stars themselves.

The category assigned to a binary has a lot to do with the orientation of the binary orbit in relation to the observer. Two stars can appear as an eclipsing binary, for example, only if their orbital plane is nearly lined up with the line of sight of our observations. In certain orientations, however, whether a system is classified as astrometric or spectro-

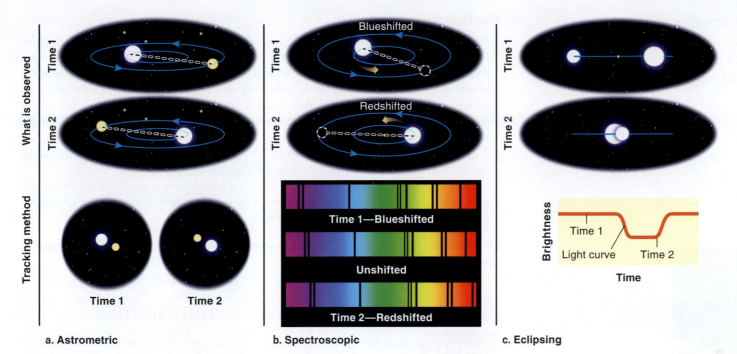

a. Astrometric   b. Spectroscopic   c. Eclipsing

Figure 11.12 Three Types of Binary Systems
a. In astrometric binaries, both stars are visible and the motion of one star around the other is directly observed. **b.** In spectroscopic binaries, Doppler shift of one star indicates its motion toward and then away from the observer as it orbits its unseen companion. **c.** In eclipsing binaries, the total light from the binary system decreases periodically as one star passes in front of the other to our line of sight.

scopic depends on whether the telescope used for viewing is powerful enough to resolve the stars' orbital motions.

Measuring Velocity

Astronomers are keenly interested in knowing a star's velocity—speed and direction—through space. (Like distance, velocity is a relational property.) Although velocity isn't used directly in modeling the lifetime of one star, measuring thousands of stellar velocities can help astronomers understand the galactic context of stellar evolution. "Our galaxy is constantly in motion," says Kaler, "and knowing stellar velocities now can sometimes help us figure out the history of galaxy-wide star formation and evolution in the past."

Two measurements are needed to determine a star's velocity: proper motion and radial velocity. If we watch a star long enough, we see it move across the sky relative to other stars. This movement, measured in arcseconds per year, is called **proper motion**. If astronomers know the distance to the star, they can use the small-angle formula (see Chapter 2) to convert proper motion to a space velocity measured in kilometers per second. But proper motion tells us how a star is moving in only the two-dimensional plane of the sky. Since space has three dimensions, we also need the star's **radial velocity**—its motion perpendicular to that plane, toward or away from us. To measure radial velocity, astronomers look for the Doppler shift of stellar absorption lines (**Figure 11.13**). If a line is blueshifted, the star is moving toward Earth; a redshifted line indicates that the star is moving away. Obtaining radial velocity requires a single measurement of Doppler shift, whereas proper motion measurements require multiple images taken over many years.

When information from proper motion (along with distance) is combined with Doppler shift data, astronomers can determine the full three-dimensional motion of the stars through space.

a. **Proper motion**

b. **Radial velocity**

Figure 11.13 Proper Motion and Radial Velocity
a. Proper motion is an object's change in position in the plane of the sky, relative to background stars, over time. **b.** Radial velocity is the velocity of the source perpendicular to the plane of the sky, measured by detecting the light's Doppler shift.

"Knowing stellar velocities now can sometimes help us figure out the history of galaxy-wide star formation and evolution in the past."

Figure 11.14 Annie Jump Cannon
As an employee of Harvard College Observatory, Annie Jump Cannon (1863–1941) created a powerful system of classifying stellar spectra that is still in use. The spectral classes are O, B, A, F, G, K, and M.

Figure 11.15 Stellar Spectra
Sample spectra (in visible wavelengths) for stars of each spectral class are shown. Because each element or molecule has a unique set of absorption lines, the spectrum of a star reveals its chemical composition (and temperature via Wien's law). The "fingerprints" of these lines dictate the star's spectral class. Spectral lines of hydrogen are identified at the top, and lines from other elements and ions, at the bottom. (Roman numerals indicate ionization levels.)

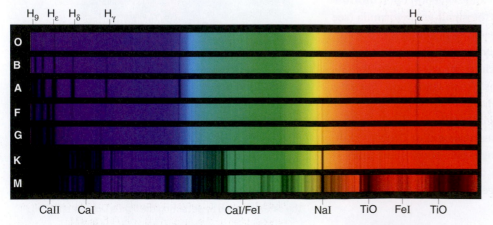

Section Summary

- An inverse square law relates brightness to distance, so astronomers can calculate the distance to objects by comparing their apparent brightness with their luminosity. Distance, a relational property, is measured for nearby stars via parallax, the apparent motion of an object against background stars. Standard candles (such as Cepheid variables), which have known luminosity, an intrinsic property, are used to measure distance to more distant objects.
- Brightness of distant objects is categorized on a magnitude scale; brighter objects are assigned lower numbers than dimmer objects.
- A star's peak spectral wavelength gives its temperature via Wien's law. A star's size can be derived from temperature and luminosity via the Stefan-Boltzmann law.
- Spectral lines are used to determine stellar mass and composition; stellar mass can also be calculated from binary stars. A star's three-dimensional velocity is measured from its proper motion and its Doppler shift, which yields its radial velocity.

CHECKPOINT Explain how a star's temperature can be measured via the peak spectral wavelength.

From Observations to Explanations

11.3 After measuring all of these stellar properties, astronomers can begin to deal with the theoretical side of stellar evolution. The data gathered from their observations provide the raw materials for identifying patterns and, ultimately, explaining those patterns.

From Data into Insight: The Hertzsprung-Russell Diagram

You can't explain the patterns in nature without finding them first. All good science begins with buckets of data; once the buckets have been gathered, the work of sifting through them begins. Facing down those difficult hours in search of patterns in the properties of stars was the task given to the indefatigable Annie Jump Cannon in 1884 (**Figure 11.14**).

While working at the Harvard College Observatory, Cannon was given the task of sorting through absorption spectra from hundreds of thousands of stars. Cannon used the presence and breadth of different absorption lines to sort her stellar spectra. In the end, she combined them into a **spectral classification** of seven categories denoted by the letters O, B, A, F, G, K, and M.

In Cannon's scheme, O stars, for example, have strong Balmer-series hydrogen lines and prominent helium, oxygen, nitrogen, and silicon lines. Category M stars show no hydrogen lines but do have lines from many molecules (most notably, titanium oxide). Eventually it was recognized that Cannon's scheme connects stellar temperature and the underlying blackbody emission. The bluish O stars are the hottest stars in her classification, and the redder M stars are the coolest (**Figure 11.15; Table 11.4**). The Sun is a class G star.

Cannon's spectral classification was a critical first step in understanding the story hidden in stellar properties. She provided a way of arranging stars into logical, property-based categories. The next step was to look for deeper patterns that would emerge from sorting stars into the categories. That task fell to Ejnar Hertzsprung and Henry Norris Russell (**Figure 11.16**), who both, separately, decided to relate Cannon's spectral classification to measurements of stellar luminosity. In 1910, the two men made a graph of spectral type (and therefore stellar temperature), plotted on the *x*-axis, versus stellar luminosity, plotted on the *y*-axis (**Figure 11.17**). What these diagrams revealed changed astronomy and our understanding of stars forever.

Table 11.4 Properties of Spectral Types

Spectral Type	Typical Mass (M_{Sun})	Typical Radius (R_{Sun})	Typical Temperature (K)	Peak Wavelength (nm)	Main Spectral Lines
O	60	12	42,000	69 (UV)	Helium II
B	5.9	3.9	15,200	191 (UV)	Helium I, hydrogen
A	2.0	1.7	8,180	355	Hydrogen, calcium II
F	1.4	1.3	6,650	436	Calcium II, hydrogen, metals
G	0.92	0.92	5,560	522	Calcium II, iron, other metals
K	0.67	0.72	4,410	658	Metals, CH (methyl-idyne radical), CN (cyanogen radical)
M	0.21	0.27	3,170	915 (IR)	Titanium dioxide

a. Ejnar Hertzsprung **b. Henry Norris Russell**

Figure 11.16 Hertzsprung and Russell
In 1910, **a.** Danish astronomer Ejnar Hertzsprung (1873–1967) and **b.** American astronomer Henry Norris Russell (1877–1957) independently graphed stellar luminosity versus temperature (though they used magnitude and stellar classification). Their work led to a powerful tool for understanding patterns in stellar evolution: the Hertzsprung-Russell or HR diagram.

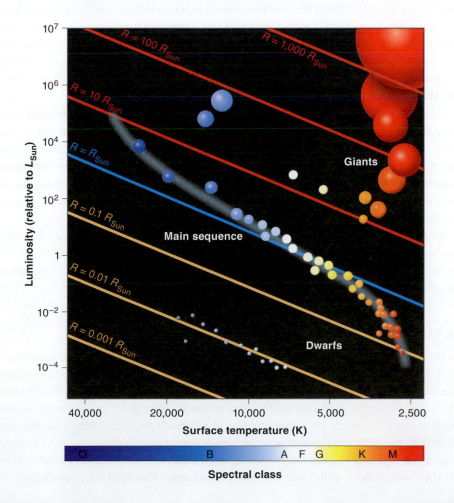

Figure 11.17 Hertzsprung-Russell Diagram
An HR diagram is a graph of luminosity versus surface temperature for many stars at many phases in their lifetimes. Notice that stars tend to "collect" in different parts of the graph. Most stars are typically found on the main sequence (the diagonal curved band running from upper left to lower right). Star radii can be found from their location relative to the diagonal lines of the diagram. The spectral classes are also shown, according to their temperatures.

spectral classification A fundamental categorization of stars, based on observed spectral lines, into seven categories denoted by the letters O, B, A, F, G, K, and M.

"Their surfaces are ringing
like a gong. The outer layers
swell and shrink in cycles."

Hertzsprung and Russell, essentially, sought a correlation between luminosity and temperature in stars. In a correlation, as one property (such as temperature T) changes, another property (such as luminosity L) changes too, in a regular and understandable way. If you start with a bunch of money in your wallet, then the number of times you take it out to buy something and the amount of cash left in it at the end of the week are going to be correlated. In contrast, the number of times you take out your wallet in a month and the amount of rainfall in Brazil that month are unlikely to be correlated. Perfectly correlated data might look like a straight line (an increase in quantity X is perfectly correlated with an increase or decrease in quantity Y). Uncorrelated data show points uniformly spread over the entire plot.

Before Hertzsprung and Russell made their graphs, no correlation between stellar properties such as temperature and luminosity was known for certain. "But they did see a correlation," says Jim Kaler. "Once they plotted spectral type versus luminosity, it was clear that the two quantities were strongly related for stars." Taking hundreds of thousands of data points and plotting them on the L-versus-T graph, Hertzsprung and Russell saw the vast majority of stars falling in a well-defined band, which astronomers call the **main sequence**, from high temperature and high luminosity to low temperature and low luminosity. Hertzsprung and Russell had no way of knowing why most stars "live" on the main sequence in their graph, now called the **HR diagram**, but they had clearly discovered a correlation that later astronomers with a better understanding of stellar physics would have to explain.

Though the main sequence is home to most stars, significant numbers of stars exist in other parts of the HR diagram. Using the Stefan-Boltzmann law, which links radius, temperature, and luminosity, astronomers categorized these stars by their physical size. In particular, astronomers group stars according to their radii. Stars above the main sequence are more luminous than their main-sequence cousins at the same temperature. These stars have large radii, and astronomers call them *giants* or *supergiants*, depending on their size. Stars below the main sequence are less luminous than main-sequence stars at the same temperature and are called *dwarfs*. (The term *dwarf* is also used for lower-mass main-sequence stars because they have low, low temperatures and, therefore, small radii. We'll explore this topic in Chapter 13.)

Astronomers realized another way to distinguish among stars, by **luminosity class**. It was known that the width of some absorption lines in a star's spectrum can be used to find stellar density in the star's outer layers. Those outer layers are expected to have lower densities in giants than in dwarfs. Using absorption lines, astronomers developed a luminosity class system in which roman numerals I to V are added to stars' designations. Giants are given low class numbers, and dwarfs have high class numbers. Thus, the supergiant Betelgeuse is in luminosity class I; the much smaller Sun is in class V.

With the HR diagram, stellar astrophysicists now had a pattern that could be used in modeling stellar evolution. What had begun as a giant stack of spectra was condensed to a few simple facts (main sequence, giants, etc.) expressed as a single graph that demanded explanation.

Making Model Stars

The HR diagram revealed the presence of distinct patterns, such as the main sequence, in stellar properties. To understand these patterns, however, astronomers began developing theoretical models of stars. The goal was to take the data they gathered from their observations and construct mathematical models of stars at any point in evolution. If these models capture the reality of stellar evolution, they should provide insight into why a pattern such as the main sequence appears in the HR diagram.

Crucial to understanding the evolution of stars is the fact that they are forever at war with their own gravity, and it is a war they cannot win. "The light that comes from the

main sequence The diagonal band on the HR diagram representing fusion of hydrogen in stars' cores. High-mass stars appear at upper left and low-mass stars at lower right.

HR diagram A graph of luminosity versus surface temperature (or spectral class) for a population of stars.

luminosity class A system for designating a star as a giant or dwarf based on absorption lines.

surface of stars tells you they are losing energy into space," says Didier Saumon, an astrophysicist who studies the stars at the end of their lives. "That energy comes from nuclear reactions in the core, but stars don't have an endless supply of fuel and that means stars have to change over time. They have to evolve."

From the moment a star begins nuclear fusion, its days are numbered. Its quantity of nuclear fuel is limited (and set at birth), even though the star may switch from fusing one kind of element to another as it ages. Eventually, the nuclear fuel runs out when no elements that can undergo fusion are left. Gravity wins the war; the star dies.

Tracking the history of a star as it fights this lifelong battle between gravity and fusion is the job of a theoretical astrophysicist. "We have a set of equations which come from the physics of stars," says Saumon. These equations describe how density, temperature, and pressure change from a star's core to its surface. They also describe how much energy is released in nuclear fusion reactions that drive the balance between gravity's pull (inward) and gas pressure's push (outward). This crucial balance of gravity and pressure at every layer of the star is *hydrostatic equilibrium* (see Chapter 10). As the star evolves, its structure changes to maintain equilibrium. "The equations can be pretty complex," explains Saumon, "but we can solve them using computers, and that's what we call a stellar model."

A *stellar model* is a description of how the properties of a star—such as density, temperature, chemical abundance, and degree of ionization—change with distance from its center. The model also calculates what the star's surface temperature and luminosity would be if astronomers could observe it on the sky.

Models are not reality, however. Mathematical equations solved on a computer are never perfect. Some effects are neglected because, for example, the computer may not be powerful enough to take them into account. The very first stellar models were simplistic and calculated by hand. They got better with early computers, but completing one model could take a whole week. Today it takes only seconds on a smartphone.

Still, the state of the art is always moving on. "The most current research is always about adding new processes that you may have ignored before for simplicity's sake. These make the calculations more realistic but also more complicated," explains Saumon. Even with the most powerful modern supercomputers, scientists cannot produce a fully three-dimensional model of a living star from core to surface that accounts for all detailed physical processes. Still, these models have revealed a great deal about the different types of stars and the deeper patterns that underlie their positions on the main sequence.

From Insight to Theory and Back: The Meaning of the Hertzsprung-Russell Diagram and Main Sequence

"If you took a random bunch of stars," says Harriet Dinerstein, a spectroscopist from the University of Texas, "you might guess that they would have an arbitrary combination of properties like brightness, radius, and temperature. After all, why not?" Why shouldn't there be just as many very large, very cool stars as there are very small, very hot stars, as well as everything in between? "Well," she explains, "it turns out that just is not how nature set things up."

The HR diagram shows astronomers that some properties of stars, such as luminosity and temperature, correlate very well. "The next question," says Dinerstein, "is what those correlations are trying to tell us. What, in other words, does the main sequence tell us about stars and their histories?"

The most important thing to remember about stars is that they evolve very slowly, over millions, billions, or in some cases even trillions of years. In general, we don't have the opportunity to watch any single star evolve. We have only snapshots of many stars that are at different points in their life histories. By assuming that we're catching stars at different points in their trek from birth to death, astronomers can turn the HR diagram into a map of stellar evolution.

"If you took a random bunch of stars, you might guess that they would have an arbitrary combination of properties like brightness, radius, and temperature."

INTERACTIVE:
Age, Mass, Luminosity

HARRIET DINERSTEIN

"I grew up in the middle of Manhattan," laughs Harriet Dinerstein. "I barely even saw the night sky." How did she catch the astronomy bug? "The Hayden Planetarium!" At Yale University, Dinerstein was one of only a handful of female students determined to find a career in science. "I guess I never looked back."

Dinerstein studies the chemical composition of gases thrown off stars at the ends of their lives. "Stars are always cooking up new mixtures of atoms through fusion," she says. Throughout her career, the possibility of discovery has always kept her tied to astronomy. "I love the excitement of identifying things that weren't known before."

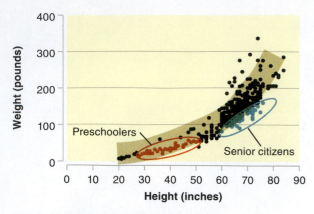

Figure 11.18 Height-Weight Correlation

Like the HR diagram, a plot of height and weight for humans shows a correlation: the points representing individuals lie along a curve rather than being randomly distributed. Extremes in age (preschoolers or senior citizens) are represented by distinct areas on the plot.

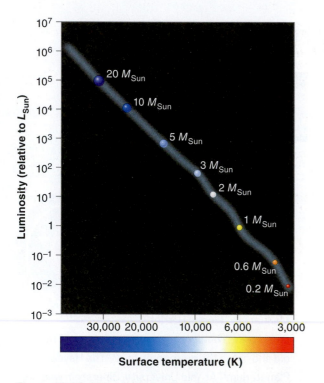

Figure 11.19 Mass and the Main Sequence

Stars spend most of their lives on the main sequence as they fuse hydrogen into helium in their cores. "Sequence" refers to stellar mass. High-mass hydrogen-fusing stars are at the top of the main sequence; they are bright and hot because their weight makes them fuse hydrogen rapidly. Low-mass hydrogen-fusing stars are at the bottom.

Construct an HR diagram for a cluster of stars as a survey contract, earning some credits for your starship while keeping an eye on the Corporation and its suspicious plans for the galaxy.

"An analogy with people can be helpful," says Dinerstein. "Imagine you went to the mall and recorded people's heights and weights. Then you made a plot of your data with height on one axis and weight on the other. Most people's height and weight go roughly together with some variation, but generally increasing or decreasing together," says Dinerstein. "That means your plot would show them to be correlated. Tall people would tend to be heavier than shorter people."

But what would happen if you repeated that experiment at a preschool? You would see only short, low-weight data points. And at a center for senior citizens, you would tend to see mid-height, low-weight data (**Figure 11.18**). Age matters.

The HR diagram is powerful for a similar reason. By taking snapshots of many stars and plotting them on the HR diagram, we get a clear picture of the patterns associated with all points in stellar evolution. Where they spend most of their lives, in terms of temperature and luminosity, is where we expect most of them to show up on the diagram. As it turns out, most stars appear on the main sequence. But the main sequence is definitely not a sequence in evolution. Instead, it's a snapshot of where most stars spend most of their lives. The main sequence shows stars of different masses appearing at their own positions along that band—high masses toward the top and low masses toward the bottom.

Our Sun is considered to fall in the low- or intermediate-mass stellar range. On the HR diagram, it plots on the main sequence. From their stellar evolution models, astronomers came to understand that the Sun is a middle-aged star that uses hydrogen fusion to support itself against gravity. As astronomers became more adept at studying the structure of stars through theoretical models, it became clear that all stars on the main sequence are also hydrogen fusers. Because hydrogen is the fusion fuel every star will rely on for most of its life, most stars plot on the main sequence. (Similarly, in the height-versus-weight analogy, most people you would encounter at the mall are between 18 and 65 years old.)

But why does the main sequence appear as an extended band from hot, bright stars to cool, dim ones? Why don't all hydrogen-fusing stars have the same luminosity and temperature and so appear in the same place on the HR diagram? The answer came when scientists learned more about the physics of nuclear fusion as applied to their model stars. "By solving equations for the structure of stars, we can calculate how fast nuclear reactions take place in the core," explains Didier Saumon. "The rate of nuclear burning is very sensitive to temperature in the core. Higher temperatures mean much faster nuclear burning." But core temperatures depend on the amount of matter squeezing down on the core. "That means the nuclear burning rates really depend on the star's mass," says Saumon. More mass means more fusion, which means both brighter stars and higher surface temperatures; that is why the "sequence" part of the main sequence reflects mass. Very luminous stars at the upper-left edge of the main sequence are very massive hydrogen-fusing stars. Stars with very low luminosity, at the lower-right edge, are hydrogen-fusing stars of very low mass (**Figure 11.19**).

The link between the main sequence and age goes beyond simply knowing that a star is in its hydrogen-fusing phase. As astrophysicists learned more about stellar nuclear fusion, they found a simple relationship between the life expectancy of a star and its mass. "Massive stars on the main sequence are burning up their fuel much faster than low-mass stars," says Saumon "That means massive stars have much shorter lifetimes. They run out of fuel faster because they burn so bright."

Astronomers developed a formula, the **life expectancy–mass relationship**, that expresses this relationship for stars (**Table 11.5**). **Going Further 11.3** explores the consequences of the life expectancy–mass relationship. It's worth noting that a massive star of 100 solar masses (100 M_{Sun}) lives for only 100,000 years (1/100,000 as long as the Sun is expected to last) and intermediate-mass stars live for 500 million to 15 billion years, while

GOING FURTHER 11.3
The Life Expectancy–Mass Relationship

A swimmer at the bottom of a deep pool feels the pressure of all the water above her. Similarly, the hydrogen plasma at a star's core feels the gravitational force—the weight—of all the matter in the layers above it. The pressure of the overlying material heats the core plasma to fusion temperatures, so hydrogen fuses into helium. The more massive the star, the higher the core temperature and the faster the hydrogen fuel fuses into helium. This is the story behind the life expectancy–mass relationship:

$$t_{ms} = 10^{10} \, \text{yr} \left(\frac{M_*}{M_{Sun}} \right)^{-2.5}$$

where t_{ms} is the amount of time a star spends on the main sequence, M_* is the star's mass, and M_{Sun} is the Sun's mass.

Let's see if we can understand what this formula tells astronomers. First, let's assume the star has a mass equal to the Sun's: $M_* = 1 \, M_{Sun}$. If we plug this value into the formula, we find that the time on the main sequence, t_{ms}, is 10^{10} years. This tells us that a Sun-like star will spend 10^{10}, or 10 billion, years on the main sequence before it uses up its hydrogen. Now let's use the formula to find the main-sequence lifetime of stars that are not Sun-like in mass.

For a star that is 10 times more massive than the Sun, $M_* = 10 \, M_{Sun}$, we get

$$t_{ms} = 10^{10} \, \text{yr} \left(\frac{10 \, M_{Sun}}{M_{Sun}} \right)^{-2.5} = 10^{10} \, \text{yr} (10)^{-2.5} = 3.2 \times 10^{7} \, \text{yr}$$

This somewhat massive star lasts only 32 million years, far less than the time elapsed since the age of the dinosaurs. For a star that is 1/10 as massive as the Sun, $M_* = 0.1 \, M_{Sun}$, we get

$$t_{ms} = 10^{10} \, \text{yr} \left(\frac{0.1 \, M_{Sun}}{M_{Sun}} \right)^{-2.5} = 10^{10} \, \text{yr} (0.1)^{-2.5} = 3.2 \times 10^{12} \, \text{yr}$$

This star will last 3 trillion years, longer than the current age of the Universe. We conclude that low-mass stars live a very long time.

a low-mass star of 0.08 M_{Sun} lives for 5.5 trillion years (more than 500 times longer than the Sun). That is quite a spread.

Section Summary

- Astronomers categorized the spectra from many stars into seven spectral classes, which also represented temperature classes.
- Hertzsprung and Russell independently correlated the spectral classes (representing temperature) with luminosity and created the HR diagram. The majority of stars plot on the HR diagram's main sequence, a diagonal band that reflects different stellar masses.
- By knowing the equations for the relevant physical processes, astronomers can build mathematical models that describe a star's structure from core to surface.
- All stars on the main sequence fuse hydrogen. Very massive, very luminous stars are much hotter, and use up their nuclear fuel much faster, than less massive stars and thus have shorter lifetimes.

CHECKPOINT How can a snapshot of stars at different points in their lives be translated into an understanding of the evolution of stars throughout their lives?

The March of Time: Stellar Evolutionary Tracks

11.4 The life expectancy–mass relationship is just one benefit of using stellar evolution models to understand a pattern, such as the main sequence, seen in the HR diagram. But other patterns in the HR diagram require explanations as well. For example, how do astronomers account for stars off the main sequence? How can the HR diagram be used to understand all of stellar evolution and the long middle age of stars?

Table 11.5 Star Masses and Life Expectancies

Mass (relative to M_{Sun})	Life Expectancy (yr)
0.08	5.5 trillion
0.25	300 billion
0.5	60 billion
0.75	20 billion
1	10 billion
2	2 billion
4	300 million
8	60 million
20	6 million
100	100,000

life expectancy–mass relationship The mathematical connection between a star's life expectancy and mass: the greater the star's mass, the shorter its life expectancy.

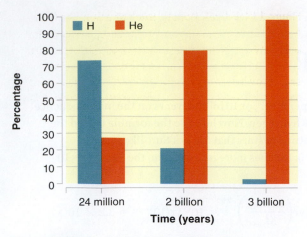

Figure 11.20 Stellar Model for a 3-M_{Sun} Star

Three points in the evolution of a star with 3 times the Sun's mass are shown. As hydrogen fusion in the core proceeds, the model describes how the relative amounts of hydrogen (H) and helium (He) change. By the time the star is 3 billion years old, its core is almost entirely composed of helium.

"We astronomers talk about the star 'moving' along the track as the star ages."

"That means globular clusters give us a natural controlled experiment to test our stellar evolutionary tracks."

stellar evolutionary track A model-generated path on the HR diagram representing the evolution of a star through time.

globular cluster A gravitationally bound cluster of 100,000 or more stars born at the same time.

Modeling Stellar Evolution

Theoretical astrophysicists are a lot like kids with Lego kits. They love to make models of things they study, turn them on, and see what happens. We've seen how models gave astronomers insights into the meaning of the main sequence, but models can go much further. Comparing the models and real data helps astronomers understand, for example, how stars originally on the main sequence end up in the giant region.

The best way to explore such issues is to take a model of a single star with specified starting conditions (mass and chemical composition) and allow that star to evolve. Astronomers then watch the changes in that star's structure and properties, such as luminosity L_* and temperature T_*, over time as the star consumes its nuclear fuel. They could also follow how the star's surface properties L_* and T_* change over time. In this way, astronomers create a **stellar evolutionary track** on the star's HR diagram from a stellar model. The track is the sequence of HR diagram locations (L_* and T_* values) that the star occupies over its lifetime.

To create a track, astronomers construct a computer model of a star by starting with an initial core composition of about 90 percent hydrogen, 10 percent helium. They follow the model as it fuses small amounts of hydrogen. For instance, they change first to 89 percent hydrogen, 11 percent helium, then run the model with these new core conditions and get a different temperature and luminosity at the star's surface. By calculating successive models with incrementally less core hydrogen (**Figure 11.20**), astronomers track the star as its position on the HR diagram changes. Connecting these positions gives a stellar evolutionary track.

Astronomers have built enough evolutionary tracks to lay out the basic story of stars—from stars of extremely low mass ($M_* \approx 0.08\ M_{Sun}$) to stars of extremely high mass ($M_* \approx 100\ M_{Sun}$). "The evolutionary track of a star on the HR diagram is basically a record of how its observable properties change with time," says Saumon. "We astronomers talk about the star 'moving' along the track as the star ages." Remember: movement on the HR diagram refers not to the star but to stellar properties.

Exit on the Right: Leaving the Main Sequence

In the next two chapters we will explore the evolution of stars, from their birth in interstellar clouds of gas to their spectacular deaths. In all cases, the evolutionary tracks computed via detailed stellar models' let us translate what we observe in different stellar populations into an understanding of the physical processes guiding the stars' changes.

For now, we focus on one important change: the stars' departure from the main sequence. We already know that stars on the main sequence are hydrogen fusers: they fuse hydrogen nuclei into helium nuclei and release energy, supporting themselves against their own weight in the process. But recall that the process cannot go on forever. The core has only so much hydrogen; once most of it has been replaced with helium, the star runs into trouble. Temperatures in the core are not high enough to start helium fusion (slamming helium nuclei together to make either carbon or oxygen). Therefore, by the time the core has become mostly helium, the star is not producing enough energy to maintain itself. Something has to give.

We will explore post-main-sequence evolution in Chapter 13, but for now notice only that there *is* post-main-sequence evolution. When a star stops getting its energy from core fusion of hydrogen, the crush of gravity forces a rearrangement of the star's distribution of density, temperature, and so forth. As a consequence, the surface properties of temperature and luminosity change, and the star leaves the main sequence (**Figure 11.21**).

While a star is on the main sequence, its mass barely changes. Although a Sun-like star loses mass in a solar wind, the amount of mass lost while the star is on the main sequence is negligible. Once the star leaves the main sequence, however, these winds become much more powerful.

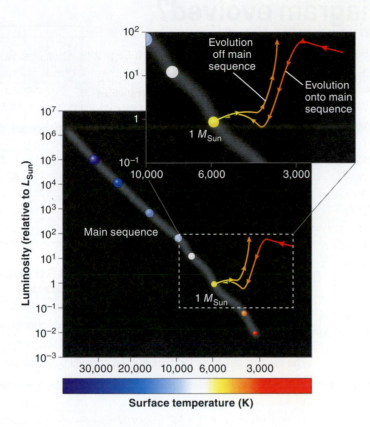

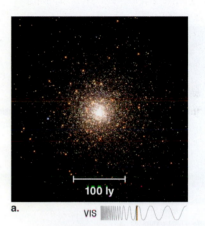

Figure 11.21 Evolutionary Track of a Sun-like Star
A star's lifetime, as seen on an HR diagram by changes in its luminosity and temperature, is shown. A Sun-like star traces such a path from its early life as protostar, onto the main sequence where hydrogen begins to fuse in the core, and off the main sequence (almost 10 billion years later) once hydrogen fusion has ended.

What is most remarkable about model-generated evolutionary tracks is that they make very specific predictions about how long it takes for a star to use up its hydrogen in the core and turn off the main sequence. That timescale—the *main-sequence turnoff time*—depends only on the star's mass, which determines the pressures, densities, and temperatures in the core. These factors, in turn, determine the star's rate of nuclear fusion.

The one-to-one relationship between a star's initial mass and the time it takes to leave the main sequence turns out to be of no small consequence. It is, in fact, how astronomers know they've gotten their models right (see **Anatomy of a Discovery** on p. 308). To elucidate that point, we must travel to the edge of the galaxy and meet some of astronomy's most beautiful players: the globular clusters.

Testing the Tracks: Globular-Cluster Hertzsprung-Russell Diagrams

"Astronomy is not really an experimental science," says Didier Saumon. "It's an observational one. We can't go into a lab and do an experiment that isolates a particular effect we are interested in. Instead we have to just go with what the Universe shows us." How can astronomers test the validity of stellar evolution models?

Since astronomers cannot watch a star evolve, they test evolutionary tracks on a group of stars that have different masses but were born at the same time. That is exactly the stellar evolution playground astronomers find in **globular clusters** (Figure 11.22a). A globular cluster consists of hundreds of thousands of stars born together. "And they are all basically at the same distance from us," explains Saumon. "That means globular clusters give us a natural controlled experiment to test our stellar evolutionary tracks."

Not only were all stars in a globular cluster born at about the same cosmic moment (billions of years ago when the galaxy was forming), but they also display a wide range of stellar masses. Having so many stars that were born together but vary greatly in mass makes globular clusters perfect for calibrating evolutionary tracks. By plotting a cluster's stars on an HR diagram, astronomers can see the full range of stellar evolution (Figure 11.22b).

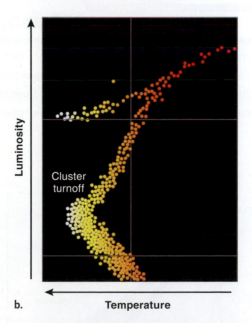

Figure 11.22 Globular Cluster M80
a. A Hubble Space Telescope image of this gravitationally bound collection of more than 100,000 stars. **b.** Astronomers use the HR diagram of a globular cluster to test models of stellar evolution. The known hydrogen-fusing lifetimes of stars at the main-sequence turnoff point indicate the cluster's age. More massive stars have already evolved off the main sequence.

How has the HR diagram evolved?

observation

In 1908, **Henrietta Swan Leavitt**, a Harvard College Observatory "computer," discovered the period-luminosity relationship of Cepheids, later used to measure distances to star clusters and galaxies.

Original HR diagram

observation

Building on Leavitt's work and Annie Jump Cannon's classification system, **Ejnar Hertzsprung** and **Henry Norris Russell** independently discovered a correlation between stellar luminosity and surface temperature. The HR diagram, circa 1910, holds the secrets to stellar evolution.

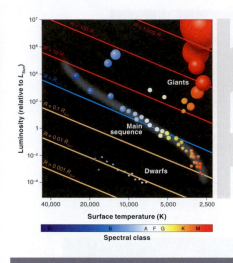

Main sequence on an HR diagram

hypothesis

Astronomers found that, on the HR diagram, most stars plot to a diagonal band: the main sequence. Russell proposed that young stars start at its lower-right corner; contract and heat, traveling to the upper left; then cool as they age, trekking back down the main sequence.

Eddington's main-sequence correlation

failed hypothesis

In 1924, **Arthur Eddington** discovered a correlation between the masses and luminosities of main-sequence stars, challenging Russell's proposal. By the next decade, astronomers accepted that all main-sequence stars fuse hydrogen in their cores. Giants—which do not fuse core hydrogen and are above the main sequence—represent key stages of stellar evolution that invalidate Russell's hypothesis.

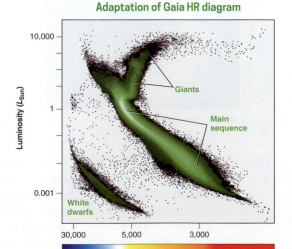

Adaptation of Gaia HR diagram

techniques

Today, astrophysicists such as **Didier Saumon** use mass, composition, and energy source(s) to develop theoretical models able to predict how a star's luminosity and surface temperature change with age. In 2018, Gaia space telescope measurements of the distances and temperatures of more than 1.3 billion stars resulted in the most detailed HR diagram ever— a snapshot of stellar evolution!

When you look at a globular cluster, because it contains stars that were born at the same time with different initial masses, you are seeing stars at very different evolutionary stages. "Remember that more massive stars have shorter lives and don't sit on the main sequence for as long as the less massive stars," says Saumon. "That means an HR diagram for a globular cluster won't have any massive stars on the main sequence since they will have already peeled away." The highest-mass stars have used up their fuel and are off the main sequence, unlike the lowest-mass stars (see Figure 11.22b). Stars of intermediate mass that have just used up their core hydrogen have started to turn off the main sequence.

This means that, once astronomers create an HR diagram using only the cluster's stars, they have a way to find the age of the cluster. Younger globular clusters should show a main sequence that still includes higher-mass stars. An older globular cluster should have lost most of its massive stars and show only intermediate- and low-mass stars on the main sequence. In this way, the HR diagram gives astronomers a *globular-cluster turnoff mass*, also called the *main-sequence turnoff point*, which yields the age of the entire cluster. The cluster age is the main-sequence lifetime of the stars that have just left the main sequence.

By gaining the measure of the age of the globular cluster, astronomers can now compare the locations of all the other stars on the HR diagram and compare them with predictions from the stellar evolution tracks. Stars with masses less than the observed main-sequence turnoff will still be on the main sequence, so they don't offer much information. But stars in the cluster with masses above the observed main-sequence turnoff will have positions on the HR diagram predicted by the stellar evolution tracks. By comparing the distribution of observed positions (temperature and luminosity) with those that come from the stellar evolution models, astronomers can gauge how well their theory for stellar evolution compares with reality.

Comparing real cluster HR diagrams with predictions from evolutionary tracks in this way has enabled astronomers to improve stellar models; now the match between data and prediction is remarkably close. "We have developed our understanding of stellar evolution over almost a hundred years of work," says Saumon. "Using these globular clusters, we can be confident about the results of all that work."

From Life Stories to Birth and Death

Now that we've seen how astronomers use measurements of stellar properties to understand the evolution of stars, we can turn to the more difficult question of stellar evolution's extremes. As strange as it may sound, we seem to understand main-sequence stars pretty well. For all of its power, a hydrogen-fusing main-sequence star is a fairly stable object whose most important secrets have been revealed through 100 years of scientific inquiry.

The birth and death of stars, however, are another story. While stars change slowly throughout the bulk of their lives, during the initial and final stages, their evolution can be very fast and very violent. Stellar birth and death are associated with such speed and chaos that their investigation requires more exotic physics and more powerful tools. In the next chapter we begin at the beginning: the formation of stars.

Section Summary

- To create a stellar evolutionary track, astronomers model the conditions within a star as it continuously converts hydrogen to helium.
- When hydrogen fusion ends in the core, a star leaves the main sequence and continues to evolve.
- The part of the HR diagram where a star leaves the main sequence depends on the star's mass.
- Globular clusters provide the means for astronomers to test their models of stellar evolution, since the stars in a cluster were born at the same time but vary in mass.

CHECKPOINT Describe the physical process that determines that the lifetime of a star will depend on its mass.

DIDIER SAUMON

"I guess I was one of those kids who got interested in science from a fairly early age," explains Didier Saumon. "I was interested in science and by that I mean all kinds of science." He was lucky to have had a neighbor who shared his interest. "We were always geeking out together." Eventually his friend got a small telescope as a gift, which started them on the path to astronomy. "We built our own telescopes as teenagers and developed a passion for observing the sky. Eventually I thought that 'yeah, this would be a cool thing to do.' So I got a bachelor's degree in physics and took the first step towards becoming an astronomer." Now a theoretical astrophysicist at New Mexico's Los Alamos National Laboratory, he studies dwarf stars, young planets, and dense plasmas.

CHAPTER SUMMARY

11.1 The Life of the Stars

Stars evolve over long timescales. Key observational properties are vital to understanding stars: distance, luminosity, temperature, radius, mass, and composition. To measure these properties, astronomers use the information that reaches Earth in the form of light. Information about star velocities helps astronomers understand stellar evolution in a galactic context.

11.2 Measuring Stars

Astronomers use the inverse square law of brightness to determine the distance to objects by comparing their apparent brightness with their luminosity. Astronomers measure distances to nearby stars via parallax. For more distant stars, they use variable stars such as Cepheids as standard candles because of their period-luminosity relationship. Brightness is measured on a magnitude scale. Astronomers apply Wien's law to observations of peak wavelength to determine the surface temperatures of stars and the Stefan-Boltzmann law to derive stellar sizes. Specific spectral lines indicate a star's composition; for some stars, the width of those lines provides measurements of mass. The motions of binary stars also enable astronomers to derive the stars' masses. A star's velocity is measured from its proper motion and Doppler shift.

11.3 From Observations to Explanations

The identification of stellar spectral classes led to the discovery that a star's luminosity and temperature are highly correlated. Most stars fall on the main sequence, a diagonal band of the HR diagram, when they're fusing hydrogen into helium in their cores (the majority of their lives). The HR diagram shows that star mass varies by evolutionary stage, leading to an understanding of stellar life cycles. To understand stellar structure and evolution, astronomers use mathematical models that describe the opposing forces of gravity and gas pressure from nuclear fusion. High-mass stars use up their fusion fuel faster, per the life expectancy–mass relationship. Low-mass stars smaller than the Sun live many billions of years. Intermediate-mass stars live 500 million to 15 billion years. Most massive stars have lifetimes on the scale of just millions of years.

11.4 The March of Time: Stellar Evolutionary Tracks

Sophisticated computer models help astrophysicists understand how stars of various masses evolve. The end of hydrogen fusion in a star leads to changes in temperature, pressure, and density that take the star off the main sequence of the HR diagram. Globular clusters contain many stars of different mass that are all born at the same time, enabling astronomers to test models of stellar evolution. By creating an HR diagram of a globular cluster, astronomers can assess the cluster's age from the point where stars turn off the main sequence.

QUESTIONS AND PROBLEMS

Narrow It Down: Multiple-Choice Questions

1. What is the primary source of information needed to make stellar measurements?
   a. cosmic rays
   b. magnetic fields
   c. electric charge
   d. light
   e. atomic structure

2. What is a standard candle?
   a. an object whose luminosity remains the same throughout its life
   b. an object whose age is known
   c. an object whose luminosity is equal to the Sun's
   d. an object whose luminosity is known
   e. a star within a cluster

3. True/False: A standard candle is useful for determining distance but not orbital velocity.

4. Wien's law tells us that
   a. wavelength is related to frequency.
   b. blackbodies are essentially black.
   c. blackbodies radiate light at all wavelengths and absorb light at all wavelengths.
   d. temperature is regulated by spectral class and luminosity.
   e. in blackbodies, temperature and peak radiation wavelength (color) are related.

5. Two stars are observed to have different spectral types. Which statement is true?
   a. Their temperatures may be equal.
   b. They cannot have the same apparent magnitude.
   c. They cannot have the same absolute magnitude.
   d. They may have some spectral lines in common.
   e. They may have equal mass.

6. Which definition of proper motion is correct?
   a. motion due to the rotation of the galaxy
   b. observed motion of an object against very distant background objects
   c. motion that does not follow Newton's laws
   d. motion around the Sun
   e. perpendicular motion that can be detected with Doppler shift

7. Astronomers must take many factors into account when modeling individual stars. Which of the following is *not* a factor?
   a. limitations of computers
   b. star mass
   c. star composition
   d. distance between stars
   e. stellar energy generation

8. Choose the correct list of spectral classes in ascending order of temperature.
   a. OBAFGKM
   b. ABFGKMO
   c. MKGFABO
   d. OAGMBFK
   e. MGAOKFB

9. A star on the upper right of the HR diagram is
   a. a giant.
   b. cool and small.
   c. hot and small.
   d. a dwarf.
   e. hot and large.

10. What does the main sequence show?
    a. the entire life cycle of a star
    b. a population of 1-M_{Sun} stars
    c. the distribution of core hydrogen-fusing stars by mass
    d. the middle third of a star's lifetime
    e. the series of stages in a star's life

11. A star's position on the main sequence does *not* tell us which of the following? Choose all that apply.
    a. its chemical composition
    b. its mass
    c. its luminosity
    d. its temperature
    e. its exact age

12. The more massive the star,
    a. the longer it remains on the HR diagram.
    b. the lower it is on the HR diagram.
    c. the longer it stays on the main sequence.
    d. the hotter it is as a main-sequence star.
    e. the higher its apparent magnitude value is.

13. A star's changing position on the HR diagram indicates
    a. its motion within the galaxy.
    b. its evolutionary track.
    c. how its mass changes as it ages.
    d. its location in space.
    e. its degree of parallax.

14. Which of the following is true of live, low-mass stars? Choose all that apply.
    a. They may be spectral class O.
    b. They are not found on the main sequence.
    c. They are hotter than intermediate-mass stars.
    d. They are positioned at the bottom of the HR diagram.
    e. Their life expectancies are in the billions or trillions of years.

15. What does the HR diagram of a cluster *not* tell astronomers?
    a. the age of the cluster
    b. the highest mass of cluster stars still on the main sequence
    c. the rotation velocity of the cluster
    d. which stars are similar to the Sun in size and mass
    e. the highest luminosity among the remaining stars on the main sequence

16. Binary stars detectable by direct observation of their orbital motion are
    a. astrometric binaries.
    b. spectroscopic binaries.
    c. eclipsing binaries.
    d. visual binaries.
    e. a symbiotic binary system.

17. Which of the following is true of stars in binary systems?
    a. They orbit a common center of mass.
    b. They will eventually merge into a single star.
    c. They are rare.
    d. They may not be gravitationally related.
    e. They almost always have identical masses.

18. Choose the statement about variable stars that is *not* true.
    a. They have unstable outer layers.
    b. They change in temperature.
    c. They change in luminosity.
    d. They may be useful as standard candles.
    e. Their luminosity may increase continuously.

19. Suppose you observe the same stars from Earth and from a fixed location in space several astronomical units away, closer to those stars. How would your observations made from space compare with those made from Earth? Choose all that apply.
    a. The stars would appear to be at greater distances.
    b. The stars would appear to be closer.
    c. Parallax for a given star would be greater.
    d. Parallax for a given star would be less.
    e. Parallax would be unchanged for all stars.

20. True/False: A star with a mass 10 times that of the Sun would have a life span one-tenth as long.

To the Point: Qualitative and Discussion Questions

21. Define the following terms: parallax, energy transport, luminosity, hydrostatic equilibrium, and correlation.

22. What are the six properties vital to studying the life stories of stars?

23. What is a standard candle, and how is it used to determine distance?

24. What information in a star's spectrum indicates the star's surface temperature?

25. What information in a star's spectrum indicates the star's chemical composition?

26. Describe the two approaches by which astronomers determine a star's mass.

27. The Doppler shift can provide information on which characteristics of a star's motion through space?

28. A star is observed with a pattern of alternating redshifted and blue-shifted spectral lines. What does this pattern tell you about the star?

29. Explain what the main sequence on the HR diagram indicates. What is the difference between a star that plots at the upper left of the main sequence and a star that plots at the lower right?

30. What is going on in the core of a main-sequence star?

31. How are globular clusters used in studying stellar evolution?

32. What is meant by globular-cluster turnoff point?

33. Name and describe three categories of binary stars.

34. Which variable stars have helped astronomers understand the scope of the Universe, and how have they done so?

35. Describe how an astronomer develops a stellar evolution model.

Going Further: Quantitative Questions

36. You observe that a star's wavelength of maximum intensity is a very reddish 770 nm. What is that star's surface temperature, in kelvins, and how does it compare with the Sun's? (See Going Further 4.1.)

37. A star has a surface temperature of 6,500 K (a bit hotter than the Sun's). What is its wavelength of maximum intensity, in nanometers? (See Going Further 4.1.)

38. Find the expected main-sequence lifetime of a 4-M_{Sun} star, in years.

39. What is the ratio of the main-sequence lifetime of a 2.5-M_{Sun} star to that of the Sun?

40. A star has a mass one-third of the Sun's. How many times longer will it live on the main sequence than the Sun will?

41. A star has a mass of 55 M_{Sun}. What is the ratio of its expected lifetime to that of the Sun?

42. What is the distance, in parsecs, to a star whose parallax is 2″? Compare this distance with that of the closest known star. Why would discovering a star with this parallax be a surprise?

43. How far from Earth, in parsecs, is a star whose parallax is 0.43″?

44. Two stars are known to have the same luminosity, but one appears one-sixteenth (1/16) as bright as the other. How many times more distant is the dimmer star?

45. Star A is 9 times as far away as star B. Both stars appear to have the same brightness. What is the ratio of the luminosity of star A to that of star B?

"A COUPLE OF TIMES THE PROJECT LOOKED AS IF IT WERE IN TROUBLE"

12

Nursery of the Stars

THE INTERSTELLAR MEDIUM AND STAR FORMATION

Seeing in the Dark

12.1 "It took a long, long time," says Judy Pipher, a professor of astronomy at the University of Rochester, telling the story of the Spitzer Space Telescope (SST). "It started in 1983. That was the year a group of us proposed that NASA build a big, space-based infrared telescope. All of us were interested in studying how stars form in interstellar gas clouds." Recall from Chapter 4 that light of any given wavelength is absorbed by objects with a size close to that wavelength, and the dust in most star-forming clouds is just the right size to absorb visible light. "The best way to see into those clouds and see what was going on with star formation was to use long-wavelength light in the infrared," says Pipher.

The project was approved, but over the next two decades Pipher and her colleagues had to ride waves of interest and antagonism from within and outside NASA. "A couple of times the project looked as if it were in trouble," she recalls. "And at one point the funding got scaled back really dramatically. All the astronomers working on the project had to sit down and decide what parts of the mission were most important and what parts we could let go. We knew we needed to get the details of what was happening in those star-forming clouds, so we focused on that as our primary mission."

Despite hurdles, on August 25, 2003, the SST's launch day arrived. Pipher and the rest of her team were there to watch the Delta 5 rocket, carrying the telescope, rise on a pillar of flame into the sky (**Figure 12.1**). "I was pretty nervous," she says. "Really, it had taken so much work I was just happy to finally make it to launch day." The launch went off without a hitch, and within a few weeks the team confirmed that the telescope was working perfectly.

At the time, the SST was the only instrument capable of seeing so deeply, and with such accuracy, into interstellar clouds, where the mysteries of star formation are hidden. A few months later, new data poured in, giving the waiting astronomers unprecedented views of giant clouds collapsing, new stars forming, and young planets hiding in surrounding disks of gas and dust. "It was definitely worth the wait," Pipher says with a wide smile.

← The Fox Fur Nebula is a vast star-forming region located 2,700 light-years from Earth. The filamentous appearance is produced by the emission of light from hydrogen, excited by ultraviolet radiation from hot young stars in the cluster.

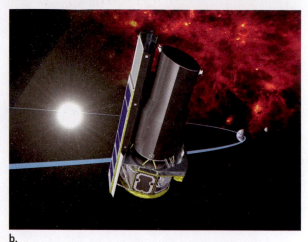

Figure 12.1 Spitzer Space Telescope (SST)

a. Launched in 2003, the SST departed Earth aboard a Delta 5 rocket. The telescope captures infrared light, which, unlike visible light, is not absorbed by dust grains. Thus, the SST observes the inner regions of dusty components of the interstellar medium. **b.** The SST trails behind Earth as it orbits the Sun, keeping its distance so that Earth's infrared emission doesn't overwhelm the emissions of objects being observed. The SST continues to drift away from us at about 0.1 AU per year.

In Space, Everyone Can Hear You Scream: The Interstellar Medium

"In space, no one can hear you scream," said the trailer for the 1978 space horror movie *Alien*. Indeed, most folks think of space as a vacuum, devoid of matter and energy. The truth, however, is a lot more interesting.

Interstellar space is far from empty. Floating between the stars is the **interstellar medium (ISM)**, a constantly moving mix of gas and dust between stars within a galaxy. The density of material in the ISM is tenuous (weak) by Earth standards: a cubic centimeter of air at Earth's surface contains 10^{19} atoms, but in the denser regions of the interstellar medium a cubic centimeter of gas contains about 10^6 atoms. That means that the ISM is one 10-million-millionth as dense as the air you're breathing. But 10^6 is not zero. Even at these low densities, if you added up all of the matter among the stars in our galaxy, it would amount to billions of times as much material as the Sun contains. On larger scales, space is full of stuff, and it is from this stuff that new stars are born (**Figure 12.2**).

The space among the stars is also not quiet or still. The ISM is in a constant state of motion called *turbulence*. A gas in turbulent motion swirls, tumbles, and crashes against itself on many size scales. As in a fast-moving river that swirls around obstacles, in the turbulent ISM there are huge "whorls" that are hundreds of parsecs across, and within those large-scale motions are smaller whorls that may extend over less than a single parsec.

It is unclear what drives the ISM into such turbulence, but many lines of evidence point to the combined action of supernovas (explosions of high-mass stars). Because the ISM is a gas (though a very tenuous one), disturbances in the ISM create sound waves like those that occur in Earth's atmosphere. If the disturbance is violent enough (as in a supernova), the extreme cousins of sound waves called *shock waves* form. These shocks can cross parsecs of space and sweep up everything in their path (**Figure 12.3**). Supernova blast waves are a specific example of astrophysical shocks.

Over millions of years, the explosions of many massive stars set local regions of gas into motion. When these supernova blast waves expand enough, they collide with one another. Collisions tear each blast wave into shreds, sending gas tumbling in all directions. Computer simulations of this process indicate that the explosion of about one supernova every 10,000 cubic parsecs (pc^3) every million years is enough to keep the ISM in its tur-

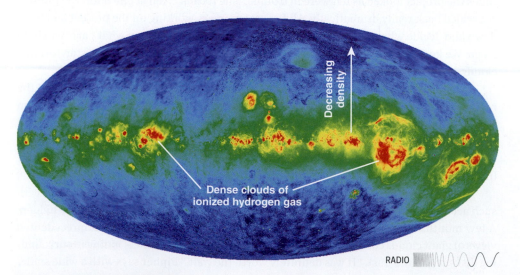

RADIO

Figure 12.2 Interstellar Medium

A whole-sky map indicating the location of ionized hydrogen gas, in which the hydrogen atoms have been stripped of their electrons. Interstellar gas is broadly distributed throughout space, but the brightest regions are highly localized and form distinct clouds.

bulent state. Winds from young and middle-aged stars, and even the rotation of the galaxy itself, may also contribute to turbulence in the ISM.

Section Summary

- The interstellar medium (ISM) is composed of a low-density mix of gas and dust but contains enough mass to enable ongoing star formation.
- The ISM is in a state of turbulent motion caused primarily by blast waves from supernovas.

CHECKPOINT Give one example of how the study of the ISM is linked to the study of stars.

Anatomy of the Interstellar Medium

12.2 To understand star formation, astronomers must first understand the different kinds of gases in the ISM. There's a sort of recycling system in the galaxy: some kinds of interstellar material can collapse and form new stars; then, as stars age, they return gas back to the ISM. Thus the ISM is a subject of great interest and an important part of our story about the lives of stars and, eventually, the lives of galaxies.

Phases of the Interstellar Medium

"The ISM is made up of different kinds of clouds," says Judy Pipher, "and they all have different conditions"—specifically, densities and temperatures. In addition, the gas takes on different microscopic states in terms of ionization or chemistry. "In some clouds," says Pipher, "the gas will be primarily composed of molecules, and in others it will be atoms that have become highly ionized."

Accordingly, astronomers have distinguished different *phases* of the interstellar medium: molecular clouds, H I clouds, H II regions, warm interstellar medium, and coronal gas (**Table 12.1**). Astronomers think of these phases the way physicists think about liquid water, ice, and water vapor. Each is a phase of water whose existence depends on conditions such as temperature and pressure. Similarly, it has proved useful for

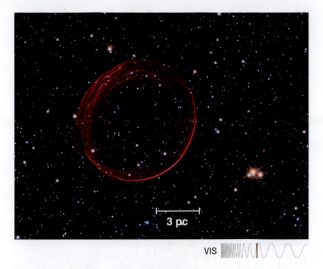

Figure 12.3 Supernova Blast Wave
SNR 0509 is the remnant of a powerful stellar explosion in the Large Magellanic Cloud, a small galaxy near the Milky Way. A supernova remnant, bounded by an outward-expanding shock or "blast" wave, results from the explosion of a massive star. As the shock expands, it encounters and compresses nearby clouds of the ISM and eventually can collide with other supernova shocks, stirring the ISM into a turbulent state.

"The ISM is made up of different kinds of clouds, and they all have different conditions."

Table 12.1   Components of the Interstellar Medium

Phase	Temperature (K)	Density (atoms/cm³)	State of Hydrogen	How Observed
Molecular clouds	10	10^2 to 10^6	Molecular (plus many other molecules)	Molecular emission and absorption lines
H I clouds	10^2	10 to 10^3	Neutral atomic	21-cm-line absorption
H II regions	10^4	10 to 10^6	Ionized	H alpha emission (656 nm, in red part of visible spectrum)
Warm interstellar medium (WISM)	8×10^3	0.2–0.5	Ionized	H alpha emission
Coronal gas	10^6	10^{-4} to 10^{-2}	Highly ionized	X-ray emission

interstellar medium (ISM) The gas and dust spread among the stars in different phases that depend on temperature and density conditions.

VIS

Figure 12.4 Star-Forming Nebulas
The Orion Nebula (M42), the region of massive star formation that is closest to Earth, displays glowing gas clouds and hot young stars. Its red color is due to the emission of H alpha photons (see Chapter 4). Astronomers have identified what seem to be numerous infant planetary systems within M42. Smaller nebulas—M43 and the dusty, bluish NGC 1977—appear nearby.

astronomers to think about the different kinds of clouds as phases of the ISM under different conditions of temperature, pressure, and chemical makeup. Note that there is no barrier between the different types of ISM. The different phases represent ISM material in different states.

MOLECULAR CLOUDS. The coldest and densest regions of the galaxy are vast clouds of gas in which stars form. These are called **molecular clouds** because their low temperatures, about 10 kelvins (K), and relatively high densities, about 10^2–10^6 atoms/cubic centimeter (cm³), allow complex molecules to form. Astronomers use emission lines from the molecules to map out these clouds (**Figure 12.4**).

H I CLOUDS. Much of the galaxy's volume is occupied by extended clouds of atomic hydrogen gas called **H I clouds**. (*H* refers to hydrogen, and the Roman numeral I indicates the neutral state.) H I clouds have temperatures of a few hundred kelvins and densities that range from on average tens of particles per cubic centimeter to as high as thousands of particles per cubic centimeter. These cold clouds tend to be found in the spiral arms of our galaxy and are confined to a relatively thin region (300 pc in height) around the midplane of the galaxy's disk (**Figure 12.5**).

H I clouds are so cold that they do not emit light through transitions of electrons in hydrogen atoms between orbits. There simply is not enough energy for collisions between particles to kick an electron up from the ground state. How, then, do astronomers know these clouds exist?

All of the particles that make up an atom have an intrinsic spin, like a top. Each charged, spinning particle (such as a proton or an electron) is like a tiny magnet. Magnetic forces between an electron and a proton tend to keep their spins oppositely oriented. When an occasional collision between particles knocks both spins into alignment, a quantum jump eventually flips them back into their lower-energy state. During this quantum "spin-flip," a photon is emitted with a wavelength of 21 centimeters (cm), which is in the radio band (**Figure 12.6**; see also Chapter 9). These long-wavelength radio photons can travel enormous distances without being absorbed. In this way, 21-cm radio observations enable astronomers not only to detect neutral hydrogen across the galaxy but also to map the entire galactic distribution of H I clouds.

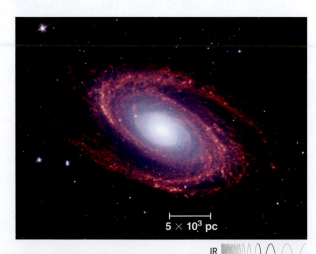

IR

Figure 12.5 H I Clouds
Spiral galaxy M81. Cool H I clouds are typically found close to the midplane in the spiral arms of such galaxies.

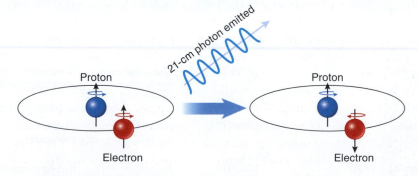

a. **Higher-energy state after collision**   b. **Lower-energy state**

Figure 12.6 The 21-Centimeter Line
H I clouds (neutral hydrogen, H) are too cold to emit light from transitions of electron energy states. They are seen instead via emission of the 21-cm line, which occurs when **a.** a collision (not shown) between two H atoms leaves one of the atoms with the spins of its electron and proton aligned. This higher-energy state eventually decays back to **b.** a configuration with opposite spins while a 21-cm photon with energy equal to the difference between the aligned and unaligned states is emitted.

H II REGIONS. "While H I clouds are neutral," says Pipher, "other regions of the ISM are defined by their ionization." **Photoionization** occurs when a photon with enough energy is absorbed and kicks an electron off an atom or ion (**Figure 12.7**). Photoionizing radiation can come from many sources, including young massive stars and evolved low- and intermediate-mass stars (even the disks of gas surrounding supermassive black holes at the centers of galaxies).

Photons with energy of, or greater than, 2.18×10^{-18} joule (J)—which corresponds to a wavelength of 91.2 nanometers (nm), in the ultraviolet range—are capable of ionizing hydrogen. An electron, once it is kicked off a hydrogen atom, will collide with other gas particles, sharing its energy and heating the gas. "That means ionized regions of the interstellar medium will be a lot hotter than neutral or molecular gas," says Pipher.

Typical temperatures in ionized interstellar gas are about 10,000 K. The density of this so-called warm component of the ISM depends on where it occurs. When a star produces the radiation-stripped hydrogen atoms, astronomers call the environment an **H II region** (or an *emission nebula* or *ionized nebula*). The Roman numeral II indicates that the hydrogen atoms have each lost one electron. H II regions have typical densities of 100–10,000 atoms/cm³.

WARM INTERSTELLAR MEDIUM. While H II regions surround individual stars (or groups of stars), another phase of ionized interstellar gas is distributed *among* the stars. These regions are far from any particular ionizing source but still experience a diffuse *background* of ionizing radiation from the sum of all ionizing stellar sources in the galaxy. Although this background radiation is relatively weak, it still produces ionizations, creating what is called the **warm interstellar medium (WISM)**. The WISM has temperatures similar to those in H II regions but lower average densities. Regions colder than the WISM, such as H I clouds, maintain their low temperatures through emission lines not available to WISM material. Recall from Chapter 4 that collisions can drive the electrons in an atom to an excited state. When an electron drops down from that state, it emits a photon. If the photon leaves the gas, it takes its energy with it, cooling the gas.

CORONAL GAS. Worming through and between the other ISM phases are regions of extremely hot, low-density gas. Temperatures in this phase can reach as high as 10^6 K. Densities in this phase can be as low as 10^{-4} atom/cm³.

"If these kinds of conditions seem familiar," says Pipher, "it's because we see them in the Sun's corona. That's why astronomers use the term 'coronal gas' to refer to this phase of the ISM." Despite the similar conditions, the source of **coronal gas** is not living stars but dead ones. The hot, low-density interstellar gas originates with supernova blast waves created at the end of a massive star's life. "When a star goes supernova, its explosion can sweep across hundreds of light-years of interstellar space," says Pipher. "All the heat released in the explosion drives swept-up ISM gas to such high temperatures that it gets strongly ionized." As she explains, an element such as iron can lose up to 13 of its 26 electrons.

At these temperatures and ionization states, the newly heated gas can't cool effectively by emitting radiation, because there are no electronic transitions that can be easily excited for the electrons that remain around the atoms. Thus, once a passing supernova blast wave drives ISM gas into the coronal state, it remains hot for billions of years. As time passes, the hot gas continues to expand, filling in regions among other phases of the ISM. Eventually, a kind of Swiss-cheese network of hot, low-density regions forms throughout the galaxy (**Figure 12.8**).

These ISM phases do not exhaust the forms of clouds floating between the stars. Astronomers have found other mixes of hot or cold, dense or tenuous gas in the ISM.

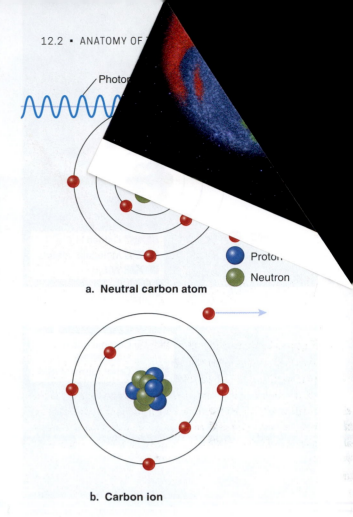

a. Neutral carbon atom

b. Carbon ion

Figure 12.7 Photoionization

a. When a photon of high enough energy collides with a neutral atom (which has as many electrons as protons), **b.** an electron is ejected from the atom. The result of this photoionization is an ion with one fewer electron than it has protons. Photoionization can also strip an ion of more electrons.

molecular clouds The low-temperature, high-density phase of the interstellar medium where molecules can form. Molecular clouds are the site of star formation.

H I clouds The phase of the interstellar medium that is composed of low-temperature, neutral hydrogen gas.

photoionization The stripping of an electron from an atom or ion via the absorption of a photon.

H II region The phase of the interstellar medium that contains ionized hydrogen gas.

warm interstellar medium (WISM) The phase of the interstellar medium that is warmed by the galaxy's diffuse ionizing radiation (the sum of all its ionizing stellar sources).

coronal gas The phase of the interstellar medium that consists of extremely hot, low-density gas resulting from explosions of high-mass stars (supernovas).

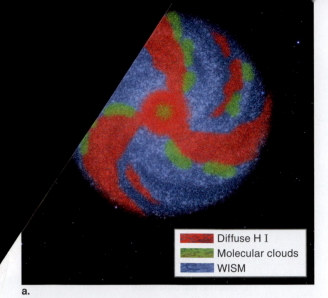

Diffuse H I
Molecular clouds
WISM

a.

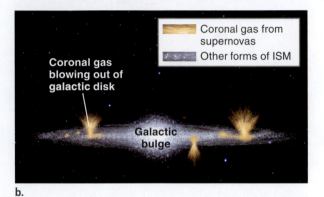

Coronal gas from supernovas
Other forms of ISM

Coronal gas blowing out of galactic disk

Galactic bulge

b.

Figure 12.8 Coronal Gas and ISM Structure
a. A top-down view of the distribution of major phases of the ISM in our galaxy. Threaded among molecular, H I, and warm phases is hot coronal gas (shown in part b) created by many supernova blast waves over galactic history. **b.** High pressures in the hot material also lead to blowouts above and below the plane of the galactic disk.

"If these kinds of conditions seem familiar, it's because we see them in the Sun's corona. That's why astronomers use the term 'coronal gas' to refer to this phase of the ISM."

"The pressure inside the balloon has dropped with the temperature. That means the balloon has to shrink."

INTERACTIVE:
Interstellar Reddening

However, the phases discussed in this section provide the main distinctions necessary for understanding the ISM and star formation.

Many Clouds, One Pressure

Remember that the ISM is a gas. Therefore it obeys the *ideal gas law*, which states that the pressure of a gas depends on its density and temperature.

Anyone who has ever played with balloons has an intuitive understanding of pressure. To blow up a balloon, you force air into it until its elastic skin expands. But if you inflate a balloon and then put it in the freezer for a couple of days, the balloon will deflate a little, even if no air escapes from it. What happened? "By cooling the trapped air, you lowered its pressure," says Pipher. "The air molecules are at a lower temperature, and that means they are no longer hitting the inside wall of the balloon with a high speed." Temperature, speed, and pressure are related. Recall from our discussion of planetary atmospheres in Going Further 6.2 that higher temperatures in a gas mean particles moving at higher thermal velocities. Since pressure is the effect of collisions with gas particles (a balloon stays inflated because atoms inside bounce off the balloon's skin), high temperatures also mean high pressures. "The pressure outside the balloon that's been inside the fridge hasn't changed," says Pipher, "but the pressure inside the balloon has dropped with the temperature. That means the balloon has to shrink."

If we think of the different ISM phases as clouds pushing against one another, what applies to balloons should also apply to them. When scientists calculated the pressure of each phase, they found that many of the phases have the same pressure even though the various types of clouds seem very different from one another. When neighboring gas clouds have the same pressure, astronomers say they are in equilibrium. A cloud that is in equilibrium with its surroundings will neither expand nor contract. From these observations, astronomers concluded that gas in the ISM naturally expands or contracts until it reaches phases that are in pressure equilibrium. The net result of these findings was a multiphase model of the ISM. There is hot, low-density gas and cold, high-density gas. The different phases can coexist indefinitely because all of them have the same pressure (see **Going Further 12.1**).

As better observations were obtained, astronomers realized that the clouds of the ISM are not in perfect pressure equilibrium. Many regions of the ISM are in motion and were found to be turbulent. "That changed how astronomers had to imagine the ISM and its history," says Pipher. The turbulent motions of a gas can be thought of as another form of pressure. The roiling motions of a turbulent gas cloud push against other material, just as the random motions of individual atoms in a hot gas do.

Interstellar Dust: It's the Little Things That Matter

Individual atoms or molecules of gas are not the only inhabitants of interstellar space. Along with the gas are the smallest examples of solids known to astronomers: small grains of matter called dust. "Actually," says Pipher, "calling it *dust* is kind of a misnomer. Most of the grains we see in space are much more like soot from automobile exhaust or the smoke in a smoke-filled room rather than the dust you find under a bed." What matters, explains Pipher, is the size of the grains.

Typical interstellar dust grains range from 0.1 to 10 microns (μm) in length (1 μm = 10^{-6} meter), though within molecular clouds the grains can be even larger. If you compare these sizes with the wavelength of optical light (0.390–0.700 μm), you will see one reason that dust is so important to astronomers. "The size of dust grains is just right to absorb or scatter optical light," says Pipher. "The grain sizes and the light wavelengths are pretty much the same [**Figure 12.9**]. That means dust limits how far we can see across the galaxy. It also keeps us from seeing inside molecular clouds, where stars form."

GOING FURTHER 12.1

Pressure, Equilibrium, and the Interstellar Medium

Astronomers have come to understand a lot about the temperature (T), number density (d)—that is, the number of particles per cubic centimeter—and pressure (P) of clouds in the ISM. These properties are related as follows:

$$P \propto dT$$

pressure is proportional to density $\times$ temperature

Let's use this proportion to learn something about the relationship of the different phases of the ISM, which have different temperatures and densities. First, we give this relationship for two cloud phases (labeled 1 and 2):

$$P_1 \propto d_1 T_1 \quad \text{and} \quad P_2 \propto d_2 T_2$$

If we divide the two proportions, we get a ratio:

$$\frac{P_1}{P_2} = \left(\frac{d_1}{d_2}\right)\left(\frac{T_1}{T_2}\right)$$

Now imagine that we observe a cloud of phase 1 embedded inside a larger cloud of phase 2. If we expect that these clouds are in pressure equilibrium, then

$$\frac{P_1}{P_2} = 1$$

That equation gives us a relationship between the densities and temperatures in the two clouds:

$$\left(\frac{d_2}{d_1}\right) = \left(\frac{T_1}{T_2}\right)$$

Thus, for pressure equilibrium to be maintained, if the temperature in cloud 1 is much higher than that in cloud 2, the density in cloud 1 must be much lower than that in cloud 2.

Imagine that we know the temperature in cloud 1 (10,000 K), the temperature in cloud 2 (100 K), and the density in cloud 1 (0.001 particle/cm³). If we also know that the clouds are in pressure equilibrium, we can calculate the density in cloud 2 by rearranging the equation as follows:

$$d_2 = d_1\left(\frac{T_1}{T_2}\right) = 0.001 \text{ cm}^{-3}\left(\frac{10,000 \text{ K}}{100 \text{ K}}\right) = 0.1 \text{ cm}^{-3}$$

The density in the second cloud must be 100 times higher than that in the first cloud to maintain equilibrium. This exercise illustrates how the concept of pressure equilibrium, when it applies, gives astronomers a powerful tool to understand the properties of ISM clouds.

When clouds are not in equilibrium, however, astronomers must also describe the motions that occur when pressure differences between two phases of the ISM set the gas in motion.

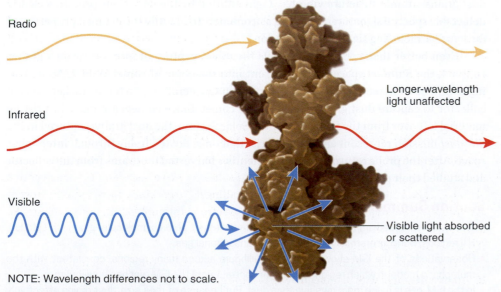

Radio

Infrared

Visible

Longer-wavelength light unaffected

Visible light absorbed or scattered

NOTE: Wavelength differences not to scale.

> "Actually, calling it *dust* is kind of a misnomer. Most of the grains we see in space are much more like soot from automobile exhaust or the smoke in a smoke-filled room rather than the dust you find under a bed."

Figure 12.9 Interstellar Dust
A dust particle absorbs light whose wavelength is less than or equal to the particle's size, typically 0.1–10 μm. Thus interstellar dust grains absorb or scatter visible light, with wavelengths of 0.390–0.700 μm. Longer-wavelength light, such as radio and infrared, can pass through dusty regions with little absorption or scattering.

"Dust limits how far we can see across the galaxy. It also keeps us from seeing inside molecular clouds, where stars form."

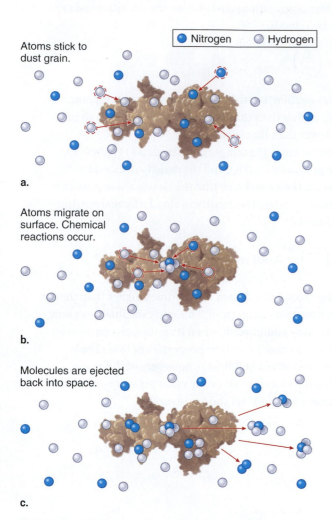

Atoms stick to dust grain.

a.

Atoms migrate on surface. Chemical reactions occur.

b.

Molecules are ejected back into space.

c.

Figure 12.10 Molecule Formation on Dust Grains

a. Atoms from interstellar gas collect on the icy surfaces of dust grains. **b.** Atoms encounter one another as they move randomly along the grain surface, allowing chemical reactions to occur. Here, hydrogen and nitrogen atoms combine, forming molecules of ammonia (NH_3). **c.** Eventually the new molecules are ejected back into the interstellar gas.

reflection nebula A dense, dusty interstellar cloud made visible by reflected starlight.

giant molecular cloud (GMC) A large-scale region of gas in molecular form, spanning tens to hundreds of parsecs and containing up to 10 million solar masses.

dark globule or **Bok globule** A small molecular cloud containing a few solar masses' worth of material whose density and dust content render it opaque to light in visible wavelengths.

On average, there is 1 kilogram (kg) of dust for every few hundred kilograms of gas in the ISM. Although dust accounts for only a small fraction of the ISM's mass, it creates two effects that astronomers must account for in their studies. First, dust grains absorb certain wavelengths of incoming visible light, making objects appear dimmer than they actually are. Second, dust tends to scatter blue light. This scattering can create beautiful blue **reflection nebulas**, relatively dense interstellar dust clouds off which starlight reflects, but it also makes objects beyond the dust appear redder than they actually are. (Closer to home, it is sunlight scattering off dust in Earth's atmosphere that creates the red color seen at sunset. We notice this effect only because of the long path through the atmosphere the light has to take when the Sun is low in the sky.) Astronomers in the early 1900s didn't understand the power of dust to dim distant stars. By miscalculating the true brightness of these stars, astronomers greatly underestimated the true size of our galaxy.

Dust is crucial to astronomers for another reason: the structure of a typical dust particle enables it to act as a chemical factory. "As the smallest bits of solid matter, dust grains act as a platform for complex chemistry," says Pipher. "The grains are also the first step in building larger solid structures like rocks, asteroids, and eventually planets."

Most interstellar dust grains form in the dense winds of dying stars (as we'll see in Chapter 13). These grains appear to be irregularly shaped crystals of silicates (silicon and oxygen) or carbon, mixed with many other elements at lower abundances. The dust grains are often surrounded by a "mantle" or covering of water ice or other frozen volatiles. "When atoms collide with dust grains, they get stuck to the grain surface," explains Pipher. "Once an atom is on the surface, it's more likely to find other elements that are being bounced around the grain." When these surface atoms combine, they form molecules (**Figure 12.10**). "Building complex molecules could never happen in free space, because out there, atoms rarely collide. The dust grains act like flypaper for the atoms." By catching atoms on their surface, the grains can build up high-enough atomic abundances for chemical reactions to occur, creating ever-more-complex molecules in relatively short times, astronomically speaking.

Astronomers can make such definitive statements about the properties of interstellar dust grains because the grains emit light in the form of blackbody radiation (they are solids), providing information about their temperature. More important, however, the dust grains' atomic structure vibrates. Light emitted by the vibrations produces clearly detectable spectral signatures, enabling astronomers to identify the detailed properties of dust grains, including size and composition.

Even better than spectral analysis is the direct capture of interstellar dust grains. In 2004, the *Stardust* space probe was sent into the coma of comet Wild 2. Approaching within 300 kilometers (km) of the comet, the spacecraft used a tennis racket–shaped collector to capture dust grains blown off the comet. Since comets are made of pristine material left over from the formation of the Solar System, the dust grains captured in the *Stardust* mission represent ancient material that once floated freely through interstellar space. After the probe returned to Earth, scientists harvested the grains from the collector and studied their properties.

Section Summary

- The interstellar medium (ISM) is composed of five distinct phases: molecular clouds, H I clouds, H II regions, warm interstellar medium (WISM), and coronal gas.
- Observations of the ISM show pressure equilibrium across many regions, consistent with the multiphase model in which regions expand or contract until the ISM reaches equilibrium.
- In the ISM, gas is much more massive than dust. But because of their size, dust grains effectively extinguish visible light, making background objects appear dimmer and redder.
- The surfaces of dust grains capture atoms and provide a platform for the formation of molecules.

CHECKPOINT Explain what astronomers mean when they say that the different kinds of interstellar clouds are in pressure equilibrium.

Molecular Clouds: The Birthplace of Stars

12.3 For the story of star formation to make sense, we need to delve more deeply into the composition and structure of molecular clouds. Only inside these dense, dusty clouds are the conditions just right for new stars to be assembled.

The Molecules in Molecular Clouds

If the molecules that give molecular clouds their name were to form on dust grains in other parts of the ISM, they would quickly be dissociated (broken apart) by passing ultraviolet photons. Within a molecular cloud, however, the gas and dust are dense enough to shield the molecules from dissociating radiation. The high densities of dust and gas in molecular clouds thus enable them to act as stellar nurseries.

Given that hydrogen is the most abundant element in the Universe, it shouldn't come as a surprise that molecular hydrogen (H_2) is the dominant molecule inside molecular clouds. However, it's difficult for H_2 to radiate at the low temperatures inside molecular clouds ($T = 10$ K). "Hydrogen just doesn't emit a lot of light at conditions within a cloud," says Judy Pipher, "unless something violent happens to raise the temperature of the gas."

Because hydrogen molecules can't be used to study these clouds, astronomers need another molecular tracer, or indicator, of cloud structure. The term *tracer* is used because these molecules may not constitute much of a cloud's mass, but they are prevalent enough to trace out the cloud's structure. Luckily, these clouds contain other kinds of molecules, such as carbon monoxide (CO), that radiate more efficiently at low temperatures. Using CO as a tracer, astronomers can map molecular clouds in detail and determine important properties such as cloud mass and internal motions.

"The chemistry in these clouds is pretty rich," says Pipher. "You get a lot more than just H_2 and CO molecules." Astronomers have identified hundreds of molecular species in star-forming clouds, including vast amounts of water (H_2O) and ethanol (CH_3CH_2OH). "One way to look at a molecular cloud is that it is like a giant martini," quips Pipher, who notes that ethanol is a form of alcohol.

Table 12.2 lists some of the molecules now known to exist in star-forming clouds. Even a number of amino acids have been discovered. This fact has provoked some astrobiologists to wonder whether life's building blocks might be produced in space first and then delivered to planets such as Earth much later.

Molecular Clouds: Shapes and Sizes

"There isn't just one kind of molecular cloud," says Pipher. "They come in a bunch of different sizes." On the largest scales are **giant molecular clouds (GMCs)**, which can span 100 pc and contain enough mass to form a million stars. On the smallest scales are **dark globules** (sometimes called **Bok globules**), which extend only a parsec or less and contain enough mass for just a few stars (**Figure 12.11**; see **Anatomy of a Discovery** on p. 323). Densities within an individual cloud can vary as well. In their central regions, molecular clouds can be quite dense by ISM standards, with up to 10^6 particles/cm³. At the edges of a cloud, however, density drops by a factor of a million as the cloud merges with other regions of the ISM.

This variety of sizes and masses may have a lot to do with the internal structure of molecular clouds. The largest star-forming clouds, GMCs, are not uniform but instead show a hierarchy of structure. "Hierarchy means there are big clouds embedded with smaller, denser clouds," says Pipher, "and within those smaller clouds are even smaller, denser structures."

Table 12.2 Some of the Molecules Found in the Interstellar Medium

Molecule	Chemical Formula
Molecular hydrogen	H_2
Molecular oxygen	O_2
Molecular nitrogen	N_2
Diatomic carbon	C_2
Ammonia	NH_3
Acetone	$(CH_3)_2CO$
Benzene	C_6H_6
Carbon monoxide	CO
Carbon dioxide	CO_2
Ethylene glycol	$HOCH_2CH_2OH$
Ethanol	CH_3CH_2OH
Formaldehyde	H_2CO
Hydrogen cyanide	HCN
Hydrogen peroxide	H_2O_2
Iron oxide	Fe_2O_3
Methanol	CH_3OH
Methane	CH_4
Nitrous oxide	N_2O
Ozone	O_3
Propanol	$CH_3CH_2CH_2OH$
Water	H_2O

"One way to look at a molecular cloud is that it is like a giant martini."

Figure 12.11 Dark (Bok) Globules

a. Barnard 68, one of the closest dark globules to Earth, is 410 light-years away. Such globules are regions of cold dense gas, molecules, and dust that block all starlight from behind. **b.** NGC 281 includes several dark globules.

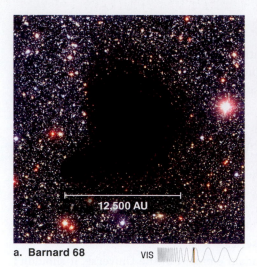

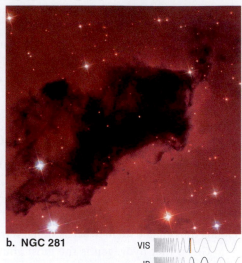

a. **Barnard 68**   VIS

b. **NGC 281**   VIS
                     IR

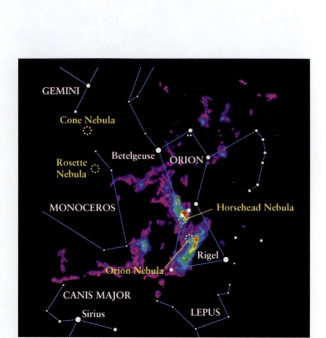

a.   RADIO

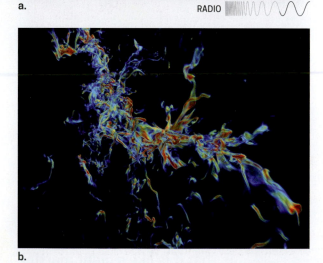

b.

Figure 12.12 Giant Molecular Clouds (GMCs)

a. A radio map of the Orion molecular cloud and **b.** a computer simulation of the formation of structure within a GMC. GMCs are not smooth spheres but irregular clouds displaying filaments and clumps.

A 50-pc-wide GMC, for example, has smaller, denser regions called *clumps*, which may extend across a few parsecs and contain enough mass for hundreds of stars. Astronomers find within clumps many denser conglomerations called *cores*, which may be 0.1 pc or less in diameter and contain enough mass for just a single star. Given these nested sets of ever-denser structures, it is likely that objects such as dark globules were once chunks of a larger cloud but then separated from the parent body.

Although astrophysicists might assume a cloud to be spherical for theoretical simplicity, the GMCs that are home to the most active sites of star formation are not even close to that (**Figure 12.12**). "Scientists like to picture things as spheres because it makes the math easier," says Pipher, "but what we see is really much messier. Many clouds appear as a network of long, dense filaments with less dense gas in between."

The shapes of clouds may give an indication of their history. For many years, astronomers believed that molecular clouds assemble slowly. The idea was that atomic gas in H I clouds would condense into molecular form once a chance bump or wiggle allowed the cloud to become dense enough to shield the interior from ultraviolet light. According to this hypothesis, once the molecular clouds formed, they would persist for many millions of years. The inward-directed gravitational weight would also be balanced by either the clouds' thermal pressure or magnetic fields. This idea implied that molecular clouds are relatively quiet structures without turbulence.

However, Doppler shift measurements have shown that many clouds are turbulent; they are filled with fast-moving, randomly directed gas, as mentioned earlier. Some astronomers hypothesize that the observed turbulence is an important clue to the formation and history of molecular clouds. They propose that clouds form from collisions between huge gas flows in the galaxy. These flows could come either from the galaxy's rotation or from some other mechanism, such as supernova blast waves. In either case, when these enormous streams collide, dense pockets of gas form, allowing molecular chemistry to get started. Clouds formed through this kind of violence still show turbulent motions. Note that, in both models, star formation takes on the order of a million years.

These two models of cloud formation—quiet coalescence versus turbulent collisions—represent the cutting edge of studies of star-forming clouds and their histories. They also represent the kind of split that often occurs at the frontiers of science. The advocates for competing models battle it out for years until new data, often obtained through new observing technologies, allow the issue to be settled.

Astronomers also understand that all molecular clouds are threaded by magnetic fields. Their strength is under debate, but in general, the fields do not appear strong

How do we see inside star-forming regions?

Dark globule Barnard 68

VIS   IR

observation

E. E. Barnard discovered and cataloged some of the coldest, densest molecular clouds—including the dark globule Barnard 68 in 1919. In visible light, these star-forming regions appear as empty patches on the night sky. Modern infrared (IR) cameras let astronomers peer through the same clouds, showing how different the Universe looks at different wavelengths of observation.

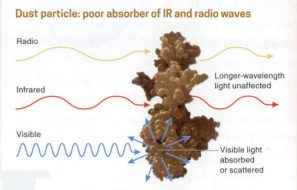

Dust particle: poor absorber of IR and radio waves

Radio

Infrared — Longer-wavelength light unaffected

Visible — Visible light absorbed or scattered

technique

Light at longer-than-visible wavelengths (IR, radio, and so on) has two important traits: (1) It passes through dense clouds of dust more readily than visible light does. (2) Cool objects, such as star-forming regions, emit light mainly at IR and longer wavelengths. So astronomers who study star formation rely on telescopes that detect long-wavelength light.

improved observations

Judy Pipher's pioneering work on IR cameras was used in the first ground-based IR telescopes in the late 1970s. With William Forrest and colleagues, Pipher also developed some of the first space-based IR instruments, including one on the **Spitzer Space Telescope**. Launched in 2003, Spitzer revolutionized the study of star formation. Spitzer even captured a burst of star formation at the dusty interface of two colliding galaxies.

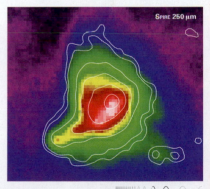

Barnard 68 (false-color 3D map)

SPIRE 250 μm

IR

Star formation in colliding galaxies

improved observations

In 2009, the **Herschel Space Observatory** was launched. Its IR telescope's image of Barnard 68 shows emission from the dust in the cloud, thought to be a future birthplace of low-mass stars in the next 200,000 years. Such images are deepening astronomers' understanding of early stages of star formation.

JUDY PIPHER

"My mom was a biologist," says Judy Pipher, "so we never grew up with any biases about girls and science. I liked physics quite a bit, and that was my major at the University of Toronto." An observational astronomy course changed her life's direction—and the field of astronomy. "That class in my junior year just let me see how physics was the basis of astronomy."

Pipher found she was good at building infrared (IR) detectors and interpreting their observations. As a professor at the University of Rochester, she became a leader in advancing generations of IR instruments. Pipher is considered the "grandmother" of IR astronomy. "I am not sure if I like being the grandmother of anything," she laughs. In 2007 she was inducted into the National Women's Hall of Fame.

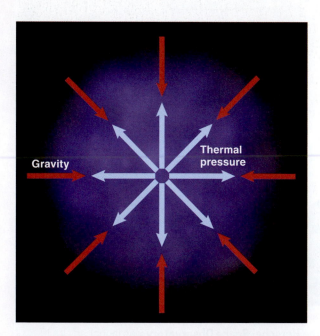

Figure 12.13 Thermal Pressure Resisting Gravity
A molecular cloud may resist collapse for millions of years by balancing the inward pull of gravity and the outward push of thermal pressure from collisions among gas particles. (See also hydrostatic equilibrium, Chapter 10.)

enough to keep most clouds from collapsing. The fields are strong enough, however, to affect the way a cloud behaves as it collapses during star formation, as we will see next.

Section Summary
- The high densities in molecular clouds enable molecules formed on dust grains to remain shielded from radiation that would break them apart.
- Molecular hydrogen dominates in molecular clouds, but many other molecules occur—notably carbon monoxide, whose abundant emission allows us to study the structure of the clouds.
- Molecular clouds are networks of long, dense filaments as opposed to spherical shapes.
- Turbulence in molecular clouds may be the result of collisions between large-scale gas flows.

CHECKPOINT What do you think is the link between molecular clouds being the densest version of the ISM and their being the site of star formation?

From Cloud to Protostar

12.4 Now that we've explored the properties of molecular clouds, we're ready to ask how stars form within these clouds. How does the mass in the clouds become so concentrated that densities and temperatures are high enough for fusion reactions to begin? The key to the story is gravity and its ability to drive the cloud, or at least a part of it, into collapse.

Collapse

Like the Sun and other stars, molecular clouds are at war with their own gravity. "The clouds, however, tend to lose that war a lot faster than stars," says Judy Pipher. Molecular clouds contain a large amount of mass in a relatively small region of space; they are often close to the point at which gravity can overwhelm gas pressure and cause the cloud to collapse on itself. Stars, in contrast, have the benefit of nuclear fusion reactions in their cores that convert mass into energy, generating enough pressure to counteract the force of gravity for most of their lives. "Molecular clouds can't generate any new energy," says Pipher. "What they can do is support themselves with thermal energy gained during their formation. If the cloud does not radiate its energy quickly, it can maintain a balance between gravity and pressure [**Figure 12.13**] for millions of years."

Molecular clouds may also get limited support from internal magnetic fields (recall the discussion in Chapter 10 of magnetic-field pressure in sunspots) or from the tumbling chaos of their own turbulence. In both cases, the fields or the turbulence provides an outward push that helps counter the inward pull of gravity. But these support mechanisms cannot last forever. Thermal energy is eventually lost as the cloud emits radiation and cools, magnetic fields dissipate with time, and turbulence eventually slows down. Without renewed energy deposition into the cloud, gravity eventually wins.

Even a cloud that reaches a balance between gravity and internal forces might not be stable. "You can balance a pencil on its point if it's in equilibrium," says Pipher, "but if you nudge the pencil, it will fall. That means the equilibrium wasn't stable. In the same way, a cloud can be in equilibrium between gravity and pressure, but a passing supernova shock can compress it enough to push it into collapse."

Gravitational collapse occurs when internal forces can no longer support some portion of a molecular cloud against its own gravity. Specifying that a *portion* of the cloud begins collapse is important. Astronomers never see an entire GMC collapsing all at once. Instead, smaller regions of the cloud—the cores—are the sites where gravity overwhelms pressure. "Once gravity starts winning," says Pipher, "the core begins falling in on itself."

To develop theories of cloud collapse, astronomers imagine following a "parcel," or blob, of gas as it falls toward the core's center. As many parcels drop inward, the density of gas increases: ever more material squeezes into an ever-smaller space. During the initial phases of collapse, the falling gas maintains its low temperature. Any heat generated as material falls inward and gets denser can escape into space as radiation. Since pressure still plays no role in slowing the collapsing gas, during this phase astronomers speak of the gas as being in free-fall.

How long does this phase of collapse last? The **free-fall time** is the amount of time it takes for a cloud of a given density to complete the free-fall phase. Remarkably, free-fall time depends on a cloud's density only: denser gas collapses faster than less dense gas. The radius of the cloud does not affect the free-fall time. For typical conditions in a cloud core, the free-fall time is 1 to 2 million years. "That is why a million years is the typical timescale astronomers think about for assembling a star," says Pipher.

Contraction to a Protostar

"Collapse can't last forever," explains Laura Arnold, a former graduate student at the University of Rochester who studied star formation. "As the density at the center of the collapsing core increases, any photons the gas emits begin to get trapped. The infalling gas starts heating up, and pressure begins to play a role again." Trapping radiation in the collapsing cloud's dense center means trapping heat, and that means pressure can begin to halt the collapse.

Where does this heat come from? Physicists have found that energy in a physical system is always conserved (see Chapter 4). That means the total energy at the beginning of a process must equal the total energy at the end. Energy can, however, be transformed from one form into another. Star formation begins with an extended cloud with a lot of gravitational potential energy (like a book on a high shelf). As collapse proceeds, that gravitational energy is converted into the kinetic energy of motion—the energy associated with gas in free-fall toward the center of the core (like a book falling off a shelf). As the density at the center of the collapsing cloud increases, particles begin to collide more frequently. The kinetic energy of motion is then transformed into heat energy. In the collapse's early phases, this heat is effectively radiated away as photons. But in the later phases, density at the collapse center becomes very high and traps the emitted photons, turning them back into heat and thus raising the temperature and pressure of the gas.

The free-falling collapse of the central regions ends when the gas pressure at the center becomes high enough to decelerate infalling material. The center is now a warm, dense, extended object called a **protostar**. Further evolution of the protostar is governed by how fast photons can radiate away from its surface. Only by losing energy in the form of light from the surface while the density at the center increases can the protostar continue to shrink. This process of contraction dominates the subsequent evolution of the protostar.

In general, a protostar has a radius of 50 million km, about the distance at which Mercury orbits the Sun. Densities within the protostar are, on average, 10^{18} particles/cm³. Temperatures at the surface are 3,000 K and at the center climb to 1 million K. Although this is much higher than the temperature in the original cloud (10 K), it is still too low to initiate nuclear fusion. Thus, a protostar is an object that shines because of escaping heat energy alone. (The photons that escape are mainly in the infrared and therefore cannot be observed with optical-wavelength telescopes.)

On a Hertzsprung-Russell (HR) diagram of luminosity versus temperature, protostars appear to the right of and above the main sequence. From this region, stars contract and get hotter, "moving" toward the main sequence and core hydrogen fusion (**Figure 12.14**). The path a star of a given mass takes to the main sequence is called a *Hayashi track* after Chushiro Hayashi, the Japanese astrophysicist who first calculated them. High-mass protostars appear at higher luminosities than their low-mass cousins, just like the main-sequence stars they eventually become.

"The clouds tend to lose that war a lot faster than stars."

"Collapse can't last forever."

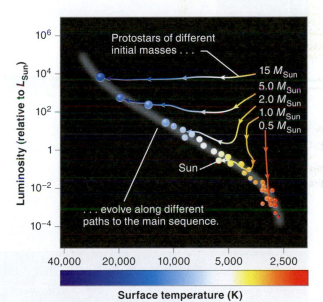

Figure 12.14 Evolution onto the Main Sequence
Protostars enter the HR diagram at upper right. They are cooler and more luminous than the stars they will eventually become. From this region, stars of different masses begin their evolutionary tracks toward the main sequence and core hydrogen fusion. Colors along the tracks indicate the temperature of the (proto)star at the time.

free-fall time The length of time required for a cloud to collapse under the force of its own gravity.

protostar A dense central object, plus an accretion disk and envelope, that forms after cloud collapse occurs but before nuclear fusion begins.

Why a Protostar Is More Than a Star

In the process of collapse, more structures than just a star are formed. "The centrally condensed object is not the whole story," says Laura Arnold. "A protostar's structure includes the thing that is going to become a star and an accretion disk spinning around it."

Since molecular clouds are turbulent, a cloud's gas rotates continuously. "Each chunk of the cloud has some spin," says Arnold. By conservation of angular momentum, the speed of rotation increases as the mass approaches the axis of rotation. This is why an Olympic skater spins faster upon pulling his or her arms inward (see Chapters 3 and 5). Similarly, as a spinning cloud core begins its gravitational collapse, its rate of spin increases. Gas along the core's spin axis, however, has little rotational speed. It can fall straight onto the forming protostar. The rest of the material takes a spiral path inward. These parts of the cloud fall inward toward the core's center while they rotate ever faster around the rotation axis (**Figure 12.15**).

For gas located at the core's "equator"—the plane perpendicular to the rotation axis—the increased rotational speed about the axis prevents that gas from falling onto the surface of the forming star. "That stuff goes into orbit around the protostar," says Arnold. In this way, a disk of gas and dust forms. Astronomers consider this disk to be part of the extended object they call the protostar.

Material in the disk, however, doesn't orbit forever. Friction slows the disk material, allowing it to spiral slowly inward. Eventually, much of the gas in the disk makes it all the way to the newly forming star, where it falls—or *accretes*—onto the central object. These *accretion disks* are also the locations of planet formation (see Chapter 5). "That's why accretion disks are so important," says Arnold. "They help build the star *and* they're the place where planets grow" (**Figure 12.16**). This dual role compels astronomers to expend enormous effort studying the properties of accretion disks, using both observations and theoretical tools such as computer simulation.

Finally, protostars—the dense central objects plus their accretion disks—are always surrounded by a third component, called the **envelope**. The envelope is made of material from the original gaseous core that is still free-falling inward. As the system evolves, the envelope continuously falls toward the center until eventually only the star and disk remain.

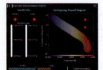

INTERACTIVE:

Evolutionary Tracks

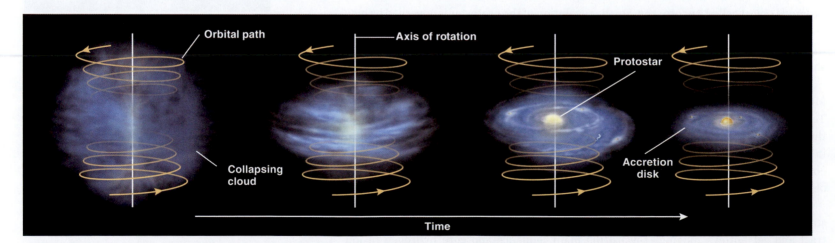

Figure 12.15 Formation of a Protostar

The spinning core of a forming star collapses along its axis of rotation, creating a central protostar surrounded by a flat accretion disk. As the cloud collapses, conservation of angular momentum leads to higher rotation rates. Material along the poles can plunge to the center; material near the equator eventually rotates fast enough to orbit the protostar. Thus, what began as an extended, slowly rotating spherical cloud ends up as a much smaller, flattened, rapidly rotating disk.

In the later stages of their evolution, protostars with mass less than 3 times the Sun's mass enter the **T Tauri star** phase. The name *T Tauri* comes from the ideal example of a young star discovered in the constellation Taurus. T Tauri stars are variable stars (see Chapter 11) that emit X-rays created by both powerful flares on the star's surface and magnetic interactions between the star and its surrounding disk.

Navigate the dangers of a star-forming accretion disk in Mission 2 and recover an artifact of unknown but dangerous origin.

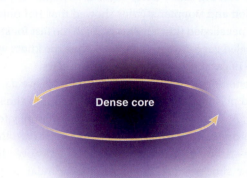

a. Molecular cloud

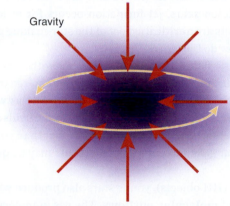

b. Gravitational collapse

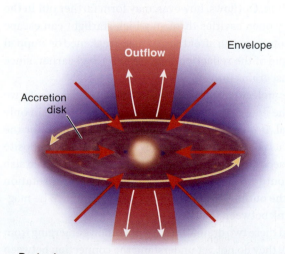

c. Protostar

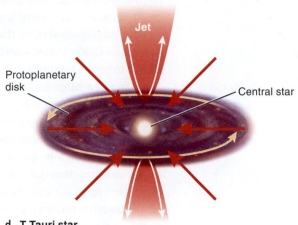

d. T Tauri star

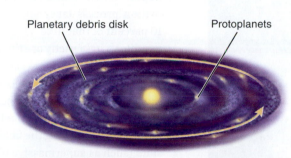

e. Pre-main-sequence star

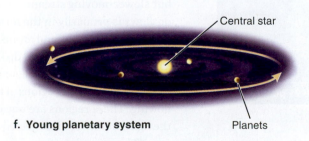

f. Young planetary system

Figure 12.16 Stages of Formation of a Star and Planetary System

A protostar consists of a centrally condensed object (which will eventually begin nuclear fusion reactions) and its surroundings, which include an accretion disk and the infalling envelope of gas and dust around it. **a.** The initial cloud **b.** collapses and **c.** forms the protostar, which includes the central object, accretion disk, and envelope. Outflows and jets also form during this stage. **d.** As the envelope thins and the central object continues to contract, the protostar enters the T Tauri phase, and **e.** planets begin to form within the disk. **f.** Evolution leads to a main-sequence star and planetary system.

envelope The region surrounding a protostar that is composed of free-falling gas.

T Tauri star A variable star of less than 3 solar masses that is in a later stage of star formation and exhibits bright X-ray flares.

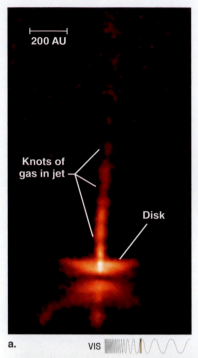

200 AU

Knots of
gas in jet

Disk

a.   VIS

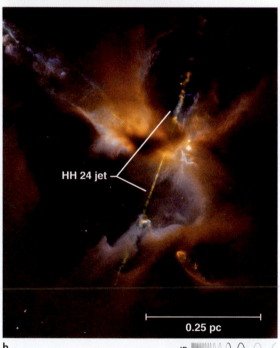

HH 24 jet

0.25 pc

b.   IR

Figure 12.17 Herbig-Haro (HH) Objects
These objects are outflows of gas created during star formation. **a.** The knots of gas in HH 30 are driven away from their
star by pulses at the source of the jet. By tracking the motion
of these knots over time, astronomers have measured the jets'
speed. **b.** Bipolar jets HH 24 emerge from a new star forming in
the Orion B molecular cloud complex.

Jets and Outflows: Star Formation Fireworks

In the 1980s, just as astronomers had started to identify newly forming stars, they were beginning to understand the role of an actor that had been discovered in the 1950s. "In almost every star-forming region they looked at," explains Arnold, "astronomers also found these bright knots moving at really high speeds." The bright knots were first seen by the Mexican astronomer Guillermo Haro and were also observed by the American astronomer George Herbig. At first these **Herbig-Haro (HH) objects** were thought to be newly forming stars. Then, proper motion and Doppler studies showed that HH objects race through space at more than 100 km per second (km/s). This is much too fast for stars. But by the 1980s, better telescopes revealed gas between the knots. Thus the knots were shown to be the brightest parts of beams of gas that are racing away from stars.

This new information helped astronomers understand that HH objects are part of **protostellar jets**—beams of high-speed plasma launched from protostars into the surrounding molecular cloud (**Figure 12.17**). At least 50 percent of all young stars may form protostellar jets. "In some cases the jets are enormous," says Arnold. "We can see them stretching across 5 pc or more." Ahead of the jet, where fast-moving jet gas overtakes slower-moving material, are bright bow shocks—curved shock waves, similar to bow waves that form in front of a boat plowing through water. The jets begin during the earliest protostellar phases and continue as long as the disk is present, even after fusion has begun in the new star.

If we take 100 km/s to be the typical speed of an HH object, then a 5-pc-long beam implies that a central star drives its jet for about 50,000 years. Since the entire star formation process takes approximately a million years, jet formation occurs for at least 10 percent of that time. This suggests that jets are a critical aspect of the star-making process. But what exactly are these jets, and how do they form?

"Most astronomers agree that magnetic fields have something to do with creating jets," says Arnold. In the most popular models, strong magnetic fields that emerge from the accretion disk act like rigid wires. As the disk rotates, the field lines rotate as well. Because charged particles feel magnetic forces, plasma becomes attached to the fields and is flung outward along the field lines. Since jets are seen in environments other than young stars (such as supermassive black holes near galactic centers), this process may be quite common.

Along with jets and their bright knots (HH objects), young stars also produce wider but slower-moving streams of matter called **molecular outflows**. The gas in molecular outflows is primarily in the form of molecules, whereas protostellar jets tend to be composed mostly of ions and atoms. Molecular outflows most likely form in the same way as jets: via magnetic fields and disk rotation. Outflows, however, may form farther out in the disk, away from the central star. They open cavities through which starlight can escape from the dusty inner regions of the protostar even after the jets have weakened or stopped flowing. These regions are one example of the reflection nebulas described earlier, since the light that we see is reflected off dust grains into our line of sight.

"There is a lot of mass in these molecular outflows," says Arnold. "Sometimes we see 0.1 to more than 1 solar mass in the flows." Protostellar jets, in contrast, contain only a fraction of a Sun's worth of material. Both jets and molecular outflows appear, for the most part, as *bipolar outflows*: material is observed moving away from the star in opposite directions along poles defined by the accretion disk's axis of rotation (**Figure 12.18**; see also Figure 12.16). "The fact that bipolar outflows are aligned with the disk's axis of rotation tells astronomers that the disks and the outflows are connected," says Arnold. "The magnetic fields in the disk are the likely link between the two."

Although astronomers have found these two dramatic forms of outflows emerging from young stars (often from the same star), they do not yet understand the connection between them. An early proposal was that jets drive the molecular gas, just as a snowplow pushes snow.

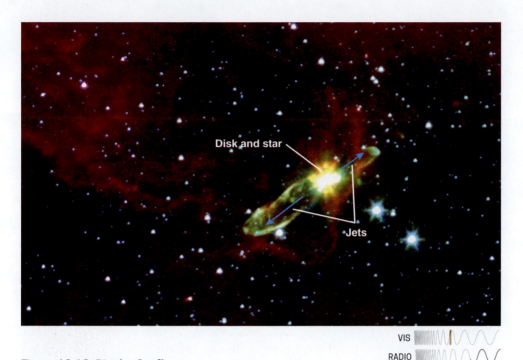

VIS
RADIO

Figure 12.18 Bipolar Outflows
A composite image of object HH 46/47, a bipolar (two-sided) outflow, which emerges from a star embedded in a molecular cloud. Here the HH jet moving to the right through the cloud creates a cavity (one lobe of the bipolar outflow). The jet streaming to the right breaks out of the dense cloud, so its swept-up cavity is not as apparent. Light is emitted by both atoms (the jets) and molecules (the cavities). Not all bipolar outflows show internal jets.

LAURA ARNOLD

Laura Arnold's story started with a love of science engendered by supportive parents—but she soon found her own ways to engage with the subject. When she started college at the University of Rochester, where her mother works, Arnold was sure she was headed toward a PhD and a research career in astronomy. But during a summer internship in Germany, she had the chance to focus on teaching. "I'm really excited and kind of passionate, and I like to talk to people. A life focused only on research wasn't going to be a good fit for me."

Arnold threw herself into learning how to teach astronomy. After graduation she looked for science education programs and began to work toward becoming an astronomy educator. "This is what I love, and I am really excited that I can spend my life doing it."

But the jets' energy and momentum are too weak to drive all of the observed molecular material into motion. (Imagine using a snowplow connected to a go-cart to try to push a mountain of snow.) For now, the relationship between jets and outflows remains a mystery.

Section Summary

- Like stars, clouds must have energy to resist the collapsing influence of gravity. If overwhelmed by gravity, a portion of a cloud will free-fall inward.
- A protostar forms when a cloud's density and temperature are high enough to trap heat and generate pressure in the interior but not so high as to initiate nuclear fusion. Further contraction radiates heat energy from the surface.
- During collapse, conservation of angular momentum amplifies a protostar's spin. Material from the resulting accretion disk eventually falls onto the protostar or may form planets.
- Fast-moving protostellar jets and slower-moving molecular outflows emanate from young stars and are thought to be caused by magnetic field lines from the disk and star.

CHECKPOINT Describe the evolutionary stages of a protostar.

"Most astronomers agree that magnetic fields have something to do with creating jets."

From Protostars to Fusion and Brown Dwarfs

12.5 A protostar is a long way from generating its own energy via nuclear fusion. The density and temperatures in its core are too low for hydrogen nuclei (protons) to overcome electromagnetic repulsion and fuse into helium. Recall from Chapter 10 that the proton-proton (P-P) chain, which occurs in stars such as our Sun, requires central temperatures of greater than 10 million K. How do

Herbig-Haro (HH) object A protostellar jet that is distinguished by knots in the jet beam.

protostellar jet A focused beam of plasma launched at high speed from a protostar.

molecular outflow The bipolar swept-up shells of molecular material, which may be driven by protostellar jets.

Figure 12.19 Carbon-Nitrogen-Oxygen Cycle

In this cycle, the dominant source of energy in stars larger than about 1.3 M_{Sun}, massive stars convert hydrogen to helium through the decay of unstable isotopes and the fusion of carbon, nitrogen, and oxygen nuclei with hydrogen nuclei. It is most effective at temperatures of 16 million K and above. Each "flash" represents a transformation—a decay or a reaction.

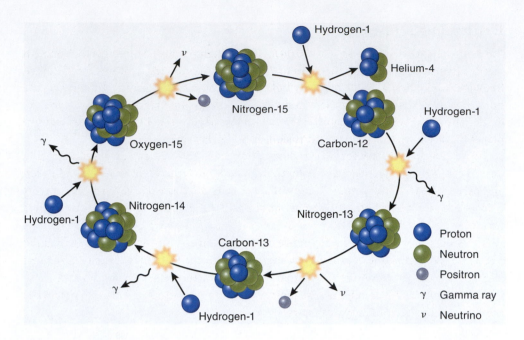

> "Brown dwarfs are essentially failed stars."

CNO cycle A chain of nuclear fusion reactions, involving carbon, nitrogen, and oxygen, that produces helium; the mechanism for powering stars of more than 1.3 solar masses.

brown dwarf A failed star that has too little mass to drive core temperatures high enough to initiate hydrogen-burning nuclear fusion.

initial mass function (IMF) The mathematical relationship describing the relative abundance of stars of different masses.

protostars reach that magic temperature and graduate into full-fledged stars? Time and contraction are the answer, and along the way a lot more evolution has to occur. That evolution is closely tied to a star's mass.

Proton-Proton Chain and Carbon-Nitrogen-Oxygen Cycle

The protostar continues to shrink as heat energy escapes from its surface. With each decrease in radius, both the central density and the central temperature increase. The protostar's surface temperature, however, does not increase very much. During this final contraction phase, the evolutionary track takes the protostar toward the main sequence on the HR diagram. The track moves in a mix of downward (lower luminosity) and leftward (higher temperature) directions depending on the protostar's mass.

As the core temperature increases, a series of fusion reactions is initiated (such as the fusing of two protons into deuterium), but these reactions are relatively ineffective at generating energy. Eventually, in protostars larger than about 0.08 solar mass (M_{Sun}), a full network of reactions begins that turns hydrogen nuclei into helium nuclei. This transformation occurs by way of the P-P chain in Sun-like stars.

In stars just 1.3 times more massive than the Sun and greater, where core temperatures exceed 16 million K, another route for fusion of hydrogen to helium dominates: the carbon-nitrogen-oxygen or **CNO cycle**. In the CNO cycle, nuclei of the elements carbon, nitrogen, and oxygen play an intermediate role. The chain of reactions begins when a hydrogen nucleus fuses with a ^{12}C nucleus, creating an unstable (radioactive) nucleus of ^{13}N. The ^{13}N then decays into ^{13}C, which combines with another hydrogen nucleus, forming ^{14}N. The chain of reactions continues until, finally, the nucleus splits into a ^{4}He and a ^{12}C. The carbon nucleus then begins the cycle anew (**Figure 12.19**). The CNO cycle is very temperature sensitive. Small increases in core temperature lead to much larger increases in energy production in stars dependent on the CNO cycle compared to (less massive) stars fueled by the P-P cycle. (Chapter 13 explains the consequences of this sensitivity for the lifespans of stars. Note that the CNO cycle can also be initiated for lower-mass stars like the Sun after the main sequence.)

For the CNO cycle to occur, the star must already contain some carbon, nitrogen, and oxygen. Those elements must have come from earlier generations of stars. When the stars died, these elements were blown back into space, eventually ending up in the cloud that formed today's protostars.

Brown Dwarfs

For protostars of very low mass ($M < 0.08\,M_{Sun}$), core temperatures never get high enough to burn hydrogen. With too little mass to drive further contraction, these objects, called **brown dwarfs** (Figure 12.20), can only initiate the fusion of deuterium. They spend billions of years slowly bleeding their stores of heat energy into space. "Brown dwarfs are essentially failed stars," says Laura Arnold. "They lie somewhere between a very massive planet like Jupiter and a full-fledged, fusion-powered, low-mass star."

The range of masses for brown dwarfs bookends the two categories that Arnold describes. Any object with a mass larger than $0.08\,M_{Sun}$ drives hydrogen fusion in its core and must be considered a star. The lower end of the mass spectrum for brown dwarfs is less certain, though some researchers put it at $0.01\,M_{Sun}$, which is about 10 Jupiter masses.

"Using fusion to define the difference between a star and a brown dwarf is very clear," says Arnold, "but the line between a low-mass brown dwarf and a high-mass planet is less clear." Generally, astronomers believe that brown dwarfs form from gravitational collapse in molecular clouds. Planets, even the most massive ones, are believed to form within the accretion disks surrounding young stars.

High-Mass Stars versus Low-Mass Stars

What brown dwarfs lack in mass they make up in number. Astronomers expect that up to 100 billion brown dwarfs may be floating freely in our galaxy; a smaller number may be members of binary or other multiple-star systems with fusion-powered stars. In total, the number of brown dwarfs is probably equal to the number of fusion-powered stars in the Milky Way.

Thus, the galaxy has far more brown dwarfs than 1-M_{Sun} stars. In turn, there are far fewer high-mass stars than 1-M_{Sun} stars. This overall negative correlation between stellar population and stellar mass is neatly captured in a relationship called the **initial mass function (IMF)**. The distribution of stellar masses is set at birth as stars form from molecular clouds. The IMF tells astronomers how many 1-M_{Sun} stars they can expect to form relative to the number of 0.5-M_{Sun} stars, how many 2-M_{Sun} stars they should expect relative to 1-M_{Sun} stars, and so on.

The shape of the IMF was discovered in the 1950s by astronomer Edwin Ernest Salpeter, who found a steep decrease in the number of stars with larger and larger mass (Figure 12.21). From Salpeter's work, astronomers expect that for every 1,000 1-M_{Sun} stars that form, only a single 10-M_{Sun} star should be created. Massive stars, which astronomers define as having a mass greater than $8\,M_{Sun}$, are thus relatively rare in the galaxy.

The reasons that stars form with such a particular distribution of masses (the IMF) is still a mystery to astronomers. They do understand that the gravitational collapse process favors fragmentation into smaller, rather than larger, protostars, but the detailed physics that sets the IMF into its exact mathematical relationship remains an open question.

Section Summary

- Protostars contract over hundreds of thousands of years, increasing their internal temperatures.
- A protostar's mass determines whether its future fusion process will be the P-P chain or the CNO cycle.
- Brown dwarfs are basically failed stars that do not have enough mass to initiate fusion as they contract.
- The initial mass function (IMF) tells astronomers the relative abundance of stars with different masses; there are many more low-mass than high-mass stars.

CHECKPOINT Why does the mass of a protostar determine what kinds of nuclear fusion (if any) can occur in its core?

VIS
IR

Figure 12.20 Brown Dwarfs

A young star cluster, NGC 1333, containing swarms of brown dwarfs (circled). In this image, taken with the Subaru Telescope, brown dwarfs were identified in two separate surveys. The arrow marks the least massive brown dwarf known in this cluster, at about 6 times Jupiter's mass.

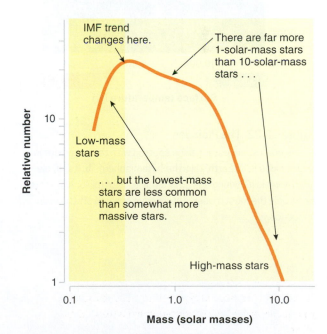

Figure 12.21 Initial Mass Function (IMF)

Not all stellar masses are equally represented in our galaxy: lower-mass stars far outnumber high-mass stars. The IMF describes the steep decrease in the numbers of stars with larger and larger masses. Only at the lowest end of the stellar-mass range do we see the trend reverse direction.

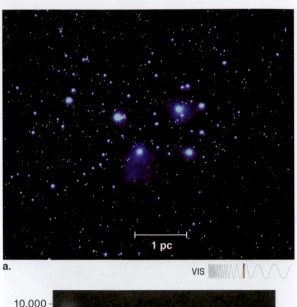

a.

VIS

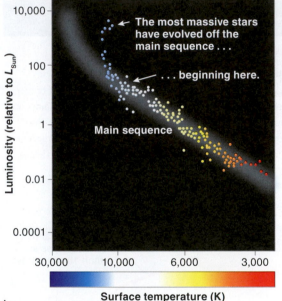

b.

Figure 12.22 The Pleiades

a. This open star cluster can be spotted with the naked eye in winter skies from Earth's Northern Hemisphere. **b.** Its HR diagram reveals its young age (only a few tens of millions of years), since only the most massive stars have evolved off the main sequence.

Stellar Interaction

12.6

So far, we have discussed star formation as though each star forms in isolation. In most cases, however, stars form in relatively close proximity to other stars, and they can affect one another's evolution.

A Family Affair: Young Star Clusters

Stars are almost never born alone. Instead, they form in families called *young star clusters*. The dense regions called clumps within giant molecular clouds, with masses that are typically in the range of 1,000 M_{Sun}, are the forerunners of star clusters. Young star clusters and the globular clusters discussed in Chapter 11 are similar in that, in both cases, all of the stars formed together. The difference, however, is that eventually the stars in a young cluster will disperse; in other words, they will not remain gravitationally bound together as the stars in a globular cluster do.

The best-known young star cluster is the Pleiades, a collection of over a thousand stars visible from Earth's Northern Hemisphere (**Figure 12.22a**). The Pleiades is an **open star cluster**, meaning that the stars are not as tightly packed together as they are in some clusters. Seven bright, massive stars called the "Seven Sisters" dominate our view of the Pleiades. (These noticeable stars became the logo of auto maker Subaru, the Japanese name for "Pleiades.") **Figure 12.22b** shows an HR diagram for all stars in the Pleiades. Notice that most of the stars are on the main sequence; only the most massive stars have begun to evolve away from hydrogen fusion. "That means the Pleiades must be young," says Laura Arnold. "All of its stars were born less than tens of millions of years ago."

Other clusters, such as NGC 1333 in the Perseus molecular cloud, contain only a few hundred stars, none belonging to the most massive class O. Some clusters, however, are far more densely populated. "The number of stars that form within a cluster," says Arnold, "will depend on the size and conditions in the clump. Massive clumps make lots of stars (and some O stars too), while low-mass clumps tend to make clusters with fewer stars and mostly low-mass stars." Thus, massive stars appear only in massive clusters where many thousands of stars (most of them low- and intermediate-mass) are born. The Orion Nebula, an H II region at the edge of the Orion molecular cloud that can be seen even with binoculars, is considered to be a rich, young star cluster because it contains so many new stars. An even more spectacular example is the Carina Nebula, visible from Earth's Southern Hemisphere (**Figure 12.23**). This nebula contains many rich clusters, including the massive binary system Eta Carinae, with a combined mass of approximately 125 M_{Sun} and at least 14,000 stars in total.

Compared with the range of ages across the rest of the galaxy, the stars in a cluster can be thought of as having the same birthday. On closer examination, these stars do have a range of ages, but it's narrow—on the order of a few million years. "Like triplets in a human family, some stars will be born before others," says Arnold, "even though on galactic timescales they all appear to be born at the same time."

The age range within a cluster may be a consequence of the turbulence occurring in the collapse process. Another contributing factor may be *feedback*, a process in which star formation in one region of a cluster affects star formation in another region. Some astronomers have argued that the jets, winds, and radiation from newborn stars can suppress or enhance local star formation. Feedback is most dramatic in **triggered star formation**, a process in which a supernova blast wave, created by the death of a nearby massive star, sweeps through a molecular cloud, compressing gas and pushing it into gravitational collapse (**Figure 12.24**). If triggered star formation occurs in a molecular cloud, astronomers would expect to see successive generations of star clusters spread across a region of the cloud.

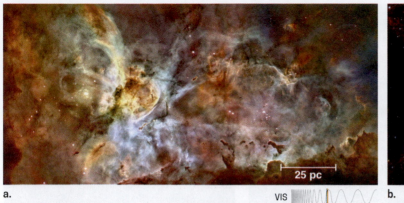

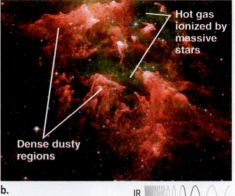

a.

b.

VIS

IR

Figure 12.23 Molecular Clouds in the Carina Nebula
a. Dark knots in this nebula are regions of dense molecular gas. **b.** The complex structure of a large region of the Carina Nebula's molecular cloud.

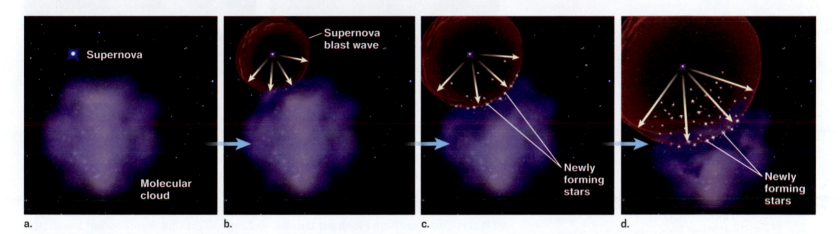

a.   b.   c.   d.

Figure 12.24 Triggered Star Formation
Star evolution in one region of a cluster may affect star formation in another region. **a, b.** The supernova blast wave from the death of a nearby massive star sweeps through a molecular cloud. **c, d.** By compressing and gravitationally collapsing the cloud, the blast wave sets off a new cycle of star formation. Successive generations can occur if massive stars form in the first cycle and later become supernovas themselves.

High-Mass Star Formation: Mess with Your Surroundings

High-mass stars—in particular, class O stars with masses of at least 25 M_{Sun}—dominate the regions of a molecular cloud where they form, and sometimes far beyond. Recall from Chapter 11 that O stars have high temperatures, above 30,000 K. "Look at a blackbody spectrum for a star like this," says Arnold, "and you'll see it generates lots and lots of ultraviolet photons." As we saw in Section 12.2, ultraviolet photons with a wavelength less than 91.2 nm can photoionize neutral H atoms. O stars, which produce a torrent of these photons, are cosmic engines of ionization. "And it's not just neutral atoms that feel the heat," says Arnold. "Molecules get dissociated by the ultraviolet light of O stars too."

A single O star can ionize material out to 100 pc or so—an enormous volume of space. When ionized, the gas surrounding an O star is also heated. "Molecular material that was 10 degrees above absolute zero before the O star 'turned on,'" explains Arnold, "gets flash-ionized and heated. Its temperature jumps to 10,000 degrees. High temperatures mean high pressures, so the ionized gas expands." Like a rapidly inflating balloon, the hot gas sweeps up surrounding, un-ionized cloud material. Eventually, the expanding bubble, now an H II region, reaches the edge of the molecular cloud and rapidly blows outward into the lower-density ISM. Thus, ionized H II regions created by O stars are agents for dispersing the clouds.

open star cluster A loosely distributed group of stars that were born from the same cloud at the same time and remained bound for some period before dispersing.

triggered star formation Star formation that is initiated by the compression of a molecular cloud by a supernova blast wave.

Figure 12.25 Elephant Trunks
a. The "Pillars of Creation," elephant trunks in the Eagle Nebula. Ionizing radiation from a massive star (out of view above this image) erodes dense molecular gas at the edges of many H II regions, forming these structures. **b.** Star formation can occur within the densest regions of elephant trunks.

a. VIS

b. IR

> "If dense knots in the elephant trunk haven't formed stars by that time, their evolution will be cut short."

As the ionizing radiation eats into cold molecular material, it sculpts some of astronomy's most astonishing and beautiful structures. The most dense knots of gas may protect the gas behind them (relative to the star), thereby shielding part of the molecular cloud from the hot star's ultraviolet photons. In this way, molecular cloud material at the edges of H II regions may form **elephant trunks**—columns of gas that seem to point toward the hot stars. Low-mass stars may form inside these trunks and can be seen only via infrared radiation (**Figure 12.25**). Eventually, however, the O stars' ultraviolet photons erode even the trunks. "If dense knots in the elephant trunk haven't formed stars by that time," says Arnold, "they get eroded too, and their evolution will be cut short."

The ionizing radiation affects more than just the molecular gas. A single newly formed O star is surrounded by thousands of lower-mass stars (recall the IMF). The histories of these newly forming low-mass stars, which are still surrounded by their accretion disks, are also altered by the O star's ultraviolet photons. When an accretion disk is irradiated by ionizing photons, its surface heats up. The pressures on the disk surface can be high enough to launch gas into space through a process called **photoevaporation**. A disk can lose so much mass via photoevaporation that any ongoing planet formation within the disk can be cut short. If the flood of ultraviolet photons from the massive star is high enough, the entire disk can be boiled away in just 10,000 years.

Demonstrating all these features of high-mass star formation is the Orion Nebula—a stellar hothouse. More than 1,000 newly formed stars are crowded together in a cluster called the Trapezium, which occupies less than a cubic parsec. (In comparison, the Sun's nearest neighbor is more than a parsec away from us.) Most of these stars are of low or intermediate mass (less than 8 M_{Sun}). The evolution of the nebula is dominated by four massive stars (one O star and three B stars; see **Figure 12.26**). The ultraviolet photons from these stars (mainly the O star) illuminate, ionize, and sculpt the Orion Nebula.

In today's Universe, there appears to be an upper limit to the mass of a newly formed star: Astronomers don't see stars above about 150 M_{Sun}. The reason for this limit is the force exerted by the very light these stars would create. High-mass stars are brighter than low-mass stars, and stars above about 125 M_{Sun} (no one knows the exact limit) may produce so much light pressure (force per area) that they cannot hold together.

elephant trunk A column of molecular gas that forms at the edge of an H II region and seems to point toward the ionizing star or stars.

photoevaporation The process by which gas is launched from the surface of an accretion disk or molecular cloud through heating by ionizing radiation from a young, massive star.

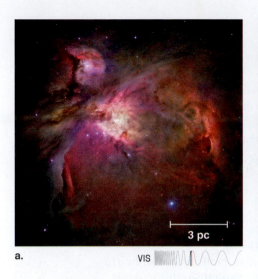 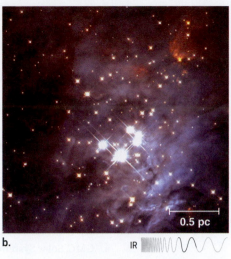

a.   VIS   b.   IR

Figure 12.26 Orion Nebula and the Trapezium
a. The Orion Nebula, with more than 1,000 stars forming, provides rich examples of the processes by which stars form from turbulent, collapsing clouds of gas and dust. **b.** The Trapezium, a cluster that includes four massive stars at the core of the Orion Nebula, provides the ultraviolet light that creates the extended H II region.

Section Summary

- Stars in clusters tend to be born at about the same time as their siblings.
- Triggered star formation results from supernova blast waves, an example of feedback in star formation.
- High-mass (very hot) stars produce enough ultraviolet photons to ionize large volumes of gas around them, creating hot H II regions that can expand into the ISM.

CHECKPOINT Why don't low-mass stars affect their surroundings as strongly as high-mass stars do?

CHAPTER SUMMARY

12.1 Seeing in the Dark

Infrared light is needed to study star-forming clouds and the star formation process, since visible light is absorbed by dust in the clouds. The Spitzer Space Telescope was one tool used in that effort.

12.2 Anatomy of the Interstellar Medium

The space between the stars (within galaxies) contains the interstellar medium (ISM), a turbulent mix of gas and dust. The ISM has different phases, including molecular clouds (where stars form), H I clouds of atomic hydrogen gas, H II regions, warm interstellar medium, and coronal gas. All phases tend to have pressures in the same range. Dust acts as a platform on which molecules can form in the ISM.

12.3 Molecular Clouds: The Birthplace of Stars

Molecular clouds are the birthplaces of stars because they are both cold and dense. They contain many molecules, such as molecular hydrogen, water, and ethanol (a form of alcohol). They vary widely in size and shape and tend to be turbulent, with dense regions and, in giant molecular clouds, a filamentous network.

12.4 From Cloud to Protostar

Gravitational collapse occurs when internal forces in a portion of a molecular cloud (a "core") can no longer support it against its own gravity. The free-fall time of a cloud core in collapse depends inversely on its density.

Gravitational potential energy inside the collapsing cloud core is transformed into infalling motions and then internal gas pressure as its core heats up. Once it is hot enough for its gas pressure to stop the collapse, the cloud core has become a protostar; further contraction requires it to radiate away energy. The original collapse of the spinning cloud core creates an accretion disk. Some of the disk's material falls onto the protostar, and some may form planets. An envelope of still-infalling gas surrounds the protostar. Bipolar outflows associated with star formation include protostellar jets and slower molecular outflows.

12.5 From Protostars to Fusion and Brown Dwarfs

Continued contraction causes the protostar to heat up enough to initiate nuclear fusion of hydrogen into helium via the P-P chain or, if the star is massive enough, the CNO cycle. If the mass is not sufficient for hydrogen fusion, a brown dwarf will result. Stars form in a broad range of masses, from 0.08 to 150 M_{Sun}. The most massive stars are very rare. The initial mass function (IMF) gives the relative proportions of stars of different masses that are expected to form.

12.6 Stellar Interaction

Stars in young star clusters are born together over a few million years. Star formation clusters are found in two sizes—high-mass and low-mass—and the most massive stars are born in high-mass star clusters. Young and fully formed stars can influence star formation around them via feedback. Supernova blast waves can cause triggered star formation.

QUESTIONS AND PROBLEMS

Narrow It Down: Multiple-Choice Questions

1. Which of the following characteristics is most likely to be the same among stars in a given cluster?
   a. age
   b. spectral type
   c. temperature
   d. mass
   e. presence of planets

2. Which phase of the ISM is most opaque to visible light?
   a. H I cloud
   b. H II region
   c. molecular cloud
   d. WISM
   e. coronal gas

3. Which of the following is/are *not* a source of turbulence in the ISM?
   a. rotation of the galaxy
   b. supernovas
   c. ultraviolet radiation from young stars
   d. jets from young stars
   e. infrared radiation from young stars

4. Given three forms of hydrogen, rank them according to increasing temperature of the environments in which they are usually found.
   a. neutral, ionized, molecular
   b. molecular, neutral, ionized
   c. ionized, neutral, molecular
   d. molecular, ionized, neutral
   e. neutral, molecular, ionized

5. Which phase of the ISM is found near hot, young stars?
   a. H I cloud
   b. H II region
   c. molecular cloud
   d. WISM
   e. coronal gas

6. Which of the following is *not* a quality that all phases of the ISM share?
   a. absence of free electrons as part of gas
   b. more gas than dust
   c. more hydrogen than other elements
   d. turbulence
   e. much lower density than that of Earth's atmosphere

7. What kinds of photons are absorbed by typical dust grains? Choose all that apply.
   a. infrared
   b. visible
   c. ultraviolet
   d. radio
   e. microwave

8. Why does interstellar dust play a vital role in the chemistry of the galaxy?
   a. It allows gamma rays to penetrate into molecular clouds.
   b. When it breaks apart, it provides new elements for the ISM.
   c. There is so much more of it than there is gas.
   d. Chemical reactions can occur on its surface.
   e. Its color changes the appearance of stars.

9. How do molecular clouds provide protection from ultraviolet photons that can dissociate molecules?
   a. Their thermal energy deflects the ultraviolet photons.
   b. Their temperature is too low for the photons to be effective.
   c. Water within the cloud absorbs the photons.
   d. Their turbulence keeps molecules in rapid motion.
   e. The dense gas and dust in their interiors shield molecules from interstellar ultraviolet radiation.

10. Which of the following is *not* a characteristic of molecular clouds?
    a. uniform structure
    b. presence of several kinds of molecules
    c. higher density than in other phases of the ISM
    d. presence of dust
    e. mass many times greater than the Sun's

11. Many astronomical objects have dense cores. Which of the following do *not*?
    a. stars
    b. protostars
    c. molecular clouds
    d. globular clusters
    e. coronal gas clouds

12. Which of the following places the objects in increasing order of radius?
    a. GMC, Solar System, Sun, protostar, brown dwarf
    b. GMC, protostar, Solar System, Sun, brown dwarf
    c. brown dwarf, Sun, protostar, Solar System, GMC
    d. Sun, brown dwarf, protostar, Solar System, GMC
    e. brown dwarf, Sun, Solar System, GMC, protostar

13. Which of the following is *not* a typical outcome for material in an accretion disk around a protostar?
    a. It accretes onto the protostar.
    b. It becomes part of a planet.
    c. It gets eroded by photoevaporation.
    d. It remains in the disk indefinitely.
    e. It returns to the ISM.

14. Two protostars have evolved to the point of nuclear fusion. One has a temperature of 12 million K; the other, 17 million K. Which of the following statements is/are true? Choose all that apply.
    a. The 17-million-K star fuses nuclear fuel faster than the 12-million-K star.
    b. The 17-million-K star is likely to use the CNO cycle.
    c. The 17-million-K star will die sooner than the 12-million-K star.
    d. The 17-million-K star can use only the P-P chain.
    e. The stars may have identical masses.

15. A star has a mass of 0.7 M_{Sun}. Which statement about it is true?
    a. It will live less than 10 billion years.
    b. Its temperature is greater than 16 million K.
    c. It is likely to use the CNO cycle.
    d. It is a brown dwarf.
    e. It is likely to use the P-P chain.

16. One or more stars form from a 1,000-M_{Sun} molecular cloud. Which of the following is likely to be true?
    a. Many stars form that are almost all 1-M_{Sun} stars.
    b. One 1,000-M_{Sun} star forms.
    c. Since molecular clouds exist near massive young stars, none of the stars will have accretion disks.
    d. A number of stars will form with a range of sizes, and most will be less-massive stars.
    e. Astronomers will be able to observe all stars that form in visible wavelengths throughout the star formation process.

17. The last supernovas in the Milky Way Galaxy were observed more than 400 years ago. Which of the following statements related to these events is/are true? Choose all that apply.
    a. The IMF explains the rarity of these events, at least in part.
    b. Each supernova may trigger the formation of new stars.
    c. Turbulence will increase in the ISM surrounding these supernovas.
    d. Coronal gas will have been created.
    e. Many more would have been observed in our galaxy with today's observation techniques.

18. Which of the following statements about interstellar dust is *not* true?
    a. It provides a platform for the formation of molecules.
    b. It emits blackbody radiation.
    c. It makes distant objects appear closer than they are.
    d. It shields star-forming regions from ultraviolet light.
    e. It is much rarer than gas in the ISM.

19. Carbon monoxide is useful as a tracer for molecular clouds because
    a. it is found in high concentrations.
    b. it can be excited to emit photons at lower temperatures than many other molecules do.
    c. it is an organic molecule.
    d. the light it emits is mostly ultraviolet light.
    e. it is found at higher temperatures than other molecules within molecular clouds.

20. An object has a mass that is less than 8 percent of the Sun's mass and an elemental composition that includes carbon, nitrogen, and oxygen. Which of the following is/are true? Choose all that apply.
    a. It could be a planet.
    b. It could be a brown dwarf.
    c. It must be on or moving toward the main sequence.
    d. It must not have reached 16 million K.
    e. It must be fusing protons into deuterium.

To the Point: Qualitative and Discussion Questions

21. Name and describe the phases of the ISM. How do these phases differ, and what do they have in common?

22. Where does the WISM's warmth come from?

23. In the Sun's structure, the corona is the outer layer of the atmosphere. Where is coronal gas found in the ISM, and what are its characteristics?

24. How does the ideal gas law relate to the ISM?

25. What two measurement techniques provide astronomers with information about the temperature and composition of interstellar dust?

26. Even though they use observations of CO emission lines to locate most molecular clouds, astronomers believe most molecular clouds are made up of molecular hydrogen primarily. Why don't they just look for the molecular hydrogen?

27. Why do molecular clouds have limited lifetimes?

28. Describe the path that protostars follow on the HR diagram as they evolve to stars.

29. Describe the components of a protostar.

30. How does a disk form around a protostar?

31. Describe the differences in the formation of a brown dwarf versus a planet; then describe the differences in the formation of a brown dwarf versus a main-sequence star.

32. How do main sequence stars affect nearby stars and clouds? What about dying massive stars?

33. Define *elephant trunks* as they relate to star formation and the ISM.

34. Describe the cause, location, and direction of jets in young stars.

35. The original mission of the Spitzer Space Telescope lasted from launch in 2003 to depletion of its helium coolant in 2009, but then the mission was redesigned to use the capabilities that remained. Do some outside reading and see whether you can identify other space missions that have done the same.

Going Further: Quantitative Questions

36. Molecular cloud A has a pressure P. Molecular cloud B has a temperature half that of cloud A and a density 3 times as great. What is the ratio of the pressure in cloud B to that in cloud A?

37. Two regions of the ISM have identical pressure, but the temperature of region 1 is 2.5 times that of region 2. How does the density of region 1 compare with that of region 2?

38. Radiation from a nearby star has heated a cloud in the ISM from 100 K to 550 K, and its density has dropped to 0.25 times its earlier density. What is the ratio of its current pressure to its pressure before these changes?

39. Region 2 of an interstellar cloud has 1.7 times the density of region 1. If the pressure of the two regions is equal, what is the ratio of the temperature in region 2 to that of region 1?

40. The temperature in a region of coronal gas is 10^6 K; a nearby H II region has a temperature of 10,000 K. If the pressure of the two regions is equal, what is the ratio of the density of the coronal gas to that of the H II region?

41. The ratio of the pressure of region A of the ISM to that of region B is 0.4. The ratio of the temperature of region A to that of region B is 1.2. What is the ratio of their densities?

42. An H II region of the ISM has a density of 10^2 atoms/cm^3 and temperature of 10^4 K. What is the ratio of its pressure to that of another H II region with a density of 10^6 atoms/cm^3 and a temperature of 10^4 K?

43. From Salpeter's projections, how many 10-M_{Sun} stars, on average, will form among 17,000 newborn stars of 1-M_{Sun}?

44. How big a population of 1-solar-mass stars would you expect if you found 52 stars, each of 10 solar masses?

45. HH 30 is 450 light-years from Earth. A knot in the jet of HH 30 moves from a distance of 1.1 pc to 1.2 pc from its protostar. What would you observe the change to be in degrees? (Hint: Use the small-angle formula from Going Further 2.1.)

"IT TURNED OUT TO BE A GOLDEN WEEKEND IN ASTRONOMY"

13

To the Graveyard of Stars

THE END POINTS OF STELLAR EVOLUTION

Fireworks in a Galaxy Not So Very Far Away

13.1 Stan Woosley had his family's ski equipment packed into the car when the call came. It was February 23, 1987, and the Universe, or at least our corner of it, had different plans for Woosley, a leading expert on the titanic stellar explosions called supernovas. "One of my graduate students called me and said, 'Check your e-mail,'" explains Woosley. "At first I thought it was a joke."

Supernovas, which mark the death of massive stars, are as rare as they are spectacular. The last one seen in our galaxy occurred four centuries ago, long before the advent of telescopes, spectrographs, and computer analysis. But when Woosley checked his e-mail that morning, he learned that a supernova had just been observed in the Large Magellanic Cloud (LMC), a satellite galaxy orbiting the Milky Way. The supernova was close enough to be seen with the naked eye (**Figure 13.1**).

After waiting 400 years, astronomers finally had a ringside seat to watch a massive star tearing itself apart. "It turned out to be a golden weekend in astronomy," Woosley explains. "An international collaboration of teams from all around the world turned every instrument they had at the supernova. There were so many discoveries happening so fast, it was extraordinary. And of course there was also just the sheer excitement and joy at this new presence in the heavens." But not everyone was overjoyed at the appearance of this supernova (SN 1987A, as it was called). "The ski trip had to be canceled," laughs Woosley. "I made it up to my family later that year."

Out, Out, Brief Candle: Why Stars Die, and Cosmic Recycling

Gravity pulls stars together, driving their central cores to extraordinary temperatures and densities at which nuclear fusion reactions become possible. In turn, only through the energy released by this nuclear fusion can stars support themselves against the crushing force of their own gravity. But sooner or later, the raw fuel for fusion runs out. When the

← The Crab Nebula is a supernova remnant—the glowing remains of a massive star 2,000 parsecs away. It was bright enough to be visible on Earth in daylight; its appearance in 1054 CE was recorded by Chinese observers. At its center is a rapidly spinning, ultradense neutron star (a pulsar), which is all that remains of the original giant.

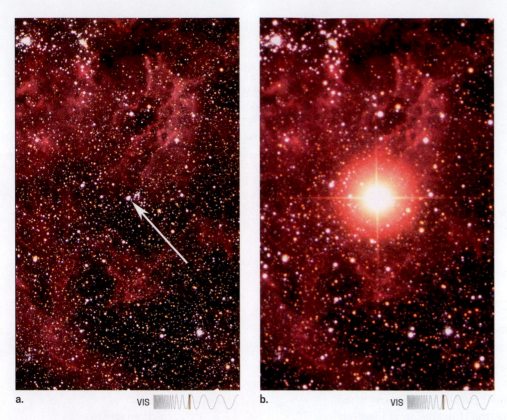

a.

VIS

b.

VIS

Figure 13.1 Supernova 1987A
Images taken **a.** before and **b.** after the dramatic explosion of a dying massive star show the enormous power of this thermonuclear explosion. Seen in 1987, it was the closest observed supernova in 400 years. It occurred in the outskirts of the Tarantula Nebula in the Large Magellanic Cloud.

"How [stars] die depends on how they lived, which depends on how they were born."

supernova The highly energetic explosion of a massive star or a star in a binary system.

fuel in the stellar core is gone, the star finds itself on the losing end of its gravity war and heads inevitably toward its death. This chapter takes up the story of stellar evolution at that point and brings the story to its conclusion.

The narrative of stellar death is also a story of renewal. In Chapter 12 we saw how gas in interstellar clouds, through gravitational collapse, is transformed into stars, which spend their lives fusing lighter elements into heavier ones. But in the long march to their end, stars return much of their mass back to the interstellar medium (ISM). This can happen through stellar winds, which become particularly strong after a star leaves the main sequence, or through powerful stellar explosions, which form much of the Universe's heavier elements. When heavy-element-rich gas is driven out into space, it eventually finds its way into new interstellar clouds. Thus a kind of *cosmic recycling* takes gas from interstellar clouds to stars and back to interstellar clouds. The cycle repeats itself as new stars form.

But not all stars die the same way. While supernovas are spectacular, only the most massive stars go out that dramatically. Lower-mass stars such as the Sun meet their fate differently. "Stars are like people," says Orsola De Marco, an Italian-born astronomer who has been working on the death of Sun-like stars for most of her career. "How they die depends on how they lived, which for a star depends on how they were born. In particular, a star's death depends on how much mass the star was born with." Recall from Chapter 10 that stars are in hydrostatic equilibrium: pressure pushing out and gravity pulling in must balance at every point in the star. The more mass a star has, the greater the gravitational weight crushing down on its core, which determines the temperature of the core. The core temperature, in turn, determines the rate at which stars fuse hydrogen. Double the core temperature, and the nuclear fusion rate increases by at least a factor of 16. The amount of time a star spends fusing hydrogen on the main sequence therefore depends on the star's mass **(Figure 13.2)**.

Stars are categorized according to their mass (although astronomers vary in how they define these categories). *Very-low-mass stars* have masses ranging between 0.08 and 0.85 times the mass of the Sun. *Low-mass stars*, including the Sun, range between 0.85 and 2 solar masses. *Intermediate-mass stars* are between 2 and 8 solar masses. *High-mass stars* include everything above 8 solar masses. Very-low-mass stars are so puny they basically fuse hydrogen forever, relative to the age of the cosmos. "The Universe hasn't been around long enough for them to have consumed their core hydrogen and evolved off the main sequence," says De Marco. Low- and intermediate-mass stars eat up their core hydrogen at a rate that lets them live on the main sequence for anywhere between half a billion years and 15 billion years. High-mass stars use up their hydrogen very quickly, spending at most a few million years on the main sequence. As you will learn in this chapter, a star's mass determines not only the length of its life on the main sequence but also the way it leaves.

Mass is not the only factor that influences a star's evolution. Recall from Chapter 11 that at least 50 percent of the stars in our galaxy exist in binary systems with another star. Having a companion can change a star's evolutionary track. The effect that being binary has on stellar evolution depends heavily on details of the two stars' orbit. "If the orbital separation between the two stars is always very large, then they don't really affect each other much," says De Marco. Astronomers call such systems with orbital separations of 1,000 astronomical units (AU) or more *wide binaries*. "At the other end of the spectrum are stars that are so close that gravity distorts their shape and they are essentially

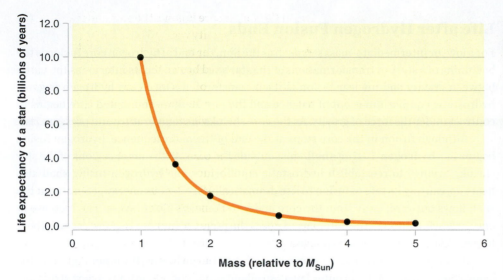

Mass (M_{Sun})	Life Expectancy (yr)
0.08	5.5 trillion
0.125	1.8 trillion
0.25	320 billion
0.5	57 billion
1	10 billion
1.5	3.6 billion
2	1.8 billion
3	642 million
4	313 million
5	179 million
6	113 million
7	77 million
8	55 million
16	10 million
32	2 million

Figure 13.2 Stellar Mass and Life Expectancy
Stars have a wide range of life expectancies, which depend on mass: the larger the star, the faster it uses up its fuel. The most massive stars live only millions of years. A 1-solar-mass (1-M_{Sun}) star will live for about 10 billion years.

touching." Such stars are *contact binaries*. In between are binaries with orbits large enough that they evolve separately for billions of years but then begin to interact when one of the stars evolves off the main sequence and becomes a giant. "The more massive of the pair will become a giant first," says De Marco. "Once it does, it can become so large that it interferes with the second star."

A number of things can happen once binary stars begin to interact. Perhaps the most important consequence is that mass from the giant star can be pulled onto the smaller star. The process in which mass is exchanged from one star to the other in this way is called *mass transfer*. Mass-transfer binaries occur in many forms, depending on the kinds of stars in the binary. For now, the important point is that the accretion disks that form around these objects often emit enough radiation to enable detailed analysis of their properties. Thus, astronomers know quite a bit about the physics of mass transfer in binary systems and the ways it can affect the ultimate fate of the two stars.

We begin our story of stellar death close to home, with low- and intermediate-mass stars like our Sun.

INTERACTIVE:
Stefan-Boltzmann Law

Section Summary

- Supernovas, the explosive deaths of massive stars or stars in binary systems, are rare events.
- Stars return much of their mass to the interstellar medium as they die.
- The time for stars to use up their core hydrogen and leave the main sequence is more than the age of the Universe for very-low-mass stars, between 5×10^8 and 1×10^{10} years for low- and intermediate-mass stars, and a few million years for massive stars.
- Many stars are born with binary companions. The first star that evolves to the giant phase can affect the subsequent evolution of its companion.

CHECKPOINT What is the lowest main-sequence mass for what astronomers categorize as high-mass stars?

How to Become a Giant: The Fate of Low- and Intermediate-Mass Stars

13.2 Stars on the main sequence are temporarily winning their battle against gravity by releasing energy through hydrogen fusion. But what happens when all of the hydrogen in the core has been converted to helium? No matter what a star's mass is, the end of core hydrogen fusion marks a crisis point. The star's time on the main sequence has come to an end.

ORSOLA DE MARCO

"I liked to mix things and make them explode," Orsola De Marco says. "That didn't make my parents very happy." To turn her attentions elsewhere, De Marco's parents bought their ambitious elementary school student a telescope. "I used it a lot to spy at my neighbor's windows."

Near the end of high school, De Marco's interests turned away from chemistry, toward astronomy. She pursued her studies at University College London, where she encountered the topic that would become her passion. "That was the time I began my love affair with planetary nebulas. They are so beautiful, and there is so much we don't understand about them."

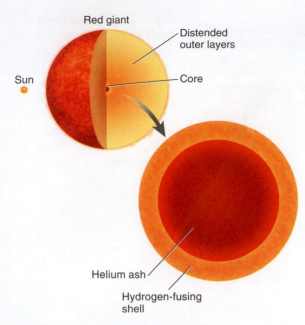

Figure 13.3 Hydrogen-Fusing Shell
Once the hydrogen in a star's core has been converted to helium, nuclear fusion there ceases. A hydrogen-fusing layer then surrounds the core, but because the energy is released outside the core it can't stop the core from contracting.

> "Once core helium burning begins, stars enter a stable phase of their lives."

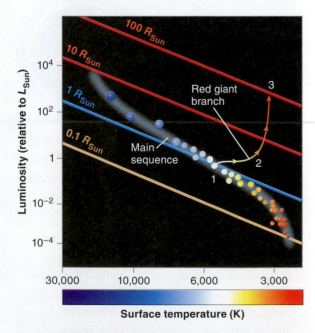

Figure 13.4 Red Giant Branch
A star that has run out of hydrogen in its core follows the evolutionary path 1→3 on the HR diagram. Such a star evolves off the main sequence and moves up the red giant branch.

Life after Hydrogen Fusion Ends

For a low- or intermediate-mass star such as the Sun, the end of hydrogen core fusion is the beginning of a story of transformations at the star's surface and in its interior as the battle between gravity and nuclear fusion is itself transformed. Once core hydrogen is gone, hydrostatic equilibrium is out of balance and the now helium-dominated core begins to contract under the force of gravity. As the core shrinks, its temperature and pressure rise.

Although fusion in the core stops at the end of the main sequence, hydrogen fusion begins in a **hydrogen-fusing shell**—just outside the core. As the inner regions of the star shrink, "trying" to reestablish hydrostatic equilibrium, the hydrogen-fusing shell also heats up, causing its rate of fusion to rise dramatically. Prodigious energy generated at the shell flows outward, away from the core and into the star's outer layers. The increase in energy flowing through the outer layers raises the pressure in these regions, causing them to swell like an oversaturated sponge.

Thus, after core hydrogen fusion ends and shell fusion begins, the inner regions of the aging star contract while its outer layers swell outward. The star attains a new structure, reestablishing hydrostatic equilibrium. The star becomes a giant, with its radius increasing by 200 times. And even though the stellar core and hydrogen-fusing shell increase in temperature, the temperature of the distended surface layers decreases by a few thousand degrees. Using Wien's law (Going Further 4.1), we find that a stellar surface at this temperature yields a blackbody spectrum that peaks in the red part of the visible spectrum (**Going Further 13.1** on p. 344). Thus the star is not just a giant but a **red giant**: a cool, evolved star with a radius much larger than that of the Sun (**Figure 13.3**).

Recall from Chapter 11 that the HR diagram serves as astronomers' principal tool for decoding the properties and life stories of stars. The evolutionary track of a star on the HR diagram is the snail's trail that an individual star makes on the diagram as it ages and its observable outer layers respond to internal changes. After hydrogen core fusion ends in low- and intermediate-mass stars, their evolutionary tracks head up and to the right on the HR diagram, a region called the *red giant branch* (**Figure 13.4**). The movement upward on the HR diagram means that a star's luminosity (energy emitted per time) is increasing. The increase in luminosity is a direct result of the increasing energy generated in the hydrogen-fusing shell. The move rightward on the HR diagram happens because the temperature of the stellar surface is decreasing—a result of the star's distended size.

For a Sun-like star, the ascent up the red giant branch on the HR diagram is not the end of the story. Eventually, the core temperature reaches the critical value of 100 million K—high enough for helium nuclei to fuse to carbon (leaving what astronomers think of as carbon ash in the core). This is a new source of energy; the star, for a while at least, gets a new lease on life.

"When helium ignites, it can do so either in a quiet way or as a kind of internal, unseen explosion, which astronomers call a core **helium flash**," explains Orsola De Marco. The difference depends on the core's mass: heavier cores lead to a helium flash. "Either way, once core helium burning begins, stars enter a stable phase of their lives." The energy released in a helium flash causes the core to expand. This brings the core temperature down slightly and leads to a decrease of the star's luminosity. The post-helium-flash drop in the star's energy output is another "attempt" to maintain hydrostatic equilibrium, with its balance of pressure and gravity.

Stars in this stage of evolution reach a region called the *horizontal branch* on the HR diagram (**Figure 13.5**). Horizontal branch stars are the core helium-fusing versions of main-sequence stars. "But eventually, the helium in the core burns out too, just like hydrogen did," says De Marco. Helium core fusion happens much faster than hydrogen main-sequence fusion, however. "A star like the Sun can spend just 50 million years on the horizontal branch," says De Marco. "Like the main sequence, more-massive stars spend less time there than lower-mass stars."

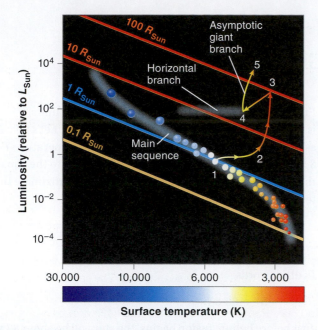

Figure 13.5 Horizontal Branch and Asymptotic Giant Branch
When stars begin a period of stable helium fusion in their cores, they move onto the horizontal branch of the HR diagram (path 3→4). Eventually, the core helium supply runs out. Then nuclear fusion occurs only in shells. The star's luminosity and radius increase again as it moves onto the asymptotic giant branch of the HR diagram (path 4→5).

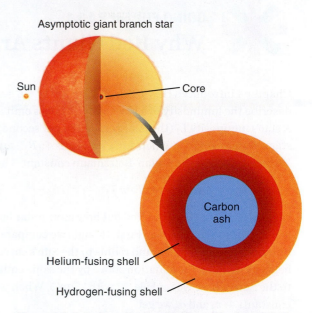

Figure 13.6 Helium-Fusing Shell and Carbon Ash Core
When a star moves onto the asymptotic giant branch, its inner structure consists of an outer hydrogen-fusing shell, an inner helium-fusing shell, and a central core of carbon ash.

When all of the helium in the core has fused to carbon ash, the star must again try to reestablish hydrostatic equilibrium. Its response mimics the first red giant transition, at the end of the main-sequence phase. The inner regions of the star contract and can now be defined as an inert carbon ash core inside two nuclear-fusing shells: a helium-fusing shell just outside the core, and a hydrogen-fusing shell above it (**Figure 13.6**). As the inner regions contract, the temperature climbs again, driving a fury of nuclear fusion in the shells. A new torrent of energy courses through the outer layers, which once again expand. The star becomes a red giant again, but this time with a vengeance. The star climbs onto the **asymptotic giant branch (AGB)**, the region of the HR diagram for highly evolved low- and intermediate-mass stars when their surfaces are cool and they have very large radii and very high luminosities (Figure 13.5). The AGB phase is the next-to-last stage of Sun-like stars. Unlike its massive cousins, a low- or intermediate-mass star does not have enough mass to raise the core's temperature as high as is needed to fuse carbon. The central engine has been shut down permanently (**Table 13.1**).

> "Eventually, the helium in the core burns out too, just like hydrogen did."

Table 13.1   Stages of Stellar Death in Low- or Intermediate-Mass Stars

Evolutionary Phase	Duration (yr)	Size (relative to R_{Sun})	Surface Temperature (K)
Post-main-sequence to red giant	100 million	3	4,500
Red giant	100,000	100	3,500
Helium fusion	10 million	10	5,000
AGB star	100,000	500	3,000

hydrogen-fusing shell The layer outside a star's core where hydrogen continues to fuse to helium after such reactions have ended in the core.

red giant A cool star with a large radius that has evolved off the main sequence because hydrogen fuel in its core is depleted.

helium flash The rapid internal ignition of helium fusion in the core of an evolved star.

asymptotic giant branch (AGB) The region of the HR diagram to which a low- or intermediate-mass star evolves when it has used up the hydrogen and helium in its core. The star then has a cool surface, large radius, and high luminosity.

GOING FURTHER 13.1
Why Red Giants Are Red and Giant

Chapter 4 introduced the Stefan-Boltzmann law, $L = A\sigma T^4$, to describe the luminosity of a blackbody. All stars emit as blackbodies. A star's luminosity L_* (the energy radiated per second) depends on the star's surface area A_* (which, for a sphere, is $4\pi R_*^2$) and temperature T_*; recall that σ is the Stefan-Boltzmann constant. Thus

$$L_* = A_*\sigma T_*^4 = 4\pi R_*^2 \sigma T_*^4$$

Using this formula, we can find out how large a star becomes after leaving the main sequence. First, though, we compare the star's luminosity, temperature, and radius to the Sun's current values. Let's begin by dividing the equation above by the same equation referring to the Sun (where we replace "*" with "Sun"). When we cancel out the constants 4, π, and σ, we get

$$\frac{L_*}{L_{Sun}} = \left(\frac{R_*}{R_{Sun}}\right)^2 \left(\frac{T_*}{T_{Sun}}\right)^4$$

If we already know the star's luminosity and temperature but want to know its radius, we can solve this formula for R_*/R_{Sun}:

$$\frac{R_*}{R_{Sun}} = \sqrt{\left(\frac{L_*}{L_{Sun}}\right)}\left(\frac{T_*}{T_{Sun}}\right)^2$$

Now we'll look at the evolutionary track on the HR diagram in Figure 13.4 to find the luminosity and temperature of a low-mass star just before helium core fusion begins (which is around point 3 in

Figure 13.4). The luminosity is already in "solar units," since the graph is plotted in terms of luminosity relative to L_{Sun}. Let's take $L_* = 10^3\ L_{Sun}$ as a representative value. We can also use a temperature T_* of about 3,500 kelvins (K). Now let's plug these values into the equation, using the values for the Sun from Chapter 10 (we approximate the Sun's temperature as 5,800 K):

$$\frac{R_*}{R_{Sun}} = \sqrt{\left(\frac{10^3\ L_{Sun}}{1\ L_{Sun}}\right)}\left(\frac{3{,}500\ \text{K}}{5{,}800\ \text{K}}\right)^2 = 86.8$$

In other words,

$$R_* = 86.8\ R_{Sun}$$

Just before helium fusion begins, the star will be about 87 times larger than the Sun is today. That is why it's called a giant.

But why is the giant red? To answer this question, we need Wien's law ($\lambda_{max} = 0.0029$ m K/T; see Chapter 4), which tells us the peak wavelength at which a blackbody radiates energy. For our star with surface temperature $T_* = 3{,}500$ K, Wien's law indicates the radiated light will have a peak wavelength λ_{max}, in nanometers (nm), of

$$\lambda_{max} = \frac{0.0029\ \text{m K}}{3{,}500\ \text{K}} = 8.29 \times 10^{-7}\ \text{m} = 829\ \text{nm}$$

This is the infrared part of the spectrum, so the giant star appears red in optical telescopes (and to the naked eye).

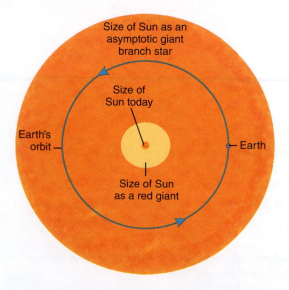

Figure 13.7 Size Comparisons of the Aging Sun
The Sun of today will be dwarfed by its future self—first as a red giant, then as an AGB star, at which point it may expand so far that it will engulf Earth.

An AGB star is a behemoth. Its outer layers can swell far beyond the radius reached during the first red giant phase. When our Sun becomes an AGB star some 5 billion years from now, it will likely become so large that Earth will, in the worst case, become engulfed by its own bloated star. At best, the Sun's surface will have expanded so close to Earth's orbit that the surface of our world will be baked to searing concrete as the oceans boil away and the atmosphere becomes too hot to breathe (though this may happen when the Sun first reaches the red giant branch). Thus the story of the end of low- and intermediate-mass stars is also the story of the end of Earth. As the Sun swells during its final AGB phase in some 5 billion years, Earth may even be engulfed by the Sun's outer layers and reduced to its constituent atoms (**Figure 13.7**).

Section Summary

- After core hydrogen fusion stops, a low- or intermediate-mass star's core contracts and hydrogen fusion begins in a shell around the core as the outer layers expand, creating a red giant.
- Temperatures rise in the star's core, and helium eventually fuses to carbon.
- When helium core fusion ends, the star's core contracts, the outer layers swell, and the star becomes an asymptotic giant branch (AGB) star.

CHECKPOINT How does the star's need to balance gravity and pressure (hydrostatic equilibrium) help explain the changes in a star at the end of the main sequence?

The Last Hurrah: Planetary Nebulas

13.3

The outer layers of a red giant are so extended that they are not tightly bound by gravity to the star. With the flood of energy (as photons) flowing outward through these layers, some fraction of the bloated stellar atmospheres is easily blown into space. Radiation impinges on tiny dust grains that form in the cool atmospheres. As a red giant, or particularly as an AGB star, pushes these grains outward, they collide with gas particles, and the entire outer atmosphere accelerates away from the star in the form of a dense wind. "A star like the Sun will lose about half of itself, half its mass, as an AGB star," explains De Marco. "For more-massive stars, the winds are even more dramatic. An intermediate-mass star that's, say, 5 times the Sun on the main sequence will lose almost 4 solar masses through its wind. That means four-fifths of the star gets blown into space."

The powerful winds of red giants contain carbon, oxygen, and nitrogen—by-products of billions of years of nuclear fusion. Deep convective flows dredge up some fraction of the processed elements from the star's core to the surface during the AGB phase (recall the discussion of stellar convection in Chapter 10). These elements become part of the stellar winds, which then return their heavy elements to the galaxy's interstellar medium in a kind of cosmic recycling. (Recall that, in astronomy, any element heavier than helium is a heavy element.) Eventually, atoms of those elements find their way into the next generation of stars (and planets), perhaps setting the stage for the evolution of life. Closer to the dying star, however, the powerful winds initiate a process that produces some of the Universe's most beautiful structures, **planetary nebulas**—glowing, sculpted gas clouds surrounding dying low- or intermediate-mass stars (**Figure 13.8**).

In 1764, the French astronomer Charles Messier attempted to catalog all nonstellar features that can be seen in the night sky by telescope (**Figure 13.9**). As his work progressed, he found a class of objects that looked like extended, spherical, glowing clouds. Because telescopes at that time were tiny (with apertures ranging from 3 to 8 inches), he was unable to resolve much detail. Later, astronomer William Herschel named the objects "planetary nebulas" because the shapes and colors of these celestial clouds reminded him of Solar System planets. It was an inaccurate choice of words, but it stuck and is still used today. "Planetary nebulas have nothing to do with planets. That much is certain," explains

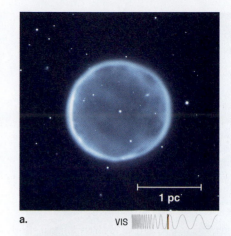

a.   1 pc   VIS

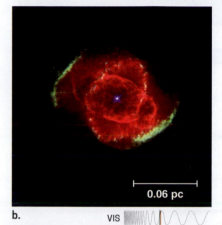

b.   0.06 pc   VIS

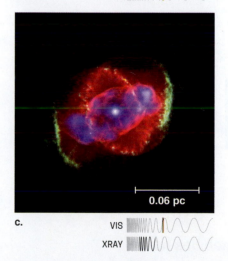

c.   0.06 pc   VIS   XRAY

Figure 13.8 Planetary Nebulas
a. Planetary nebula Abell 39. A dense shell of gas was swept up when fast winds from its hot central star expanded into material ejected earlier during the AGB phase. **b., c.** Planetary nebula NGC 6543. The dense inner bubble of a planetary nebula, unlike the spectacular outer layers, is not visible in optical light but glows in the X-ray part of the spectrum.

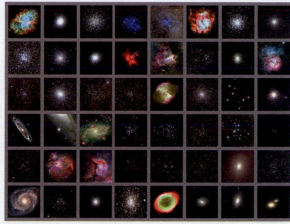

a.   b.

Figure 13.9
Charles Messier's Catalog
a. Charles Messier (1730–1817), a comet hunter who cataloged all the fuzzy things he saw on the sky to distinguish them from comets. **b.** A portion of his catalog, which is now known to contain galaxies, star clusters, double stars, and nebulas. Prized for its usefulness by amateur and professional astronomers, it was later expanded by others to include 110 objects.

planetary nebula In low- and intermediate-mass stars, the evolutionary stage (between the AGB and white dwarf phases) in which material is driven off the star in a stellar wind, forming a luminous cloud around the star.

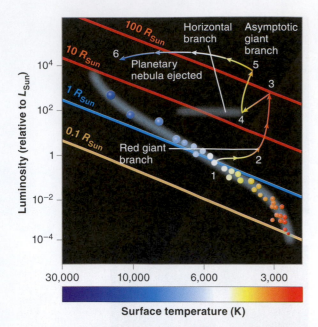

Figure 13.10 Planetary Nebulas on the HR Diagram
After low- and intermediate-mass stars reach the top of the AGB on the HR diagram, the giant star loses its outer surface as stellar wind. Its hotter inner layers are then exposed, and it moves leftward on the diagram to higher temperatures at approximately constant luminosity (path 5→6). Ultraviolet photons from the hot star ionize the ejected gas and allow it to glow as a planetary nebula.

INTERACTIVE:
Expansion of a Planetary Nebula

"Planetary nebulas have nothing to do with planets."

De Marco. "But while we do know they are connected with the death of low- and intermediate-mass stars, much of our understanding of these beautiful objects is in a state of flux right now."

Astronomers do know that after a star has spent 100,000 years on the AGB, its stellar winds have stripped it down almost to its nuclear-fusing shells. The "surface" above the shells and core is a violent scene of high-temperature gas releasing a torrent of energetic photons, mostly in the ultraviolet range. The increase in surface temperature takes the star off the AGB in the HR diagram and moves it leftward (**Figure 13.10**). The photons punch much of the remaining atmosphere into space, creating a tenuous, high-speed wind. This "fast" wind, with speeds up to 1,000 kilometers per second (km/s), quickly overtakes the dusty wind from the star's AGB phase; that slower wind moves at only 10 km/s. The expanding fast wind slams into the inner edge of the AGB wind with the force of a trillion 1-megaton hydrogen bombs. Then the fireworks begin.

A shock wave forms when gas is pushed faster than it can react by getting out of the way. As a shock wave moves through a medium, it quickly and violently smashes together atoms of gas like cars in a highway pileup. When fast stellar wind reaches slower AGB wind, the collision of the two winds produces two powerful shock waves. First, a shock wave moves outward, accelerating and compressing the AGB wind, squeezing it into a dense layer of atoms and ions. At the same time, another shock wave rebounds off the AGB material, moving back through the fast wind toward the star. This rebound shock jerks the fast wind to a near stop, and the violent deceleration heats the material to more than 10 million K, creating a hot bubble of gas. Ultimately, a kind of shock wave layer cake results. The inner shock wave is closest to the star, surrounded by the hot bubble, which in turn is surrounded by the dense shell of swept-up AGB gas and its outer boundary, the outer shock wave (**Figure 13.11**).

As the shock waves heat and compress the gas, the ultraviolet light from the hot central star ionizes the surrounding material and causes it to emit light (as occurs in H II

Figure 13.11 Shock Waves
The structure of a planetary nebula is driven by a sequence of spherical winds. An inner shock wave forms closest to the star, heating the fast-wind gas and creating a hot bubble. The expansion of this bubble sweeps up a dense shell of previously ejected AGB gas and is bounded by an outer shock wave.

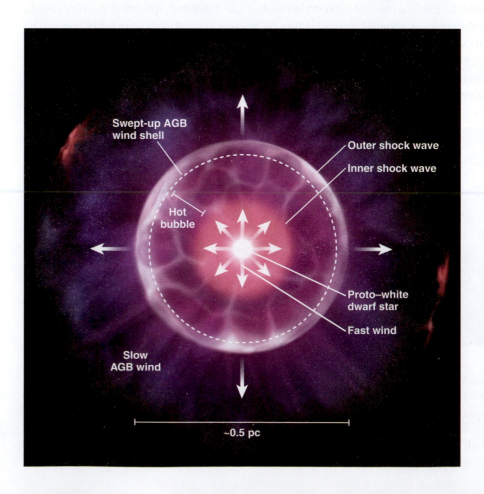

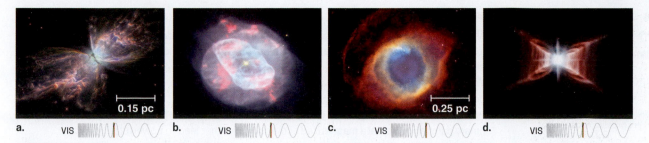

Figure 13.12 Planetary Nebula Shapes
a. NGC 6302 (Butterfly Nebula), **b.** NGC 5882, **c.** NGC 7293 (Helix Nebula), and **d.** HD 44179 (Red Rectangle). The incredible variety of planetary nebulas led astronomers to broaden their models of nebula formation. Notice the bipolar (two-lobed) structure and symmetry sometimes seen between the two lobes.

regions around young massive stars; see Chapter 12). The high densities in the shell make it emit most intensely, creating the bright rims seen in planetary nebulas (Figure 13.8). The gas in the hot bubble is not dense enough to produce much optical light. It does, however, produce enough X-rays to be seen with orbiting X-ray telescopes, such as the Chandra X-ray Observatory. The shell and the outer shock wave make the beautiful glowing forms we see when we view a planetary nebula from Earth.

New Data, New Stories

Spherical winds produce spherical shock waves, so we might expect all planetary nebulas to be round. But in the 1990s, astronomers who turned the Hubble Space Telescope (HST) toward planetary nebulas got a shock of their own. "It's an absolute zoo," says De Marco, referring to the bizarre assortment of shapes of planetary nebulas (**Figure 13.12**). To explain the HST images, more players were needed in the planetary nebula drama. As often happens in science, new data forced astronomers to go beyond their old ideas.

"We reached a crisis," says De Marco. "The paradigm we had all been using just wasn't working anymore." To explain the amazing variety of planetary nebula shapes, astronomers now have advanced theories that link back to the jets and disk we encountered in star formation (see Chapter 12). "We are pretty sure that the outflowing gas in nonspherical planetary nebulas is shaped by jets from a magnetized accretion disk that forms around a star," says De Marco. "It's just like the jets you see with young stars. The spinning accretion disk whips magnetic-field lines around, and gas tied to those field lines gets flung out into space. On the large scale of the nebula, the twisting magnetic-field lines can explain why we also get those amazing shapes" (**Figure 13.13a**).

Where do the accretion disks come from? A dying Sun-like star whose cloud dispersed 10 billion years ago can create an accretion disk only if it has a companion—another star, a brown dwarf, or even a giant planet. De Marco explains, "If you have a companion to your giant and the companion is close enough, then you can make an accretion disk."

When two objects orbit each other, each one defines a region of gravitational influence. A **Roche lobe** is the extent of a star's domain of gravitational influence. As one star in a binary system becomes a giant, its outer layers may expand past its own Roche lobe. Some of its gas then cascades across to the second star. Because the two stars orbit each other, by conservation of angular momentum the transferring gas can't just fall onto the second star; instead, it goes into orbit. New material piles up, forming an extended accretion disk—a process called *Roche lobe overflow* (**Figure 13.13b**).

This new paradigm for planetary nebulas is controversial. Astronomers who favor this **binary-accretion-disk-magnetic-field model** have much work ahead of them to convince skeptical colleagues. But progress in science can involve conflicts—out of which come new advances.

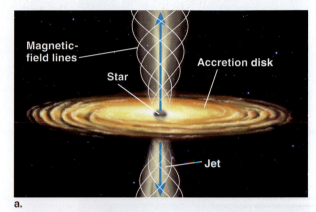

a.

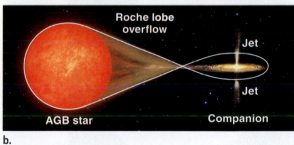

b.

Figure 13.13 The Binary-Accretion-Disk-Magnetic-Field Model
An AGB star provides material that is drawn in by a companion as an accretion disk. **a.** Magnetic-field lines threading the disk push material back into space as a jet that drives through the previously ejected AGB wind. **b.** The disk forms by Roche lobe overflow (or else by capture of the AGB wind).

Roche lobe The region around a star within which orbiting material is gravitationally bound to that star.

binary-accretion-disk-magnetic-field model A theory of planetary nebula formation that involves a binary system and accretion disks. Jets from the disk sculpt the shape of the planetary nebula.

"We reached a crisis. The paradigm we had all been using just wasn't working anymore."

AT PLAY IN THE COSMOS THE VIDEOGAME Analyze the properties of the "Ant Nebula's" central star to discover an ancient and dangerous secret in Mission 12.

Earth

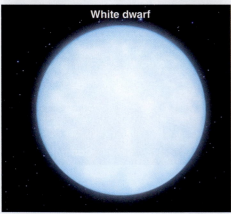

White dwarf

Figure 13.14 White Dwarf versus Earth
Although the mass of a white dwarf is equivalent to that of our Sun, the radius of a white dwarf is at most 1/100 that of the Sun.

Heisenberg uncertainty principle The quantum mechanical principle stating that as the position of a particle is specified to higher accuracy, its speed becomes increasingly uncertain, and vice versa.

degeneracy pressure Pressure exerted because of quantum mechanical motions of particles resulting from confinement to a small volume.

white dwarf The compact remnant of a low- or intermediate-mass star, supported by electron degeneracy pressure.

Section Summary

- The outer layers of gas and dust are blown off an AGB star as a dense, slow wind. Once the star is stripped almost to its core, the surface heats up and a fast wind is created. When the fast and slow winds collide, shock waves form.
- We observe a planetary nebula when ultraviolet light from the central star ionizes the wind-sculpted surrounding gas, making it glow.
- Astronomers are debating a new binary-accretion-disk-magnetic-field model, according to which binary companions create magnetized accretion disks that lead to powerful jets.

CHECKPOINT Describe the process by which an AGB star evolves to a planetary nebula.

White Dwarfs: Stellar Corpses of the First Kind

13.4 You might expect that once nuclear fusion in the core had stopped, gravity would have its final say and crush the star completely. But what does "crushed completely" mean? Can all of a star's mass be crushed out of existence? Doesn't that mass have to exist somewhere? Are there forces that can stop a dead star from being completely crushed by gravity? Physicists and astronomers wrestled with such questions through the middle of the 20th century. Their answers, which relied on the most advanced theories of atomic and subatomic matter, led them into the bizarre world of stellar cinders—the forms a stellar corpse can take.

Just as the final years of a star depend on its original mass, so does its final fate. During the planetary nebula stage of a low- to intermediate-mass (Sun-like) star, the hydrogen-fusing shell is just below the surface; most of what little mass the star has above the core and the nuclear-fusing shells is blown into space in the final fast wind. What's left after core helium fusion is the inert carbon ash core. With no nuclear fusion to create thermal pressure, why doesn't this inert core get crushed?

The answer comes from quantum mechanics, the remarkable understanding of the microscopic world (molecules, atoms, and subatomic particles), which you first encountered in Chapter 4. One of the most important conclusions of quantum mechanics is the **Heisenberg uncertainty principle**, which tells us that particle properties in the micro world are essentially uncertain in very specific ways. For example, this principle tells us that if we could specify the position of a particle to great accuracy, then its speed would be highly uncertain. Conversely, if we could specify the particle's speed to great accuracy, then its position would be highly uncertain. These uncertainties have nothing to do with our measuring instruments but are built into the very structure of matter. If you think this is weird, you're in good company. The world at the atomic scale of atoms behaves in very different ways than does the world at the human scale.

How does this principle relate to dying stars? As gravity crushes carbon nuclei and their electrons in a star's core, each particle occupies less space. At this scale, quantum physics rules and the Heisenberg uncertainty principle determines the reaction of the nuclei and electrons to the big squeeze. As the range of possible electron positions becomes more confined, the uncertainty principle leads them to take on an ever-larger range of velocities. The rapid oscillation of confined electrons corresponds to a pressure—called electron **degeneracy pressure**—that supports the almost-naked core against further collapse. Unlike the thermal pressure created by nuclear fusion, the degeneracy pressure of these electrons does not require any material to be consumed and does not depend on temperature, so the star's core has won a permanent victory over gravity.

The almost-naked core, supported by electron degeneracy pressure, is now called a **white dwarf**—the compact remains of a dead low- to intermediate-mass star. The *dwarf* part of the name is easy to understand, since a typical white dwarf, containing about half of the Sun's mass, has a radius equal to about 1 Earth radius (**Figure 13.14**). (The exact relationship between core mass and the mass the star began with depends on the history of its winds, which strip material away over the star's lifetime.) It is called a *white* dwarf because of its surface temperature. A very thin atmosphere sits atop the carbon ash core, and it's kept very hot because of the intense gravitational field associated with cramming so much matter into so small a sphere. Temperatures on the surface of young or newly formed white dwarfs are typically 100,000 K or more. Wien's law shows that a blackbody at this temperature would radiate most of its energy in the ultraviolet part of the electromagnetic spectrum. In the visible band, all wavelengths of the blackbody spectrum are almost equally represented, so the star appears white to us.

As time goes on, the white dwarf's atmosphere radiates away its energy and cools down. There is a well-defined white dwarf cooling track on the HR diagram (**Figure 13.15**), which astronomers have painstakingly confirmed through observational studies. After billions of years, white dwarfs cool enough to emit barely any visible radiation. The final fate of a Sun-like star is to become a *black dwarf*, spending eternity as a dark cinder.

Electron degeneracy has its limits, however. If you could pack even more mass into a white dwarf, you would find that it *shrinks* as its mass increases. This makes sense in terms of the Heisenberg uncertainty principle. Electron degeneracy pressure depends on squeezing particles into less space to increase their range of velocities. If more matter and hence more gravity are added, the star must shrink for its electrons to be squeezed further and increase in degeneracy pressure. Thus, more-massive white dwarfs have smaller radii than their lower-mass cousins (**Figure 13.16**).

The Chandrasekhar Limit

Matter cannot be packed onto a white dwarf forever, and understanding the limitations required the innovations of both relativity and quantum physics. In July 1930, the young physicist Subrahmanyan Chandrasekhar (**Figure 13.17**) boarded a boat from his native India to England once he was awarded a scholarship to attend graduate school at the University of Cambridge. After overcoming seasickness, he used his weeklong journey to work on the unresolved problem of the fate of stars. He started his calculations by trying to combine the new science of quantum physics with Einstein's theory of relativity. As the ship made its way to England, Chandrasekhar developed the basic outline of how degeneracy pressure could support a dead star and the limits that pressure would set in terms of the final mass of the stellar cinder. According to his calculations, electron degeneracy

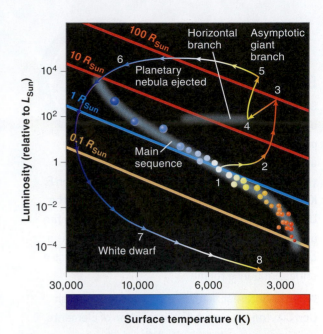

Figure 13.15 White Dwarfs on the HR Diagram
After the winds of the AGB and planetary nebula phases, the star has been stripped down to its core. Nuclear fusion has ended, and the star, now a white dwarf, dims and cools over time as it moves downward and to the right on the HR diagram (path 6→8).

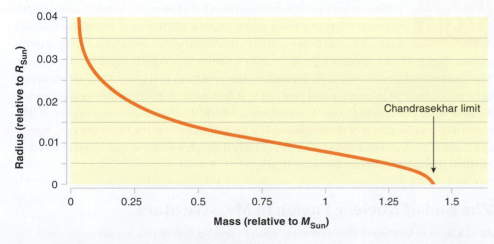

Figure 13.16 Relative Size of a White Dwarf
For the degeneracy pressure of more-massive white dwarfs to increase, they must shrink. Thus they have smaller radii than their less-massive counterparts.

Figure 13.17 Subrahmanyan Chandrasekhar
As a young man, Chandrasekhar (1910–1995) described the physics of degeneracy pressure in stars and discovered the maximum allowable mass that pressure can support, now called the *Chandrasekhar limit*.

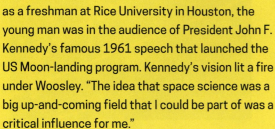

STAN WOOSLEY

Stan Woosley started college without a strong passion for astronomy. But as a freshman at Rice University in Houston, the young man was in the audience of President John F. Kennedy's famous 1961 speech that launched the US Moon-landing program. Kennedy's vision lit a fire under Woosley. "The idea that space science was a big up-and-coming field that I could be part of was a critical influence for me."

When his interest flagged, a chance meeting with Don Clayton, an expert on the creation of elements and supernovas, again ignited Woosley's interest. "It's amazing to me looking back now that I almost lost interest, because I love what I do," says Woosley. A professor at UC Santa Cruz since 1975, he has been one of the world's leaders in the study of supernovas.

Chandrasekhar limit The maximum mass (1.4 M_{Sun}) of a white dwarf; the mass that electron degeneracy pressure can support.

pressure could not support a core of more than about 1.4 solar masses (M_{Sun}) against its own gravity. One way to think about this limit is the understanding that came from Einstein that nothing can travel faster than the speed of light. That means there is a limit to the electron speeds being used to support ever-more-massive white dwarfs.

When Chandrasekhar began working out the details of his theory in England, he often discussed his results with Sir Arthur Eddington, who was the most famous astrophysicist at the time. Eddington did not believe the new results but encouraged Chandrasekhar and invited him to present the results at a highly prestigious congress of astronomers.

With some trepidation, Chandrasekhar delivered his talk on electron degeneracy pressure and the 1.4-M_{Sun} limit. In one of the most famous betrayals in the history of astronomy, Eddington immediately followed with a talk of his own in which he savaged Chandrasekhar's work. To have a scientist of Eddington's stature take so strident a position against one's work could end a young scientist's career. But though the experience was initially devastating to Chandrasekhar, Eddington was wrong. History and astronomical data eventually proved the truth of degeneracy pressure and its limits. Chandrasekhar earned the 1983 Nobel Prize in Physics for his research on stellar evolution. Astronomers now use the term **Chandrasekhar limit** to describe the 1.4-M_{Sun} limit on dead cores supported by electron degeneracy pressure.

For cores with mass higher than the Chandrasekhar limit, electron degeneracy pressure is not strong enough to counteract the inward crush of gravity. Thus, more massive cores from more massive stars must rely on a different form of support after fusion ends. We turn to that story next.

Section Summary

- Once the planetary nebula phase is over, the inert carbon ash core supports itself as a white dwarf via electron degeneracy pressure.
- White dwarfs begin with very hot atmospheres but cool down over time to black dwarfs.
- The more massive a white dwarf is, the smaller its radius. White dwarfs cannot be more massive than 1.4 M_{Sun}, the Chandrasekhar limit.

CHECKPOINT   How does electron degeneracy fit with our ideas of hydrostatic equilibrium and stability of stars?

Living Fast, Dying Young: The End of Massive Stars

13.5   It takes more than 100 million years for a low- to intermediate-mass star to evolve from the main sequence to a white dwarf, but more-massive stars evolve far faster and with more dramatic consequences. Once a star greater than 8 M_{Sun} completes its core hydrogen fusion, it proceeds rapidly to its demise. The timescale of a star's evolution depends on temperatures in its core. Core temperature determines both what can fuse and how fast that fuel is spent. Even on the main sequence, massive stars produce high core temperatures, which shorten their main-sequence lifetimes to a small fraction of those achieved by their smaller cousins. Once the core hydrogen is exhausted, a high-mass star moves quickly through a series of transformations that hurtle it toward its eventual dramatic death.

The End of Nuclear Fusion in Massive Stars

As in a low- to intermediate-mass star, when hydrogen fusion in a massive star ends, the core starts to contract, producing large amounts of energy that flow through the star and cause its outer layers to swell. The surface temperatures drop, and the star moves to the red supergiant branch of the HR diagram. However, the internal transformations happen more rapidly and take the high-mass star into domains its smaller cousins can never enter. At the end of their lifetimes, the most-massive stars swell to become *supergiants*—large enough to swallow the orbit of Mars.

Let's consider a 25-M_{Sun} star. After the main sequence, the core of this massive star contracts and rapidly switches to helium fusion. After only hundreds of thousands of years, all of the helium is consumed and turned into carbon. Carbon is created through the *triple-alpha process* (**Figure 13.18**), in which two helium nuclei (which, for historical reasons, are known as alpha particles) combine, forming a nucleus of beryllium (four protons and four neutrons). The beryllium nucleus then fuses with another helium nucleus, creating a nucleus of carbon with six protons and six neutrons.

Intermediate-mass stars never achieve temperatures high enough to fuse carbon, but the core of a high-mass star is squeezed by a tremendous force of gravity. Eventually, the temperature in a massive star reaches 600 million K; at that point, carbon begins fusing to neon. That fusion lasts only about a thousand years, however. After carbon is exhausted, neon begins fusing to oxygen; this stage lasts about a year. After the neon is spent, oxygen begins fusing to silicon, which also lasts about a year. Things happen fast now because each new form of fusion requires a higher temperature. As we've seen, higher temperatures mean that the star uses up its store of nuclear fuel rapidly.

Eventually, silicon in the core fuses to iron while the other elements are still fusing in their surrounding shells (**Figure 13.19**). But after a timescale on the order of Earth days (depending on the star's mass), all of the silicon in the core is exhausted. This is the beginning of the star's crisis point.

Up to the creation of an iron ash core, the star was able to extract energy from the fusion reactions in the various shells, which provided enough thermal pressure to resist gravitational collapse. But iron fusion requires more energy input than it gives off. (Consider the curve of binding energy in Chapter 10, especially Figure 10.8.) Once the core is full of iron, the star cannot support itself against gravity. The 24-hour period of silicon fusion is a day of reckoning. "That's when the star has to die," says Stan Woosley, who has studied the fate of massive stars for more than 40 years. "It burns through carbon, neon, oxygen, and silicon until it finally ends up having sort of a spherical layer cake of elements, with iron down in the core and the untouched hydrogen on the surface. When a star gets to a point where it has made an iron core, you can't get any more energy out of fusing it to heavier elements, because iron is the tightest-bound state of matter. The neutrons and protons inside the iron nucleus are packed as tightly to one another as possible. If you try to release more energy by putting two iron nuclei together, it doesn't work. So the creation of iron is the end point of stellar evolution—the end point of core nuclear fusion" (**Table 13.2**).

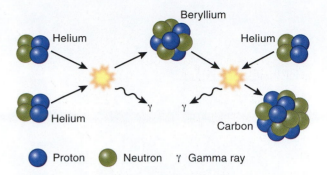

Figure 13.18 Triple-Alpha Process

This set of nuclear fusion reactions occurs in massive stars as they become red giants. Three helium-4 nuclei (also called alpha particles) are transformed into carbon; beryllium forms in an intermediate step. The triple-alpha process is similar to the P-P chain in Sun-like stars (see Figure 10.10).

"That's when the star has to die."

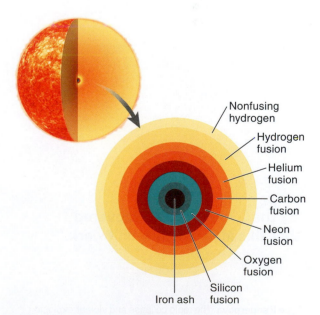

Figure 13.19 Structure of a Massive Star before Death

The succession of nuclear-fusing shells above the core is depicted. Because the fusion of iron into heavier elements takes more energy than fusion gives back, formation of the iron ash core is the beginning of the end of a massive star's life.

Table 13.2	Stages of Nuclear Fusion in Massive Stars		
Fusion Phase	**Required Temperature (K)**	**Required Mean Density (kg/m³)**	**Duration**
Hydrogen	4×10^7	5,000	7 million years
Helium	2×10^8	7×10^5	700,000 years
Carbon	6×10^8	2×10^8	600 years
Neon	1.2×10^9	4×10^9	1 year
Oxygen	1.5×10^9	10^{10}	6 months
Silicon	2.7×10^9	3×10^{10}	1 day

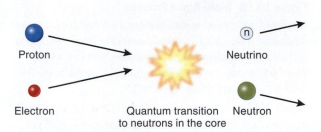

Figure 13.20 Transformation to Neutrons and Neutrino Release

In the core collapse of a star, the quantum mechanical transformation of electrons and protons to neutrons is accompanied by the release of a neutrino.

"By the 1990s we all discovered that this idea simply didn't work."

Type II supernova The rapid collapse and violent explosion of a massive star.

stellar ejecta Material expelled in a stellar explosion such as a supernova.

supernova remnant An expanding shock wave that contains ejected material and swept-up interstellar material resulting from the explosion of a star.

Section Summary

- Because of their high core temperatures, massive stars can fuse elements heavier than helium after the main sequence, from carbon to neon to oxygen to silicon to iron.
- Once silicon fuses to iron, nuclear fusion in the core ends because energy cannot be extracted through the fusion of iron nuclei.

CHECKPOINT How does a massive star get to the point at which carbon can be fused in its core?

Big and Bigger Bangs: Novas and Supernovas

13.6 Once the iron ash core is complete and the nuclear engine dies, gravity works quickly. At this point, the core of a massive star contains at least 1.5 M_{Sun} of iron. Within a few milliseconds, it begins its catastrophic collapse.

The density in the core rockets to 10^{17} kilograms per cubic meter (kg/m^3) at the center, and the violence of the infall drives the temperature up to 10 billion K. Under these extreme conditions, the core is so dense and hot that neutrinos can be created from interactions within the iron nuclei and other particles in the core. In particular, protons and electrons are squeezed so tightly that they undergo a quantum transition to neutrons. In the process, they emit a neutrino (**Figure 13.20**). Few protons and electrons remain, and the core is made almost entirely of neutrons.

Recall from Chapter 10 that neutrinos are ghostly high-energy particles that rarely interact with matter. If you absolutely had to capture just one neutrino, you would need to put up a wall of lead 22 light-years long in its path. But during the core collapse, neutrinos are produced prodigiously (including through thermal processes associated with the ultrahot gas). The core is so dense that the neutrinos cannot just escape; they have to worm or "diffuse" their way out, so in a sense, the core becomes a neutrino star. "The star is emitting a huge amount of energy in neutrinos," says Woosley. "In fact, the energy coming out in neutrinos exceeds the energy radiated by the rest of the entire visible Universe. All that energy has to come from somewhere, and the core gets it by shrinking, becoming more and more tightly bound. It's the gravitational collapse that powers the huge neutrino losses."

The core is now collapsing at about one-third the speed of light. "The core would collapse forever if it didn't reach a point where the strong nuclear force actually began to be repulsive," explains Woosley. "We think of a nuclear force as being attractive—and on the scale of ordinary nuclei, it is—but if you squeeze enough, a nucleus will actually push back, and if you try to compress matter beyond the density of nuclear matter, it also pushes back."

Supernova!

Once the repulsive nuclear force kicks in, the core suddenly stops collapsing and becomes rigid like a hard rubber ball. Stellar layers above the core, however, continue to come crashing down. They are instantly halted when they reach the core, creating a powerful shock wave that begins to travel outward. Astronomers were confident that this rebound-generated shock wave would travel all the way to the stellar surface and blow the star apart, resulting in the energetic explosion of a massive star called a *supernova*. Unfortunately, that simple idea met with an ugly reality. "The shock gets stalled in the collapsing layers above the core," says Woosley. "By the 1990s we all discovered that this idea simply didn't work." Like their colleagues studying planetary nebulas, astronomers studying the death of massive stars were forced to look to other ideas.

To explain supernovas, astrophysicists returned to a theory from the late 1960s that the wild rush of neutrinos emanating from the core could be the power source. "If just a little bit of the energy the core radiates in neutrinos could be deposited in the material above it," says Woosley, "that would be enough to blow away the rest of the star." For nearly 20 years, scientists have used the world's most powerful supercomputers to build complex mathematical models of neutrino-driven explosions of massive stars (**Figure 13.21**). The problem is immensely difficult: to get the physics right, the whole star must be realistically modeled in three dimensions with many processes—the flow of neutrinos, the collapse of the stellar gas, the nuclear physics of the core—happening at once. Although scientists think the neutrino idea is promising, they still do not know the exact mechanism that blows a massive star apart. "People have been trying to make neutrino-driven supernova explosions with mixed results," explains Woosley. "Sometimes the star blows up this way, but more often it doesn't."

While astronomers may not know exactly how a star rips itself apart, they do know that eventually, some process within the star forms a shock wave powerful enough to blow everything above the core into space. These supernovas, called **Type II supernovas** (**Figure 13.22**), are so bright that they can be seen in distant galaxies. Lasting for roughly a few months, Type II supernovas shine with the energy of a billion Suns—as much energy as an average-sized galaxy has. Along with this intense radiation, the explosions create **stellar ejecta**, material above the core that is blown outward at speeds of about 10,000 km/s, or about one-third the speed of light. The stellar ejecta slams into the surrounding interstellar medium, sweeping it up and creating a **supernova remnant**—a vast, bubble-shaped shock wave that can grow to sizes of 100 light-years or more.

As the stellar ejecta and swept-up interstellar material move outward, they cool and decelerate, causing the supernova remnant to fragment. With their multicolored glowing networks of filaments and knots, these supernova remnants constitute some of the most beautiful and striking objects on the sky (**Figure 13.23**). Astronomers have become adept at using the remnants' observed structures to piece together the evolution of the explosion and its interaction with the surrounding gas.

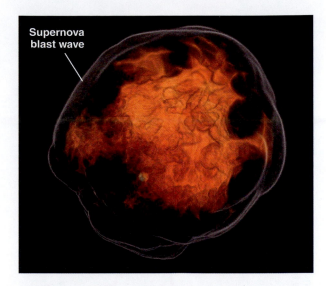

Figure 13.21 Computer Model of a Supernova
A 3D simulation of a neutrino-driven core-collapse supernova, 850 milliseconds after the explosion began. Only the star's inner regions are shown, as some of the neutrinos produced in the contracting core are absorbed in the dense gas above it, providing extra pressure to drive the supernova blast wave. The process is not purely spherical, as plumes of neutrino-driven gas rise upward.

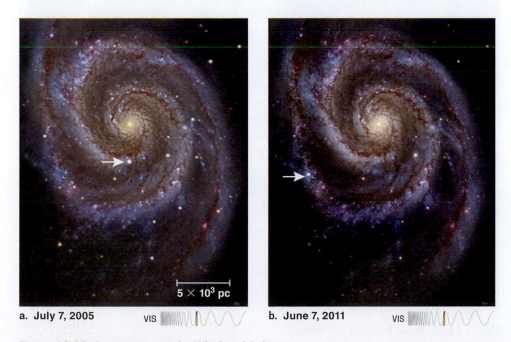

Figure 13.22 Supernovas in the Whirlpool Galaxy
Occurrences in **a.** 2005 and **b.** 2011 of Type II supernovas in the nearby spiral galaxy M51, the Whirlpool Galaxy. Supernovas are rare, but M51 has yielded three observable ones since 1995. Such massive stars have short life spans, so they die close to where they were born—in the galaxy's blue spiral arms.

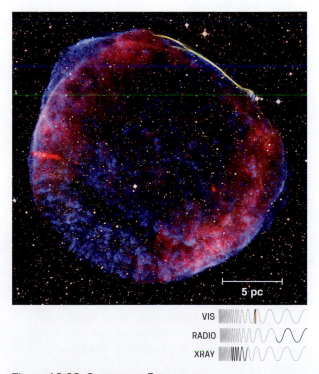

Figure 13.23 Supernova Remnants
A composite image of the bubble-shaped shock wave of SN 1006. This Type Ia supernova exploded more than a thousand years ago, driving stellar material into space and sweeping up the interstellar medium around it.

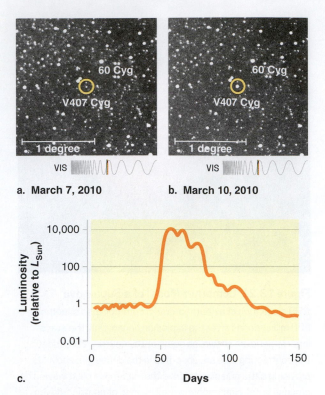

a. **March 7, 2010**   b. **March 10, 2010**

c.

Figure 13.24 Nova

a. Before and **b.** after images of Nova Cygni ("V407 Cyg"), as captured by amateur astronomers. It was 10 times brighter in the span of just a few days. **c.** Changes in a nova's brightness over time. A nova can increase in brightness by a factor of 10,000 in just a few days.

nova A nonterminal explosive nuclear fusion occurring on the surface of a white dwarf after it has accreted matter from a companion star.

thermonuclear runaway Nuclear fusion that ignites and spreads catastrophically, consuming all available fuel.

Type Ia supernova The thermonuclear explosion of a white dwarf that has accreted mass from a binary companion, causing it to exceed the Chandrasekhar limit and experience thermonuclear runaway.

One aspect of supernovas and their remnants that astronomers understand in great detail is the way these explosions seed the Universe with heavy elements. "We are all made out of heavy elements," says Woosley. "Earth is made out of heavy elements too. That makes supernovas really important because most of the Universe's heavy elements are born in a supernova." We have seen that a massive star just before its death is a kind of many-layered sphere of processed elements. The outer atmosphere is hydrogen, surrounding shells of nuclear-fusing hydrogen, helium, carbon, neon, oxygen, and silicon, all around an iron ash core. In a supernova, each layer is blown into space. The violent conditions in the explosion are so extreme that elements heavier than iron form. Supernovas create heavy elements such as nickel, lead, copper, and so forth, which enable life as we know it. As the astronomer Carl Sagan once said, "We are all starstuff looking at starstuff."

By combining the action of supernovas with the main-sequence fusion of low-mass, intermediate-mass, and massive stars, astronomers can account for a lot of the elemental abundances in the Universe. We will soon discuss other modes of *nucleosynthesis*— that is, formation of atomic nuclei and therefore elements. But we can now answer the question of why carbon is relatively common in the cosmos, compared with an element such as scandium, by tracking where and when each is fused in the course of stellar evolution.

From Nova to Supernova: Exploding Stars in Binaries

From careful observations, astronomers studying supernovas have found distinct flavors of these stellar explosions. For there to be *super*novas, there first had to be "regular" novas. For hundreds of years, astronomers cataloged the recurrent flaring of certain stars, which they called **novas** (*nova* is Latin for "new"). A typical nova increases in brightness by a factor of 10,000 to 100,000 for a few days to a few weeks (**Figure 13.24**). The time between nova outbursts can vary from a few decades (which can be observed) to 100,000 years (which

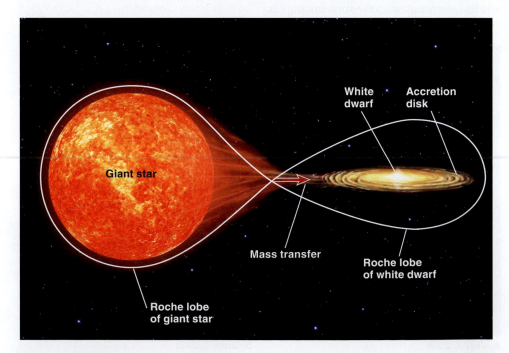

Figure 13.25 Mass Transfer and Nova

The Roche lobes of the primary and secondary (white dwarf) stars in a binary system. Mass transfer through the Roche lobes triggers a nova. When the primary star becomes a giant and overflows its Roche lobe, the excess material crosses over, forms a disk around the white dwarf, and eventually spirals onto the white dwarf's surface.

can only be hypothesized). By the mid-1960s, astronomers had worked out a basic model explaining these periodic flarings of novas.

If the orbit of two stars in a binary system brings them close enough together that one star extends outside its Roche lobe, their evolutionary pathways can be dramatically altered. "A nova happens if a white dwarf is in a binary star system and accumulates a little bit of matter from the companion star," says Woosley. A thermonuclear explosion of the accumulated mass on the white dwarf's surface produces a nova (**Figure 13.25**) when the temperatures and densities on that surface reach a **thermonuclear runaway**. In a thermonuclear runaway, the fusion happens so fast that the matter has no time to expand. "In the case of a regular nova," says Woosley, "you're building up only about 10^{-5} of a solar mass of hydrogen and helium on the surface of the white dwarf before the nuclear runaway happens. So if the accretion rate from the companion is quite low like this, meaning that it's a solar mass every billion years instead of a solar mass every 10 million years, then you get into the case where the surface layer explodes before the white dwarf grows very much in mass."

Although the explosion is powerful enough to increase the brightness of the white dwarf 100,000-fold or more, it is not powerful enough to damage the binary star system permanently. "When this layer, this 10^{-5} of a solar mass, explodes," says Woosley, "it has enough energy to push off everything that had fallen onto the white dwarf and makes the star very bright for weeks, but it doesn't destroy the white dwarf." Once the explosion is over, the companion begins dumping matter onto the white dwarf again, setting the stage for another nova event in the future.

If the accretion rate onto the white dwarf is more dramatic, however, the white dwarf can be entirely consumed. This may be the origin of **Type Ia supernovas**, explosions that involve binary stars and white dwarfs. Type Ia supernovas tell a very different story than their single-star cousins, Type II supernovas, in which a massive star collapses (see **Anatomy of a Discovery** on p. 357; and **Table 13.3**). "If you put a white dwarf into a binary system and let the star next to it donate enough matter," says Woosley, "the mass of the white dwarf can grow to the Chandrasekhar limit. At that point, its central density rises and allows it to finally ignite carbon fusion. The star couldn't do this by itself, but the matter being dumped on it by the companion makes it happen."

In a Type Ia explosion, a "flame front" of carbon fusion begins in the center and races through the rest of the white dwarf. Once the front gets going, it takes only a second for the carbon to fuse all the way through the elements to iron. Unlike the steady fusion in a massive star, here the fusion is part of an explosive detonation. The released energy tears the white dwarf and its companion apart (**Figure 13.26**). "A Type II supernova is powered by

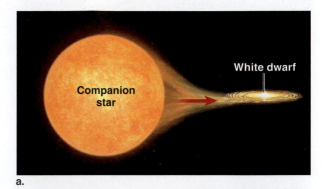

a.

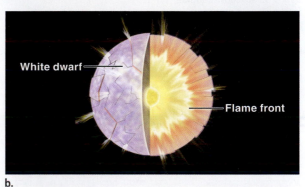

b.

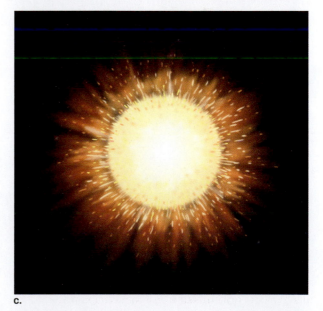

c.

Figure 13.26 Type Ia Supernova
The stages of evolution of a binary system that lead to a Type Ia supernova: **a.** Through mass transfer, matter is dumped onto the white dwarf. **b.** The additional mass ignites a wave of nuclear fusion (the "flame front") within the white dwarf. **c.** The white dwarf then explodes.

Table 13.3	Characteristics of Supernova Types	
	Type II (core collapse)	**Type Ia (white dwarf binary)**
Type of Star(s)	Single massive star (above 8 solar masses)	Binary system (white dwarf plus another star)
Causal Event	Nuclear fusion ends as fuel runs out; then gravitational collapse and explosion	Accretion of mass onto white dwarf; explosion when it exceeds Chandrasekhar limit
Luminosity	Billions of solar luminosities	Billions of solar luminosities
End Results	Destruction of star; supernova remnant; neutron star or black hole	Destruction of both stars; supernova remnant
Other		Useful as standard candle

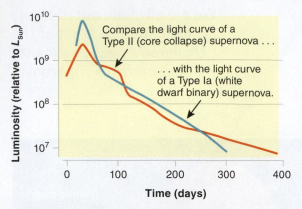

Figure 13.27 Light Curves of Supernovas
A light curve is the pattern of changes in an object's luminosity over time. The characteristic light curves of Type Ia and Type II supernovas are distinct. The predictable light curves and incredible brightness of Type Ia supernovas, visible from across the galaxy, make them ideal for determining distances and thus useful as standard candles.

gravitational collapse. I call it a gravity bomb," says Woosley. "But a Type Ia—that is really all about fusion. It's basically a tremendous thermonuclear bomb."

Some researchers argue that the merger of two white dwarfs, rather than mass transfer, may be the source of Type Ia supernovas. In the last two decades or so, Type Ia supernovas have become extremely important in astronomy because their luminosity changes with time in a characteristic way. When astronomers plot any object's energy output versus time, the resulting plot is a called a **light curve**. The shapes of light curves are often quite different for different classes of objects whose light output changes. Recall from Chapter 11 that the light curve of a Cepheid variable can be used to infer the distance to the Cepheid.

The light we see from a Type Ia supernova comes from the radioactive decay of elements that were abundantly built up during the explosion. After looking at many Type Ia light curves, astronomers found similarities in shape—for instance, how long the light curve remains at its peak (**Figure 13.27**). They recognized that the curve's shape is tied to the total energy release in the supernova, which is equal to the total luminosity. This discovery meant that astronomers had discovered a way to use Type Ia supernovas as standard candles, similar to Cepheid variables. Comparing that luminosity with the apparent brightness of the supernova on the sky then enabled astronomers to find the supernova's distance by using the inverse square law for brightness (see Chapter 11).

The fact that Type Ia supernovas can be used as standard candles was an unexpected boon for astronomers, who are always hungry for new distance determination methods. Supernovas are bright enough that they can be seen across significant fractions of the visible Universe, so Type Ia supernovas are extremely valuable standard candles.

Finally, it's worth noting that explosions more than 100 times as powerful as a typical supernova have also been observed. Sometimes called *hypernovas*, they represent a frontier in astrophysical research. The model many astronomers favor for these objects is the collapse of a massive star with the brief formation of a powerful jet near the core that explosively plows through the infalling stellar layers.

Section Summary

- Once an iron ash core forms in a massive star, core fusion ends and the star quickly collapses. Repulsive forces in the core halt further collapse while surrounding layers continue to fall inward.
- Type II supernovas are explosions that mark the death of massive stars. The exploding stellar ejecta sweeps up surrounding interstellar material, forming a supernova remnant and dispersing heavy elements across the galaxy.
- Novas occur when a white dwarf in a binary system pulls matter from its companion, causing periodic thermonuclear runaways on the white dwarf's surface.
- If too much matter falls onto the white dwarf, carbon fusion begins in the core, and the star may be blown apart in a Type Ia supernova (whose light curve makes it useful as a standard candle).

CHECKPOINT   a. Of the two kinds of supernova discussed in this chapter, which is better understood in terms of how the explosion is initiated? **b.** Explain how that type of supernova proceeds.

Neutron Stars: Stellar Corpses of the Second Kind

13.7   Type Ia supernovas completely destroy the white dwarf stars that gave them birth. But what remains at the center of the once-massive star after a Type II supernova, when the torrent of light fades and the blast wave blows the bulk of the stellar material far into space? The answer, as with almost all questions related to stellar evolution, hinges on mass. The mass of the already ultradense core determines its final fate.

light curve A graph of an astronomical object's light output versus time.

What happens when a star explodes?

Tycho's supernova remnant

Kepler's supernova remnant

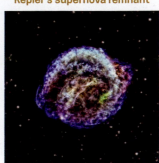

observations

Tycho Brahe observed a "new star" in 1572. Johannes Kepler observed one in 1604. To Tycho and Kepler, the transient objects tested the immutable nature of Aristotle's model of the Universe. Tycho determined that the object resides in the heavens, far beyond the Moon's orbit and Earth's atmosphere. Thus, he proved that the heavens cannot be unchanging.

Tycho viewing a "new star"

observation

The "new stars" were supernovas—representing the final moment of a massive star's life (Type II) or the instant a white dwarf accretes enough mass to cross the Chandrasekhar limit (Type Ia). While visible light from Tycho's and Kepler's supernovas has faded, space-based telescopes detect both X-rays from hot gas and IR emission from dust grains in their remnants. A Type II occurs just once every 50–100 years in the Milky Way; a Type Ia (like Tycho's and Kepler's) is even rarer.

observation

In 1987, nearly 400 years after Kepler's supernova and the invention of the telescope, light from a Type II supernova in the nearby Large Magellanic Cloud (LMC) reached Earth. At Las Campanas Observatory in Chile, **Ian Shelton** noticed a "new star" in an image of the LMC and reported his discovery. The proximity of this supernova, named SN 1987A, gave astronomers a superb chance to study the explosion in great detail and at all wavelengths.

HST image of SN 1987A

SN 1987A (before)   SN 1987A (after)

hypothesis

SN 1987A was the spectacular death of a supergiant star. In subsequent decades, **Stan Woosley** and his team used observations of this nearby supernova to constrain theoretical models of the mechanism driving these explosions. Such work has vastly expanded our understanding of the energy supernovas produce and the elements they synthesize, seeding interstellar space with the building blocks of the next generation of stars.

"The end of a big star's life is the beginning of something else," says Woosley. When the iron ash core collapses just before the explosion, quantum physics again comes into play, allowing one kind of subatomic particle to transform into another. Recall that as the core collapses, the charged electrons and protons are combined into neutrons, and this rapid creation of neutrons helps drive the intense production of neutrinos that may power the supernova. Turning so many charged electrons and protons into neutral neutrons enables the core to get rid of repulsive electromagnetic forces and draws its components together ever tighter. The core continues to shrink; in the end, only neutron degeneracy pressure can save it from complete collapse.

Neutron degeneracy pressure is similar to the quantum process of electron degeneracy pressure (Section 13.4), which occurs in white dwarfs. In both cases, the squeezing of particles into a smaller space forces them to move at higher speeds. The high-speed particles rattling around inside the dead star generate a pressure force that supports the stellar corpse against its own gravity. In white dwarfs, electrons provide the degeneracy pressure. In a neutron star, neutrons play that role.

"At the end of the core's collapse, it has nearly the same density as matter in the nucleus of an atom," says Woosley. "Basically, the core *is* a giant nucleus, and that's the beginning of the life of a neutron star." What Woosley describes is one of two forms of a stellar "corpse" left over from the death of massive stars. **Neutron stars**—the compact remains of massive stars—have masses between 1.4 and 2.1 M_{Sun}, but they pack that matter into a sphere just 10 km in radius (**Figure 13.28**). So much material is stuffed into such a small space that a neutron star's gravitational field is almost 200 billion times stronger than Earth's. A mere teaspoon of neutron star material would weigh about 1 billion tons, which is about 10 times the weight of the entire human population.

These extremes of gravity and density in neutron stars make them perfect laboratories for scientific investigation of the frontiers of physics. For example, physicists believe that entirely new states of matter may exist in a neutron star's core, where strange particles are created in such large numbers that they might form an ultradense kind of nuclear liquid. These calculations have proved extremely useful (though in some cases inconclusive) for ascertaining the nature of neutron stars.

When neutron stars appear as part of a binary system with a normal star, they can also help reveal how accretion works under extreme conditions. Binary systems known as **X-ray bursters** show intense outbursts of energy in the X-ray band of the electromag-

neutron star A stellar remnant, supported by neutron degeneracy pressure, that results from the gravitational collapse of a massive star.

X-ray burster A binary system that shows intense X-ray outbursts caused by the pileup of material on the surface of a neutron star.

pulsar A rotating, magnetized neutron star whose radio waves produce periodic emission patterns that are observed on Earth as energy bursts.

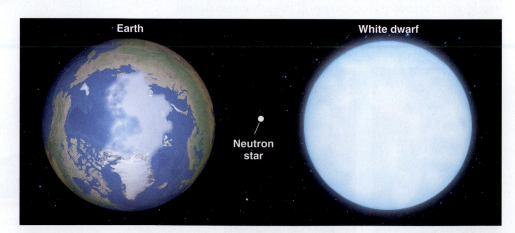

Figure 13.28 Relative Size of a Neutron Star
Matter is so compressed in a neutron star that a teaspoon would weigh about 1 billion tons. A white dwarf is about the size of Earth, but the mass of both is much less than that of a neutron star, which is many times smaller—close to the size of one city.

netic spectrum. The outbursts appear to be linked to a pileup of material on the neutron star's surface. As in a nova, when enough material builds up on the neutron star's surface, nuclear fusion is triggered, releasing a burst of X-ray photons.

These kinds of observations have helped astronomers unpack the remarkable physics of neutron stars. Nature, however, has provided scientists with a most unusual and unexpected laboratory for studying neutron stars: the enigmatic cosmic lighthouses called pulsars.

Pulsars

"A neutron star starts out very hot," explains Woosley, "but it doesn't stay that way for very long. Since hot gas radiates as a blackbody, the star produces lots of radiation in the visible part of the spectrum, just like a hot white dwarf. But as the star cools, it quickly becomes invisible in optical light." If that were the whole story, you might wonder how astronomers have studied neutron stars so extensively. In many cases, the conditions that create neutron stars also provide them with a means to be recognized far across the galaxy.

"Neutron stars are born rotating very rapidly," says Woosley. "That's because even if the massive star was rotating slowly on the main sequence, when it collapses and shrinks to such high density, conservation of angular momentum forces it to spin faster and faster." Recall that the principle of conservation of angular momentum applies to star and planetary system formation (see Chapters 5 and 12). This principle works the same way for a massive collapsing star as it does for giant star-forming clouds.

The reduction in size of the rotating core is so extreme that newborn neutron star rotation rates are mind-numbingly fast. A typical young neutron star—a 20-km sphere packing a Sun's worth of mass—completes more than 600 rotations every second, giving it a rotation period of about 2 milliseconds (ms). That incredible speed powers **pulsars**—rotating, magnetized neutron stars whose radio waves periodically sweep across Earth, producing emission patterns that are observed as bursts of energy. "On top of their outrageously fast rotation rates," Woosley explains, "neutron stars are also born with incredibly powerful magnetic fields." Most stars have some degree of magnetism, but just as the collapse amplifies the spin of the massive star's core, it also drives even weak stellar magnetic fields to extreme values. Thus, newborn neutron stars also host hyperstrong dipole magnetic fields whose shapes are like Earth's field, only a trillion times more intense.

The axis of a neutron star's magnetic field need not align with the star's axis of rotation. Significant misalignment of these axes leads to dramatic, far-reaching effects, as the star's rotation sweeps the magnetic poles around. Electrons in the neutron star's atmosphere react to the magnetic field's movement, becoming rapidly accelerated and radiating energy. The dipole magnetic-field lines funnel the electrons and the radiation along the field's axis. The resulting beams of radiation, which are intense enough to be detected across many light-years, emanate from the north and south magnetic poles **(Figure 13.29)**. The radio beams rotate with the pulsar much like the rotating light from a lighthouse, swinging around so they appear to pulse. Pulsars have also been observed through space-based X-ray and gamma-ray telescopes. When electrons in the region surrounding a pulsar interact directly with photons or the magnetic field, they produce emissions in these high-energy, short-wavelength bands of the spectrum.

In 1967, Jocelyn Bell, a University of Cambridge graduate student, used a specially built radio telescope to look for new radio sources. While scanning the sky, Bell found what she first thought was "a bit of scruff" on the paper readout. Observing again, she found, to her amazement, beautifully timed repeating pulses of radio emission coming

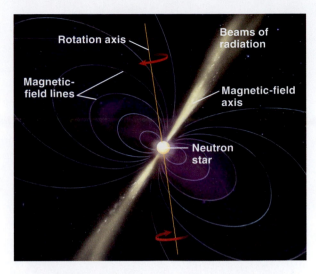

Figure 13.29 Pulsar

The rapid rotation of a neutron star with misaligned magnetic-field and rotation axes produces a pulsar. As the neutron star rotates, beams of radio waves sweep around with the magnetic field. A pulsar can be observed from Earth only if a beam of radio waves passes across the planet.

INTERACTIVE:
Pulsar Observation:
Beam vs. Viewing Angle

"On top of their outrageously fast rotation rates, neutron stars are also born with incredibly powerful magnetic fields."

Figure 13.30
Jocelyn Bell and the Discovery of Pulsars
a. Jocelyn Bell was the astronomer who found the perfectly timed radio pulses of neutron stars.
b. The nonrandom cadence of the pulses initially caused Bell and her colleagues to ponder the possibility of extraterrestrials as the source, before they identified the true nature of the phenomenon.

a.

b.

"When a massive star dies and the iron core collapses, there are only two things it can do. It can make a neutron star, or it can make a black hole."

Table 13.4	Origin of the Elements
Source	**Elements**
Big Bang nucleosynthesis	Hydrogen, helium
Stellar nucleosynthesis	Helium, carbon, neon, oxygen, silicon, iron (plus all other elements heavier than hydrogen and lighter than iron)
Explosive nucleosynthesis	All naturally occurring elements heavier than iron

at 1.3-second intervals (**Figure 13.30**). She brought the stunning results to her graduate adviser, Antony Hewish. For a moment the pair considered the idea that the pulses might be evidence of extraterrestrial intelligence, naming the object LGM-1 for "Little Green Men-1." But they soon decided the phenomenon must have a natural origin and named it a pulsar. A year later, the American astrophysicist Thomas Gold and others provided the theoretical explanation of pulsars in terms of rotating, magnetized neutron stars. Later Hewish alone was awarded the Nobel Prize in Physics for the discovery, although many astronomers argue that Bell deserved to share the prestigious prize.

Since then, more than 2,600 pulsars have been discovered, having a range of periods from milliseconds to seconds. Astronomers now know that pulsars slow down as they age. Conservation of energy demands that the energy leaving a pulsar as radiation be balanced through a reduction in another form of energy held by the dead star. That balance comes from a reduction in the pulsar's spin energy. The radiation pulsars emit gradually slows their spin over millions of years.

Pulsars have a remarkable property: they act as precise "cosmic timepieces." Once a pulsar's spin and its rate of spin decay are determined, astronomers can use the steady pulses as extremely accurate stopwatches to measure properties of the pulsar and its environment.

Stellar Evolution and the Abundance of Elements

Why is gold rare while carbon is easy to find? Now that we know how stars evolve from their birth until their demise as planetary nebula or supernovas, we can make a connection between stellar evolution and the *abundance of elements*. What sets the relative numbers of atoms of the different kinds of elements? The answer to this question takes us from the heart of stars and stretches throughout the entire history of the cosmos.

Beginning with Chapter 10, we saw how Sun-like stars on the main sequence fuse hydrogen into helium via nuclear reactions in their cores. Later, on the horizontal branch and AGB, they fuse helium into carbon. In this chapter we've seen how fusion in the core of more massive stars can create heavier elements such as oxygen and neon, all the way up to iron. Overall, this process is called *stellar nucleosynthesis*. We've also seen that when massive stars explode as supernovas, conditions are extreme enough to drive fusion reactions. This process, called *explosive nucleosynthesis*, can create elements heavier than iron. For many years, astronomers conjectured that the merger of orbiting binary neutron stars produces *kilonovas*, explosions so intense that, like supernovas, they might produce some of the Universe's heavy elements. This theory received some support in 2017 when the first neutron-star merger was detected and spectral signatures of elements such as gold were detected. As we will see in Chapter 18, some fraction of the Universe's lighter elements, such as helium and lithium, formed in the first few minutes of cosmic history in what's called the *Big Bang nucleosynthesis*.

Taken together, these forms of nucleosynthesis can account for all of the elements we see today (**Table 13.4**) as well as their different abundances in the Universe. Our ability to describe the origin, and census, of every kind of atom in the Universe represents an incredible achievement of science. Ancient thinkers wondered about the origin of the elements and their relative abundances; it is extraordinary that you live among the first few generations to have solved that puzzle.

The Final Collapse

"When a massive star dies and the iron core collapses, there are only two things it can do," explains Woosley. "It can make a neutron star, or it can make a black hole. Which kind of stellar corpse results depends on the mass of the thing that is finally left behind. If the core is more than about 2.1 times the mass of the Sun, then the strong nuclear force, the crowding of neutrons, and neutron degeneracy pressure just can't hold it up."

Scientists still debate which massive stars fit into this category. Although astronomers are certain that a 10-M_{Sun} star on the main sequence will leave a neutron star (or pulsar) behind, the fate of a 30-M_{Sun} star is not so clear. That depends on how much mass is blown off the star by stellar winds as it evolves throughout its life, as well as how much mass is blown away in the supernova explosion. One thing is clear: If there is too much mass in the core, then nothing can stop gravity from winning a complete victory. The core collapses all the way until it becomes that most extreme and remarkable denizen of the stellar graveyard—a black hole.

Section Summary

- Neutron stars are the compact remnants of massive stars that have undergone Type II supernovas and have masses between 1.4 and 2.1 M_{Sun}.
- Pulsars are neutron stars that emit periodic bursts of radio emission along the star's magnetic poles. Pulsar periods vary from milliseconds to seconds; the period slows over millions of years.
- Explosive nucleosynthesis creates elements heavier than iron.

CHECKPOINT Are we more likely to have an accurate accounting of all of the pulsars in the galaxy than all of the white dwarfs? Explain your answer.

CHAPTER SUMMARY

13.1 Fireworks in a Galaxy Not So Very Far Away

Stars live on the main sequence until they can no longer counteract their own gravity by producing energy through nuclear fusion of hydrogen. Once core hydrogen is used up, stars begin a series of changes that eventually returns much of their mass to the interstellar medium. A star's mass determines how, and how quickly, these processes proceed and what the end state will be. Another important factor is the presence of a binary companion and its proximity to the dying star.

13.2 How to Become a Giant: The Fate of Low- and Intermediate-Mass Stars

Once the core hydrogen is depleted in a low- or intermediate-mass star, the core contracts and heats up. A shell outside the core heats up and begins fusing hydrogen, causing the outer layers to swell; the star becomes more luminous and its surface cools down. This phase corresponds to the red giant branch of the HR diagram. Core contraction creates a temperature high enough to begin fusing helium into carbon in the core; the star is then on the horizontal branch of the HR diagram. Once helium is depleted, the carbon ash core contracts again, and a helium-fusing shell above the core is created. The star, now very large and luminous, moves to the asymptotic giant branch of the HR diagram.

13.3 The Last Hurrah: Planetary Nebulas

Dense, relatively slow stellar winds blow off the outer layers of the dying star, causing it to lose much of its mass. Eventually, a fast wind blown from the now high-temperature dwarf star overtakes the slower wind, creating violent shock waves. These shells are visible to us as planetary nebulas.

13.4 White Dwarfs: Stellar Corpses of the First Kind

Stars with core masses below the Chandrasekhar limit of 1.4 M_{Sun} collapse into white dwarfs, which are smaller high-temperature stars able to withstand

gravity through electron degeneracy pressure. White dwarfs cool over billions of years until they fade as black dwarfs.

13.5 Living Fast, Dying Young: The End of Massive Stars

In a massive star, higher temperatures accelerate nuclear fusion and enable it to proceed to heavier elements. The star is then a red giant and begins fusing helium to carbon in the core by the triple-alpha process. It then proceeds through a series of contraction and increased-temperature stages in the core, fusing neon, oxygen, silicon, and finally iron. Each transition is accompanied by the creation of a new shell fusing the previous element.

13.6 Big and Bigger Bangs: Novas and Supernovas

Since fusing iron cannot release energy, the once-massive star gravitationally collapses into an ultradense state consisting almost entirely of neutrons. Many neutrinos are produced. A strong shock wave explosively blows away the outer layers of the star in a Type II supernova. Material blown from above the core sweeps through the interstellar medium and becomes a supernova remnant. First appearing as a bubble, it will finally fragment. Elements heavier than iron are created through fusion in supernovas and are ejected into the interstellar medium. In contrast, Type Ia supernovas involve a white dwarf with a binary companion that accumulates additional material until it exceeds the Chandrasekhar limit and violently explodes. Type Ia supernovas are used as standard candles to determine distance.

13.7 Neutron Stars: Stellar Corpses of the Second Kind

The neutron stars formed in Type II supernovas have strong magnetic fields and rotate many times per second initially, slowing as they age. The orientation of their magnetic fields with respect to their spin axes can create radio emission, seen as pulsars. Massive stars whose cores exceed the limit for neutron stars (which may be a core mass of about 2.1 M_{Sun}) collapse into black holes.

QUESTIONS AND PROBLEMS

Narrow It Down: Multiple-Choice Questions

1. A star is 10 billion years old. What final form may it take when it dies?
   a. white dwarf or black hole, but not neutron star
   b. neutron star or black hole, but not white dwarf
   c. only white dwarf
   d. only black hole
   e. There's not enough information to answer the question.

2. How does the mass of a star relate to its life expectancy?
   a. They are known to be unrelated.
   b. They may be related but the relationship has not yet been established.
   c. The greater the mass, the longer the life expectancy.
   d. The greater the mass, the shorter the life expectancy.
   e. While star masses are known, star life expectancies are not.

3. Just after a star has used up the hydrogen in its core, the core properties differ from those of the shell directly above it. Which of the following statements regarding the differences between the two layers is *not* true?
   a. The elemental composition differs.
   b. The density in the core is greater than in the shell.
   c. Gravity increases in the core only.
   d. The hydrogen fraction is lower in the core.
   e. Nuclear fusion of hydrogen to helium occurs in the shell only.

4. As the hydrogen in the core of a star runs out, which of the following does *not* start to occur?
   a. Hydrogen fuses in the shell above the core.
   b. Helium immediately fuses in the core.
   c. The star's total energy output increases.
   d. The core contracts.
   e. The shell heats up.

5. Which of the following describes nuclear fusion within a star on the horizontal branch of the HR diagram?
   a. only hydrogen to helium in the core
   b. helium to carbon in the core; none in shells above the core
   c. no fusion in the core; hydrogen to helium in the first shell
   d. helium to carbon in the core; hydrogen to helium in the first shell
   e. no fusion in the core; helium to carbon in the first shell; hydrogen to helium in the second shell

6. Which of the following accurately describes changes to a star when it first moves off the main sequence of the HR diagram?
   a. increased radius and increased surface temperature
   b. increased radius and decreased surface temperature
   c. shorter wavelength of peak radiation
   d. decreased luminosity
   e. absence of nuclear fusion

7. A nearby star is fusing helium into carbon in a shell above the core. What mass can the star be?
   a. only low-mass
   b. only intermediate-mass
   c. only high-mass
   d. either intermediate-mass or massive
   e. any size

8. True/False: If you discovered a gas cloud that is composed only of carbon and elements heavier than carbon, you could conclude with certainty that it did not originate from a supernova.

9. Which of the following is/are true about an accretion disk around a star? Choose all that apply.
   a. Its star must be massive.
   b. Its star must be low-mass.
   c. It may be evidence of a companion star.
   d. It may be composed of material left over from the birth of the star.
   e. It is almost certain to contain only the same elements as the star.

10. Which of the following is *not* a possible source of an accretion disk?
   a. the birth of a new star
   b. one star pulling material from a binary companion
   c. a larger star pulling a smaller binary companion apart
   d. a slowly spinning single main-sequence star
   e. All are possible.

11. What does the Heisenberg uncertainty principle imply about the behavior of an electron?
   a. The more confined it is, the greater will be the range of its velocity.
   b. When its velocity is greater, it requires more space.
   c. The greater the uncertainty in its position, the higher will be the range of its velocity.
   d. The electron can be squeezed out when matter becomes degenerate.
   e. The charge of the electron is reduced when it is confined.

12. Which of the following has the highest density?
   a. intermediate-mass main-sequence star
   b. white dwarf
   c. neutron star
   d. planetary nebula
   e. masssive main-sequence star

13. Which of the following statements about white dwarfs is true?
   a. They can be the remains of massive stars.
   b. They don't allow light to escape.
   c. They first appear at the center of a planetary nebula.
   d. They are smaller than neutron stars.
   e. The larger their mass, the larger their radius.

14. The Chandrasekhar limit is
   a. the minimum mass of a very-low-mass main-sequence star.
   b. the minimum mass of a white dwarf or neutron star.
   c. the maximum mass of a white dwarf.
   d. the maximum mass of a neutron star.
   e. the maximum mass of a black hole.

15. Which statement about a star that is creating the element silicon is true?
   a. It must be a low-mass star.
   b. It must be an intermediate-mass star on the main sequence.
   c. It must be an intermediate-mass star on the red giant branch.
   d. It must be a high-mass star on the main sequence.
   e. It must be a high-mass star on the supergiant branch.

16. Which of the following statements about the sequential nuclear fusion that occurs in a high-mass star is true?
    a. The duration of fusion increases as increasingly heavy elements are produced.
    b. Lighter end products are created in each successive stage.
    c. The last element that serves as input to nuclear fusion is iron.
    d. Higher temperatures are required for nuclear fusion of each successively heavier element.
    e. Fusion can occur in only one shell at a time.

17. What characteristic do the processes that produce planetary nebulas and Type II supernovas share?
    a. electron degeneracy pressure
    b. density of the remaining star
    c. shape of the outflows
    d. mass of the dying star
    e. the outer layers of the star being driven into space

18. Which of the following is/are *not* characteristics that Type Ia and Type II supernovas have in common? Choose all that apply.
    a. mass of the dying star
    b. fusion of elements up through iron prior to the stellar explosion
    c. creation of elements heavier than iron
    d. enormous shock waves
    e. usefulness as standard candles

19. Which statement accurately describes a neutron star?
    a. It is created from the outer layers of a high-mass star.
    b. It is a star fusing iron in its core.
    c. It results from the death of a massive star.
    d. It is created by electron degeneracy pressure.
    e. It is the end product of a very-low-mass star.

20. A pulsar is also
    a. a neutron star.
    b. a white dwarf.
    c. a strong source of visible light.
    d. a dead low-mass star.
    e. the site of a nova.

To the Point: Qualitative and Discussion Questions

21. Explain the relationship between mass and core temperature in stars and how it determines a star's longevity.

22. Describe how different elements are created and recycled by stellar life cycles.

23. What event signals the beginning of the death process for all stars?

24. Why does a midsized star move up and to the right on the HR diagram relative to the main sequence as it approaches death?

25. In Sun-like low- and intermediate-mass stars, what keeps the inner core from collapsing after the star's hydrogen fuel is depleted and fusion stops?

26. Describe the correlation between the masses and radii of white dwarfs.

27. Describe the most current explanation of how planetary nebulas form.

28. Subrahmanyan Chandrasekhar determined that the limit for a dead star to be supported by electron degeneracy pressure is $1.4\,M_{Sun}$. What happens to stars more massive than that?

29. Describe the end-of-life stages in a massive star.

30. Why was the Hubble Space Telescope so important to the development of a new theory of planetary nebula formation? What does this tell you about the role of new technologies in science?

31. Why are massive stars unable to continue fusing elements heavier than iron?

32. What causes the core of an intermediate-mass star to cease collapsing? How about that of a massive star?

33. Describe the differences in the causes and observations of Type Ia and Type II supernovas. Which type serves as a standard candle, and why?

34. Would you expect to observe more planetary nebula remnants or supernova remnants? Why?

35. How does the conservation of angular momentum explain the initial spin rate of neutron stars and the subsequent changes as they age?

Going Further: Quantitative Questions

36. A star's radius and temperature are twice those of the Sun. How does its luminosity compare with the Sun's? Which had a greater effect on the final luminosity: the doubling of temperature or radius?

37. A star's temperature is 3 times as high as the Sun's, and its luminosity is 48 times that of the Sun. What is the ratio of the star's radius to the Sun's radius?

38. The temperature of a red giant is 3,300 K, and its radius is 60 times that of the Sun. What is its luminosity, in L_{Sun}? Does this result make sense, given the cooler surface temperature of the red giant?

39. A red giant has a temperature of 3,700 K and luminosity of $1.8 \times 10^3\,L_{Sun}$. What is its radius in solar radii, R_{Sun}?

40. What is the wavelength of peak radiation, in meters, of a white dwarf with a temperature of 35,000 K? What kind of light is this?

41. A neutron star has a temperature of 50,000 K. What is the wavelength of its peak radiation, in meters? What kind of light is this?

42. What is the temperature, in kelvins, of a star with a peak wavelength of 6.7×10^{-7} meter?

43. A white dwarf has a peak radiation of 3.3×10^{-8} meter. What is its temperature, in kelvins?

44. A planetary nebula with a radius of 0.1 parsec (pc) was created during the death of its star 3,490 years ago. At what rate has it been expanding, in kilometers per second?

45. A supernova remnant has expanded at the rate of 6,000 km/s and now measures 4.1 pc in diameter. How many years ago did the supernova occur?

"PEOPLE WERE GETTING DESPERATE"

14

Down the Rabbit Hole

RELATIVITY AND BLACK HOLES

The Black-Hole Shuffle

14.1 Nothing was working. In fact, nothing had worked for almost a decade. "People were getting desperate," says Manuela Campanelli, a renowned expert in exploring the physics of black holes by computer. "The idea should have been relatively straightforward: take the equations Einstein gave us that describe how matter bends space-time and use those to simulate something like two black holes orbiting each other." Einstein's theory of relativity—the theory behind black holes—is notorious for its difficulty, in terms of both its mind-bending consequences and its complex mathematics. Still, in principle it should have been possible to convert it into a form that computers could use for simulations. After all, many other difficult astrophysical phenomena, including the turbulent interiors of stars, have been simulated with supercomputers. "There were some initial successes in the 1970s," says Campanelli, "and that gave everyone hope that you could do black-hole simulations. But then things started grinding to a halt."

For years, astrophysicists struggled to simulate even a single, isolated black hole moving through space. The computers would take a few steps but then spew errors and fail. "People even began to think that it could not be done," says Campanelli. "Maybe Einstein's equations governing black holes really were just too hard."

Then, in August 2005 at a conference in Aspen, Colorado, one lone researcher took the stage and shocked the world. "His name was Frans Pretorius, and he was the one who first made it all work," says Campanelli. "He showed these beautiful simulations of binary black holes spiraling inward toward each other. It was amazing." Within months, two other groups, including Campanelli's, were simulating the entire process of black-hole mergers (**Figure 14.1**). "It was like a dam had broken," says Campanelli. "Now we can get a detailed understanding of these strangest of nature's creatures. It's a black-hole revolution happening right before our eyes."

← An artist's conception of a black hole and an accretion disk of gas surrounding it. As material spirals inward through the disk, jets of matter and energy are driven outward at close to the speed of light along the axis of rotation.

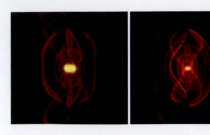 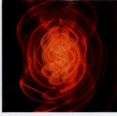 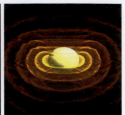

Figure 14.1 Black Holes Merging
Computer simulations of the merger of two black holes are shown. In the process, the structure of space and time around the black holes is distorted as predicted by Einstein's general theory of relativity. Left to right: emission of gravitational waves increases in intensity as the two black holes spiral inwards toward each other. Eventually (far right), the holes merge into a single object.

"The patent office job paid [Einstein's] bills and gave him a lot of time to think about physics."

black hole A region of space-time whose strong gravitational distortion prevents anything, including light, from escaping.

Gravity's Final Victory

All stars must fight with their own gravity, and sooner or later they must lose that fight. As you have learned in previous chapters, stars support themselves against their own gravitational weight by generating energy and pressure through nuclear fusion. Eventually, however, the fuel for nuclear fusion runs out. With nothing to balance gravity, dying stars must contract in on themselves. For stars with low enough mass, quantum mechanical properties associated with subatomic particles, such as electrons and neutrons, can ultimately halt the collapse. White dwarfs the size of Earth and neutron stars the size of Manhattan give testimony to how tightly more than 10^{30} kilograms (kg) of matter can be squeezed, as quantum effects support these burned-out stars against their own crushing gravity.

But white dwarfs form from stars with core masses of up to 1.4 solar masses (M_{Sun})—the *Chandrasekhar limit* (see Chapter 13). These correspond to main-sequence stars with a total mass of 8 M_{Sun} or less. For neutron stars, the exact upper limit on progenitor main-sequence core mass is not yet known. The best estimates, however, suggest that only stars with cores of 2–3 M_{Sun} end their lives as ultradense neutron stars. These may correspond to main-sequence stars with masses of about 8–20 M_{Sun}.

What fate awaits stars more massive than this limit? What can keep a 30-, 50-, or 70-M_{Sun} behemoth from infinite collapse? The remarkable answer is *nothing*. No known force can balance the titanic gravity of the most massive stars when they exhaust their fusion fuel. These stars collapse all the way down, their matter crushing into apparently infinite densities and zero radii. It is stars of such enormous mass that become the most enigmatic of astrophysical objects: **black holes**.

"Black holes are objects with gravity so powerful that even light cannot escape their pull," says Sean Carroll, a theoretical physicist at the California Institute of Technology. Black holes are so strange that they push our understanding of physics to new limits that might appear to be more science fiction than fact. Using Albert Einstein's theory of relativity, scientists have come to understand and model black holes, but even Einstein was unsure whether such outrageous objects could occur in nature. Only through the most intense study have astronomers built a convincing case for the existence of black holes. The efforts poured into their study have shown that black holes exist and provided new insights into their bizarre nature.

Before delving into a detailed discussion of black holes, we must first explore the theory that led to their discovery: Einstein's radical reimagining of the physical world, known as the theory of relativity.

Section Summary

- When the most massive stars run out of fuel, they collapse to black holes because there is no force to stop their gravitational collapse.

CHECKPOINT **a.** How do stars that have not run out of fuel keep from collapsing? **b.** Why do only more massive stars collapse?

Special Theory of Relativity

14.2 To understand a black hole, we must first understand how Einstein reimagined gravity. But to do that, we must first understand how he reimagined the very nature of space, time, and motion. This first step was the subject of his famous 1905 paper introducing what he called the *special theory of relativity*.

Electromagnetic Waves and the Speed of Light

Einstein was only 26 years old in 1905 when he submitted his manuscript on special relativity to the most famous physics journal of his era. At the time he wasn't a physics professor, a researcher at a major laboratory, or even a graduate student but a clerk working at the Swiss patent office (**Figure 14.2**). This was the best job he could find after he got his degree in physics. "The patent office job," says Sean Carroll, "paid his bills and gave him a lot of time to think about physics."

Einstein was tasked with evaluating patents for new devices that could synchronize clocks across giant factories or immense cities. Time and space were the meat and potatoes of his daily work life. Ironically, he came to the job already engaged in a puzzle deeply linked to ideas of time and space that had been bothering him since he was a child. By the time he reached the patent office, he was determined to think his way to a solution.

Eighteen years before Einstein was born, the British theoretical physicist James Clerk Maxwell devised a set of equations linking electric phenomena (such as charge and current) with magnetic phenomena (bar magnets, magnetic fields from moving charges, and so forth). An explanation of light as an electromagnetic wave also emerged from these equations. According to this explanation, the visible light to which our eyes respond is nothing more than waves of oscillating electric and magnetic fields traveling through space at the tremendous speed of 300,000 kilometers per second (km/s), or 670 million miles per hour (see Chapter 4).

But what are these waves made of? "Scientists in Einstein's era knew about water and sound waves," says Carroll. "They also knew these waves needed a medium through which to propagate. So they assumed that the existence of electromagnetic waves also implied a medium for their propagation."

Since light crosses vast distances between stars, physicists imagined a "luminiferous aether" filling all of space and supporting the passage of light waves. Although there was no direct evidence of the aether, physicists were certain it existed. "Once light was recognized as an electromagnetic wave," says Carroll, "the race was on to try to reconcile its behavior with the known principles of mechanics as Newton had laid them down." Researchers started designing experiments to study the propagation of light through the aether, expecting to detect the speed of light changing as Earth moves against the background aether.

"Before Einstein came along, adding up different velocities was pretty straightforward," says Carroll. "Imagine you are standing on the shore watching a boat move up a river. The water is flowing past you at some fixed velocity, so the velocity of the boat that you measure is the velocity of the water *compared to you* plus the velocity of the boat *compared to the water*." Viewed from the shore, the boat *appears* to move slowly because its engines are fighting a strong current. Physicists expected that light (like the boat) and the aether (like the river) would behave in the same way (**Figure 14.3**). "If you're trying to measure the velocity of an object like a light beam moving through the aether, then you'd expect the velocity you measure to be that of a light beam moving with respect to the luminiferous aether plus the velocity of the aether moving with respect to you."

But what physicists observed was not what they expected. No matter which way they set up their experiments, the observed speed of light always stayed the same. No effect of an "aether current"—the relative flow of the aether past a moving observer—was ever found. In response to the experiments, physicists began inventing new ideas to save the aether. Einstein, however, declined

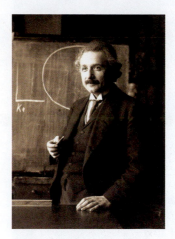

Figure 14.2 Albert Einstein
After graduating with a degree in physics, Einstein (1879–1955), shown in 1921, worked as a Swiss patent clerk. This work, involving patents dealing with the transmission of electric signals and the electrical-mechanical synchronization of time, served as a backdrop for the development of the special theory of relativity and its dramatic reimagining of the role of space and time.

Figure 14.3 Relative Motion
From the shore, you would measure the velocity of a boat moving on a river to be the velocity of the water relative to you (7 kilometers per hour, km/h) *plus* the velocity of the boat relative to the water (12 km/h). Because the boat moves in the opposite direction of the river's flow, the velocity of the water relative to you must be considered negative—so, the boat's velocity relative to you is $(-7 \text{ km/h}) + (12 \text{ km/h}) = 5 \text{ km/h}$.

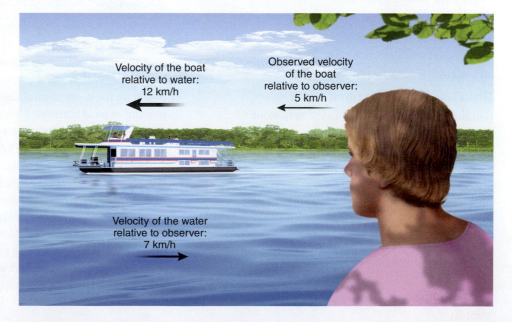

Velocity of the boat relative to water: 12 km/h

Observed velocity of the boat relative to observer: 5 km/h

Velocity of the water relative to observer: 7 km/h

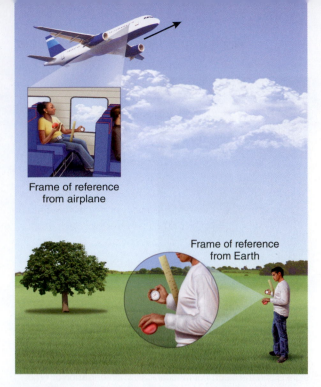

Figure 14.4 Frame of Reference
A frame of reference is the individual perspective of a person making observations—for example, using a clock to measure time, using rulers to measure distance, or watching a ball fall to the ground. The frame of reference of a person standing on Earth while an airplane passes differs from that of a passenger in that plane.

Frame of reference from airplane

Frame of reference from Earth

SEAN CARROLL

When directors must get the physics right for a super-hero movie, Sean Carroll is one of the scientists they call. He has advised on films such as *Thor* and *The Avengers*. He says with a smile, "I don't get paid much, but it's really cool to hang out on the set."

"There was a book I found in my town's library," Carroll recalls of his childhood. "It . . . explained everything they knew at the time about the fundamental structure of matter." That emphasis on fundamentals became the heart of his scientific career in the nature of relativity. "I still get really excited thinking about the basic nature of time." Carroll was one of the first scientists to begin blogging and has appeared on television in venues from the History Channel to *The Colbert Report*.

to play by the rules. He did not believe in the aether, so he did not care about solving the aether problem. Instead, he changed the entire discussion.

"It was Einstein who took the leap," says Carroll. "It was Einstein who said, 'You know what? Forget the aether and just assume that the speed of light is always constant.' That changed everything."

Frames of Reference and Special Relativity

Einstein recognized that the problem lay in describing phenomena from different frames of reference—a key concept in physics. A **frame of reference** is the perspective that constitutes an observer's account of the world. (An observer is anyone making a measurement or viewing the situation.) "It's the way we measure length and time intervals according to a point of view," says Carroll. "A frame of reference is just the individual perspective of a person extended throughout the whole Universe." If you're standing in a field and watch clouds pass overhead, you have one frame of reference. If you're sitting in an airplane that's flying at 500 miles per hour as you watch the same clouds, you have a different frame of reference (**Figure 14.4**).

Since the age of Galileo and Newton, physicists had understood that the motion of a frame of reference could affect the outcome of experiments performed inside that frame. They learned how to reconcile the different ways that observers in different frames of reference would describe what they saw. In doing so, physicists developed a coherent account of Newtonian physics. For Newton, frames of reference that are *not* accelerating—called *inertial frames*—played an important role because his laws were simplest to describe in such frames. A description of an object in a *noninertial frame* (that is, an accelerating frame) must include the apparent force that results from the acceleration of everything in the frame of reference. An accelerating car is a noninertial frame, and the feeling you get of being pushed back into the car seat as the car speeds up is the resulting apparent force. (The seat forces you to accelerate with the car.)

Although Einstein called his theory "relativity," he was really after invariants—the parts of physics that do *not* change from one frame of reference to another. To find nature's true invariants, Einstein first abandoned the aether and then let go of Newton's version of space and time as separate entities that are unchanging from one frame of reference to another. He arrived at the **special theory of relativity**, which rests on two postulates:

1. All motion is relative: no special frame of reference exists from which an absolute standard of motion can be judged.

2. The speed of light must be the same for all observers, no matter what their state of motion.

Taken together, the consequences of these two postulates are stunning and very much against our intuition. "If someone shines a flashlight in our direction, and we measure the beam of light as it passes us," explains Carroll, "then we will get 300,000 km/s. Now imagine that same person is on a spaceship zipping toward us at 200,000 km/s. It would be the most natural thing in the world to believe that now we would see the light passing us by at 300,000 + 200,000 = 500,000 km/s. We would expect to add the velocity of the spaceship to the velocity of the light with respect to that spaceship. But Einstein and his postulates of special relativity say, 'Nope.' Despite the fact that the person on the spaceship sees the light leave them at 300,000 km/s and the spaceship is moving with respect to us at 200,000 km/s, we see the light pass by us at 300,000 km/s."

To see how strange this behavior of light is relative to our common experience, imagine two workers on a fast-moving train. Both workers are in a moving boxcar with all doors closed. The worker at the back end of the car heaves a bulky bag of mail forward to the other. The velocity of the bag as it flies through the air is, from the perspective of the other worker, the velocity the first worker gave it when he threw it. The train's velocity does not affect the workers' experience. But now imagine that you're standing beside the tracks

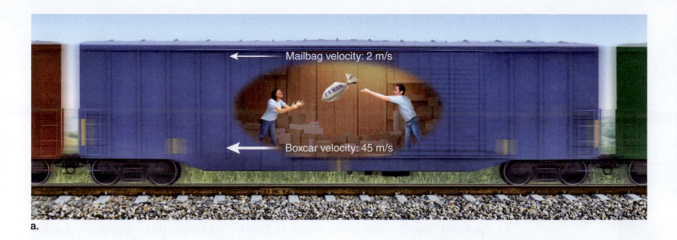

Mailbag velocity: 2 m/s

Boxcar velocity: 45 m/s

a.

Mailbag velocity: 47 m/s

b.

Figure 14.5 Addition of Velocities
a. Inside the train, the heavy mailbag thrown at a low velocity is easy to catch. **b.** In the reference frame of an observer beside the tracks, however, the velocity of the train must be added to the velocity at which the worker throws the mailbag, making the catch much riskier.

as the train roars by and a worker tosses the heavy mailbag to you from an open boxcar. Would you want to catch the mailbag? Not likely. The velocity of the bag would now be the velocity at which the worker released it plus the velocity of the train. From your frame of reference, the bag's velocity and the train's velocity must be added together (**Figure 14.5**).

This *addition of velocities* is what physicists would have expected for light as well. Einstein, however, had a deeper vision. The second postulate of relativity is analogous to demanding that the mailbag thrown from the moving train travels toward you at the same velocity at which it leaves the worker's hands—as if the train's velocity does not exist. This is how light behaves.

Relativity and Space-Time

Einstein further hypothesized that the speed of light is an upper bound on all cosmic motion. This fact alone changes the meaning of time and space. Einstein realized that if the Universe has a speed limit and light runs at this maximum, then something else must give way in order for light to achieve its constancy from one frame to another.

Every measurement of velocity V is a mix of two other measurements: a change in position x and a duration or change in time t. In the language of math, this means

$$V = \frac{x_2 - x_1}{t_2 - t_1} = \frac{\Delta x}{\Delta t}$$

"If you need a constant light speed," says Carroll, "then something else needs to change— either your definition of space and the way you measure it, or your definition of time and the way you measure it, or both." If the speed of light is a constant (c), independent of its frame of reference, then measurements of length (Δx) and time (Δt) can't be independent

"A frame of reference is just the individual perspective of a person extended throughout the whole Universe."

"If you need a constant light speed, then something else needs to change."

frame of reference An individual perspective from which observations are made.

special theory of relativity Albert Einstein's first theory of relativity; it states that the laws of physics are the same for all observers in inertial (nonaccelerating) frames of reference and that the speed of light in a vacuum is the same for all observers, regardless of their relative motion.

Twin 1:
Age =
20 years

Twin 2:
Age =
20 years

a. Before launch

The returned space traveler is . . .

. . . younger than her Earth-based twin.

Twin 1:
Age =
23 years

Twin 2:
Age =
80 years

b. After return

Figure 14.6 The Twin Paradox

Relative ages of twins are shown **a.** before launch and **b.** after return. For the twin who travels on a spaceship at a substantial portion of the speed of light, time flows more slowly relative to the twin who stays on Earth. However, each twin feels her own time passing at a normal rate. Only the relative flow of time between the two frames of reference changes. This paradox illustrates the effect of "time dilation" in special relativity.

space-time A unified, four-dimensional geometry of the Universe that incorporates both space and time.

time dilation The effect of relativity in which time runs slower for observers who are moving at relativistic speeds relative to that of distant observers.

of each other. Length (space) and duration (time) must be flexible, changing from one frame of reference to another. Thus, if you and I are in motion relative to each other, the Δx that *you* measure and the Δx that *I* measure for the same event will be different. Our two measurements of the time between those events (Δt) will also be different.

For the practical purposes of daily life, these differences are so small as to be imperceptible. Only when objects are in motion at a significant fraction of the speed of light do the predictions of relativity become noticeable. (An object must travel at about 10 percent or more of the speed of light, or $V = 0.1c$, before physicists consider it to be moving at "relativistic" speeds.) That's why Newton's physics works fine in everyday life, where speeds are much less than c. What Einstein saw was that Newton wasn't wrong; it was just that Newton's physics described reality for motion at a low-speed limit.

Thus, the heart of Einstein's new vision for physics was the recognition that space by itself and time by itself are illusions. In Newton's physics, three-dimensional space is a kind of empty stage on which the drama of physics plays out. Time flows uniformly through this empty stage; that is, the rate of time's flow is always the same for all observers. Einstein's special theory of relativity did away with Newton's separate and unchanging time and space. In their place, Einstein gave physics a new actor—a single unified **space-time**, which is four-dimensional (three dimensions of space and one dimension of time).

Einstein's conceptual revolution led, as well, to the merger of another long-standing distinction: energy and matter. The famous equation $E = mc^2$ is a statement that mass is not something unchanging and unchangeable. Einstein's theory of relativity showed that mass is a form of energy. Both the terrible power of atomic weapons and the gift of light from the Sun are testaments to how, through $E = mc^2$, Einstein's theory of relativity captured new fundamental truths about the world.

THE TWIN PARADOX. A good example of relativity's remarkable consequences is the *twin paradox*. "The twin paradox," says Carroll, "brings home how different the world of relativity is from the world of Newton, which is the world of our learned 'common sense.'"

Consider a pair of identical twins. At age 20, one twin climbs into a spaceship and rockets away to a star 30 light-years away, always traveling at 99.9 percent of light speed. When she reaches the star, she turns around and returns to Earth at the same speed. The twin on Earth has waited 60 years (30 years + 30 years) for her sister to return and is now 80 years old. She is stunned, when the spaceship lands, to see that her space-traveling twin is still a young woman, only 23 years old (**Figure 14.6**).

How is that possible? According to Einstein, the stay-at-home twin recorded 60 birthdays, but the space-traveling twin experienced only 3. Time did not flow at the same rate for the sisters. Time flowed faster for the stay-at-home twin, as measured by everything from clocks in the town square to the pulse in her chest; thus, duration (Δt) was different for the two twins. The fact that time flows more slowly in some reference frames than in others is an effect called relativistic **time dilation** (**Going Further 14.1**). "Neither twin perceives her own time to be moving faster or slower," says Carroll. "It's only the measurements of duration each makes about the other's reference frame that show something interesting happening."

Why is there a difference between the two twins if all motion is relative? Couldn't we say that the twin on Earth was moving and the twin on the spaceship was standing still? Because the twin in the spaceship had to accelerate and decelerate during her voyage to the star and back, she is the one who experiences the slower relative clock ticks. The acceleration (no matter how brief) is what breaks the symmetry between the sisters.

The important point about the twin paradox and relativity is that both twins have made proper and accurate measurements of duration. Einstein's fundamental insight was that no "right" answer exists for these kinds of questions about time—or space—because there is no *absolute* frame of reference in which to judge the answer. Understanding the physics requires abandoning the notion that space and time are separate entities. The problem is that the time

GOING FURTHER 14.1
How Time Dilation Works

Remarkably, you can prove time dilation yourself. Imagine you're riding in a spaceship at a significant fraction of the speed of light, and suddenly you sneeze. Your head snaps forward with a loud "achoo!" The ship has a special kind of clock that shoots a laser from near the floor to the ceiling (at height H), where a mirror bounces the laser pulse back down to the floor. The time it takes for the laser to make a round-trip and the duration of your sneeze happen to be the same from your perspective (your frame of reference).

Let's calculate how long your sneeze took. As **Figure 14.7** illustrates, because time = (distance covered)/speed, the total round-trip time Δt_{you} of the laser pulse—and of your sneeze—from your perspective must be

$$\Delta t_{you} = 2H/c$$

(Given the speed of light c and the height of your spaceship H, this would be a remarkably fast sneeze, but let's ignore that.)

Would someone on a planet watching you whiz by measure the same time for the sneeze (let's call it $\Delta t_{observer}$) as the time you experience? Let's figure out what $\Delta t_{observer}$ is in terms of Δt_{you}.

Einstein said that the speed of light is the same for all observers, no matter whether or how they are moving. To see how this works in terms of $\Delta t_{observer}$ and Δt_{you}, let's draw a picture of what the observer on the planet sees as your spaceship shoots past at speed V (**Figure 14.8a**). From the observer's perspective, the light beam doesn't go straight up and down; we must now take into account the sideways motion of the spaceship. We use geometry to calculate how far the light beam travels in its round-trip, as seen from the planet-based observer's frame of reference:

$$(c\Delta t_{observer}/2)^2 = (V\Delta t_{observer}/2)^2 + H^2$$

$$(\text{hypotenuse})^2 = (\text{horizontal side})^2 + (\text{vertical side})^2$$

This is Pythagoras's theorem for the triangle made by the laser beam, as seen from the planet (**Figure 14.8b**).

How do we get your sneeze duration (Δt_{you}) into the picture (and equation)? We take the first equation in this box, $\Delta t_{you} = 2H/c$, solve it for H, and get $H = c\Delta t_{you}/2$. Then we substitute that into the second equation, Pythagoras's theorem:

$$(c\Delta t_{observer}/2)^2 = (V\Delta t_{observer}/2)^2 + (c\Delta t_{you}/2)^2$$

Now we solve for $\Delta t_{observer}$ to turn this into an equation that gives us $\Delta t_{observer}$ in terms of Δt_{you}. When the smoke clears, we get

$$\Delta t_{observer} = \sqrt{1/(1 - V^2/c^2)}\,\Delta t_{you}$$

(continued)

Figure 14.7 Sneeze as Seen by the Sneezer
The duration of your sneeze, from your frame of reference on a moving spaceship.

What you see:
$\Delta t_{you} = 2H/c$

H

"achoo"

$V = 0.99c$

$\Delta t_{observer} = \sqrt{1/(1 - V^2/c^2)}\,\Delta t_{you}$
$= 7 \times \Delta t_{you}$

"a-"   "-aa-"   "-choo"

What the observer sees

a.

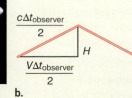

$\dfrac{c\Delta t_{observer}}{2}$

H

$\dfrac{V\Delta t_{observer}}{2}$

b.

Figure 14.8 Sneeze as Seen by a Planet-Based Observer
a. The duration of the same sneeze, from the frame of reference of the observer as your spaceship speeds by at 99 percent of the speed of light c.
b. Geometrical calculation of distance according to Pythagoras's theorem.

What does this equation tell us? If the speed of your spaceship were a lot less than the speed of light ($V \ll c$), then the factor in front of Δt_{you}, which is $\sqrt{1/(1 - V^2/c^2)}$, would be about 1. The time you would experience for the light pulse to bounce up and down (which is also the duration of your sneeze) is pretty much what the person on the planet would observe. But if your spaceship is moving close to light speed, as in Figure 14.8a, something amazing happens. Let's say you're moving at 86 percent of light speed (that is, $V = 0.86c$). In this case, the calculation works out to

$$\Delta t_{observer} = 2.0 \Delta t_{you}$$

In other words, your sneeze takes twice as long as seen by the planet-based observer. If you were instead traveling at 99 percent of light speed, then

$$\Delta t_{observer} = 7.1 \Delta t_{you}$$

Now your sneeze takes seven times as long. That would be really slow motion.

This is not magic; it's a consequence of the constant speed of light. But those consequences are profound. There is no universal time throughout the Universe. Your experience of time is literally your own and differs, even if by just a little, from that of anyone moving relative to you.

INTERACTIVE:

Time Dilation/Twin Paradox

we recognize is the one our brains evolved for us to experience. Since we never move close to the speed of light, we never experience these domains of the real world directly.

Section Summary

- Experiments consistently showed that the speed of light remains constant regardless of an observer's motion.
- Frames of reference must be taken into account for an accurate understanding of physics.
- Einstein created a new paradigm, called special relativity, in which (1) there is no special reference frame from which to measure motion and (2) the speed of light is the same for all observers, regardless of their motion.
- Since the speed of light is fixed, space and time must be flexible. Traveling at speeds close to the speed of light creates the effects of time dilation.

CHECKPOINT **a.** Give an example of a frame of reference in ordinary life. **b.** Why does the constant speed of light require different observers moving relative to each other to experience different flows of time?

> "The twin paradox brings home how different the world of relativity is from the world of Newton, which is the world of our learned 'common sense.'"

General Theory of Relativity

14.3 Einstein's first version of relativity is called "special" because it does not consider the role of acceleration produced by all possible forces. One force in particular, *gravity*, is so general and so prevalent that Einstein spent 7 years trying to understand how to encompass it in his new vision of physics. The result is called the **general theory of relativity**, and in its wake, the concept of gravity and the human understanding of space and time were forever altered.

Space-Time and the New View of Gravity

Einstein often used thought experiments to work out his next steps beyond special relativity. In conducting a thought experiment, you imagine a set of circumstances and then use the laws of physics to work through the logical consequences. To understand the effect of gravity in relativity, Einstein needed a thought experiment that would help him grasp how accelerations, gravity, and different frames of reference are related.

You can roughly replicate Einstein's thought experiment by imagining yourself as the sole occupant of a windowless enclosure, such as a rocket ship floating in space. When

general theory of relativity Albert Einstein's second theory of relativity, which extends the special theory of relativity by incorporating gravity; it states that matter and energy dictate how space-time curves, while space-time dictates how matter and energy move.

principle of equivalence The equivalence between the "force" experienced in an accelerating frame of reference and the force of gravity in an inertial (non-accelerating) frame.

the rocket motor is turned on and the ship begins to accelerate, the "floor" of the capsule (where the rockets are located) rushes up to meet you, as the entire enclosure accelerates in response to the force of the motor. The floor hits you and the equipment on board, transmitting the rocket's incessant push. You feel as if you're being pulled toward the floor. At this moment in the thought experiment, Einstein came to a realization: A person inside an accelerating rocket has the same experience as a person in a rocket at rest sitting on a planet's surface. From a physics perspective, the two situations are equivalent.

Now think about what would happen if the capsule's motor was turned off. You would have no way to know whether the capsule was in motion. It might be stationary with respect to the stars, or it might be moving at constant velocity. In either case, you and your equipment would float freely inside the enclosure. No experiment you performed would reveal any difference between the ship's standing still and moving at constant velocity, because both frames of reference are inertial. (Recall that an inertial reference frame is a nonaccelerating frame.)

As a final step in the thought experiment, imagine that the capsule is in gravitational free-fall (Chapter 3); for example, it is in a stable orbit around a planet. Einstein realized that in this situation, too, a person would float freely in the capsule, and the conditions would be indistinguishable from a state of constant motion. Therefore, gravitational free-fall must also be equivalent to an inertial state, even though the capsule and its inhabitant are accelerating (**Figure 14.9**). This is one statement of what Einstein called the **principle of equivalence**, an essential step on the road to his new conception of gravity. But how can an object be in an inertial state while still experiencing acceleration?

This question inspired Einstein to make a bold conceptual leap. He did away with gravity as a separate force and substituted the *geometry of space-time*. Einstein saw that if the geometry of four-dimensional space-time were malleable in response to matter, it could play the same role as gravity.

a.   **b.**   **c.**

Figure 14.9 Gravity and Acceleration

Einstein realized that moving in an accelerating frame of reference feels equivalent to the presence of gravity. **a.** The astronaut and tools in a spacecraft moving at constant velocity seem to be weightless. But the astronaut and his tools react similarly when they are **b.** in an accelerating spacecraft and **c.** experiencing gravity on a planet's surface. In both cases the magnitude of the force pushing the astronaut down on the scale is the same.

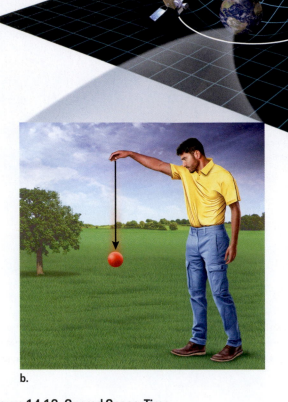

Figure 14.10 Curved Space-Time

a. Think of space-time as stretched or curved toward the center of Earth by Earth's mass. (The curvature of three-dimensional space in four-dimensional space-time is represented by the curvature of a gridded two-dimensional sheet in three-dimensional space.) A satellite in space is in free-fall: it follows a path dictated by the curvature of space-time around Earth. **b.** A ball dropped to Earth is also in free-fall and moves according to the shape of space-time.

"Like all good scientists, Einstein thought about a number of different ways of testing general relativity."

muon An elementary particle that is extremely short-lived and heavier than an electron.

cosmic ray A high-speed subatomic particle (typically, an atomic nucleus) from space.

gravitational lensing A change in the path of light passing a massive object due to the curvature of space-time.

For example, if you drop a ball on Earth, the ball accelerates toward the ground. Using his new vision of a flexible fabric of space-time, Einstein translated this acceleration into the ball's unforced movement through a curved space-time. According to Einstein, Earth's mass does not create a gravitational force that pulls on the ball. Instead, it distorts geometry, the very shape of space-time, around the ball. Remove all imposed forces (such as the support of your hand), and the ball is free to fall along the curve of space-time—the way water flows down a ramp. Thus, both a ball falling from your hand and a spaceship orbiting Earth are in free-fall, moving according to the shape of space-time around them (**Figure 14.10**).

This general theory of relativity extended Einstein's earlier special theory. Now, individual frames of reference are thought to move on a flexible fabric of reality—four-dimensional space-time—that can be bent, stretched, and folded. Two new postulates of general relativity can be described in the following way:

1. Matter-energy determines how space-time can bend.
2. Space-time determines how matter-energy can move.

In special relativity, measurements of space or time depend on an individual frame of reference and its motion. But in general relativity, an observer's location relative to a large body of matter can also affect space and time measurements through the massive body's distortion of the fabric of space-time. Like special relativity, general relativity has implications that challenge our everyday understanding of space and time. For example, clocks close to a planet's surface run slower than those far away (the phenomenon of gravitational time dilation, as we saw earlier), and lengths measured close to a planet's surface yield smaller values than do measurements taken far out in space (a phenomenon called *length contraction*). Time and space are relative, but this malleable geometry of space-time and the distributions of matter-energy provide the framework for understanding all of them as a single, unified whole. In this way, our entire understanding of reality changed through Einstein's efforts.

How Do We Know? Tests of Relativity

Einstein's ideas may seem so unintuitive that your initial response is "No way!" How can 1 year for someone on a rocket ship moving close to light speed be 20 years for me? How can empty space bend and warp? You would be right to be skeptical, because good scientists seek experimental verification of theories before accepting them. As it turns out, both the special and general theories of relativity have been verified many times. Let's look at a few ways in which they've been shown to be accurate descriptions of the world we inhabit.

PARTICLE LIFETIMES: A TEST OF SPECIAL RELATIVITY. Physicists studying the structure of matter have used certain short-lived particles to provide evidence of relativistic time dilation. These particles, which move at speeds close to the speed of light, quickly undergo radioactive decay into other types of matter (see Chapter 6) and thus serve as very exact clocks. One such particle is the **muon**. "The muon," says Sean Carroll, "is an elementary particle very much like the electron, but it's heavier and it decays in about one-millionth of one second. If you have a bag of muons, then one-millionth of a second later, half of them are gone." Muons are created in Earth's upper atmosphere when **cosmic rays**, fast-moving particles from space, slam into atmospheric atoms. The muons are sent flying toward the surface at speeds close to the speed of light and caught by detectors on the ground.

How are muons used to test special relativity? "If we plug in how much time it takes a muon to travel from the upper atmosphere toward detectors here on Earth," says Carroll, "it is a lot longer than a millionth of a second." If you do the calculation, you'll find that the trip from atmosphere to ground takes 3×10^{-5} second. Because muons tend to live only about one-millionth (10^{-6}) of a second, they should decay before reaching the ground. But scientists do find muons in ground-based detectors, suggesting that these particles are not decaying at their normal rate.

The reason for this surprising finding is relativistic time dilation. Time is flowing more slowly for the high-speed muons relative to time measured on the ground. Thus, in the Earth-based frame of reference, muons from the top of the atmosphere can reach the ground before they decay (**Figure 14.11**). The observed behavior of the muons exactly matches the predictions of special relativity.

BENDING STARLIGHT: A TEST OF GENERAL RELATIVITY. Other experiments have confirmed the predictions of general relativity. "Like all good scientists," says Carroll, "Einstein thought about a number of different ways of testing general relativity." One of these involved comparing the positions of stars in the night sky (which can be measured very accurately) with their apparent positions during a solar eclipse. According to general relativity, when starlight passes near the position of the Sun, a massive object, the light's path should be deflected by the Sun's curvature of space—an effect called **gravitational lensing**. However, scientists can observe this deflection only during a solar eclipse, when the Moon obscures the Sun.

"By bending space-time, the Sun acts as a distorting lens," says Carroll. "It bends the path of the light rays. A picture of the stars' positions when they were near the Sun during a total solar eclipse would look different from their relative positions at night," when those stars are far from the Sun's position on the sky. General relativity gives a definitive way to predict the change in the stars' positions, and astronomers were looking for a way to measure that change (**Figure 14.12**).

A solar eclipse is not an everyday event, but in 1919, just 5 years after Einstein finished his general theory, scientists were able to observe an eclipse and demonstrate that even light is subject to the bending of space-time by a massive object. Separate teams of astronomers were sent to Brazil and Principe, off the coast of Africa, to observe the eclipse. Although the sky was cloudy in Brazil, the eclipse was visible from Principe. The deflection of starlight was observed, just as Einstein had predicted. General relativity was experimentally confirmed; the eclipse measurements were seen as a triumph for relativity. "That was really the event that catapulted Einstein into worldwide celebrity," says Carroll. More recently, there has been some dispute about the degree to which the data taken that day matched the theory, as scientists insist on repeating tests. Later experiments, however, have provided many confirmations of the bending of starlight by curved space-time.

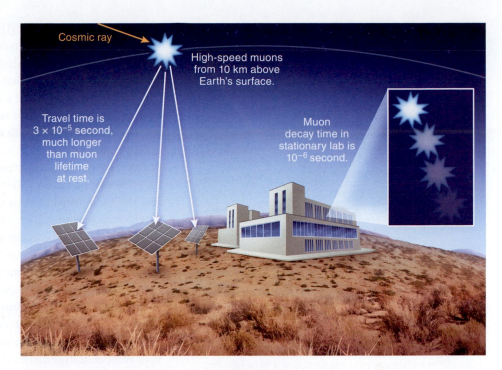

Figure 14.11 Time Dilation and Particle Lifetimes
Muons live only about one-millionth of a second, yet those created 10 km up in the atmosphere reach Earth's surface—a trip that requires several muon lifetimes. According to special relativity, the muon's time relative to observers on Earth is dilated (slowed) because of its high speed. Thus, even though the muon should have decayed during the trip, in the muon's frame of reference less time has passed.

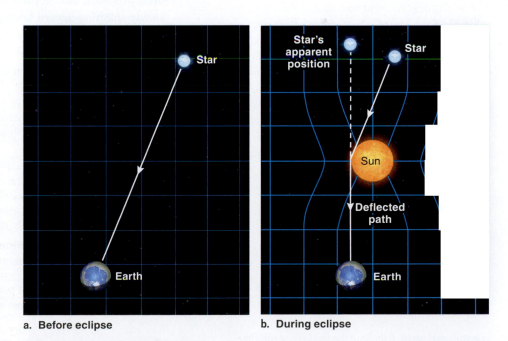

a. **Before eclipse**   b. **During eclipse**

Figure 14.12 Gravitational Lensing
The total solar eclipse of 1919 provided a way to test general relativity. **a.** The expected position of a distant star before eclipse is shown. **b.** During eclipse, the star's observed position changes because of the Sun's gravitational distortion of space. As light from the star passes the Sun, the curvature of space-time bends the light's path, altering the star's position as perceived by an observer on Earth.

Figure 14.13 Time Dilation and Global Positioning Systems
A global positioning system compares time signals from orbiting satellites and the ground-based receiver. It must take into account effects of both special relativity (motion) and general relativity (location relative to a massive object). According to general relativity, the clocks on the satellites run faster than those on Earth's surface (gravitational time dilation).

GLOBAL POSITIONING SYSTEMS: A GENERAL RELATIVISTIC TECHNOLOGY. When you use a *global positioning system* (*GPS*) to find your position on Earth, you're conducting an experiment in general relativity. The GPS uses a network of about 30 orbiting satellites. Each has an atomic clock that keeps hyperaccurate time and continuously broadcasts the exact time to a billionth of a second. Ground-based receivers (such as cell phones) pick up the signals from four of the satellites and compare signal travel times. A cell phone compares its own clock with the time measured by the satellites. It then uses the difference between the times to calculate the distance to each satellite. Since light travels at 300,000 km/s, a time difference of one-thousandth of a second implies that the ground-based receiver must be 300 km from the satellite. This comparison of the receiver's time and the time signal sent by the four satellites enables accurate determinations of the receiver's position.

This process may seem straightforward, but a hyperaccurate measurement of time is not possible without an understanding of general relativity, in which space-time is distorted by the presence of matter. Time passes more slowly for observers on the ground than for observers in space (**Figure 14.13**). General relativity predicts that GPS satellite clocks will run ahead of ground-based clocks by about 45 millionths of a second per day. While this difference might seem tiny, if relativistic time dilation weren't accounted for in GPS technology, after only a week or so your GPS would locate you hundreds of miles from your real position and a couple of miles above the ground.

PULSARS AND GRAVITATIONAL WAVES. In a famous test of general relativity, the binary pulsar PSR 1913+16—two pulsars orbiting each other (see Chapter 13)—was used to verify Einstein's theory of gravity. The theory predicted that the two orbiting pulsars should produce **gravitational waves**—ripples in the fabric of space-time—as they orbit each other. Gravitational waves would not be directly detected until 2015, so in 1974 it was still unclear whether Einstein got this part of his theory right. Using the pulsar pulses to measure the orbit precisely, astronomers demonstrated conclusively in 1974 that the two pulsars are slowly spinning inward toward each other. (We'll discuss gravitational waves further in Section 14.5.) Their orbit is shrinking as energy is lost to the elusive gravitational waves. The exact amount of orbital decay matched perfectly the predictions of gravitational wave generation. Einstein had been proved correct again.

Section Summary

- The general theory of relativity, which includes gravity and the motions it produces, states that matter-energy determines how space-time bends and space-time determines how matter-energy moves.
- Because of relativistic time dilation, muons detected on Earth's surface have traveled farther than is consistent with their short lifetimes.
- As light travels, its path is bent by the curvature of space via gravitational lensing.
- A global positioning system (GPS) works with high accuracy only when relativistic effects are factored in.
- Astronomers inferred the existence of gravitational waves, ripples in the fabric of space-time, by measuring pulses from a binary pulsar.

CHECKPOINT How did the principle of equivalence help Einstein see that gravity should be considered to be more than just a force?

gravitational wave A ripple in the fabric of space-time, produced whenever matter moves.

singularity The central point of a black hole, where density and space-time curvature is infinite.

Schwarzschild radius The distance from a black-hole singularity to points where the escape velocity equals the speed of light.

Anatomy of a Black Hole

14.4   The British geologist John Michell, while contemplating Newton's formula of escape velocities in 1783, noticed something peculiar. If the mass M of a central object is large enough—or, more important, if its radius R is small enough—then the escape velocity should be greater than the speed of light:

$V_{escape} > c$. This means it is theoretically possible to have a "black star" from which light cannot escape.

Like a cannonball blown straight upward with less than escape velocity, a light beam shot from such a black star would fall backward and never reach outside observers. "All light emitted from such a body would be made to return toward it by its own proper gravity," Michell wrote to a colleague. Over the next century or so, only a few astronomers paid attention to the possibility of these black stars; the idea was thought to be nothing more than a theoretical oddity.

But when the new physics of general relativity replaced Newton's vision of gravity, the concept of such black stars was taken up again and pushed to its limit. These black stars were eventually reconceptualized as black holes, and the Universe as we understand it was forever changed.

How to Make a Black Hole: Massive Stellar Cinders

Chapters 12 and 13 tracked the evolution of stars through their main-sequence phases and beyond. When the hydrogen-burning phase ends for intermediate-mass stars ($M_* < 8\, M_{Sun}$), the cores contract enough to initiate helium fusion. Once the helium is gone, core fusion ends. From that point on, only the quantum mechanical degeneracy pressure of electrons keeps the burned-out stellar cinder alive as a white dwarf.

Higher-mass stars ($M_* > 8\, M_{Sun}$) burn through a sequence of elements up to iron after their core hydrogen is exhausted. But since iron requires more energy to fuse than it gives back, it represents the end of the line for the star. The star collapses on itself, creating a solid neutron core in some cases (see Chapter 13). The ultradense neutron star is another kind of stellar cinder that, like white dwarfs, is supported against gravity by the quantum physics of degeneracy pressure, though in this case the neutrons are degenerate.

"There's a limit to what quantum mechanics can do in terms of supporting a dead star against its own weight," says Manuela Campanelli. "If the core mass is too big, then even degeneracy pressure can't fight gravity's pull." Thus there can be no salvation for the highest-mass stars. With nothing to restrain the inward crush of gravity, the star collapses further and further toward an ever-smaller radius. Eventually the star becomes so small that its escape velocity is greater than the speed of light, as John Michell had imagined. But the star's infall doesn't stop there. "There's no known physical force that can resist gravity for these massive objects," says Campanelli, "so on the face of it, the star collapses all the way down toward zero radius."

A zero radius implies a zero volume, so the stellar density (mass divided by volume) shoots toward infinity as the radius shrinks to nothing. Physicists call this kind of point a **singularity**, and its creation is the hallmark of the beast we call a black hole.

Astronomers are not certain how massive a star must be to end up as a black hole. Many factors, including the rate of rotation, determine whether a newly formed neutron star will be stable or will survive for just a brief time before collapsing into a black hole. In truly massive stars, a black hole may even form at the center before a neutron core can form. Astronomers currently estimate that stars greater than 20–30 M_{Sun} become black holes, but future research may revise that limit.

Schwarzschild Radius and Event Horizon

You don't need Einstein's general theory of relativity to see that a massive enough star will eventually collapse completely under its own weight. You do, however, need general relativity to understand what is left after the total collapse is over.

One of the important features of a black hole is its **Schwarzschild radius** (R_s), the distance from the center to locations where the escape velocity V_{escape} equals the speed of

MANUELA CAMPANELLI

Manuela Campanelli grew up in Switzerland. "When I was a little girl, I read about everything scientific I could get my hands on. I was interested in biology, but I didn't want to do experiments with animals, so I knew this was not the field for me. Then I tried chemistry. . . . But then one day I did an experiment in my room and everything caught fire," she says. "This is when I realized that I needed to work on the theory side of science."

Campanelli says that even from her youngest days, she was interested in answering "big" questions. "Those questions never left me, and I still get to think about them in my work with black holes." She went on to become a leader in the field of gravitational astrophysics.

"There's a limit to what quantum mechanics can do in terms of supporting a dead star against its own weight."

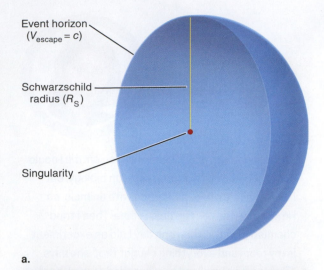

Event horizon ($V_{escape} = c$)

Schwarzschild radius (R_S)

Singularity

a.

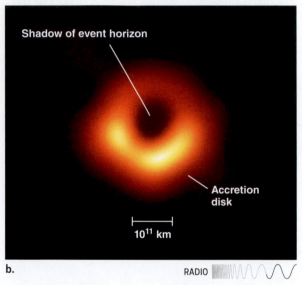

Shadow of event horizon

Accretion disk

10^{11} km

b.   RADIO

Figure 14.14 Structure of a Black Hole

a. A black hole has a simple structure: the event horizon, as defined by the Schwarzschild radius (where the escape velocity is the speed of light), and a dimensionless singularity at the center (where all mass appears to be concentrated). **b.** The first image ever obtained of a black hole—the supermassive black hole at the center of galaxy M87, 55 million light-years from Earth—surrounded by hot gas is shown. What's visible is not the event horizon but its shadow, outlined by light that has been curved around it.

INTERACTIVE:

Schwarzschild Radius

light *c*, just as Michell described. This radius defines a sphere called the **event horizon** (**Figure 14.14**). What Michell could not see, however, is that the distance that is the Schwarzschild radius defines a point of no return for all objects falling into a black hole.

"Even Newton could have figured out that a massive and compact enough star would have an escape velocity equal to light speed," says Campanelli. "But general relativity tells us much more. It also shows us that once you cross the Schwarzschild radius, no amount of force can get you back out. Newtonian gravity would not have given you that conclusion."

Once an object drops below the Schwarzschild radius, the structure of space-time demands that it continue its fall inward toward the center. That is why the sphere defined by R_s is called an event horizon. Just as you can't see distant objects that lie over the horizon on Earth, objects closer to the center of a black hole than the Schwarzschild radius lie over a "horizon" for observers in the rest of the Universe. Inside a black hole—meaning inside the event horizon—is a region of space-time that is entirely cut off from the rest of the cosmos.

A simple formula gives us the Schwarzschild radius (see **Going Further 14.2** on p. 382):

$$R_s = \frac{2GM}{c^2}$$

$$\text{Schwarzschild radius} = \frac{2(\text{gravitational constant})(\text{mass of black hole})}{(\text{speed of light})^2}$$

Using this equation, we can see that the event horizon for a 1-M_{Sun} black hole would be only about 3 km in radius! That is basically the distance from the US Capitol to the Lincoln Memorial. Imagine trying to hide 2,000 billion billion billion kg of matter (1 M_{Sun} = 2 × 10^{30} kg) on the National Mall!

For larger masses, the Schwarzschild radius of a billion-solar-mass black hole, like those at the center of some galaxies, would be only 20 AU, about equivalent to Uranus's distance from the Sun. For smaller masses, if all of the mass in your body were squeezed down to the point where a black hole formed, it would have an event horizon smaller than an atom (**Table 14.1**). "While it might sound crazy," says Campanelli, "when the Universe was born in the Big Bang, the conditions were so violent that these kinds of micro–black holes may have formed quite easily."

The Singularity and Quantum Gravity

General relativity provides precise descriptions of the physics of the event horizon, but the singularity is an unknown, untamed monster. The problem for astrophysicists trying to build models of black holes is that the density, temperature, and curvature of space-time become infinite at the singularity. "The singularity is essentially a geometric point," says Campanelli. "That's where all the mass and energy of the black hole live." But geometric points have no volume. How can the entire mass of a star occupy zero volume? There must be more to the story.

"Of course, the real density can't actually go all the way to infinity, just as the radius of the star can't shrink all the way to zero," Campanelli continues. "There must be some form of new physics which occurs before the radius reaches zero and the density goes to infinity. Unfortunately, we don't yet know what that new physics is, so we work within the limit of what Einstein's theory of relativity tells us is possible."

Physicists know from studies of matter that, on small enough scales, all of reality must be described by quantum physics. Its central lesson is that, at the smallest scales, nature is granular. As Campanelli puts it, "Nature breaks up into chunks on small enough scales. But the fabric of space-time as Einstein described it is smooth and continuous like a bedsheet. There is nothing quantum about it, and that's the problem."

Table 14.1	The Schwarzschild Radius		
Object	**Mass**	**Schwarzschild Radius (R_S)**	**R_s Comparison**
Human being as a black hole	70 kg	10^{-28} km	Radius of an atom
Sun-like star as a black hole	$1\,M_{Sun}$	2.9 km	Width of Manhattan Island, NY
Massive star as a black hole	$10\,M_{Sun}$	29 km	Length of Long Beach Island, NJ
Supermassive black hole progenitor	1 billion M_{Sun}	20 AU	Uranus's orbital radius

To describe what happens inside a singularity, physicists need a theory of **quantum gravity** to explain how space-time itself breaks up into discrete bits. The size scale at which space-time should begin showing its quantum nature is the **Planck length**, estimated to be around 10^{-44} meter. "That's about a hundred million billion billion billion billionth of a meter," says Campanelli. "It's beyond small, and much, much smaller than the nuclei of atoms." Although physicists know when this new physics should kick in, the true theory of quantum gravity remains unknown—defining the frontiers of our knowledge of the Universe.

Since the known laws of physics break down at singularities, some physicists see them as dangerous abominations. The great cosmologists Stephen Hawking and Roger Penrose proposed a principle of **cosmic censorship**, in which singularities would never occur without an event horizon around them. Singularities with no event horizon are called **naked singularities**. In a sense, the event horizon forms a protective membrane for the Universe, since nothing can ever cross the event horizon *going outward*. "Hawking never explained the physics controlling cosmic censorship," Campanelli explains. "He simply stated that nature will find a way to keep naked singularities from forming. Only time is going to tell us if he was correct."

Science fiction movies may suggest that a black hole has infinite gravitational power and will inexorably suck in everything across the cosmos. The reality is more benign. "If the Sun turned into a black hole tomorrow," explains Campanelli, "Earth would still happily go on in its orbit (though in darkness, of course). It's only relatively close to the event horizon that the strange general relativistic effects appear." (Our Sun, however, does not have enough mass to become a black hole and will end its life as a white dwarf.)

What are these "strange relativistic effects"? To find out, we must embark on a very dangerous and, happily, imaginary plunge into a black hole.

The Longest Fall: Plunging into a Black Hole

The nature of a black hole will change if it is rotating or carries an electric charge. For our journey, we'll keep things simple by making the plunge into a black hole that has no spin and no charge.

Imagine you're in a spaceship orbiting a black hole at a safe distance, thousands of Schwarzschild radii away from the hole. Space-time is gently curved, and you feel no unusual effects of the black hole, except for its gravitational attraction (as described by Newton's version of gravity).

A friend in her own ship orbits with you as you prepare for your trip into the black hole. The two of you routinely exchange signals by radio, ensuring that your clocks are

> "That's about a hundred million billion billion billion billionth of a meter. It's beyond small, and much, much smaller than the nuclei of atoms."

event horizon The boundary of a black hole, located one Schwarzschild radius away from the singularity.

quantum gravity A theory (yet to be fully worked out) that unifies general relativity and quantum mechanics, describing how space-time becomes granular ("quantized") at the smallest scales.

Planck length The size scale at which space-time should begin showing its quantum nature, about 10^{-44} meter.

cosmic censorship The conjecture that a singularity cannot exist without an event horizon.

naked singularity A singularity that lacks an event horizon, which cannot exist if cosmic censorship is correct.

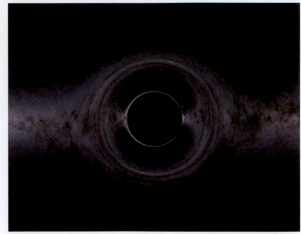

a.

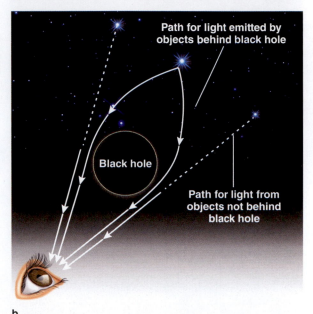

Path for light emitted by objects behind black hole

Black hole

Path for light from objects not behind black hole

b.

Figure 14.15 Gravitational Lensing by Black Holes
a. A simulation of what you would see if you were near a black hole. **b.** The path that light would take to reach you. Because of the strong curvature of space-time, the light from stars behind a black hole (relative to your line of sight) bends into a ring around the event horizon via gravitational lensing.

"Near a black hole, space-time is so distorted that your feet are being stretched toward the black hole much more strongly than your head."

supermassive black hole (SMBH) A black hole of millions or billions of solar masses that exists at the center of a galaxy.

perfectly synchronized. Since the two ships are close together and far from the event horizon, your clocks appear to run at the same rate. The only thing that looks strange is the view of space in the direction of the black hole (via a telescope, since you are so far away).

"Because the black hole distorts space-time so strongly," Campanelli explains, "you can actually see objects directly behind the black hole. The space-time around the hole acts like a lens bending the path of starlight. Objects like stars and galaxies that are directly behind the hole would appear to a distant observer as a bright band that runs around the edge of the event horizon" (**Figure 14.15**). This is an example of gravitational lensing.

To start your plunge, you fire the ship's retro-rockets in the opposite direction of your orbital motion to slow down. The black hole's gravity begins to pull you inward; your path becomes a long spiral that will end at the event horizon. By firing the rockets *to speed up*, you could still stop your inward spiral and get back into a stable orbit.

But you are brave and continue your spiral inward, as the black maw of the event horizon and its gravitationally lensed rim of stars grow in your field of view. From a safe distance, your friend notices that the closer you get to the hole, the slower your clock appears to run relative to hers. The radio signals that you send her are also getting redshifted; to receive them, she must keep tuning her receiver to longer and longer wavelengths. Both of these effects become stronger as you drop deeper into the black hole's well of stretched space-time (**Figure 14.16**).

"The slowing of your clock relative to your friend's is called *gravitational time dilation*," says Campanelli. "Time in your frame of reference will run slower relative to that of someone far from the black hole. The second effect is called a *gravitational redshift*. Radiation emitted close to a black hole gets shifted to longer wavelengths as it climbs out of the hole's gravitational well." Both effects occur around any mass, not just black holes. Gravitational time dilation and redshift occur around Earth too—but they are so weak that only extremely precise instruments (such as a GPS) can detect them.

As you get closer, suddenly your ship lurches and begins to plunge directly toward the blackness of the event horizon. You ramp up your rockets to speed up, but you can't get back into orbit. You're falling fast. "What you just discovered," says Campanelli, "is that no stable orbits are possible around a black hole past a certain point. The shape of space-time is such that a balance between orbital motion and gravity is not possible anymore." The location of this last stable orbit is at $R = 3R_s$ for a nonrotating black hole; for a rotating hole, the position is farther out.

Time dilation is very strong now. Even though you feel time passing at its normal rate, your friend sees you slowing down to a crawl. You feel strange in a different way, though. Your feet seem to be pulling away from your head as if you are being stretched on a medieval rack. At the same time, your insides are getting squeezed, and you feel like you're being crushed in a giant fist. It's becoming too painful to bear.

You're experiencing an extreme form of tidal forces (Chapter 6), caused by the fact that space-time curves more strongly on the side of an object closer to a large mass than on the side farther away. "From Newton, we know that tidal forces result because the effect of gravity decreases with distance from a central object," explains Campanelli. "In general relativity, we think of tidal forces as differences in the stretching of space-time from one point to another. Near a black hole, space-time is so distorted that your feet are being stretched toward the black hole much more strongly than your head." The effect you're feeling is called *spaghettification*. Objects falling into a black hole are pulled into spaghetti-like strands. When intermolecular forces can't hold up, the strands are pulled apart. This is your eventual fate (**Figure 14.17**).

Even after you and your spaghettified spaceship are pulled apart, the remains fall toward the event horizon. From your friend's perspective, however, your time stops entirely when you reach the Schwarzschild radius. All signals become redshifted to infinity, so your friend can't see an image of you anymore; if she could, you would appear to be forever frozen at the lip of the event horizon.

Your remains, which are now just a spaghettified cloud of subatomic particles, pass right through the event horizon. Notice that the horizon is not a physical boundary like the surface of a lake. It's just a location where the stretching of space-time becomes so extreme that light and everything else are permanently bound to the black hole.

What happens inside the horizon? "According to Einstein's equations, space and time reverse roles in a very strange way," explains Campanelli. "Just as we can only travel into the future in normal space-time, now there is only one direction in space you can travel and that is *into* the singularity. You can't help but move toward it."

Thus, your journey ends as you are drawn toward that giant question mark that is the singularity. What happens when you reach it is more than anyone can say right now.

Section Summary

- Very massive stars (probably greater than 20–30 M_{Sun}) collapse to black holes when their fusion fuel is gone.
- The Schwarzschild radius (R_s) of a black hole defines the event horizon, where the escape velocity is c and inside of which nothing can be seen or can escape. For each solar mass in the black hole, R_s is about 3 km.
- A black hole has the same gravitational pull as a star or other object of the same mass (although a much smaller radius).
- The space-time distortion near a black hole causes outbound light to be redshifted and the flow of time relative to distant observers to be dilated (slowed down). Tidal forces pull apart (spaghettify) infalling objects.

CHECKPOINT What is the relationship between the curvature of space and the tidal effects that would cause spaghettification if you fell into a black hole?

Real Black Holes in Astronomy

14.5 The formation of black holes through stellar collapse was inferred from Einstein's general theory of relativity as far back as 1939. But at that point most scientists didn't imagine that black holes were real. "Even Einstein believed that event horizons were too strange to exist," says Manuela Campanelli. "He thought the star would somehow stop collapsing before one formed."

As the decades passed, however, researchers explored general relativity's description of event horizons and other aspects of black-hole physics in ever-greater detail. "All that work built up confidence that nature might actually create something like black holes," says Campanelli. "But until you go out and find one through observations, you can't know for sure."

From the 1960s to the 1990s, advances in observational techniques helped build the case that black holes exist. Astronomers generally group black holes into two categories: *stellar-mass black holes*, with masses between 1 and (approximately) 100 M_{Sun}, and **supermassive black holes (SMBHs)**, with masses between 10^6 and 10^9 M_{Sun}. In principle, there is no reason why an intermediate-mass black hole of, say, 10^4 M_{Sun} can't exist. At this point, however, astronomers do not think that a large population of such objects has naturally formed.

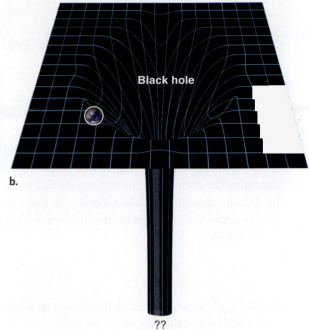

a.

b.

??

Figure 14.16 Gravitational Well of a Black Hole
a. The gravitational well of a normal star differs significantly from **b.** the gravitational well of a black hole. The space-time surrounding a black hole is stretched far more near the singularity at its center. If the Sun were replaced by a black hole of equivalent mass, Earth would continue to orbit (without heat or light).

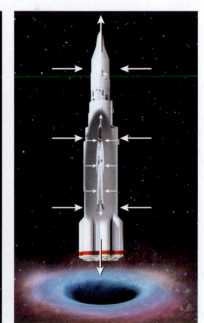

a. Normal   b. Near a black hole

Figure 14.17 Spaghettification
Tidal forces are greatly amplified near a black hole. An unfortunate person approaching a black hole would be pulled apart ("spaghettified") as his feet were pulled away from his head and his body compressed, or squeezed, perpendicular to the pull.

GOING FURTHER 14.2
Black-Hole Edges: The Event Horizon

One of the strangest attributes of black holes comes from one of the simplest ideas in physics. The event horizon represents the point of no return for an object falling toward a black hole. Mathematically, the event horizon is defined as the point where the black hole's escape velocity equals the speed of light. For a black hole of mass M, we can determine the event horizon's radius—the Schwarzschild radius (R_s), the distance from the black hole's center to the event horizon.

To derive the Schwarzschild radius, we write the equation for escape velocity:

$$V_{escape} = \sqrt{\frac{2GM}{R}}$$

We let $V_{escape} = c$ (which defines the Schwarzschild radius). If we then square both sides of the equation, we get

$$c^2 = \frac{2GM}{R_s}$$

Next we solve for R_s (because $R = R_s$ when $V_{escape} = c$):

$$R_s = \frac{2GM}{c^2}$$

Since the speed of light (c) and Newton's gravitational constant (G) don't change, we can write this equation in a way that shows us how R_s changes (or *scales*, as astronomers say) with mass:

$$R_s = (2.9 \text{ km})(M/M_{Sun})$$

Note how the equation is written so that mass is expressed in units of the Sun. Thus a 1-M_{Sun} black hole has a 2.9-km event horizon, a 10-M_{Sun} black hole has a 29-km event horizon, a 100-M_{Sun} black hole has a 290-km event horizon, and so on. We now know that a black hole exists at the center of the Milky Way Galaxy; how large is its event horizon? One place astronomers find evidence for (big) black holes is at the center of galaxies. If they find one with a mass of 3.7 million M_{Sun}, we can calculate its size as

$$R_s = (2.9 \text{ km})(3.7 \times 10^6 \, M_{Sun}/M_{Sun}) = 1.1 \times 10^7 \text{ km}$$

Converting to astronomical units (AU), we find

$$R_s = (1.1 \times 10^7 \text{ km})/(1.5 \times 10^8 \text{ km/AU}) = 0.07 \text{ AU}$$

Thus, the size of a black hole containing as much mass as 3.7 million Suns is much less than the distance from Mercury to the Sun.

Black Holes and X-ray Binaries

The search for stellar-mass black holes got a strong push forward when NASA launched the satellite Uhuru in 1970, one of its first X-ray space telescopes. Scouring the sky for sources of bright X-rays, Uhuru scientists detected strong signals emerging from a region associated with a blue giant star (of spectral class B) located in the constellation Cygnus. The radiating object, christened Cygnus X-1, was tagged as a candidate for part of a black-hole binary system when analysis of the B star's motion showed the star in orbit with a massive but unseen object. The binary orbital period was just 5.6 days. Using Newton's laws, astronomers quickly determined the mass of the invisible companion to be at least 10 M_{Sun}. "That was big," says Campanelli, "and way too massive for the object to be a 'normal' compact object like a white dwarf or a neutron star."

The X-rays that Uhuru detected were also considered strong evidence of a black hole, which would make the star and its companion the first example of an *X-ray binary*. Astronomers predicted that material pulled off the companion B star by the black hole would spiral inward toward the event horizon, forming an accretion disk. "Gas in any accretion disk gets heated as it drops deeper into the gravitational well of the central object," says Campanelli, "and the deeper the well, the higher the temperature of the gas as it spirals around." Since black holes represent the deepest gravitational well possible, astronomers expected temperatures in the inner regions of a black-hole disk to reach many millions of degrees. At these temperatures, dense gas radiating as a blackbody would hit peak emission in the X-ray region of the spectrum. Thus, any accretion disk around a black hole was expected to be glowing in the X-ray region of the spectrum (**Figure 14.18**).

Given the mass of the unseen companion and the X-ray emission from the orbiting gas, you might think the identification of Cygnus X-1 as a black hole would have been an

"The mass was so large and was all packed into such a small region of space that a black hole was really the only explanation."

gamma-ray burst (GRB) A brief, intense outburst of gamma rays at relativistic speeds during a hypernova or kilonova.

open-and-shut case. Scientists are skeptical, however, and explore every explanation for a given set of data—not just the one that gets everyone excited. Throughout the 1970s and 1980s, alternative explanations were suggested that didn't involve a black hole. In the meantime, other X-ray binaries were discovered. Some of these had companions that turned out to be neutron stars. For others, as with Cygnus X-1, a black hole was still considered a viable possibility. Only later did new data from advanced X-ray telescopes reveal accretion disks spinning around objects so small but massive that they had to be black holes. Astronomers finally obtained the evidence needed to convince even the most hardened skeptics that stellar-mass black holes had been discovered.

Supermassive Black Holes and Galaxies

A galaxy is a vast collection of stars, gas, and dust with a total average mass of about 10^{12} M_{Sun}. Beginning in the 1960s, it became clear that the centers of many galaxies are home to wildly energetic phenomena (**Figure 14.19**). Torrents of light pour from the central regions of these galaxies, and many of them drive energetic jets of gas into space. Over time, astronomers came to realize that only a black hole of titanic proportions could account for these observations. They hypothesized that SMBHs, with masses more than a billion times that of the Sun, must exist at the core of these galaxies.

These supermassive monsters turned out to be good targets for black-hole research. "Using the Hubble Space Telescope and other high-resolution instruments," Campanelli explains, "astronomers could zoom in on material spiraling into these larger black holes." The light from this material was emitted by gas orbiting far enough from the event hori-

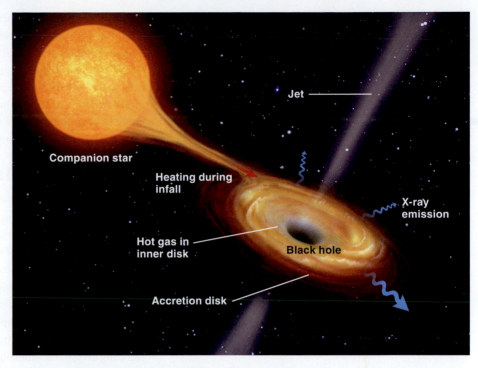

Figure 14.18 X-ray Binaries

Astronomers expected temperatures in the inner regions of a black-hole accretion disk to reach many millions of degrees, as gas in the disk would be drawn deeper into the black hole's gravitational well. At such temperatures, any gas that radiates as a blackbody would reach peak emission in the X-ray region of the spectrum.

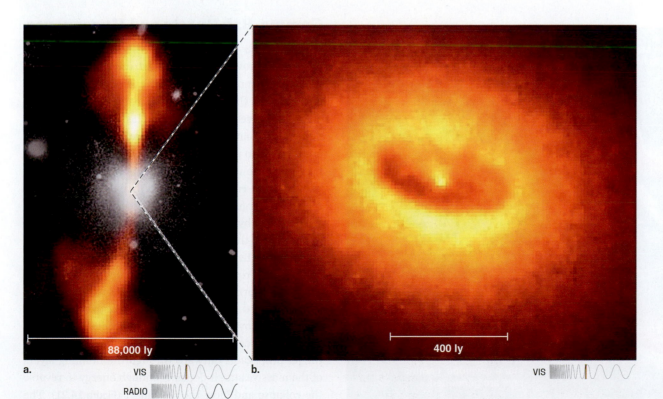

Figure 14.19 Supermassive Black Holes (SMBHs)

Supermassive black holes (SMBHs) exist at the center of many galaxies. **a.** The galaxy NGC 4261, with its brilliant radio jets (driven by the unseen black hole at the center) aligned perpendicular to the disk's spin axis and extending well beyond the galaxy's limits. **b.** A Hubble Space Telescope image of the dusty disk surrounding NGC 4261's black hole. Spectral observations of gas suggest a SMBH of 5×10^8 M_{Sun}.

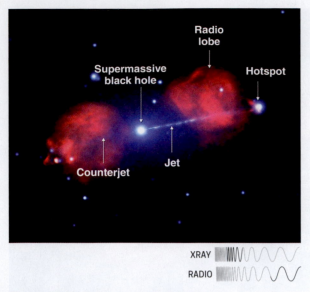

Figure 14.20 Jets from a Black Hole
A Chandra X-ray image of Pictor A, a radio galaxy with a SMBH at its center. A spectacular jet emanates from the SMBH, and a "counterjet" ejects from the opposite side.

Figure 14.21 Gamma-Ray Burst (long type)
When a star of more than 20 M_{Sun} has lost its outer layers, a compact iron-ash core surrounded by lighter nuclei remains. After gravitational core collapse, a rapidly rotating black hole and a surrounding accretion disk form. Twin jets emerge, bursting through the stellar envelope and producing gamma rays. The collapsed star soon explodes in an energetic hypernova. An observer can detect this rare, violent outburst only when in the direct path of a jet.

zon that it could easily escape and reach Earth-based telescopes. By analyzing the light's Doppler shift, astronomers determined the speed of the orbiting gas and therefore could derive the mass of the central object (see Going Further 3.4).

"The mass was so large and was all packed into such a small region of space that a black hole was really the only explanation," says Campanelli. The discovery of these SMBHs ushered in a new era of black-hole studies. It is perhaps ironic that some of the black holes we've come to know best are the ones Einstein and the early pioneers of general relativity never imagined. In fact, astronomers now think that *all* galaxies have black holes at their centers.

Jets from Black Holes

While you may think of black holes as giant cosmic vacuum cleaners, the truth is more complicated. Many environments hosting black holes also show material flowing outward from the central object. In particular, many black holes show evidence of powerful jets blasting across space at speeds just shy of light speed. We encountered jets in discussing star formation (Chapter 12) and evolution of low- and intermediate-mass stars (Chapter 13).

Microquasars are examples of stellar-mass systems in which jets are associated with black holes. (Quasars, phenomena associated with SMBHs at the center of galaxies, are discussed in detail in Chapter 16.) A microquasar is a type of X-ray binary, so it includes a black hole or a neutron star. Jets of material moving at 99 percent of light speed are seen hurtling away from microquasars and their larger galactic cousins.

Galactic jets are larger-scale beams of relativistic matter created by a black hole. The engines of galactic jets, which can extend many thousands of parsecs into space, are SMBHs in galactic centers (**Figure 14.20**). These black holes are also surrounded by accretion disks, which form as material falls in toward the central region of a galaxy.

In both cases, the relativistic jets are composed of material that has been lifted off the inner regions of an accretion disk surrounding a black hole and launched into space by magnetic fields. The jets' relativistic speeds indicate that the launching must occur in the inner regions of the disk, where the orbital speeds are also relativistic (roughly 10 percent of the speed of light or greater).

Gamma-Ray Bursts and Hypernovas

In Chapter 13 we encountered the stellar explosions called supernovas. For a time, astronomers thought of typical supernovas as the most energetic eruptions of energy in the Universe. Over the last few decades, however, they have discovered new classes of related phenomena that are even more energetic, and black holes are thought to play an essential role in getting some of these explosions to such extremes.

Gamma-ray bursts (GRBs) are brief outbursts of the most energetic forms of light (see Chapter 4). GRBs are classified by duration: long (greater than 2 seconds) or short (less than 2 seconds). The most accepted models of long GRBs invoke a form of hypernova (see Chapter 13): a rapidly rotating core-collapse supernova in which a black hole forms at the star's center. As material spirals inward toward the black hole, a magnetized accretion disk forms a powerful bipolar jet. The speeds in the jet are close to the speed of light, giving it tremendous energy. As the jet propagates outward through the infalling layers of the massive star, it deposits enough energy to reverse the collapse and blow the star apart (**Figure 14.21**). The

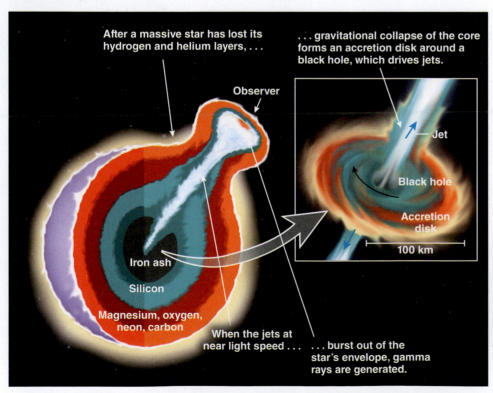

Are there ripples in space-time?

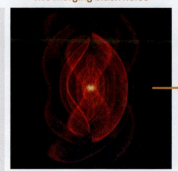

Simulation of two merging black holes

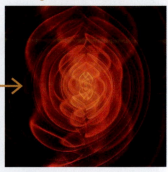

Resulting gravitational waves

theory

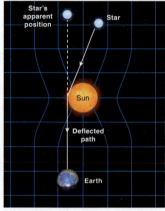

Star's apparent position

Star

Sun

Deflected path

Earth

Gravitational lensing: test of general relativity

In 1915, **Albert Einstein** published his general theory of relativity. It revolutionized our understanding of the interactions of massive bodies with the fabric of space and time. In it he predicted that the path of even a massless wave of light is altered when passing near a massive object.

hypothesis

Einstein's theory also predicts the generation of gravitational radiation by objects accelerating through space-time. The merger of very massive, dense objects such as neutron stars or black holes is predicted to radiate brief bursts of the Universe's most energetic gravitational waves (ripples in space-time), which may be detectable on Earth.

confirmed by observation

Using the Arecibo radio telescope, **Russell Hulse** and **Joseph Taylor** discovered the first binary pulsar in 1974. They noticed the system's orbital period was decreasing, as though the two stars were losing energy via gravitational waves—the mechanism Einstein had predicted 60 years earlier. Although seen as a confirmation of one prediction of Einstein's theory, this evidence of gravitational waves was indirect.

LIGO's Hanford Observatory

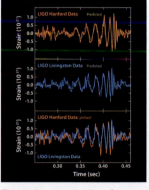

Detecting gravitational waves

confirmed by experiment

In 2015, an international team including **Manuela Campanelli** used the Laser Interferometer Gravitational-Wave Observatory (LIGO) to make the first direct detection of gravitational waves from the merger of two massive black holes. The "ripples" LIGO measured have amplitudes on the order of one-thousandth of the diameter of an atomic nucleus!

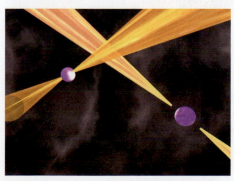

Artist's conception of binary pulsar

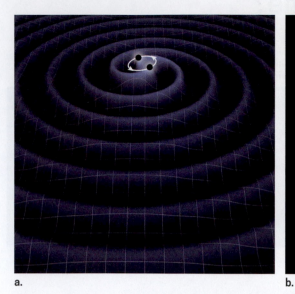

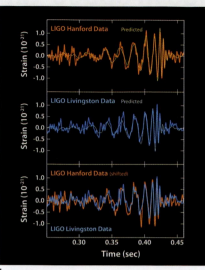

Figure 14.22 Gravitational Waves from Merging Black Holes

a. The merger of two massive and very distant black holes provided the event needed to capture gravitational waves. Each of the two LIGO locations independently recorded the phenomenon. **b.** The oscillation space in each LIGO detector as the waves pass are here seen to be strikingly similar, via overlay of the two sets of results, which were offset exactly by the additional time delay due to distance the waves had to travel to the Livingston facility relative to the Hanford detector.

gamma rays in GRBs are generated as the jet plows through dense material.

Short GRBs are believed to come from merging binary neutron stars and, therefore, are related to the resulting explosions called kilonovas (see Chapter 13). Thus while long GRBs involve black holes, both kinds of GRBs involve matter moving at relativistic speeds and gravitational behavior that requires Einstein's general theory of relativity to describe.

Gravitational Waves and Black Holes

In Section 14.3 we saw how the timing of binary pulsars infers the presence of unseen gravitational waves. Einstein's general theory of relativity predicted the existence of these ripples in the fabric of space-time. Gravitational waves are produced any time matter moves through space—but are generally quite weak. When tremendous amounts of matter are squeezed into tiny volumes of space and move at really high speeds, however, the gravitational waves emitted can carry a huge amount of energy. That's why the search for gravitational waves has targeted binary black holes.

In the 1980s, work began on the Laser Interferometer Gravitational-Wave Observatory (LIGO). The plan was to use the precision of laser light to look for distortions in space caused by gravitational waves originating from distant merging binary black holes (and other sources). To do this, LIGO would bounce lasers back and forth down two 4-km-long tunnels intersecting at a 90° angle. The laser light would then be combined to detect small changes in the length of the tunnels caused by a passing gravitational wave. The expected difference in tunnel length was, however, less than the width of an atomic nucleus. To detect such a tiny signal, the LIGO team—including Manuela Campanelli—needed measurements with unprecedented accuracy.

To ensure signal validity, the LIGO team built two laser-based gravitational wave detectors—one in Hanford, Washington, and one in Livingston, Louisiana. (A similar detector, called VIRGO, was later constructed in Italy.) After almost 30 years of work, the LIGO instruments reached the necessary level of sensitivity.

On September 14, 2015, both the Hanford and Livingston detectors registered a wiggling change in tunnel length, signaling the passage of gravitational waves from the merger of two black holes (**Figure 14.22**). Almost 100 years since their prediction, gravitational waves had finally been directly observed (see **Anatomy of a Discovery**). Within just a year, representatives of LIGO were awarded the Nobel Prize in Physics for this stunning achievement.

With the detection of gravitational waves, a new window on the Universe had opened. Astronomers learned surprising details about binary black holes, as the masses of the dead stars (29 and 36 M_{Sun}) were far greater than predicted. In 2017, LIGO detected the gravitational wave signal of merging neutron stars. Astronomers, quickly turning their telescopes on the region of the sky from which the waves had originated, were delighted to observe a kilonova in mid-explosion. Gravitational wave astronomy, still in its infancy, has already extended the frontiers of our knowledge of the cosmos. We can expect to learn much more from this field in years to come.

Section Summary

- The deeper a gravitational well is, the hotter the gas around it. Accretion disks in black-hole binaries can be observed via X-ray telescopes, indicating their extremely high temperatures and thus providing strong evidence of the existence of stellar-mass black holes.
- Some of the first black holes discovered were supermassive black holes in the centers of galaxies, observed via the light emitted from orbiting and infalling material. Some black holes power jets that move at relativistic speeds.
- Gamma-ray bursts are brief, intense bursts of high-energy light associated with the massive explosion during a hypernova or kilonova.
- LIGO confirmed the existence of gravitational waves, detecting them from mergers of black holes and of neutron stars.

CHECKPOINT   Compare the 1974 "detection" of gravitational waves from a binary pulsar (discussed in Section 14.3) with LIGO's 2015 detection of gravitational waves.

The End of Stellar Death

With our exploration of black-hole physics, we have now come to the end of our exploration of stars and their lifetimes. Stars live by the billions in vast, swarming structures called galaxies, and the Universe, it turns out, is teeming with galaxies. Now that we've learned about planets and stars, it is time to take the last step on our journey outward and view the cosmos first through its multitude of galaxies and then through its own fundamental history. It is time to go to the edges and beginnings of the Universe itself.

CHAPTER SUMMARY

14.1 The Black-Hole Shuffle

Black holes are objects whose properties defy our everyday experience. Dying massive stars collapse into black holes when thermonuclear fusion no longer offsets gravity's inward pull. Not even light can escape from a black hole.

14.2 Special Theory of Relativity

Einstein proposed (and experiments later confirmed) his special theory of relativity, which considers how measurements of space and time differ in different frames of reference. Two important consequences of special relativity are as follows: (1) The speed of light must be the same for all observers, so measurements of space and time must be different for different frames of reference. (2) If a person is moving close to the speed of light relative to an observer, the observer will see the moving person's time dilated (flowing more slowly), although the person in motion will perceive no time change.

14.3 General Theory of Relativity

Albert Einstein reached some of his most radical conclusions through thought experiments, realizing, for example, that the physics of a body at rest in a gravitational field is equivalent to that of a body moving at constant acceleration. His general theory of relativity implies that (1) matter-energy determines how space-time bends and (2) space-time determines how matter-energy moves. According to general relativity, gravity is the effect of curved space-time. In gravitational time dilation, clocks run more slowly when they are closer to large masses. Observations of the extended lifetimes of muons confirmed special relativity, and general relativity was initially confirmed by the bending of starlight. Today's global positioning systems must account for the effects of special and general relativity.

14.4 Anatomy of a Black Hole

Black holes are extremely compact and massive. They can result from the deaths of very massive stars or may be located at the centers of galaxies. At the singularity, a central point in all black holes, our current understanding of physics breaks down. A theory of quantum gravity, not yet fully developed, is expected to be the key to understanding what happens there. Black holes are bounded by an event horizon at the distance of the Schwarzschild radius from the center, where the escape velocity equals the speed of light. An object falling into a black hole is spaghettified and torn apart by strong tidal forces. Light from the object is greatly redshifted. The flow of time near the event horizon, relative to a distant observer, is strongly dilated.

14.5 Real Black Holes in Astronomy

At least two classes of black holes are observed to exist: stellar-mass black holes and supermassive black holes (SMBHs). The rapid motions of nearby stars and gas are evidence of SMBHs at galactic centers. Black holes resulting from stellar death were first observed as part of a binary pair with a visible companion star. Accretion disks around black holes have high temperatures and radiate in the X-ray region of the spectrum. Jets are frequently observed flowing at relativistic speeds from stellar-remnant black holes. Dying stars that collapse to black holes may end as massive explosions, generating gamma-ray bursts. The recent detection of gravitational waves from interactions of binary black holes and of binary neutron stars confirmed their existence.

QUESTIONS AND PROBLEMS

Narrow It Down: Multiple-Choice Questions

1. Which of the following statements about time dilation is true?
   a. Time dilation is only theoretical and not proven.
   b. Time dilation increases when moving at slower speeds.
   c. Both an individual traveling at close to the speed of light and one moving much more slowly perceive time to pass at a normal rate, although the rates are actually quite different.
   d. Time is more dilated the farther an observer is from a black hole.
   e. Time is dilated only when an object is moving faster than the speed of light.

2. Which force or process is primarily responsible for creating a black hole?
   a. strong nuclear force
   b. inertia
   c. gravity
   d. centrifugal force
   e. gas pressure

3. An observer moving at 90 percent of the speed of light ($0.9c$) would observe light around him to move at
   a. $0.9c$.
   b. $0.1c$.
   c. c.
   d. $0.81c$.
   e. $-0.1c$.

4. Which of the following shares your reference frame while you're reading this book in a room?
   a. an astronaut on the Moon
   b. a geosynchronous satellite directly over your location
   c. an airplane 30,000 feet up in the atmosphere traveling 400 miles per hour
   d. a racecar driver speeding on a track
   e. a person in an adjacent room

5. Spaceship A is traveling at $0.9c$, and spaceship B is traveling at $0.95c$ in the opposite direction. Which of the following statements is true?
   a. B measures A moving at the speed of light.
   b. A measures B moving at the speed of light.
   c. Both measure light passing them at c.
   d. Only A is actually moving.
   e. Only B is actually moving.

6. In the twin paradox, the twin who travels in the spaceship experiences
   a. time passing slowly.
   b. time passing quickly.
   c. a normal flow of time in her frame of reference.
   d. no sense of time passing.
   e. time passing at the same rate as her twin experiences it, as measured by the same distant observer.

7. Which of the following is true?
   a. General relativity alone deals with gravity.
   b. Special relativity alone has c as a constant.
   c. Special relativity incorporates the effects of gravity.
   d. In special relativity alone, space and time are flexible.
   e. "General relativity" refers to objects in uniform motion.

8. Which of the following is true about gravitationally induced motion, according to general relativity?
   a. It is a result of an object's spin.
   b. It occurs as objects fall freely, following the curvature of space-time.
   c. It is identical to gravitational motion according to Newton's laws.
   d. It is the same as uniform motion.
   e. It exists only near extremely massive objects.

9. Which of the following does/do *not* represent a successful test of relativity? Choose all that apply.
   a. the detection of muons created by cosmic rays
   b. the inferred existence of gravitational waves via binary pulsars
   c. the discovery of the wave nature of light
   d. the weightlessness experienced by astronauts in space
   e. the accuracy of GPS devices

10. If GPS satellites orbited higher, the general relativistic adjustments currently required for accuracy of positioning on Earth's surface would be
    a. greater because of the difference in the satellites' speed.
    b. lesser because of the difference in the satellites' speed.
    c. the same, since the surface positions remain the same.
    d. greater because of the greater difference in the curvature of space-time between the ground and satellite.
    e. the same because general relativity can be ignored in the near-Earth environment.

11. Which of the following is the densest?
    a. core of a 1-M_{Sun} star
    b. core of a 100-M_{Sun} star
    c. white dwarf
    d. neutron star
    e. singularity of a black hole

12. Which statement about gravitational waves is correct?
    a. They are a form of light.
    b. The first detection of gravitational waves showed they differ greatly from predictions about them.
    c. Gravitational waves from the motion of a planet in orbit should be readily detectable with current technology.
    d. X-ray telescopes can detect them.
    e. Their strength relates directly to the mass of the object whose motion generates them.

13. A patent clerk in a spaceship observes that time on the clock of an astronaut on a spaceship passing at $0.25c$ runs slower than does time on his own clock. This phenomenon is called
    a. relativistic time dilation.
    b. time travel.
    c. gravitational time dilation.
    d. curvature of space-time.
    e. time warp.

14. As an object approaches the event horizon of a black hole, the light from it is observed to become
    a. longer in wavelength.
    b. shorter in wavelength.
    c. slower and slower.
    d. faster and faster.
    e. unchanged.

15. If you could be 475 km from the center of a 200-M_{Sun} black hole, which of the following would be true?
    a. You would see an accretion disk.
    b. You would be orbiting at a constant speed.
    c. You could no longer communicate with the rest of the Universe.
    d. You would be outside the event horizon.
    e. You could speed up to move away from the black hole.

16. Spacecrafts *X101* and *X102*, with different velocities relative to each other, pass in interstellar space. Which of the following *cannot* be true?
    a. *X101* measures its own velocity as zero and measures a nonzero velocity for *X102*.
    b. *X102* measures its own velocity as zero and measures a nonzero velocity for *X101*.
    c. Each measures both its own velocity and the velocity of the other as nonzero values.
    d. Both measure light passing them at exactly *c*.
    e. A distant observer measures zero velocities for both.

17. Which of the following statements about Einstein's theory of gravity, compared with Newton's law of gravity, is true?
    a. It concluded that the effect of mass is less than expected, while that of distance is more than expected.
    b. It concluded that the effect of distance is less than expected, while that of mass is more than expected.
    c. It proved Newton's law entirely wrong.
    d. It made the use of Newton's law unnecessary.
    e. It showed that Newton's law is a "limiting case" of general relativity that applies only when velocity and gravity are relatively low.

18. Which of the following statements is/are consistent with Einstein's special theory of relativity? Choose all that apply.
    a. There is no privileged frame of reference.
    b. The speed of light is constant, except when affected by relativistic time dilation.
    c. The speed of light is an upper bound to motion in the Universe.
    d. Time is not the same in all frames of reference.
    e. Space and time exist independent of each other.

19. True/False: Spaghettification upon approaching the event horizon of a black hole occurs because gravity varies with distance from its source.

20. True/False: The 1919 solar eclipse provided support for relativistic time dilation.

To the Point: Qualitative and Discussion Questions

21. Explain the difference between core mass and main-sequence mass in determining the end product of star death.

22. Describe two of Einstein's thought experiments.

23. Describe the evidence of the existence of black holes.

24. What was the "luminiferous aether" believed to be, and why did physicists propose that it existed?

25. What is a frame of reference? What and who currently share your frame of reference?

26. Describe a situation in which the "addition of velocities" gives a correct result.

27. Why did Einstein conclude that space and time must be unified and that measurements of each depend on an observer's frame of reference?

28. What is the twin paradox?

29. For what kinds of situations is special relativity relevant? For what kinds is general relativity relevant?

30. What are some of the effects that occur near black holes?

31. How does the detection of muons on the surface of Earth, after they are created high in Earth's atmosphere, provide evidence of special relativity?

32. Describe the method used to detect gravitational waves. Why were they so difficult to detect?

33. What are the differences between supermassive black holes and stellar-mass black holes, in terms of where they reside?

34. What evidence suggests that massive black holes exist in galaxies?

35. Think about any movies or TV shows in which black holes have been portrayed. According to what you now know about black holes, what did the writers get right and what did they get wrong?

Going Further: Quantitative Questions

36. What is the Schwarzschild radius, in kilometers, of a 13-M_{Sun} black hole? Of a 2,500-M_{Sun} black hole?

37. What is the mass, in M_{Sun}, of a black hole whose Schwarzschild radius is 990 km?

38. A black hole has a Schwarzschild radius of 7×10^6 km. What is its mass, in kilograms?

39. What is the Schwarzschild radius, in meters, of a black hole whose mass is 4×10^{22} kg?

40. A planet's orbital radius is 43 AU. If the star were a black hole instead, at what radius would the planet orbit?

41. Sophia (on Earth) sees her twin sister, Stella, passing on her spacecraft at 0.4*c* and observes that Stella's clock runs slower than her own. Sophia's favorite movie runs 2 hours. By Sophia's clock, how long, in hours, would the movie run for Stella?

42. While traveling from a distant space station at 0.9*c*, the occupants of a spaceship take 35 minutes to prepare their dinner. How long, in minutes, does their preparation take, according to the clock of a NASA scientist on Earth who is observing the activity and the clock of the spaceship?

43. An American football game runs 60 minutes (not counting time-outs and commercials). You are on Earth observing a game being played aboard a starship traveling at 0.79*c*. By your clock, how many minutes does the game last?

44. An event takes 2 hours on the clock of a spacecraft at relativistic speed but 8.4 hours on the clock of an observer on a nearby planet. At what speed is the craft traveling? (State your answer in terms of *c*.)

45. What is the escape velocity from a location that is 92 km from the singularity of a black hole with mass 25 M_{Sun}? (State your answer in terms of *c*.)

"I WAS REALLY EXCITED BECAUSE I THOUGHT I HAD A NEAT, AND PRETTY SIMPLE, IDEA"

15

Our City of Stars

THE MILKY WAY

A Hard Rain: A New Vision of Galactic Studies

15.1 Students don't usually get calls from world-famous scientists, but that is what happened to Bob Benjamin after he submitted his first scientific paper for review. "I was a graduate student and had written a short paper on clouds of gas falling down onto the disk of the Milky Way Galaxy," says Benjamin, now a professor at the University of Wisconsin. "I was really excited about the idea because I thought I had a neat, and pretty simple, idea for explaining them."

Benjamin hypothesized that supernovas could blow material out of the plane of the Milky Way, but the gas would move too slowly to escape our galaxy's gravitational pull. This phenomenon, called a *galactic fountain*, explicitly links the death of stars and the evolution of galaxies. The gas would rise but then fall back (**Figure 15.1**). "I had written down some simple equations that compared the falling gas clouds to raindrops," says Benjamin. "The speed of the cloud is determined by the balance between gravity and aerodynamic drag, just like rain." If correct, Benjamin's idea would help explain a piece of the puzzle of galactic recycling: how gas from one part of the galaxy ends up in another.

Then Lyman Spitzer (**Figure 15.2**), for whom the Spitzer Space Telescope was named, called Benjamin. "I've been given your paper to referee, and it's very interesting," said Spitzer. "However, I have some concerns about equation 1." Benjamin was stunned. "Oh dear!" he said to himself. "I'm in for it now." Equation 1 was just a version of $F = ma$.

What Benjamin was in for was a long, fruitful discussion with the great scientist. Eventually, Benjamin's work was published with Spitzer's blessing. "This paper presents a new and potentially important theory of interstellar dynamics," the older scientist wrote. Benjamin's research on our galaxy, the Milky Way, had begun.

← Like our own Milky Way Galaxy, NGC 1309 is a dramatic spiral galaxy with a central bulge and star-forming spiral arms. Unseen—but pivotal to its formation, structure, and motion—are its central black hole and dark matter halo. NGC 1309 resides 100 million light-years from Earth.

**Figure 15.1
Galactic Fountain**

a. A composite image of the Milky Way shows clouds of gas above the plane of the galaxy (in visible light). Particles in the cloud "rain" down on the Milky Way, seeding it with material for star birth. **b.** An artist's rendering of the galactic fountain concept. Supernovas drive material rich in heavy elements out of the plane of galaxy, which then rains down elsewhere.

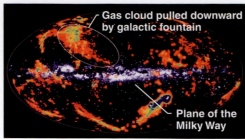

a. **All-sky map**

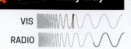

VIS
RADIO

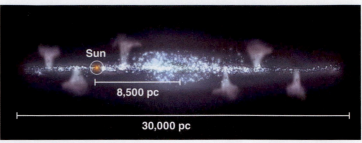

b. **Edge-on view of Milky Way**

Sun

8,500 pc

30,000 pc

Figure 15.2 Lyman Spitzer

Lyman Spitzer (1914–1997), an American theoretical astrophysicist known for his research in many areas of astronomy and for conceiving of telescopes operating in outer space. NASA's infrared Spitzer Space Telescope is named in his honor.

"I guess in that way you could call Galileo the first galactic astronomer."

Life in a Universe of Galaxies

Every day the Sun rises and sets, and every night the stars appear and create their patterns of illumination against the backdrop of empty space. But where are all the stars located? Is there any structure to their distribution in space, or is the cosmos just an infinite array of stars? Does the night sky provide any clues about where we reside in the architecture of the Universe?

For most of human history, a hint to these giant questions was seen overhead every night in an arc of light reaching from one horizon to the other (**Figure 15.3**). Today most of us see the Milky Way only when we leave cities and suburbs. "If you get far away from city lights, you'll see many, many, many more stars than you ever realized were up there," says Benjamin. "And in particular, you'll see a sort of a smooth band of light stretching across the sky. That band is the Milky Way, and it's where most of the stars in the galaxy are located."

For hundreds of millennia, humans looked at the Milky Way in wonder, creating explanations first in myth and later in science. Through the efforts of generations of astronomers, the **Milky Way** was gradually recognized to be a galaxy—a teeming "city" of a few hundred billion stars, all gravitationally bound together. Moreover, astronomers have shown just how vast the Milky Way is. If we could shrink the Sun to a grain of sand and Jupiter's orbit to just a few centimeters, in proportion, the Milky Way would still have a size of about 150,000 km.

Figure 15.3 The Milky Way on the Sky

VIS

Our galaxy appears as an arc of diffuse light reaching from one horizon to the other. The widespread use of electric lights has made it a rare sight for anyone living in highly populated areas.

Milky Way   The spiral galaxy in which our Sun is located.

The last step outward in building our story of the Universe is essentially a story of galaxies, the building blocks of the Universe. What better way to understand the structure, formation, and history of galaxies than to start with our own vast system of stars, the Milky Way?

The Long Road to the Milky Way

Ancient astronomers marveled at the nightly spectacle of the Milky Way, but telescopes were needed to determine its true nature. "It was Galileo with his telescope who first realized this Milky Way was not just some fuzziness in the sky," says Benjamin. "He saw it was a collection of thousands of stars whose light was blurred together. I guess in that way you could call Galileo the first galactic astronomer."

After Galileo surmised the starry nature of the Milky Way, the race was on to find its true shape. That required an understanding of the distribution of visible matter (which at that time meant stars) in the Universe. By Isaac Newton's time, many astronomers believed that an infinite number of stars is distributed smoothly across infinite space. But as telescopes became more powerful, some astronomers resolved to determine the Milky Way's size and structure by using the best tool at their disposal: direct observation.

By the late 1700s, sibling astronomers William and Caroline Herschel had set out to measure the distribution of stars in the Milky Way directly by counting them (**Figure 15.4**). "William Herschel was already an expert at building big telescopes, and he had used these instruments to make amazing discoveries," says Benjamin. "He was the first person in history to discover a planet, Uranus, and that made him an instant celebrity."

In time, the Herschels turned their attention to the Milky Way. They assumed that the Sun is at its center and spent hours each night at their telescope counting stars in different directions of the sky. "Pointing the telescope at a certain location in the sky," explains Benjamin, William would "count up the number of stars he could see through the eyepiece and call the numbers out to Caroline at the base of the big instrument. Then he'd reset the telescope, point it in another direction, count up those stars, and call them out again." After many months of meticulous, exhausting work, the Herschels built up a map of star counts.

The effort bore fruit: the Herschels accurately inferred that the Milky Way is shaped like a grindstone or disk (**Figure 15.5**). Their assumption that the Sun is located at the grindstone's center was, however, flawed for reasons they could not have suspected, as we will see soon.

By the early 20th century, astronomers not only had a sense of the Milky Way's shape but also had begun determining its size limits. "By then, people had photographic plates to make their star counts," says Benjamin. "That was a huge advance over staring into an eyepiece." In 1906, the astronomer Jacobus Kapteyn began using the star count method, together with maps of stellar motions, to deduce that the Milky Way is a flat-

a.

b.

Figure 15.4 Star Counters
a. Caroline (1750–1848) and **b.** William Herschel (1738–1822) made labor-intensive counts of the number of stars in all directions of the sky to ascertain the structure of the Milky Way Galaxy.

INTERACTIVE:
Milky Way

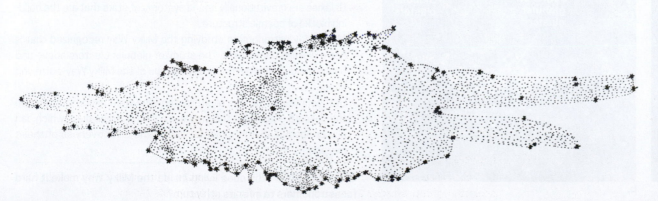

Figure 15.5 Herschel Grindstone Diagram
With a remarkable ability to interpret data, the Herschels correctly deduced from the distribution of stars in their counts that the Milky Way has a disklike shape. William Herschel likened it to a grindstone, a stone wheel used to sharpen metal blades.

a.

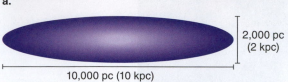

b.

2,000 pc
(2 kpc)

10,000 pc (10 kpc)

Figure 15.6 Kapteyn and the Size of the Milky Way

a. Jacobus Kapteyn (1851–1922) counted stars, as the Herschels had before him, but he used photographic plates instead of observations through an eyepiece. **b.** Kapteyn inferred that the Milky Way is a flattened disk 10,000 pc across and 2,000 pc high, much greater than previous estimates.

Figure 15.7 Finding the Galactic Center

Harlow Shapley (1885–1972) studied globular clusters in our galaxy, such as **a.** M3, and determined that they are distributed spherically around the galaxy's center. **b.** Using globular clusters, he estimated the Sun to be at least 16 kpc from that center, though we now know that distance to be 8 kpc.

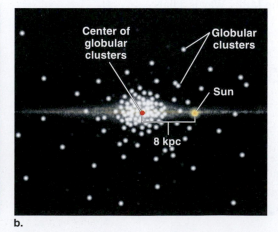

a.   VIS

Center of globular clusters

Globular clusters

Sun

8 kpc

b.

tened disk 10,000 parsecs (pc; equal to 10 kiloparsecs, kpc) across and 2,000 pc (2 kpc) high (**Figure 15.6**). These distances would have been unimaginably large a few centuries earlier. The study of the Milky Way pushed researchers ever farther out into the cosmos. The need for a new distance unit, the *kiloparsec*, equal to 1,000 pc and to 3,262 light-years, is testament to the fact that they had left the realms of stars for far larger cosmic structures.

Ironically, further progress determining the Milky Way's true (enormous) size was blocked by the tiniest pieces of solid matter: interstellar dust. Recall from Chapter 12 that these flecks of material affect observations in two ways. First, some of the starlight that passes through dust clouds is scattered or absorbed, reducing the stars' apparent brightness. Since astronomers were not taking dust into account, they thought the stars were farther from Earth than they really are.

Second, dust clouds can completely obscure light from some regions of the Milky Way, so star counts carried out in those directions are highly inaccurate. "Dust really limits how far you can see," says Benjamin. "If you're looking toward the inner galaxy through the Milky Way's disk, dust blocks out almost everything more than a few kiloparsecs away. If you don't know about dust, then you are going to get a lot of aspects of the galaxy dead wrong."

It was some time before astronomers understood how to take interstellar dust into account in calculations of the Milky Way's size. A crucial step came in 1914 when Harlow Shapley, an astronomer at Harvard University, focused not on counts of individual stars but on the jewel-like collections of stars called *globular clusters*. Shapley used the largest telescope of his day to resolve the type of variable star called Cepheid variables that can be used as standard candles (see Chapter 11). He mapped out the distances to and positions of globular clusters, which he found to be spherical, not disk-shaped. Just as important, he determined that the center of the sphere lies more than 16 kpc away from the Sun, in the constellation Sagittarius as seen from Earth. (The actual distance is about 8 kpc.) This, he concluded, is the galaxy's true center (**Figure 15.7**). It was interstellar dust that made Shapley's distance to the galactic center wrong.

"What made Shapley famous was he managed to go outside the plane of the Milky Way's disk," says Benjamin. "When Herschel and Kapteyn did their star counts, they weren't seeing very far because dust in the disk blocked their view. That meant they were getting a very biased picture of the galaxy." Shapley trained his telescope away from the dust and toward clusters that lay above and below the disk. Doing so enabled him to see farther, find the Milky Way's center, and discover that the disk is not the only component of the galaxy's structure—setting the stage for the modern era of Milky Way studies. In the decades that followed, the true structure of the Milky Way was mapped out in exquisite detail, revealing a complex anatomy.

Section Summary

- All of the stars that we see at night exist within the Milky Way Galaxy.
- Galaxies are gravitationally bound systems of stars that are the building blocks of cosmic structure.
- Many early astronomers studying the Milky Way recognized that it has a disklike component. The study of globular clusters above and below the disk allowed for better models of the Milky Way's size and shape.
- Star counts and studies of stellar motions were used to determine the galaxy's shape, but the presence of interstellar dust, which can dim or obscure distant objects, limited the effectiveness of these methods.

CHECKPOINT Why does living inside the Milky Way make it hard for astronomers to infer its structure?

Anatomy of the Milky Way

15.2 Even though we can see the Milky Way only from the inside, years of study reveal that it has three basic directly observable structural components: the galactic disk, the stellar halo, and the bulge. **Figure 15.8** shows the Milky Way's structure, and **Table 15.1** gives the basic properties of each component.

The Galactic Disk

The **galactic disk** is the most recognizable of the three directly observable components, since the Milky Way appears to be a classic example of a **spiral galaxy**, similar to NGC 1309. The disk is a thin, flattened, circular distribution of stars and interstellar material with a radius of 15 kpc and a height of approximately 0.3 kpc above the midplane (Figure 15.8). There are between 200 billion and 400 billion stars in the disk—implying a mass of approximately 10^{11} solar masses, M_{Sun}. The amount of interstellar material is lower but still significant (in the tens of billions of solar masses). The Sun is located approximately 8 kpc from the center of the galaxy and 0.03 kpc above the midplane of the disk.

Although material is distributed throughout the disk, there is strong evidence of the existence of **spiral arms**, where the density of stars and gas increases. The exact number of distinct spiral arms remains in dispute, but most astronomers agree that the Milky Way has up to four major spiral arms, which tend to be the location of molecular clouds where stars are forming. Later in this chapter, we will discuss the nature of spiral arms. With the exception of the arms, however, the density of stars and gas clouds decreases fairly smoothly with both radius and height in the disk. The fraction of heavy elements in both the stars and the stellar clouds also decreases with distance from the disk's center.

Table 15.1	Visible Components of the Milky Way Galaxy		
	Galactic Disk	**Stellar Halo**	**Bulge**
Location in the Galaxy	Equatorial plane	Surrounding the disk	Central region
Size	Radius ~15 kpc Height ~0.3 kpc	Radius ~15 kpc	Radius ~5 kpc
Shape	Flat disk; may have spiral arms	Sphere	Flattened sphere
Features	Star-forming regions in spiral arms	Old, widely dispersed globular clusters and stars	Very dense stellar population

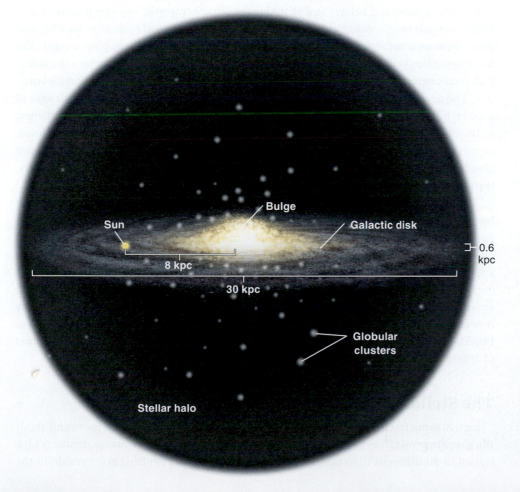

Figure 15.8 The Milky Way

A schematic view of the galaxy shows its three primary luminous components (components made of "normal"—nondark—matter): the galactic disk, stellar halo, and bulge.

galactic disk A flattened distribution of stars and interstellar material within a galaxy.

spiral galaxy A pinwheel-shaped galaxy that may have spiral arms.

spiral arm A pinwheel-shaped distribution of stars and interstellar material, typically extending from the bulge outward.

Figure 15.9 Orbital Motions in the Milky Way

Stars in the galactic disk move on circular orbits in the plane of the disk around the galactic center. In addition, gravitational encounters between stars and clouds lead to peculiar motions, up-and-down oscillations of some disk stars above and below the disk's midplane. Halo stars and the globular clusters that contain them travel far from the plane of the galaxy on highly elliptical, "plunging" orbits, while stars in the bulge move with a mix of circular and plunging orbits.

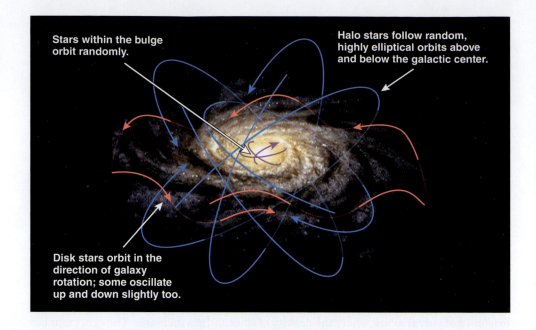

Stars within the bulge orbit randomly.

Halo stars follow random, highly elliptical orbits above and below the galactic center.

Disk stars orbit in the direction of galaxy rotation; some oscillate up and down slightly too.

"As time goes on, the amount of metals [in] interstellar gas increases as each generation of stars dumps processed heavy elements back into the galaxy."

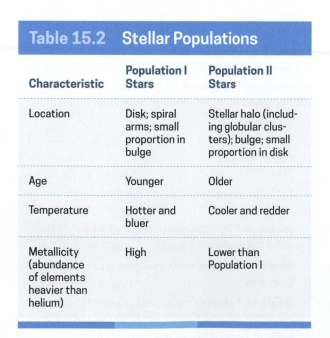

Table 15.2	Stellar Populations	
Characteristic	Population I Stars	Population II Stars
Location	Disk; spiral arms; small proportion in bulge	Stellar halo (including globular clusters); bulge; small proportion in disk
Age	Younger	Older
Temperature	Hotter and bluer	Cooler and redder
Metallicity (abundance of elements heavier than helium)	High	Lower than Population I

Material in the disk revolves around the galactic center on orbits that are generally circular. At the Sun's position, the galactic revolution period of this overall circular orbit is about 233 million years. As stars go around the Milky Way, they also move in what are called **peculiar motions**. Gravitational encounters between stars and clouds cause stars to "scatter," similar to the way planets can scatter within solar systems (see Chapter 5). The scattered disk material bobs up and down in an oscillating vertical motion that takes stars and gas above and below the disk midplane. "In other words," says Benjamin, "if you look at a star, it may be going slightly upward or slightly downward relative to the disk, and it may be somewhat above or below the disk" (**Figure 15.9**). In general, the older a star is, the more opportunity it has had for encounters and thus for scattering. That means the oldest disk stars are found at the greatest distances above and below the Milky Way's midplane.

The age distribution of stars in the disk is mixed. Many stars are older, with ages of about 8 billion years. Stars tend to become redder as they evolve off the main sequence, so the old stars contribute red light to the overall spectrum of the galaxy. Newly formed, extremely massive young stars are found close to where they were born, in molecular clouds. These clouds tend to be located in the spiral arms, which thus have higher concentrations of the massive blue stars.

"As time goes on, the amount of metals, or the metallicity, of interstellar gas increases as each generation of stars dumps processed heavy elements back into the galaxy," Benjamin reminds us. **Metallicity** is the fraction of elements heavier than helium produced by stellar fusion, as a fraction of all elements in a star (see Chapter 11). Stars are classified as either metal-poor or metal-rich. Overall, the fraction of heavy elements decreases with distance from the disk center. Most stars in the disk, however, have a high enough metal concentration to indicate that they formed from material that was already processed through earlier generations of stars. Thus, astronomers categorize disk stars as **Population I** stars. This classification distinguishes them from other populations of stars of different ages found in other locations in the galaxy (**Table 15.2**).

The Stellar Halo

"There is some fraction of stars like the Sun that are on this nice merry-go-round about the galactic center," says Benjamin. "Then there are these stars on more extreme orbits, tipped in all different directions." Most of these randomly orbiting stars reside in the

stellar halo, an extended spherical distribution of globular clusters surrounding the Milky Way. There are approximately 150 globular clusters in the halo. As we saw in Chapter 11, globular clusters are gravitationally bound collections of up to a million stars, all born at the same time. These clusters in the stellar halo are what enabled Shapley to determine the galaxy's true center. The radius of the halo is about 15 kpc—about the same size as the galactic disk, although the halo is spherical. A few outlying globular clusters, such as Pal 14, lie at much greater distances, extending out as far as 72 kpc from the galactic center.

X-ray observations indicate that the stellar halo may be full of gas at a temperature of more than a million degrees. If this is true, the mass in the halo would be comparable to that of the galactic disk. The density of the Milky Way's hot gas (like that of the Sun's corona) appears to be quite low, so it escaped detection for some time. Studies to try to understand the properties of this hot halo are ongoing.

As Benjamin emphasizes, the motion of the halo's globular clusters is very different from the stellar motions in the disk. Disk stars (and gas clouds) revolve around the galactic center on roughly circular orbits, but globular clusters tend to dive inward and then swing around the galactic center on highly elliptical paths (Figure 15.9). These *plunging orbits* provide evidence that whatever formed the globular clusters was not rotating very quickly; if it had been, halo stars would have a far more significant component of circular motion. Just as important, to the degree that the globular-cluster orbits do show rotational motion around the galactic center, though on highly elliptical orbits, many of them have a sense of rotation that is opposite that of disk material. Thus, astronomers consider these globular clusters to be on retrograde galactic orbits—a significant clue for modeling the galaxy's history.

The composition of the stellar-halo stars and interstellar medium is also very different from that of disk stars. A key difference is the lack of star-forming gas in the halo. "The disk shows a rich interstellar medium of gas and dust, including gas clouds that are forming new stars," says Benjamin. "But the amount of gas that could form stars in globular clusters is comparably pretty small." Given the absence of star-forming regions, it should come as no surprise that the halo does not contain newborn stars. "Every now and then when I am observing, I will look for star formation in the halo," says Benjamin. "But really, there's nothing." All of these findings suggest that the halo is an older system of stars than the disk.

The conclusion that the stellar halo is older than the galactic disk is also strengthened by studies of the chemical compositions of halo stars. While disk stars are relatively abundant in heavy elements, halo stars show distinctly lower processed-element compositions. Because of the low metal abundances, astronomers categorize halo stars as **Population II** stars (Table 15.2). Population II stars must have formed at an earlier epoch of cosmic history, before many generations of star formation and nuclear processing. The material from which they formed was relatively low in metallicity compared with the metal-rich material found in Population I stars.

Astronomers recently created a third category of stars, *Population III*, to describe the first generation of stars born after the creation of the Universe. These stars would have had no or extremely low metallicity, and all of them are expected to have met their stellar death in one form or another by now. We have no confirmed examples of these stars yet.

The Bulge

The central region of the Milky Way is dominated by the aptly named **bulge**, a roughly spherical distribution of stars and dust about 5 kpc in radius (Figure 15.8). Stars are packed tightly together in the bulge. Typical stellar densities there are as high as 1,600 stars per cubic parsec (pc^3); in comparison, the density of stars found near the Sun is much less than 1 star/pc^3.

"Every now and then when I am observing, I will look for star formation in the halo. But really, there's nothing."

peculiar motion The motion of stars in the Milky Way disk that is a departure from pure revolution around the galactic center.

metallicity The abundance of elements heavier than helium in an object.

Population I A category of stars, found primarily in the disk or bulge, that have high metallicity and therefore must have been born relatively recently in cosmic history.

stellar halo The spherical distribution of Population II stars that are found mainly in globular clusters.

Population II Stars found primarily in the stellar halo, which have lower metallicity and therefore must have been born relatively early in cosmic history.

bulge The roughly spherical distribution of stars and interstellar material in the central region of a galaxy.

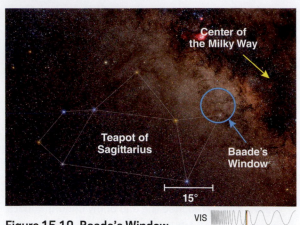

Figure 15.10 Baade's Window

The density of gas and dust near the galactic center makes the bulge very difficult to study. Baade's Window, an opening in the clouds in the constellation Sagittarius, affords astronomers a view into the bulge's central regions.

The bulge is composed mainly of Population II stars. Thus, stars in the bulge are, on average, older and redder than the Population I stars composing the disk. Do not think of the bulge as a dense version of the stellar halo, however. The bulge is a separate component of the galaxy with its own unique features and history; it includes some Population I stars as well. The gas and dust content of the bulge is quite high, making it very different from the stellar halo. The presence of the gas allows new stars to form, and there are scattered examples of hot, blue stars and supernovas in the bulge.

The presence of so much gas and dust also makes the bulge hard to study. There is so much interstellar material in this innermost component of the galaxy that seeing inside is almost impossible without using very-long-wavelength telescopes. In a few places, such as the region called *Baade's Window*, a path through the gas clouds has opened up, affording astronomers a valuable optical view into the bulge's central regions (**Figure 15.10**).

Determining the true shape of the Milky Way's bulge is not the only reason astronomers are keen to see deep into the galactic center. A bright radio source was discovered in a region close to the center of the bulge in 1974, and it appears to be associated with a supermassive black hole (see Chapter 14). The presence of an enormous, relatively nearby black hole quickened the hearts of many astronomers, and the central region of the bulge has remained a topic of intense study ever since.

Our Neighborhood of Galaxies

If we could step outside the Milky Way and look at it from afar, we would see that it resides in a collection of more than 50 galaxies, called the **Local Group** (**Figure 15.11**). Although these other galaxies are not part of the Milky Way's "anatomy" per se, many of them may

**Figure 15.11
The Local Group**

Galaxies rarely appear in isolation but occur in groups (or in larger clusters). The Milky Way is in the Local Group, which encompasses two other spiral galaxies and a host of smaller satellite galaxies. The Milky Way's satellite galaxies include the Large Magellanic Cloud (LMC) and Small Magellanic Cloud (SMC). Distances from the Milky Way are given in kiloparsecs (kpc). The vertical bars indicate distance above or below the plane of the Milky Way.

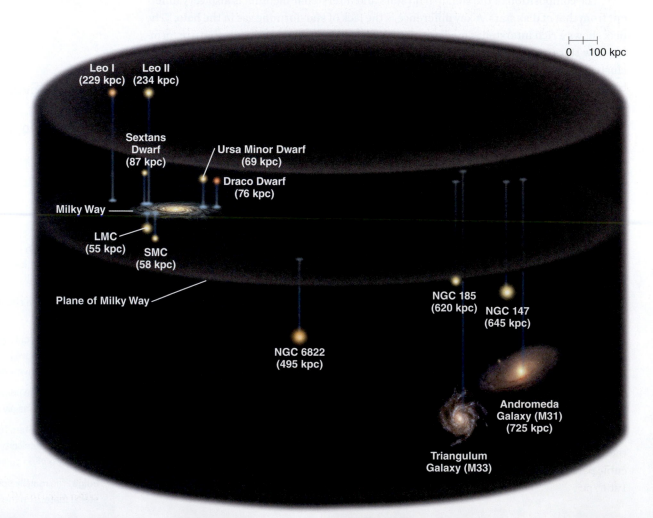

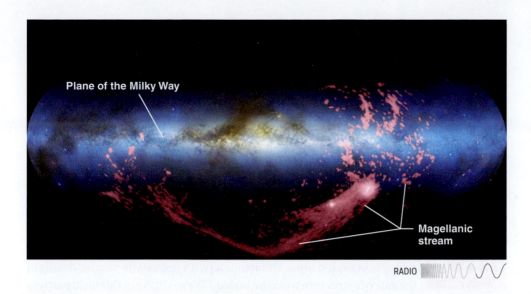

Figure 15.12 Galactic Interaction
A composite image of the Magellanic Stream, a tidal stream formed billions of years ago, when the Large and Small Magellanic Clouds passed close to one another. The bright stream is evidence of an ancient galactic interaction.

be influencing our galaxy's structure and evolution. In fact, the current understanding of the Universe suggests that every large galaxy is surrounded by a multitude of smaller "satellite" stellar systems, which are involved in ongoing interactions, collisions, and even cannibalism.

The Milky Way is orbited by at least 14 confirmed satellites. The largest and most massive are the Large Magellanic Cloud (LMC) and the Small Magellanic Cloud (SMC), two irregularly shaped galaxies with diameters of 4 kpc and 2 kpc, respectively. (They are named for the Portuguese explorer Ferdinand Magellan, who noted them on his historic circumnavigation of the globe in 1519–1522. They are visible only from Earth's Southern Hemisphere.) Many of the smaller galaxies are on orbits that will lead to a collision with the Milky Way in the distant future.

These galaxy motions imply that collisions with the Milky Way have also happened in the past. Evidence of such collisions is seen in *tidal streams*, arcs of stars and hydrogen gas that form as an orbiting satellite galaxy is pulled apart by the Milky Way's much larger gravity. For example, astronomers have found two tidal streams associated with the Sagittarius Dwarf Galaxy, torn off by the Milky Way's huge gravitational pull. Sagittarius may once have been one of the brightest of the Milky Way's satellite galaxies, but the Milky Way's immense gravity has ripped it apart, dispersing half of Sagittarius's stars and virtually all of its gas over the last billion years.

While the larger gravitational pull of the Milky Way tends to dominate all satellite interactions, the satellites themselves sometimes collide. The so-called Magellanic Stream signifies a near-collision between the LMC and SMC. This stream is composed of high-speed neutral hydrogen (H I) clouds stretching 180° across the sky (**Figure 15.12**). Detailed computer simulations indicate that the stream formed almost 2.5 billion years ago when the LMC and SMC passed close enough to each other for tidal forces to become significant. The near-collision not only pulled out the Magellanic Stream but also triggered a burst of star formation in both galaxies and deformed them into their current shapes.

AT PLAY IN THE **COSMOS** THE VIDEOGAME In Mission 16, travel to the galactic halo in search of a power source for your ship's new reactor.

Section Summary

- The basic components of the Milky Way are the galactic disk, stellar halo, and bulge. Most of the observable matter is in the disk, a thin circular distribution of several hundred billion stars with tens of billions of solar masses of dust and gas.
- The disk contains a mix of young, blue, metal-rich stars and older, redder stars. Disk stars (Population I) primarily orbit the galactic center, but some have additional peculiar motions.
- The spherical stellar halo consists of old, metal-poor (Population II) stars, primarily in globular clusters that move on highly elliptical orbits and lack star-forming gas.

Local Group A gravitationally bound collection of galaxies within which the Milky Way resides.

- The bulge at our galaxy's center includes dense star-forming gas and dust in (mostly Population II) stars packed tightly together.
- Large galaxies have smaller satellite galaxies that interact, changing galaxy properties. Gravitational interactions between the Milky Way and its most significant satellites, the Large and Small Magellanic Clouds, have produced tidal streams.

CHECKPOINT Describe the characteristics of each of the three basic components of the Milky Way.

Spiral Arms: Does the Milky Way Have Them?

15.3 One of the mysteries 20th-century astronomers faced was whether our galaxy hosts spiral arms. Unfortunately, it's not easy to resolve details of the galaxy's structure from the inside. "We're sitting in this disk, partway out from the center," says Debra Elmegreen, a galactic astronomer at Vassar College. "When we look up at that band of light in the sky we call the Milky Way, how are we supposed to see spiral structure in that?" Because we always view the Milky Way from the inside, any picture of the whole galaxy (including spiral arms), as shown from the outside, must be a computer rendering or an artist's depiction.

In other spiral galaxies considered to be similar to ours, spiral arms are often observed as part of the disk. The spirals are some of the most beautiful structures the night sky has to offer, and their origin was poorly understood for more than a century. Finding spiral arms from inside the Milky Way meant having to obtain detailed maps of its stars, star clusters, and gas clouds.

Evidence of Spiral Arms

The first evidence of spiral arms in the Milky Way came from studies of young star clusters with bright, massive stars. Astronomers found that the most massive stars (those with spectral classifications O and B) do not occur alone but in groupings. More important, their distribution is not uniform in space. O and B stars often appear in clusters, and when these clusters were mapped, their positions were inferred to be segments of spiral arms, with our Sun as part of the Orion-Cygnus Arm (**Figure 15.13**). Farther from the Milky Way's center, the O and B groupings trace out the Perseus Arm; astronomers find evidence of the Carina-Sagittarius Arm closer to the center. These names are based on the directions on the sky and the constellations in which the O and B stars appear.

The fact that spiral arms can be traced out by the massive stars they contain provides clues about their nature. Because they do not live very long compared with low-mass stars, massive stars have not had time to migrate far from where they formed. Thus, the presence of massive stars in the arms suggests that these stars were born *within* the arms. Whatever process creates the arms must trigger star formation.

Direct evidence of the existence of spiral arms in the Milky Way came with the advent of radio astronomy in the 1950s. Because of their very long wavelengths, radio waves emitted by interstellar gas can propagate unobstructed by dust from one end of the galaxy to the other. Radio astronomers' attempts to map the galaxy were aided enormously by the fact that cold, neutral hydrogen atoms naturally emit at a radio wavelength of 21 centimeters (cm). Since hydrogen is the most abundant material in the cosmos, the 21-cm emission line became the go-to tool for mapping galactic structure (see Chapter 9). Maps of the galaxy in the 21-cm line confirmed the presence of spiral arms in the Milky Way.

The link that the presence of O and B groupings suggests between star formation and spiral arms is strengthened when radio astronomers use emission lines of the carbon

DEBRA ELMEGREEN

"My husband and I were on a train coming down from Alaska," says Debra Elmegreen. "Venus was about to pass in front of the Sun [a rare transit]. We didn't want to miss it, so we got out a piece of cardboard and stuck a pinhole in it. I had binoculars, which we turned backward and held up so we could project the Sun's image." Do-it-yourself enthusiasm has been a hallmark of hers. "I grew up in Indiana, where the skies were very dark." In school, she built her own telescope. Then came local science fairs and the Westinghouse Science Talent Search.

Now president-elect of the International Astronomical Union, Elmegreen admits the details of doing science can wear her down. "Every now and then I have to step outside. Once I look up and remember why I'm doing this, it's all okay."

a.

VIS

b.

Far 3kpc Arm

Carina-Sagittarius Arm

Scutum-Centaurus Arm

Orion-Cygnus Arm

Sun

Near 3kpc Arm

Perseus Arm

Figure 15.13 Spiral Arms in the Milky Way

In our galaxy, young star clusters with bright, massive O and B stars tend to cluster in long strips, now seen as evidence of spiral arms. **a.** The cluster NGC 6231 is found in a region called the Scorpius O and B group of the Carina-Sagittarius Arm. **b.** In this schematic diagram of the Milky Way's spiral arms, the dots show the locations of O and B groupings.

monoxide (CO) molecule to map the galaxy. Recall from Chapter 12 that molecules form only in molecular clouds—both the densest, coldest regions of the interstellar medium and the sites of star formation. CO maps of the Milky Way show that molecular clouds occur not randomly but instead in bands or strips separated by wide cloudless regions. Like O and B stars, molecular clouds appear to trace out spiral arms. Again, star formation and spiral arms appear linked. Any working model of the physics of spiral arms must explain this connection.

Spiral Arms: From Windings to Waves

Tie a string to a stick. Now roll the stick between your fingers in one direction. With each rotation, the string winds around the stick once. Is this a model of what happens with spiral arms? Are the arms composed of material that winds up as the galaxy rotates?

Measurements of the galaxy's rotation rate soon debunked this early idea of the origin of spiral arms. Observations of stellar orbits about the galactic center show that the galaxy's period of rotation varies with distance from the center. At the distance of the Sun, the galaxy rotates about once every 233 million years. Measurements of the age of the oldest stars in the galactic disk show, however, that the galaxy is nearly 13 billion years old. If we divide its total age by the time one orbit takes, we find that the Milky Way has spun completely around approximately 50 times. If the spiral arms were material that winds up like a string tied to a rolling stick, then we would expect to see about as many windings as there have been galactic rotations. In other words, we would expect to see very tightly wound spiral arms. "Over the lifetime of our galaxy there should have been many, many, many windings. And so we ought to see 40 or more, but that is not what we observe in our galaxy or others," says Elmegreen. Observations suggest that the Milky Way hosts up to four well-defined spiral arms.

Astronomers now have strong evidence that spiral arms form by means of **spiral density waves**. Instead of material winding up as the galaxy rotates, the arms are seen to be patterns moving through the disk at a speed different from that of the disk's physical material (stars and interstellar gas). Like sound waves on Earth, spiral density waves are

"Over the lifetime of our galaxy there should have been many, many, many windings. . . . [B]ut that is not what we observe in our galaxy or others."

spiral density wave A wavelike pattern of compression that moves through the physical material of a galaxy's disk. Spiral density waves are seen in spiral arms.

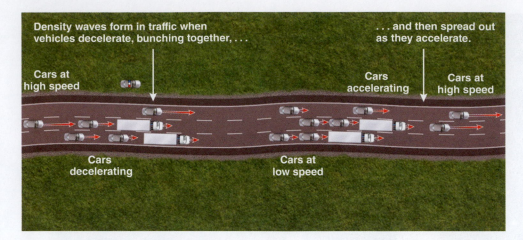

Figure 15.14 Spiral Density Waves

The spiral density waves believed to form a galaxy's spiral arms are a type of compression wave. Gas and dust passing through the spiral arms are compressed, thus increasing the density. Highway traffic is a similar wavelike phenomenon—cars bunch together as they slow down and then spread out as they resume cruising speed.

compression waves—they move through the medium that supports them. For sound waves, the medium is atmospheric gas molecules. For spiral density waves, it's gas molecules of the interstellar gas and the distribution of disk stars. Material piles up and is squeezed inside the "crest" of the compression wave and then spreads out in the wave's trough.

Highway traffic demonstrates this kind of compression wave. Cars initially traveling at high speed are forced to slow down for a time as they move through a congested region. When the traffic eases, the cars can accelerate back up to their cruising speed. To an observer on a helicopter overhead, the congestion looks the same even as the speed of individual cars passing through it continuously changes. The congestion is a stable pattern of compression waves on the highway. The density of cars (number of cars per kilometer) is highest at the peak of the wave and lower on either side of the peak (**Figure 15.14**).

Regarding spiral arms as compression waves explains many of the features observed in the Milky Way, including the association with star formation. The spiral arms rotate more slowly than the stars and gas clouds. Imagine a low-density gas cloud orbiting the galactic center. As the cloud overtakes the spiral arm and is swept into the arm's gravitational well, the gas in the cloud is compressed. Molecules then form in the dense, cold interior regions that are shielded from interstellar ionizing radiation. Thus, if spiral density waves exist, molecular clouds should occur more frequently in spiral arms than elsewhere in the galaxy. That is exactly what astronomers have observed (**Figure 15.15**).

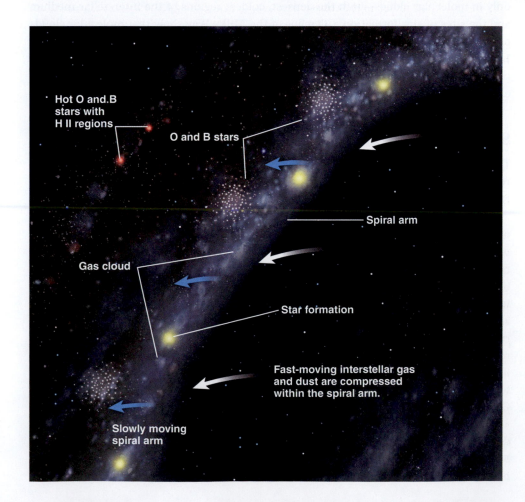

Figure 15.15 Star Formation in Spiral Arms

These regions of high density rotate more slowly than the stars and low-density gas clouds that sweep through them. When a fast-moving, low-density cloud reaches a spiral arm, the gas and dust in the cloud are compressed, setting the stage for star formation. Thus, spiral density waves explain the molecular clouds we see in spiral arms and the star formation within the clouds. O and B groupings occur in or downstream of the spiral arms.

Because massive stars live for only a few million years, we should also expect to see them close to where they formed. The compression produced in spiral arms should trigger the formation of new stars, and therefore we expect to see O and B stars preferentially in the arms. That is again what is observed.

The theory of spiral density waves provides a robust answer to the question, "What are spiral arms?" It also provides a reasonable explanation of how spiral arms form, because the spiral wave pattern is the result of what are called **instabilities** in the galactic disk. Any small disturbance of the smooth pattern of rotation will naturally organize itself into spiral density waves around the disk. These disruptions might come from a globular cluster passing through the disk or even from a collision with a satellite galaxy. Just a small nudge is enough to get the spiral pattern going.

Section Summary

- Evidence of spiral arms in the Milky Way includes the positions of young, hot (O and B) stars, the 21-cm emission line of neutral hydrogen, and the mapping of molecular clouds via carbon monoxide.
- Spiral arms cannot be caused by the winding action of galactic rotation.
- Spiral arms appear to be compression waves that move through the disk, gravitationally attract and compress gas clouds, and were created as a result of instability. This theory is consistent with star formation occurring in the arms.

CHECKPOINT Why can't spiral arms be the result of the winding action of rotation?

The Galactic Center

15.4

Located in the constellation Sagittarius, the galactic center is a place like no other in the Milky Way. It's an environment dense with stars and giant arcs of magnetic fields stretching across parsecs. Most remarkably, it is home to a black hole with millions of times the Sun's mass.

The Galactic Hub

We have seen that the bulge contains a higher density of stars than the galaxy's other components do. More than 10 million stars swarm within the central cubic parsec of the galaxy. Compare this value of 10^7 stars/pc^3 with our local value of 0.14 star/pc^3: the galactic center has a stellar density that is 100 million times greater than that of the Sun's neighborhood.

The galactic center also hosts an enormous quantity of gas, most of which resides in a dense ring of molecular material. This **molecular ring**, or **molecular disk**, orbits the galactic center at a distance of about 2 pc. While some stars are forming in the ring now, astronomers predict that within 200 million years, gravitational forces in the ring may trigger a **starburst**, a rapid increase in star formation, when a substantial amount of the molecular gas is quickly converted into stars. Starbursts are seen in many other galaxies and may be the way the majority of stars in the Universe have formed. The star formation rate in our galaxy now is puny compared with what happens in the brief "pulse" of a starburst.

The nature of stars in the galactic center is surprising. From observations of the rest of the bulge, astronomers expected to find high densities of older, redder stars. Instead, the density of red stars drops in the central regions, forming a kind of "hole" in the distribution of highly evolved stellar objects. Astronomers do not yet understand why there are so few older stars and refer to this problem as the *conundrum of old age*. Equally surprising is the number of young stars, including a large population of massive O and B stars. Since star formation is not prevalent in the rest of the bulge, the presence of these short-lived young stars came as a surprise; astronomers refer to this as the *paradox of youth*. One

INTERACTIVE:
Stellar Orbits around Galactic Black Hole

compression wave A wave of increased density that passes through background material; the density returns to normal after the wave passes.

instability A complex pattern of gas flow resulting from a disruption in an originally smooth flow.

molecular ring or **molecular disk** A dense ring of molecular material orbiting the galactic center.

starburst A rapid conversion of an appreciable fraction of molecular gas into stars.

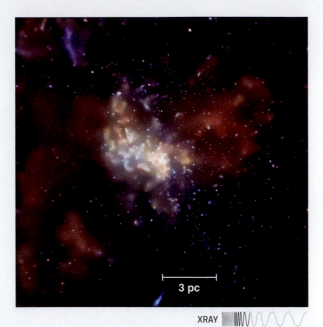

XRAY

Figure 15.16 Black Hole at the Center of the Milky Way
A Chandra X-ray Observatory image of an explosion around the galactic center's supermassive black hole, named Sagittarius A* (Sgr A*). Lobes of gas at 20 million °C (at upper right and lower left) are dozens of light-years in extent.

> "There are lots of stars near that black hole that are going around very rapidly."

explanation for the presence of young stars may be the collapse of a massive gas cloud as it orbited the central black hole.

Along with gas, dust, and stars, the galactic center hosts highly energetic phenomena, some of which are connected with the black hole. Radio observations of the center show vast arcs of magnetic fields strung across the region. Their structure is similar to that of the prominences seen on the solar surface, but the galactic center has no surface on which the fields can anchor. That means these great arcs are generated by large-scale flows of charged gas swirling around and through the center.

Monster at the Center: The Galactic Black Hole

In 1974, two graduate students, Bruce Balick and Robert Brown, were using the National Radio Astronomy Observatory's big antenna in West Virginia for their thesis project. They were looking at a complex region of radio emission in the constellation Sagittarius that includes a supernova remnant. As the radio antenna accumulated signals from the distant region, the two scientists saw a bright object pouring out radio energy off to the side of the supernova remnant.

Balick and Brown dubbed their new object Sagittarius A* (abbreviated Sgr A*). As recognition of its strange properties grew, the object became a subject of intense study, and astronomers slowly built the case that this bright radio source is none other than the Milky Way's own supermassive black hole (SMBH; **Figure 15.16**).

Twenty-eight years later, a swarm of stars that orbit Sgr A* provided the proof that it is a black hole. "There are lots of stars near that black hole that are going around very rapidly," explains Debra Elmegreen. "They allow astronomers to deduce the mass of the black hole, which of course we don't see." Using hyperprecise radio observations, astronomers accurately mapped the microarcsecond paths that these stars make on the sky (**Figure 15.17**). After tracking the stars' complete orbits for 10 years, astronomers used Newton's laws to measure the mass of the central object binding the stars to their orbits. Our galaxy's own SMBH has a mass of more than 4 million Suns: $4.3 \times 10^6 \, M_{\mathrm{Sun}}$! Since many of the stars orbit the central object at distances of just 45 astronomical units (AU), only one interpretation made sense. Only in a black hole can millions of Suns' worth of material be crammed into a volume smaller than our Solar System (see **Going Further 15.1** on p. 405).

The existence of a SMBH in the center of the Milky Way explained a wide range of high-energy phenomena in the region, such as X-ray flares in Sgr A* that come from clouds of gas as they are torn apart by the black hole. Astronomers have also seen evidence of

Figure 15.17 Star Tracking
Simulation of orbits of stars near the Milky Way's supermassive black hole, as compiled from a series of high-resolution observations taken over time. Tracking the stars that orbit Sgr A* confirmed its existence. After constructing the stars' orbits, astronomers used Newtonian physics to calculate the mass of the central object. Given the small size of the region, that mass could only exist in the form of a black hole.

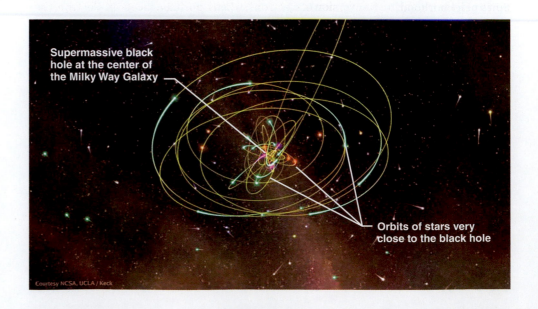

Supermassive black hole at the center of the Milky Way Galaxy

Orbits of stars very close to the black hole

Courtesy NCSA, UCLA / Keck

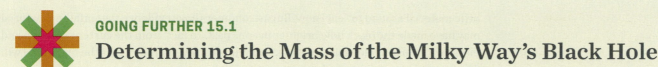

GOING FURTHER 15.1
Determining the Mass of the Milky Way's Black Hole

How did astronomers use the orbits of stars in the Milky Way's center to determine that a supermassive black hole is there? As we've seen several times (for example, in Going Further 3.4), Newton's laws can be used to determine the mass of an object from the orbital properties of its satellite. We can make a rough calculation to see how close we get to astronomers' results from more detailed and sophisticated calculations.

We'll focus on a star named S2 whose orbit around the SMBH is quite elliptical ($e = 0.88$). Its average orbital radius projected on the sky (let's call it apparent size a), as measured in arcseconds, is found to be approximately 0.1″.

First, recall the small-angle formula (see Going Further 2.1), which tells us that, for an object projected on the sky,

$$a = 206,265'' \times \frac{R}{D}$$

$$\text{apparent size of object} = 206,265'' \times \frac{\text{radius of object's orbit}}{\text{object's distance from us}}$$

(Here we use R, the orbital radius, rather than physical size d as in Going Further 2.1.) The distance D to star S2 at the galactic center has been determined to be 8.5 kpc. Since both a and D have been measured, we can rearrange the small-angle formula to find R:

$$R = a \times \frac{D}{206,265''}$$

Now we put in the known values and express D in meters (1 kpc = 3.08×10^{19} m):

$$R = 0.1 \times \frac{8.5 \text{ kpc} \times (3.08 \times 10^{19} \text{ m/kpc})}{206,265''} = 1.3 \times 10^{14} \text{ m}$$

To find the mass of the black hole M_{BH}, we also need to know the period P of the star's orbit; it has been measured to be 15.78 years = 4.976×10^8 seconds. Now we can use Newton's version of Kepler's third law (with G, the gravitational constant; see Chapter 3) to determine M_{BH}:

$$M_{BH} = \frac{4\pi^2}{G} \frac{R^3}{P^2} = \frac{4\pi^2}{6.67 \times 10^{-11} \text{ m}^3\text{kg}^{-1}\text{s}^{-2}} \frac{(1.3 \times 10^{14} \text{ m})^3}{(4.98 \times 10^8 \text{ s})^2}$$

$$= \frac{5.24 \times 10^{36} \text{ kg}}{(2 \times 10^{30} \text{ kg})/M_{Sun}} = 2.62 \times 10^6 \, M_{Sun}$$

This is about half the value found from a more detailed analysis using the details of the star's orbit around the black hole: $4.3 \times 10^6 \, M_{Sun}$. Thus, using Newton's powerful law, we got pretty close to the right answer.

---

higher-energy gamma-ray emission, possibly caused by positrons created outside the black hole's event horizon and funneled into relativistic jets. Recall from Chapter 10 that positrons are the antimatter version of electrons; when a positron meets an electron, they annihilate in a flash of gamma rays. Thus, the detection of such high-energy radiation from Sgr A* implies exotic processes that can occur only near a black hole.

The confirmation of a SMBH at the galaxy's center closed the books on one question, but it raised another that astronomers have yet to answer. Compared with monster black holes in other galaxies, which spew vast amounts of energy into space, our black hole seems relatively inactive. "The region around the black hole at the center of our galaxy does not emit much radiation," says Elmegreen. "Usually, we expect these kinds of systems (a black hole and matter surrounding it in an accretion disk) to be extremely luminous. The ones in other galaxies can be seen from enormous distances because they are so bright. But our Milky Way black hole puts out about one-billionth of their energy. No one is really sure why."

Our SMBH is also smaller than that of many other galaxies. "The really bright ones appear to have masses that are billions of times the mass of the Sun," continues Elmegreen. Because it's got relatively low mass, our black hole in the Milky Way may never have gone through a really luminous phase, as we see in other galaxies."

However, the black hole at the center of the Milky Way may be able to flare a bit brighter every once in a while. In a sense, our black hole may be on a temporary diet with little material around to "eat" now. But astronomers have evidence that infalling material may have made the black hole brighter in eons past. In fact, from the current position and motion of nearby molecular clouds, astronomers can even predict that the black hole will flare up again in a few centuries as new material drops onto it.

Along with the supermassive black hole, astronomers had long suspected the existence of many black holes formed from the death of massive stars. These black holes may have formed in the galactic center or been caught and dragged inwards over time. In 2018, astronomers using years of X-ray data found evidence that the galactic center may host a population of 20,000 stellar-mass black holes.

Section Summary

- The galactic center has the Milky Way's highest density of stars, a dense molecular ring (expected to give rise to a future starburst event), and large-scale magnetic arcs.
- Near the center, there are fewer old stars and more young stars than expected.
- Observations of stellar orbits confirmed the enormous mass (4.3 million M_{Sun}) and small radius of the radio source Sgr A*, consistent with a black hole.
- Sgr A* emits less radiation than do supermassive black holes observed in other galaxies, possibly because of its smaller mass and lack of available infalling material.

CHECKPOINT **a.** What are starburst events, and why are they important? **b.** What evidence indicates the presence of a black hole at the center of the Milky Way?

Dark Matter and the Milky Way

15.5 When you look out into the night sky, you see a lot of empty space punctuated by bright dots. With a telescope, you can see those dots for what they are: stars. You can also see other objects among the stars, such as interstellar clouds and distant galaxies. Using techniques such as spectroscopy, astronomers have been able to probe the light from these distant objects, determining that they are made up of the same kinds of physical constituents as we are—protons, neutrons, electrons, and other subatomic particles. But what if there's stuff out there that we cannot see? Are there other kinds of matter that simply do not interact with light—neither emitting nor absorbing it? How can we tell whether such stuff is out there?

The Footprints of Dark Matter

These were exactly the questions astronomers studying our own galaxy found themselves facing as they began mapping the orbits of stars in the Milky Way's disk. The idea was to measure the mass of the Milky Way by first measuring the orbital speed of stars at ever-larger distances from the galactic center. Recall from Going Further 3.4 (and Going Further 15.1) that Newton's law of gravity can be used to determine the mass M of a central object by measuring the period P and radius R of satellites orbiting that object:

$$M = \frac{4\pi^2 R^3}{GP^2}$$

$$\text{mass of central object} = \frac{4\pi^2 (\text{orbital radius of satellite})^3}{(\text{gravitational constant})(\text{orbital period of satellite})^2}$$

In the 1970s and 1980s, astronomers hoped to use the stars in the outer regions of the disk to find the mass of the Milky Way (M_{MW}). The expectation was that most of the galaxy's mass is close to the center, so measuring ever-more-distant stars would yield better and better estimates of the galaxy's total mass.

How could astronomers be sure that their assumption about the distribution of mass in the galaxy was correct? Recall that most of the mass in our Solar System resides in the Sun. Because gravity gets weaker as the distance from a massive object increases, a graph

"It really comes down to Occam's razor. You try to go for the simplest explanation."

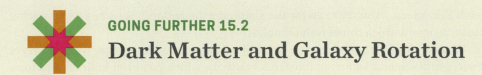

GOING FURTHER 15.2
Dark Matter and Galaxy Rotation

Astronomers measuring how the orbital velocities of stars in the Milky Way's disk change with distance didn't get the result they expected. The idea was simple: For stars orbiting the galactic center, Newton's laws show that only matter inside the star's orbit can affect its motion. That means stars orbiting at large distances from the galactic center should feel the gravitational force from the bulk of the galaxy's mass. The rotation curve of these stars should look a lot like the curve we see for objects orbiting a central mass (the Sun) in the Solar System. How, exactly, do orbital velocities change with distance for objects in the Solar System?

Chapter 3 introduced a compact formula describing the circular orbital velocity V_c of a satellite orbiting a central object of mass M:

$$V_c = \sqrt{GM/R}$$

This equation predicts the shape of the rotation curve for planets in our Solar System. Since neither the mass of the central object (the Sun) nor gravitational constant G changes, the Solar System's rotation curve looks like this:

$$V_c = \frac{C}{\sqrt{R}}$$

Here C (not to be confused with c, the speed of light) is a constant: $C = \sqrt{GM}$.

This equation says that the farther out you go, the slower the orbital (or rotational) speed of the planets should be (see **Figure 15.18**,

"Expected" curve). Astronomers expect that pattern for any system of orbits, once most of the mass has been enclosed within those orbits. But this is not what astronomers get for the Milky Way's rotation curve (see Figure 15.18, "Observed" curve).

The fact that the observed rotation speed of our galaxy never decreases, even at large distances from the center, tells us something very important. If Newton's laws hold at these distances (and we have every reason to assume they do), then there must be more matter in the Milky Way than we can see, and it must be very extensive. Scientists therefore think that the increasing mass with distance comes from a spherical distribution of dark matter that extends far beyond the visible limits of our galaxy.

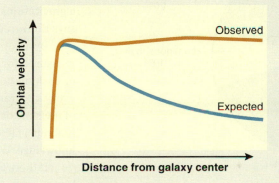

Figure 15.18 Rotation Curves

of orbital speed versus orbital radius (distance from the center)—called the **rotation curve**—within the Solar System shows velocities of orbiting objects (planets, asteroids, and so forth) decreasing with increasing distance from the Sun. Astronomers expected to see the same effect in the disk of the galaxy. Stars at the outer edges of the disk should be orbiting much more slowly than those closer to the center. This is not, however, what observations showed.

Instead of seeing orbital velocities drop with orbital radius, astronomers found that the orbital velocities of stars are constant across most of the disk, even out to the greatest distances they could observe (see Figure 15.18 in **Going Further 15.2**). The galaxy does not behave as expected. The rotation curve makes it look as though the galaxy's matter is far more spread out than astronomers had thought, with enough material at larger radii to tug on the distant stars and keep their velocities high. But no matter how closely astronomers looked by telescope, this extra matter was nowhere to be found.

At first, astronomers were certain they had made a mistake. But as more and more studies showed the same result, it became clear that something truly unexpected was going on. Faced with an anomaly—something that does not make sense in terms of the best theories available—scientists have two choices: they can abandon the theory, or they can imagine that some process is occurring that was not part of the original assumptions of the calculation supporting that theory. "It really comes down to Occam's razor," says

rotation curve The graph of rotation speed as a function of orbital radius for objects orbiting a common center, such as stars orbiting a galaxy.

Figure 15.19
Vera Rubin

The work of Vera Rubin (1928–2016), a pioneer in the field of rotation curves in galaxies, provided evidence of the existence of dark matter.

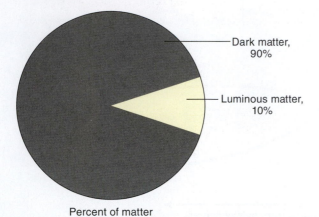

Dark matter, 90%

Luminous matter, 10%

Percent of matter

Figure 15.20 Dark versus Luminous Matter in the Milky Way

Luminous matter—which we can see through its interaction with light—constitutes only 10 percent of all matter in our galaxy. The rest is dark matter, which is observed only through its gravitational effect on luminous matter.

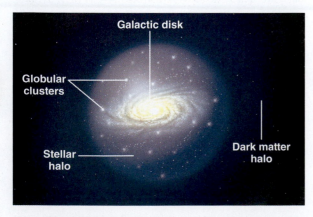

Galactic disk

Globular clusters

Stellar halo

Dark matter halo

Figure 15.21 Dark Matter Halo

The dark matter halo is believed to be a spherical region of decreasing density with an estimated radius of 100 kpc, extending several times past the galactic disk and the stellar halo.

Debra Elmegreen. "You try to go for the simplest explanation." But when we're staring into an anomaly, which choice is the simplest?

Abandoning Newtonian gravity—the first option—meant doing away with 400 years of successful science, including Einstein's relativity, which had been built on Newton's work. For the most part, scientists were unwilling to choose that option to explain the rotation curve anomaly. "The astronomer Vera Rubin was one of the first people to work on rotation curves in galaxies," explains Elmegreen. "A student once asked Vera about the idea of modifying gravity, and she didn't scoff. But she did say we don't need to do that to explain what we see." What Rubin (**Figure 15.19**) meant was that there was a second option to account for the observed anomaly: dark matter.

Rubin and others imagined that there is far more mass in the galaxy than we can directly observe. While this was a remarkable claim, it did less violence to our understanding of the Universe than abandoning Newton's laws would have. The unseen mass would produce no light but would, if distributed differently than luminous matter, produce gravity. In this way, the extra mass would provide the needed gravitational force on the orbiting stars, raising their orbital speed and producing the observed rotation curve. Because Newton's laws allow the distribution of this material around the galaxy to be inferred, astronomers soon embraced the idea, calling the new mass **dark matter**. The label *dark* means simply that they knew nothing of its properties, other than its ability to generate a gravitational force. "The only thing missing is we don't know what this dark matter is," says Elmegreen, "but as an explanation for the rotation curves, it makes sense."

By assuming the existence of dark matter, astronomers could use the velocities they measured for disk stars and work backward. Newton's laws gave them both the amount of dark matter and its distribution in space. The astonishing results suggested that for every kilogram of "normal," luminous stuff in the Milky Way, there are 9 kilograms (kg) of invisible dark material (**Figure 15.20**). As Rubin put it, "The ratio of dark-to-light matter is about a factor of 10. That's probably a good number for the ratio of our ignorance to our knowledge. We're out of kindergarten, but only in about third grade."

If there is so much dark matter in the galaxy, where is it? "Most of the dark mass can't be concentrated at the center of the galaxy," says Elmegreen, explaining that if it were, the rotation curves would still decline with distance. "It's got to be in a big extended halo around the disk of our galaxy in order to account for the rotation curves." This halo is theorized to be a spherical region whose density decreases with distance from the center. The *dark matter halo* does not have a sharp edge, but astronomers estimate its radius (based on its gravitational influence) to be approximately 100 kpc. "That means the dark matter halo extends several times past the stellar halo and the disk," says Elmegreen (**Figure 15.21**; see **Anatomy of a Discovery** on p. 410).

What Is Dark Matter? The Four Forces

If the Milky Way is constructed primarily of dark matter, a nagging little question remains: What *is* dark matter?

Astronomers once thought that dark matter might be made of ordinary matter that is extremely dim—like faded white dwarfs, neutron stars, and black holes—called **massive compact halo objects (MACHOs)**. Astronomers carried out exhaustive searches for MACHOs by looking for so-called *microlensing*, the gravitational focusing of light that would occur as a MACHO passes in front of a more distant star. These microlensing surveys did find some candidates, but far too few to account for the total amount of dark matter. That meant dark matter must be made of particles that do not respond to electromagnetic force (light).

If dark matter doesn't interact with light, how can we determine whether it exists? To answer that question, we first have to ask an even broader one: How does *anything* in the Universe interact with anything else in the Universe? A more explicit way of posing this question is to ask what kinds of forces there are in the Universe. After all, Newton taught us that only through forces does matter accelerate or decelerate. Thus, without forces there can be no interaction among bits of matter.

From their studies of subatomic particles, physicists have constructed an extremely detailed understanding of matter and forces. What they've discovered is that only four **fundamental forces** exist in the cosmos: gravity, the electromagnetic force, the strong nuclear force, and the weak nuclear force (**Table 15.3**). "You need all of the forces to account for structures we see in the Universe on different size scales," says Elmegreen. Gravity holds stars and planets together and keeps your feet on the floor. Electromagnetism holds atoms together and makes chemical reactions possible; you are able to read this book because of light, which is an electromagnetic phenomenon. The strong nuclear force binds protons and neutrons together inside atomic nuclei, and the weak nuclear force drives certain kinds of radioactive decay.

We already know that dark matter must be made of particles that do not interact with the electromagnetic force. These particles also don't respond to the strong nuclear force; if they did, astronomers would have seen traces of their strong interactions. Thus, two of the four forces were ruled out. Dark matter was assumed to feel gravity, so all that was left was the weak nuclear force.

Neutrinos are a classic example of a particle that responds to other particles mainly through the weak nuclear force. If neutrinos were fired into solid lead 1 light-year thick, particle interactions within the lead would stop only half of the neutrinos; the rest would pass through unimpeded. The weak force must be weak indeed. Therefore, dark matter might well be *weakly interacting*. Thus, astronomers and physicists began focusing on **weakly interacting massive particles (WIMPs)** as their best candidate for the mysterious unseen stuff composing most of the Milky Way. (The acronym *WIMPs* was a direct response to the acronym *MACHOs*.)

Physicists are hard at work looking for direct evidence of WIMPs. If dark matter is composed of these kinds of particles, then with large enough detectors and long enough time frames, astronomers should be able to capture at least a few WIMPs in the act of interacting with normal (luminous) matter. Extremely sensitive dark matter detectors must be located far below Earth's surface to shield them from cosmic rays and other forms of "noise" that could confuse the detectors. A number of these detectors exist, such as the Large Underground Xenon detector, 1.5 kilometers (km) below the surface in a former gold mine (**Figure 15.22**). In addition to the underground searches, astrophysicists have launched satellites such as the Fermi Gamma-ray Space Telescope, which is designed to see photons resulting from the very rare interaction of a dark matter particle slamming into a particle of normal stuff.

To date, however, dark matter has not been convincingly detected. Whatever makes up the bulk of the Milky Way Galaxy remains invisible and unknown. We can only infer its properties from the gravitational effects produced on that small fraction of the Milky Way whose matter we can see. The story of dark matter goes beyond galaxies, however; we'll encounter this important actor in the cosmic drama again in the next few chapters.

Table 15.3 Fundamental Forces

Force	Range of Influence	Effects
Weak nuclear force	10^{-18} m	Decay of nuclei and fundamental particles
Strong nuclear force	10^{-15} m	Holds nuclei together
Gravity	Infinite (declines with square of distance)	Creates stars and planets; holds us on Earth's surface
Electromagnetic force	Infinite (diminished by shielding)	Holds atoms together; includes light

Figure 15.22 Large Underground Xenon (LUX) Detector

This detector, installed in a former gold mine in South Dakota and activated in 2013, seeks evidence of dark matter. LUX uses ultrasensitive photon detectors that watch for possible collisions between liquid xenon and WIMPs. The underground location offers protection from surface noise and cosmic rays.

Section Summary

- Attempts to calculate the Milky Way's mass from stellar orbits unexpectedly revealed that stars farther from the galaxy center do not orbit more slowly than stars nearer the center.
- Dark matter, a nonluminous form of mass accounting for roughly 90 percent of the galaxy's mass, may be the cause of the unexpectedly high orbital speeds.
- Dark matter, if it exists, must interact via gravity but not via the electromagnetic force or the strong nuclear force.
- Astronomers think dark matter may be made of particles that feel the weak nuclear force (WIMPs), but no dark matter particles have yet been detected.

CHECKPOINT Why did astronomers feel the need to propose the existence of dark matter?

dark matter Nonluminous mass that is evident only by its gravitational influence.

massive compact halo object (MACHO) A proposed form of dark matter that includes objects such as black holes and extremely faint neutron stars.

fundamental forces The four basic ways matter can interact: gravity, electromagnetic force, the strong nuclear force, and the weak nuclear force.

weakly interacting massive particle (WIMP) A proposed form for dark matter consisting of particles that interact only via gravity and the weak nuclear force.

What holds the Milky Way together?

observations & theory

Newton's and Einstein's theories of gravity are rooted in observations of the ways massive objects move and interact. Using those theories and observing such motions and interactions, we can determine the masses of distant objects. For example, we use rotation speeds of stars orbiting the Milky Way's center to compute the mass of the galaxy.

Artist's conception of Milky Way

Coma Glaxy Cluster

observation & hypothesis

In 1933, **Fritz Zwicky** estimated the total mass of the Coma Cluster (via light emitted by its galaxies) and the mass required to hold the cluster together (via orbital speeds and Newton's law of gravity). The latter mass is several hundred times more than the total mass, implying the presence of "unseen" stuff, dubbed *dunkle Materie* (dark matter). Few colleagues took his findings seriously.

supported by observation

Galactic rotation curves

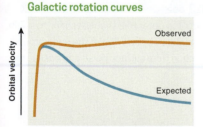

Observed

Expected

Orbital velocity

Distance from galaxy center

In 1970, **Vera Rubin** and Kent Ford measured rotation velocities of star-forming regions in the arms of the Andromeda Galaxy— a spiral galaxy like the Milky Way. Visible mass (stars and dust) is concentrated near Andromeda's center, so astronomers expected objects closest to the center to orbit fastest, as for planets orbiting our Sun. Instead, Rubin and Ford observed a roughly flat rotation curve and, like Zwicky, found evidence of unseen matter.

Andromeda Galaxy

Dark versus luminous matter in Milky Way

Dark matter, 90%

Luminous matter, 10%

Percent of matter

supported by observation

Rubin observed flat rotation curves for additional spirals. By the 1990s, astronomers had confidently measured a flat rotation curve for the Milky Way and inferred that 90 percent of its mass is dark matter. Dark matter is distributed fairly evenly in all directions and extends well beyond our galaxy's visible edges. Despite its abundance, dark matter has yet to be detected directly; its existence is only inferred from its gravitational effects.

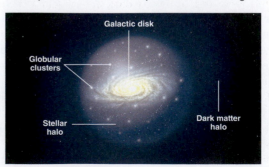

Galactic disk

Globular clusters

Stellar halo

Dark matter halo

Milky Way components

Constructing a Galaxy: Evolution of the Milky Way

15.6 To piece together the history of the galaxy, like good detectives we have to stick with the facts we can gather now and see what they tell us about the past. But which facts matter? Which features of the galaxy shed light on its evolution in time and space? A good start might be to focus on the different populations of stars in the Milky Way and use their locations to piece together the galaxy's story.

Early Stages of Cloud Collapse

To begin, recall from Section 15.2 that there are metal-rich Population I stars and metal-poor Population II stars. The abundance of processed elements (referred to as metals) in any population is a measure of age. Stars with large quantities of metals must have formed relatively recently, so Population I stars are young. Stars with few metals must have formed long ago, so Population II stars are old. The locations of these populations tell us that the disk, which contains mainly Population I stars, is young, while the halo, containing mainly Population II stars, is old. This is our first key piece of the history: the stellar halo must have formed earlier than the disk.

Stellar populations are not the only evidence indicating that the halo is older than the disk; gas and dust offer another clue. The disk is rich with gas and dust, but significant, dense distributions of interstellar material are rare in the stellar halo. Whatever gas and dust existed once in the halo appear to have been converted into stars or dispersed long ago. It also makes sense that we don't see massive stars in the halo: if star formation stopped in the halo billions of years ago, each massive star that formed would have blown up as a supernova long ago, leaving only a population of low-mass, low-metal Population II stars.

How can we turn the relative ages of the halo and disk into a realistic model of the formation of the Milky Way? As usual, everything begins with primordial clouds and gravity. "We can go back to the very early Universe and see how all structures got started," says Debra Elmegreen, who explains that after the Big Bang, gravity began to pull small concentrations of luminous mass together, creating ever-denser structures. Filaments of gas began to form; in those cosmic-scale filaments, the seeds of galaxies originated. "The buildup of galaxies over time occurs primarily through this inflow of gas," says Elmegreen.

Using the oldest stars in the stellar halo, astronomers place the birth of the Milky Way at about 12.6 billion years ago. Our galaxy most likely began as a denser-than-average region within a filament. In this sense, the Milky Way formed as a large, extended gas cloud, similar to the denser regions *within* galaxies that become centers of star formation (see Chapter 12).

Unlike the molecular clouds where star formation occurs, however, the bulk of the proto–Milky Way would have been made of dark matter. From the initial large-scale distribution of dark matter, a region of higher-than-average density occurred via random fluctuations. This dense mix of mostly dark matter and a smaller fraction of luminous matter acted as a seed, allowing local gravity to begin pulling material inward and eventually form into a well-defined dark matter halo. This structure began collapsing in on itself through its own gravitational force.

There is, however, an important distinction between the dark and luminous components of this proto–Milky Way. Because dark matter is so weakly interacting, it cannot experience the friction that slows down particles of normal matter and leads them to clump together. Thus, the galaxy's dark matter collapsed to the point at which it formed an extended halo. Any further collapse was impossible.

The luminous matter suffered no such restriction. Since luminous material experiences all four fundamental forces, the infalling, pregalactic gas could collapse to the point at

> "We can go back to the very early Universe and see how all structures got started."

> "Our galaxy might have eaten many, many tiny neighbors that were a tenth or a hundredth of the mass of the Milky Way."

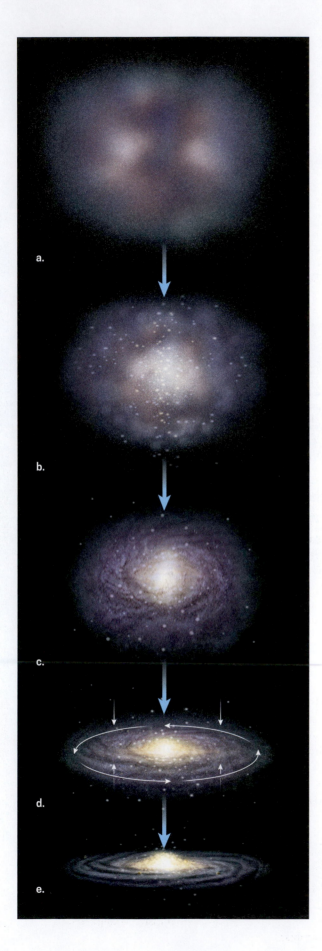

which star formation began. This process didn't occur all at once or all in one place. Instead, early on, as the extended cloud of gas began falling inward, small fragments or cloudlets must have formed. These collapsed on themselves, forming globular clusters. These early clusters were the birthplace of the Population II stars that we see today in the stellar halo.

Formation of the Disk, Bulge, and Black Hole

The gas that was not incorporated into clusters in the infalling protogalaxy continued to fall toward the cloud's center, forming a disk in much the same way that an accretion disk forms around a young star. By conservation of angular momentum, the rotation speeds around the new galaxy's center increased, while collisions within the material at the cloud's midplane flattened the mass distribution into a disk. Although the rotation speeds of the stars and globular clusters that had already formed also would have increased, the mass distributions of those structures would not have been flattened by collisions. Stars rarely collide the way gas clouds do, because the distance between stars is much larger than the typical radius of a star. (Like clouds, stars do feel each other's gravitational force from large distances.) In contrast, the distance separating interstellar clouds can be of the same size scale as the clouds. This is why astronomers expect (and observe) cloud collisions.

The material in the newly formed disk eventually began its own sequences of star formation, leading to the formation of the younger Population I stars. Spiral density waves that were generated in the disk material sped up the process by providing regions of extra compression, where clouds were more likely to collapse into stars.

In the central regions of the newly formed Milky Way, the bulge built up as material that fell straighter inward piled up at the center. The gas in the bulge still had some rotational motion. As central regions grew, material streamed inward, and mass built up there until it collapsed under its weight, forming the central SMBH. In this way, over time, the dark matter halo, stellar halo, disk, and bulge all formed, creating the Milky Way Galaxy we know today (**Figure 15.23**). The SMBH at the galactic center may have formed before the disk did, or it may have built up through inward migration of massive star-forming clumps that contained black holes. The whole process of formation took just 5 billion years.

The preceding narrative fits the classic version of the story of galaxy formation. A more modern update would include the role of smaller galaxies that formed and then collided and merged with the Milky Way. "Our galaxy might have eaten many, many tiny neighbors that were a tenth or a hundredth of the mass of the Milky Way," says Elmegreen. "Those probably became part of the stellar halo."

We must add a very important detail to the Milky Way's story. A few billion years after the galaxy's formation, a small gas cloud about 8 kpc from the galactic center collapsed and formed a very average yellow G-type star. After the chaos of the star's initial planet formation period, eight worlds settled into stable orbits. On the third planet from that star, eventually a blue-green world rich in liquid water, life, and oxygen emerged and then . . . well, you know how that story goes.

Figure 15.23 How the Milky Way Formed
a. Our galaxy began at the intersection of filaments of collapsing gas where an extended cloud, composed of more dark matter than luminous matter, gathered via gravity. **b.** By random fluctuation, a region of higher-than-average density emerged, acting as a seed for gravity to pull material inward and start to form a dark matter halo. **c.** Luminous material collapsed and stars began to form—starting with globular clusters and the stellar halo. **d.** The rest of the gas continued to fall toward the cloud's center, forming a disk via conservation of angular momentum. **e.** Star formation began in the disk, accelerated by compression waves of the spiral arms. In the central regions, the bulge and black hole formed.

Section Summary

- In the early Universe, clouds of gas and dark matter collapsed due to gravity, forming the Milky Way.
- The dark matter remained in a spherical distribution, but some star clusters also formed and created the stellar halo.
- The galactic disk formed when collisions at the rotating protogalactic cloud's midplane caused luminous matter to collapse.
- In central regions, the bulge and black hole formed.

CHECKPOINT Describe the steps involved in the formation of the Milky Way.

CHAPTER SUMMARY

15.1 A Hard Rain: A New Vision of Galactic Studies

Galileo was the first to realize that the Milky Way is made up of thousands of stars. William and Caroline Herschel used star counts to determine that the Milky Way is disk-shaped. Harlow Shapley's observations of globular clusters helped pinpoint the galactic center.

15.2 Anatomy of the Milky Way

Our galaxy's structure includes a disk that contains several hundred billion stars, moving on roughly circular orbits around the galactic center, and up to four spiral arms. The stars are mostly younger Population I stars, which are rich in heavy elements. The stellar halo has a much smaller number of primarily old, red Population II stars. Most of those stars are in globular clusters with highly elliptical paths that are randomly oriented relative to the disk. The bulge is a roughly spherical central region with extremely high densities of primarily old Population II stars.

15.3 Spiral Arms: Does the Milky Way Have Them?

Astronomers used various methods, such as mapping massive O and B stars, to determine whether spiral arms are present. Radio emission from cold, neutral hydrogen via the 21-cm emission line and emission lines from carbon monoxide provided additional evidence. Spiral arms are created by spiral density waves—compression waves that propagate through the disk. Stars then form in the compressed regions.

15.4 The Galactic Center

The extreme density of stars, gas, and obscuring dust in the galactic center makes this region particularly difficult to study. A molecular ring orbiting near the very center may be the site of future starburst activity. The center bears a supermassive black hole, whose existence has been demonstrated by the motions of orbiting stars, and giant arcs of magnetic fields.

15.5 Dark Matter and the Milky Way

From its gravitational influence on the velocities of orbiting stars, astronomers deduce that an unseen halo of dark matter surrounds the galaxy and represents the majority of the galaxy's mass. Dark matter responds to gravity but does not interact with the electromagnetic force or the strong nuclear force. Astronomers are conducting experiments to see whether dark matter particles respond to the weak nuclear force.

15.6 Constructing a Galaxy: Evolution of the Milky Way

The structure and motions of the galaxy provide insight into its formation. The stellar halo formed from an initial large sphere of gas and dust. Infalling luminous matter formed the disk and bulge. The dark matter did not collapse, instead remaining as an extended halo. First generations of stars in the disk were metal-poor (Population II), but later generations were increasingly metal-rich (Population I). The Milky Way may have consumed smaller galaxies through collisions and mergers.

QUESTIONS AND PROBLEMS

Narrow It Down: Multiple-Choice Questions

1. Using star counts, William and Caroline Herschel concluded that
   a. the Milky Way is made of stars.
   b. stars differ in mass.
   c. stars are not evenly distributed in our galaxy.
   d. the Milky Way Galaxy is disk-shaped.
   e. there are fewer massive stars than smaller stars.

2. What role did interstellar dust play in the quest to determine the shape and size of our galaxy?
   a. It magnified the light from stars, making them appear closer.
   b. It obscured some regions and made others appear dimmer and redder.
   c. It outlined the Milky Way's structure.
   d. It played no role; dust is not important on a galactic scale.
   e. It made stars appear bluer.

3. Globular clusters
   a. orbit in the plane of the galactic disk.
   b. belong to Population I.
   c. contain primarily hot, young stars.
   d. were discovered after the dimensions of the galaxy were known.
   e. have low metallicity.

4. With which of the four forces is dark matter known *not* to interact? Choose all that apply.
   a. gravity
   b. electromagnetism
   c. weak nuclear force
   d. strong nuclear force
   e. all

5. The galactic disk is *not* characterized by
   a. spiral arms.
   b. a flat shape.
   c. stars, gas, and dust.
   d. mostly old, red stars.
   e. massive, young stars.

6. In which component do gas and stars bob slightly above and below the midplane of the Milky Way?
   a. disk
   b. bulge
   c. stellar halo
   d. dark matter halo
   e. central black hole

7. Which of the following statements is true?
   a. Population I stars are lower in metallicity than Population II stars.
   b. Population I stars are found in the halo.
   c. Population I stars are older than Population II stars.
   d. Population II stars may be supermassive.
   e. Population II stars are older than Population I stars.

8. A 5.0-M_{Sun} red giant star with high metallicity
   a. may belong to Population III.
   b. is composed of primordial material.
   c. is probably a member of Population I.
   d. is likely to be found in a globular cluster.
   e. could not be located on a spiral arm.

9. Which of the following is *not* evidence that globular clusters formed very early in the history of the Milky Way?
   a. All of them formed at the same time.
   b. They have low metallicity.
   c. Some have young, blue stars.
   d. Some have retrograde galactic orbits.
   e. They are nearly as old as the Universe.

10. What can be seen through Baade's Window?
    a. the nearest large spiral galaxy
    b. a spiral arm
    c. the central black hole
    d. the galactic center
    e. a globular cluster

11. Which statement about dark matter is true?
    a. It is denser toward the outer regions of the Milky Way.
    b. It represents 10 percent of total galactic matter.
    c. It does not respond to gravity.
    d. It ends at the average orbital radius of the globular clusters.
    e. Its nature is not yet understood.

12. Which of the following is *not* evidence of spiral arms in the Milky Way?
    a. the path of the Sun on the sky
    b. patterns of O and B stars
    c. CO maps of the galaxy
    d. the distribution of hydrogen seen via the 21-cm emission line
    e. the appearance of spiral arms in similar galaxies

13. Our Solar System orbits the center of the Milky Way with approximately what period?
    a. 1 year
    b. 24 hours
    c. 200 million years
    d. 4.5 billion years
    e. 13 billion years

14. If there were no dark matter, the orbital period of the Sun around the galactic center would be
    a. shorter.
    b. longer.
    c. the same.
    d. impossible to predict.
    e. the same as that of every other star in the galaxy.

15. A starburst is
    a. the death of a massive star.
    b. a collision between two stars that annihilates both.
    c. a rapid increase in star formation.
    d. the merger of many stars into a black hole.
    e. the implosion of a Sun-like star.

16. Evidence of a central black hole in our galaxy includes all of the following *except*
    a. strong radio emission.
    b. stars orbiting rapidly around the potential black hole's position.
    c. the scarcity of brown dwarfs.
    d. a large apparent concentration of mass in a small volume of space.
    e. X-ray flares.

17. The age of the Milky Way Galaxy is about 13 billion years, but our Solar System is less than 5 billion years old. If the Sun had been one of the first stars to form, how different would it, and its evolution, have been? Choose all that apply.
    a. The Sun would have lower metallicity.
    b. The Sun would already have reached the end of its fusion lifetime.
    c. The Sun might be in the bulge.
    d. The Sun might be in a globular cluster.
    e. The Sun would be a Population I star.

18. Compare two Milky Way disk stars—one 3 billion years old and one 5 billion years old. The younger one is more likely to
    a. be farther from the midplane of the Milky Way.
    b. belong to Population II than to Population I.
    c. have lower metallicity.
    d. reach smaller heights as it oscillates above and below the disk.
    e. be redder.

19. Which statement(s) about galaxy groups is/are true? Choose all that apply.
    a. The largest galaxies tend to have satellite galaxies.
    b. Interactions can occur between massive galaxies and their satellites, as well as between satellite galaxies.
    c. Galaxy mergers are rare in galactic evolution.
    d. Tidal streams can result from galaxy interactions.
    e. All galaxies within a group tend to lie in the same plane.

20. True/False: Stars near the galactic center of the Milky Way, but not in the molecular ring, are generally Population I stars.

To the Point: Qualitative and Discussion Questions

21. What observations and techniques did each of the following astronomers use to study the galaxy?
    a. Galileo
    b. the Herschels
    c. Kapteyn
    d. Shapley

22. How did Kapteyn's estimate of the galaxy's size compare with the estimate made by today's astronomers? Why did his estimate differ from modern ones?

23. In what ways did dust inhibit early astronomers from obtaining an accurate picture of our galaxy?

24. How is dark matter believed to be similar to luminous matter, and how does it differ? How do we know this?

25. Where in the Milky Way is our Solar System located?

26. What are peculiar motions of disk stars?

27. Describe the differences in age, color, metallicity, and location of Population I and Population II stars. Why would we expect the age, color, and metallicity of stars to be interrelated?

28. How does the arrangement of globular clusters in our galaxy provide a clue to the timing of their formation?

29. How did the study of the galactic rotation curve help lead to the concept of dark matter?

30. How does the study of dark matter's interaction with the four fundamental forces provide evidence of the nature of dark matter?

31. What would our galaxy look like if spiral arms were solely the result of winding due to rotation?

32. How do O and B groupings trace out the locations of spiral arms? What other methods have astronomers used to verify the presence of spiral arms?

33. Describe how star orbits near the center of the galaxy provide evidence of the existence of a SMBH.

34. "You can't see the forest for the trees." How is this saying relevant to the study of the Milky Way? What are some other situations you've experienced when it has been difficult to discern the big picture because you were immersed in it?

35. The nomenclature for Populations I and II is somewhat counterintuitive, since Population II consists of stars that were formed earlier. Describe at least one other example of astronomical nomenclature you have learned that is counterintuitive.

Going Further: Quantitative Questions

36. The average orbital radius of a star around a galactic black hole has an angular size of 0.25″ when observed from a distance of 6.2 kpc. What is the orbital radius in kilometers? In astronomical units?

37. The orbital radius of a star orbiting Sgr A* is 3.45×10^{11} km. Observed from a distance of 7.46 kpc, what is its angular size of the orbit in arcseconds?

38. Calculate the mass of a black hole, in solar masses, based on the orbit of a star with period 7.8 years and orbital radius 2.8×10^{11} km.

39. A star orbiting Sgr A* has a semimajor axis of 1.4×10^{15} m and a semiminor axis of 7.6×10^{14} m. If the SMBH has mass $4.3 \times 10^6\ M_{Sun}$, find the star's orbital period, in years.

40. Calculate the orbital speed, in kilometers per second, of a star orbiting 15,000 pc from the center of a galaxy whose mass is 130 billion M_{Sun}.

41. Star A orbits at 5,000 pc from the Milky Way's center; Star B orbits at 7,500 pc from the center. In the absence of dark matter, what would be the expected ratio of A's speed to that of B? Assume that all of the luminous matter resides in the galactic center.

42. How many times faster would a star's orbital speed be at the Sun's position, compared with that of a star orbiting at 3.7 times the distance from the galactic center? Assume that dark matter is not a factor and that all of the mass of the luminous matter resides in the galactic center.

43. A star orbits in its galaxy at the same orbital radius as the Sun, but the mass of the galaxy is 2.6 times that of the Milky Way Galaxy. Assuming the ratio of the galactic mass lying within its orbit is the same as that of the Sun, what would be the ratio of the star's orbital speed to the Sun's?

44. A globular cluster has an orbital radius of 25,000 pc. Using a galactic mass of $1.0 \times 10^{12}\ M_{Sun}$ for both luminous and dark matter combined (and assuming that all of the mass lies within the cluster's orbit), what is the orbital speed of the globular cluster, in kilometers per second?

45. What is the Schwarzschild radius, in kilometers, of a black hole whose mass is $6.7 \times 10^7\ M_{Sun}$? (See Going Further 14.2.)

"BUT IN ALL THAT TIME, THE NATURE OF GALAXIES HADN'T BEEN RESOLVED"

16

A Universe of Galaxies

BEYOND THE MILKY WAY

The Great Debate and the Scale of the Universe

16.1 The April night was blustery in Washington, DC, as astronomers in spats and hats filed into the auditorium of the Smithsonian Institution's US National Museum (now known as the National Museum of Natural History). The year was 1920; the United States had recently emerged triumphant from World War I. The astronomers were concerned with a different kind of conflict: a contentious debate over the nature of the cosmos. The fate of galaxies, or at least our view of them, hung in the balance.

The event was called the Great Debate. For more than 100 years, astronomers had been cataloging pinwheel-shaped objects called spiral nebulas. "People had observed these so-called spiral nebulas in the sky for decades," explains David Law, a galactic astronomer at the University of Toronto. "But in all that time, their nature hadn't been resolved. That is what the Great Debate was really all about."

Some astronomers suggested that these spiral clouds are vast, complete systems of stars similar to our Milky Way—but located at what were then inconceivable distances from us. Spiral nebulas, they claimed, are island universes strung throughout the vast emptiness of space. Other scientists maintained that the Milky Way is all there is to the Universe. The debate pitted two eminent astronomers against each other. Arguing against the "island universe theory" was Harlow Shapley, then at California's Mount Wilson Observatory and a leader in Milky Way research. He claimed the Universe is a single "enormous all-comprehending galactic system" within the Milky Way. Heber Curtis, from Lick Observatory in California, defended the island universe theory, presenting multiple lines of evidence that spiral nebulas are distant, separate galaxies **(Figure 16.1)**.

Each researcher fought hard for his vision of the Universe, but the debate that April night ended without a knockout. "No one really won the Great Debate," explains Law. "Each side had points in its favor which ultimately turned out to be true. But each side had biases and incorrect assumptions." What makes the Great Debate so remarkable is

← A dramatic image of galaxy M51 (Whirlpool Galaxy), interacting with its companion, galaxy NGC 5195 (at top right). Atomic hydrogen, marked by jewel-like dots along the spiral arms, identifies many H II regions, where massive stars are forming.

a.   b.

Figure 16.1 The Great Debate
a. Harlow Shapley and **b.** Heber Curtis were combatants in the 1920 Great Debate. Their evidence focused on the nature and distance of spiral nebulas, but the deeper issue was the scale of the Universe.

that as late as 1920, when airplanes and radios were becoming commonplace, science had yet to determine the nature of our own galaxy or to prove the existence of others. The true dimensions of galactic space had to wait a few more years to be resolved.

The Island Universe Theory Confirmed

After serving in World War I and then studying at Oxford University on a Rhodes Scholarship, Edwin Hubble arrived at Mount Wilson Observatory in 1919, just a year before the Great Debate (**Figure 16.2a**). Hubble, a star athlete and boxer who once knocked out the German heavyweight champion, favored the island universe theory and mounted his own attack on the issue.

Mount Wilson was the home of the 100-inch Hooker Telescope, the largest astronomical instrument in the world at the time (**Figure 16.2b**). With it, Hubble could make out individual stars in the larger spiral nebulas. On October 5, 1923, Hubble spent the night searching the great spiral nebula called Andromeda. The next day he compared the night's work with previous observations. To his surprise and delight, he found the Great Debate's resolution staring at him on fresh photographic plates (**Figure 16.3**).

Hubble found an all-important Cepheid variable star in Andromeda. Recall from Chapter 11 that Cepheids are standard candles that brighten and dim in a regular pattern. "There is the famous period-luminosity relationship for Cepheids," says Law. "Just by observing how long the star takes to cycle from dim to bright and dim again, you can read off its intrinsic luminosity. Compare that with how bright it appears on the sky, and a distance measurement just falls right out." Through his discovery, Hubble had found astronomical gold. After taking more images over subsequent nights, Hubble determined the Cepheid's period and thus its intrinsic energy output (luminosity). Then Hubble calculated the distance to the newly discovered star and swept away 100 years of debate. "Isolating that Cepheid in Andromeda," says Law, "is what allowed him to end the Great Debate."

Using the Cepheid variable's period, Hubble calculated that the Andromeda Galaxy is almost a million light-years from Earth. (We now know it to be about 2.5 million light-years, or 770,000 parsecs, away.) This was far larger than any other astronomer's estimate for the Milky Way's outer boundary and implied that the spiral nebula in Andromeda could not be located within the Milky Way. It was, without doubt, a spiral *galaxy*, not a spiral nebula. Most important, Hubble's result showed that the Universe is far larger than anyone had imagined.

grand design spiral galaxy A spiral galaxy with well-defined spiral arms that may be traced all the way from the bulge to the outer regions of the disk.

a.   b.

Figure 16.2 Edwin Hubble and the Hooker Telescope
a. Edwin Hubble (1889–1953) used **b.** the 100-inch Hooker Telescope at Mount Wilson Observatory to resolve the Great Debate.

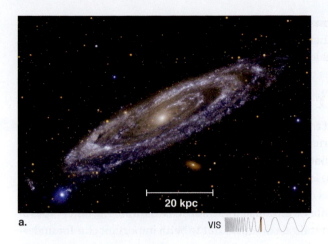

Figure 16.3 Confirmation of the Island Theory
a. Hubble observed the Andromeda Galaxy (then known only as a spiral nebula), establishing its true distance by identifying Cepheid variables in it via **b.** photographic plates. Given the great distance between the Milky Way and its neighbor, Hubble identified Andromeda as another galaxy, similar to our own.

By then Shapley was at the Harvard College Observatory, but he had not given up on his hypothesis that spiral nebulas are part of the Milky Way. On receiving news of Hubble's discovery, the defeated Shapley grimly told a student, "Here is the letter that has destroyed my Universe."

Section Summary

- In the Great Debate, scientists expressed conflicting views about the distance to "spiral nebulas" and its implications for the size of the Universe.
- Edwin Hubble used his observations of Cepheid variables to resolve the debate and determine that ours is one of many galaxies.

CHECKPOINT   Describe the positions that Shapley and Curtis took in the Great Debate.

Galactic Zoology

16.2   With Hubble's discovery, galaxy studies began in earnest. Astronomers started building extensive catalogs of galaxy observations—the kind of pure data gathering that marks the first step in any new science. "There were all these different kinds of galaxies, and no one really knew what was going on," says David Law. "So astronomers started by dividing the galaxies into bins. They created classifications based on the galaxies' observed properties and hoped those categories would expose underlying patterns in the physics of galaxy evolution and structure."

In galaxy classification, Hubble again led the pack, helping to create the basic classification scheme still used today. Hubble sorted galaxies into four main types: spiral, barred spiral, elliptical, and irregular.

Pinwheels on the Sky: Spiral Galaxies

"The main thing that sets spirals apart," laughs Law, "is how pretty they look in Hubble Space Telescope images." But spiral galaxies are not important for their looks alone; their beauty belies a wealth of physics. The Milky Way is a spiral galaxy, and studies of it indicate that the basic (normal matter) components of a spiral galaxy are a disk, a bulge, and a stellar halo with globular clusters (see Chapter 15).

Fortuitously, one of the closest large galaxies to us, Andromeda (M31), is also a spiral galaxy, so it provides us with an easily observed model for this class of galaxy. Both the Milky Way and Andromeda are called **grand design spiral galaxies** because they contain well-defined spiral arms—sites of star formation. Other examples include the Whirlpool Galaxy (M51) and M81 (**Figure 16.4**). Law reminds us that whenever a galaxy has spiral

"There were all these different kinds of galaxies, and no one really knew what was going on. So astronomers started by dividing [them] into bins."

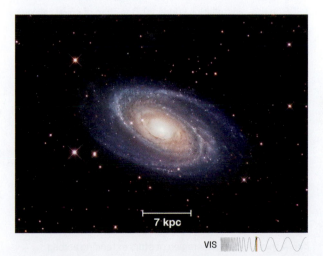

Figure 16.4 Grand Design Spiral Galaxies
Prominent spiral arms characterize the classic grand design spiral galaxies, such as M81 (in the constellation Ursa Major).

VIS

Figure 16.5 Flocculent Spiral Galaxies
Patchy, poorly organized spiral arms typify flocculent spiral galaxies, such as NGC 4414 (in the constellation Coma Berenices).

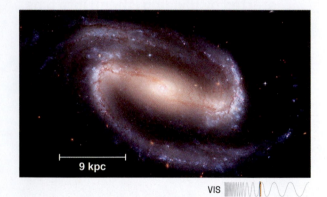

VIS

Figure 16.6 Barred Spiral Galaxies
A central bar distinguishes barred spiral galaxies, such as NGC 1300 (in the constellation Eridanus). One-third of spiral galaxies show a bar, which plays a major role in star formation.

flocculent spiral galaxy A spiral galaxy with poorly defined spiral arms.

barred spiral galaxy A spiral galaxy with a central rectangular bar.

bar A thick, rectangular region of stars extending across the center of some spiral galaxies.

elliptical galaxy A galaxy that appears elliptical on the sky. They are generally triaxial, meaning that the width is different in each dimension.

arms, how tightly bound the arms appear is a fundamental characteristic revealing the physics of spiral density waves that occur in the galactic disk (Chapter 15).

Not all spiral galaxies have such well-defined arms. **Flocculent spiral galaxies**, such as NGC 4414, show only patches of spiral arms in the disk (**Figure 16.5**). "In a flocculent spiral, sometimes it's difficult to see the spiral pattern at all," explains Law. "The spiral pattern is often fairly loosely wound, and it's broken up into lots of individual pieces. It's a very chunky spiral." In some cases, no spiral arms are seen—the galactic disk appears entirely smooth, as in the Sombrero Galaxy (M104).

Spiral galaxies come in a wide range of sizes, with disk diameters as small as 5 kiloparsecs (kpc) and as large as 100 kpc. (The Milky Way has a diameter of 30 kpc.) In mass they range from 10^9 solar masses (M_{Sun}) to 10^{12} M_{Sun}. Today, "dwarf" spiral galaxies (very small, low-mass versions) are quite rare, though dwarfs are common among other types of galaxies. The lack of true dwarf spiral galaxies turns out to be an important clue for understanding the history of all galaxy classes.

Pinwheels with a Twist: Barred Spiral Galaxies

"When you look at the central regions of a spiral galaxy," says Law, "you basically see one of two things. Either the spiral arms extend pretty much directly into the center of the galaxy where they end in the bulge, or the arms don't curve smoothly into the very center but join onto a linear feature that cuts across the central bulge." The spiral galaxies we have discussed thus far are the first type; **barred spiral galaxies** are the second type. They appear to have a central rectangular **bar** composed of stars. Bars are bright because the rate of star formation within them is quite high, as in the spiral arms themselves (**Figure 16.6**). Fully 33 percent of all galaxies classified as spiral include a bar.

The fact that the spiral arms in a barred spiral galaxy begin at the bar, as Law describes, is an important clue to the bars' origin. This strong morphological link tells astronomers that bars form from the same kinds of instabilities that form spiral arms. (*Morphology* is the study of structure.) That means bars are also density waves—moving patterns of high density that flow through the background stars and gas. "The bars are a feature reflecting some instability within the galaxy itself," says Law. "A bar will evolve over time in terms of its appearance. They are often a fairly transient feature. In computer simulations, they may stick around for many billions of years and then disappear and then reappear."

Those computer simulations show that bars form in spiral galaxies when the orbits of stars near the bulge are perturbed enough that their paths around the galactic center become more elliptical. The bar pattern thus represents an increase in the local density of stars and gas as the entire galaxy rotates (**Figure 16.7**). The extra density in a bar can, in

Figure 16.7 Bar Formation
A computer simulation of the sequence of changes leading to the formation of a bar across the center of a spiral galaxy. Bars appear to form because of instabilities in stellar orbits. The Milky Way may host such a bar.

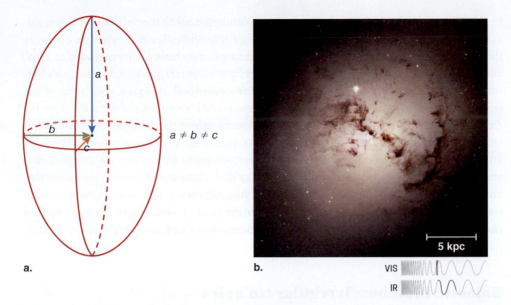

a.

b.

VIS
IR

Figure 16.8 Elliptical Galaxies
Depending on their orientation relative to our line of sight, elliptical galaxies may appear to be spheroidal or elliptical, but many are **a.** triaxial, meaning no two axes are the same length. **b.** NGC 1316 is a triaxial galaxy.

turn, pull interstellar gas from larger distances inward. As the bar "funnels" gas inward, conditions become ripe for enhanced star formation.

Footballs in the Sky: Elliptical Galaxies

As you would expect, **elliptical galaxies** are elliptical in shape. "Elliptical galaxies have much less in the way of visible structure in contrast to the spiral galaxies," says Law. "They don't have the dominant disk with all of its star formation." Elliptical galaxies appear on the sky as fairly smooth distributions of stars. One axis tends to be longer than the other, and the density of stars increases from edge to center in these galaxies. Detailed studies of elliptical galaxies show that they are *triaxial*, meaning that the size of each of the three perpendicular dimensions can be different (**Figure 16.8**). A football, for example, is biaxial, but a lima bean—which is flattened by different amounts in each dimension—is triaxial.

Elliptical galaxies show very little interstellar medium. Radio studies using the 21-cm emission line of neutral hydrogen, for example, find almost no H I clouds in elliptical galaxies. While spiral galaxies host active star formation, in elliptical galaxies stellar nurseries are practically nonexistent. "In general, ellipticals seem devoid of gas," explains Law. "That means they don't have much in the way of newly formed stars either." Most of the stars in elliptical galaxies, therefore, are old and red, whereas spiral galaxies have both red, older stars and blue, young, massive stars.

Most elliptical galaxies do exhibit a halo of low-density, high-temperature gas, however. Space-based X-ray observations using instruments such as the Chandra space telescope show that these extended halos (**Figure 16.9**) are filled with gas as hot as 10^7 kelvins (K). The presence of hot halos indicates that elliptical galaxies must have experienced violent events in their past that could drive the gas to such high temperatures. According to many scientists, it also indicates that collisions with other galaxies must be an important part of the evolution of elliptical galaxies.

Stellar motions within elliptical galaxies are also quite different from those in their spiral cousins. "Instead of stars orbiting on nice, circular orbits in a disk," says Law, "the stars in an elliptical galaxy are fairly chaotic. They are all plunging toward the galaxy center with randomly oriented directions." We saw in Chapter 15 that the globular clusters in the Milky Way move on similar plunging orbits. In fact, globular clusters and elliptical galaxies show a number of important similarities, including the old, red stellar population and the lack of an interstellar medium. Elliptical galaxies also tend to contain more globular clusters than spiral galaxies do.

> "Elliptical galaxies have much less in the way of visible structure in contrast to the spiral galaxies."

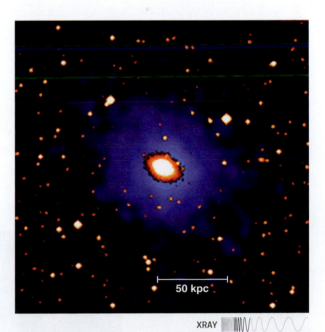

XRAY

Figure 16.9 Galactic Halos
A Chandra space telescope image of the large elliptical galaxy NGC 6482 (the bright central region) at a distance of 55 million parsecs (pc). The galaxy is embedded in a cloud of hot, X-ray-emitting gas extending multiple times the diameter of the core galaxy.

The range of sizes and masses is much wider for elliptical galaxies than for spiral galaxies. The smallest elliptical galaxies, called *dwarf elliptical galaxies*, are not much larger than the largest globular clusters. Dwarf elliptical galaxies have an average size of 1 kpc and a mass of 10^6 M_{Sun}. They represent the majority of elliptical galaxies. The largest elliptical galaxies in size and mass are known as **giant elliptical galaxies**, which live at the center of dense galaxy clusters (explored in Chapter 17). Giant elliptical galaxies have a mass of at least 10^{13} M_{Sun} and may extend across hundreds of kiloparsecs.

According to Law, there is a lesson to be learned in the difference between the mass and size ranges of spiral and elliptical galaxies. "The spiral galaxies are more limited in size," says Law. "They tend to be on the middle and lower ends of the mass spectrum. You can certainly have reasonably high-mass spiral galaxies, but these are not as massive as the most massive elliptical galaxies. The range of masses and sizes shows us that the different kinds of galaxies must form in different ways and have different star formation histories."

Galactic Blotches: Irregular Galaxies

Spiral, barred spiral, and elliptical galaxies represent relatively clean morphological "bins" into which astronomers can categorize galaxies. But some galaxies don't fit into any of these categories; called **irregular galaxies**, they have amorphous shapes. "They are, essentially, train wrecks of galaxies," says Law. At the simplest level, irregular galaxies look like messy blobs with no easily categorized geometry. Taken a step further, though, astronomers see enough pattern to distinguish two types: the truly amorphous Irr I (pronounced "Irregular One") galaxies and the Irr II (pronounced "Irregular Two") galaxies, which seem to retain some features of a disk.

In their range of sizes and masses, irregular galaxies fall between dwarf elliptical and normal spiral galaxies. The smallest irregular galaxies have masses of 10^7 M_{Sun}; the largest, 10^{10} M_{Sun}. As with elliptical galaxies, dwarf irregular galaxies represent the bulk of the class: most irregular galaxies have low mass and small diameter. In addition, the rate of star formation varies greatly among irregular galaxies.

"The irregulars are a population that's reasonably uncommon in the modern Universe," says Law. Just as important, many irregular galaxies are companions to other, larger galaxies. Because they are typically low in mass, these irregular galaxies can be considered satellites of the larger galaxies. The best-known irregular galaxies are, in fact, satellites of our Milky Way. The Large Magellanic Cloud (LMC) and Small Magellanic Cloud (SMC) are a pair of dwarf irregular galaxies (type Irr I) that can be seen from Earth's Southern Hemisphere (**Figure 16.10**). The LMC contains 10^{10} M_{Sun} of normal (nondark) matter and is about 4.3 kpc across, while the SMC contains only 7×10^9 M_{Sun} and is about 2 kpc in diameter. Stars continue to form in both systems. The fact that large galaxies can

"They are, essentially, train wrecks of galaxies."

Figure 16.10
Irregular Galaxies

Irregular galaxies have amorphous shapes and lack common galactic structures. They are typically small and are often found as satellites of spiral or elliptical galaxies. **a.** The Large Magellanic Cloud. **b.** The Small Magellanic Cloud. **c.** NGC 1427.

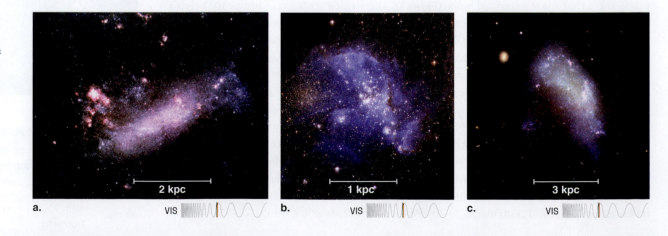

a.   2 kpc   VIS

b.   1 kpc   VIS

c.   3 kpc   VIS

Table 16.1   Galaxy Types

Type	Shape	Structures	Star Formation	Special Classes	"Tuning Fork" Designations	Notes
Spiral	Disk	• Disk • Bulge • Stellar halo with globular clusters • Most have spiral arms	Prevalent in the spiral arms	• Flocculent • Grand design • Lenticular	Sa, Sb, Sc	Dwarfs are rare
Barred spiral	Disk	• Disk • Bulge • Bar • Stellar halo with globular clusters • Most have spiral arms	Prevalent in the spiral arms and bar	• Flocculent • Grand design • Lenticular	SBa, SBb, SBc	Dwarfs are rare
Elliptical	Spheroid to triaxial ellipsoid	Stellar halo with globular clusters	Very limited	• Dwarf • Giant • Lenticular	E0 through E7	• Giants are found at the center of some galaxy clusters • Widest range of radii
Irregular	Amorphous	May have some disklike features	Varies; active in some	• Irregular I • Irregular II	Irr I, Irr II	Often found as satellites to larger galaxies

have smaller galaxies such as these trapped within their orbital influence is central to our understanding of galactic evolution.

The key features of each of the galaxy types are summarized in **Table 16.1**.

The Hubble Tuning-Fork Diagram

After establishing that galaxies can be sorted into four basic categories (spiral, barred spiral, elliptical, and irregular), the next step for astronomers was to try to find an underlying relationship among the categories. Recall how the stellar spectral classification OBAF-GKM was given physical meaning once stars were plotted on the Hertzsprung-Russell (HR) diagram. The diagram revealed how stellar temperature and luminosity combine (via

giant elliptical galaxy The largest, most massive type of elliptical galaxy.

irregular galaxy An amorphous galaxy lacking common structures.

Figure 16.11 Hubble Tuning-Fork Diagram
Categorizing the galaxies he had observed, Edwin Hubble developed a classification scheme in the shape of a tuning fork. Hubble assumed that his sequence, from left to right, represented the evolution of galaxies from young to old. That assumption has been disproved, but his classification scheme remains useful.

> "Once the [galaxy types] were defined, someone needed to work out the relationship between them. That someone, again, was Edwin Hubble."

lenticular galaxy A galaxy with a bulge, limited evidence of a disk, and no spiral arms.

Hubble tuning-fork diagram Edwin Hubble's scheme for representing galaxy types along a progression from highly elliptical to spiral or barred spiral.

megaparsec (Mpc) A unit of distance equal to 1 million parsecs.

the deeper principles of stellar structure and nuclear burning) to order stars by mass on the main sequence. "Galaxies were no different," says Law. "Once the categories (the galaxy types) were defined, someone needed to work out the relationship between them." That someone, again, was Edwin Hubble.

After providing the basic galaxy characterizations, Hubble refined the categories and organized them in a way that he believed brought order to the galaxy classification. The fruit of his effort did enable astronomers to make significant progress in understanding galaxies—but not for the reasons Hubble imagined.

For Hubble, the shapes of elliptical galaxies could be lined up on a spectrum, from circular to extremely elliptical (cigar-shaped). He designated the roundest elliptical galaxies E0. The more elliptical the galaxy, the higher the integer following the letter *E*. Hubble labeled the cigar-shaped galaxies E7.

Hubble needed to account for two properties in spiral galaxies. The first property was the size of the central bulge; the second was how tightly the spiral arms wrap around the galaxy. Luckily, the size of the bulge appeared to be correlated with the winding of the arms, enabling Hubble to use a single classification scheme for both. Hubble designated spiral galaxies with the largest bulges and the tightest, most circularly wrapped spiral arms as *Sa galaxies*. In contrast, *Sc galaxies* have the smallest bulges and spiral arms that are fairly open, and *Sb galaxies* have intermediate bulges and intermediate spiral arm windings.

Hubble used the label *SB* to denote a spiral galaxy with the presence of a bar. He distinguished three subcategories of barred spiral galaxies by the size of the bulge and angle of the spiral arms: *SBa*, *SBb*, and *SBc*.

Hubble found galaxies that seem intermediate between the last elliptical galaxies (E7) and the first spiral galaxies (Sa). These objects, called **lenticular galaxies**, appear to be almost all bulges with just a hint of a disk. "Lenticulars can get a little complicated," says Law. "They look elliptical, but you can often see they've got a disk. Still, those disks aren't necessarily forming a great deal of new stars." Hubble classified these objects as *S0*.

With this inventory of galaxies, Hubble created what became known as the Hubble sequence, which could be represented on what is known as the **Hubble tuning-fork diagram** (Figure 16.11). "Hubble placed the ellipticals on a line running from E0 to E7," explains Law. "The line ends with the S0 lenticulars and then splits off into two branches." The spiral galaxies run along one branch from Sa to Sc. The barred spiral galaxies run along the other branch from SBa to SBc. "So the entire sequence looks like a tuning fork" (a common instrument of Hubble's day), Law continues. "The irregulars don't get included because they are just a mess."

To Hubble, the sequence exhibited on the tuning-fork diagram appeared to be an *evolution in time*. He believed his classification scheme laid galaxies out in their different stages from birth onward. Hubble was suggesting that an individual galaxy would begin with an amorphous elliptical shape and then evolve into the grand pattern of spiral arms, perhaps with a central bar as well.

Two aspects of the Hubble sequence stand out. The first is that it appears *Hubble was entirely wrong about the evolution of galaxies*: there is no evidence that elliptical galaxies ever turn into spiral galaxies, and in fact, the opposite seems to be true. "We now know elliptical-type galaxies tend to be bigger and older than the spiral galaxies," says Law, "which tend to be both lower-mass and more actively forming stars." So, rather than elliptical galaxies turning into spiral galaxies, astronomers now understand that spiral galaxies may turn into elliptical galaxies through galaxy collisions (**Figure 16.12**). It's important to note, however, that the processes that shape galaxies remain at the frontier of research.

VIS

Figure 16.12 Galactic Collisions
Collisions between galaxies can drive galactic evolution. Each image **a.–f.** shows a different pair of interacting spiral galaxies. Taken together, however, the composite is similar to what is seen in simulations of a single pair of colliding galaxies as gravitational interactions result in the loss of spiral structure and evolution toward elliptical shapes.

The second remarkable aspect of Hubble's scheme is how useful it turned out to be, despite being wrong in terms of the path of evolution. Categorizing galaxies by their main morphological types, as Hubble suggested, proved to be a good way for astronomers to link observed properties and important physical aspects such as gas content and star formation. In this way, the Hubble sequence, as a classification scheme, has outlived the organizing idea (galaxy age) underlying its creation.

Section Summary

- Spiral galaxies have a disk, bulge, and halo; they may have spiral arms that result from spiral density waves. Dwarf spiral galaxies are rare.
- Bars form across the center of one-third of spiral galaxies, apparently due to spiral density waves. The presence of a bar drives enhanced star formation.
- Elliptical galaxies have triaxial shapes, minimal star formation, a stellar density that smoothly increases toward the center, the broadest range of sizes of all galaxy types, and stars that plunge in highly elliptical orbits toward the galactic center.
- Irregular galaxies are amorphous, although some have features of a disk. Many dwarf irregular galaxies are satellites of larger galaxies.
- Hubble's tuning-fork diagram categorizes elliptical galaxies by degree of ellipticity, and spiral and barred spiral galaxies by bulge size and how tightly wound the spiral arms are. Hubble believed galaxies evolve from elliptical to spiral, but astronomers later discovered that the reverse is more likely because of collisions.

CHECKPOINT   Why does the Hubble tuning-fork diagram have a fork?

The Cosmic Distance Ladder

16.3   Studying galaxies beyond our own Milky Way means dealing with an enormous increase in size scale. Distances are measured in parsecs for stars and in kiloparsecs for the structure of the Milky Way. The space between galaxies, however, takes us to size scales that are thousands to millions of times greater. The distance between the Milky Way and Andromeda, the nearest spiral galaxy, is about 770 kpc, or 770,000 pc. The separation between galaxies must therefore be measured in units of **megaparsecs (Mpc)**, equal to 1 million pc.

DAVID LAW

Some discoveries are accidents. "My colleague and I were just looking through some of the Hubble Space Telescope data we had recently taken," David Law says of very distant galaxies. "Most of them looked weird, blobby, and irregular. But then we just saw one that looked totally different. It looked like a regular spiral galaxy. We figured we must have gotten the distance wrong." But they confirmed it was 10 billion light-years away. It was the oldest spiral galaxy ever seen.

Law wasn't a kid determined to be an astronaut or astronomer; his fascination grew slowly. What pushed him over to the dark side (astronomical observing) was a summer program at the Haystack Observatory in Westford, Massachusetts. "I learned that I really, really liked astronomy."

"What we are doing is pulling ourselves up by our own bootstraps."

verification The use of a proven measurement method to validate a new measurement method.

cosmic distance ladder The sequence of methods for determining the distance to astronomical objects; arranged by increasing distance over which each method is effective.

spectroscopic parallax A method of determining distance to an astronomical object that first relies on the width of absorption lines in the object's spectrum.

As you think about astronomical distances, keep in mind that light has a finite speed of 300,000 kilometers per second (km/s). Although this is fast enough to seem instantaneous in our everyday life, the light from objects at distances of megaparsecs takes millions of years to reach Earth. Thus, when we look at a galaxy that is 65 million light-years (19.9 Mpc) away, we see it as it looked 65 million years ago. That was when the age of the dinosaurs ended.

Given these staggering astronomical distances, the first question must be, How do we know? How do we measure distances on these immense scales? "The answer to that question turns out to be remarkably hard," says David Law. "We're stuck here on Earth, and we can't take out a ruler and measure how far away an object is. We're really limited in terms of how we can determine astronomical distances. No one method for determining distances works over all ranges. So we have to come up with lots of different methods and hopefully get enough overlap between the ranges in which they're valid that we can validate them all against one another."

Verification means using a known method to tune and prove a new method. When astronomers discover a new method of distance determination, they first test it and validate it against proven methods to ensure its usefulness and accuracy. They can then use the new method to survey even greater volumes of cosmic space. Each new method of determining astronomical distances essentially represents a rung on a ladder extending out to ever-greater distances, so the entire set of techniques is known as the **cosmic distance ladder** (Figure 16.13). "What we are doing," says Law, "is pulling ourselves up by our own bootstraps. We combine lots of different methods, one after the other, to get to ever-larger distances."

Lowest Rungs of the Ladder

The lowest rung of the ladder could be considered the use of radar, which works for objects within the Solar System. But other techniques are required to measure distances beyond the planets. "For the very nearest stars, we can use parallax," says Law. "Parallax is just the yearly back-and-forth shift in the apparent positions of nearby stars on the sky." Parallax is the most direct method for determining astronomical distances. It relies on only basic trigonometry and the observed displacement of a star on the sky as Earth moves from one side of the Sun to the other during its orbit (see Chapter 11). "The problem is, for traditional ground-based telescopes, parallax works only out to about 100 pc or so," says Law. Space-based observatories give astronomers the ability to use parallax for even larger distances. The Gaia satellite can measure stellar positions to distances as far as the galactic center, many kiloparsecs away. Such sophisticated technologies have only recently become available. Before satellite technology was available, astronomers had to find a different method to measure greater distances.

In the 1940s, such a method was found: **spectroscopic parallax**. With spectroscopic parallax, astronomers use spectral classification to determine a star's horizontal placement on the HR diagram. If the object is a hydrogen-burning star, locating it on the main sequence gives an estimate of its luminosity. If, however, the star is a giant that is no longer fusing hydrogen in its core, the widths of absorption lines in the spectrum provide a good measure of its luminosity. In all cases, measurements of spectra yield values of luminosity that can be compared with apparent brightness to find distance. (Note that parallax is not involved in this method, but since astronomers had a geometric distance measure called parallax, they added *parallax* to the name of the new method as well.)

"But you can't use any new distance determination method until you validate it," Law reminds us. Thus, astronomers compared stellar distances that had already been determined through geometric parallax with their findings from the new spectroscopic parallax method—and confirmed that it could, in fact, produce accurate distances. With this

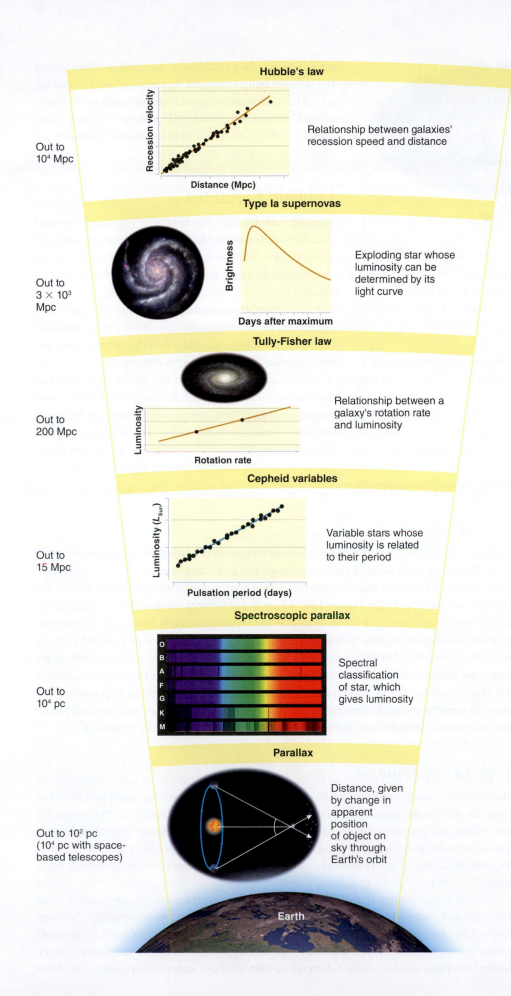

Hubble's law

Recession velocity

Distance (Mpc)

Out to
10^4 Mpc

Relationship between galaxies'
recession speed and distance

Type Ia supernovas

Brightness

Days after maximum

Out to
3×10^3
Mpc

Exploding star whose
luminosity can be
determined by its
light curve

Tully-Fisher law

Luminosity

Rotation rate

Out to
200 Mpc

Relationship between a
galaxy's rotation rate
and luminosity

Cepheid variables

Luminosity (L_{Sun})

Pulsation period (days)

Out to
15 Mpc

Variable stars whose
luminosity is related
to their period

Spectroscopic parallax

O
B
A
F
G
K
M

Out to
10^4 pc

Spectral
classification
of star, which
gives luminosity

Parallax

Out to 10^2 pc
(10^4 pc with space-
based telescopes)

Distance, given
by change in
apparent
position
of object on
sky through
Earth's orbit

Earth

Figure 16.13 Cosmic Distance Ladder
A series of methods ("rungs," represented by curved horizontal
bars) are used for determining distance to faraway astronomical objects. Each higher rung extends to greater distances.
Astronomers validate the accuracy and results of new methods
by using already-proven and overlapping methods.

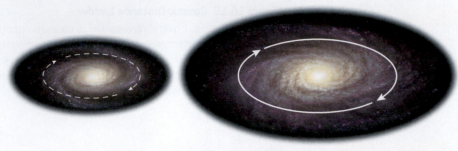

a. Galaxy A: Slower rotation rate   b. Galaxy B: Faster rotation rate

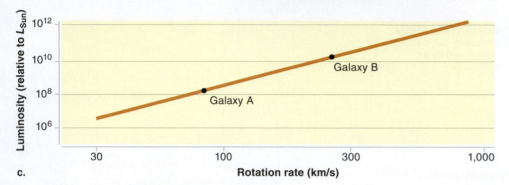

c.

Figure 16.14 Tully-Fisher Law
a. Stars in the disk of a low-mass galaxy rotate more slowly than **b.** stars in the disk of a more massive galaxy. Greater mass also means more stars, which means higher luminosity. **c.** The Tully-Fisher law defines the linear relationship between galactic rotation speed (related to galactic mass) and a galaxy's luminosity. Luminosity compared with observed brightness provides a distance measurement.

"The discovery of supernova standard candles was fantastic."

Tully-Fisher law A relationship of galactic rotation speed to galaxy mass that is used to infer the galaxy's luminosity.

Hubble's law The linear relationship between a galaxy's recession velocity and its distance.

Hubble's constant The slope of Hubble's law relating recession velocity and distance; expressed in (kilometers per second) per megaparsec [(km/s)/Mpc].

confirmation, they could push outward in space, using the new method to measure objects at greater distances.

The usefulness of spectroscopic parallax does not extend indefinitely, however. It is accurate out to only about 10,000 pc. To measure even greater distances, yet another new method was needed.

Cepheid Variables

Cepheid variables are the next step higher on the ladder. As we have seen, they are the classic example of a standard candle—an object whose luminosity can be inferred through some basic measurement of the object's properties. Measure the pulsation period of a Cepheid, and you can find its luminosity. Compare this luminosity to the brightness you see on the sky, and you can use the inverse square law for brightness (see Chapter 11) to determine the object's distance.

If an individual Cepheid can be resolved in a distant object, such as a galaxy, so that the star's pulsation period can be tracked, the period-luminosity relationship can be used to calculate the distance to that object. In this way, Hubble used Cepheid variables to measure the correct distance to the Andromeda Galaxy and prove that spiral nebulas are, in fact, spiral galaxies. Current telescope technologies allow Cepheids to be used for distance determination out to 15 Mpc.

Tully-Fisher Law

Higher-mass galaxies are expected to have higher energy outputs than galaxies with lower mass. "That's because most of the visible mass will come in the form of stars," says Law. Thus, knowing a galaxy's mass could provide a way to estimate its luminosity. This is the principle behind the **Tully-Fisher law**, which is a relatively quick method of determining distances for galaxies as far away as 200 Mpc. Its application, however, is limited to galaxies like edge-on spirals. While astronomers can't directly measure a distant galaxy's mass, they can indirectly determine mass from the galaxy's average rotation speed. As we have seen, orbital revolution speeds provide some measure of a galaxy's mass via Newton's law of gravitation. "If you measure the rotation speed of a galaxy using a spectrograph," explains Law, "then you can use the Tully-Fisher law to calculate how intrinsically bright the galaxy itself must be, and that gives you an estimate of its distance" (**Figure 16.14**).

Type Ia Supernovas

The brighter a standard candle is, the farther away it can be seen and used to determine distance. (Recall the inverse square law for brightness from Chapter 11.) Therefore, it's easy to imagine why astronomers became giddy when, in the 1990s, one type of supernova was found to be useful as a standard candle. "The discovery of supernova standard candles was fantastic," says Law. "It gave us a method of getting distances out to incredibly large scales."

Recall from Chapter 13 that Type Ia supernovas are created in binary systems when a companion dumps so much mass onto a white dwarf that thermonuclear runaway occurs, causing catastrophic nuclear fusion and blowing the white dwarf apart. In the mid-1990s, researchers discovered that the shape of a Type Ia supernova's light curve can be used to infer the total intrinsic light-energy output of the explosion (**Figure 16.15**). "The period over which the Type Ia supernovas brighten and fade turns out to be very closely linked to the intrinsic brightness during the peak of the explosion," says Law. "That means the light curve alone tells you precisely how much energy they put out. That makes them a standard candle and allows you to get their distances."

Since supernovas are some of the most powerful explosions known in the Universe (the Big Bang, which created the cosmos, was *the* most powerful), they can be seen across enormous distances—about 3,000 Mpc. Thus, the recognition that Type Ia supernovas can be used as standard candles revolutionized astronomy.

Hubble's Law

Early in the 20th century, American astronomer Vesto Slipher measured the spectra of a small sample of distant galaxies. He found that most of them showed strong redshifts, meaning they appeared to be receding from us. Later, in the mid-1920s, Edwin Hubble also began exploring the motions of galaxies through space. To determine their velocities, he too made measurements of galactic spectral lines. "These come from the combined starlight of each galaxy," says Paul Green, an astronomer at the Harvard-Smithsonian Center for Astrophysics. "The spectra gave him velocities, but he also needed distance measurements so he could understand if there was a pattern in the way different galaxies were moving."

To get each galaxy's position in space, Hubble continued using Cepheid variables. Accumulating the measurements (velocity and distance) required many frigid nights sitting in the "cage" atop Mount Wilson Observatory's 100-inch telescope, keeping the 100-ton instrument trained on the right position as the faint light from the distant galaxy slowly exposed photographic plates at the telescope's focus. "I did this kind of thing as an undergraduate," says Green, "and it's really, really cold in the cage."

Working with Hubble was Milton Humason, a former mule-train driver. After finding work at the observatory as a mechanic, Humason soon showed himself to be an exceptionally gifted astronomical observer (**Figure 16.16**). Together, Humason and Hubble painstakingly gathered the needed data. Putting it all together, the unlikely pair of colleagues stumbled across one of the greatest discoveries in the history of astronomy.

First, Hubble discovered that nearly every one of the 500 distant galaxies he had observed showed redshifts. That meant all the galaxies are moving away from us; "shunning us like the plague" was the phrase Hubble used. This, by itself, was a remarkable result, but when Hubble plotted the galaxies' speed versus their distance, he found a clear pattern. The farther away a galaxy was from Earth, the faster it was receding. In other words, Hubble found a *linear relationship* between recession velocity V (as measured in units of kilometers per second, km/s) and distance D (in megaparsecs, Mpc). In mathematical form, the new relationship, called **Hubble's law**, says

$$V = H_0 D$$

recession velocity of a galaxy = Hubble's constant × distance to galaxy

The term H_0, **Hubble's constant**, is what sets the slope of the linear relationship.

Although the exact value of H_0 is a subject of ongoing research, the best current estimate is 69 (km/s)/Mpc (**Figure 16.17**). "You should pay attention to the funny units on Hubble's constant [(km/s)/Mpc]," says Green. "That combination comes because Hubble's law can take measurements of galactic recession velocity (measured in kilometers per

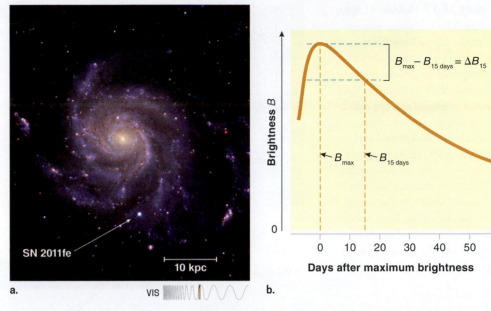

a.   VIS   b.

Figure 16.15 Type Ia Supernovas
a. Supernova SN 2011fe, in the Pinwheel Galaxy, is a Type Ia supernova. **b.** A light curve of such an event. The measured difference between a supernova's peak brightness and the brightness 15 days later (ΔB_{15}) yields a value of the explosion's luminosity at its peak. In this way, astronomers use Type Ia supernovas as standard candles to determine accurate distances to objects out to 3,000 Mpc.

Figure 16.16 Milton Humason
Humason (1891–1972), working with Edwin Hubble, made the remarkable discovery that galaxies are receding from us at speeds proportional to their distances. The implication of this observation is that the Universe itself is expanding.

Figure 16.17 Hubble's Law

Hubble's law correlates galactic distance with recession speed. **a.** Hubble's original plot of this relationship, for galaxies out to a distance of about 2 Mpc. **b.** More recent studies of galaxies out to distances of hundreds of megaparsecs give an estimate of $H_0 = 69$ (km/s)/Mpc. **c.** Astronomers determine recession speed from the redshift of absorption lines in a galaxy's spectrum. The farther a known absorption line is shifted from its rest wavelength, the faster the galaxy is moving away and the greater its distance.

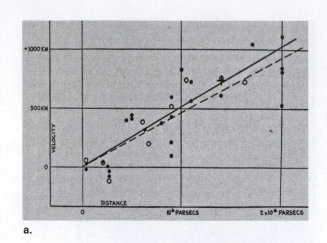

a.

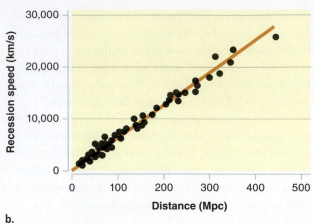

b.

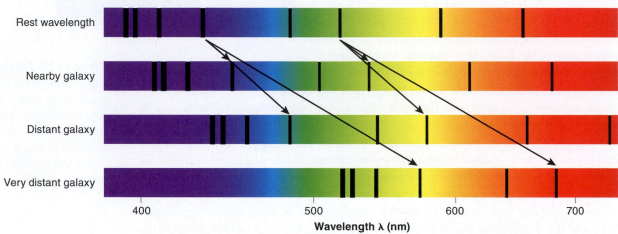

c.

second) and turn them into distances (in megaparsecs), or the other way around" (see **Going Further 16.1**).

Hubble's discovery that almost every galaxy is moving away from us at a recession speed that increases with distance had profound theoretical implications. It was the first direct evidence that the Universe had a beginning and is not infinite in time, as Aristotle and others had suggested. A full discussion of that topic will be the focus of Chapters 17 and 18; for now, it's enough to see how Hubble's law functions as a rung on the cosmic distance ladder with an upper limit of about 10^4 Mpc.

"Using Hubble's law, distance comes straight from velocity," says Green. "All an astronomer needs to do is make a measurement of a galaxy's redshift (which is relatively easy). Then the redshift gets converted into velocity and the velocity then becomes a distance via Hubble's law." Hubble's discovery opened up the vast distances between galaxies to astronomical exploration. The known size of the Universe, once again, was exploding to ever-larger scales.

INTERACTIVE:
Hubble's Law Generator

Section Summary

- The cosmic distance ladder consists of methods of determining astronomical distances, with each method calibrating the next to greater distances. Parallax, an object's apparent shift of position against the background of stars, yields the distance to the nearest objects beyond the Solar System. Spectroscopic parallax, which relates stellar spectral class and luminosity, is used for more distant objects.
- Astronomical distances can be measured using standard candles such as Cepheid variables and the application of the inverse square law for brightness.
- The Tully-Fisher law infers a galaxy's mass and luminosity from its rotation speed.

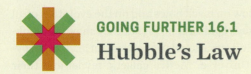

Hubble's Law

Hubble's finding that galaxies are receding from one another was one of the most significant scientific discoveries in history. His conclusion is cleanly summed up in Hubble's law, which shows how galaxy velocity (as measured with redshift) and galaxy distance go together.

Using the modern value of Hubble's constant $H_0 = 69$ (km/s)/Mpc, Hubble's law—recession velocity V (measured with redshift) = $H_0 \times$ galaxy distance D (in megaparsecs)—takes the form

$$V = 69 \text{ (km/s)/Mpc} \times D$$

Thus, if two galaxies have recession velocities of 6,900 and 13,800 km/s, we calculate their distances from us as 100 and 200 Mpc, respectively.

Notice that if the distance doubles, the velocity also doubles, because Hubble's law expresses a linear relationship between distance and velocity. This is why Hubble's law is so useful to astronomers: they can translate a measure of redshift (velocity) into a measure of distance by rearranging the equation:

$$D = \frac{V}{69 \text{ (km/s)/Mpc}}$$

Therefore, if a galaxy is moving away from us at 3,000 km/s, we can calculate its distance as follows:

$$D = \frac{3,000 \text{ km/s}}{69 \text{ (km/s)/Mpc}} = 43 \text{ Mpc}$$

Again, doubling the velocity would double the distance, as expected for a linear relationship.

How do we get from an actual measurement of redshift to velocity? Imagine that, for a certain galaxy, a spectral line of a particular element is measured in the laboratory (at rest with respect to us—as though the source were not moving) at a wavelength $\lambda_{lab} = 500$ nanometers (nm). In the galaxy's observed spectrum, however, the same spectral line appears at $\lambda_{galaxy} = 505$ nm. To convert from spectrum to speed, astronomers use the formula

$$V = \frac{\lambda_{galaxy} - \lambda_{lab}}{\lambda_{lab}} \times c$$

where c is the speed of light. (This equation is valid only when the velocity V is much less than the speed of light c.) Plugging in our values, we find

$$V = \left(\frac{505 \text{ nm} - 500 \text{ nm}}{500 \text{ nm}} \right) \times (3 \times 10^5 \text{ km/s}) = 3,000 \text{ km/s}$$

That's the power of Hubble's law. All you need is a galaxy's spectrum to determine its velocity, and from there you get distance. At distances the size of the Universe itself, however, Hubble's law is not accurate. That story will have to wait until Chapter 18.

- The light curves of Type Ia supernovas correlate with their luminosity.
- Combining measurements of 500 galaxies, Hubble and Humason found a velocity-distance relationship. Hubble's law shows that almost all galaxies are receding from us.

CHECKPOINT Why is it crucial to use an existing distance determination method to validate a new one?

Monsters in the Deep: Active Galactic Nuclei

16.4 Hubble used only optical telescopes to identify the basic categories of galaxies (elliptical, spiral, barred spiral, and irregular). Once other wavelength windows opened on the Universe, beginning with radio in the 1950s, the study of galaxies went in entirely new directions. Radio, infrared, X-ray, and other wavelength bands yielded a treasure trove of new insights. The most surprising discovery was the presence of extremely energetic and violent activity in the centers of some galaxies. Studies of these **active galactic nuclei (AGNs)** eventually led to the recognition that a supermassive black hole (SMBH) resides in the deep gravitational well at the center of almost all galaxies (see Chapter 14). Although only 1 percent of galaxies have AGNs,

active galactic nucleus (AGN) A highly luminous, energetic galactic nucleus powered by a supermassive black hole.

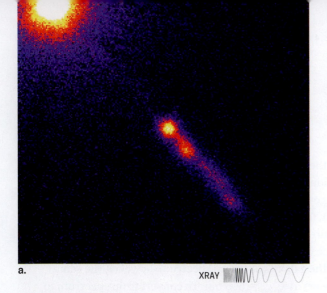

a.

XRAY

Laboratory spectrum (for comparison)

Hδ Hγ   Hβ

Spectrum of 3C 273

b.

Figure 16.18 Quasars
a. Light from quasar 3C 273—the first one discovered—and a relativistic jet driven from its nucleus are shown. Astronomer Maarten Schmidt determined the nature of quasars in 1963.
b. Hydrogen (H) emission lines observed in the lab (no redshift) are compared with the same lines in the observed spectrum of 3C 273. Its emission lines were redshifted so much as to be almost unrecognizable, implying great distance, and its brightness exceeded expectations for an object so far away.

"Looking at objects that have extreme, even bizarre, observational properties can lead to discovering entirely new kinds of physics in the Universe."

Seyfert galaxy A galaxy whose nucleus shows broad emission lines, indicating highly ionized material moving at great speed.

quasar A quasi-stellar radio source; an extremely luminous and distant active galactic nucleus.

radio galaxy A galaxy whose nucleus is the origin of relativistic bipolar jets.

blazar A form of radio-loud AGN that shows rapid variability in radio, optical, and X-ray wavelengths but few emission lines.

these active galaxies have proved to be a window on black holes and on the evolution of all galaxies. "What's so cool about these AGNs," says Paul Green, "is that they show how looking at objects that have extreme, even bizarre, observational properties can lead to discovering entirely new kinds of physics in the Universe." Active galactic nuclei and their black-hole engines remain at one of the most vibrant frontiers of astrophysical research.

Early Findings

The study of active galactic nuclei began even before new instruments such as radio and X-ray telescopes were available. But the development of these technologies pushed the field forward, leading to the detection of quasars and radio galaxies in the 1960s. As is typical, the process of discovery and the process of recognizing the meaning of discoveries did not move in a straight line. Scientists sometimes find that a new discovery sheds light on phenomena that may have been seen many years before.

SEYFERT GALAXIES. The earliest recognition of what would eventually be called AGNs came in the 1940s, when optical studies by astronomer Carl Seyfert unveiled a class of galaxies, now called **Seyfert galaxies**, whose spectra show a distinct pattern of broad emission lines. "These lines show up as tall peaks in the spectrum from elements like hydrogen, oxygen, neon, sulfur, and nitrogen," says Green. "The thing that's unusual about Seyfert galaxies is that they show lines from highly ionized versions of these elements. That means there must be very intense ultraviolet or X-ray radiation stripping the elements of their electrons."

The fact that the emission lines are so broad tells us that they come from gas moving in the galaxy core at high speed. "Whatever is going on in there must include clouds with orbital speeds of up to 10,000 km/s," says Green. This is about 1/30th the speed of light. In addition, Seyfert galaxies show rapid changes in brightness not seen in any other object. "Taken together, the high ionizations, high velocities, and rapid variability all tell you something remarkable is happening in the nuclei of these galaxies." Seyfert galaxies are categorized as Type I or Type II. Type I Seyferts show the broadest emission lines, indicating fast-moving gas. Type II Seyferts show narrower emission lines, indicating that the material is moving less rapidly.

QUASARS. In 1963, Maarten Schmidt, a professor of astrophysics at the California Institute of Technology, was working with radio astronomers to identify optical counterparts of newly discovered radio sources. "Radio astronomy was still a relatively new science at this point," says Green. "Stars don't normally give off huge amounts of radio waves in the first place, and only one in 10,000 optical objects in the sky will give off radio emission that you could detect back then." Just as important, radio telescopes at the time had lower resolution than optical ones, so it was difficult to match an object on a photographic plate to an object detected via radio telescope. Schmidt, however, had become quite good at his work, and he compiled an impressive list of galaxies that could be linked to specific radio observations.

Then along came 3C 273 (whose name stands for the 273rd object in the third Cambridge University radio survey). As well as Schmidt could determine, 3C 273 was associated with an unremarkable, bluish-looking star. When Schmidt looked at the object's spectrum, however, he was shocked. Its emission lines made no sense. No matter how hard he tried to fit the spectral lines to the usual library of elements, he failed. Either he had discovered new forms of matter (an unlikely option) or something very strange was going on. After some months, Schmidt finally came up with the answer: the spectral lines were just normal hydrogen being redshifted to an enormous degree (**Figure 16.18**).

Racing through his analysis, Schmidt found that 3C 273 is receding from Earth at a whopping 16 percent of light speed. Using Hubble's law, he calculated the object's distance to be 750 Mpc, making it the farthest object ever seen up to that point. Schmidt then calculated the luminosity of 3C 273. "That is when the real shock hit him," says Green.

At a luminosity of approximately 10^{40} watts (W), 3C 273 pumps out more than a trillion times the Sun's energy output and is hundreds of times more luminous than the entire Milky Way Galaxy. While supernovas produce more light, they shine for only a few weeks at most. The energy output from 3C 273 stayed high no matter how long astronomers looked. Whatever was powering the distant object was something new and unknown to astrophysics at the time.

Other sources like 3C 273 were quickly discovered and given the name **quasar**, for *quasi-stellar radio source*. That name meant that although these objects looked like stars, there must be more to them. Their great distance and power made quasars an enigma for more than 20 years, but by the 1980s a scientific consensus had emerged that quasars are a type of active galactic nucleus.

RADIO GALAXIES. As extragalactic studies matured, a new phenomenon closely related to quasars was recognized to be living in our own neck of the woods (on relative cosmic scales). Radio astronomical studies in the 1960s found another new class of galaxies, called **radio galaxies**. They have powerful, focused jets of gas from their central regions pushing into the intergalactic medium (these jets are, in principle, similar to those discussed in other contexts like star formation and the late stages of stellar evolution: see Chapters 12, 13, and 14). "Radio galaxies are spectacular," says Green. "They show huge, bipolar lobes extending thousands of parsecs in both directions away from the galactic nucleus" (Figure 16.19). The shape of the radio spectrum tells astronomers that the lobes must be filled with energetic electrons spiraling around strong magnetic fields.

With the advent of interferometric methods (many radio telescopes working in concert), the resolution of radio observations vastly improved. The jets and their lobes were mapped in exquisite detail, and astronomers were able to trace the source of the jets to central regions of their host galaxies just a few light-years across. Step by step, astronomers were zeroing in on the mysterious sources of AGNs that we consider the "engines" driving AGN activity.

Active Galactic Nuclei: Shake, Rattle, and Roll

With the discovery of an ever-larger census of galaxies, it became clear that a small fraction of the total galactic population shows remarkably strange behavior. Over time, astronomers recognized that Seyfert galaxies, quasars, and radio galaxies display essentially similar behavior. By the 1980s, astronomers had lumped the driving engines of all three together as AGNs.

Other subclasses of AGN activity were also discovered; they tended to be distinguished by the amount of radio emission coming from the AGN. Seyfert galaxies, for example, are considered "radio-quiet" active galaxies. Low-ionization nuclear emission line regions, or *LINERs*, are another type of radio-quiet active galaxy; the spectra of LINERs show gas with weak emission lines at the galactic cores. Radio galaxies, with their jet emissions, are "radio-loud" active galaxies, as are quasars. **Blazars** are also radio-loud; these AGNs show rapid variability in their brightness in radio, optical, and X-ray wavelengths but also show few emission lines.

In general, AGNs have some combination of six principal characteristics:

1. Compact size
2. High luminosity
3. Strong emission from the core (continuum radiation emitted at a wide range of wavelengths, from radio to X-ray)
4. Strong emission lines
5. Variability in both continuous and spectral-line emission
6. Strong emission at radio wavelengths

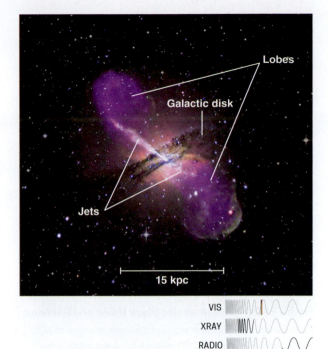

Figure 16.19 Radio Galaxies
The lobes above and below the dusty disk of radio galaxy Centaurus A are created by the powerful radio jets driven from the galaxy's nucleus.

"Radio galaxies are spectacular."

PAUL GREEN

"I started out wanting to be in a rock band," says Paul Green. He recalls his shift to astronomy. "During the winter term at Oberlin, they let you get out of town and do something different for a month. I decided to go to the National Observatory in Arizona." Green tested photographic plates for astronomical images. He sat 20 feet above ground in a cage atop the giant Kitt Peak telescope. "It was dead quiet, except for the clanking of the machinery and the buzz of the electronics. And the sky . . . was perfect, just the black of space and the sharp light of the stars. I was in heaven."

At the Harvard-Smithsonian Center for Astrophysics, Green uses the Chandra X-ray space telescope to study quasars. "But I still have a band, okay?" he says.

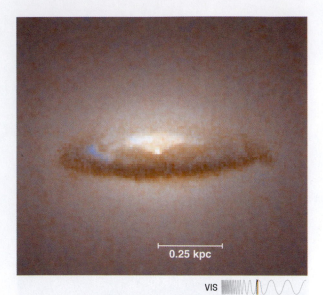

0.25 kpc

VIS

Figure 16.20 Supermassive Black Holes and Galaxies
A Hubble Space Telescope image of the central region of NGC 7052, an elliptical galaxy 200 million light-years away, in the constellation Vulpecula. The disk surrounds a 300-million-M_{Sun} black hole. Such SMBHs are the source of the behaviors associated with AGNs.

"Originally everyone was just cataloging these as different kinds of objects," says leading AGN astronomer Meg Urry, a former president of the American Astronomical Society. "It was kind of a mess." Despite the differences among the types of AGNs (radio-loud, radio-quiet, strong emission lines, weak emission lines), astronomers were motivated to find one model that could unite all of the behaviors they observed. The model they eventually found is based on the assumption that the principal actors driving all of the remarkable behavior of AGNs are the enigmatic SMBHs. "When people recognized that it all could not be powered by stars and that black holes must be the ultimate source of all the different kinds of activity," explains Urry, "that was the key to making progress understanding AGNs."

Linking Active Galactic Nuclei and Black Holes

Chapter 14 introduced supermassive black holes, with masses between 10^6 and 10^{11} M_{Sun}, which reside at the centers of galaxies. Even before firm proof could be obtained for the existence of SMBHs, astronomers were convinced they are the engines of AGNs. The jets discovered in radio galaxies were one of the early findings that led astronomers to suspect that black holes exist. "Early on, people knew the jets were traveling at close to the speed of light," says Meg Urry. "They also knew there was a tremendous amount of energy coming from AGNs where the jets were launched. Finally, they knew that the source of the jets was compact, meaning it was really small." Urry points to a paper written by British astronomer Donald Lynden-Bell in 1969 showing how these properties imply that the AGN source must be a black hole. "These kinds of conditions actually give you a black hole," says Urry, "whether you want one or not" (**Figure 16.20**).

The relativistic speeds of AGN jets are a compelling piece of evidence. (At relativistic speeds, an object is moving faster than 10 percent of light speed, or $V = 0.1c$.) They imply that whatever launches the jet must itself be relativistic. If a jet formed just outside the event horizon of a black hole is to escape into space, the jet must have a speed nearly the speed of light. In some cases, astronomers calculate jet speeds that appear to be even faster than the speed of light. These **superluminal motions**, however, are optical illusions that occur when jets are oriented such that they're moving toward us. However, superluminal motion will appear only for relativistic jets moving close to light speed.

Perhaps the most important early evidence of a connection between AGNs and SMBHs is the timescale of AGN luminosity variations. Depending on the wavelength, X-ray observations show that AGNs can have increased light output by a factor of 2 or more on timescales of hours to days. How do these rapid changes in luminosity offer a clue to the size of AGN central engines? "To illustrate how this works, I do a demo in class," says Urry. "I whisper something to a student in the front row and then ask them to whisper it to the person behind them. When each person is done whispering, they stand up. It takes a certain amount of time for the person in the back row to stand up. That shows nicely that it takes time for a signal to propagate." Since nothing can travel faster than light speed, a process that causes an AGN's luminosity to increase can't propagate any faster than the time it takes for light to make it across the entire AGN.

Urry's reasoning means that if an AGN's luminosity increases in a span of 3 days, then the size of the AGN must be 3 light-days, or just 80 billion km (see **Going Further 16.2**). That is a remarkably small region of space, when you consider how much energy an AGN releases. Only a black hole could account for that much energy being unleashed in such a small volume.

Understanding that AGNs and black holes are linked did not, however, explain the bewildering array of behavior from the different kinds of AGNs. Why are some radio-loud and others radio-quiet, for example? For astronomers, the next step was to come up with a model that used black holes and other actors to develop a single unified model of all AGNs.

superluminal motion The illusion that an object's speed exceeds the speed of light; created by the object's motion toward the observer.

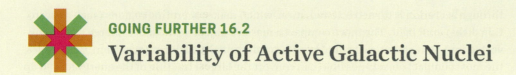

GOING FURTHER 16.2
Variability of Active Galactic Nuclei

The rapid timescales of AGN variability provided a critical clue for astronomers trying to pin down the nature of the engines that drive AGN activity. Using the time it takes for an AGN to brighten significantly and some geometry, astronomers can calculate the central engine's size.

Suppose that the AGN engine is a spherical region (shown in cross-section in **Figure 16.21**). We could think of the time it takes for a physical process that causes brightening to propagate from one side of the engine to the other (the diameter of the circle). Another approach is to imagine that the whole engine suddenly brightens at the same time. Let's explore that case in more detail.

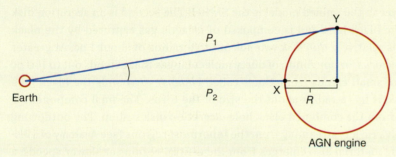

Figure 16.21 Calculating the Size of the Central Engine of an Active Galactic Nucleus

If the entire spherical engine brightened all at once, then observers on Earth would detect the change from a point X, closest to Earth on the sphere, before they would detect it from point Y, which is farther from Earth. We use trigonometry to express the path difference between point X to Earth and point Y to Earth. Take R to be the sphere's radius, P_1 to be the path length from point Y to Earth, and P_2 to be the path length from point X to Earth.

Earth is so far from the AGN that the two paths are pretty much parallel. Thus their lengths are related by this approximation:

$$P_1 \approx (P_2 + R)$$

So we take R to be

$$R = P_1 - P_2$$

The difference in arrival time ΔT of the signal of light detected from X versus that of the signal from Y is the difference in the paths $(P_1 - P_2)$ divided by the speed of light c:

$$\Delta T = (P_1 - P_2)/c = R/c$$

Thus,

$$R = \Delta T \times c$$

Since astronomers see AGN variations within a matter of hours (1 hour = 3,600 seconds), this last equation tells us that the engine must have a radius of

$$R = 3{,}600 \text{ s} \times (3 \times 10^5 \text{ km/s}) = 1.1 \times 10^9 \text{ km}$$

That is a very small region, equivalent to 7.4 AU.

Now we can ask what the mass M of the sphere would be if R_s were the Schwarzschild radius of a black hole. Recall from Section 14.4 of Chapter 14 that

$$R_s = \frac{2GM}{c^2}$$

where G is Newton's gravitational constant. Thus,

$$M = c^2 R_s/(2G)$$

In using this formula, we must convert radius R_s from astronomical units to meters (1 AU = 1.5×10^{11} meters). We get

$$M = \frac{(3 \times 10^8 \text{ m/s})^2 \times 7.4 \text{ AU} \times (1.5 \times 10^{11} \text{ m/AU})}{2 \times (6.67 \times 10^{-11} \text{ m}^3/\text{kg s}^2)}$$

$$= (7.5 \times 10^{38} \text{ kg}) \times \left(\frac{1 \, M_{Sun}}{2.0 \times 10^{30} \text{ kg}}\right) = 3.7 \times 10^8 \, M_{Sun}$$

This is an appropriate mass for a SMBH.

The Power of One: A Unified Model of Active Galactic Nuclei

The deep gravitational well of a black hole is the most effective means known to convert matter into radiation. Recall from Chapter 12 that when an object falls onto a gravitating body, astronomers say the falling material has accreted. During accretion, the initial gravitational potential energy of an object (when it is far away) is converted into kinetic energy. Thermal energy (heat) is also generated if the infalling material experiences friction on its way down (as can happen when something is pulled apart by tidal forces). Heat released

MEG URRY

Meg Urry says she owes it all to Miss Crawley, her high school chemistry teacher. "I did not always have astronomy in mind [as a career]," says Urry, a professor at Yale University and a former president of the American Astronomical Society. "I liked everything—English, history, math, languages." But Miss Crawley's class (and Urry's parents, both scientists) let Urry find her own excitement for physics and astronomy. "When I got to college, I . . . found Electricity and Magnetism class challenging, decided to conquer it, and then discovered it was beautiful." And she realized that she thought like a scientist.

Urry has been at the forefront of studying active galactic nuclei. Nevertheless, she says, "Probably the happiest I've been is teaching students physics—either explaining something they hadn't yet understood, or getting them excited about what the Universe is like. That's definitely the most satisfying part of my career."

"It was like a zoo. These were lions, but these were bears, and over here were tigers."

unified model of AGNs A consolidated depiction of the morphology of the various types of active galactic nuclei.

AGN outflow Material driven away from the central regions of an active galactic nucleus.

through accretion is turned into radiation, which makes accreting regions (such as accretion disks) emit light. The more compact a massive object is, the more energy can be converted into light as material falls down its gravitational well during accretion. Black holes, the most compact objects possible, can convert the largest fraction of the energy locked up in an accreting mass m into light. "Active accretion onto a black hole fit so much of what we saw when we looked at active galactic nuclei," says Meg Urry. "But that, by itself, wasn't enough."

The black-hole scenario on its own could not yet account for all of the behaviors seen in different types of AGNs. "For a long time in the study of AGNs, it was just murky," says Urry. "It seemed like every other galaxy was something different. There were the ones that are bright in the radio and the ones that weren't, the ones with broad lines, the ones with narrow lines. It was like a zoo. These were lions, but these were bears, and over here were tigers." A simple, efficient mechanism was needed to explain the many differences. It took years of study, but eventually astronomers—including Urry—developed a coherent framework for understanding all AGNs through what is called the **unified model of AGNs**. The model begins with a few basic players and then adds the concept of viewing angle, a critical factor.

The first player in the unified model is the SMBH. The second is an accretion disk that surrounds the black hole and is composed of galactic gas captured by the black hole's gravity. This relatively *thin disk* extends out to a radius of about 1 pc. At greater distances is a thick *torus*, or doughnut, of dusty molecular gas that extends out to 100 pc or more. A halo of small clouds (sometimes called *clumps*) surrounds the black hole and accretion disk; it has a radius about the size of the torus. The final component is an **AGN outflow** from the combined black hole–accretion disk system. The outflowing gas includes relativistic jets streaming from the innermost regions (see **Anatomy of a Discovery**). Outflowing material also emerges from the central regions at slower speeds, a "mere" 1,000–10,000 km/s.

With these players in place, the angle of observation—the inclination of the entire system relative to our telescopes—can account for the differences among AGNs. "All of a sudden we had a geometry that could help us explain the different classes of AGNs," says Urry. "The idea was that because of the geometry of the torus and the disk, the central engine can be hidden if you're viewing from certain directions."

Astronomers whose view of the area nearest the black hole is blocked by the dense molecular and gas clouds will see one type of AGN. But if that view is not blocked, then astronomers see a different class of AGN. If, for example, we view the torus–disk–black hole system from the side, we can clearly see the jets, but our view (even using radio telescopes) of the innermost regions is obscured by the dense torus. Such a system would be considered radio-quiet, so this viewing angle explains the Seyfert galaxies. If we see the system more from the poleward direction, looking down the jet, then the system would be deemed radio-loud. "When we get to see the jets coming right at us," says Urry, "the AGNs are incredibly bright in the radio." Looking "down the throat" of the AGN also lets astronomers detect its rapid variability because they're seeing down into the central regions. This viewing angle would reveal blazars. Depending on the viewing angle, other objects will show a mix of properties, such as those seen in classic quasars (**Figure 16.22**).

"There are still differences between different AGNs," Urry reminds us. "They can have different black hole masses, or some may have more dust than others. But by far the big observed differences have to do with viewing angle. That's what we had to figure out first."

The success of the unified model teaches us about more than just the nature of AGNs and the central regions of galaxies. At the center of science's mission is the idea of unity in diversity—the belief that phenomena that appear to be distinct may be attributable to a few basic principles operating together. Once those principles are understood, the bewildering buzz of experience becomes a symphony of order and unity.

What lurks within the heart of a galaxy?

Hydrogen lines with extreme redshift

Laboratory spectrum (for comparison)

Hδ   Hγ   Hβ

Spectrum of 3C 273

Quasars are indistinguishable from stars

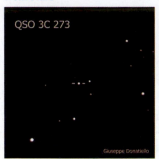

QSO 3C 273

Giuseppe Donatiello

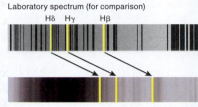

observation

Robert Vickrey's 1966 portrait of **Maarten Schmidt** appeared on a *Time* cover 3 years after Schmidt used an extreme redshift and Hubble's law to find the distance to object 3C 273. His discovery placed quasars near the edge of the observable Universe as some of the most powerful objects ever observed. *Time's* publisher Bernhard Auer wrote, "Perhaps not since Galileo has an astronomer so jolted the specialists whose field of study is the Universe."

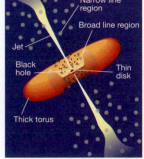

Galaxy M87

Relativistic jet

hypothesis

The brightness of quasars varies on short timescales, indicating that the source of their energy is confined to a "small" region of space. Radio observations suggested that quasars must generate jets of matter outflowing at near light speed. What compact object could contain the mass of millions of Suns and power relativistic jets? Early ideas focused on supermassive black holes (SMBHs).

Galaxy Centaurus A and jets

confirmed by observation

The hunt was on for SMBHs. In the 1970s, Bruce Balick and Robert Brown detected strong radio emission from the Milky Way's center; others observed stars quickly orbiting the centers of some elliptical galaxies. Both findings indirectly point to very massive compact objects at these galactic centers. By 2010, teams led by **Andrea Ghez** and **Reinhard Genzel**, monitoring stellar orbits, had measured the mass of the "unseen object" at our galaxy's center to be roughly 4 million solar masses.

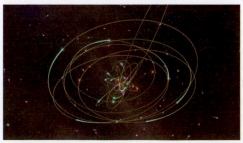

Stars orbiting galactic center

Narrow line region

Broad line region

Jet

Black hole

Thin disk

Thick torus

Unified model of AGNs

theory

We now know quasars are a type of active galactic nucleus (AGN). Further studies place SMBHs at the heart of most galaxies, indicating that black holes power quasars and other AGNs. Active galaxy studies led by **Meg Urry** produced a unified model of AGNs. It features an accretion disk and a gas torus surrounding a central black hole that powers two jets—and predicts many of the observed traits of active galaxies.

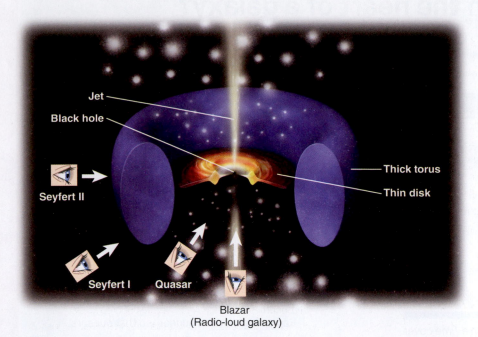

Jet

Black hole

Seyfert II

Seyfert I

Quasar

Blazar
(Radio-loud galaxy)

Thick torus

Thin disk

Figure 16.22 Unified Model of Active Galactic Nuclei
Every AGN has a central black hole, a thin disk, a thick torus, small clumps of clouds, and an AGN outflow, including relativistic jets. Astronomers now realize that the variety of AGNs they observe is a result of orientation. If we look down the barrel of the jet, we see a blazar. If we look directly from the side, the torus obscures the innermost regions, so we see a Type II Seyfert galaxy. An AGN seen from an intermediate position will appear as a quasar or a Type I Seyfert galaxy.

Explore the AGN of galaxy Centaurus A and use its supermassive black hole to power the last hyperspace jump in search of the aliens and their advanced technology.

velocity dispersion The characteristic range of velocities of bodies in an astronomical object, such as stars in an elliptical galaxy.

Black Holes and Bulges

While only a tiny fraction of galaxies host an AGN, astronomers now believe that all galaxies host a SMBH. The reason not all SMBHs create AGNs is that if there is not enough matter falling into the black hole, then luminosity will be relatively weak, as appears to be the case for the black hole at the center of our own Milky Way.

But there is more to the black hole–galaxy relationship than simply having a black hole in every galactic center. As astronomers became more adept at "weighing" black holes via the motion of surrounding material, they were able to build a fairly complete comparison of black-hole mass to the host galaxy's mass. In particular, spiral galaxies show a tight correlation between the mass of the bulge and the mass of the black hole. Bulges are almost always approximately 700 times more massive than the central black hole. This suggests that galaxy evolution is connected to black hole evolution.

Section Summary

- Studies revealed energetic activity in many galactic centers. Seyfert galaxies exhibit high ionizations, high speeds, and rapid variability. Quasars are enigmatic, extremely bright, very distant objects. Radio galaxies have powerful jets that emanate from a compact center.
- Seyfert galaxies, quasars, radio galaxies, LINERs, and blazars are types of active galactic nuclei (AGNs), which are driven by supermassive black holes (SMBHs).
- Only the presence of a black hole can explain the compact size of the core of AGNs, their relativistic jets, and their rapid variability.
- The unified model of AGNs attributes observed differences among AGN types to our viewing position relative to the system's components: black hole, accretion disk (thin disk), torus, cloud halo, and AGN outflow. Not all galaxies are observed to have AGNs, but astronomers believe all galaxies have a SMBH.

CHECKPOINT a. Describe the components of an active galactic nucleus, according to the unified model of AGNs. **b.** Explain why a wide variety of AGNs were identified before the unified model was developed.

Galaxies and Dark Matter

16.5 Chapter 15 described how studies of the Milky Way's rotation curve led astronomers to conclude that a significant component of our galaxy's mass is in the form of dark matter. Further evidence of the existence of dark matter came from explorations of other galaxies. "It's easier to study galaxies from the outside," says Paul Green. "The strongest early evidence for dark matter came from extragalactic studies—studies of galaxies other than our own."

Extragalactic Evidence of Dark Matter

In the late 1960s, Vera Rubin began using a new generation of high-resolution spectroscopes to graph the rotation curves of relatively nearby spiral galaxies. Her spectroscope was sensitive enough to get Doppler shifts of stars at different radii within the galactic disk. When Rubin and her collaborator, Kent Ford, plotted the Doppler-measured rotation velocities versus the stars' distance from their galactic centers, the researchers found that each galaxy showed the kind of rotation curve that the Milky Way has. "For every galaxy Rubin observed, she saw the same thing," says Meg Urry. "The velocity of the stars stayed remarkably constant out to quite large radii" (**Figure 16.23**). According to Newtonian physics and the distribution of normal (luminous) matter, the rotational velocities of stars farther from the center should have decreased (see Going Further 15.2). "The fact that the velocities stayed constant was really interesting," says Urry. "It implied that mass in each galaxy is rising the farther out you go. It was a major discovery."

Rubin's work forced astronomers to see that the combined masses of a galaxy's stars, gas, and dust do not account for all of the gravitational force the galaxy as a whole exerts on its stars. The dark (unseen) component of mass in every galaxy is greater than the luminous mass. Astronomers now generally accept that every spiral galaxy, including the Milky Way, has an enormous spherical "halo" of dark matter and that stars, gas, and dust inhabit only the inner regions. While dark matter halos vary in mass and size from one galaxy to another (just as the luminous components do), other halo properties inferred from the rotation curves are remarkably similar. For example, the shape of the dark matter distribution (the way it varies as a function of distance) seems to be common to all spiral galaxies.

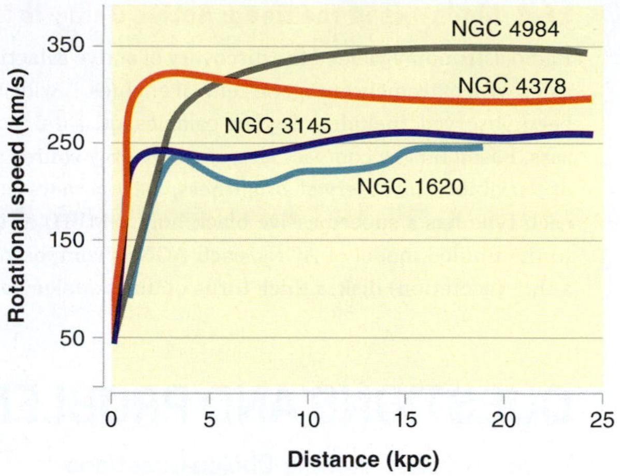

Dark Matter in Elliptical Galaxies

So far, we have discussed dark matter only in terms of spiral galaxies. Is there evidence of dark matter in elliptical galaxies too? Definitively, yes. In fact, it's easier to determine the presence of dark matter in elliptical galaxies than in spiral galaxies, precisely because the orbits in the former are highly elliptical. Instead of a uniform circular motion, the orbits in an elliptical galaxy make up a kind of buzzing swarm of stars. Rather than mapping an entire rotation curve from center to exterior, astronomers can look for the average orbital velocity V_{avg} in elliptical galaxies and compare that with the escape velocity V_{escape} deduced from all of the luminous matter. In all cases, V_{avg} is greater than V_{escape}, meaning that there is not enough luminous matter to hold the elliptical galaxy together. "You just measure what we call the velocity dispersion of the stars (basically V_{avg})," says Paul Green, "and you can derive the mass from there." **Velocity dispersion** is the entire range of stellar velocities. From these kinds of studies, astronomers infer that in elliptical galaxies, as in their spiral cousins, dark matter composes more than 90 percent of the galaxy's mass.

Section Summary

- The mapping of galactic rotation curves unexpectedly showed that stars farther from the galactic center do not rotate more slowly than stars near the center, indicating that a halo of unseen matter is present.
- Unseen dark matter is present in both spiral and elliptical galaxies.

CHECKPOINT   How did the evidence of dark matter in the Milky Way compare with evidence of dark matter in other spiral galaxies?

Figure 16.23 Galactic Rotation Curves and Dark Matter

Vera Rubin's work with galactic rotation curves showed that rotational speeds in spiral galaxies do not drop off with distance from the center, as Newtonian physics would suggest. Additional unseen mass (dark matter) must be present, keeping the stars rotating faster than expected from the total mass of luminous material alone.

"The fact that the velocities stayed constant was really interesting. It was a major discovery."

CHAPTER SUMMARY

16.1 The Great Debate and the Scale of the Universe

The true nature and scale of galaxies were not known until well into the 20th century. The Great Debate of 1920, on the distance to observed spiral nebulas, left the issue unresolved. Soon after that, Edwin Hubble used Cepheid variables in determining that our galaxy is one of many in a vast Universe of galaxies.

16.2 Galactic Zoology

Hubble classified galaxies into four main types. Spiral galaxies exhibit a disk, bulge, and stellar halo, and many also have some form of spiral arms. Barred spiral galaxies show all of these characteristics as well as a central bar, where star formation is accelerated. Elliptical galaxies vary widely in ellipticity and size. Irregular galaxies are amorphous; some have disk-like features. Hubble arrayed the various galaxy types into a tuning fork–shaped diagram and defined subclasses. He incorrectly assumed that the arrangement represents an evolutionary track from elliptical to spiral morphology.

16.3 The Cosmic Distance Ladder

The cosmic distance ladder enables astronomers to use various methods to determine ever-greater distances to objects, and each rung (method) is used to validate the rung that follows. Most steps rely on comparing the observed brightness with the known luminosity of a class of objects. In order of increasing distance, the rungs are parallax, spectroscopic parallax, Cepheid variables, the Tully-Fisher law, Type Ia supernovas, and Hubble's law. Using Cepheid variables to measure brightness, combined with distance measurements, Hubble determined that the faster a galaxy appears to be receding from us, the more distant it is. This discovery implied that the Universe is expanding and may have evolved from a different original state.

16.4 Monsters in the Deep: Active Galactic Nuclei

Radio astronomy yielded the discovery of active galactic nuclei (AGNs), galaxies with extremely energetic central engines. A wide variety of AGNs have been observed, including Seyfert galaxies, quasars, radio galaxies, and blazars. Based on the compact size of the energy source and the short period of variability in observed brightness, astronomers have determined that each type has a supermassive black hole (SMBH) at its center. According to the unified model of AGNs, each AGN is composed of a central SMBH, a thin (accretion) disk, a thick torus of dusty molecular gas, a halo of small clouds, and an AGN outflow, including a relativistic jet and a slower wind. Differences in AGN characteristics are due to the galaxy's orientation relative to the observer. All galaxies are believed to have a SMBH, but not all have an AGN.

16.5 Galaxies and Dark Matter

The presence of dark matter is implied by the rotation curve of spiral galaxies, as discovered by Vera Rubin, and by the velocity dispersion observed in elliptical galaxies.

QUESTIONS AND PROBLEMS

Narrow It Down: Multiple-Choice Questions

1. The factor that is most important in making Cepheid variables as useful as standard candles is
   a. their distance.
   b. their spectral class.
   c. the inverse square law for brightness.
   d. their position on the HR diagram.
   e. the period-luminosity relationship.

2. Which of the following characteristics is/are observed in *all* spiral galaxies? Choose all that apply.
   a. disk
   b. bulge
   c. jets
   d. bar
   e. spiral arms

3. A galaxy is observed to be spheroidal and to have increasing stellar density toward its center. What is its galaxy classification?
   a. spiral
   b. barred spiral
   c. lenticular
   d. elliptical
   e. irregular

4. A galaxy seen edge-on is observed to have a disk and a bulge. What is its galaxy classification?
   a. spiral
   b. barred spiral
   c. either elliptical or barred spiral
   d. either spiral or elliptical
   e. either spiral or barred spiral

5. Which of the following galaxy types is most likely to be clearly identifiable, regardless of orientation?
   a. E0
   b. Irr
   c. SBa
   d. E6
   e. SBc

6. Which of the following is/are *not* a subcategory or special class of spiral galaxies? Choose all that apply.
   a. elliptical
   b. halo
   c. flocculent
   d. barred
   e. grand design

7. Which of the following galaxy types has no discernible structure?
   a. Irr II
   b. E7
   c. Irr I
   d. lenticular
   e. SBb

8. Which of the following cosmic distance ladder methods use(s) actual or apparent motion of objects to determine distance? Choose all that apply.
   a. Tully-Fisher law
   b. spectroscopic parallax
   c. Type Ia supernovas
   d. parallax
   e. Cepheid variables

9. Which of the following cosmic distance ladder methods uses cyclical changes in a star's brightness over time to determine distance?
   a. Tully-Fisher law
   b. spectroscopic parallax
   c. Type Ia supernovas
   d. parallax
   e. Cepheid variables

10. Which method of determining distance would be most appropriate for a ground-based observer trying to find accurate distances out to approximately 100 pc?
    a. Tully-Fisher law
    b. spectroscopic parallax
    c. Type Ia supernovas
    d. parallax
    e. Cepheid variables

11. Hubble's law states that
    a. all galaxies are expanding.
    b. the more distant the galaxy is, the faster it appears to be receding from us.
    c. larger galaxies rotate faster.
    d. the larger the galaxy is, the faster it is receding.
    e. all galaxies that will ever exist already do exist.

12. Galaxy A is receding from us at x km/s, while galaxy B's recession speed is $3x$ km/s. Based on Hubble's law, which statement is true?
    a. B is 1/3 as far from us as A.
    b. A is 1/3 as far from us as B.
    c. B is 9 times farther from us than A.
    d. The relative positions of A and B depend on their galaxy classifications.
    e. The relative positions of A and B depend on their size.

13. Which of the following is *not* true of galaxy 3C 273?
    a. It is very distant.
    b. It is very luminous.
    c. It is receding from the Milky Way.
    d. It is believed to have a black hole at its core.
    e. Its spectrum reveals no hydrogen.

14. Which of the following is/are designated as an AGN? Choose all that apply.
    a. Seyfert galaxies
    b. radio galaxies
    c. quasars
    d. lenticular galaxies
    e. blazars

15. Which of the following describe(s) characteristics of AGNs that point to a black hole as the central engine? Choose all that apply.
    a. compact size of the core
    b. abundance of dark matter
    c. relativistic speed of jets
    d. short timescale for variability in brightness
    e. rapid star formation

16. Superluminal motions of the jets of AGNs are the result of
    a. jets moving faster than the speed of light.
    b. jets that appear to move faster than the speed of light because of their orientation.
    c. optical illusions due to interference from dust.
    d. light from the jets that has been accelerated to greater than light speed by the energy of the black hole.
    e. measurement error from our own galactic motion.

17. The unified model of AGNs
    a. says that the types of AGNs have different physical characteristics.
    b. says that the observer's orientation accounts for the differences among the types of AGNs.
    c. says that all galaxies have an AGN.
    d. explains all variations in observations of the various types of AGNs, except those of quasars.
    e. is based on observations of the Milky Way's AGN.

18. True/False: Hubble's subclasses of spiral and barred spiral galaxies (a, b, c) are based on the relative thickness of the bulge.

19. What led astronomers to infer the existence of dark matter?
    a. infrared images showing galactic halos
    b. the presence of far more satellite galaxies than expected
    c. observations of dark nebulas
    d. unexpected results in extragalactic rotation curves
    e. the discovery of black holes

20. Astronomers determined that elliptical galaxies have dark matter by
    a. calculating that the average stellar orbital velocity is greater than the expected escape velocity based on luminous matter.
    b. calculating that the expected escape velocity based on luminous matter is greater than the average stellar orbital speed.
    c. measuring star rotation speeds near the axis of the galaxy.
    d. measuring the galaxies' rotation curves.
    e. measuring the observed brightness versus the luminosity.

To the Point: Qualitative and Discussion Questions

21. What was the controversial topic debated in the Great Debate, and what position did each participant argue?

22. Speculate on the role that dust played in determining the distance to faraway objects such as the spiral nebulas.

23. What are the four main classes of galaxies? Describe the characteristics of each type.

24. Which of the four galaxy types commonly occur as dwarfs?

25. Which types of galaxies show active star formation?

26. Describe the stellar orbits in spiral versus elliptical galaxies.

27. What does the Hubble tuning-fork diagram describe?

28. What is the cosmic distance ladder, and what does it mean that each rung has been used to validate the next-higher rung?

29. Describe each of the following methods of determining astronomical distances: parallax, spectroscopic parallax, Cepheid variables, Tully-Fisher law, Type Ia supernovas. How did each technique rely on the previous one for validation?

30. Hubble's law describes a correlation between what two quantities?

31. What did the spectrum of 3C 273 indicate? Why was it a surprise?

32. Define the following terms: AGN, Seyfert galaxy, radio galaxy, quasar.

33. What evidence do astronomers have that AGNs are associated with supermassive black holes?

34. What is the evidence for dark matter in spiral galaxies? In elliptical galaxies?

35. The orientation of AGNs to the observer's line of sight (see Figure 16.22) impacts the characteristics that can be observed, and thus their classification. How does the orientation of spiral galaxies affect their observed characteristics and classification? (See Figure 16.11.)

Going Further: Quantitative Questions

36. A galaxy is observed moving away from ours at 2,415 km/s. What is its distance from us, in megaparsecs?

37. A galaxy is measured to be receding at 5,800 km/s. How far away is it, in megaparsecs?

38. A galaxy is determined to be 162 Mpc from us. At what speed, in kilometers per second, is it receding?

39. What is the recession velocity, in kilometers per second, of a galaxy 47 Mpc from us?

40. A spectral line whose wavelength in the laboratory is 500 nm appears at 540 nm in the spectrum of a galaxy. What is the distance to the galaxy, in megaparsecs?

41. The wavelength of a particular spectral line is 580 nm in the laboratory. It is observed in the spectrum of a galaxy at 660 nm. What is the recession velocity of the galaxy, in kilometers per second?

42. An AGN brightens significantly in a period of 5.5 hours. What is the maximum radius of the black hole, in kilometers? Estimate the mass of the black hole if this is the maximum size.

43. An astronomer has calculated the radius of a black hole to be 68 AU by observing its period of variability in brightness. What period, in hours, did she observe?

44. What is the mass, in kilograms, of a black hole whose Schwarzschild radius is 4.3×10^9 km?

45. Calculate the mass, in M_{Sun} units, of a black hole with Schwarzschild radius 7.0×10^{11} km.

"EXACTLY WHAT WE EXPECTED BASED ON OUR UNDERSTANDING OF HOW THESE LARGE-SCALE STRUCTURES FORM"

17

The Cosmic Web

THE LARGE-SCALE STRUCTURE OF THE UNIVERSE

Bright Lights, Big Universe

17.1 "When one door closes, somewhere a window opens," the old saying goes. When it came to measuring the density of gas between distant galaxies, astronomer Eric Wilcots was pretty sure he had found the portal he needed. "It's often really hard to directly measure gas density between galaxies," says Wilcots, a professor of astronomy at the University of Wisconsin. Intergalactic gas usually does not produce emission lines that can be used to make density measurements. This difficulty in directly measuring density posed a problem for Wilcots, since he was interested in seeing how the density between galaxies changes in different environments. "Some galaxies are packed fairly close together in big structures called galaxy clusters," says Wilcots, "and others come in less dense structures called *galaxy groups*. What we wanted was an *indirect* way of finding the background gas density between the galaxies in these different structures."

That's when Wilcots had his good idea. "Some galaxies produce jets," he explains, "and when those galaxies and their jets move through the background intergalactic medium, the jets get blown backward" (**Figure 17.1**). You get the same effect when you spray water from a garden hose into a strong wind. The galactic jets feel the intergalactic medium as they fly through it, and its pressure bends the jets away from the direction their host galaxy is traveling.

By measuring the deflection of the galactic jets and the speed of the host galaxy, Wilcots and his team could infer the density of the intergalactic gas. "It worked out really well," says Wilcots with a broad smile. "We were able to see that the density between galaxies in the groups was lower than the density between galaxies in the clusters—exactly what we expected based on our understanding of how these large-scale structures form across cosmic history."

← Over 1 million galaxies are shown by the infrared Two Micron All Sky Survey (2MASS). They form a web of structures over a range of size scales, from galaxy clusters to superclusters. The distribution of galaxies (their large-scale structure) provides important clues to the evolution of the Universe. The vertical slash near the center is created by very bright stars inside our disk-shaped Milky Way.

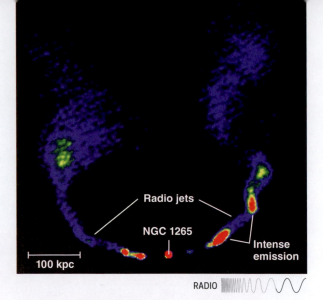

Figure 17.1 Galactic Jets

Jets emanating from galaxy NGC 1265 interact with and are deflected by the intergalactic medium.

Figure 17.2 An Expanding Universe

a. Philosophers and astronomers alike had long assumed that the Universe is unchanging and has existed in its current state for eternity. **b.** The discovery that all galaxies are moving apart disproved that "static" view and raised the possibility that the Universe is evolving—that it was different early in its history than it is today.

Wilcots's discovery was one thread in a tapestry of studies focusing on the distribution of matter across the Universe at immense distances. That distribution is called the **large-scale structure of the Universe**.

Large-Scale Structure and the History of Everything

When Edwin Hubble wrote his paper on the relationship between galaxy redshifts and distance (what we now call Hubble's law; Chapter 16), he never tried to interpret the monumental discovery's meaning. His paper reported that galaxies are receding from us with a speed that depends on their distance—end of story. But he and almost every other astronomer knew what the galaxy redshifts implied: if all galaxies are flying away from each other, then the entire Universe must be expanding.

Cosmic expansion was a radical idea that suddenly had solid observational support. At the time of Hubble's discovery, most astronomers assumed the Universe is static, stationary, and eternal—an idea that found its roots in the writings of Aristotle. Individual stars, planets, and people come and go, but the Universe in its entirety is immune to movement and change, or so it was thought. The discovery of Hubble's law changed those assumptions overnight (**Figure 17.2**). In time, the recognition that the Universe is expanding led to the idea that the Universe had a beginning. In other words,

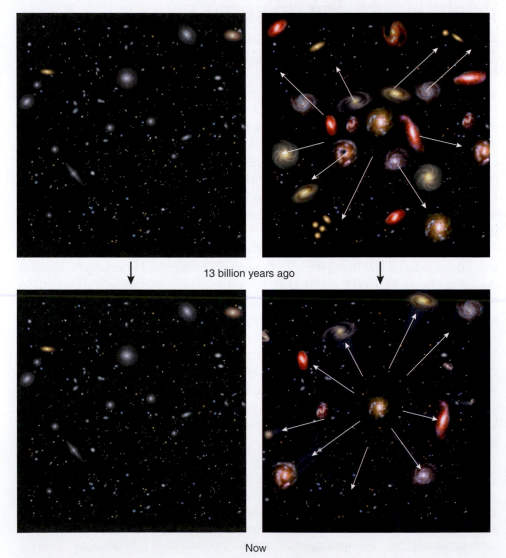

13 billion years ago

Now

a. Static Universe   **b. Expanding Universe**

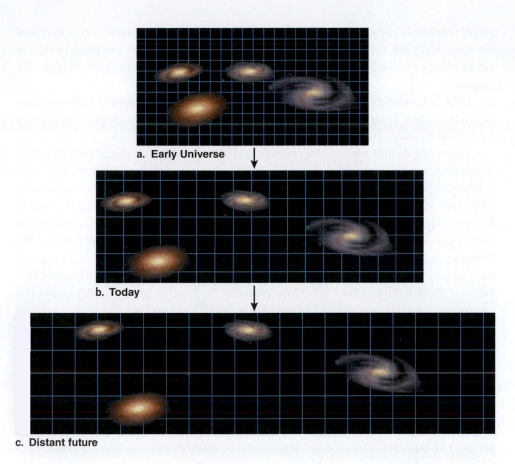

a. Early Universe

b. Today

c. Distant future

Figure 17.3 Expansion of Space-Time
We can visualize the expanding Universe as a vast rubber sheet being stretched in all directions, and the galaxies are swept along with it as it expands.

Hubble's law was the first major piece of evidence for what astronomers would come to call the **Big Bang**.

Today, after nearly a century of study, astronomers have compiled an overwhelming case that the Universe began in a single event—the Big Bang—that sent all space, time, matter, and energy flying apart. This is the standard model of **cosmology**, the field of astronomy that studies the Universe as a whole. Cosmology is such a rich branch of astrophysics that we will spend this chapter and the next exploring its ideas and consequences. This chapter focuses on the later stages of cosmic history and the ways that cosmic expansion affected the evolution of the large-scale structure of the Universe.

Using the world's largest telescopes and most powerful supercomputers, astronomers have mapped out the evolution of matter across space and time as it formed not only galaxies but also much larger structures called groups, clusters, and superclusters of galaxies. By studying these maps, astronomers have gained insights into the Universe's history going back billions of years. Hidden in these maps are clues to the Universe's earliest moments. We will use those clues in our study of the Big Bang itself in Chapter 18, the final chapter of our cosmic exploration.

Cosmic Expansion: Looking Out, Looking Back

In terms of the Universe's history, Hubble's law tells us that the Universe as a whole is expanding. And according to Einstein's general theory of relativity (see Chapter 14), that means the fabric of space-time itself is expanding. As Wilcots puts it, "It's not that all the galaxies in the Universe are rushing away from each other through space. Instead, it is space (or space-time itself) which is stretching. It's all the points of space that are pulling away from each other."

One way to visualize this cosmic expansion is to imagine a grid of galaxies glued to a vast rubber sheet (**Figure 17.3**). If the sheet is stretched uniformly in all directions, each

large-scale structure of the Universe The hierarchy of cosmic structures on ever-larger scales from galaxy groups up to superclusters and voids (and beyond).

cosmic expansion The expansion of space-time since the Big Bang.

Big Bang The event that initiated cosmic expansion.

cosmology The field of astronomy that studies the Universe as a whole.

galaxy moves away from all the others, even though no galaxy ever moves across the sheet or expands. It's the viewpoint that's key. As Wilcots explains, "From the point of view of any randomly chosen galaxy, all the neighbors appear to be speeding away from it like the plague."

Just as important, the speed at which the neighbor galaxies appear to be receding depends on their distance from the observer galaxy. That means every observer in every galaxy sees the same Hubble's law for the motion of surrounding galaxies.

Now imagine that you film this cosmic expansion and then run the movie backwards. All galaxies would draw closer together as time wound backward. "Eventually," says Wilcots, "a point would be reached where each galaxy (or each point on our imaginary rubber sheet) would be squashed against its neighbors." Thus, we can imagine going so far back that the distance between points on the sheet would *approach zero*. Everything everywhere would be infinitely compressed together. In this way, the expansion of the Universe that we see today can, in principle, imply a cosmic beginning: the Big Bang. For now, we focus on the amazing *long-term* consequences of the Big Bang: the origin and evolution of the structure we see in the Universe today. To do so, we need to understand a few key cosmological ideas, which can be stated in terms of Hubble's law.

Recall the equation for Hubble's law (Chapter 16):

$$V = H_0 D$$

where recession velocity V (redshift) and distance D are related by Hubble's constant H_0. That constant is the slope of the relationship between recession velocity and distance. The first critical idea hiding in Hubble's law is the (approximate) age of the Universe: H_0 tells us how fast space is expanding now. But we can flip that information around and use H_0 to get an idea of how long ago the points in space were almost on top of each other.

Using Hubble's law and our best observational value for Hubble's constant, approximately 69 kilometers per second per megaparsec [(km/s)/Mpc], we estimate that the Universe is about 14 billion years old (see **Going Further 17.1**). This number is roughly consistent with estimates based on other, more accurate, measures. As we will see in Chapter 18, however, this estimate ignores any acceleration or deceleration that may have occurred during the evolution of the Universe. Taking those factors into account, most scientists now agree that the age of the Universe is closer to 13.7 or 13.8 billion years.

The second critical idea implied by Hubble's law (and the Big Bang) is that by looking out into space, we're looking back in time toward earlier epochs of cosmic history. Even if the Universe were not evolving, the fact that the speed of light is not infinite means that when we look outward, we always look back in time. So when we observe a star 1,000 light-years away, we are seeing that star as it was 1,000 years ago. The same relationship is true of any galaxy.

The Big Bang and Hubble's law added something new to this picture. If the Universe has evolved, when we look into the past we're looking at something different from the present. In a static Universe, we would see a galaxy that is 7 billion light-years distant as it existed 7 billion years ago, but the Universe would be the same then as it is now. "The cosmos itself would not have changed in any significant way," says Wilcots. If, however, the Universe began about 14 billion years ago and has been evolving ever since, seeing a galaxy 7 billion light-years away means looking a significant fraction of the way back to the beginning of everything! In an evolving Universe that began from a Big Bang, the cosmic conditions 7 billion years ago would have been different from what exists today.

Thus, because Hubble's law describes how a galaxy's recession velocity increases with its increasing distance from us, redshift (in cosmology, denoted as z) becomes a kind of clock. An object with redshift $z = 0.1$ is seen as it was when the Universe was 90 percent of its current age (about 1.3 billion years ago). An object with redshift $z = 1$ is seen as it was when the Universe was a little more than half as old as it is today (about 7.7 billion years

> "From the point of view of any randomly chosen galaxy, all the neighbors appear to be speeding away from it like the plague."

INTERACTIVE:
Look-Back Time

cosmological look-back time A measure of time elapsed between the moment photons are detected on Earth and the moment at which they were emitted by a distant object.

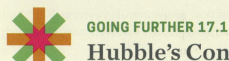

GOING FURTHER 17.1

Hubble's Constant and the Age of the Universe

The exact value of Hubble's constant is enormously important to astronomers because it controls the derivation of many quantities in the study of the Universe—most notably, the age of the Universe. To see how this works, we start by rearranging the equation $V = H_0 D$:

$$V/D = H_0$$

so

$$D/V = 1/H_0 = 1 \left| \left(\frac{69 \text{ km/s}}{\text{Mpc}} \right) \right. = 1 \text{ s} \times 1 \text{ Mpc}/69 \text{ km}$$

Now recall from Chapter 3 that velocity is a measure of distance per unit time. That is, $V = D/T$, and therefore time is a measure of distance over velocity: $T = D/V$. We can replace D/V in the equation above with T:

$$T = 1 \text{ s} \times 1 \text{ Mpc}/69 \text{ km}$$

Here, T represents the time it has taken any galaxy to reach its current distance from Earth, given its current velocity. In other words, T is equal to the age of the Universe, or T_U.

Next we cancel out the units of distance by converting megaparsecs to kilometers (1 Mpc = 3.08×10^{19} km) and then converting seconds to years (1 year = 3.16×10^7 seconds):

$$T_U = 1 \text{ s} \times \frac{3.08 \times 10^{19} \text{ km}}{69 \text{ km}} \times \frac{1}{3.16 \times 10^7 \text{ s/yr}} = 1.41 \times 10^{10} \text{ yr}$$

The current value of Hubble's constant tells us that cosmic expansion began about 14 billion years ago.

ago). If you're looking at an object with a redshift of $z = 4$, you're seeing the Universe when it was just 10 percent of its current age (about 12.4 billion years ago). In this way, we associate every redshift with a **cosmological look-back time**—the time measured from where we are now back to the moment in cosmic history when the photons we see today were radiated (**Figure 17.4**; **Table 17.1**).

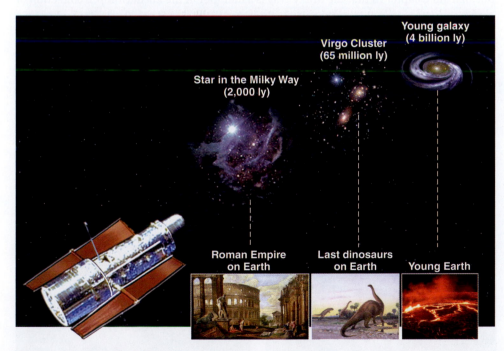

Figure 17.4 Look-Back Time
The light travel times to a few classes of objects and the events occurring on Earth when light from those sources was emitted are depicted. Looking back in space means looking back in time. When we see an object 1 million light-years (ly) away, we are viewing it as it was when its photons were emitted 1 million years ago.

Table 17.1   Redshifts (z) and Look-Back Times

Value of z	Look-Back Time (billions of years ago)
0	0
0.01	0.14
0.03	0.40
0.1	1.3
0.3	3.5
0.5	5.1
1.0	7.7
2.0	10.6
3.0	11.7
4.0	12.4
10.0	13.5

For the Universe as a whole, the mathematical connection between redshift and look-back time is not simple, because light from distant objects has been moving through a space-time that has also been expanding.

Section Summary

- Hubble's discovery that all galaxies appear to be receding from us was the first step in understanding that the Universe has been expanding ever since its beginning at the Big Bang.
- The largest-scale structures that make up the Universe—galaxies, galaxy groups, galaxy clusters, and superclusters—have been evolving since the Big Bang.
- All galaxies recede from other galaxies as space-time expands. Astronomers imagined that expansion in reverse to trace the evolution of the Universe to the Big Bang, the timing of which is estimated by Hubble's law.
- Measurement of an object's recession velocity (redshift) indicates the object's look-back time.

CHECKPOINT Why does the fact that light travels at a finite speed mean that looking at distant objects also mean you are seeing them as they were in the distant past?

Cosmic Neighborhoods: Galaxy Groups and Galaxy Clusters

17.2 Climbing up the ladder of cosmic structure is equivalent to climbing outward on the cosmic distance scale. Beyond the neighborhood of the Milky Way and its small satellite galaxies, the next rungs we encounter are galaxy groups and galaxy clusters.

Galaxy Groups

Recall from Chapter 15 that the Milky Way is a member of the *Local Group*, a collection of gravitationally bound objects that also includes two other large spiral galaxies and a number of dwarf galaxies (**Figure 17.5**). The smallest spiral galaxy in the Local Group is M33, the Triangulum Galaxy. The Milky Way is next in size, and the Andromeda Galaxy, M31, is the largest galaxy in the Local Group. All told, the Local Group contains at least 50 galaxies and more than 700 billion stars.

"All of these galaxies are tied together by their mutual gravity," says Eric Wilcots. "They orbit around a common center of mass, and over time, many of them experience collisions with their neighbors." We've already seen that many of the Milky Way's satellites have interacted with each other or the Milky Way. Those interactions extend even to the largest members of the Local Group. "From their current relative motions toward each other," says Wilcots, "astronomers predict that the Milky Way and Andromeda will

INTERACTIVE:
Scale of the Universe

"Astronomers predict that the Milky Way and Andromeda will collide in 4 billion years."

Figure 17.5 The Local Group
Like other galaxy groups, the Local Group includes a mix of galaxy types, gravitationally bound as a whole, with dwarf galaxies clustered around the more massive spiral galaxies. The Milky Way is one of only three spiral galaxies in the Local Group; the other 50 or so galaxies are elliptical or irregular. **a.** Triangulum Galaxy (M33), the smallest spiral galaxy in the group. **b.** NGC 205 (M110), a dwarf elliptical galaxy that is a satellite to the Andromeda Galaxy (M31).

5 kpc

a. VIS

3 kpc

b. VIS

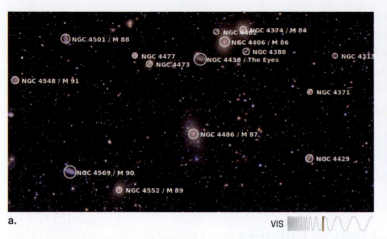

a.

VIS

NGC 4889

NGC 4874

Dwarf galaxies in Coma Cluster

10"

b.

VIS

IR

Figure 17.6 Galaxy Clusters
a. The central region of the Virgo Galaxy Cluster, some 16 Mpc from the Milky Way. Dominant among its 1,000-plus galaxies is the giant elliptical galaxy M87.
b. A false-color mosaic of the central region of the Coma Galaxy Cluster, revealing thousands of faint dwarf galaxies. Two large elliptical galaxies, NGC 4889 and NGC 4874, dominate the cluster's center.

collide in 4 billion years." Thus, even at the largest scales, galaxy collisions are an important part of galaxy evolution.

The Local Group is not unusual. Many, if not most, galaxies can be found in gravitationally bound aggregates called **galaxy groups**. The typical group consists of fewer than 50 galaxies and has a diameter of about 2 megaparsecs (Mpc). The total mass in such groups is approximately 10^{13} solar masses, or M_{Sun} (which would equal 10 Milky Way galaxies). Like the motion of stars within galaxies, the motion of galaxies within groups indicates the presence of unseen material whose gravity binds the visible matter together. That means galaxy groups show evidence of dark matter. Typically, galaxy groups show ratios of dark to luminous matter of 5.5:1, providing more evidence that dark matter dominates the Universe's mass.

Galaxy Clusters

Groups are the smallest association of galaxies that have formed in the Universe. Astronomers who make larger three-dimensional maps of galaxies and their distribution in space, however, find larger and far richer structures. **Galaxy clusters** are the next step upward in cosmic structure. "Galaxy clusters contain anywhere from hundreds to thousands of galaxies in aggregates that stretch across 2–10 Mpc of space," says Wilcots. The first indication that some galaxies appeared to be clustered goes back to the 18th century, but it was in the 1930s that photographic surveys showed the first strong evidence for clusters.

The Virgo Galaxy Cluster, in the constellation Virgo, located 16 Mpc from the Milky Way, is a relatively nearby example of a **regular rich cluster**. Astronomers group clusters by the number of galaxies they contain and their shape. Regular clusters are spherical and centrally condensed, meaning that the density of galaxies (the number per volume) increases toward the cluster's center. Rich clusters contain more than 1,000 galaxies. With between 1,300 and 2,000 members, Virgo is a moderately rich cluster (**Figure 17.6a**). The Virgo Galaxy Cluster has a diameter of about 5 Mpc, about the same size as the Local Group. "That means the cluster packs almost 50 times as many galaxies into the same space as is occupied by the Milky Way, Andromeda, and their 50 neighbors," says Wilcots.

The outer parts of the Virgo Galaxy Cluster host a fairly even mix of galaxy types: spiral, elliptical, and irregular. "In the inner, denser parts of Virgo, it's a different story," says Wilcots. Virgo's core shows far more elliptical than spiral galaxies. "Ninety percent of the galaxies in Virgo are dwarfs, meaning it's the little guys that really make up the cluster," Wilcots explains.

Another well-studied galaxy cluster occurs in the constellation Coma Berenices. The Coma Galaxy Cluster is a spectacular regular rich cluster about 99 Mpc from the Milky Way. This cluster contains at least 2,000 galaxies (perhaps as many as 10,000) in a volume

galaxy group A gravitationally bound collection of galaxies, usually containing fewer than 50 members.

galaxy cluster A gravitationally bound collection of galaxies containing thousands of members.

regular rich cluster A spherical, centrally condensed galaxy cluster with more than 1,000 galaxies.

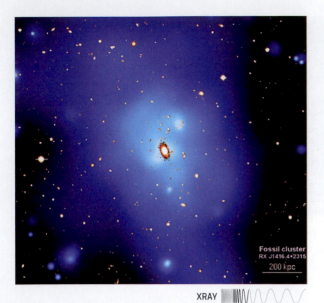

Figure 17.7 Cluster X-ray Halos
Galaxy cluster RX J1416.5+2315 (captured by the XMM-Newton satellite). Its hot halos, which appear as a vast cloud, have temperatures that reach tens of millions of kelvins and constitute a substantial fraction of the cluster's total mass.

"Density matters because that is what drives collisions."

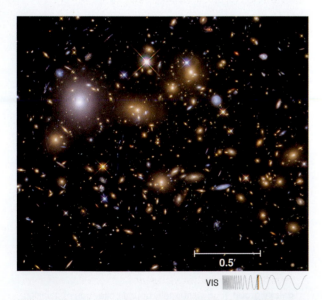

Figure 17.8 Galaxy Cluster MACSJ0717.5+3745
A Hubble Space Telescope image of one of the densest, most massive clusters known (and the largest gravitational lens known). The galaxies in a cluster are gravitationally bound and orbit a common center of mass.

of space more than 6 Mpc across (**Figure 17.6b**). At the heart of the cluster are two giant elliptical galaxies, each with mass more than $2 \times 10^{13} M_{Sun}$ (more than 10 times the Milky Way's mass). Beyond these monsters, most of the bright galaxies in the central regions of the Coma Galaxy Cluster are elliptical or S0-type galaxies.

Unlike the situation in the Virgo Galaxy Cluster, only 15 percent of the brightest 1,000 galaxies in Coma are spirals. "The deficit of spirals in a rich cluster like Coma, compared with less dense collections of galaxies like the galaxy groups, points to an important feature of galaxy evolution," says Wilcots. "Density matters because that is what drives collisions. The higher the density of galaxies in a region, the higher the likelihood of galaxies running into each other in ways that disrupt spirals and turn them into ellipticals or irregulars."

X-ray Halos of Galaxy Clusters

Galaxy clusters are made of more than just galaxies. As X-ray telescopes were lifted above the atmosphere with increasing success, astronomers used these space-based instruments to discover that every galaxy cluster has an **X-ray halo**—a halo of hot, X-ray-emitting gas.

The X-ray halos surrounding galaxy clusters are large, roughly spherical regions of gas at temperatures of 30–100 million kelvins (K). The halos' origin is a subject of intense debate, but they are composed mainly of ionized hydrogen (that is, free protons and electrons). Detailed observations using high-resolution space-based X-ray instruments, such as the Chandra space telescope, reveal that each galaxy in a cluster has its own bright X-ray halo (as the Milky Way does). "The space in between the galaxies is also filled with hot, X-ray-emitting gas," explains Wilcots, and this gas forms an **intracluster medium** (**Figure 17.7**). The cluster halos contain at least 9 times as much mass as all of the galaxies in the cluster.

One of the most important characteristics of the cluster X-ray halos is their pressure. The halo is like a giant balloon. Without the cluster's gravity, the hot, X-ray-emitting gas would expand into space. Astronomers can determine the gas pressure of the X-ray gas from direct measurements of its density and temperature and find that the halos have far too much pressure to be held by the visible mass of the cluster. "All of the galaxies and the X-ray-emitting gas itself do not produce enough inward gravitational pull to constrain the outward push from the hot X-ray-emitting gas," says Wilcots. There must be more mass to keep the X-ray gas in equilibrium—that is, to keep it from expanding out to space. The extra mass must be in the form of dark matter. "That means cluster halos provide yet more indirect evidence for the existence of dark matter." Only by adding 10 times more mass in the form of dark matter can astronomers explain why the X-ray halos remain confined.

Thus, as in galaxy groups, both the X-ray gas and the galaxies themselves are gravitationally bound within galaxy clusters. That means the collective gravity of all of the mass in the cluster is large enough to keep the galaxies and the intracluster medium from escaping. Hence clusters are *long-lived structures*.

Further evidence that clusters are not just random, short-lived configurations comes from a general astronomical principle: the present moment we live in, the moment when we're making our observations, is not in any way special. Thus, if we see galaxies in a cluster (which means we're seeing them as they were hundreds of millions of years ago if the cluster is hundreds of millions of light-years away), it is improbable that they are just a random collection of objects passing close enough at the moment of observation to *appear* to be a cluster (**Figure 17.8**). Instead, the galaxies must orbit a common center of mass created by the gravitational pull of the total mass of the cluster, including its dark matter (see **Going Further 17.2**). The fact that we see similar structures at different distances, which means different look-back times, also convinces astronomers that there is nothing particularly special about our time. If we saw certain structures (such as clusters) only at certain times, then we could indeed say that those times must have been special.

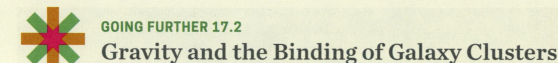

Gravity and the Binding of Galaxy Clusters

Imagine you have a bunch of marbles in a transparent box and take photos as you shake the box hard. One image might show that most of the marbles were bunched up at the center at the moment that image was taken. If that were your only image, how would you know whether the "cluster" of marbles was a random occurrence or a more permanent structure formed by forces pulling the marbles together? Astronomers face that type of question when they try to understand galaxy clusters. To answer it, they apply a bit of physics and common sense.

Using Doppler shift measurements, astronomers can find the speed of any galaxy within a cluster. For the Virgo Galaxy Cluster, typical galaxies move at V_{galaxy} = 900 km/s. At this speed, how long would it take a galaxy to cross from one end of the cluster to the other, over a diameter D of about 5 Mpc? Using the formula relationship time = distance/velocity, where the diameter is the distance, we calculate the crossing time t_{cross} as follows (including unit conversions):

$$t_{cross} = D/V_{galaxy} = (5 \text{ Mpc})/(900 \text{ km/s})$$

$$= \frac{(5.0 \times 10^6 \text{ pc}) \times (3.1 \times 10^{13} \text{ km/pc})}{(900 \text{ km/s}) \times (3.15 \times 10^7 \text{ s/yr})}$$

$$= 5.5 \times 10^9 \text{ yr}$$

Next we compare this crossing time t_{cross} with the age of the Universe T_U, which (from Section 17.1) is 1.37×10^{10} years. If t_{cross} is sufficiently less than T_U, then the galaxy has had plenty of time to go from one end of the cluster to the other in the history of the Universe. Therefore, if the galaxy were not bound to the cluster and you looked at a random time (which is essentially what happens when an astronomer makes an observation), you would not expect to see that galaxy bunched up with others in the cluster. Since you *do* see a cluster and find that $t_{cross} \ll T_U$, the cluster must be held together by the force of gravity. Thus, the cluster is a bound structure, not a random occurrence.

We can also use the average galaxy velocity in a cluster, along with cluster size, to determine a cluster's mass. If we assume, for convenience, that the galaxy is on a circular orbit around the cluster center, we can begin with the circular-velocity formula (see Going Further 5.1):

$$V_{galaxy} = V_c = \sqrt{\frac{GM_{cluster}}{R}} = \sqrt{\frac{GM_{cluster}}{(D/2)}}$$

where G is the gravitational constant. Rearranging, we get

$$M_{cluster} = (V_{galaxy})^2 \frac{D}{2G}$$

Using the values V_{galaxy} = 900 km/s and D = 5 Mpc, we have

$$M_{cluster} = (900 \text{ km/s})^2 \times 5 \text{ Mpc}/2(6.67 \times 10^{-11} \text{ m}^3/\text{kg/s}^2)$$

Carrying out the unit conversions and performing the calculation, we find that $M_{cluster} = 4.7 \times 10^{14} \, M_{Sun}$. That is far more than what is observed from the luminous material in the galaxies and X-ray halos. Thus, our calculation points to the existence of dark matter in clusters.

Galaxy Clusters and Gravitational Lenses

Unlike individual galaxies, galaxy clusters provide a means for astronomers to investigate the amount and distribution of dark matter through gravitational lenses—a remarkable effect predicted by Einstein's general theory of relativity (see Chapter 14). General relativity predicts that space-time is curved by the presence of matter. Any object moving freely through space (not being acted on by a force such as a rocket motor) must move along paths defined by the shape (the curvature) of space-time. Photons traveling through empty space are perfect "sensors" of this curvature. Photon paths are deflected from straight lines to the degree that the geometry of space-time is distorted by mass.

"Just like a glass lens in a telescope can bend the path of light, collecting it into an image," explains Eric Wilcots, "the distortion of space-time by matter can deflect the path of light, creating an image of background objects." The dark matter within a galaxy cluster creates a large-scale deformation of space-time. Consider an object (such as a quasar or young galaxy) that is more distant than a galaxy cluster, relative to an observer (**Figure 17.9a**). Light from that object is deflected from its original path as it passes near the cluster. Like a lens, the curved space gathers the light into an image. When the cluster's dark matter is arranged in a highly symmetric distribution, **Einstein rings** form,

X-ray halo A large, quasi-spherical region of hot gas surrounding a galaxy or galaxy cluster.

intracluster medium Thin gas in the space between galaxies in a galaxy cluster.

Einstein ring The ring-shaped appearance of a distant background object that is caused by a gravitational lens formed via symmetric distribution of dark matter in the foreground.

Figure 17.9 Gravitational Lensing
a. The general theory of relativity predicts that the presence of mass bends light's path through space-time. In this way, a massive galaxy cluster (including its dark matter) acts like a lens, bending the light of a background quasar, and may form Einstein rings or multiple arcs. **b.** A foreground galaxy cluster, Abell 2218, has "lensed" the light of an extremely young galaxy in the background, 13.4 billion light-years away, into many arcs (marked by arrows).

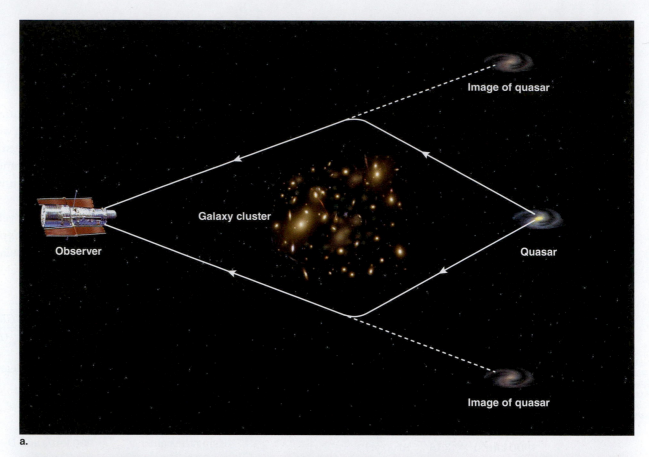

a.

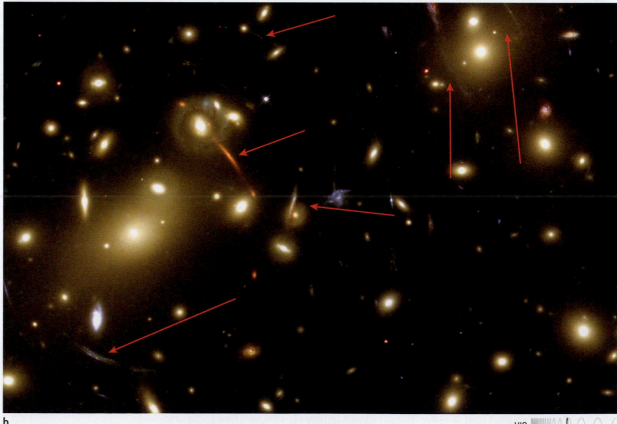

b.

VIS

in which the image of a background object appears in a thin, circular region around the cluster (**Figure 17.9b**). But since a cluster's dark matter tends to be asymmetric, the lens can rarely gather the light from the background object into a single image ring. Instead, multiple images of the background object usually form and appear as arcs.

"Since the properties of the image depend on the shape of space-time in and around the cluster, and since space-time is distorted by mass," explains Wilcots, "gravitational lenses give astronomers a powerful tool for exploring dark matter in clusters." Using the shape of the images and their brightness, astronomers can determine the amount of dark matter as well as probe its distribution. In this way, gravitational lenses have yielded some of the most compelling evidence of the existence of dark matter. As with other methods (galaxy rotation curves, X-ray halos, and so forth), studies of lenses show that the ratio of dark to luminous material is about 10:1. "Thus, for every kilogram of stuff like the kind you are made of," says Wilcots, "there are 9 kilograms of the mysterious dark stuff." However, the ratio of dark to normal matter is not the same for different kinds of structures. Galaxy groups show ratios of about 5.5:1, while clusters show ratios of 10:1. The reason for these differences is a point of active research.

Section Summary

- Galaxies are typically found in gravitationally bound groups with up to 50 interacting members. Galaxy movement within groups provides evidence of dark matter.
- Larger in scale than galaxy groups are galaxy clusters, which contain hundreds to thousands of members. The denser the cluster, the more collisions occur and the higher the proportion of elliptical galaxies near cluster centers.
- Galaxy clusters have large X-ray halos that may form when the gaseous halos of member galaxies merge. Halos and galaxies are gravitationally bound within a cluster, whose total mass is great enough to indicate the presence of dark matter. We see similar structures at different look-back times.
- Gravitational lensing occurs when a large concentration of mass distorts space-time, bending the light from background objects. Lensing also provides further evidence of dark matter in galaxy clusters.

CHECKPOINT Why would there be more galaxy collisions in a cluster than in a group?

ERIC WILCOTS

Eric Wilcots was born in Philadelphia—not exactly dark-sky territory. He was 8 when his parents gave him a telescope. "You could see the rings of Saturn!" he recalls. "I mean, how cool is that?" The arrival of that telescope coincided with the *Voyager* mission passing by Jupiter. "The guys at JPL [Jet Propulsion Laboratory]," says Wilcots, "they just looked like they were having fun. I mean, there was Io and its volcanoes."

After earning an undergraduate degree at Princeton University and a PhD at the University of Washington, Wilcots joined the University of Wisconsin's astronomy department, where he studies galaxies and their environments. "It's still fun," he says. "I still get excited thinking about getting new data and figuring out ways to answer a question no one else has asked."

Superclusters and Large-Scale Structure

17.3 Before we can get to the broadest perspective on cosmic structure, we must understand the difficulty involved in gaining that perspective. "Galaxy groups and galaxy clusters can both be found in relatively nearby regions of space," explains Eric Wilcots. "For example, for all its majesty, the Coma Cluster is 'only' about 100 Mpc away. While that might seem like an enormous distance, it constitutes only 1 percent of the distance to the edge of the visible Universe."

To get a sense of structure on the largest scales, astronomers need to map the distribution of galaxies in space out to billions of parsecs. Creating accurate maps means establishing the three-dimensional positions—and therefore distances—of perhaps millions of galaxies. To make these maps, astronomers must again turn to Hubble's law and measurements of redshift.

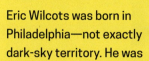

"For every kilogram of stuff like the kind you are made of, there are 9 kilograms of the mysterious dark stuff."

Redshift Surveys

"If we have a telescope, especially a big one, we can open the shutter for a long time, collect the light, and see that the Universe is full of galaxies," says Margaret Geller, the astronomer who led the way in producing a three-dimensional map of the distribution of galaxies.

Figure 17.10 Sloan Digital Sky Survey
The SDSS used this dedicated 2.5-meter telescope at Apache Point Observatory, New Mexico, to survey a large portion of the sky.

"The Universe is big and life is short."

redshift survey Measurements of redshift (recessional velocity) for many galaxies across a region of the sky, from which astronomers produce three-dimensional maps of galaxies.

"So the first thing we do," she continues, "is to take pictures of the sky where you can actually see these concentrations of galaxies like the Coma Cluster. You can see them on the sky, but the question is, are they really all together in space? What's missing is we don't have the distance."

That's where redshift and Hubble's law come in. Astronomers focus on redshift when studying cosmic structure because it is, in principle, easy to measure. Just take a spectrum of a galaxy, locate familiar spectral lines, and determine how much they have shifted to longer wavelengths. Using that measured redshift, Hubble's law then enables a determination of distance. By measuring hundreds, thousands, or even millions of galaxy redshifts, astronomers can take the galaxies' positions on the two-dimensional bowl of the night sky and create a three-dimensional map of matter in the Universe over a tremendous scale. In this way, redshifts reveal the Universe's large-scale structure. That is why, in the 1980s, astronomers began the difficult task of compiling **redshift surveys**.

A redshift survey is a compilation of spectra from many, many galaxies. It's not an easy task. "The Universe is big and life is short," says Geller. "Part of what determines whether you can detect the galaxy and take a good redshift is how much light you can detect." The inverse square law for brightness shows that even very bright objects appear faint if they are far from us. Long exposures are needed to catch enough light to make a high-quality spectrum for accurate redshift determination. When Geller began her pioneering work in redshift surveys, even the detectors were a problem. "When I started doing this, the light detectors were very inefficient. Now they are much better and you catch 90 percent of the incoming photons." Even today, getting long exposures for hundreds of thousands of galaxies requires lots of telescope time. Because the point of a survey is to map the three-dimensional locations of galaxies across space, astronomers must make observations over as wide a patch of sky as possible. That requires even more telescope time. All told, redshift surveys are some of the astronomical community's most complicated and time-consuming undertakings.

One of the most famous redshift surveys was the Sloan Digital Sky Survey (SDSS), which ran from 2000 to 2005 in its initial phase. To accomplish its goal of mapping the three-dimensional structure of a sizable portion of the sky, a special-purpose 2.5-meter telescope was built in the mountains of New Mexico (**Figure 17.10**). To facilitate rapid data acquisition, special filters were constructed for each region of the sky to be observed by the SDSS. Holes were drilled into a metal plate at the exact locations where galaxy images would appear. Fiber-optic lines were attached to the holes so that the light from the galaxies could be quickly and efficiently collected and analyzed. "So the same kind of fibers that carry everyone's telecommunications also are used to carry light to our detectors," says Geller. "That means people can easily capture 1,000 redshifts at 10 times the distance in one night." Using such technology, the SDSS imaged more than 350 million celestial objects during its operation. Additional spectra were then taken from 930,000 galaxies, 120,000 quasars, and 460,000 stars included in the full sample. Other surveys, such as the Australian Two-degree Field (2dF) project, collected data from almost 250,000 objects, none of which were covered in the SDSS survey.

Redshift surveys tell us a great deal about the future of astronomy, as well as our ever-expanding narrative of cosmic history. These surveys are large-scale projects requiring the collaboration of many scientists over many years. Typically, a scientific paper by the SDSS team can have 30 or more authors. Modern astronomical research can involve small armies of researchers and technicians—a trend sometimes called *big science*.

Redshift surveys also demonstrate the importance of large-scale computation in modern astronomy. They accumulate so much information that special software for data storage and analysis had to be developed. "When we did our first redshift survey, just one slice of space would mean 1,000 galaxies worth of data," explains Geller. "Back then it took a half a year for the computers to process and reduce that data." Even with faster computers, today's surveys are so extensive that the largest ones represent the incursion of

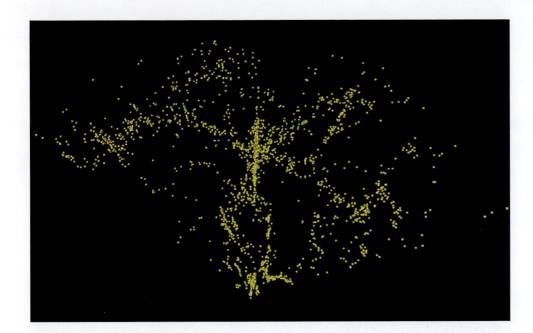

Figure 17.11 "Stickman"
Margaret Geller and her team conducted early redshift surveys and stunned the astronomy community by showing that, contrary to earlier assumptions, the distribution of matter in the Universe is not smooth at the large scales they studied. Instead, it has filaments and vast voids. Their map was dubbed "Stickman" for the distinctive shape of large-scale structure it revealed.

big data into astronomy. All in all, redshift surveys, by demanding very large collaborative efforts and very large sets of data, are creating new ways of doing astronomical science.

What emerged from the redshift surveys were, however, not just new models for doing science but fundamentally new results. The three-dimensional maps of the Universe that SDSS and other surveys produced showed just how far away and how large cosmic structures can be. The maps also provided a story of those structures across more than 13 billion years of cosmic history.

Galaxy Superclusters

Even the first attempts to produce large-scale redshift surveys, the ones created by Margaret Geller and others, stunned astronomers of the time.

The distribution of matter in the Universe, on large enough scales, had always been assumed to be uniform—what astronomers call "smooth." Surveys out to hundreds of millions of parsecs, astronomers thought, would illustrate such uniformity. Geller's first redshift survey, however, showed astronomers how wrong they could be. With just 1,100 galaxies observed in a thin strip of sky, Geller's team was able to probe the mass distribution out to more than 200 Mpc. Although the map was sparse, it showed two important features: First, the galaxies were arranged into long filaments, not distributed uniformly. Second, the filaments were separated by vast voids where no galaxies appeared (Figure 17.11). Even on these enormous scales, matter has structure. The team had discovered **galaxy superclusters**.

A galaxy supercluster is a chain of clusters and groups that are being shaped by their mutual gravitation. Superclusters are so large that not all of the material may be gravitationally bound. Some of the matter is still in the process of collapsing into the supercluster. That means supercluster formation is still ongoing even after more than 13 billion years of cosmic evolution. The Milky Way lies near the edge of the Local Supercluster, an oblong structure containing at least 100 groups and clusters. It is also called the Virgo Supercluster because the Virgo Galaxy Cluster makes up the bulk of its mass. ("Local Supercluster" is sometimes used to refer to the vast Laniakea Supercluster as well, which includes the Virgo Supercluster and a number of others. *Laniakea* is Hawaiian for *immense heaven*.) The Local Supercluster is typical of most superclusters in mass and luminosity, at $10^{15} M_{Sun}$ and $10^{12} L_{Sun}$. The Local Supercluster differs from other superclusters, however, in the way its galaxies are distributed from the center to the outer regions. The Local

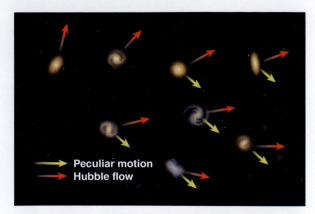

Figure 17.12 Hubble Flow versus Peculiar Motions
Galaxy motions based on Hubble flow occur because of the expansion of space-time associated with the expanding Universe. Peculiar motions are responses to the gravitational pull of large masses nearby (on cosmic scales). Here peculiar motions are influenced by an unseen large mass in the downward direction.

"The voids are huge, . . . and it's a whole lot of nothing."

AT PLAY IN THE **COSMOS** THE VIDEOGAME

In Mission 18 your xFLT reactor fails as you travel back to the early Universe. Find a region of colliding galaxies and look for starbursts to get the heavy elements needed to make repairs.

galaxy supercluster A chain of clusters and groups that are being shaped by their mutual gravitation.

void A region between superclusters, spanning tens to hundreds of megaparsecs, where galaxy density is much lower than average.

Hubble flow The pattern of large-scale galaxy motion associated with the expansion of space-time.

peculiar motion The motion of a galaxy through space that is due to gravitational attraction, distinct from motion due to the expansion of space-time.

Great Attractor An unseen concentration of mass whose gravity affects the motion of galaxies in a large volume of space that includes the Milky Way.

Supercluster is considered to be "poor" in the sense that it does not have a central condensation of galaxies. Notice that the definitions of *poor* and *rich* are the same for superclusters and clusters—"rich" structures have a high density of galaxies at their centers.

The Perseus-Pisces Supercluster is a rich supercluster. More than 76.7 Mpc away, it appears as a particularly dense, filamentary concentration of galaxies 45 Mpc in length. It is dominated by a centrally condensed region containing the Perseus Cluster, which is one of the most massive galaxy clusters known.

While the Perseus-Pisces Supercluster is a notable example of a supercluster on redshift survey maps, it borders a similarly notable example of a **void**. The Taurus Void is a spherical region 33 Mpc in diameter that is bounded on all sides by galaxies (including those of Perseus-Pisces). Across the vast expanse of the Taurus Void, astronomers find only a handful of galaxies; it appears almost completely devoid of clusters, with only a few isolated galaxies. "The voids are huge," says Geller. "They can be 150 million light-years across, and it's a whole lot of nothing." Voids are essentially the complements of superclusters. "We use the galaxies to trace where the matter is," says Geller, "and they give us a fairly good picture that there really is less material in them than there is in these dense structures that surround them."

After completing their redshift surveys, astronomers were faced with the reality that structure still exists on hundred-million-parsec to billion-parsec scales. More important, such structure took the remarkable form of alternating nearly empty voids and denser supercluster filaments. That discovery rocked the astronomical world. The next crucial step in explaining this large-scale structure could not be completed without adding motion to the story.

Rivers of Galaxies

Redshift surveys did more than tell astronomers where galaxies are located in space; they also showed astronomers how those galaxies move. Since redshifts represent velocities (which are then converted to distance via Hubble's law), astronomers could see that in some cases the galactic motions include components that differ from simple expansion. By Newton's laws of gravity, if there is motion, then mass must be present to lead to that motion.

Astronomers call the basic pattern of galactic expansion the **Hubble flow**. Remember from Chapter 14 that the motion of expansion is not that of the galaxies through space but the expansion of space itself. Galaxies are simply carried along with space-time as it expands in the aftermath of the Big Bang. Galaxy clusters also do not expand along with the Hubble flow, because their gravity holds them together. Redshift surveys can, however, detect deviations from the pure expansion pattern of Hubble flow, called **peculiar motions** (Figure 17.12). The presence of peculiar motions tells astronomers that galaxies do not just flow *with* space as a result of the Big Bang but also flow *through* space. Thus, it is gravitational attraction that causes galaxies to stream toward each other.

One of the most dramatic examples of peculiar motion is caused in part by a concentration of mass called the **Great Attractor** (Figure 17.13). Deviations from Hubble expansion of as much as 700 km/s are seen in galaxies in a region of sky centered on the constellation Centaurus. That means astronomers have detected a vast river of galaxies flowing in the same direction.

Although astronomers originally associated the river of galaxies with the Great Attractor, it soon became apparent that an even larger concentration of mass at even greater distances also plays a role. The Shapley Supercluster, the most massive supercluster known, lies 500 Mpc away, beyond the Great Attractor. The Shapley Supercluster packs the equivalent of 20 Virgo Clusters into the same volume as the Virgo Cluster.

In all such rivers of galaxies flowing through space, the mass needed to account for the motions of the material we see (the galaxies) is far too large to be accounted for by other collections of observable luminous matter. Thus, peculiar motions represent another strand of evidence for dark matter in the cosmos.

The Cosmic Web

As we've seen, exploring the Universe on ever-larger scales has meant *discovering structure* on ever-larger scales. Astronomers found galaxy groups first (on the scale of megaparsecs), then galaxy clusters (tens of megaparsecs), and then galaxy superclusters (hundreds of megaparsecs). "There is an obvious question," says Avi Loeb, a theoretical astrophysicist at Harvard University. "Does it ever stop? What we are looking for is the length scale at which one patch of the Universe looks the same, on average, as any other patch." "We know there's structure on the scale of hundreds of millions of light-years," says Margaret Geller. "But is there structure on a scale to a billion light-years or more?"

What Loeb and Geller mean is this: if we pull back far enough, does the distribution of matter in the Universe ever smooth out? "You compare different regions of the Universe and see if they are the same in a statistical sense," explains Geller. "Think of these regions as boxes. The question is, what is the scale where you can take this-sized box and compare it with another, similar-sized box somewhere else in the Universe and have them look the same in terms of the average properties? So, what's the size of that box? We know it's not the size of a house, it's not the size of Earth, it's not the size of the Solar System, it's not the size of the galaxy. How big is it? Have we found it?"

The answer for some astronomers is yes. As they mapped the large-scale Universe in greater and greater detail, it seemed that no new distinct structures appear beyond a few hundreds of megaparsecs. **Figure 17.14** shows one of SDSS's deepest and most complete

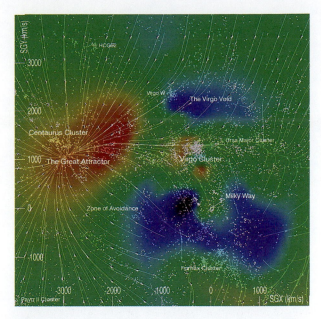

Figure 17.13 The Great Attractor
The enormous mass concentrated in the region of space occupied by the Centaurus Cluster produces so much gravitational attraction that it dominates the motions of our local region of the Universe, leading astronomers to call the region the Great Attractor.

Figure 17.14 Cosmic Web
A cosmic-scale, 3D SDSS map showing the interconnected filaments of superclusters and the voids between them. Earth is at the center, and each point outward represents a galaxy. The outer circle is at a distance of 2 billion light-years. The region between the wedges was not mapped; dust in our own galaxy obscures the view of the distant Universe in these directions.

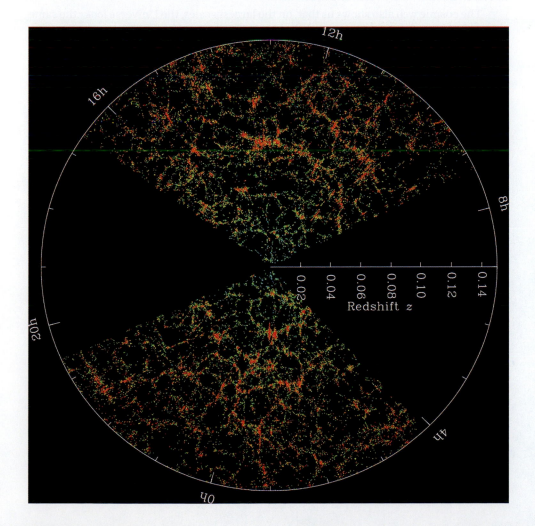

"We know there's structure on the scale of hundreds of millions of light-years. But is there structure on a scale to a billion light-years or more?"

three-dimensional maps of the Universe. This stunning image appears to show that, on billion-parsec scales, the pattern of superclusters and voids begins to repeat itself. At such scales, individual superclusters look like a network of interconnected filaments joined at the densest regions (the dense clusters). Together, these filaments become walls surrounding the voids, and the entire pattern takes on the appearance of a dense network of structure that some astronomers call the **cosmic web**. "I also like to think of it as if the galaxies are on the surfaces of giant bubbles," says Geller, "where the surface is kind of a weblike structure defining where the clusters and groups are."

The idea that the hierarchy of structure ends with the cosmic web has important consequences for the field of cosmology. It means that, at the largest scales, the Universe is **homogeneous**—the distribution of matter is the same, on average, in every part of space. For the cosmic web, that means, statistically, the overall pattern of voids and filaments remains constant across different regions of space. Consider the example of a city at different scales. From one city block to another the number of, say, grocery stores, clothing stores, and playgrounds may vary quite a bit. If you pulled back and viewed the city as on larger scales you might find that patches of city 10 blocks across always had the same average number of grocery stores, clothing stores, and playgrounds. That means the distribution of stores and playgrounds becomes homogeneous at large enough scales.

Returning to the Universe and its structure, even though any one void might be larger than another, the *average* properties of the cosmic web—such as density or void size—are the same from one place to another. "This means you must consider large enough volumes of space for such averages to make sense," says Loeb. Calculations of the average properties must include many voids and filaments.

The cosmic web also shows us that, on the largest scales, the Universe *looks* the same in all directions: the same average pattern of voids and filaments occurs, wherever you point your telescope. The Universe is therefore **isotropic** (see **Anatomy of a Discovery** on p. 462).

The fact that, on these largest scales, the Universe is both isotropic and homogeneous makes the development of grand, all-inclusive cosmological theories much easier. "If every region of the Universe looked different at even the largest scales, it would be impossible, or at least very difficult, to embrace the whole Universe as a simple system that is statistically predictable," says Loeb.

It is remarkable that when Einstein and others began developing the first modern cosmological models 100 years ago, they assumed isotropy and homogeneity because they had to. Any other choice made the mathematics too difficult to solve. (From these early models, the theory of the Big Bang eventually emerged.) A century later, astronomers have discovered that the Universe may have exactly the right large-scale properties to make models like those of Einstein and his followers possible. Cosmologists, perhaps, got lucky.

However, the scientist who was responsible for the first maps of large-scale structure remains unconvinced that we've found the scales at which the Universe is truly homogeneous and isotropic. "There aren't surveys large enough and dense enough to really limit these issues," says Margaret Geller, who believes more work must be done. Her reticence demonstrates an important aspect of how science works. It's sometimes the insistence of members of the research community to reexamine claims that helps scientists avoid complacency.

Section Summary

- Redshift surveys that cover large areas of the sky capture the dim light of at least hundreds of thousands of galaxies. These surveys provide the largest-scale three-dimensional maps of the Universe and clear evidence of the evolution of cosmic structure.
- Redshift surveys reveal superclusters separated by vast, nearly empty voids. On the size scale of superclusters, cosmic structure is not seen to smooth out as expected.

"We know it's not the size of a house, it's not the size of Earth, it's not the size of the Solar System, it's not the size of the galaxy."

cosmic web Structure on the largest cosmic scales, appearing as a network of interconnected filaments (joined where dense clusters are found) and surrounding voids.

homogeneous Uniformly distributed. A homogeneous distribution of matter has the same average properties at every point in space.

isotropic Uniform in appearance when viewed from different angles. An isotropic distribution of matter looks the same in all directions.

- The Hubble flow is the pattern of galactic expansion of space-time. Deviations from that flow, known as peculiar motions, are galaxy motions caused by the presence of large-scale concentrations of material (mostly dark matter), creating "rivers of galaxies" that flow through space.
- On the largest scales, the Universe's structure takes the form of a homogeneous and isotropic cosmic web, built from filaments of superclusters surrounding voids.

CHECKPOINT   Describe the large-scale structure of the Universe.

Large-Scale Cosmic Structure and Cosmic History

17.4

As discussed in Section 17.1, when astronomers look outward far enough to see large-scale cosmic structure, they are also looking backward into the depths of cosmic evolution. Thus, the vast superclusters of galaxies and the cosmic-scale flows associated with them give us one thread in the narrative of the Universe's history. Since the dawn of time, the Universe has been building ever-larger structures (galaxy superclusters) out of smaller ones (galaxy clusters, galaxy groups, and galaxies themselves).

To piece together the full story of the Universe from the Big Bang to the present (and into the future), we need to understand which mechanisms can take the Universe from small to large structures across cosmic time. But to do that we must, once again, rely on observations that look as far back in time as possible. In addition to mapping the entire sky, we need to map backward along *individual lines of sight*, constructing a time line of the most basic building blocks of cosmic structure. In this way, we come to one of astronomy's holy grails: the quest to understand when the first stars and galaxies assembled.

The Era of Quasars

One of the earliest clues to the history of galaxies came from the study of quasars. Recall from Chapter 16 that quasars are a highly luminous form of active galactic nuclei (AGNs). The prodigious light pouring out of a quasar is directly connected with material pouring into a supermassive black hole (SMBH) at the center of a galaxy. Quasars are so far away that, in general, only the superabundant light from regions around the black hole is visible. Light from the rest of the galaxy is overwhelmed by emissions from the nucleus. But the enormous distances to quasars mean that we're seeing these galaxies in the earliest epochs of cosmic history (**Figure 17.15a**).

Since the discovery of quasars in 1963, astronomers have pushed to find ever-more-distant examples. With a redshift of $z = 0.3$, the first quasars observed went back about 25 percent of the way to the Big Bang. The farthest quasars observed thus far have $z = 5.7$ or more, placing them 95 percent of the way to the Big Bang. Seeing that far back into cosmic history is a remarkable achievement. Astronomers also made an unexpected discovery as they pushed back the frontiers of quasar observation: a cosmic era of quasars had come and gone.

Figure 17.15b shows the density of quasars in the Universe (meaning how many quasars existed per unit of volume of space) as a function of time since the Universe evolved, according to the most likely scenario. The graph tells a simple story of quasar history. The number of quasars begins low, rises to a peak, and then drops off sharply. As of a few billion years ago, no quasars were left; therefore, there are no quasars in the nearby Universe. The epoch of quasars ended long ago.

But why was there a quasar era in the first place? The answer to that question gives us two key pieces of information for understanding the history of galaxies in their proper cosmic context. "On the early side of the quasar era, it must have taken some time for

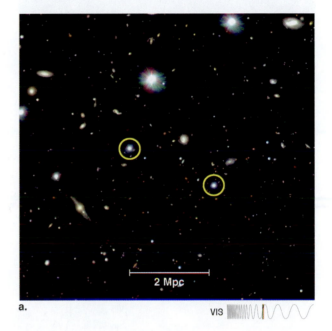

a.   VIS

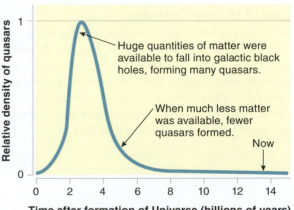

b.

Figure 17.15 The Age of Quasars
a. A pair of quasars (denoted QP0110-0219) at redshift $z \approx 1$, which implies a look-back time of 7.7 billion years ago. **b.** A graph of the relative number of quasars versus the age of the Universe shows that no quasars are forming now. Their birth required the availability of huge amounts of material flowing into the center of a newly forming galaxy.

a. VIS

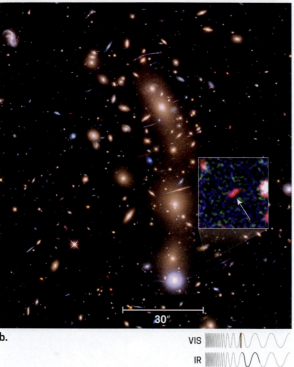

b. VIS

IR

Figure 17.16 Deep Field Images
The Hubble Deep Field project carried out long-time exposures on a small patch of dark sky to image very faint, very distant galaxies. The resulting images provide a direct observation of galaxies in various stages of their early evolution. **a.** Part of a Hubble Extreme Deep Field image that contains about 5,500 galaxies. **b.** An HST image revealing a compact galaxy (inset) forming just 400 million years after the Big Bang.

the black holes (and the galaxies that host them) to form," says Avi Loeb. "The end of the quasar era tells us that the massive infall of material into the supermassive black holes at these galaxies' centers could not last forever. There simply was not enough dense material in the center of the young galaxies to keep feeding the monsters at such a high rate." Thus, quasar history and galaxy history are closely tied, and the fact that quasars came and went gives us an important glimpse into the evolution of galaxies.

Hubble Deep Field

To study the evolution of galaxies across cosmic history, astronomers need to see the galaxies directly. Gaining such a direct view means taking the best instruments, including the Hubble Space Telescope (HST), to extremes.

There is always fierce competition for even a few hours of observing time on the HST. Given the level of demand, it was a shock when, in 1995, HST's director, Robert Williams, announced that a whole week of telescope time was being dedicated to staring at a tiny region of the sky (about the size of a tennis ball seen from across a football field). Even more incredible, the part of the sky he had chosen was empty! At least, no *known* objects were located at those coordinates. But by staring at this region of space for a week, the HST slowly gathered photons from the faintest objects in that patch of sky. This was the origin of the **Hubble Deep Field (HDF)**. All of the objects discovered by the HST that week were indeed very faint—that is, very, very far away. The HDF had captured the first portrait of galaxies still in the agonies of birth.

The HDF was soon followed by other, similar projects, including the Hubble Ultra Deep Field, from images captured in 2003 and 2004, and the Hubble Extreme Deep Field, which was created in 2012 from overlaying 10 years' worth of images (**Figure 17.16**). Working in concert with the Spitzer (infrared) and Chandra (X-ray) space telescopes, the HST also produced the Great Observatories Origins Deep Survey (GOODS). Together, these projects enabled astronomers to see galaxies from even earlier eras of the evolution of the Universe—back to 13.2 billion years ago. That is a redshift of 11.9 and constitutes some of the earliest cosmic structures ever resolved with a telescope.

What astronomers find in their deep fields are galaxies that look like they're still being put together. "There are many irregular-looking objects with hints of spiral structure," says Loeb. "Also, any of the recognizable spiral galaxies appear remarkably blue, and that means they are being overwhelmed by the formation of bright, massive stars." Close cousins of such starbursts (see Chapter 15) have been seen in the local Universe near the Milky Way, but the deep fields make it clear that in the early stages of assembly, entire galaxies can be gripped by the starburst phenomenon. In addition, clearly recognizable in the images are interactions among new galaxies. Everywhere, galaxies appear to run into other galaxies, merge with them, or tear their neighbors apart. It's stark evidence that space was so much more crowded when the Universe was young and that collisions were an essential process in making the galaxies we see today (**Figure 17.17**).

The Dark Ages

Using data such as the Hubble Extreme Deep Field, astronomers can see galaxies already in the process of forming, but going much farther back than about 13 billion years presents its own problems. For galaxies, the most important milepost of cosmic history comes about 380,000 years after the Big Bang. That's when most of the normal matter was neutral hydrogen (the rest was mostly helium). Chapter 18 will explore this important transition in detail. For now, the critical point is that because the Universe had become filled with neutral hydrogen, it was also dark, and in this darkness the first stars and galaxies were born.

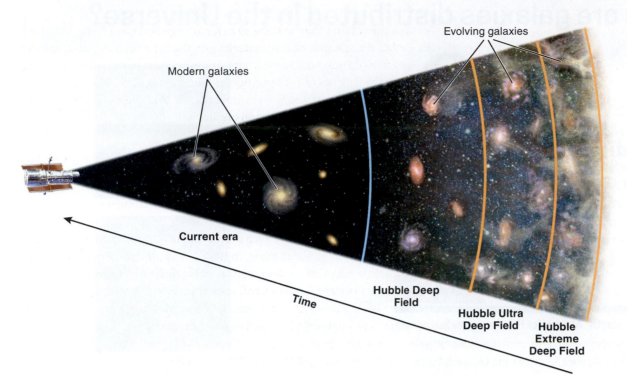

Modern galaxies

Evolving galaxies

Current era

Time

Hubble Deep Field

Hubble Ultra Deep Field

Hubble Extreme Deep Field

Figure 17.17 Evolution of the Universe

To understand the evolution of the Universe, astronomers work backward along lines of sight, constructing a time line of the most basic building blocks of cosmic structure and the stages that created them. Images from the Hubble Deep Field series represent the farthest (and therefore most ancient) galaxies directly observed to date.

"Neutral hydrogen is very good at absorbing photons in the optical [visible] and UV bands," explains Loeb. "That means UV and optical light could not travel very far without being absorbed. If you were around back then, you couldn't have seen very far out into space before everything was lost in a gray fog." The situation that Loeb describes is very different from today's cosmos, in which a red or blue photon can cross billions of parsecs of intergalactic space without being absorbed. Astronomers therefore call the era when space was full of neutral hydrogen the **Dark Ages** (Figure 17.18).

But how did the Universe go from the Dark Ages of early cosmic history to the transparent, well-lit conditions we see today? "It's the formation of stars and galaxies themselves that ended the Dark Ages," says Loeb. "That is why we are so interested in this epoch of cosmic history. It's where stars and galaxies begin."

> "The end of the quasar era tells us that the massive infall of material into the supermassive black holes at these galaxies' centers could not last forever."

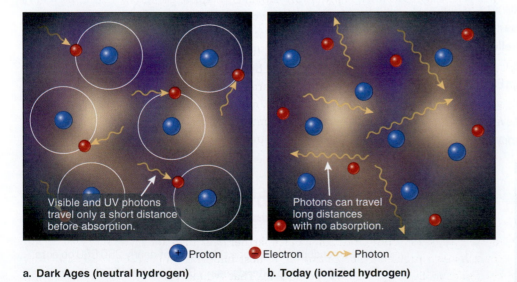

Visible and UV photons travel only a short distance before absorption.

Photons can travel long distances with no absorption.

+ Proton   – Electron   〰 Photon

a. Dark Ages (neutral hydrogen)   **b. Today (ionized hydrogen)**

Figure 17.18 The Dark Ages

a. During this era, dense clouds of hydrogen gas reabsorbed visible and ultraviolet photons as soon as they were emitted, making the Universe opaque then. **b.** Today, most intergalactic hydrogen is ionized, so the Universe is transparent to visible photons.

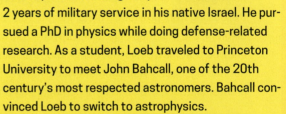

AVI LOEB

After high school, Avi Loeb continued his studies as he completed the obligatory 2 years of military service in his native Israel. He pursued a PhD in physics while doing defense-related research. As a student, Loeb traveled to Princeton University to meet John Bahcall, one of the 20th century's most respected astronomers. Bahcall convinced Loeb to switch to astrophysics.

As Loeb's research came to focus on the early Universe, he found he was returning to his first passion, philosophy. "In our work we address questions touching the realm of philosophy. For example, we consider how the first light was produced, which appears in the very first chapter of the Bible," he says. "I can actually address the big questions I started with, but this time I'm using scientific tools and I feel like I'm making progress."

How are galaxies distributed in the Universe?

Artist's conception of expanding Universe

predictions

In 1927, **Georges Lemaître** used Albert Einstein's equations of the general theory of relativity to predict that the Universe is expanding.

Independently, **Alexander Friedmann** did so as well. An imaginary rewinding of cosmic expansion leads to the logical conclusion that the Universe had a very dense, hot beginning—now called the Big Bang.

Distribution of matter: homogeneous

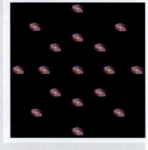

Appearance: isotropic

hypothesis

Lemaître and Friedmann assumed that matter is distributed evenly, especially on large scales (think Mpc). As a result of this homogeneous distribution of matter, the Universe should, on average, be isotropic—it should appear the same in all directions when observed out to great distances.

observations

In 1929, Edwin Hubble published observations of cosmic expansion—but his survey did not measure redshifts of galaxies distant enough to test whether the Universe is homogeneous and isotropic. In the 1980s, **Margaret Geller**'s trailblazing work provided a major insight. Her "Stickman" diagram, the first map of galaxies beyond 200 Mpc from Earth, showed that matter distribution at these distances is not homogeneous! Astronomers would have to map at even greater distances to search for homogeneity and isotropy.

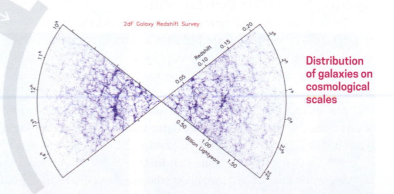

Distribution of galaxies on cosmological scales

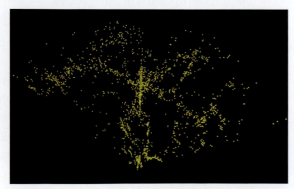

Stickman map

confirmed by observation

In 2003, the Australian 2dF project released the results of a redshift survey, conducted by robotic telescope, of nearly 250,000 objects. The 2dF galaxy map extends nearly 850 Mpc from Earth. It confirms that the Universe's large-scale structure is a cosmic web—a pattern of filaments rich in clusters upon clusters of galaxies bordering vast voids of space. The data also confirm that on these *larger scales*, the Universe "smooths" out, appearing both homogenous and isotropic.

Section Summary

- The era of quasars came early in cosmic history and ended when the abundance of available material that fed their supermassive black holes at galactic centers was used up.
- The Hubble Deep Field images are ultralong exposures of small sections of sky that reveal myriad galaxies far back into cosmic history. These images show early phases of galaxy evolution, including details such as collisions and starbursts.
- The emergence of stars and the first large black holes (associated with galaxies) marks the end of the Dark Ages, a period following the Big Bang when neutral hydrogen had already formed and absorbed all visible-wavelength photons.

CHECKPOINT **a.** Compare the era of quasars and the Dark Ages of cosmic history. **b.** What have astronomers learned from deep-field images?

Building Cosmic Structure

17.5

Using powerful computer simulations, astronomers have been zeroing in on the earliest stages of cosmic structure formation. Their conclusion can be summarized: All cosmic structures began forming in earnest at the outset of the Dark Ages. But even these beginnings hinged on holdovers from the truly "early" Universe: ripples in the sea of matter and energy called perturbations.

The First Stages: The Rich Get Richer

Density perturbations are small regions where the density of hydrogen gas is slightly higher or lower than that of the surroundings. (As we will see in Chapter 18, these perturbations formed randomly in the barest instants after the Big Bang.) "Gravity takes the regions of slightly denser-than-average gas and makes them into the seed of all the structure we see today," says Avi Loeb. Since these perturbed regions have a little more density (and more mass) than their surroundings, they lie at the bottom of shallow "gravitational wells." Once neutral hydrogen has formed, gravity pulls surrounding material inward, toward the center of the perturbations. As more mass is pulled in, what started as a small ripple soon becomes a distinct cloud of gas.

In contrast, regions that started with slightly lower-than-average densities of hydrogen gas become evacuated (that is, all contents are lost). "It's a case of the rich getting richer and the poor getting poorer," says Loeb. "The low-density regions eventually become voids, and the high-density regions are where stars, galaxies, and galaxy clusters form."

But which structures form first? According to computer simulations, the evolution of large-scale structure represents a process of **hierarchical formation**. The smallest structures are the first to emerge via gravity. Only later do the larger structures form.

The first generation of stars that formed out of the Dark Ages was, however, unlike anything we see today. Astronomers who simulate the formation of the first stars have found monsters. Somewhere around a redshift of $z = 20$, clouds of a million or so solar masses have already formed from the sea of neutral hydrogen. Within these clouds, smaller pockets of gas collapse under their own gravity, becoming stars. But instead of the stars we see today, with an average mass less than the Sun's, astronomers believe that the first generation of stars was dominated by behemoths with masses of 100–300 M_{Sun}.

In the current epoch, astronomers have never measured a star mass greater than 150 M_{Sun}, and massive stars are relatively rare. Why were the first-generation stars so big? "There were no heavy elements around because there had been no stars to make them," says Loeb. "Heavy elements like carbon and oxygen can be much more effective at emitting radiation that allows gas to cool than lighter elements." This cooling is important in the evolution of star-forming clouds (see Chapter 12). Today there are enough heavy elements

density perturbation A small region where the density of hydrogen gas is slightly higher or lower than that of the surrounding regions.

hierarchical formation An ordered sequence of formation, from smallest to largest structures.

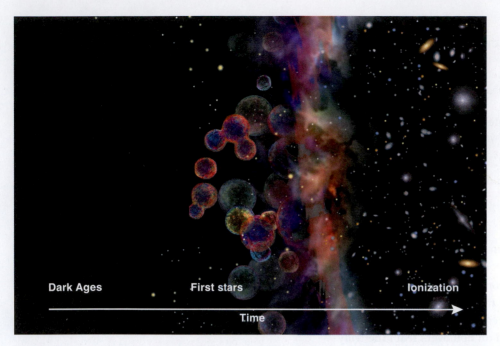

Dark Ages   First stars   Ionization

Time

Figure 17.19 The End of the Dark Ages
Ionizing bubbles were created first from massive stars, then from those stars' supernova blasts, and finally from young quasars. As these bubbles overlapped, the vast stores of neutral hydrogen clouds were eroded until at last the Universe became transparent and the Dark Ages ended.

to let star-forming clouds cool, which leads them to fragment into smaller pieces. The cooling, says Loeb, "produces smaller-mass stars. Without those heavy elements, the first generation of stars ended up really massive."

The monster first-generation stars produced powerful fluxes of ultraviolet light, which then created ionized bubbles in the otherwise neutral sea of hydrogen atoms. And when the tremendous stars ended their lives as supernovas, their powerful blasts created more ionizing radiation that further eroded neutral hydrogen. When these stars were close enough together, their surrounding ionized regions merged, creating more space free of neutral hydrogen. In this way, the first-generation stars helped end the cosmic Dark Ages. But they weren't alone; the young black holes also did their part.

As larger and larger clouds were gathering, gravity pulled them together and formed the first protogalaxies. While stars formed in some regions, in others the progenitors of SMBHs were created with 10,000 M_{Sun} or more in mass. Their formation is an active topic of research, but hypotheses include the direct infall of gas and the mergers of remnant black holes from the first stars. Once these intermediate-mass black holes formed, they quickly became surrounded by accretion disks that continued to feed them new matter from the surrounding gas. Thus, the first black holes became the first protoquasars.

Enough ionizing radiation was created by these young quasars that they, too, became surrounded by bubbles of ionized gas. As the number of quasars grew, the ionized bubbles overlapped (in analogy with the first-generation stars). "About a billion years after the Big Bang, all the neutral hydrogen was gone," says Loeb. "The entire Universe became, for the most part, transparent to optical photons." Only within the galaxies themselves could gas recombine into clouds of neutral hydrogen. Thus, the cosmic Dark Ages finally ended (**Figure 17.19**).

Thus, there is an explicit link between hierarchical formation and galaxy formation. Smaller structures such as stars and black holes form first. Later these objects, along with surrounding clouds of gas, are pulled together, creating galaxies. As more galaxies form, neighbors begin to interact and collide, leading to further changes in galactic structure and the creation of satellite galaxies and groups.

"It's a case of the rich getting richer and the poor getting poorer."

Dark Matter, Galaxy Formation, and Large-Scale Structure

As we've climbed up the ladder of cosmic structure—from galaxies to galaxy superclusters—we've seen that at every rung, dark matter dominates over luminous matter. Taking a giant step back and trying to simulate the evolution of all this structure from just after the formation of neutral hydrogen ($z = 2,000$) up to the present ($z = 0$), astronomers find that this dominance of dark matter extends to almost every feature of the Universe's evolution. More important, using these simulations as a guide, astronomers have been able to infer how the properties of dark matter shape the history of cosmic structure formation.

Modern observations enable astronomers to place fairly strict constraints on the composition of matter in the Universe. Taken as a whole (as opposed to what is seen in specific individual structures), 87 percent of all the matter in the Universe is composed of the dark component, while only 13 percent is the luminous kind that makes up stars

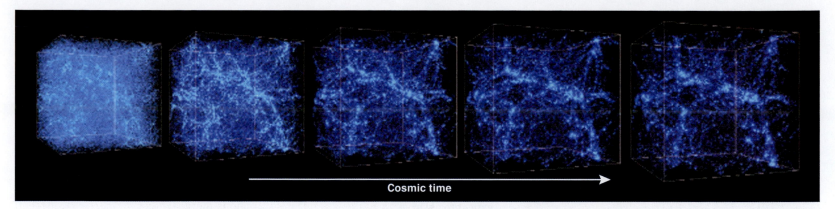

Cosmic time

Figure 17.20 Evolution of Cosmic Structures
Computer simulations show the emergence of the filamentary structure of matter associated with superclusters and voids. Such simulations of cosmic history show how structures emerge from small-scale perturbations that were imposed moments after the Big Bang. Gravity from the small, overly dense regions allows the perturbations to be amplified as more mass from surrounding regions is drawn in. Eventually they form the hierarchy of large-scale structure we see today.

and galaxies and human beings. (Most of the Universe's luminous matter is in the form of interstellar gas.) This ratio represents the starting condition for simulations of cosmic structure formation. But since nothing is known from observation about the form that dark matter takes, astronomers had to make a choice: Should they treat dark matter as particles moving at high speeds (even close to the speed of light), as neutrinos are known to travel? Or should they treat dark matter as slow-moving particles, like the majority of luminous matter particles (protons, neutrons, electrons, and so forth)?

"It turns out that the difference between such 'hot' and 'cold' dark matter scenarios really matters," says Loeb. When astronomers begin their simulations with **hot dark matter**, gravity has a hard time getting the fast-moving particles to collapse into structures like those seen on the sky. Only by assuming a **cold dark matter (CDM)** scenario do simulations show gravity grabbing hold of the initial perturbations and drawing sufficient material together to re-create the beautiful patterns of the cosmic web observed in redshift survey maps. "Whatever dark matter is made of, it must be moving slowly," explains Loeb. "That's an extremely important bit of information." Knowing that dark matter moves relatively slowly limits the kinds of candidates that particle physicists (who study the nature of matter) can propose to make up the invisible but dominant form of cosmic mass. Relativistic particles, for example, can't make up dark matter.

With the question of cold versus hot dark matter resolved, astronomers have turned to building ever-more-detailed and highly resolved simulation-based explorations of structure formation. These supercomputer studies show an initially smooth cosmic gas (other than tiny perturbations) emerging from the era when neutral hydrogen forms. Gas slowly collects in and around the perturbations, creating clouds that become ever denser and ever more massive. Long, dense filaments grow and link the densest regions. As we have seen, as the gas pours into these deepening gravitational wells, it fragments into clouds that form the first stars and the progenitors of SMBHs (and hence quasars). As gas is drawn together, any rotational motion it had on large scales is amplified as it collapses, leading to the disks of spiral galaxies.

With time, more material is drawn together, lowering the density in the voids and increasing the mass of proto–galaxy groups and proto–galaxy clusters (**Figure 17.20**). Galaxies within these structures are gravitationally bound together but don't necessarily interact gracefully. "The orbits of galaxies through the clusters are often highly elliptical," says Loeb. "They dive through the cluster centers, rise to the cluster edges, and then dive back again like bees swirling around a hive." Collisions between young galaxies are inevitable. Spiral galaxies often run into each other, stripping away the organized disks of stars and gas. Although the stars rarely hit each other in these galaxy-galaxy collisions, vast clouds of molecular gas are driven together, creating tremendous starbursts. These spikes in the stellar birth rate last only a few tens of millions of years. In that brief time, however,

"They dive through the cluster centers, rise to the cluster edges, and then dive back again like bees swirling around a hive."

hot dark matter Nonluminous matter moving at a relativistic speed.

cold dark matter (CDM) Nonluminous matter moving at a slower-than-relativistic speed.

the rate of star formation can be 100–1,000 M_{Sun} per year. (The current rate in our galaxy is about 1 M_{Sun} per year.)

In the aftermath of the collision of two spiral galaxies, existing stars are tossed about but not lost. They eventually fall back, creating a new, larger galaxy that is distinctly elliptical. Thus the history of hierarchical large-scale structure formation explains the higher percentage of elliptical galaxies in the centers of rich clusters. As gravity draws galaxies and their groups together into the clusters, the density of galaxies (the number per volume) increases, thereby increasing the likelihood of collisions. A collision between elliptical galaxies produces another elliptical galaxy (or perhaps an irregular one). "Since collisions between spirals also produce ellipticals," explains Loeb, "you can see why the more galaxy collisions that occur, the larger the fraction of ellipticals we astronomers expect to observe."

Section Summary

- Density perturbations, which are small, regional differences in mass, formed the seeds around which all larger structures could be created by gravity.
- The first stars lacked heavier elements to limit their growth and so were many times larger than present-day stars. The first black holes were the progenitors of quasars. Ionizing radiation from the first stars and black holes ended the Dark Ages.
- Dark matter played a fundamental role in cosmic evolution; astronomers determined it must be "cold" (moving at nonrelativistic speeds).
- In galactic evolution, spiral galaxies evolve into elliptical galaxies through collisions.

CHECKPOINT **a.** Describe hierarchical formation. **b.** Why did astronomers conclude that dark matter must be "cold"?

Final Questions for the Beginning of Everything

It's important to distinguish which parts of these stories are widely accepted and which parts are still being worked out or contain speculative elements. The study of large-scale structure is still very much cutting-edge science, and astronomers are still wrestling with aspects of the models of cold dark matter and hierarchical formation to see whether those models can explain all observations. This is a natural and exciting part of the scientific process. It takes a lot of effort and time to see just how far a theory can go in explaining the world. Sometimes ideas that seem the most promising have to be pruned or even rejected to make progress.

Still, models of the evolution of cosmic structure that are based on cold dark matter have, overall, been remarkably successful at predicting everything from the distribution of galaxy types to the statistics of supercluster density (how much more dense superclusters are than their surroundings). They give us a coherent and consistent story of how the Universe went from a relatively structureless gas of neutral hydrogen 300,000 years after the Big Bang to the rich and diverse structures we see today, 13.7 billion years after the Big Bang.

But where did the initial perturbations that form such an important part of this story come from? And what happened before the formation of neutral hydrogen? Finally, and perhaps most important, what *was* this event called the Big Bang and how did the whole shebang we call the Universe get set into motion?

Attempting to answer these questions takes us to the end of our cosmic journey by bringing us to the beginning and, perhaps, the deepest questions science can ask. How did time begin, and why is there something rather than nothing? Chapter 18 explores these ultimate questions of cosmology: the birth of the Universe and its eventual fate.

CHAPTER SUMMARY

17.1 Bright Lights, Big Universe

The discovery that the Universe is expanding was the first step in refuting the long-held belief in a static cosmos. It eventually led to the conclusion that the Universe has evolved: it is different today than it was in the past. The expansion of the Universe is apparent from observed receding motions of all galaxies, and it suggests that the Universe was once extremely compact. Using Hubble's law alone, astronomers estimate the Universe is about 14 billion years old. More detailed methods indicate the Universe is 13.7 billion years old. An object's recession velocity is measured in terms of the redshift of its spectrum, which correlates with look-back time—how far back in time we see the object.

17.2 Cosmic Neighborhoods: Galaxy Groups and Galaxy Clusters

The Milky Way is part of the Local Group of galaxies. Galaxy groups are sometimes members of galaxy clusters. Galaxy clusters often have more than 1,000 members. Clusters differ in their relative abundances of specific galaxy types. Galaxy clusters include X-ray halos of hot gas, which appear to result from the merging halos of individual galaxies. Evidence of dark matter comes from the fact that more mass is required to hold galaxy clusters (including their halos) together than is observed as luminous matter. Gravitational lensing—in which large collections of matter cause space-time to bend and focus light from distant background objects—also demonstrates the presence of dark matter in galaxy clusters.

17.3 Superclusters and Large-Scale Structure

Redshift surveys provide a three-dimensional picture of galaxy distribution on the largest scales. Through these surveys, vast superclusters of galaxies have been discovered, with similarly vast voids between them, showing that structure exists even on very large scales. The surveys have also identified that, along with the Hubble flow of cosmic expansion, galaxies show peculiar motions driven by the gravitational pull of other large collections of mass, giving further evidence of dark matter. On the largest scales, the Universe's structure smooths out into a homogeneous and isotropic cosmic web, consisting of interconnected filaments of superclusters and the voids between them.

17.4 Large-Scale Cosmic Structure and Cosmic History

Studying the history of the Universe requires moving backward in time along individual lines of sight. Quasars came into existence relatively early in cosmic history as large quantities of matter were funneled into supermassive black holes at galactic centers. As the reservoirs of matter ran out, the quasars dimmed.

17.5 Building Cosmic Structure

Deep fields, in which astronomers collect light from very distant (faint) cosmic objects, have advanced the study of the Universe's history by revealing galaxies in their early stages of construction. Visible in these deep fields are many galaxies in various stages of evolution and evidence of frequent collisions. Stars and galaxies first started forming during the Dark Ages, when normal matter was mostly neutral hydrogen that absorbed ultraviolet and visible photons. Massive neutral hydrogen clouds formed first, then massive stars, and then intermediate-mass black holes that formed the cores of galaxies. By 1 billion years after the Big Bang, almost all neutral hydrogen had been ionized, and the Universe became transparent to visible and ultraviolet light. The dominance of dark matter over luminous matter is seen throughout cosmic evolution and is central to the formation of galaxies, galaxy clusters, and superclusters. Simulations agree with observations in showing that dark matter must be "cold" (moving at nonrelativistic speeds).

QUESTIONS AND PROBLEMS

Narrow It Down: Multiple-Choice Questions

1. Select the correct sequence for cosmic structures, from smallest to largest.
   a. galaxy, galaxy group, galaxy cluster, supercluster, cosmic web
   b. cosmic web, galaxy, galaxy cluster, galaxy group, supercluster
   c. galaxy, galaxy cluster, galaxy group, cosmic web, supercluster
   d. galaxy, galaxy cluster, galaxy group, supercluster, cosmic web
   e. cosmic web, galaxy, galaxy group, galaxy cluster, supercluster

2. Which of the following provide(s) evidence for dark matter? Choose all that apply.
   a. the Solar System
   b. galaxies
   c. galaxy clusters
   d. superclusters
   e. black holes

3. Which statement about the observed expansion of space-time is true?
   a. Observers in all galaxies should see farther galaxies receding from them faster than nearer galaxies.
   b. Since we see all galaxies receding from the Milky Way, we must be near the center of the Universe.
   c. On average, all galaxies appear to be receding at the same speed.
   d. The farther away galaxies are, the more slowly they appear to recede.
   e. Scientists have observed an outer edge of space-time at the boundary of the Universe.

4. The Local Group is a group of
   a. planets.
   b. stars.
   c. galaxies.
   d. galaxy clusters.
   e. astronomers working on a redshift survey.

5. If H_0 were constant at 90 (km/s)/Mpc, how would the estimated age of the Universe change from current estimates?
   a. The value would be the same.
   b. The value would be larger.
   c. The value would be smaller.
   d. The value can't be determined from the information given.
   e. The value would depend on the density of the Universe.

6. From what you've learned about the stages of evolution of the Universe, what would be the earliest time frame or event that could be seen in visible wavelengths?
   a. the instant of the Big Bang
   b. within 300,000 years after the Big Bang
   c. the birth of the first galaxy
   d. the end of the Dark Ages
   e. none of the above

7. What does a redshift of $z = 0$ mean?
   a. The object is almost at the edge of the Universe.
   b. The object is nearly 14 billion years old.
   c. The object's motion cannot be determined.
   d. The object shows no redshift.
   e. The object is obscured from our view by the plane of the Milky Way.

8. Which of the following is *not* characteristic of rich galaxy clusters?
   a. Their diameters are measured in megaparsecs.
   b. They can have more than 1,000 members.
   c. They contain many galaxy groups.
   d. They are gravitationally bound.
   e. They contain mostly spiral galaxies.

9. Which of the following is *not* indicated by the observation of an Einstein ring?
   a. the presence of dark matter
   b. a large mass between the observer and the source object for the ring
   c. a source object with a circular shape
   d. dark matter with a symmetric distribution
   e. curved space-time

10. Redshift surveys are difficult and time-consuming to obtain. Which of the following factors does *not* contribute to that difficulty?
    a. lack of evidence showing that redshifts imply expansion
    b. dim light from galaxies observed
    c. number of objects to be observed
    d. need for specialized equipment
    e. need for a great deal of telescope time

11. Which of the following statements about the cosmic web is *not* true?
    a. It is filamentary.
    b. It is isotropic.
    c. It is homogeneous.
    d. It shows no voids.
    e. It has interconnected superclusters.

12. Which of the following is true of an object with a redshift of $z = 5$?
    a. Its spectrum shows longer wavelengths for specific elements than are measured in a laboratory.
    b. It is receding faster than an object with $z = 6$.
    c. It is only a few light-years away.
    d. It may be significantly closer than a second object with $z = 5$.
    e. It is farther away than an object with $z = 7$.

13. Which of the following were the earliest to form in the evolution of the Universe?
    a. 1-M_{Sun} stars
    b. galaxies
    c. supermassive stars
    d. protoquasars
    e. intermediate-mass black holes

14. Which of the following is/are expanding as the Universe expands? Choose all that apply.
    a. the Solar System
    b. galaxies
    c. galaxy clusters
    d. galaxy superclusters
    e. space-time

15. X-ray halos of galaxy clusters consist primarily of which of the following?
    a. neutral hydrogen gas
    b. hot, ionized gas
    c. brown dwarfs
    d. supernova remnants
    e. X-ray radiation

16. Imagine that when the Local Group began to form, it was more massive and denser than astronomers believe it actually was. Which of the following would be true?
    a. Fewer galaxy mergers would likely have occurred.
    b. Fewer large galaxies would likely have formed.
    c. Star collisions would likely have frequently occurred.
    d. The galaxies would have started moving relative to each other at lower speeds.
    e. The ratio of luminous matter to dark matter would likely have been approximately equal to their current ratio.

17. If the Universe has been expanding at a constant rate since the Big Bang, how would the calculated age of the Universe differ from the age obtained by using the current value of H_0?
    a. Spiral galaxies form from the merger of other spirals.
    b. Elliptical galaxies can form from the merger of spirals.
    c. Elliptical galaxies may form from the merger of ellipticals.
    d. Irregular galaxies may form from the merger of ellipticals.
    e. The outcome of galaxy mergers is not predictable.

18. What specific piece(s) of evidence for dark matter have astronomers found? Choose all that apply.
    a. More mass lies in various large-scale structures than is visible as luminous matter.
    b. Large dark regions exist between galaxies.
    c. Galactic rotation curves do not match expectations based on mass and radius.
    d. X-ray halos in galaxy clusters have too much pressure to be gravitationally bound by the observed luminous matter in the clusters.
    e. Galaxies appear dimmer than expected.

19. Which statement(s) about redshift and look-back time is/are true? Choose all that apply.

    a. The greater the redshift, the longer the look-back time.
    b. The smaller the value of z, the shorter the look-back time.
    c. Objects with greater redshift appear younger than they actually are today.
    d. Look-back time is limited by the speed of light and the age of the Universe.
    e. The greater the redshift, the closer the look-back time approaches the age of the Universe.

20. True/False: Gravitational lensing can cause objects to appear shifted from their actual positions.

To the Point: Qualitative and Discussion Questions

21. Explain what redshift means as it applies to the study of galaxies.

22. What galaxy types are found in the Local Group, and how are they arranged?

23. The Virgo Galaxy Cluster is a regular rich cluster. What do the terms *regular* and *rich* tell you about its structure?

24. How do astronomers explain the formation of X-ray halos in galaxy clusters?

25. What is required for the formation of an Einstein ring?

26. What are redshift surveys? Name some notable examples. How are their results used?

27. Explain the meanings of *big science* and *big data*.

28. How do peculiar motions differ from Hubble flow? How does dark matter affect peculiar motions?

29. Define the terms *homogeneous* and *isotropic*.

30. Explain why there are no quasars near the Milky Way.

31. How do deep-field images contribute to our understanding of the evolution of the Universe?

32. How do small perturbations form the seeds of larger structures?

33. Explain the term *cold dark matter*.

34. The first *Star Wars* movie series began with the phrase "A long time ago in a galaxy far, far away . . ." On the basis of your current understanding of the size and time scales of the observable Universe, how long ago and how far away could that have been?

35. Astronomers describe the Universe as homogeneous and isotropic. Describe the scales on which this description is observed to be true.

Going Further: Quantitative Questions

36. Estimate the age of the Universe, in years, using a value of 55.4 (km/s)/Mpc for Hubble's constant and assuming that the rate of expansion has remained constant.

37. Estimate the age of the Universe, in years, using a value of 81.5 (km/s)/Mpc for Hubble's constant and assuming that the rate of expansion has remained constant.

38. An early estimate found the age of the Universe to be 7.5 billion years old. What value of Hubble's constant would be consistent with this age?

39. The average speed of galaxies in a galaxy cluster is 850 km/s. The cluster's diameter is 3.2 Mpc. How long, in years, would it take a galaxy to cross the cluster?

40. Calculate the mass, in M_{Sun}, of a galaxy cluster that has an average galactic speed of 970 km/s and a diameter of 8.5 Mpc.

41. What is the average galactic speed, in kilometers per second, within a cluster that has mass $2.5 \times 10^{14}\ M_{Sun}$ and diameter 2.7 Mpc?

42. A star orbits a black hole whose mass is $4.7 \times 10^6\ M_{Sun}$ at a radius of 1.9×10^{13} meters. What is its speed, in meters per second? (See Going Further 15.1.)

43. What is the recession velocity, in kilometers per second, of a galaxy that is 790 Mpc from Earth? (See Going Further 16.1 and use the value of H_0 given there.)

44. In the spectrum of a distant galaxy, a spectral line with expected value 510 nm is observed at 710 nm. If you ignore relativistic effects, at what speed, in kilometers per second, is the galaxy receding from the observer? (See Going Further 16.1.)

45. An astronomer calculated a galaxy's recession velocity to be 65,000 km/s, by noting the redshift of a spectral line whose wavelength is 420 nm. (Ignore relativistic effects.) At what wavelength, in nanometers, was the observed spectral line? (See Going Further 16.1.)

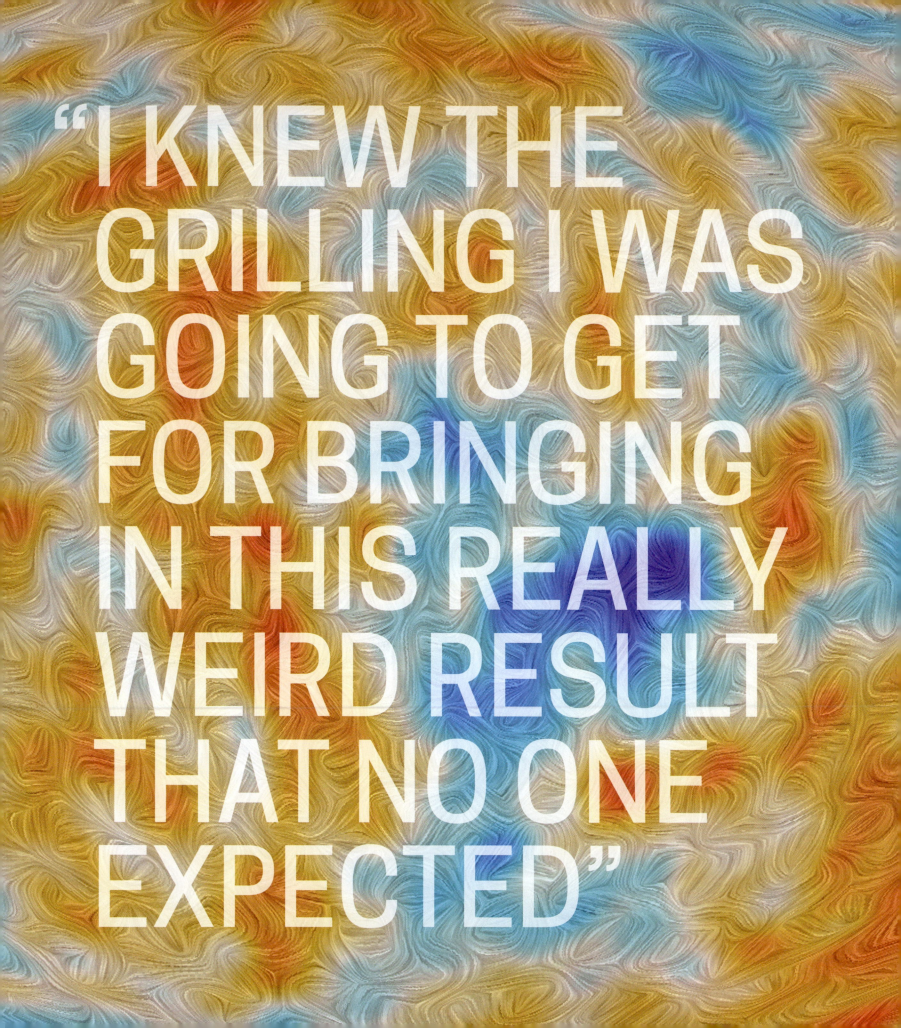

18

Endings and Beginnings

COSMOLOGY

How to Win a Nobel Prize: The Accelerating Universe

18.1 Adam Riess knew he might have a winning hand, and it scared him. The year was 1998, and Riess had a postdoctoral fellowship at UC Berkeley. With a team of other astronomers, he was trying to use distant supernovas to find evidence that the Universe is slowing down.

Ever since the Big Bang (the event that initiated the Universe) was first proposed, astronomers assumed that the mutual gravitational pull of all cosmic matter would slow down the cosmic expansion discovered by Edwin Hubble. (Recall our discussion of cosmic expansion from Chapter 17.) That made the rate of "cosmic deceleration" a key factor in cosmology. Everyone wanted to know its value.

Riess and his collaborators were using a new approach. "Supernovas are so bright," says Riess, "you can see them from very far away, which also means looking far back in time. Using supernovas, we see more of the history of the Universe's expansion" (**Figure 18.1**). (We will explore the role of supernovas in cosmology later in this chapter.)

After many nights at the world's most powerful telescopes, Riess and his colleagues had all of the data they needed to calculate the exact value of cosmic deceleration. There was only one problem: "I found just the opposite," says Riess. "The data said the Universe wasn't slowing down in its expansion. It was speeding up. It was accelerating."

Riess knew he was sitting on the scientific version of a huge grenade. "I remember being at my desk at UC Berkeley, looking at my results and feeling kind of sweaty-anxious about it," he says. "I was imagining having to explain this to the whole research group I was part of. I knew the grilling I was going to get for bringing in this really weird result that no one expected."

Finding out that the Universe is accelerating meant Riess had also found evidence of a new and invisible form of energy—"dark" energy—that powers the cosmic acceleration. If he was right, he was in Nobel Prize territory. If he was wrong, he would make a fool of himself very publicly. "It was like being in a big poker game and thinking you're going to

---

← Part of an all-sky survey made by the European Space Agency's Planck satellite, showing the polarization of the light that constitutes the cosmic microwave background (CMB) in one region of the sky. The CMB represents light waves emitted just 380,000 years after the Big Bang. Probing its polarization gives astronomers clues to the conditions in the early Universe.

Figure 18.1 Distant Supernovas

These Type Ia supernovas were imaged by the Hubble Space Telescope. When astronomers learned how to use distant Type Ia supernovas as standard candles, these supernovas became the key to discovering that the Universe's expansion is accelerating.

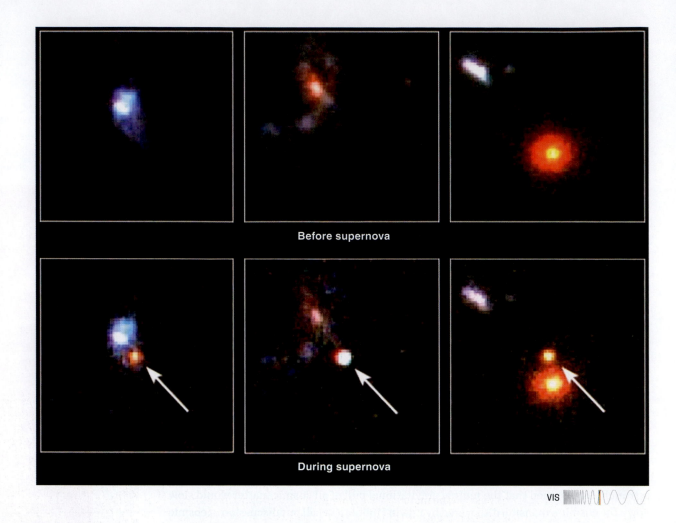

Before supernova

During supernova

VIS

"As you go backward in time, the Universe must have been much denser than what we see today."

steady-state model A theory, advanced by Fred Hoyle, stating that although the Universe is expanding, new matter is slowly created between galaxies, maintaining a constant average density.

play for relatively small stakes," he says. "But then you get dealt the most amazing hand ever, and you realize you can't just bet with pennies and nickels. You need to go all in."

Riess and his collaborators rechecked their calculations many times as they prepared to announce their results. "Brian Schmidt, who led the supernova search, and I were planning what we would do if we had to leave the field in disgrace because we were proved wrong," Riess chuckles. But they weren't wrong. Another group, led by Saul Perlmutter, used a similar method and found the same result. The two teams announced their discoveries at the same time. Dark energy was here to stay. Fourteen years after that night of worrying at his desk, Riess (along with Schmidt and Perlmutter) picked up his Nobel Prize in Physics.

The Last Step in the Cosmic Journey

Seventeen chapters ago, we began a journey through time and space. We started by considering our own human evolution and the evolution of science from Galileo to Newton's laws and on through to the properties of light and the structure of matter. Using these laws of physics, we pushed outward, exploring planets, stars, and galaxies. Now we are at the end of the journey. With everything we've learned, we can finally ask the biggest and most all-embracing questions possible. We are finally ready to consider *cosmology* and *the Universe as a whole*.

Beginnings have always held a special place in human curiosity, but endings are equally important. In this chapter we will use the laws of physics to ask not only about the beginning of our Universe but also about the ultimate cosmic ending. What is the long-term future of the Universe? Will it end or continue forever?

Time

Figure 18.2 Running Expansion Backward
Reversing the expansion of space-time seen today in the recession of galaxies leads first to an ever-higher density of galaxies. Eventually, the stars and gas making up the galaxies must merge and mix. As we go further back in time and increasing density, even all of the matter making up stars must have been crushed together. This "thought experiment" shows how the expansion we see today can imply very different cosmic conditions earlier in cosmic history.

More than almost any other science, cosmology takes us to the limits of what it means to ask scientific questions. Does the theory of the Big Bang really tell us why there is something rather than nothing? Does it tell us what time is or how time began? Does it make infinite space easier to understand? As we tell this final story in the narrative of humanity's understanding of astronomy, we need to pay close attention not only to what we know, but also to what we don't know and the difficulties we face in asking the deepest kinds of questions.

What the Big Bang Isn't

We have already seen how Hubble's discovery of galaxy redshifts led some astronomers to the idea that the Universe must have looked much different in the past than it does today. "If you imagine running the 'movie' of cosmic expansion backward," says Riess, "then it would seem obvious that as you go backward in time, the Universe must have been much denser than what we see today."

If we could run the "movie" of cosmic expansion backward as far as possible, we would find that at some point in the past, all the matter in the Universe (which is now expanding) must have been infinitely squeezed together (**Figure 18.2**). This scenario implies that the Universe began with an infinite density. However, an infinite density of matter makes no sense from a physics standpoint. An infinite amount of matter can't be squeezed into a cubic meter. So while cosmic expansion might seem to imply that the Universe had a beginning—a time when time itself started—most physicists and astronomers were not willing to go that far when Hubble made his discovery.

The modern theory of the Big Bang began in the late 1940s and 1950s with Russian-American physicist George Gamow and his graduate student Ralph Alpher (**Figure 18.3**). Their calculations suggested that the early Universe might have begun very dense and very hot and then cooled and rarefied (thinned out) with cosmic expansion. As we will see, this idea led to specific predictions that could be tested against observations (though 12 years passed before that happened), becoming the basis for today's cosmological models. Gamow and Alpher did not explain, however, where all of this matter, energy, space, and time came from in the first place.

Also in the 1950s, astrophysicist Fred Hoyle proposed a **steady-state model** of cosmology, in which the Universe has always been expanding and will always expand. In Hoyle's theory, new particles of matter slowly appear between galaxies, such that the average cosmic density remains the same (**Figure 18.4**). When other scientists objected to Hoyle's continuous creation of matter, he asked whether having everything created in a "big bang" was any better. And so it was that Hoyle invented the name *Big Bang*, using the term to distinguish it from his own ideas.

The eventual fate of the steady-state model is a wonderful example of the scientific process at work. Many scientists at the time viewed Hoyle's idea favorably. But by the mid-1960s, new discoveries made clear that the Universe looked very different in the past than it does today. Thus the evidence did not back up Hoyle's steady-state model (that the

a.

b.

Figure 18.3 Gamow and Alpher
In 1948, **a.** George Gamow (1904–1968) and his graduate student **b.** Ralph Alpher (1921–2007) published "The Origin of Chemical Elements," a paper on the Big Bang theory. In this paper and others, they introduced the idea that the Big Bang could have produced only specific light elements, and they predicted the discoveries that now support the theory.

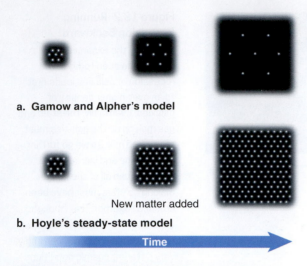

a. Gamow and Alpher's model

New matter added

b. Hoyle's steady-state model

Time

Figure 18.4 Big Bang Model versus Steady-State Model
a. In Gamow and Alpher's Big Bang model, expansion increases the separation between galaxies over time. **b.** In Hoyle's steady-state model, new galaxies form slowly and fill in the space left by the expansion of older ones, so the density of galaxies never changes.

"Knowing what kind of a stage you're on—meaning knowing the geometry of cosmic space-time—really changes how you interpret what you're seeing."

past should look the same as the present). In fact, the new data pointed to a Universe that must have been much hotter and denser in the past than it is today, as Gamow and Alpher had predicted.

Still, although the theory of the Big Bang prevailed, it is important to remember that it is not a theory of the origin of the cosmos; it's a theory of what happened after the Universe began. The Big Bang never tells us how the cosmos, or existence itself, came into being. For now, that question remains unanswered by science.

Section Summary

- Astronomers now have answers to many—but not all—important questions in cosmology, the study of the Universe and its evolution.
- Early scientific cosmological theories included the steady-state model, in which the Universe did not evolve but instead always looked as it does now.
- Today, the generally accepted model is the Big Bang, which describes what happened *after* cosmic expansion began.

CHECKPOINT Contrast the Big Bang model with the steady-state model of cosmology.

Our Cosmology: The Big Bang

18.2

Now that we know that the Big Bang *isn't* a theory of the origin of the Universe, we are ready to explore what the Big Bang *is*: a theory of what happened from just after expansion began up to the present.

Part I. The Stage and the Actors

To lay out our modern narrative of cosmic evolution, we must set the stage for this scientific drama and introduce the actors, some of whom will be new while others are old friends. Let's start with the stage.

SPACE-TIME. The Big Bang is a general relativistic cosmological theory. It begins with the flexible space-time at the heart of Einstein's general theory of relativity (explored in Chapter 14). In a sense, cosmology was waiting for Einstein. There had been early attempts to build scientific models of the entire Universe. Newton had thought about it, but previous efforts were hobbled without Einstein's great insight into the nature of space-time as a four-dimensional continuum (meaning a smooth fabric of reality). By allowing physics to describe space-time in terms of geometry and by assuming that the Universe is highly symmetric (that is, homogeneous and isotropic on large scales, as described in Chapter 17), Einstein was able to formulate equations describing the Universe *as a whole*.

With general relativity, describing cosmic history meant describing the geometry (or curvature) of space-time and the evolution of that geometry over time. "Geometry is the stage," says Adam Riess, "and the play must take place on the stage. Knowing what kind of a stage you're on—meaning knowing the geometry of cosmic space-time—really changes how you interpret what you're seeing."

The geometry of the Universe is determined by the matter and energy within it. The only way to distinguish which geometry the Universe actually has is by making measurements that are either direct (looking for effects of the geometry) or indirect (measuring the Universe's store of matter and energy). Remarkably, these experiments have been carried out successfully and, as we will see, current evidence points to our living in a Universe with a flat geometry. But future research could indicate that the situation is more complex.

The Big Bang can't just describe the expanding stage, meaning the space part of space-time, however. It must also describe what's happening *on* that stage. Modern

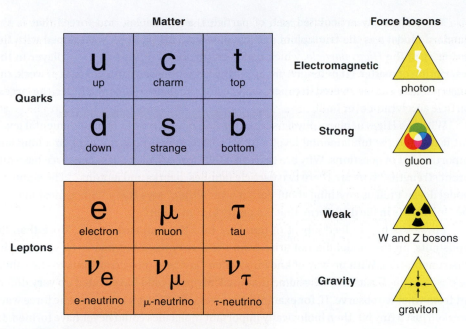

Figure 18.5 Standard Model of Particle Physics
This model accounts for all known subatomic particles making up normal (nondark) matter and for the forces with which they interact. Leptons and quarks are the basic classes of matter particles, each of which occurs in six forms. Force bosons are force-carrying particles. The graviton remains a hypothetical particle and has not been directly detected.

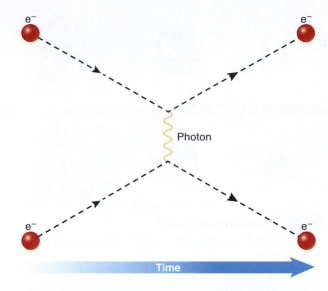

Figure 18.6 Force Bosons
All of the forces between matter are due to an exchange of force-carrier particles. When two electrons collide, the point of impact is really the exchange of a photon—a light particle—which is the carrier of the electromagnetic force. The photon exchange changes the electrons' direction as they move through space.

science has a theory to do just that, and it is a crowning achievement of the human intellect: the standard model of particle physics.

THE STANDARD MODEL OF PARTICLE PHYSICS. During the 1950s and 1960s, physicists expended enormous effort and resources to develop giant **particle accelerators**, machines designed to bring the smallest specks of matter to the highest speeds (and therefore the highest energies) possible. By smashing the particles into each other and scanning the remains, physicists gleaned hints of the particles' internal makeup. The American physicist Richard Feynman described the effort this way: "It's like smashing two watches together and using the debris to figure out how they worked." By the mid-1960s, accelerator-based studies had produced a remarkably successful description of all known subatomic particles, known as the **standard model of particle physics**.

Physicists found that there are just two classes of fundamental matter particles: **leptons** and **quarks** (Figure 18.5). The electron is the most familiar lepton, but two other lepton "generations"—the *muon* and the *tau*—fill out the family. Quarks, which make up protons and neutrons, constitute the other family of matter. Every particle has an *antimatter* twin, such that the "anti" and "normal" particles take opposite electric charges. Particles that don't have an electric charge can be their own antiparticles.

Taking a census of all of the Universe's particles and antiparticles was not the only job of the standard model. Particles "feel" each other by exerting forces. Recall from Chapter 15 that only four forces are at work in the Universe: gravity, electromagnetism, the strong nuclear force, and the weak nuclear force. The two classes of particles differ significantly in that only quarks can feel the strong nuclear force, whereas both quarks and leptons feel the weak and electromagnetic forces. And both quarks and leptons feel the force of gravity.

The standard model describes three of these forces and their effects on subatomic particles. The forces themselves are mediated by a separate class of particles called **force bosons**. The photon, for example, is the particle that carries the electromagnetic force. Exchanging photons is the way charged particles "feel" the electromagnetic force (**Figure 18.6**).

INTERACTIVE:
Explorable List of Particles and Forces

INTERACTIVE:
Big Bang Timeline

particle accelerator An experimental apparatus used to study the structure of matter by colliding atoms and subatomic particles together at very high speeds (high energies).

standard model of particle physics A description of subatomic particles and forces.

lepton A class of fundamental particles of matter that includes electrons, muons, tau particles, and neutrinos.

quark A class of fundamental particles of matter that make up other particles, such as protons and neutrons.

force boson A particle that carries one of the four fundamental forces.

The exquisitely articulated web of particles, antiparticles, and forces that is the standard model was the triumph of particle physics. This triumph was capped with the discovery of the *Higgs particle*, which had been predicted in 1964 as a key player in the description of matter. (It helps give many particles their mass.) After decades of work, the Higgs particle was discovered in experiments using the Large Hadron Collider, the largest particle accelerator ever built.

With the Higgs in place, physicists had a grand map of matter at a fundamental level. But it was not *the* fundamental level. The standard model could not explain a long and important list of questions. Why are there just four forces? Why does each force have different strengths? Why are there two particle families: quarks and leptons? "The standard model doesn't tell us anything about dark matter either," says Riess. "Since most mass in the Universe is in the dark form, that is a pretty big omission."

Physicists had to feed a host of numbers into the standard model. More than 20 "constants" in the model did not arise from the theory itself but had to be specified by experimentation. With no way of knowing which "law" set those constants to the values we observe, physicists had to assume that the Universe could have ended up very different from what we observe. If, for example, the strength of the electromagnetic force were tweaked just a tiny bit, then biologically important molecules could never have formed. In general, any small change to the constants of nature in the standard model would lead to a Universe unlike what exists today. Physicists call this the "problem of fine-tuning," since it appears that the constants in the standard model must be tuned to the exact values we have or we could not exist.

Problems such as fine-tuning are unresolved and constitute the exciting frontiers of particle physics. Even with open questions, however, enough has been learned from the standard model to enable a coherent description of cosmic evolution all the way back to fractions of a second after the Big Bang.

We must consider one last point before we begin telling the story of cosmic history. In order to treat the entire Universe as an entity that can be studied by using the mathematical laws of physics, astronomers invoke what they call the **cosmological principle**. It states that, on average, the conditions in one part of the Universe do not differ from those in any other part. This concept greatly simplifies the treatment of cosmological problems. In the end, however, astronomers must observationally test the truth of the cosmological principle. But recall from Chapter 17 that, at the largest scales, matter does appear to become uniformly distributed in the Universe.

Part II. The Story

Having set the stage and introduced the actors in our cosmic drama, we are ready to begin the story. Scientists break up the narrative of the Big Bang into a sequence of periods based on the dominant physical processes that operated during each period. Remember that as soon as we step forward from the "moment of creation," or $t = 0$, the Universe is expanding, and the matter in that space thins out. Thus, as we progress from one era of cosmic evolution to the next, we are also moving through a sequence of ever-decreasing average temperatures and densities.

Let's begin, then, just after the beginning. **Table 18.1** previews the various periods of the early Universe.

SINGULARITY. Tracing the history of the expanding Universe backward suggests that the Universe may have begun with a state of infinite density, infinite temperature, and infinite space curvature. This is what physicists call a **cosmic singularity**. It is similar to the situation in the center of a black hole. To understand what happens in a singularity, however, physicists need a theory of quantum gravity, which they don't have yet. Thus, the current version of the Big Bang applies only *after* the Big Bang. "That," says Riess, "is what

> "The standard model doesn't tell us anything about dark matter either. Since most mass in the Universe is in the dark form, that is a pretty big omission."

cosmological principle The idea that, on average, the conditions in one part of the Universe are the same as those in any other part.

cosmic singularity The initial condition of the expanding Universe as a state of infinite density, temperature, and space curvature from which the Big Bang was initiated.

grand unified theory (GUT) A yet-to-be-discovered theory to explain how the strong and weak nuclear forces and the electromagnetic force emerged from a unified force moments after the Big Bang.

quantum field The distribution of energy that fills space, within which fundamental particles appear as excited states.

Table 18.1	Eras of Early Cosmic Evolution							
Era or Event	Singularity	Planck era	GUTs era	Inflation	Electroweak era	Particle era	Big Bang nucleosynthesis	Recombination era
Time from "Origin"	"Origin" of Big Bang	From $t = 0$ s to $t = 10^{-43}$ s	From $t = 10^{-43}$ s to $t = 10^{-34}$ s	$t = 10^{-34}$ s	From $t = 10^{-34}$ s to $t = 10^{-12}$ s	From $t = 10^{-12}$ s to $t = 10^{-3}$ s	From $t = 10^{-3}$ s to $t =$ a few minutes	$t \approx 380{,}000$ years
Forces	Unified force: quantum gravity	Unified force: quantum gravity	Classical gravity and unified strong + electroweak force	Classical gravity and unified strong + electroweak force	Classical gravity, strong force, and unified electroweak force	Four forces of standard model	Four forces of standard model	Four forces of standard model
Important Transitions		Big Bang at $T = 0$ s		Hyperrapid expansion of space-time		All particles of standard model had formed	Nuclear reactions created H nuclei, He-4 nuclei, deuterium, He-3, and Li	Neutral H atoms formed
Temperature (K)	∞	$>10^{32}$	$>10^{27}$	10^{22}	$>10^{13}$	$>10^{10}$	10^9	3,000–4,000

GUTs, grand unified theories; K, kelvins. Classical gravity is nonquantum gravity. The four forces of the standard model are classical gravity, strong force, weak force, and electromagnetism.

we mean when we say our theory of the Big Bang really starts a short interval after all the players are already in place."

PLANCK ERA ($t = 0$ SECOND TO $t = 10^{-43}$ SECOND). Physicists expect that at the incredible temperatures just after the Big Bang, all of the forces were unified into a single force via what they call a "theory of everything." As the Universe expanded and cooled, these forces separated into the four we experience today. Gravity is thought to have separated first, at approximately $t = 10^{-43}$ second, which is called the *Planck era*. Before that time, gravity can be described only by using quantum mechanics. Thus the Planck era is the realm of quantum gravity, which physicists still don't know how to describe. We will explore quantum gravity in Section 18.5.

GRAND UNIFIED THEORIES ERA ($t = 10^{-43}$ SECOND TO $t = 10^{-34}$ SECOND). After gravity separated out, the strong, weak, and electromagnetic forces eventually became separately acting forces one by one as the Universe cooled. The Planck era was followed by the *GUTs era*—the realm of **grand unified theories** (or **GUTs**), which focus on the unification of these last three forces and aim to explain how all of the particle interactions we see today emerged from the early history of the Big Bang. There are, however, no generally accepted grand unified theories yet, and the ones that have been proposed are highly speculative. As a result, the "start" and "end" times of the GUTs era (and the next few that follow it) should be seen as approximate.

During the GUTs era, the Universe was filled with **quantum fields** (quantum energy fields). These are distributions of energy that fill all space but appear as different kinds of fundamental particles. (Recall from Chapter 4 that, in quantum mechanics, particles may behave as either particles or waves.)

INFLATION ($t = 10^{-34}$ SECOND). Many theorists think that when the Universe was about $t = 10^{-34}$ second old, something remarkable happened in post–Big Bang space as a quantum field found itself in an unstable, excited state (much like an electron in an outer orbit

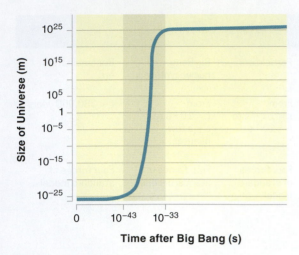

Figure 18.7 Inflation

During inflation, post–Big Bang space-time expanded very quickly, such that the distance between any two points grew rapidly. Inflation began just a fraction of a second after the Big Bang and lasted even less time. After inflation ended, the Universe settled into the rate of expansion we see today, and cosmic evolution continued.

"In Einstein's general relativity, gravity can actually be repulsive if there is the right kind of energy around. This is what happens during inflation."

Figure 18.8 Emergence of the Four Fundamental Forces

The fundamental forces emerged (separated out) from one unified force within the first second after the Big Bang. Physicists continue to search for a grand unified theory explaining how the three forces that were still unified after gravity emerged form the basis of all particle interactions we see today.

of an atom). The energy locked up in that elevated state was enormous. Then, suddenly, the quantum field dropped back down to its ground state. As the field made its transition, the energy released pushed space into a kind of expansion on steroids. This period is called **inflation** (Figure 18.7).

"In Newton's gravity, there were only attractive forces," says Riess, explaining how inflation works. "But in Einstein's general relativity, gravity can actually be repulsive if there is the right kind of energy around. This is what happens during inflation when the quantum field drops back to its ground state."

During inflation, the scale of the post–Big Bang Universe blew up from the size of an atom to the size of a softball. By just a bit after 10^{-34} second following the Big Bang, the scale of the inflated region had increased by a factor of about 10^{60}. But inflation ended as quickly as it started, leaving behind a vastly enlarged space, which settled back into an expansion that was much slower (on the order of what we see today). Keep in mind that the Big Bang and inflation represent a stretching of space-time itself, not an expansion into a void. The phrase "the scale of the Universe" therefore refers to the part of it undergoing inflation that would become the observable Universe we see today. (The observable Universe is that part from which light can reach us after 13.7 billion years of evolution.) It's worth noting when you think about inflation that general relativity allows the expansion of space to be faster than light, even though no particle can travel faster than light *through* space.

Among the inflation era's most important legacies for the future of cosmic history were the small bumps and wiggles it left in the cosmic bath of matter-energy. These perturbations were the result of random quantum fluctuations in the energy fields driving inflation. They would persist through all the eras that followed.

ELECTROWEAK ERA ($t = 10^{-34}$ SECOND TO $t = 10^{-12}$ SECOND). Immediately after inflation, both gravity and the strong force separated out—but the weak and electromagnetic forces were still combined. During this period, the *electroweak era*, the temperature of the Universe was above 10^{13} K. While previous eras can only be studied with computer models, temperatures in the electroweak era can be created in particle accelerators, so scientists can study this era experimentally.

PARTICLE ERA ($t = 10^{-12}$ SECOND TO $t = 0.001$ SECOND). Around 10^{-12} second, the electromagnetic and weak forces separated by *symmetry breaking*. That process marked the beginning of the **particle era**, in which cosmic density and temperature continued to fall with the expansion of space. **Figure 18.8** shows each era of the early Universe and the emergence of the four fundamental forces.

In the particle era, the matter and energy filling the Universe went through a sequence of transitions with cooling and rarefaction. Many of these changes, particularly those at the earliest times and highest temperatures, are not fully understood today. What physicists are sure of, however, is that whatever emerged eventually gave rise to the particles that make up the standard model (quarks, leptons, and force bosons).

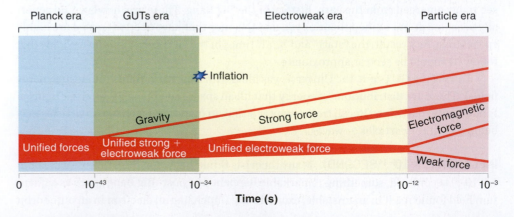

To understand how these particles formed, physicists tend to think in terms of **phase transitions**, changes that occur when matter moves from one form to another as temperature changes. The most familiar phase transition is the switch from liquid water to ice. As the temperature of liquid water drops, freely moving H_2O molecules become locked into ice crystals. Physicists think of the early Universe as a soup of particles (and the quantum fields associated with them). In this phase of matter, particles could interact and switch from one form to another in specific ways. As the temperature dropped during the particle era, some kinds of particles could no longer participate in the back-and-forth transformations; those particles were locked in their current form and number for the rest of cosmic history. For example, when temperatures were high enough, quarks could move freely through the early Universe. But as cosmic temperatures dropped, all quarks became bound together by the strong nuclear force into particles called *baryons* (such as protons and neutrons), consisting of three quarks, and *mesons*, consisting of a quark and an antiquark. This phase transition from free quarks at high temperatures to bound quarks at lower temperatures was essential to the emergence of the Universe we experience today (**Figure 18.9**).

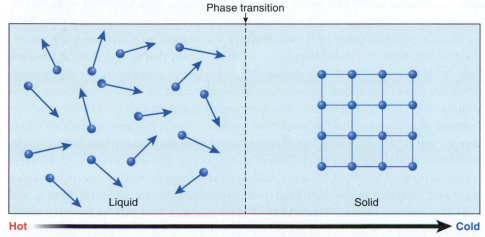

a. Phases of water

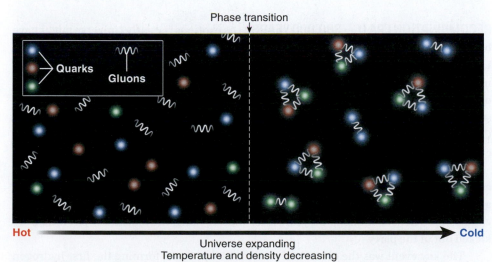

b. Particle era

Figure 18.9 Phase Transitions

The early Universe is described as a sequence of phases with particles of matter interacting and changing state. **a.** A familiar phase transition is the cooling of liquid water enough that it freezes into a solid. **b.** In the early Universe, at high temperatures, quarks and gluons (bosons that carry the strong force) moved freely. As temperatures dropped, a phase transition occurred: all quarks became bound into composite particles (3 quarks = 1 baryon; 1 quark + 1 antiquark = 1 meson).

inflation A hyperrapid expansion of space-time that occurred extremely soon after the Big Bang.

particle era A period in cosmic evolution when the particles described by the standard model of particle physics emerged.

phase transition A change in the state of matter based on changes in temperature (or another quantity).

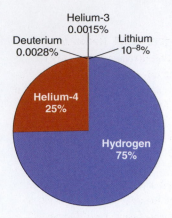

Figure 18.10 Big Bang Nucleosynthesis
During BBN, falling temperatures and densities in the expanding Universe allowed fusion of protons (hydrogen nuclei) into light nuclei such as deuterium, helium, and lithium. The proportions of each nucleus are collectively referred to as the "primordial abundance of elements."

"That's what makes the BBN era so remarkable. It only lasts about 3 minutes."

Big Bang nucleosynthesis (BBN) The nuclear fusion of light elements in the first few minutes after the Big Bang.

recombination era The era about 380,000 years after the Big Bang in which protons captured electrons, forming the first hydrogen atoms and thus decoupling matter from cosmic blackbody radiation.

BIG BANG NUCLEOSYNTHESIS (t = 0.001 SECOND TO t = A FEW MINUTES). When the temperature of the Universe fell to 10^9 K and the density fell to 10^4 kilograms per cubic meter (kg/m^3), conditions were just right to turn the entire Universe into a nuclear reactor. In this epoch, known as **Big Bang nucleosynthesis (BBN)**, light elements, such as hydrogen, helium, and lithium, were first forged.

Big Bang nucleosynthesis must be distinguished from stellar nucleosynthesis (discussed in Chapter 13). Elements could not be "cooked" in the center of a star until the first stars formed, about 10 billion or so years after the Big Bang. Big Bang nucleosynthesis, however, began when the mix of protons, neutrons, electrons, and photons emerging from the particle era had the right range of temperatures and densities to allow fusion to begin. "It's important to distinguish between the relatively constant density and temperature in the core of a star and the rapidly dropping density and temperature in the Universe in the early moments after the Big Bang," says Riess. "That's what makes the BBN era so remarkable. It only lasts about 3 minutes."

Collisions between protons and neutrons led to deuterons, the same heavy form of hydrogen nuclei that is produced by fusion reactions in the Sun. As cosmic deuterium built up, continued collisions added neutrons and protons until helium-4—with its two protons and two neutrons—became abundant.

Nuclear reactions depend on temperature and density, so as the cosmic temperature dropped, some reactions slowed to a crawl, like a cake that stops baking as the oven cools. In this way, the Universe's expansion carried the soup of matter through successive stages where again some nuclei stopped transforming via nuclear reactions. A few minutes after t = 0, temperatures and densities dropped enough that further nuclear reactions were impossible. (The exact time nucleosynthesis stopped depends on which reaction you are interested in, but all were done within minutes.) When fusion ended, the elemental abundances of the lightest elements and isotopes were set. Just a few minutes after the Universe began, the nuclear content was 75 percent hydrogen nuclei, 25 percent helium-4 nuclei, 0.0028 percent deuterium, 0.0015 percent helium-3, and about 10^{-8} percent lithium (**Figure 18.10**). (The close match of this "primordial abundance of elements" with theoretical predictions is a primary piece of evidence for the Big Bang.)

RECOMBINATION ERA ($t \approx$ 380,000 YEARS). As exotic as the early Universe might have been, it was still a collection of hot, dense matter (plasma). In this way, the entire early Universe was a blackbody. We have encountered blackbodies and blackbody radiation many times in our journey from planets to stars and on to galaxies. When matter in an object is dense enough that the distance between a photon's emission and its absorption is small (compared with the object's size), then that object behaves as a blackbody. We saw in Chapter 4 that blackbody radiation has a distinct spectral energy distribution, with peak energy occurring at a wavelength determined by Wien's law: λ_{max} = (0.0029 m K)$/T$.

This interplay of matter and radiation was the rule during the Universe's entire early history. Matter particles jostled with photons that were absorbed and spat out again in endless reactions. All the while, the young Universe remained in blackbody conditions. Then, about 380,000 years after the Big Bang, the blackbody photons were kicked out of the party.

The key event was the capture of electrons by protons, forming the first hydrogen atoms. This period is called the **recombination era**. "This is a terrible misnomer," says Riess, "because it's the *first* time protons and electrons could form long-lived hydrogen atoms. We really should call it *combination*."

One of the most important changes that came with recombination was the fate of the blackbody photons. The newly formed hydrogen atoms could not absorb most of the still-present blackbody electromagnetic radiation. The physics of hydrogen's electron

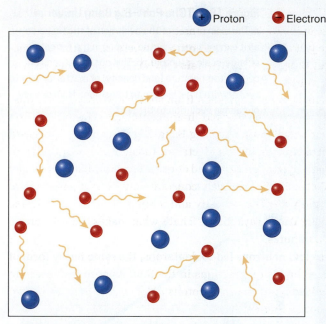

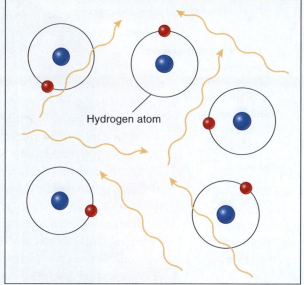

+ Proton   − Electron   ⌇⟶ Blackbody photon

Hydrogen atom

a. **Before recombination: photons are absorbed after traveling a short distance**

b. **After recombination: most photons can travel freely without being absorbed**

Figure 18.11 Recombination
a. Before recombination, matter and photons were in equilibrium because of their rapid interactions, giving the radiation a blackbody spectrum.
b. Due to lower temperatures in the recombination era, protons and electrons moved slowly enough to bind together, forming the first hydrogen atoms. The atomic hydrogen could not absorb the blackbody photons, leaving the radiation "decoupled" from matter. The blackbody photons became "fossils" that we observe today as the cosmic microwave background.

orbits made interactions between hydrogen atoms and the blackbody photons all but impossible. Thus, recombination also meant a *decoupling* of matter (the now ubiquitous hydrogen atoms) from the bath of blackbody photons (**Figure 18.11**).

The blackbody photons were left without dance partners. Just a few thousand years earlier, they could not travel more than a tiny distance through the Universe without being absorbed. After decoupling, they were free to wander space almost unimpeded. "It's like when you are stuck in a fog," says Riess. "If you turn on your headlights, you can't see anything. The light just scatters around. But once the fog clears, the headlight beams can travel a long distance through the air without interruption." As time marched forward, the only change this "fossil" light would experience was a stretching in wavelength directly tied to the expansion of space itself. Note that the recombination era represents the beginning of the Dark Ages described in Chapter 17. While the Universe became transparent to longer wavelength blackbody photons created before recombination, afterward any short wavelength photons (like UV) would not be able to propagate very far until the Universe became mostly ionized again.

The decoupling of matter and radiation via recombination had other consequences, the most important of which are related to cosmic structure. As long as radiation and matter were strongly linked, the tiny perturbations left over from inflation could not grow. Before recombination, if gravity tried to draw material into a region of higher density, the region simply grew hotter. Increased heat meant increased pressure, which pushed back against gravity, smoothing out the perturbations via sound waves. After matter and blackbody radiation went their separate ways, gravity could begin its work because contracting bodies could cool by emitting radiation that escaped from the infalling gas. Thus, dense regions could become denser and build up the large-scale structure we explored in Chapter 17. In this way, recombination marked the end of the Universe's early history. **Figure 18.12** summarizes the periods following the Big Bang (which were previewed in Table 18.1).

INTERACTIVE:
Radius of the Visible Universe versus Age

"If you turn on your headlights, you can't see anything. But once the fog clears, the headlight beams can travel a long distance through the air without interruption."

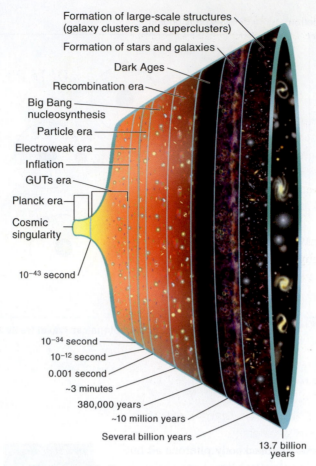

Formation of large-scale structures
(galaxy clusters and superclusters)

Formation of stars and galaxies

Dark Ages

Recombination era

Big Bang
nucleosynthesis

Particle era

Electroweak era

Inflation

GUTs era

Planck era

Cosmic
singularity

10⁻⁴³ second

10⁻³⁴ second
10⁻¹² second
0.001 second
~3 minutes
380,000 years
~10 million years
Several billion years

13.7 billion
years

a. **Time since Big Bang**

Figure 18.12 The Post–Big Bang Universe
a. The main stages of the evolution of the Universe as it expands are depicted, ending in the formation of large-scale structure. **b.** This cosmic expansion causes temperature (and density) of matter-energy to decline with time, leading to transitions from one era of cosmic evolution to the next.

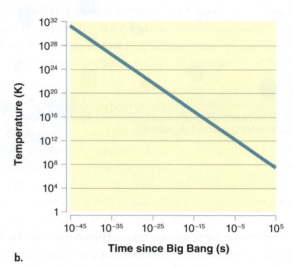

b.

Section Summary

- The Big Bang theory is built on the concept from general relativity of flexible space-time. The standard model of particle physics describes matter and energy at a fundamental level.
- Cosmic history is traced backward to a cosmic singularity, which implies infinite density, temperature, and space-time curvature. That was followed by the Big Bang and then the Planck era, when temperatures were extremely high and the four forces were unified.
- The grand unified theories (GUTs) era began when gravity emerged as a separate force and ended with inflation (a hyperrapid expansion of space-time). Next, in the electroweak era, the strong nuclear force separated from the electromagnetic and weak forces.
- In the particle era, the electromagnetic and weak forces separated and the particles of the standard model formed as the Universe's temperature and density dropped.
- During Big Bang nucleosynthesis (BBN), the nuclei of the lightest elements formed by fusion and their elemental abundances became fixed. Then, in the recombination era, matter and blackbody photons decoupled, allowing the photons to travel unimpeded through space.

CHECKPOINT Describe the different eras in the post–Big Bang history of the Universe.

> "It's kind of like seeing a cake that is rising as it bakes. It's natural to infer that it was all much smaller and more compact when it started out."

How Do We Know? The Three Observational Pillars of Big Bang Theory

18.3

It's quite a story: the Universe begins in an ultradense, ultrahot state and immediately begins expanding. What observations, however, convince astronomers that the Big Bang actually happened?

primordial abundance of light elements The proportions of hydrogen, helium, deuterium, and other light elements created in the first few minutes after the Big Bang.

Three key pieces of evidence support the classic narrative of the Big Bang. Here, *classic* means everything except inflation, which is a more recent addition to Big Bang theory. Let's deal with the classic part of the story first.

Pillar 1: Cosmic Expansion

Edwin Hubble's discovery of galaxy recession (see Chapter 16) was the first evidence of a Universe that was not static. However, if cosmic expansion were our *only* evidence, it would still be possible to develop a model like Fred Hoyle's steady-state Universe, in which space always expanded but always looked the same. The Big Bang model, in contrast, tells us that the Universe looked very different in the past than it does today. "It's kind of like seeing a cake that is rising as it bakes," says Adam Riess. "It's natural to infer that it was all much smaller and more compact when it started out."

Although Hubble discovered cosmic expansion in 1929, he did not claim it to be evidence of a cosmic beginning. It was the great Belgian physicist Georges Lemaître who, in 1927, had proposed the concept that subsequently became known as the Big Bang.

Pillar 2: Big Bang Nucleosynthesis (BBN)

The brief period lasting a few minutes after the Big Bang, when the Universe was hot enough to drive fusion reactions, left a mark on the cosmos that we can see today. The relative abundance of nuclei such as hydrogen, deuterium, helium-3, helium-4, lithium-6, and lithium-7 can be directly measured and compared against detailed models of the Universe during the BBN era. In this way, the light elements act as a kind of fossil record of the Big Bang, enabling us to infer conditions in the early Universe.

Astronomers measure the **primordial abundances of light elements** (the abundances established just after the Big Bang) in a number of ways. For example, the primordial helium-4 abundance is measured in relatively "primitive" systems—astronomical environments that haven't been tainted by helium-4 formed within stars. Dwarf galaxies are useful as primitive systems because they are especially poor in oxygen and nitrogen (which must be processed within stars). Thus, measurements of helium-4 in dwarf galaxy gas clouds give astronomers a measure that is close to the abundance it exhibited when the BBN era ended.

The measured values of light-element abundances correspond well with the predictions from BBN theory. This match between theory and observations makes it clear to astronomers that the Universe must have been hot enough and dense enough early in its history for nucleosynthesis to occur. In other words, the Universe *then* looked very different from the Universe *now*.

Pillar 3: The Cosmic Microwave Background (CMB)

In the spring of 1964, Bell Laboratory scientists Arno Penzias and Robert Wilson tripped over one of the greatest cosmological discoveries ever made, for which they won a Nobel Prize (**Figure 18.13**). They were working on then-new technology of satellite telecommunications via a massive, horn-shaped microwave antenna for Bell Labs in Holmdel, New Jersey, which was then a powerhouse of both industrial and pure scientific research. (Recall from Chapter 4 that microwaves are electromagnetic waves with wavelengths between 1 meter and 1 millimeter [mm].) When the two astronomers had a chance to use the antenna for basic astronomy, they planned to look at distant galaxies in microwave radiation. But that plan kept failing.

The problem was an annoying, low-level "noise"—a microwave hiss—that persisted regardless of which direction the antenna was pointed. For weeks, Wilson and Penzias struggled to root out the problem. They rebuilt the electronics. They cleaned layers of

Figure 18.13 Arno Penzias and Robert Wilson
The scientists stand in front of the radio telescope they used to discover the cosmic microwave background—fossil photons left over from the early Universe.

pigeon poop from the antenna surface. Nothing changed. Through painstaking work, they came to understand the problem: there was no problem.

The signal was real, and it was another "fossil" left over from the Big Bang. The light waves flooding the microwave antenna were radiation released more than 13 billion years ago. When the full spectrum of Wilson and Penzias's mystery microwaves was measured, it turned out to be a *perfect blackbody*. That meant the electromagnetic waves they discovered were the blackbody light left over from the recombination era 380,000 years after the Big Bang.

The peak wavelength λ_{max} of their blackbody radiation occurred in the microwave portion of the spectrum at 1.9 mm (1.9×10^{-3} meters). This was almost exactly what Gamow and Alpher's theory had predicted back in 1954. With some subsequent fine-tuning, the temperature of the **cosmic microwave background (CMB)** was found to be just 2.7 K. It was so cold because the wavelength of the blackbody radiation had stretched along with the expansion of space since recombination had decoupled it from the cosmic gas. This spatial stretching translated into an ever-decreasing temperature of the waves. Thus, although the blackbody's temperature had been about 3,000–4,000 K at the time the radiation was released more than 13 billion years earlier, now it was just barely above absolute zero (**Figure 18.14**; **Going Further 18.1**). In 1978, Penzias and Wilson were awarded the Nobel Prize in Physics for their monumental discovery of the CMB.

"The CMB tells us that the Universe had to have been much hotter than it is now," says Riess, "because you still see the glowing radiation. It's as though you step into a room a moment after somebody pulled a grenade pin, and you see shrapnel flying apart and you feel heat. So even though you didn't see somebody trigger the grenade—you didn't actually see the explosion—you can infer what's going on in the room." A bit of the static you see on a TV screen (when it's connected to an antenna) comes directly from the CMB. Some of those little flashes on the screen are actually signals from the infancy of the Universe.

> "Even though you didn't see somebody trigger the grenade—you didn't actually see the explosion—you can infer what's going on in the room."

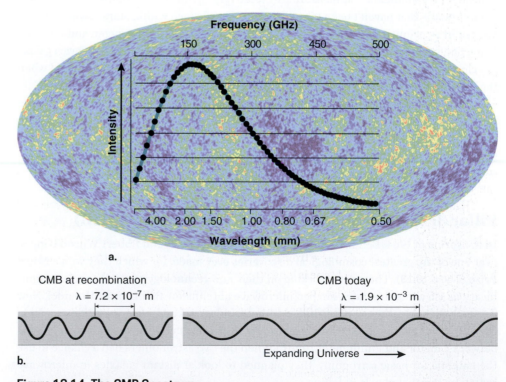

a.

b.

Figure 18.14 The CMB Spectrum
a. The spectrum is that of a blackbody that has a peak radiation consistent with a temperature of 2.7 K.
b. As the Universe expanded, the wavelength λ of the electromagnetic waves making up the CMB was stretched. Correspondingly, the CMB's temperature dropped.

cosmic microwave background (CMB) Blackbody radiation released during recombination, with a peak wavelength in the red part of the electromagnetic spectrum. The expansion of the Universe now puts the peak of CMB in the microwave range.

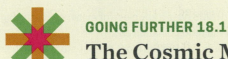

GOING FURTHER 18.1

The Cosmic Microwave Background and the Expansion of the Universe

How much has the Universe expanded since the cosmic microwave background (CMB) radiation decoupled from matter during the recombination era?

Recall that the temperature of the plasma filling the Universe had dropped enough that electrons and protons were finally moving slowly enough for electromagnetic attraction to allow them to find each other and bind together into neutral hydrogen atoms. At the start of recombination the Universe was, essentially, a blackbody with a temperature of about $T = 4,000$ K. Let's calculate the wavelength at which this blackbody radiation peaked. We can use Wien's law (Chapter 4):

$$\lambda_{max} = (0.0029 \text{ m K})/T = (0.0029 \text{ m K})/(4,000 \text{ K})$$
$$= 7.2 \times 10^{-7} \text{ m} = 720 \text{ nm}$$

Thus at the moment of decoupling, the whole Universe was glowing with a peak at 720 nm, in the red part of the visible spectrum.

But how does this number tell us anything about the Universe's expansion? Recall that after decoupling, the electromagnetic waves making up this blackbody radiation stopped interacting with matter. The only thing affecting the radiation was the expansion of space, which explains why the CMB seen today has a peak wavelength that is so much longer than 720 nm. Thus, the expansion of the Universe is encoded in the CMB radiation.

Let's call the size scale of the Universe at decoupling $R_{decoupling}$ (think of it as the distance to a very, very distant galaxy) and the size scale of today's Universe $R_{current}$. Because the peak wavelength of the CMB is directly proportional to the size scale, we have

$$\frac{R_{current}}{R_{decoupling}} = \frac{(\lambda_{max})_{current}}{(\lambda_{max})_{decoupling}}$$

The peak of microwave radiation in the current cosmic epoch is $(\lambda_{max})_{current} = 1.9 \text{ mm} = 1.9 \times 10^{-3}$ meter. Plugging these values in, we see that

$$\frac{R_{current}}{R_{decoupling}} = \frac{1.9 \times 10^{-3} \text{ m}}{7.2 \times 10^{-7} \text{ m}} = 2.6 \times 10^{3}$$

Our estimate is good enough to tell us that the Universe has expanded by a factor of about 1,000 since the recombination era. A lot of new space has been added to the cosmos.

An Old Question Answered: Olbers's Paradox

If the Universe is infinite and full of an infinite number of stars, then when you look at the sky, your line of sight should always cross a star somewhere. Then why is the sky dark at night—why doesn't the light of infinite stars fill the night sky? Heinrich Olbers asked this question in the early 1800s.

How does the Big Bang theory answer that question? By positing that the Universe had a beginning and is expanding, the Big Bang theory tells us that light from distant objects is redshifted; its wavelength has been stretched beyond the visible, making it undetectable by eye. The expansion of the Universe from a beginning also means that light from some stars has yet to reach us, if its travel time is longer than the current age of the Universe.

The role of cosmic redshifts is important for more than Olbers's paradox. For example, UV light that would be blocked at the era of recombination gets significantly redshifted by the time we observe it.

In Missions 19 and 20, use the properties of the CMB radiation to find the alien technology you need to save the galaxy from the clutches of the Corporation.

Section Summary

- The first evidence of the Big Bang came from the expansion of the Universe, as seen in the recession of galaxies described by Hubble's law.
- Additional evidence comes from models predicting the relative abundances of the lightest elements that Big Bang nucleosynthesis would have produced, which match the abundances observed in the Universe today.

- The discovery of the cosmic microwave background (CMB) was a key piece of evidence for the Big Bang, as the CMB constitutes fossil radiation from the recombination era.
- Olbers's paradox is explained by cosmic expansion, as light from the most distant stars either has been redshifted out of the visible range or hasn't reached us yet.

CHECKPOINT Explain the three pillars of evidence for the Big Bang theory.

INTERACTIVE:

Recombination of the Cosmic Microwave Background

Beyond the Classic Big Bang Model

18.4 One of the most important consequences of the CMB studies was the development of support for the existence of inflation as part of the Big Bang. The idea of inflation was first fully developed by Alan Guth of MIT at the start of the 1980s. It was soon picked up by others and became a major addition to the story of cosmic evolution. But astrophysicists needed to find solid evidence before they could accept the possibility that it had, in fact, occurred.

> "It's like firing a missile that doesn't reach escape velocity. The rocket eventually falls back to Earth. With the Universe, if the initial Big Bang isn't strong enough, then the Universe collapses back on itself."

Inflation, the Cosmic Microwave Background, and the Origin of Structure

When Wilson and Penzias stumbled on the CMB, they found that every region of the sky is filled with blackbody radiation at a temperature of 2.7 K. But as we saw in Chapter 17, astrophysicists knew that the large-scale structure such as today's galaxy clusters must have originated in small density perturbations during cosmic evolution. Those fluctuations were present in the recombination era, when the CMB radiation was released. According to the physics governing the plasma during recombination, regions that were slightly denser would be slightly hotter, while regions that were less dense would be cooler. Thus, small perturbations in density implied that small perturbations in temperature must have existed as well.

These temperature fluctuations were expected to show up in maps of the CMB. But when physicists estimated the expected changes in temperature, the differences were tiny—only one part in 100,000. Taking these measurements required overcoming enormous technical difficulties; the observations had be made from space. After years of effort, the Cosmic Background Explorer (COBE) satellite, launched in 1989, successfully mapped the CMB across the entire sky with enough sensitivity to find the cosmic temperature fluctuations (**Figure 18.15**). It was yet another triumph for the Big Bang theory. Other CMB satellites followed, such as the Wilkinson Microwave Anisotropy Probe (WMAP), active 2001 to 2010, and Planck in 2009, each mapping the sky at ever-higher resolution.

One significant prediction of inflationary Big Bang cosmology was the **spectrum of perturbations**. Quantum fluctuations imposed on matter during inflation became the tiny bumps and ripples that persisted into the recombination era as the seeds of all current large-scale structure. Inflation made very definite predictions for the distribu-

Figure 18.15 Mapping the Cosmic Microwave Background

Fluctuations in the CMB are shown, as mapped with ever-higher resolution by successive satellites: **a.** COBE (1989), **b.** Planck (2009), and **c.** WMAP (2010). The fluctuations correspond to small regions in the early Universe with different temperature (and density) conditions—which result in the large-scale structure of today's Universe.

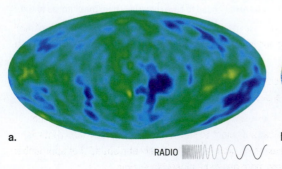

a.

RADIO

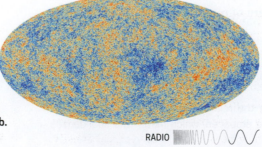

b.

RADIO

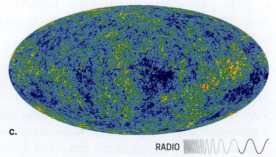

c.

RADIO

tion, or spectrum, of those perturbations in terms of their size and strength. With higher-resolution maps of the CMB and its temperature fluctuations, astronomers were able to show how closely inflation's prediction for the spectrum of perturbations matched what is observed. This was the first real, data-based test of inflation theory, and it passed that test with flying colors.

The Paradox of Universal Uniformity (the Horizon Problem)

Why were astronomers so keen on the idea of inflation? "Inflation ended up solving a number of problems or paradoxes that were appearing in the classic version of the Big Bang," says Clifford Johnson, a theoretical physicist at the University of Southern California.

One of these paradoxes was the recognition that the Universe, as shown by the CMB, is remarkably uniform. Statistically, temperature fluctuations on one side of the sky are just like those on the other side of the sky. (It is as though every city on Earth has exactly the same temperature.) Since these different regions correspond to seeing different parts of the Universe at early times, the uniformity implies that all regions of the entire Universe at recombination must have already shared information, smoothing out any large-scale differences. Astronomers call this the *horizon problem* because, just as we can't see past the horizon when we stand on Earth, the Universe's distant regions should not have been able to "see" each other because the travel time of light between them was greater than the age of the Universe at that time.

"They should have been in contact with each other," says Johnson, "but that just wasn't possible with the classic version of the Big Bang because the expansion of the Universe was too slow." According to classic Big Bang theory, there had not been enough time from $t = 0$ to the recombination era for light waves (the fastest signals possible) to cross from one side of the cosmos to the other, bringing different regions into contact and hence smoothing out their differences (**Figure 18.16**).

By assuming there was an early period of inflation, cosmologists were able to resolve this and other paradoxes in classic Big Bang theory. The brief period of inflation implies that the different parts of the Universe we now see as widely separated were once close enough together to be connected by light signals before inflation occurred.

Density and Cosmic Destiny (the Flatness Problem)

Inflation solved another important problem for cosmologists. In the classic Big Bang theory, Einstein's equations imply a critical link between the density of the Universe and its geometry. If the Universe began with a high enough density of matter-energy, then gravitational attraction should, eventually, overwhelm the expansion initiated at the Big Bang. The gravitational attraction of everything to everything else should eventually stop cosmic expansion and turn it around into an overall cosmic contraction, leading to the opposite of the Big Bang: a Universe ending in a **Big Crunch**. "It's like firing a missile that doesn't reach escape velocity," says Johnson. "The rocket eventually falls back to Earth. With the Universe, if the initial Big Bang isn't strong enough, then the Universe collapses back on itself."

In another scenario, the matter-energy density of the Universe is *too* low. In this case, gravity could never decelerate cosmic expansion enough to halt the stretching of space, and the Universe would expand forever.

The final possibility is that the Universe has just the right value of matter-energy density—what cosmologists call the **critical density**. The Universe would slow down in such a way that even though the expansion would proceed indefinitely, the expansion speed would approach zero more and more closely as time passes. Thus, which scenario is most realistic depends on the density of the Universe relative to the critical density.

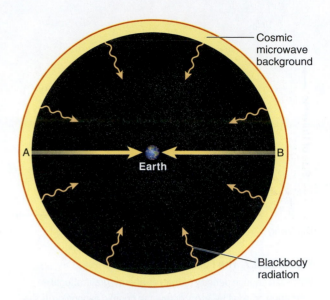

Figure 18.16 The Paradox of Universal Uniformity
As shown by the CMB, temperature fluctuations on one side of the sky statistically match those on the other side. If light from points A and B on the CMB is reaching Earth only now, then there has not been enough time since recombination for light waves to smooth out the different parts of the cosmos. The brief period of inflation implies that the parts we now see as widely separated were close enough together before inflation occurred to be connected by light signals.

spectrum of perturbations The range and strength of variations in density imposed on cosmic matter during inflation.

Big Crunch A theoretical contraction of the Universe that would occur if the combined matter-energy exceeded critical density.

critical density The matter-energy per unit volume that is just sufficient to slow cosmic expansion such that the expansion proceeds indefinitely at a speed that approaches zero more and more closely over time.

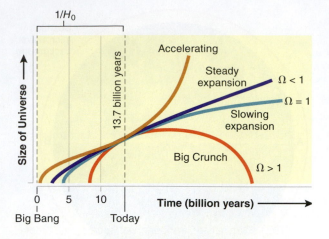

Figure 18.17 The Fate of the Universe

With high enough density of matter-energy, gravitational attraction could slow, stop, or reverse cosmic expansion. If the density matches the critical density ($\Omega = 1$), the expansion will slow, proceeding indefinitely at a speed that approaches zero as time passes. A density above the critical value ($\Omega > 1$) will result in a Big Crunch; at a density below that value ($\Omega < 1$), the Universe will expand forever.

Figure 18.18 The Curvature of Space-Time and the Cosmic Microwave Background

Three possible geometries of the Universe: each is tied to an Ω value, the CMB's appearance, and the path of light rays from CMB "granules" (the seeds from which galaxies formed). **a.** Positive curvature ($\Omega > 1$), like that on a sphere's surface: converging light rays would seem to magnify the CMB. **b.** A flat Universe with no curvature ($\Omega = 1$): unbent light rays would show the CMB correctly. **c.** Negative curvature ($\Omega < 1$), yielding a saddle-shaped geometry: diverging light rays would seem to reduce the CMB. Thus the CMB's appearance provides evidence that our Universe has a flat geometry.

Astronomers express this relationship in terms of a parameter they call **omega** (Ω): the ratio of the measured density to its critical value (**Figure 18.17**). Each scenario is associated with its own form of geometry.

Einstein's equations also predict that the Universe can have different kinds of geometry, meaning different kinds of curvature, depending on density (**Figure 18.18**). If the Universe has a density greater than the critical value ($\Omega > 1$), then it has positive curvature, with three-dimensional spherical geometry similar to the surface of a beach ball. A Universe with positive curvature would be finite (only so much space) but have no boundaries. If instead the Universe has a cosmic density that equals the critical value ($\Omega = 1$), then the Universe is flat—it has no curvature. With a density below the critical value ($\Omega < 1$), the Universe would have negative curvature, yielding a saddle-shaped geometry. Note that the fate of the Universe depends on its geometry too, with positive curvature leading to a Big Crunch and negative curvature leading to a Universe that expands (and thins out) forever. A flat universe expands "asymptotically," meaning it stops expanding only when its age equals infinity. That means, as time goes on, a flat Universe gets ever closer and closer to halting its expansion but never actually reaches such a state. Finally, it's worth noting that this discussion does not include the effect of dark energy, which we will consider shortly.

Before the inflation theory was developed, one problem cosmologists faced was that observations implied a cosmic density close to the critical value of 1. (Typical Ω values were about 0.2 or 0.3.) But according to the equations governing cosmic evolution, if the density started even a little below the critical density (say, $\Omega = 0.9999$), it would evolve into incredibly low values by our current epoch (on the order of 10^{-22} or less). Likewise, if the density at the Big Bang were slightly above the critical value (say, $\Omega = 1.0001$), it should now be greater than 1 by many orders of magnitude. In other words, unless density at the Big Bang was *exactly* the critical density, cosmic history should have

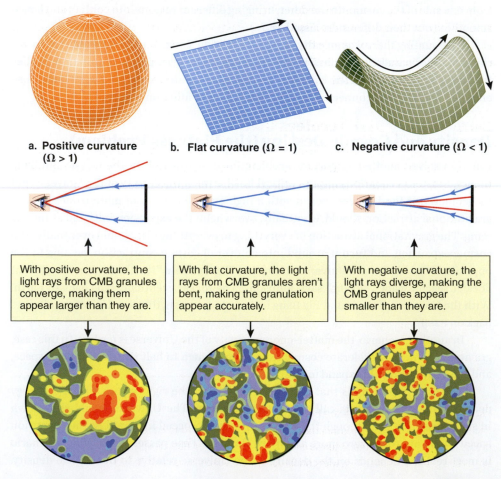

a. Positive curvature ($\Omega > 1$) **b. Flat curvature ($\Omega = 1$)** **c. Negative curvature ($\Omega < 1$)**

With positive curvature, the light rays from CMB granules converge, making them appear larger than they are.

With flat curvature, the light rays from CMB granules aren't bent, making the granulation appear accurately.

With negative curvature, the light rays diverge, making the CMB granules appear smaller than they are.

pushed it very far from 1 as the Universe evolved. But observations showed density close to but not exactly critical, so either astronomers were missing something or somehow the Universe had been fine-tuned early on to give this nearly-but-not-quite-1 value that we see today.

Inflation solved this dilemma, called the *flatness problem*. By pushing space-time into hyperexpansion, inflation flattened the geometry of space, driving it toward rather than away from critical density (**Figure 18.19**). Because the rapid expansion stretched space, in turn reducing the curvature, inflation took the part of space we can see and drove its density toward the critical value. Thus, inflation theory allowed cosmologists to breathe a sigh of relief, since finding densities close to the critical density no longer seemed like a miracle.

But this result presented another dilemma. As cosmologists got a better look at the CMB, their best models predicted a density not *close* to critical but *exactly* the critical density. If the matter-energy density detected (including the dark matter) added up to only $\Omega = 0.3$, then what made up the rest of the Universe that would give a total density of $\Omega = 1$? The answer came as a surprise and represented one of the most important cosmological discoveries of the 20th century, after the discovery of the CMB.

Although the geometry of space-time might seem abstract, it has observable consequences. In a flat space, the sum of the angles in a triangle must equal 180°. But in a space with positive curvature, the three angles would sum to more than 180°, while a space with negative curvature would have angles that sum to less than 180°. Remarkably, the fluctuations in the CMB provide a direct way of distinguishing such effects. Rather than providing giant triangles on the sky, it's the size of the CMB fluctuations that is sensitive to the kind of curvature our Universe has. Measurements of the size of the CMB temperature variations carried out with the satellites WMAP and Planck have confirmed that the Universe has a geometry that is very nearly flat, perhaps exactly flat.

Finally, it is important to understand that the space-time of our Universe (as represented by the geometries discussed in this section) does not expand *into* anything. It can be very hard to think about the space-time of the Universe as a whole, but the two-dimensional curved surfaces in Figures 18.18 and 18.19 represent the whole of physical reality, which is itself expanding. There is no other area for it to expand into.

Dark Energy and Acceleration

By the 1990s, astronomers had learned how to use Type Ia supernovas as standard candles for measuring distance (see Chapter 11). These bright supernovas made it possible to determine accurate distances to objects across the entire observable Universe. One

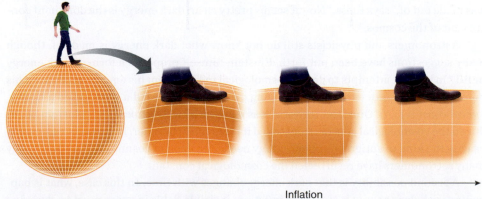

Inflation

Figure 18.19 Inflation Explains Flatness
Each subsection of the sphere is inflated by a factor of 3 from the one in the previous panel. By the third iteration, the sphere's curvature is undetectable. Similarly, inflation explains the flattening of the Universe's geometry, consistent with critical density (an observed value of $\Omega = 1$).

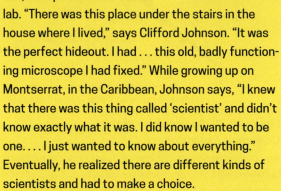

CLIFFORD JOHNSON

If you're going to be a scientist, it helps to have a secret lab. "There was this place under the stairs in the house where I lived," says Clifford Johnson. "It was the perfect hideout. I had . . . this old, badly functioning microscope I had fixed." While growing up on Montserrat, in the Caribbean, Johnson says, "I knew that there was this thing called 'scientist' and didn't know exactly what it was. I did know I wanted to be one. . . . I just wanted to know about everything." Eventually, he realized there are different kinds of scientists and had to make a choice.

After moving to England, Johnson specialized in theoretical physics. He is now an expert in string theory. Strings are considered a good candidate for a theory of everything. Perhaps Johnson has not strayed far from his dreams in the secret lab.

omega (Ω) The ratio of the observed density of matter-energy in the Universe to the critical density.

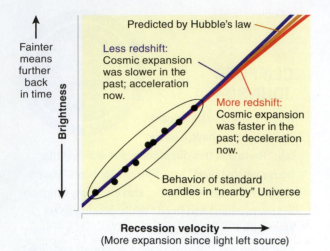

Figure 18.20 Acceleration of the Universe's Expansion

How cosmic expansion has changed over time: the relationship between brightness and recession velocity of standard candles in the nearby and distant Universe is depicted. All standard candles have the same luminosity, so fainter objects are farther away and hence further back in time. Near us cosmologically, the plot is linear, consistent with Hubble's law. But at great distances sampled by Type Ia supernovas as standard candles, the curve will bend either downward, signifying faster expansion in the past (hence, deceleration now), or upward, signifying slower expansion in the past (hence, acceleration now). The Type Ia data favor acceleration and imply the existence of dark energy.

"Objects far away (deeper in the past) would be moving faster than the linear Hubble's law predicted because they had not yet been decelerated."

project that used Type Ia supernovas set out to determine how Hubble's law changes at very large distances. "That is how I got involved with the project," says Riess. "I was a grad student at Harvard, and I heard that Robert Kirshner had a project using Type Ia supernovas to test Hubble's law at huge distances. So I asked him if I could join his team." The rest is history.

Recall that astronomers had always expected that the mutual gravitational attraction of the Universe's matter-energy would slow cosmic expansion to some degree. Lots of deceleration (with $\Omega > 1$) would lead to a Big Crunch, whereas almost no deceleration (with $\Omega < 1$) would mean infinite expansion. Using Hubble's law and measuring distances (D) and velocities (V) of objects very far away, astronomers could determine the deceleration by plotting V versus D on a graph and seeing how their relationship departed from a straight line. "Objects far away (deeper in the past)," explains Riess, "would be moving faster than the linear Hubble's law predicted because they had not yet been decelerated."

But Riess and his colleagues' painstaking multiyear project to push the V-versus-D plot deeper into space and further back in time found acceleration rather than deceleration, as we saw in Section 18.1. The expansion of the Universe isn't slowing down; it is speeding up (**Figure 18.20**).

Cosmic acceleration implies that some kind of force is pushing space apart. Since forces require energy, astronomers began speaking in terms of a **dark energy** that pervades all space. The term *dark* simply means that we could not measure it directly but could only measure its effects on luminous matter (as with dark matter). This new form of dark energy, however, must be different from the "normal" matter-energy we usually encounter. Whereas normal matter-energy always leads to gravitational attraction, dark energy has an "antigravity" effect in pushing space apart (see **Anatomy of a Discovery** on p. 492). Dark energy leads to gravitational repulsion.

Because of this difference between normal matter-energy and dark energy, astronomers now specify different forms of the density parameter Ω. There's Ω_m for normal matter, Ω_{dm} for dark matter, and Ω_{de} for dark energy. The total matter-energy density of the Universe, Ω_T, is the sum of the three forms: $\Omega_T = \Omega_m + \Omega_{dm} + \Omega_{de}$. Astronomers soon gathered other evidence for the existence of dark energy, including direct measures of the Universe's geometry via the size of CMB temperature fluctuations. And recall that the matter-energy density detected (matter plus dark matter, $\Omega_m + \Omega_{dm}$) adds up to only $\Omega = 0.3$, while the total matter-energy density $\Omega_T = 1$. All of this led to the remarkable discovery that the matter-energy density of dark energy $\Omega_{de} = 0.7$, making it the Universe's major component. **Figure 18.21** shows the relative amounts of each matter-energy type. "It's pretty remarkable that as late as 1998, we still didn't know what most of the Universe was made out of," says Riess. "Now it seems pretty clear; dark energy is the dominant constituent of the cosmos."

Astronomers and physicists still do not know what dark energy is made of, though many suggestions have been put forth. Einstein himself proposed a form of dark energy in 1917 in his first attempts to derive cosmological models from his equations. To keep his model Universe static—that is, unchanging over time—he added a term called a *cosmological constant* (Λ). Once the Universe was discovered to be expanding, Einstein abandoned the cosmological constant and called it his greatest blunder. But since it represents exactly the kind of antigravitational energy needed to explain cosmic acceleration, today many astronomers have resurrected the cosmological constant.

A second possibility is that dark energy changes with time. In this case, what is happening with dark energy is similar to inflation: invisible fields in space must be decaying (very slowly), with their energy driving cosmic acceleration.

Many other ideas about the nature of dark energy have been proposed and are being actively explored. Without a doubt, learning the origin and nature of dark energy is one of the greatest challenges for modern physics and cosmology. Its discovery shows just how much there still is to learn about our Universe.

Cosmology and Accuracy

"Even as recently as when I was an undergraduate, cosmology used to be the butt of a lot of jokes," says Clifford Johnson. "People made fun of how poorly known most of its basic measured quantities were. The accuracy was pretty bad."

That's not the situation anymore. From the CMB to Big Bang nucleosynthesis to the statistics of large-scale structures, astronomers now have many ways to test the modern narrative of cosmology. With the addition of inflation and dark energy, the story of the Universe's evolution from the Big Bang has become richer and more detailed than the early version of the classic Big Bang. Most important, our ability to test key features of that story has grown. Just 50 years ago, cosmological parameters such as Hubble's constant H_0 or the density parameter Ω were poorly known, if known at all. Today, astronomers can pin these numbers down to within less than a few percentage points. This is what scientists sometimes mean when they talk about cosmology as a "precision" science (though an "accurate" science might be a better label). "We've also got a lot more statistics now," says Johnson. "That allows us to really bring down the uncertainties. We know what we know pretty accurately."

One measure of precision is the ability to use the observed data to distinguish among variations of a cosmological model. For example, all versions of Big Bang models make predictions of the number of temperature fluctuations of a given size that should be imprinted in the CMB. As we've seen, these fluctuations come from perturbations imposed by inflation—but they were also modified by sound waves running through the cosmic gas during recombination.

The number of large-scale temperature fluctuations (in terms of how much space they occupy) versus small-scale fluctuations differs for different models of the Big Bang. For example, more large-scale fluctuations occur in models with more dark energy relative to dark matter. Astronomers have computed many "model" universes in this way, each with its own proportions of dark energy, dark matter, and normal matter. Each model yields different predictions of the statistics of temperature fluctuation in the CMB. "By comparing these predictions with the actual data from something like the Planck satellite," explains Johnson, "we've been able to eliminate a lot of possibilities for the constitution and history of the Universe."

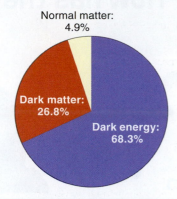

Figure 18.21 Relative Abundance of Dark Energy
Atoms (normal luminous matter, of which we, planets, stars, and cell phones are made) constitute only about 4.9 percent of the matter-energy of the Universe. Dark matter makes up 26.8 percent, and dark energy represents the majority, 68.3 percent.

> "We've also got a lot more statistics now. That allows us to really bring down the uncertainties. We know what we know pretty accurately."

Section Summary

- Large-scale structures must have formed from density fluctuations in the early Universe that left fingerprints found in temperature variations in the CMB. Inflation explains how the Big Bang produced statistically uniform structure from once-close parts of the Universe.
- The fate of the Universe depends on the density of matter-energy and the resulting gravitational effects.
- The Universe has three possible geometries: (1) finite but unbounded, with expansion that will eventually stop and yield to contraction; (2) saddle-shaped, with space whose expansion continues indefinitely; or (3) infinitely flat (no curvature), with unending expansion whose speed approaches zero over time.
- Observations support a flat geometry as a result of the hyperexpansion that occurred during inflation, which was crucial to reaching the critical density.
- The discovery that the expansion of the Universe is accelerating led to the identification of dark energy, which represents 70 percent of all matter-energy.

CHECKPOINT **a.** Identify the three possible geometries of the Universe. **b.** Which geometry does current evidence support, and what is that evidence?

dark energy Energy driving the acceleration of the expansion of space-time.

How has the Universe evolved over time?

Supernova SN 1994D

observation

In 1990 (26 years after Arno Penzias and Robert Wilson discovered radiation from the Big Bang), the **Hubble Space Telescope** (HST) was launched. **Wendy Freedman** led the HST Key Project, which sought to make a precise measurement of Hubble's constant (H_0), an indicator of the Universe's expansion rate. Like Edwin Hubble, her team used Cepheid variables to obtain distances to remote galaxies.

improved technique

In 1992, **David Branch** and colleagues identified a new standard candle: Type Ia supernovas. Much brighter than Cepheids, this type of exploding star allows astronomers to measure distances hundreds of times greater than before and to calculate H_0 when the Universe was only a couple of billion years old.

Hubble Ultra Deep Field

observation

Two teams—the High-Z Supernova Search Team and the Supernova Cosmology Project—hunted for Type Ia supernovas in distant galaxies. In 1998, both teams concluded that the expansion rate of the Universe is *increasing*, as though something is pushing galaxies apart ever faster! The accelerating expansion of the Universe remains an incredible and unexpected, yet widely accepted, result.

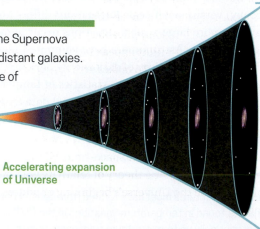

Accelerating expansion of Universe

theory

Independently, the teams had discovered the main component of today's Universe: dark energy, a mysterious form of energy needed to power the acceleration. The finding fundamentally altered theories of cosmic evolution. In 2011, team leaders **Saul Perlmutter**, **Adam Riess**, and **Brian Schmidt** received the Nobel Prize in Physics.

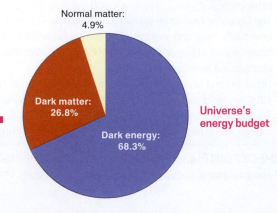

Normal matter: 4.9%

Dark matter: 26.8%

Dark energy: 68.3%

Universe's energy budget

Questions about "Before" and "Everything"

18.5 As we've seen, the Big Bang is a theory not about "creation" but about what happens "after time began." Most scientists, however, were never quite satisfied with the idea of the Universe just popping into existence. All theories led back to questions of what happened before the first instants of the Big Bang and the possible existence of more than one universe.

Before the Big Bang: The Hunt for Quantum Gravity

As mountains of data piled up showing the early Universe to be hotter and denser than it is today, researchers assumed that, one day, someone would figure out what $t = 0$ really means. "People always knew there were several options for the beginning of space and time," explains Clifford Johnson. "One idea is that the Universe always has been. It just seems to start at zero at the beginning of the Universe, but really there was no beginning. Another option is that there was no 'pre-Universe' in the sense of there being a large, smooth space in which one could move around with everything widely separated from everything else." What Johnson means is that before the Big Bang, the kind of space and time we're familiar with may not have existed. Scientists contemplating the Universe's very earliest moments must begin with this kind of speculative theorizing. "Maybe both space and time are just something completely different fundamentally from what we think of," says Johnson. "Maybe they require a different language to describe them, such that creation of the Universe means taking some kind of protospace and prototime and allowing them to evolve into the smooth space and time we know today."

The problem with describing the Big Bang's true beginning is that the further back in time scientists push, the hotter and denser the Universe gets, until space-time becomes so compressed that infinities appear in Einstein's general relativity. Scientists believe that in this cosmic singularity, a space-time that's so strongly curved on such tiny scales, general relativity is no longer applicable. "The equations themselves become singular," says Johnson, "meaning infinities appear in particular ways that tell us that we just can't use those equations anymore." To make further progress, scientists must switch to a different theory—a *quantum description* of space-time.

Quantum mechanics describes nature at the smallest scales. At the Universe's very beginning, even space-time itself would have been quantized, appearing as a kind of "foam" of location and duration. Unfortunately, no one has ever developed a full working theory for these "atoms" of space-time. Without such a theory of *quantum gravity*—a merger of Einstein's general theory of relativity with quantum mechanics—scientists cannot really describe the Universe's beginning or address whether something preceded the Big Bang.

> "Maybe both space and time are just something completely different fundamentally from what we think of."

Section Summary

- Discovering what happened at $t = 0$ requires a theory of quantum gravity, or quantized space-time, which scientists have not developed yet.

CHECKPOINT Why does a complete theory of cosmology require a theory of quantum gravity?

Journey's End

We have come a long way on our journey. We began learning about small bands of hunter-gatherers looking in wonder at the night sky and followed our species' progress as we initiated the agricultural, city-building, industrial, and digital revolutions. With each step, we humans learned more about the celestial drama occurring over our heads—including the facts that we live on a spherical planet and the drama plays out below our feet as well.

As human culture matured, our understanding of the sky grew richer. Our journey through the modern story of astronomy in this book reflects that richness. From the birth and evolution of planets to the birth and evolution of stars, we have seen that the cosmos is in a constant state of change and evolution. In these last chapters we have seen how, even at galactic, extragalactic, and cosmological scales, change and evolution are the rule as well. It is a grand and sweeping story playing out across billions of years and billions of parsecs.

What makes this narrative of cosmic history even more remarkable is that it's a story we have learned for ourselves. No book was dropped by aliens telling us what happens inside a star or how to interpret the fossil cosmic microwave photons left over from the Universe's early history. We figured all of it out for ourselves—using our capacity to reason and to measure, as well as the remarkable tools that, taken together, are called science. It's a remarkable accomplishment, requiring endless hours spent working with telescopes or hunched over lab benches or computers. For scientists of the past—and those working today—there is no better work than to hear the world's own voice speak to us through our investigations. That effort and adventure is ongoing because the story of the Universe and our place in it is not yet complete. We remain at play in the cosmos.

CHAPTER SUMMARY

18.1 How to Win a Nobel Prize: The Accelerating Universe

Cosmology is the study of the Universe as a whole. Early attempts at cosmological theories include Fred Hoyle's steady-state model, which said that the Universe has always expanded and will do so forever. After new data and observations were gathered, this model and others were replaced by the Big Bang theory, but even that does not explain how the Universe originated.

18.2 Our Cosmology: The Big Bang

Classic Big Bang theory includes a flexible space-time from general relativity, along with the standard model of particle physics, which defines the subatomic particles and the forces they experience. The classic model begins from a cosmic singularity, the infinitely dense and hot source of all matter and energy, that began expanding and cooling 13.7 billion years ago. The post–Big Bang space-time underwent inflation, an accelerated expansion, and this region became the observable Universe. Key epochs in cosmic evolution were the Planck era, when forces were unified; the grand unified theories (GUTs) era, when gravity emerged; inflation, when the strong force emerged; the electroweak era, when the weak and electromagnetic forces were still combined; the particle era, when those two forces separated and the currently dominant particles emerged; Big Bang nucleosynthesis (BBN), when the original proportions of the lightest elements formed; and the recombination era, at about 380,000 years after the Big Bang, when matter and blackbody photons separated.

18.3 How Do We Know? The Three Observational Pillars of Big Bang Theory

Three primary pieces of evidence support the Big Bang theory: the observed expansion of the Universe, the primordial abundance of light elements produced during the BBN, and the cosmic microwave background (CMB, blackbody radiation emitted by photons during recombination).

18.4 Beyond the Classic Big Bang Model

Several aspects of the classic Big Bang model, including the uniformity of temperature fluctuations in the CMB, were problematic but are addressed by the addition of inflation to the theory. These temperature fluctuations in the CMB have been mapped to ever-greater spatial resolutions by satellite. Temperature fluctuations represent density differences, which themselves represent the seeds of today's large-scale structure. The relationship between cosmic expansion and the density of matter-energy is key to understanding the Universe's fate. On its own, the gravity of normal matter-energy might have eventually slowed or reversed the expansion. Experiments indicated an accelerated expansion and therefore a repulsive force powered by dark energy. Dark energy makes up 70 percent of the Universe; luminous and dark matter make up the other 30 percent.

18.5 Questions about "Before" and "Everything"

To understand what may have preceded the Big Bang, astronomers are working to formulate a theory of quantum gravity—quantized space-time. It would describe how space-time behaved in the ultrahot, ultradense conditions of the cosmic singularity.

QUESTIONS AND PROBLEMS

Narrow It Down: Multiple-Choice Questions

1. Which of the following is/are considered direct evidence of the Big Bang? Choose all that apply.
   a. the existence of black holes
   b. the expansion of the Universe
   c. the primordial abundance of light elements
   d. the presence of life in the Universe
   e. cosmic microwave background radiation

2. In which ways did Big Bang nucleosynthesis differ from nuclear fusion in stars? Choose all that apply.
   a. Its duration was a tiny fraction of time over which nuclear fusion has occurred in stars.
   b. It produced fewer elements and isotopes.
   c. It was incapable of producing carbon.
   d. It resulted in the production of hydrogen.
   e. The temperature was falling as nuclear fusion began.

3. Which of the following statements about space-time is *not* correct?
   a. Evidence suggests that it is finite and bounded.
   b. It constitutes a four-dimensional continuum.
   c. It was first recognized as a unified quantity within special relativity.
   d. The current extent of its expansion is limited by the age of the Universe.
   e. There is strong evidence to suggest its current geometry.

4. Which of the following was/were "steady" according to Hoyle's steady-state model? Choose all that apply.
   a. the amount of matter
   b. the amount of space
   c. the average cosmic density
   d. the distance between any two galaxies
   e. all quantities and measurements

5. Which of the following statements is true?
   a. Only quarks feel the force of gravity.
   b. Both quarks and leptons feel the strong nuclear force.
   c. Quarks, including electrons, feel the strong nuclear force.
   d. Only leptons, including protons and neutrons, feel the force of gravity.
   e. Both quarks and leptons feel the force of gravity.

6. Which of the following statements is false?
   a. The tau lepton carries the electromagnetic force.
   b. The photon is a force boson.
   c. W and Z bosons carry the weak nuclear force.
   d. The muon is a lepton.
   e. Gravitons are hypothesized to carry the gravitational force.

7. Which of the following does *not* describe conditions since shortly after the Big Bang?
   a. Temperature and density have decreased.
   b. The volume of space-time has increased.
   c. The overall size of the observable Universe has increased.
   d. Inflation occurred and ended.
   e. Initial perturbations have been mostly smoothed away.

8. Cosmic inflation refers to
   a. the Universe filling up with hydrogen gas.
   b. the ongoing expansion of the Universe.
   c. a brief early period of hyperrapid expansion of space-time.
   d. the current period of cosmic acceleration.
   e. stellar winds creating bubbles of gas.

9. How long after the Big Bang was the primordial abundance of elements established?
   a. It is still changing today.
   b. 10^{-33} second
   c. 0.001 second
   d. a few minutes
   e. 380,000 years

10. What temperature can the CMB reasonably be expected to have 5 billion years in our future?
    a. 0 K
    b. less than 2.7 K
    c. 2.7 K
    d. There is no way to know.
    e. The CMB will not exist then.

11. Which of the following statements about the CMB is/are *not* true? Choose all that apply.
    a. Its photons originated at the moment of the Big Bang.
    b. Its photons have been redshifted dramatically since their creation.
    c. Its current temperature is 2.7 K.
    d. It is a blackbody.
    e. It is completely homogeneous in temperature.

12. Inflation theory helps explain many current properties of the Universe. Which of the following statements about inflation is/are *not* true?
    a. It drives the density of the Universe toward the critical density.
    b. It resulted in a rapid stretching of space-time.
    c. It explains why the Big Bang occurred.
    d. It accounts for a major drop in temperature and density early in cosmic evolution.
    e. It explains why different sides of the sky show the same CMB fluctuations.

13. Dark energy accounts for
    a. 30 percent of the matter-energy in the Universe.
    b. less of the matter-energy in the Universe than does dark matter but more than luminous matter.
    c. more of the matter-energy in the Universe than does dark matter but less than luminous matter.
    d. more of the matter-energy in the Universe than do dark matter and luminous matter combined.
    e. inflation.

14. If we ignore the effect of dark energy, a density value greater than the critical density
    a. would be impossible.
    b. would mean the Universe will eventually contract.
    c. would mean there is more luminous matter than dark matter.
    d. would mean the Universe will expand at an ever-increasing rate forever.
    e. is what astronomers believe was the actual value at the Big Bang.

15. Which of the following is true about the recombination era?
    a. It was the period when photons could not escape.
    b. It occurred within seconds of the Big Bang.
    c. It was marked by the re-creation of H atoms from component parts that had been separated long before.
    d. It marked the release of the CMB.
    e. It smoothed out perturbations in the dense matter-energy.

16. Which of the following statements about dark matter and dark energy is/are true? Choose all that apply.
    a. While the nature of dark matter has been determined, the nature of dark energy has not.
    b. The effect of dark energy is observed only on cosmological scales, while the effect of dark matter has been observed on the scale of galaxies.
    c. Both far exceed the matter-energy equivalent of luminous matter.
    d. Evidence of the existence of dark energy was identified first.
    e. Dark matter has only an attractive gravitational effect, while dark energy is repulsive.

17. Which of the following describes commonalities between Big Bang nucleosynthesis and the recombination era?
    a. average temperature
    b. duration of the period
    c. size of the Universe
    d. initial formation of a new configuration of matter
    e. presence of the CMB

18. True/False: The Universe is continuing to cool today.

19. True/False: The primordial abundance of light elements provides evidence of the Big Bang by indicating that the Universe was briefly hot enough for nucleosynthesis to occur.

20. True/False: A fully developed theory uniting all four fundamental forces has yet to be defined.

To the Point: Qualitative and Discussion Questions

21. How did the term *Big Bang* come into being?

22. What did Gamow and Alpher contribute to the search for a theory of cosmic evolution?

23. In what sense is the Big Bang not a theory of origin?

24. Describe what is meant by *positive*, *negative*, and *flat curvature*.

25. What are the two classes of fundamental matter particles? What distinguishes the two classes?

26. Which of the three pillars of evidence for the Big Bang theory is most convincing to you, and why?

27. What is the difference between the "stage" and the "actors" in Big Bang theory, as described in the chapter?

28. Describe the process used to discover the smallest elemental particles, such as the Higgs particle.

29. What are the proportions of hydrogen and helium in the primordial abundance of elements? Why were heavier elements not formed during Big Bang nucleosynthesis?

30. Describe the differences between the steady-state model and the Big Bang model.

31. How did observations of the Universe's expansion point to the presence of dark energy?

32. What role did phase transitions play in leading to the contents of the Universe that we see today?

33. How is dark energy similar to Einstein's cosmological constant?

34. Explain the concept of quantum gravity.

35. Recall your view of the origin, scope, and age of the Universe before you began this course. How has it changed (if at all) during this course?

Going Further: Quantitative Questions

In the questions that follow, ignore any acceleration of the expansion rate due to dark energy.

36. What value would you expect for the peak wavelength of the CMB if the age of the Universe were double its current value?

37. What value would you expect for the peak wavelength of the CMB if the Universe had expanded by a factor of 800 since recombination?

38. By what factor had the Universe expanded since recombination when the peak wavelength of the CMB was 0.00038 meter?

39. By what factor will the Universe have expanded when the current wavelength of the CMB is 0.0076 meter?

40. What do you expect the temperature of the CMB to be when the Universe has expanded to 1.5 times its current size?

41. What was the temperature of the CMB when the Universe had expanded to one-third of its current size?

42. What is the recession speed, in kilometers per second, of a galaxy that is 420 megaparsecs (Mpc) from Earth? (See Going Further 16.1).

43. The discovery of dark energy was based on the difference of the observed brightness of Type Ia supernovas from their expected brightness. Comparing two such supernovas, A and B, what would be the expected ratio of A's brightness to that of B, if A is 1/7th as far from Earth as B?

44. If cosmic evolution were at a point where the temperature of the CMB was 4.5 K, how would the Universe's size compare to its current size?

45. If cosmic evolution were at a point where the temperature of the CMB was 1.3 K, how would the Universe's size compare to its current size?

APPENDIX 1

PHYSICAL PROPERTIES OF PLANETS AND DWARF PLANETS

	Mass (M_E)	Mass ($\times 10^{24}$ kg)	Radius (km)	Density (kg/m³)	Average Surface Temperature (°C)	Average Surface Temperature (K)	Escape Velocity (km/s)
Mercury	0.055	0.33	2,440	5,430	167	440	4.3
Venus	0.82	4.87	6,052	5,240	464	737	10.4
Earth	1	5.97	6,378	5,520	15	288	11.2
Mars	0.11	0.642	3,397	3,940	−65	208	5.05
Jupiter	317.82	1,898	69,911	1,330	−110	165	59.5
Saturn	95.16	568	58,232	690	−140	124	35.5
Uranus	14.54	86.8	25,362	1,300	−195	73	21.3
Neptune	17.15	102	24,622	1,760	−200	73	23.5
Vesta (asteroid)	4.47×10^{-5}	0.00026	251	3,456	−93	180	0.36
Ceres (dwarf planet)	1.5×10^{-4}	0.00087	476	2,170	−105	168	0.5
Pluto (dwarf planet)	2.2×10^{-3}	0.013	1,185	1,879	−225	48	1.1
Haumea (dwarf planet)	6.6×10^{-4}	0.004	1,700	2,550	−241	32	0.91
Makemake (dwarf planet)	6.7×10^{-4}	0.004	715	2,300	−239	34	0.8
Eris (dwarf planet)	2.8×10^{-3}	0.017	1,163	2,520	−231	42	1.38

APPENDIX 2

ORBITAL PROPERTIES OF PLANETS AND DWARF PLANETS

	Orbital Inclination to Ecliptic (deg)	Rotation Period (h)	Orbital Radius (AU)	Orbital Period (yr)	Orbital Speed (km/s)	Eccentricity
Mercury	7	1,407.6	0.39	0.24	47.9	0.205
Venus	3.39	5,832.5	0.72	0.62	35	0.007
Earth	7.16	23.9	1	1	29.8	0.017
Mars	1.85	24.6	1.52	1.88	24.1	0.093
Jupiter	1.31	9.9	5.2	11.86	13.1	0.048
Saturn	2.49	10.7	9.54	29.42	9.7	0.054
Uranus	0.77	17.2	19.19	83.75	6.8	0.047
Neptune	1.77	16.1	30.10	163.72	5.4	0.009
Vesta	7.14	5.4	2.36	3.63	19.34	0.089
Ceres	10.59	9.1	2.77	4.6	17.9	0.076
Pluto	17.15	153.3	39.48	248.02	4.7	0.249
Haumea	28.2	3.9	43.22	284.12	4.5	0.191
Makemake	29	7.77	45.79	309.09	4.4	0.156
Eris	44.04	25.9	68.05	561.40	3.4	0.441

APPENDIX 3

PROPERTIES OF MAJOR SOLAR SYSTEM MOONS

Planet	Moon	Radius (km)	Mass (kg)	Density (kg/m³)	Orbital Radius (10³ km)	Orbital Period (days)	Eccentricity
Earth	Moon	1,737	7.35×10^{22}	3,341	384.4	27.32	0.055
Mars	Phobos	11.2	1.06×10^{16}	1,876	9.38	0.32	0.015
	Deimos	6.2	2.4×10^{15}	1,471	23.46	1.26	0.00033
Jupiter	Callisto	2,403	1.08×10^{23}	1,834	1,883	16.69	0.0074
	Europa	1,565	4.80×10^{22}	3,013	671	3.55	0.009
	Ganymede	2,634	1.48×10^{23}	1,936	1,070	7.15	0.0013
	Io	1,821	8.93×10^{22}	3,528	422	1.77	0.0041
Saturn	Pan	14.1	4.95×10^{15}	420	133.58	0.58	0
	Prometheus	43	1.60×10^{17}	480	139.35	0.61	0.0022
	Pandora	40.7	1.37×10^{17}	490	141.7	0.63	0.0042
	Mimas	198	3.74×10^{19}	1,148	185.52	0.94	0.0196
	Enceladus	252	1.08×10^{20}	1,610	238.02	1.37	0.0047
	Titan	2,576	1.35×10^{23}	1,881	1,222	15.95	0.0288
Uranus	Miranda	236	6.59×10^{19}	1,200	129.39	1.41	0.0013
	Ariel	579	1.35×10^{21}	1,660	191.02	2.52	0.0012
	Umbriel	585	1.17×10^{21}	1,390	266.3	4.14	0.0039
	Titania	789	3.53×10^{21}	1,711	435.91	8.71	0.0011
	Oberon	761	3.01×10^{21}	1,630	583.52	13.46	0.0014
Neptune	Proteus	210	5.04×10^{19}	1,300	117.65	1.12	0
	Triton	1,353	2.14×10^{22}	2,061	355.76	5.88	0
	Nereid	170	2.70×10^{19}	—	5,513.4	360.13	0.751
Pluto	Charon	593	1.62×10^{21}	1,650	19.6	6.39	0

APPENDIX 4

PROPERTIES OF THE NEAREST STARS TO EARTH

Common Name	Constellation	Scientific Name	Distance (ly)	Distance (pc)	Spectral Type	Apparent Magnitude	Absolute Magnitude	Luminosity (L_{Sun})
Sun	N/A	Sol	0.000016	0.000005	G2	−26.74	4.83	1
Alpha Centauri	Centaurus	Alpha Cen	4.2	1.3	G2	−0.27	4.38	1.50
Barnard's Star	Ophiuchus	Gliese 699	5.9	1.8	M4	9.53	13.22	0.0004
Wolf 359	Leo	Gliese 409	7.8	2.4	M6	13.44	16.55	0.00002
Lalande 21185	Ursa Major	Gliese 411	8.3	2.5	M2	7.47	10.44	0.021
Sirius	Canis Major	Alpha CMa	8.6	2.6	A1	−1.44	1.45	25
Luyten 726	Cetus	Gliese 65b	8.6	2.6	M5.5	12.99	15.85	0.00006
Ross 154	Sagittarius	Gliese 729	9.7	3.0	M3	10.43	13.07	0.0038
Ross 248	Andromeda	Gliese 905	10.3	3.2	M6	12.29	14.79	0.0018
Epsilon Eridani	Eridanus	Gliese 144	10.5	3.2	K2	3.73	6.19	0.34
Lacaille 9352	Piscis Austrinus	Gliese 887	10.7	3.3	M2	7.34	9.75	0.011
Luyten 789	Aquarius	Gliese 866	10.9	3.4	M5.5	11.06	13.30	0.00001
Ross 128	Virgo	Gliese 447	10.9	3.4	M4	11.13	13.51	0.00036
Struve 2398	Draco	Gliese 725	11.3	3.5	M3	8.90	11.16	0.039
61 Cygni	Cygnus	Gliese 820	11.4	3.5	K5	5.21	7.49	0.15
Procyon	Canis Minor	Alpha CMi	11.4	3.5	F5	0.4	2.68	6.93
Groombridge 34	Andromeda	Gliese 15	11.7	3.6	M2	8.08	10.32	0.0064
Epsilon Indi	Indus	Gliese 845	11.8	3.6	K5	4.69	6.89	0.15
Tau Ceti	Cetus	Gliese 71.0	11.9	3.7	G8	3.49	5.68	0.52
YZ Ceti	Cetus	Gliese 54.1	12	3.7	M4	12.02	14.17	0.0002
Luyten's Star	Canis Minor	Gliese 273	12.4	3.8	M3.5	9.86	11.97	0.0004

APPENDIX 5

PROPERTIES OF THE BRIGHTEST STARS

Common Name	Constellation	Scientific Name	Distance (ly)	Distance (pc)	Spectral Type	Apparent Magnitude	Absolute Magnitude	Luminosity (L_{Sun})
Sun	N/A	Sol	0.000016	0.000005	G2	−26.72	4.8	1
Sirius	Canis Major	Alpha CMa	8.6	2.6	A1	−1.44	1.45	25
Canopus	Carina	Alpha Car	309	95	A9	−0.62	−5.53	15,100
Alpha Centauri	Centaurus	Alpha Cen	4.2	1.3	G2	−0.27	4.38	1.5
Arcturus	Bootes	Alpha Boo	37	11	K0	−0.05	−0.31	170
Vega	Lyra	Alpha Lyr	25	8	A0	0.03	0.58	40
Capella	Auriga	Alpha Aur	43	13	G1	0.08	−0.48	79
Rigel	Orion	Beta Ori	862	265	B8	0.18	−6.69	120,000
Procyon	Canis Minor	Alpha CMi	11.4	3.5	F5	0.4	2.68	6.93
Betelgeuse	Orion	Alpha Ori	498	153	M1	0.45	−5.14	150,000
Achernar	Eridanus	Alpha Eri	139	43	B6	0.45	−2.77	3,150
Hadar/Agena	Centaurus	Beta Cen	392	120	B1	0.60	−5.42	41,700
Altair	Aquila	Alpha Aql	17	5	A7	0.77	2.21	11
Acrux	Crux	Alpha Cru	322	99	B1	0.77	−4.14	25,000
Aldebaran	Taurus	Alpha Tau	67	20	K5	0.85	−0.63	518
Spica	Virgo	Alpha Vir	250	77	B1	1.04	−3.55	12,100
Antares	Scorpius	Alpha Sco	553	170	M0.5	1.09	−7.20	57,500
Pollux	Gemini	Beta Gem	34	10	K0	1.15	1.08	43
Fomalhaut	Piscis Austrinus	Alpha PsA	25	8	A4	1.16	1.72	17
Deneb	Cygnus	Alpha Cyg	1,411	433	A2	1.25	−8.38	196,000
Mimosa	Crux	Beta Cru	278	85	B0.5	1.30	−3.92	34,000

APPENDIX 6

PERIODIC TABLE

PERIODIC TABLE OF THE ELEMENTS

Legend:
- 1 — Atomic number
- H — Symbol
- Hydrogen — Name
- 1.0079 — Average atomic mass
- Metals
- Metalloids
- Nonmetals

1 / 1A	2 / 2A	3 / 3B	4 / 4B	5 / 5B	6 / 6B	7 / 7B	8	9 (8B)	10	11 / 1B	12 / 2B	13 / 3A	14 / 4A	15 / 5A	16 / 6A	17 / 7A	18 / 8A
1 **H** Hydrogen 1.0079																	2 **He** Helium 4.0026
3 **Li** Lithium 6.941	4 **Be** Beryllium 9.0122											5 **B** Boron 10.811	6 **C** Carbon 12.011	7 **N** Nitrogen 14.007	8 **O** Oxygen 15.999	9 **F** Fluorine 18.998	10 **Ne** Neon 20.180
11 **Na** Sodium 22.990	12 **Mg** Magnesium 24.305											13 **Al** Aluminum 26.982	14 **Si** Silicon 28.086	15 **P** Phosphorus 30.974	16 **S** Sulfur 32.065	17 **Cl** Chlorine 35.453	18 **Ar** Argon 39.948
19 **K** Potassium 39.098	20 **Ca** Calcium 40.078	21 **Sc** Scandium 44.956	22 **Ti** Titanium 47.867	23 **V** Vanadium 50.942	24 **Cr** Chromium 51.996	25 **Mn** Manganese 54.938	26 **Fe** Iron 55.845	27 **Co** Cobalt 58.933	28 **Ni** Nickel 58.693	29 **Cu** Copper 63.546	30 **Zn** Zinc 65.38	31 **Ga** Gallium 69.723	32 **Ge** Germanium 72.63	33 **As** Arsenic 74.922	34 **Se** Selenium 78.96	35 **Br** Bromine 79.904	36 **Kr** Krypton 83.798
37 **Rb** Rubidium 85.468	38 **Sr** Strontium 87.62	39 **Y** Yttrium 88.906	40 **Zr** Zirconium 91.224	41 **Nb** Niobium 92.906	42 **Mo** Molybdenum 95.96	43 **Tc** Technetium [98]	44 **Ru** Ruthenium 101.07	45 **Rh** Rhodium 102.91	46 **Pd** Palladium 106.42	47 **Ag** Silver 107.87	48 **Cd** Cadmium 112.41	49 **In** Indium 114.82	50 **Sn** Tin 118.71	51 **Sb** Antimony 121.76	52 **Te** Tellurium 127.60	53 **I** Iodine 126.90	54 **Xe** Xenon 131.29
55 **Cs** Cesium 132.91	56 **Ba** Barium 137.33	57 **La** Lanthanum 138.91	72 **Hf** Hafnium 178.49	73 **Ta** Tantalum 180.95	74 **W** Tungsten 183.84	75 **Re** Rhenium 186.21	76 **Os** Osmium 190.23	77 **Ir** Iridium 192.22	78 **Pt** Platinum 195.08	79 **Au** Gold 196.97	80 **Hg** Mercury 200.59	81 **Tl** Thallium 204.38	82 **Pb** Lead 207.2	83 **Bi** Bismuth 208.98	84 **Po** Polonium [209]	85 **At** Astatine [210]	86 **Rn** Radon [222]
87 **Fr** Francium [223]	88 **Ra** Radium [226]	89 **Ac** Actinium [227]	104 **Rf** Rutherfordium [265]	105 **Db** Dubnium [268]	106 **Sg** Seaborgium [271]	107 **Bh** Bohrium [270]	108 **Hs** Hassium [277]	109 **Mt** Meitnerium [276]	110 **Ds** Darmstadtium [281]	111 **Rg** Roentgenium [280]	112 **Cn** Copernicium [285]	113 **Uut** Ununtrium [284]	114 **Fl** Flerovium [289]	115 **Uup** Ununpentium [288]	116 **Lv** Livermorium [293]	117 **Uus** Ununseptium [294]	118 **Uuo** Ununoctium [294]

6 Lanthanides

58 **Ce** Cerium 140.12	59 **Pr** Praseodymium 140.91	60 **Nd** Neodymium 144.24	61 **Pm** Promethium [145]	62 **Sm** Samarium 150.36	63 **Eu** Europium 151.96	64 **Gd** Gadolinium 157.25	65 **Tb** Terbium 158.93	66 **Dy** Dysprosium 162.50	67 **Ho** Holmium 164.93	68 **Er** Erbium 167.26	69 **Tm** Thulium 168.93	70 **Yb** Ytterbium 173.05	71 **Lu** Lutetium 174.97

7 Actinides

90 **Th** Thorium 232.04	91 **Pa** Protactinium 231.04	92 **U** Uranium 238.03	93 **Np** Neptunium [237]	94 **Pu** Plutonium [244]	95 **Am** Americium [243]	96 **Cm** Curium [247]	97 **Bk** Berkelium [247]	98 **Cf** Californium [251]	99 **Es** Einsteinium [252]	100 **Fm** Fermium [257]	101 **Md** Mendelevium [258]	102 **No** Nobelium [259]	103 **Lr** Lawrencium [262]

We have used the U.S. system as well as the system recommended by the International Union of Pure and Applied Chemistry (IUPAC) to label the groups in this periodic table. The system used in the United States includes a letter and a number (1A, 2A, 3B, 4B, etc.), which is close to the system developed by Mendeleev. The IUPAC system uses numbers 1–18 and has been recommended by the American Chemical Society (ACS). While we show both numbering systems here, we use the IUPAC system exclusively in the book.

APPENDIX 7

STAR MAPS

Northern Hemisphere Night Sky in Spring

1 A.M. on March 1
11 P.M. on April 1
9 P.M. on May 1
(Add 1 hour for daylight saving time)

Northern horizon

Eastern horizon

Western horizon

Southern horizon

Northern Hemisphere Night Sky in Summer

1 A.M. on June 1
11 P.M. on July 1
9 P.M. on August 1
(Add 1 hour for daylight saving time)

Northern Hemisphere Night Sky in Autumn

1 A.M. on September 1
11 P.M. on October 1
9 P.M. on November 1
(Add 1 hour for daylight saving time)

Northern Hemisphere Night Sky in Winter

2 A.M. on December 1
Midnight on January 1
10 P.M. on February 1

Northern horizon

Eastern horizon

Western horizon

Southern horizon

APPENDIX 8

INTERVIEWEE BIOGRAPHIES

LAURA ARNOLD

Laura Arnold's story, like that of many other scientists, started with a love of science engendered by supportive parents. "My father's an engineer who helped design Gorilla Glass for electronics like iPhones. My mother works at the University of Rochester. When I was a kid, she enrolled me in a bunch of science, math, and computer science programs they had for young girls." Soon Arnold found her own ways to engage with the subject. "My parents tell stories of me doing math on the sidewalk in chalk as a little girl, just adding and subtracting numbers while the other kids were drawing ponies."

Arnold's interest in physics began in high school with an AP class that let her carry forward her own research project. Astronomy, however, came from the night sky. "I grew up right on the beach, where the sky can be really thrilling. Later, I got to see Jupiter's moons through a telescope at the planetarium. I thought that was pretty cool, but when I came back later in the night, they had moved. They were swinging around in their orbit, and I could actually see it happening. That really blew me away."

When she started college at the University of Rochester, Arnold was sure she was headed toward a PhD and a research career in astronomy. But during a summer internship in Germany, she had the chance to focus on teaching rather than research. "I realized that research just wasn't really right for me. I'm really excited and kind of passionate, and I like to talk to people. A life focused only on research wasn't going to be a good fit for me."

On her return from Germany, Arnold threw herself into learning how to teach astronomy. "I helped teach some of the intro courses," she says, "and I loved it." After graduation she looked for science education programs and began to work in earnest toward becoming an astronomy educator. "This is what I love, and I am really excited that I can spend my life doing it." (See Chapter 12, p. 329.)

JIM BELL

"I was definitely one of those people who always knew they wanted to be an astronomer," says Jim Bell. "One of the great things about growing up in the early 1970s was being around to watch guys drive a car on the Moon. That was very, very cool, and it changed how I looked at the world and at space exploration."

Another astronomical landmark in Bell's early life also came to him from TV. "There was this program called *Cosmos* by the Cornell University astronomer Carl Sagan," Bell explains. "Kids today have no idea how influential it was. I try to describe to my own students how different things were back then for someone interested in science. I ask them to imagine a world where there are only three big networks on the television and pretty much none of them ever covered science." Sagan took viewers on a tour of everything from the theory of relativity to the birth of stars and on to the origins of the Universe. "Sagan was something no one had seen before—a charismatic scientist (in his signature turtleneck) who could translate science into plain English," recalls Bell. After Sagan's grand tour, Bell became a convert, determined to make a place for himself as a professional scientist.

After years of study, Bell's dream of exploring the Solar System became reality through his involvement in a series of major missions. Meanwhile, he became a professor at Cornell himself and then later at Arizona State University. For all of his success, Bell has not forgotten the power of his early inspirations. "When eventually I got on the faculty at Cornell and was able to meet and talk with Carl Sagan, it was just a dream come true," he says. "In many ways it still is." (See Chapter 7, p. 196.)

BOB BENJAMIN

Bob Benjamin's journey into astrophysics began with a single high school math teacher. "She was always telling me, 'You've got to take more math,'" he recalls. Once Benjamin had arrived at Carleton College, he not only took his teacher's advice but found he had real talent in the subject. He also took a single astronomy class. That drew Benjamin's interest but was still not enough to set the fire roaring. "When I reached the end of my undergraduate career, I really didn't know what I wanted to do with my life. I had a pretty intense debate with myself the summer before senior year about graduate school in physics versus astronomy. On impulse, I decided astronomy."

After starting the astronomy program at the University of Minnesota, Benjamin soon knew he'd made the right choice. "There was this moment that really stayed with me. This other graduate student named Kim and I were sitting in our office talking. First we started comparing our favorite bands and where they came from. Then we started talking about O stars, which are the biggest and brightest stars in the sky. Both of us were interested in them in our research. At some point I said, 'Kim, do you realize that we just talked for 20 minutes about O stars, and it was as interesting as the stuff about the bands?' She smiled and said, 'Yeah. Wasn't that weird?'"

Now a professor of astrophysics, Benjamin has become an expert on the structure and evolution of galaxies. "I remember in the beginning of graduate school wondering, 'Well, how do you think of a research project?' Then somewhere in the middle of graduate school I was thinking of my own exciting research projects, and I realized I couldn't stop thinking about new research projects I wanted to do." And Benjamin still can't stop thinking up new research projects. "I know that shouldn't be a problem," he laughs. "But I already have way too many things on my to-do list." (See Chapter 15, p. 397.)

CRYSTAL BROGAN

"When I was much younger, I wanted to be an archaeologist," says Crystal Brogan. "I loved *Raiders of the Lost Ark*, and I thought finding clues and assembling ideas about how stuff happens in the past was really exciting." Eventually she concluded that all of the biggest mysteries on that front had been solved and, based on a healthy diet of TV science and science fiction, she decided that assembling clues about the Universe would be just as exciting.

It took some time, though. "I started out interested in engineering and a bunch of other stuff too," she says. "It wasn't really until I was an undergraduate that I really got fixed on astronomy." She graduated from James Madison University, earned a PhD from the University of Kentucky, and did postdoctoral work in both New Mexico and Hawaii. Brogan's research focuses on star formation and protoplanetary disks. Currently an astronomer at the National Radio Astronomy Observatory in Charlottesville, Virginia, she also teaches at the University of Virginia.

One of Brogan's most exciting moments in science came when she and her colleagues used the Atacama Large Millimeter Array (ALMA) telescope to get one of the highest-resolution views ever of a young planet-forming disk, HL Tau. "It was the gaps that were really amazing," she says of the image, which shows concentric dark bands in the disk that likely are created by young planets. "The gaps were a very telling sign of planet formation in a disk that is only a million or a couple million years old. That was really different from what people expected for such a young disk. It was spectacular on so many fronts." (See Chapter 5, p. 140.)

MANUELA CAMPANELLI

Manuela Campanelli grew up in Switzerland. "I was always interested in science," she says. "When I was a little girl, I read about everything scientific I could get my hands on. I was interested in biology, but I didn't want to do experiments with animals, so I knew this was not the field for me. Then I tried chemistry. I liked chemistry a lot. But then one day I did an experiment in my room and everything caught fire. I stopped doing chemistry after that," she explains. "This is when I realized that I needed to work on the theory side of science."

Campanelli says that, even from her youngest days, she was interested in answering "big" questions. "I remember watching all astronomy shows and documentaries. They were so fascinating. The questions about the nature of time and of space really fired my imagination. Those questions never left me, and I still get to think about them in my work with black holes." Campanelli went on to become one of the leaders in the field of gravitational astrophysics. Her team, along with another led by Joan Centrella of Drexel University, helped open the door to computer simulations of binary black holes.

"My parents were not scientists," Campanelli says. "They work in fields aligned with economy and business. My brother is an economist too. My whole family is made up of very pragmatic people. I guess I am the exception to the rule." (See Chapter 14, p. 377.)

SEAN CARROLL

When Hollywood film directors must get the physics right for a superhero movie, Sean Carroll is one of the working scientists they call. Carroll has acted as adviser on films such as *Thor* and *The Avengers*. "I see my role as helping them at least understand the limits of the science they want to include in the films." He says with a smile, "I don't get paid much, but it's really cool to hang out on the set."

Carroll picked up his passion for science in childhood. "There was a book I found in my town's library," he recalls. "It was called *High-Energy Physics*, and it explained everything they knew at the time about the fundamental structure of matter." That emphasis on fundamentals became the heart of Carroll's scientific career. After earning a BS at Villanova University, Carroll was accepted to Harvard University for graduate school and began looking into the nature of relativity. "The structure of relativity as a set of physical ideas has always interested me," he says. "In particular, I still get really excited thinking about the basic nature of time."

Carroll is the author of a number of trade books on physics and cosmology, along with a graduate textbook on general relativity. In addition, he was one of the first scientists to begin blogging, and he has appeared numerous times on television in venues as diverse as the History Channel and *The Colbert Report*.

For all his interactions with the public sphere, Carroll still loves his work as a scientist. "I have some of my best moments doing science on airplanes," he says. "Since there is nothing to do but work, I sometimes have these experiences where I am working on something and it all just falls together. Those moments of insight make everything else pale in comparison." (See Chapter 14, p. 368.)

ORSOLA DE MARCO

Orsola De Marco's initial scientific inclinations leaned more toward chemistry than astronomy. "I liked to mix things and make them explode," she says. "That didn't make my parents very happy." In an effort to turn her attentions elsewhere, De Marco's parents bought their ambitious elementary school student a telescope. "I used it a lot to spy at my neighbor's windows."

Near the end of high school, De Marco's interests did turn away from chemistry, toward astronomy. She traveled to England to pursue her studies at University College London, where she encountered the topic that would become her passion. "That was the time I began my love affair with planetary nebulas. They are so beautiful, and there is so much we don't understand about them."

De Marco became one of the leading astronomers advocating for the new binary-accretion-disk-magnetic-field paradigm in planetary nebula studies. Pushing the envelope of what is known has provided some of her most difficult and exciting moments as a scientist. "At a meeting in Hawaii on planetary nebulas, I gave a review talk outlining the arguments for the new binary model. After the talk, I remember being pinned by some older scientist who kept telling me I was discrediting myself. I remember thinking 'Okay, if this is what it's like to push against the establishment, then I don't want to do this.' But later in the meeting, there was this amazing moment when someone stood up and made a key point about the need for binary stars and magnetic fields. All of a sudden you could feel people in the room get it. You could feel the tide turning. People were starting to be convinced, and I knew it. It was really quite incredible." (See Chapter 13, p. 341.)

HARRIET DINERSTEIN

"I grew up in the middle of Manhattan," laughs Harriet Dinerstein. "I barely even saw the night sky. I did look through binoculars at the Moon when I was a senior in high school." How did the young Dinerstein, living in the middle of a large city, ever catch the astronomy bug? "The Hayden Planetarium!" she says. "Every time there was a new show at the Hayden, my mom would take us."

The interest that began at the planetarium deepened in the library. "I remember reading book after book by Isaac Asimov," Dinerstein says. "One chapter he wrote about the Solar System's weirdest objects—like how Uranus rotates on its side. Then, when I was 14 came the *Apollo* Moon landing."

In college, Dinerstein was one of only a handful of female students determined to find a career in science. "I went to Yale when it had barely turned coed," she says. "I plunged right into physics, chemistry, linear algebra, and astronomy my freshman year. I guess I never looked back."

Dinerstein eventually ended up at the University of Texas at Austin, where she studies the chemical composition of gases thrown off stars at the end of their lives, to see how fusion transformed them. "Stars are always cooking up new mixtures of atoms through fusion. Then they spew those newly formed elements back into space," says Dinerstein. "What I see coming off the stars now will eventually end up in the next generation of stars or even some future planet."

Throughout her career, the possibility of discovery has always kept Dinerstein tied to astronomy. "I love the excitement of identifying things that weren't known before," she says. "Two emission lines were discovered long ago and remained unidentified for 25 years, until I recognized that they came from the elements selenium and krypton. It was really a thrill to understand something that had never been understood before." (See Chapter 11, p. 303.)

DEBRA ELMEGREEN

If you're ever on a train and you see a fellow passenger building a telescope out of cardboard, you might have just run into Debra Elmegreen. "A little while ago my husband and I were on a train coming down from Alaska," she explains. "Venus was about to pass in front of the Sun [a rare event called a transit]. We didn't want to miss it, so we got out a piece of cardboard and stuck a pinhole in it. I had binoculars, which we turned backward and held up so we could project the Sun's image. A couple of people on the train got really excited. 'This is crazy!' they were saying. 'What is it?'"

That kind of do-it-yourself enthusiasm has been a hallmark of Elmegreen's astronomical career, going all the way back to when she was a kid. "I grew up in Indiana, where the skies were very dark. We lived along the river, and there were no other houses around. My parents got me a telescope when I was 5, one of those 2-inch refractors. I was hooked from that moment on." In school, Elmegreen joined the science club and built her own telescope. Then came local science fairs and the Westinghouse Science Talent Search.

Now president-elect of the International Astronomical Union, Elmegreen admits that the responsibilities of teaching, grant-proposal writing, serving on national committees (including recently having been president of the American Astronomical Society, AAS), and the nuts-and-bolts details of doing science can wear her down. "You know, the day-to-day computer drudgery can get to you sometimes," she says. But a view of the sky is all she needs to fire up her imagination again. "Every now and then I have to step outside. Once I look up and remember why I'm doing this, it's all okay." That sense of excitement can also come from new data. "You get this new plot and you realize no one else has ever seen what you are finding before. I'm living my dreams. It's fun. I still love it. I can't imagine doing anything else." (See Chapter 15, p. 400.)

MARGARET GELLER

Margaret Geller is a trailblazer. You can tell that from the long, long list of awards she's been honored with, including the MacArthur Fellowship (also known as the "Genius" award). What earned her all those prizes and medals was a willingness to take on a seemingly impossible project lying at the frontiers of her field and to apply a stubborn resistance to difficulties. That's how she produced the first maps of the Universe's large-scale structure.

Born in Ithaca, New York, Geller got her bachelor's degree in physics at UC Berkeley. Although she began her studies with down-to-earth concerns in the field of solid-state physics, a mentor suggested that she study physical cosmology at Princeton University, where she got her PhD in physics in 1975. Geller then headed to the Harvard-Smithsonian Center for Astrophysics, which is where she carried out that groundbreaking research on the large-scale structure of the Universe.

Throughout her storied career, while seeking to understand the Universe as a whole, Geller has always kept a keen eye on how science works in the rest of society. She has collaborated on films and TV shows that bring astronomy to people who may have never thought about it before. In 2004 she wrote an essay called "The Black Ribbon," which explored how poverty keeps many bright and talented students from being able to pursue careers in science. As she puts it, "The differences between the population of physical scientists and the population of the United States—or of the world—are not subtle." (See Chapter 17, p. 455.)

OWEN GINGERICH

"I'm a lapsed astrophysicist," says Owen Gingerich, a professor emeritus at Harvard University. One of the world's foremost experts on the Copernican revolution, Gingerich never expected to become a historian. Astronomy was his first love, and it was a romance that began early. As far back as his childhood in Iowa, Gingerich knew with certainty that he wanted to be an astronomer. His own turn to history was more accident than intention. "There were all these wonderful anniversaries coming up," says Gingerich, referring to the 400th anniversary of Kepler's birth in 1971 and the 500th anniversary of Copernicus's birth in 1973. "That is what got me started working on the history of science as a way to understand how science works."

"I remember one very hot summer night," Gingerich recalls. "My mother took cots out in the backyard for us to sleep." Looking up into the sky, he asked, "Mommy, what's that?" "Those are the stars," she answered. "You've often seen them." It was a moment when the world opened up for the young boy. "I didn't know they stayed up all night," he recalls, laughing. His future path was set. "I've been at it ever since."

Gingerich went on to study astronomy at Harvard University, where he eventually became a professor of astrophysics. "I was trained in the pioneering days of high-speed computing," he says, "and worked on studies of stellar atmospheres." (See Chapter 3, p. 57.)

MARCELO GLEISER

The tragic loss of his mother when he was just a child left Marcelo Gleiser searching for answers even as a boy. Growing up in Rio de Janeiro, Gleiser found those answers in tales of the supernatural. "I was really fascinated with the possibilities that there might be greater intelligences living somewhere in the vastness of space," he says.

Gleiser earned his PhD in theoretical astrophysics from King's College London and eventually joined the faculty of Dartmouth College. One of the most popular science writers in Brazil, Gleiser has written two books on the intersection between science and human culture: *The Prophet and the Astronomer* and *The Dancing Universe*.

Marcelo Gleiser was in his teens when he discovered science. "I gawked at the TV when Neil Armstrong and Buzz Aldrin planted the US flag on the Moon," he says. "And the destructive power of the H-bomb really terrified me. I was amazed that, for the first time in history, we could obliterate civilization at the push of a few buttons." Stanley Kubrick's famous movie *2001: A Space Odyssey* left Gleiser with questions about greater intelligences living in the Universe. "Could *they* have been our creators?"

Such questions propelled Gleiser toward a career in science. After earning his bachelor's and master's degrees in Brazil and his PhD in England, he spent a few years as a researcher in theoretical particle physics and cosmology at both Fermilab and the Kavli Institute for Theoretical Physics. Teaching at Dartmouth since 1991, Gleiser was awarded the Appleton Professorship of Natural Philosophy in 1999.

"Science is more than a way to make sense of the material world," Gleiser says. "It is a way of living, a very human creation and the product of the same curiosity that has moved our collective imagination for thousands of years. It is an expression of our deeply ingrained desire to make sense of the world and to know our place in the big scheme of things." (See Chapter 2, p. 41.)

ALYSSA GOODMAN

Not every astronomer is featured on a public-television game show for kids. But then again, Alyssa Goodman is not your everyday astronomer, and it's more than her appearance on *Fetch! with Ruff Ruffman* that gives her those credentials.

Goodman's interest in astronomy has taken her to telescopes around the world, though the Boston area has remained her home base. After completing her undergraduate degree in physics at MIT, she earned a PhD in physics from Harvard University. She won a President's Fellowship at UC Berkeley but returned to Harvard in 1992 as a professor of astronomy.

Goodman loves the process of observing best. "It's amazing to be up at 14,000 feet working all night under the stars," she says. "You get tired, but working with your collaborators while the data comes in is just one of the most exhilarating experiences you can have."

Goodman's interest in imaging—making pictures from her radio astronomy data—has led her to seek a deeper understanding of how other fields represent and visualize data. "We would like to see what the clouds we study look like in 3D," she explains, "but of course we can't do that in astronomy. So we need a way of putting our images together into something that looks more like a 3D picture." Searching for the tools she needed led her into an unusual collaboration with Harvard Medical School. "People in other fields, like health science, figured out how to do what we needed with our radio telescope data. We call the project astronomical medicine; it gives people 3D views of what these very large regions of space look like." The beauty of astronomy is just one location where she has found a passion for visualization. Goodman's interest in the image extends to her work as an artist. (See Chapter 4, p. 85.)

PAUL GREEN

"I started out wanting to be in a rock band," says Paul Green. "Astronomy came much later." Green recalls the exact moment of his shift to astronomy. "There was this one night when I pretty much decided that astronomy was the coolest thing out there. During the winter term at Oberlin, they let you get out of town and do something different for a month," he explains. "I decided to go to the National Observatory in Arizona. I got hooked up on this project using what was then a big telescope: the 4-meter at Kitt Peak."

Green tested different kinds of photographic plates for astronomical images. For him to carry out his tests, the giant telescope had to be tipped over so that Green could climb into a cage mounted at the top. Then the telescope was pointed back at the sky as Green sat 20 feet above the ground, packed into the cage. "I would be up there all night," he says. "First I'd put in a photographic plate. Then I'd take an image of the Andromeda Galaxy with the enormous telescope. Then I'd change out the plates again." It wasn't easy work, and it definitely wasn't comfortable. "The observatory is 7,000 feet up on a mountain," says Green. "You needed very warm clothes if you were going to sit there all night."

Despite the cold, the project was a revelation for Green. "It was dead quiet, except for the clanking of the machinery and the buzz of the electronics. And the sky . . . oh, man! The sky was perfect, just the black of space and the sharp light of the stars. I was in heaven."

Those nights at Kitt Peak put Green on a track that led him to an undergraduate degree in physics at Oberlin College, a PhD in physics from the University of Washington, and a Hubble Fellowship. Green finally landed at the Harvard-Smithsonian Center for Astrophysics, where he uses the Chandra X-ray space telescope to study quasars. "But I still have a band, okay?" he says. "Make sure you get that in." (See Chapter 16, p. 433.)

LIKA GUHATHAKURTA

Neither of Lika Guhathakurta's parents were scientists, but that did not stop them from instilling in their young daughter a scientist's most important quality: curiosity. "There was lots of very general discussion in the family, and I think both my parents were very open-minded individuals."

Born in Kolkata, India, Guhathakurta moved with her parents to Mumbai when she was 8. From an early age she was introduced to astronomy through books. "I remember these wonderful picture books about stars, comets, and planets," she says with a smile. Guhathakurta's early fascination deepened as she learned to read and began absorbing the story behind the pictures. "I remember poring through all those books in my native Bengali. I never lost that fascination."

Eventually, Guhathakurta applied to the University of Colorado to study astrophysics. There, in the Laboratory for Atmospheric and Solar Physics, she began her concentration in the study of the solar atmosphere. "There was so much going on then: eclipse observations and analyses of new satellite data. One of the images we took in X-rays of the solar corona made it onto the cover of *Time* magazine."

But Guhathakurta was always interested in seeing the big picture. As her career progressed, she became involved in one major solar observation program after another. The different interests of each project made her determined to keep the big picture of the Sun in mind. "I really like bringing people together from the different subdisciplines because it's getting the overview that makes science really fun! People should always try to remember that gaining a grasp of the whole is why we do what we do as researchers." (See Chapter 10, p. 279.)

PAT HARTIGAN

Pat Hartigan got his first subscription to *Sky & Telescope* in 1969 when he was just 9 years old. He has never lost his focus since. At 10 years of age he received his first telescope. "I went outside, swatted mosquitoes (this was Minnesota), and spent the whole night trying to find things in the sky," he explains. "Since then I just basically continued on doing the same thing (minus the mosquitoes)." Hartigan pauses for a moment to correct himself. "Well, I did want to be a rock star or a professional baseball player, but neither one of those things panned out."

Hartigan, now a professor at Rice University in Houston, Texas, is an observational astronomer whose work has focused on studies of star-forming regions. He has traveled to some of the world's most remote observatories for his work. "I remember being on top of a mountain in Chile, the top of the Andes," he says. "It was spectacularly clear up there, and the Sun had just gone down. Up in orbit, the shuttle was about to rendezvous with the space station, and you could see them both in the sky. They were about a degree apart and brilliantly bright. I watched them tracking across the night sky in a perfectly smooth motion. I really had a sense of what the ancients were imagining with their celestial spheres and the perfect kinds of motions they wanted for the heavens."

"Another time, back when I was a graduate student, I was working with the 4-meter telescope in Chile. I was in the 'cage' at the top of the telescope. I remember looking down into the mirror—this huge, 4-meter-diameter block of glass. There, reflecting back at me, were all the brilliant stars of the Southern Hemisphere. It was like looking down into this ocean of starlight. That was really quite something." (See Chapter 4, p. 92.)

AMANDA HENDRIX

It can take decades for a project such as *Cassini* to go from proposal to design to assembly to launch and then, finally, arrival. Amanda Hendrix, now a senior scientist at the Planetary Science Institute, was with the project from its launch in 1997. Seven years later, Hendrix's most cherished moment in an impressive career came when she viewed the *Cassini* data.

"We were having a little image-viewing celebration as the data were coming back from *Cassini*," says Hendrix. "We were looking at images of Saturn's outer moon Rhea. There was so much detail I was stunned. I am a moon person, so this was really exciting to me. Finally I was seeing one of those moons up close!" The *Cassini* mission has been wildly successful, mapping out Saturn, its moons, and its spectacular ring system.

Hendrix is a planetary scientist whose own journey began even earlier than *Cassini*'s. "I was in second grade when I learned about the Solar System," she recalls. "I loved it." As she grew older, Hendrix added astronaut to her list of ambitions. "I thought being an astronaut and an astronomer would tie in pretty naturally."

The mix of Hendrix's interests in the science of planets and in the engineering aspects of being an astronaut led her to major in aerospace engineering as an undergraduate. The University of Colorado, where Hendrix did her PhD studies, also had a large planetary sciences program. "I was doing planetary science through Colorado's aerospace engineering department, and that worked out fine," she says. As an aerospace graduate student, Hendrix also analyzed data collected on the Moon and asteroids from the ultraviolet instrument on the *Galileo* Jupiter mission. (See Chapter 8, p. 212.)

FALK HERWIG

It wasn't the lure of the night sky and its mysteries that led Falk Herwig into astronomy; it was the simple desire to find a field of study that was as far as possible from anything practical. "When I was younger, I think I must have had a big chip on my shoulder or something," he says. "I just couldn't imagine using my time to do anything applied to real life. I wanted to find the most abstract endeavor I could and dedicate myself to it. Now, of course, I am not so resolute about the importance of doing something useless."

When he started his PhD studies in Germany, Herwig explored a number of disciplines. "Areas like geophysics were interesting, but they still seemed too applied," he says. "Really abstract fields like string theory sounded great, but I had a feeling I would never be able to find a job if I made that my specialty. Astrophysics, however, seemed to have just the right mix of distance from day-to-day reality and a reality that included job prospects," he says, laughing.

Whatever his motivation, Herwig's skill as a scientist led to a successful career modeling the evolution of stars like the Sun. His work has placed him at the frontiers of understanding what happens deep within stars as they reach the end of life. From the vantage point of this success, Herwig now realizes that it's the methods he loves most. The specific field to which those methods are applied comes second. "Now I think I could have gone into geophysics and it would have all been fine. It's the process of science and the tools like the computer simulations I use that are exciting. That's what I really love doing. The exact application doesn't matter that much." (See Chapter 10, p. 275.)

DAVID JEWITT

The skies of London may not seem like the best place for a young boy to find astronomical inspiration, but in the 1960s the hazy skies of England's urban capital were dark enough to let David Jewitt see his future.

"I grew up in an industrial part of London, where the city lights were pretty bright," he explains. "One night I was riding my bike home from the park. I looked up and saw a bunch of meteors crossing the sky. I didn't really know what they were, but I must have seen 10 or 15. The main thing that got me was they all came from the same direction." With this celestial vision in his mind's eye, Jewitt pedaled his bike straight home and asked his mother the simplest of questions: "What are those things?"

"She told me they were shooting stars," recalls Jewitt. With a stubborn curiosity that is the hallmark of scientists, budding or otherwise, Jewitt replied, "How can a star shoot?"

The meteor shower Jewitt saw that night set the course for his life. "The thing I really remember was I saw something that my parents didn't understand. Nobody could tell me what it was. That really impressed me." Eventually, Jewitt's grandparents bought him a small telescope. "It was about as big as Galileo's," he recalls. "I watched the phases of the Moon. I watched sunrise and sunset on various lunar craters. I thought it was all amazing. I hadn't imagined that a person in North London and of my age and lack of knowledge could see things like that."

The years of backyard observing paid off for Jewitt. After attending the University of London, Jewitt went on to earn a PhD at the California Institute of Technology, ending up as a professor in the Earth, Planetary, and Space Sciences Department at UCLA. Jewitt's discovery of the first Kuiper Belt objects in 1992 marked a turning point in studies of our Solar System. "There are some things we do know," he says, "and there are lots of things we don't know. And then there are things we thought we knew, but we're learning we have to go back and rethink them. It's that last aspect of science that keeps it all interesting." (See Chapter 5, p. 122.)

CLIFFORD JOHNSON

If you're going to be a scientist, it helps to have a secret lab. "There was this place under the stairs in the house where I lived," says Clifford Johnson. "The house was still being built, so you had to sort of climb up into this spot. It was the perfect hideout. I had my little bottles of pond water and this old, badly functioning microscope I had fixed."

While growing up on the island of Montserrat, in the Caribbean, Johnson was vaguely aware of where he was going with his work. He says, "I knew that there was this thing called 'scientist' and didn't know exactly what it was. I did know I wanted to be one. In fact, I remember a family friend once asking me, 'What kind of a scientist do you want to be?' and me being really angered by the question. I just wanted to know about everything, so I didn't want to choose." Eventually, Johnson realized that there are different kinds of scientists, and sooner or later he would have to make a choice. "I actually just got a dictionary and tried to find all the '-ists' and '-ologists' to see what they studied. In some sense, I chose to be a physicist because whoever wrote that entry said it was the study of everything."

After moving to England, Johnson excelled in his high school studies. Despite his continuing interests in everything, however, he ended up specializing in theoretical physics. He is one of the world's experts in string theory. Strings are considered a good candidate for a TOE, or theory of everything. Perhaps Johnson has not strayed too far from his dreams in the secret lab under the stairs. (See Chapter 18, p. 489.)

JIM KALER

When Jim Kaler was 8 years old, he asked his grandmother a simple question: "Grandma, why do stars all have points?" Kaler's grandmother told the young boy that stars don't have points on them. "Just look up and see for yourself," she told him (like a good scientist). "I did look up," says Kaler. "But this time I really saw! I was looking into the sky at what turned out to be the giant star Arcturus, and sure enough, stars didn't have points at all."

This first direct experience with astronomical observations set Kaler ablaze. "I remember Arcturus was a slightly orange shade of red. The more I stared at it, the more I was fascinated by it. At that moment I fell in love with astronomy and, as crazy as it sounds, I never looked back."

Kaler eventually became one of the world's experts in the late stages of evolution of stars like the Sun. "I always viewed astronomy as a very aesthetic science," he says. "It's as much an art form to me as it is a science. I remember just falling in love with the images I saw of what are called planetary nebulae—clouds of gas blown off dying solar-type stars."

But studying stars was not enough for Kaler. "Somewhere in middle life I decided I was having more fun writing about astronomy than actually doing it," he says. "I did a sort of a career change which was unheard of back in those days. I went into public education." Kaler is now a well-recognized author of popular books on stars. (See Chapter 11, p. 298.)

HEATHER KNUTSON

For Heather Knutson, a life in astronomy began with accidents of location. "I was an undergraduate at Johns Hopkins University," she explains. "I wasn't sure what I wanted to study, but right across the street from the physics department was the Space Telescope Science Institute." STScI, as it is known, is the operations center for the Hubble Space Telescope (HST). From inside the futuristic building, technicians and astronomers oversee the orbiting telescope, changing where it points for different observations and maintaining its overall health.

"That turned out to be a wonderful opportunity for me," says Knutson. "I was able to work there and got to meet real astronomers. I got my hands on real data from the HST and had a chance to see how science really works. From that experience I saw how I could take the physics I was learning in class and apply it to the unsolved astronomical problems. After a while I could tell those problems were far more interesting to me than some of the other paths that I could go down within physics."

Knutson's time at STScI showed her another aspect of life in astronomy that piqued her interest. "I noticed these conference flyers everywhere for astronomical meetings. Looking at their locations, I noticed a theme: they were all in really interesting places." Knutson saw meetings on galaxy formation being held in Hawaii, meetings on cosmology held in Italy, and meetings on exoplanets held in Brazil. She was hooked. "Looking back, one of the things I have thoroughly enjoyed about doing astronomy, in addition to the science, has been the opportunity to travel."

After completing a postdoctoral fellowship at UC Berkeley, Knutson had to look for a permanent position—something not easy to find in astronomy.

"It is kind of chancy, but I'm definitely glad that I did this," she said. "I always felt like even if I never find a permanent job, the time I spent studying extrasolar planets while in grad school was so enjoyable that it was really worth it to me." Knutson is now a professor at Caltech. (See Chapter 5, p. 129.)

CAROL LATTA

It's hard not to catch Carol Latta's enthusiasm. After a career working in management at a major international corporation and raising two children, Latta found herself ready to take on something new. Returning to a local university, she rediscovered her enjoyment in learning science and then, unexpectedly, found a passion for astronomy. "I started out with math and science, then went to linguistics, and ended up working in business. It was only after I took my early retirement that I realized how much I missed math and science." Latta now conducts youth education through the Astronomy Section of the Rochester Academy of Science.

"At one point I took a course in astronomy at the Museum & Science Center in Rochester," Carol Latta recalls. "Then my husband gave me a trip to astronomy camp in Arizona as a birthday present." That trip sealed Latta's fate. She returned to upstate New York determined to find a way to become actively involved in astronomy education. Joining the local amateur astronomy club and subsequently serving as its president for 5 years, Latta became an expert at using a telescope. Night after night she learned how to find objects like galaxies and nebulas on the sky, and her love for astronomy grew.

These days Latta divides her time among her 14 grandchildren, acting as director of the University of Rochester's public observatory program each summer, and her work with the Rochester Academy of Science. Latta's excitement and passion for astronomy remains as keen as that of any professional researcher. "I found that astronomy is something broad enough and deep enough and has so many different aspects that I could learn, be productive, and have so much fun all at the same time. In the end, becoming serious about astronomy was an obvious choice." (See Chapter 2, p. 39.)

DAVID LAW

Some discoveries, including one of David Law's best memories of astronomical research, begin by accident. "My colleague and I were just looking through some of the Hubble Space Telescope data we had recently taken," Law explains. These were images of very distant galaxies 10 billion light-years or so away. "Most of them looked weird, blobby, and irregular."

Since Law and his collaborator were seeing so far back in time, these galaxies were caught in the act of their formation, so it made sense that they looked like a mess. "But then we just saw one that looked totally different," Law continues. "It looked like a regular spiral galaxy. It had a bulge and two or three nice spiral arms. We figured we must have gotten the distance wrong." But checking and rechecking, they confirmed that the galaxy was clearly 10 billion light-years away—more than two-thirds of the way back to the dawn of time. They had found the oldest spiral galaxy anyone had ever seen.

Law wasn't the kind of star-happy kid who was determined to be an astronaut or astronomer when he grew up. Instead, his fascination with the field grew slowly. "I took a physics course in high school, and I really liked the part about planetary orbits and gravity," he explains. Once Law started his undergraduate studies, he signed up for more physics courses. But the event that really pushed him over to the dark side (astronomical observing) was a summer program at the Haystack Observatory in Westford, Massachusetts. "Even though I found out I didn't want to have anything to do with radio antenna design, I did get a lot of really useful experience and, most important, I learned that I really, really liked astronomy." (See Chapter 16, p. 426.)

AVI LOEB

It wasn't astronomy or even physics that got Avi Loeb started on his career studying the earliest cosmic structure. It was philosophy.

"Originally, I was mostly interested in philosophy because philosophy addresses the most fundamental questions we have in life," says Loeb. "It doesn't provide unique answers," he explains, "so it's difficult to feel like progress is being made." In addition, after high school, Loeb was eager to continue his studies while he was completing the obligatory 2 years of military service in his native Israel. "They wouldn't let people pursuing the humanities forgo the service," he explains. "But by being admitted to the special Talpiot program, which allowed me to pursue a PhD in physics while working in defense-related research, I could engage in intellectual work. I preferred to do that rather than run around in the field with a heavy pack on."

As a student, Loeb had the chance to travel to Princeton University to meet John Bahcall, one of the 20th century's most respected astronomers. "Bahcall said he would offer me a postdoc position after I graduated," Loeb recalls, "but only under one condition. I had to convert from the physics I was doing at the time to astrophysics."

That was a fateful step for Loeb. As his research became more enmeshed with questions about the

early Universe, he found he was returning to his first passion, philosophy. "After some months I found that I had really closed the circle. In our work we address questions touching the realm of philosophy. For example, we consider how the first light was produced, which appears in the very first chapter of the Bible," he says. "For me, it's a fulfillment of my childhood dream. I can actually address the big questions I started with, but this time I'm using scientific tools and I feel like I'm making progress." (See Chapter 17, p. 461.)

JONATHAN LUNINE

For Jonathan Lunine, a career in astronomy and planetary sciences was an accident of geography: his family lived in New York City just a few blocks from the Hayden Planetarium. "The planetarium was definitely part of it. My parents took me there all the time." Tragedy, however, almost steered Lunine away from astronomy. "I was 14 or 15 when my father died, and I began to get this guilt complex that I needed to help save the world. I decided I was not going to become an astronomer but instead would become a medical doctor."

In his senior year of high school, Lunine asked his doctor for some medical school catalogs to see what courses he would have to take. "He talked me out of it," Lunine recalls. Instead, the doctor offered him a ticket to see astronomer Frank Drake talk about the giant radio astronomy dish in Arecibo, Puerto Rico. That was all the push toward his childhood love that Lunine needed. "After Frank Drake was done lecturing, I remember coming back down in the elevator and thinking, 'I really want to be an astronomer, not a doctor.'"

Drake wasn't the only scientist to make an impression on the young Lunine. "I read Carl Sagan's *The Cosmic Connection* at the age of 14," Lunine explains. "I was so enchanted by it that my mother encouraged me to write to him. I did, and then he wrote back a two-page letter on Cornell stationery. Having a famous scientist like Sagan actually write a letter to this little junior high school student made a big impression on me. So now if a kid writes to me, I try to respond because I remember how much Sagan's letter changed my life." (See Chapter 6, p. 168.)

CHRIS MCKAY

There was a time when Chris McKay liked to go winter camping for fun. Now, as an astrobiologist who spends months at a time living in a tent in the remotest parts of the world, McKay likes to stay at home in his off-hours. "After you've camped for a month in Antarctica, there is no attraction to go winter camping for fun anymore." Nevertheless, McKay loves his science. "To do the kind of research that interests me, you have to get pretty extreme. If you are going to go out and find extremophiles, it means you're not going to be sitting in an office somewhere wearing a white lab coat."

McKay was always an adventurer. And, he says, "I was always interested in science. Sometimes I'd skip school to ride my motorcycle, come back for a physics lecture, and then leave again. The next day my physics teacher would come up to me and say, 'You know, I am sure you were in class yesterday, but according to the roll you were absent.'"

Despite his less-than-stellar attendance record in high school, McKay went to Florida Atlantic University as an undergraduate and then did graduate work at the University of Colorado. The timing with one of astrobiology's founding events was fortuitous. "It was 1976," he says. "That was when the two *Viking* probes landed on Mars." Watching those missions propelled McKay into astrobiology, a field that had yet even to be named. "It was fate in a lot of ways that got me into this," says McKay. "There are a thousand interesting questions to ask about the world. I happened to be at the right place and the right time to end up spending my life asking astrobiological ones. But I am pretty happy with the way it worked out." (See Chapter 9, p. 245.)

MIKI NAKAJIMA

When she was a little girl, Miki Nakajima didn't want to study the Moon; she wanted to draw comics. "I really like manga [a popular form of Japanese comics or graphic novels], and I wanted to be a comic artist," she says. "But I thought it would be really competitive."

Though her love for the martial-arts-themed manga series *Dragon Ball* was firm, once Nakajima reached college her real decision became a career in physics versus one in Earth science. "I kind of chose both," she explains. Her interdisciplinary approach to research eventually took her to geophysics, which uses the methods of physics to study how planets evolve. "I was really happy with my choice," says Nakajima. "It's not just the science, but also the people that I work with that really makes it fun."

She followed up degrees in Earth and planetary sciences from the Tokyo Institute of Technology with a PhD at the California Institute of Technology. Now an assistant professor at the University of Rochester, Nakajima and her colleague Dave Stevenson, of Caltech, have used computer simulations to study the Moon's formation by impact with the proto-Earth and the effect of impacts on planetary composition. Nakajima is also at the forefront of modeling the origin of other moons and of exoplanets. (See Chapter 6, p. 172.)

JUDY PIPHER

"My mom was a biologist," says Judy Pipher, "so we never grew up with any biases about girls and science. I liked physics quite a bit, and that was my major at the University of Toronto." But in Pipher's junior year, an observational astronomy course changed her life's direction and changed the field of astronomy. "Even as a kid I can remember lying on the beach at my family's cottage and using a book to find different stars. That class in my junior year just let me see how physics was the basis of astronomy."

For graduate school, Pipher went to Cornell University, where she became involved in the young science of infrared astronomy. "We used to fly very primitive infrared detectors on rockets out at the White Sands Proving Ground in New Mexico," she explains. "The data would get collected on these long strips of graph paper. It was a nightmare trying to interpret what the detectors saw during their short flights. Every night the professors would tell us to have a report ready in the morning, and then they'd head off to bed."

Pipher found she was good at building the detectors and good at interpreting their observations. Eventually, she worked her way to a professorship at the University of Rochester, where she became a leader in advancing generation after generation of infrared instruments. When the time came to plan the construction of a large-scale infrared space telescope, Pipher was ready to lead one of the teams. Twenty years later, that effort paid off with the successful launch of the Spitzer Space Telescope.

Pipher is now considered by some the "grandmother" of infrared astronomy. "I am not sure if I like being the grandmother of anything," she laughs. "I just went with what I loved to do, what interested me. I guess it worked out okay." In 2007 she was inducted into the National Women's Hall of Fame for her pioneering efforts in infrared astronomy. (See Chapter 12, p. 324.)

ADAM RIESS

Adam Riess was not that interested in astronomy, though he won the Nobel Prize for work he did at a telescope (a lot of telescopes, actually). "I was just not one of those kids who was into it," he says. "I always liked science. At first I liked biology. Then I did a summer program in high school that convinced me it was really physics that I liked." Riess majored in physics in college, but it wasn't until graduate school that astronomical science appeared in his personal universe. "That was the point when I discovered astrophysics and cosmology. It was cosmology that really got me excited."

Riess was born with the kind of curiosity that lands a kid either in science or in the hospital. "I was always blowing stuff up. I would put a wire across those two prongs on an electrical outlet to see what would happen," he recalls. "But I did less dangerous stuff too, like taking radios apart." Along with this practical experimental side, Riess held deeper philosophical questions close to his heart. It was the possibility of pursuing those questions that sent him to the "dark side" of astrophysical studies.

"When I started learning about cosmology," Riess explains, "I thought, 'Really? You can really ask basic questions like, How old is the Universe? or What's its ultimate fate?' I was even more amazed that you could actually design quantitative experiments that answer those questions. When I stumbled on this aspect of cosmology, I was just thrilled!"

Since then, Riess has never lost interest in the big questions. "I would say the questions that I have—or the questions that seem important to me, that I want to answer—are all physics questions. But I have to use astronomical tools to answer them." (See Chapter 18, p. 483.)

DIDIER SAUMON

"I guess I was one of those kids who got interested in science from a fairly early age," explains Didier Saumon. "I was interested in science, and by that I mean all kinds of science." He considers himself lucky to have had a neighbor who shared his interest. "We were always geeking out together."

Eventually his friend got a small telescope as a gift, and that started them on the path to astronomy. "We built our own telescopes as teenagers and developed a passion for observing the sky. Eventually I thought that 'yeah, this would be a cool thing to do.' So I got a bachelor's degree in physics and took the first step towards becoming an astronomer." The rest, as they say, is history.

Now a theoretical astrophysicist who has worked at New Mexico's Los Alamos National Laboratory for decades, he develops models—thousands of them—to study dwarf stars, young planets, and dense plasmas. He has done computer modeling and data analysis for the Gemini Planet Imager project, which found 51 Eridani b, the first Jupiter-like extrasolar planet ever to be discovered. (See Chapter 11, p. 309.)

LINDA SPILKER

"Girls don't go into science," Linda Spilker's high school guidance counselor told her. It was the mid-1970s, and Spilker had already caught the science bug. "I grew up with the space race. I remember watching all the astronaut missions from *Mercury* to *Gemini*

and then *Apollo*. I think that really fixed my attention on space." When Spilker was 9, her parents got her a telescope. "I was just so thrilled. I remember taking it out, looking at Jupiter and seeing the little moons as pinpoints of light. From that point, I was hooked."

Despite the guidance counselor's discouraging words, Spilker's parents were supportive and knew their daughter could do anything she set her mind to. "My mom loved math," recalls Spilker. "She took algebra when she was a girl. She loved it but got teased so much she dropped the course. I know she always regretted that."

Completing her degree in physics at California State University, Fullerton, Spilker worked for NASA's Jet Propulsion Laboratory (JPL) and later received a PhD from UCLA in geophysics and space physics. As the *Cassini* project scientist, she was responsible for the science operations of the entire billion-dollar mission. (See Chapter 8, p. 229.)

WOODY SULLIVAN

A 9-pound Russian satellite was all it took to ignite Woody Sullivan's scientific ambitions. "I was in the eighth grade on October 4, 1957," says Sullivan, now a professor of astronomy at the University of Washington. "That was the day Sputnik was launched." Sputnik was the first artificial satellite to be lofted into orbit. While many Americans worried about the Russians dropping bombs from space, for Sullivan the whole enterprise was thrilling. "I thought, 'Wow, people have launched into space!'" he recalls. "I had been interested in science before that, but Sputnik really pushed me over the edge to astronomy."

Sullivan earned an undergraduate degree in physics from MIT, where he developed a passion for radio astronomy. But only while working on his PhD at the University of Maryland did he begin thinking about the Search for Extraterrestrial Intelligence (SETI) project. "I did my graduate thesis on water vapor in the interstellar medium," he explains. "It turns out water vapor has a spectral line with a wavelength of 1.35 cm (which is in the radio band). This was a new discovery at the time, and I ended up doing much of the early research on the subject. That's what led me to think about life in terms of water and radio signals."

Later, while Sullivan was reading about Project Cyclops, a NASA plan to build 1,000 large radio dishes to scan the sky for extraterrestrial signals, he got to thinking about something more than direct signals. "I started thinking about 'leakage,'" Sullivan says. Radio leakage means signals that a technological culture does not intend to send but could be detected anyway. "We are constantly leaking radio emission from our radar and other transmitting operations," Sullivan explains. "I started to think about what detection of leaked radio signals could reveal about a civilization."

His work on the topic has become a classic SETI study. (See Chapter 9, p. 254.)

JOHN TARDUNO

John Tarduno's research as a geophysicist has taken him to some of the remotest corners of Earth, from frozen wastelands of the Arctic to the mountains of New Zealand. But his globe-trotting for science was not something he could have predicted when he was young.

"Geology had always been a hobby when I was a kid, but I was mainly interested in physics," says Tarduno. As an undergraduate, he saw a job opening advertised in a campus physics lab, and a merger of hobby and profession became possible. "The guy I was working for ran a solid-state physics lab, meaning they studied properties of solids like rocks. They were looking for ways to extract measurements of Earth's magnetic field when the rocks had first formed. That was my first experience with paleomagnetism and how I learned about the whole field of geophysics. It was just a natural fit."

Over the years, Tarduno's work in the field, looking for new paleomagnetic samples, has taken him to some extraordinary places and, a few times, has gotten him into some dangerous situations. "One time working in the Arctic, there was this narrow, steep-walled ravine that I wanted to explore," Tarduno recalls. "So I had the chopper put me down at one end and told him to pick me up on the other at the end of the day. But I had no way to know if I could make it to the other side, and if I didn't find a path I would be stuck in there because there would be no way to land a helicopter in the ravine. I was lucky because I found a natural terrace that I could move along, but it was 500 feet down and if things hadn't worked out I would have been in trouble." The hard-won samples Tarduno retrieved that day turned out to be key to recovering the record of a particularly turbulent part of Earth's magnetic history. "I guess it was worth it," he says, smiling. (See Chapter 6, p. 160.)

MEG URRY

Meg Urry says she owes it all to Miss Crawley, her high school chemistry teacher in Massachusetts. "I did not always have astronomy in mind [as a career]," says Urry, now a professor at Yale University and a former president of the American Astronomical Society. "I liked everything—English, history, math, languages. In fact, science was never the top thing." But it was Miss Crawley's class (and Urry's parents, both of whom are scientists) that let Urry eventually find her own excitement for physics and astronomy. She attended Tufts University as an undergraduate. "When I got to college, I took physics (which I had dis-

liked in high school), found Electricity and Magnetism class challenging, decided to conquer it, and then discovered it was beautiful." Most important, Urry realized that she thought like a scientist.

Urry did research at the National Radio Astronomy Observatory in Charlottesville, Virginia. There she found the astronomers to be generally nice and fun people to be around. Later she completed her graduate work at Johns Hopkins University and postdoctoral research at MIT and the Space Telescope Science Institute.

Across her career, Urry has been at the forefront of studying active galactic nuclei. And while her research has been a source of great fun and excitement, it has not been the best part of her time as an astronomer. "Probably the happiest I've been is teaching students physics—either explaining something they hadn't yet understood, or getting them excited about what the Universe is like. That's definitely the most satisfying part of my career," she says. Urry has also been particularly interested in the status of women in science and in explaining the shortage of women scientists at the highest levels. (See Chapter 16, p. 436.)

LUCIANNE WALKOWICZ

Lucianne Walkowicz likes to do science, but she likes to do a lot of things. Her journey to astronomy began in high school with new interests in physics and chemistry. "I wanted to find a way to combine them," Walkowicz says. A summer program with the New York Academy of Sciences gave her the opportunity to explore the kinds of chemical reactions that can occur in space. From that point on, she was hooked. "I got connected to an astrochemist at New York University, and I fell for astronomy after that!" Since then, Walkowicz has had a chance to explore science from an unusual number of angles, including attending launches of space shuttles that carried instruments she'd worked on.

Along with falling for astronomy, Walkowicz also developed a passion for communicating science to broader audiences. In her day job as an astronomer at the Adler Planetarium, she does both science and science communication. "The best thing about working in a planetarium is definitely the questions people ask—people are so often afraid that they will sound stupid, but honestly, people ask great questions! I even find that there are questions people ask that make me think about astronomy differently—it's so great to get to hear everyone's perspective and what's on people's minds." Walkowicz has also served in the unusual position of Library of Congress Chair of Astrobiology. In that role, she has had the chance to open new conversations into how we view the possibilities of life in the Universe. (See Chapter 1, p. 13.)

ERIC WILCOTS

Eric Wilcots was born in Philadelphia, which is not exactly dark-sky territory. He was 8 years old when his parents gave him a telescope, and he knew something big had happened. "You could see the rings of Saturn!" he recalls. "I mean, how cool is that?"

The arrival of that new telescope just happened to coincide with the *Voyager* mission passing by Jupiter. "The guys at JPL [Jet Propulsion Laboratory]," says Wilcots, "they just looked like they were having fun. I mean there was Io and its volcanoes. You can't argue with that kind of thing." The images from the space probe, combined with his own backyard observing, set Wilcots on a course to a career in astronomy from which he never wavered.

"But the thing is, I didn't know anything about astronomy. I was taking all these classes, and they were great, but I didn't know what astronomers really did." That gap in knowledge led Wilcots to ask his astronomy professor at Princeton University, "So how much do they pay you to do this?" It may have been an impolite question, but Wilcots asked it innocently. "I guess I wasn't thinking it might be rude to ask how much someone makes. I was just a kid and really wanted to know." Without missing a beat, however, Wilcots's professor answered, "Enough to go to the opera."

That was good enough for Wilcots. After earning his undergraduate degree at Princeton and a PhD at the University of Washington in Seattle, Wilcots joined the astronomy department at the University of Wisconsin, where he studies galaxies and their environments. "It's still fun," he says. "I wanted to do this when I was 8, and now here I am. Doing science, doing astronomy, thinking about galaxies . . . I still get excited thinking about getting new data and figuring out ways to answer a question no one else has asked." (See Chapter 17, p. 453.)

LILIYA WILLIAMS

Liliya Williams knew early on she was headed toward the abstract side of science. "I was born with an interest that steered me towards physics," she says, "preferably something that has very little to do with our everyday experience. I knew I wasn't going to do tabletop experiments."

Telescopes never worked for her. "I grew up in Moscow," she says. "When I was 9 or 10, somebody gave me a book on planets with colored pictures in it. My dad noticed my interest in astronomy and got me a telescope. I didn't enjoy it at all. But later I got hold of another book that was more mathematically inclined. That one I liked a lot."

Now a professor at the University of Minnesota, Williams is an expert on gravitational lensing—the use of Einstein's theory of relativity to predict how light is bent and focused as it passes a massive object in space.

Williams still loves the beauty of mathematics and its applications in astrophysics after all these years. "It's when something just clicks that you feel great," she explains. "When one of those moments of realization comes and you understand something, oh boy! Even if it's something that has been understood for decades but you understand it for yourself the first time—that is a really wonderful feeling. Of course, it's even better when what you realized is something no one else has ever explored before, when it's something new in your own research." (See Chapter 3, p. 69.)

STAN WOOSLEY

Stan Woosley started college without a particularly strong passion for astronomy. "I was taking physics, and it was really hard. I would have changed majors, but I was just stubborn enough to refuse to quit," he explains. Then something happened that changed both his life and the course of history. It was 1961, and the Texas-born Woosley was a freshman studying at Rice University in Houston. By chance or fate, the young man ended up sitting in the audience of President John F. Kennedy's famous speech at Rice that launched the US Moon-landing program. Kennedy's vision of astronomical science without boundaries lit a fire under the young Woosley. "The idea that space science was a big up-and-coming field that I could be part of was a critical influence for me."

Woosley stayed at Rice through graduate school but found his interest flagging. A chance meeting, however, with Don Clayton, an expert on the creation of elements and supernovas, once again ignited Woosley's interest and passion. "I always loved blowing things up," Woosley explains. Whatever doubts he had vanished as he plunged into the work at the fron-

tiers of physics and astronomy. "It's amazing to me looking back now that I almost lost interest, because I love what I do now so much," he says. "Sometimes I can't sleep, I am so excited about working on a problem. It is such a privilege to be able to directly explore the nature of the world and learn about how the Universe is constructed."

Woosley went on to study at Caltech before taking a professorship at UC Santa Cruz in 1975. Across his career, he has been one of the world's leaders in the study of supernovas, with more than 300 published papers to his credit. After all of those accomplishments, he has earned the right to relax now and then. "I like margaritas and playing guitar by the pool these days," he confesses. (See Chapter 13, p. 350.)

JIM ZIMBELMAN

Jim Zimbelman's interest in planets began early. "When I was in middle school, I was interested in all sorts of 'sciency' kinds of things," says Zimbelman. "But the turning point came when my dad got a small telescope. That's what really got me hooked specifically in astronomical things. But what I really loved best of all were the planets."

Zimbelman's childhood passion stayed with him from middle school through his undergraduate years. But by the time he started graduate work, he was in for a surprise. "When I got to UCLA to start a master's program in astronomy, I discovered that all of the classes I wanted were taught in a different department," Zimbelman recalls. In the late 1970s, when he began his graduate education, astronomy and planets were parting ways. Astronomers studied stars and even more distant objects, such as galaxies. If students wanted to study something as nearby as the planets, as Zimbelman did, they needed to take classes in geology. "I had to transfer my application from an astronomy PhD to an earth and space sciences PhD." After finishing his degree in 1984, Zimbelman went on to play a significant role in the study of volcanism on all the terrestrial planets.

Zimbelman, a former chair of the Smithsonian Center for Earth and Planetary Studies, attributes his broad perspective on celestial science to his family background. "One thing I appreciated about my father was, while he was a minister, he was open to listening to discussions of things." This open-mindedness has always been an encouragement to Zimbelman in his scientific studies of the sky. (See Chapter 7, p. 184.)

SELECTED ANSWERS

Chapter 1

Checkpoint Questions

1.1. Answers may vary. **a.** Canals on the Martian surface; Martian rovers currently scouring the surface of the planet. **b.** Unending quest for greater understanding, evolution of prevailing scientific descriptions of nature, role of technology in new discoveries.

1.2. Answers may vary. **a.** The number of times a value is to be multiplied (or divided) by the number 10. **b.** Divide the coefficients and subtract the exponents.

1.3. Answers may vary. Using our Sun and the Milky Way, 12 orders of magnitude.

1.4. Answers may vary. **a.** Taller candidates in presidential elections have tended to be the winner, which may be evidence that people have bias for taller leaders. Some claim cell phones are linked with brain cancer, but there is no evidence for that association. **b.** The insistence of science on testable hypotheses and its reliance on disconfirming instances to signal fundamental problems with a given model.

1.5. Answers may vary. **a.** Big data and personal privacy. **b.** Predicting eclipses; precision astronomy.

Narrow It Down: Multiple-Choice Questions

2. c

8. a

10. d

To the Point: Qualitative and Discussion Questions

22. Flashes of insight, accidental discovery.

30. Time keeping, agriculture, navigation, religious observances.

34. Selection bias altering perceived frequencies of various outcomes.

Going Further: Quantitative Questions

36. 5.0×10^4 years.

39. Nine orders of magnitude.

40. 3.75×10^4 years.

Chapter 2

Checkpoint Questions

2.2. Answers may vary. The tilt of the Earth's axis of rotation with respect to the plane of the ecliptic gives rise to the seasons, since it results in the Earth receiving more or less directly concentrated sunlight at different points in its orbit. For example, in the Northern Hemisphere, it will be summer when the rotation axis points most directly toward the Sun—the concentrated sunlight being maximally effective in warming its surface at this time of year. Similarly, 6 months later, on the opposite side of its orbit and with the axis now pointing away from the Sun, it will be winter in the Northern Hemisphere. At the same time, the axial tilt's relationship to the direction of the Sun also determines the maximum height of the Sun above the horizon—an observation that varies both from day to day and, more noticeably, over the course of the year. This explains why, for example, in the summer the Sun will pass most directly overhead—giving rise to long days and short nights—while in winter the converse is true.

2.3. Answers may vary. See Figure 2.18. Paint one half of a ball, hold it at arm's length with the painted side face-on, and spin around in a circle.

2.4. Answers may vary. Planets will exhibit "retrograde loops."

2.5. Answers may vary. Agriculture made it necessary to plant, tend, harvest, and store crops in ways appropriate to the changing cycle of the seasons.

2.6. Answers may vary. Eratosthenes reasoned that shadow lengths cast at the same time on the same day vary by location because the Earth is spherical rather than flat.

Narrow It Down: Multiple-Choice Questions

2. e

12. c, d

16. b, c

To the Point: Qualitative and Discussion Questions

24. Full.

28. The equator.

32. They would be more extreme.

Going Further: Quantitative Questions

36. $14,400''$.

40. $12.2°$.

42. 7.78×10^7 km.

Chapter 3

Checkpoint Questions

3.1. Answers may vary. **a.** The Copernican system put the Sun at the center, did away with the equant, and postulated both that the Earth spins on its axis and that the closer a planet is to the Sun, the faster it moves in its orbit. **b.** As a catch-and-pass effect.

3.2. Answers may vary. He provided evidence that the heavens were not perfect and unchanging.

3.3. Answers may vary. Kepler's laws provide a mathematically precise Sun-centered description of planetary orbits.

3.4. Answers may vary. **a.** Mountains on the Moon and sunspots show that celestial bodies are not featureless and perfect. Jupiter having moons means that not everything orbits the Earth. The phases of Venus are inconsistent with a geocentric model. **b.** He formulated the concept of inertia and demonstrated that all objects accelerate at the same rate in Earth's gravity.

3.5. Answers may vary. Newton's laws accurately explain Galileo's law of falling bodies, Kepler's three laws, and orbital and escape velocities.

Narrow It Down: Multiple-Choice Questions

6. e

8. b

10. d

To the Point: Qualitative and Discussion Questions

27. Conservation of angular momentum.

30. The strength of gravity is related to one over the square of the separation distance.

31. It would decrease since the acceleration due to gravity is inversely proportional to the square of the radius ($a = GM/R^2$).

Going Further: Quantitative Questions

36. 5.65 AU.

38. 132.6 years.

40. 16 times greater.

Chapter 4

Checkpoint Questions

4.1. Decreases.

4.2. Answers may vary. Continuous/blackbody spectra: temperature of a hot, opaque body. Emission spectra: composition of a hot, low-density gas. Absorption spectra: composition of a cold, low-density gas.

4.3. Answers may vary. **a.** Photon energies must match the energy difference between two allowed levels. **b.** Atoms of different chemical elements have different energy-level structures.

4.4. Answers may vary. **a.** It increases light-gathering power and sharpens angular resolution. **b.** The smaller the wavelength, the better the resolution. This feature of the light spectrum, coupled with the effect of the Doppler shift on light emitted by a source moving with respect to the point of observation, also permits information about the speed and direction of the motion to be extracted: redshifted (blueshifted) spectral lines indicate that the object is moving away from (toward) the observation point, while speed can be calculated from the magnitude of the shift.

4.5. Answers may vary. Ground-based: poor angular resolution but easily accessible for repair and enhancement and less expensive. Space-based: higher cost and difficult to access for repair or enhancement but operate outside the blurring and filtering effects of our atmosphere.

Narrow It Down: Multiple-Choice Questions

4. d

6. d

12. d

To the Point: Qualitative and Discussion Questions

24. It is mostly transparent to visible light and radio waves and mostly opaque across the remainder of the spectrum.

28. Emission: electron drops to lower energy state. Absorption: electron jumps to higher energy state.

32. Lessening of the atmosphere's obscuring effects.

Going Further: Quantitative Questions

38. 3,766 K.

40. Surface area of cooler star is 16 times greater.

42. Larger aperture collects 2.78 times more light.

Chapter 5

Checkpoint Questions

5.1. Answers may vary. **a.** Added a new category of objects to and increased the extent of the known Solar System. **b.** Aids in the testing of theories about planet and Solar System formation. Bears heavily on the question of the possible existence of extraterrestrial life.

5.2. Answers may vary. Terrestrial planet, Earth; gas giant, Jupiter; ice giant, Neptune.

5.3. Answers may vary. Asteroids, comets, and meteoroids are too small to be spherical. Dwarf planets are spherical but have not cleared their orbital paths. These bodies give astronomers a way of probing the conditions that prevailed in the early Solar System.

5.4. Answers may vary. Similarities: Earth-like planets and exoplanets orbiting in the habitable zones of other stars. Differences: super-Earths and hot Jupiters.

5.5. Answers may vary. Collapse of a molecular cloud, formation of a protoplanetary disk, condensation, accretion.

Narrow It Down:
Multiple-Choice Questions

6. d

10. d

18. a

To the Point:
Qualitative and Discussion Questions

26. Long-period comets.

28. Redshifted, blueshifted.

32. It accurately accounts for variations in planet composition as distance to the host star increases.

Going Further: Quantitative Questions

36. 2.

42. 23.

45. 375 years.

Chapter 6

Checkpoint Questions

6.1. Answers may vary. Gravitational interactions with the Moon stirred the primordial soup of the early oceans, continually slowed Earth's rotation rate, and continue to stabilize its axis of rotation.

6.2. Answers may vary. Heat trapped in the interior of Earth drives convection in the mantle, frictionally dragging tectonic plates across its surface.

6.3. Answers may vary. **a.** Layers of the atmosphere from lowest to highest: troposphere, stratosphere, mesosphere, ionosphere. **b.** As Earth rotates, its metal-rich core produces a magnetic field via a dynamo effect.

6.4. The fission hypothesis has been ruled out since Earth was never rotating fast enough to eject its molten material. The capture hypothesis has been shown to be so unlikely to have occurred that it has also fallen out of favor. The Moon's small iron core and lack of iron in its crust support the impact hypothesis.

Narrow It Down:
Multiple-Choice Questions

8. a, b, c, d

10. b

12. d

To the Point:
Qualitative and Discussion Questions

26. It deflects charged particles from the solar wind.

28. Stratosphere and ionosphere. The stratosphere is warmed by absorption of UV solar rays by its ozone; the ionosphere is warmed by ionization of its gas by collisions with solar particles from the solar wind.

34. Capture, fission, and impact; the impact model—in which a Mars-sized object is hypothesized to have collided with the still-forming Earth, resulting in the ejection of molten material from which the Moon formed—is currently favored.

Going Further: Quantitative Questions

36. The force of gravity is 6.9% stronger on the closer side.

40. 3,232 lead-206 nuclei.

44. Yes, oxygen would be retained, because $V_t < V_e$.

Chapter 7

Checkpoint Questions

7.1. Answers may vary. Mainly molten, smooth crust, cratered crust, lava flows, eroded landscape.

7.2. Since it would cool more slowly and shrink less, there would be fewer scarps.

7.3. Answers may vary. Atmospheric surface pressure of Venus is 90 times that of Earth.

7.4. Answers may vary. **a.** It rapidly vaporizes under low atmospheric pressure. **b.** Mars once had a much more substantial atmosphere.

Narrow It Down:
Multiple-Choice Questions

2. a, b, e

6. a, b, d

10. b

To the Point:
Qualitative and Discussion Questions

24. Gravitational separation of chemical constituents in a molten body. Heavier elements predominate in the center, lighter ones nearer the surface.

25. Greatest to least cratering: Mercury, Mars, Earth, Venus.

28. An interior layer of molten metal that is undergoing convection, and a sufficiently rapid rotation rate.

Going Further: Quantitative Questions

36. 1.1 times as great.

38. 5.05 km/s.

40. 0.33.

Chapter 8

Checkpoint Questions

8.1. It would move outward.

8.2. Answers may vary. Outer planets have thick outer atmospheres, immense internal pressures, and a layer of liquid metallic hydrogen (gas giants) or a layer of liquid hydrogen, frozen water, and ammonia and methane ices (ice giants).

8.3. Answers may vary. Temperature and pressure both decrease as the probe descends into the outermost atmosphere but increase as it passes through cloud layers of ammonia, ammonium hydrosulfide, and water, then an ocean of molecular hydrogen and a layer of liquid metallic hydrogen, and finally reaches the core.

8.4. Answers may vary. Saturn is less massive, so it has less metallic hydrogen and therefore a smaller magnetic field. Saturn's more rapid rotation makes it much more oblate. Unlike Jupiter, Saturn has a moon (Titan) with a thick atmosphere.

8.5. Answers may vary. Ice giants have only one cloud layer, lack a layer of metallic hydrogen, and have magnetic fields with axes that are highly inclined from their spin axes and shifted away from their centers.

Narrow It Down:
Multiple-Choice Questions

2. c

10. c

20. True

To the Point:
Qualitative and Discussion Questions

26. Giant planets experience faster rotation. This causes stronger and longer-lived storms, cloud bands, and oblateness.

30. Tidal forces cause a squeezing and a stretching of the moon's interior, which increases internal temperature. This also results in compression and stretching of the surface and a lack of craters due to volcanism.

34. *Cassini* and *Voyager*.

Going Further: Quantitative Questions

36. 88.82 K.

38. 79.5 times greater.

40. 6.85×10^{29} kg.

Chapter 9

Checkpoint Questions

9.1. Answers may vary. Develop an understanding of the causes and conditions that led to life on Earth.

9.2. Answers may vary. **a.** Too close in: water on the surface will boil away. Too far out: water will remain permanently frozen. **b.** Extremophiles make use of unexpected sources of energy, such as radioactive decay and the thermal energy associated with undersea vents. Liquid water exists beneath the surface of some of the moons of the giant planets.

9.3. Answers may vary. Mutation involves genetic changes at the level of DNA; an example is exposure to X-rays. Migration refers to movement of individuals with a specific trait into new regions; an example is invasive plant species. Genetic drift occurs when a random event causes some members of a species to die, leaving the surviving members to pass on their genes; an example is the meteorite that wiped out the dinosaurs. Natural selection is reproduction of the individuals more likely to survive in a given environment; an example is camouflage.

9.4. Answers may vary. The Drake equation shifts focus to the question, How many technologically advanced civilizations are out there? It directs attention to how to answer each of its subquestions.

Narrow It Down:
Multiple-Choice Questions

8. e
10. d
16. c, d

To the Point:
Qualitative and Discussion Questions

24. Geothermal energy, radioactivity.
30. Radio waves are not absorbed by interstellar dust, and hydrogen (the most abundant element) radiates strongly in the radio band.
32. No direct evidence to date.

Going Further: Quantitative Questions

36. 2.4×10^6.
40. $6^{138} = 2.43 \times 10^{107}$.
44. 0.39 AU inner and 0.72 AU outer; yes.

Chapter 10

Checkpoint Questions

10.1. Answers may vary. Photosphere, the layer of the Sun from which visible light is emitted.
10.2. Answers may vary. A small fraction of the mass of the hydrogen nuclei is converted into energy.
10.3. Answers may vary. Radiation: light from a lightbulb. Convection: heat rising from a fire. Conduction: burning your hand on a hot piece of metal.
10.4. Answers may vary. They all manifest from the interplay of charged particles in the Sun and its magnetic field.

Narrow It Down:
Multiple-Choice Questions

4. d
8. a
10. e

To the Point:
Qualitative and Discussion Questions

22. An ionized gas.
26. Photosphere, because it emits visible light.
28. Convection, in the mantle layer.

Going Further: Quantitative Questions

38. 21.3 years.
40. 1.89×10^{19} joules.
42. 33.18 times brighter.

Chapter 11

Checkpoint Questions

11.1. Answers may vary. Stars burn hydrogen fuel in their cores, creating helium and converting a fraction of the initial mass into energy.

11.2. Answers may vary. Wien's law can be applied to calculate the surface temperature of the star from the peak wavelength.
11.3. Answers may vary. Observations of multiple stars at various points in their lifetimes can be temporally ordered and used to infer the evolutionary trajectory of individual stars.
11.4. Answers may vary. Greater mass implies greater core compression, higher temperatures, more rapid nuclear burning, and shorter lifetimes.

Narrow It Down:
Multiple-Choice Questions

2. d
10. c
12. d

To the Point:
Qualitative and Discussion Questions

22. Distance, luminosity, temperature, size, mass, and composition.
24. The peak wavelength of its blackbody spectrum.
30. Hydrogen fusion.

Going Further: Quantitative Questions

36. 3,776 K, cooler.
40. 15.59 times longer.
42. 0.5 pc, closer. The closest star other than the Sun (Proxima Centauri) is at a distance of 1.3 pc, so it would be a big surprise to learn of a closer star!

Chapter 12

Checkpoint Questions

12.1. Answers may vary. The study of turbulence in the ISM can provide insights into various aspects of the supernova explosions by which intermediate- and high-mass stars end their lives.
12.2. Answers may vary. Gas in the ISM expands or contracts until the pressures in each phase of the ISM equalize.
12.3. Answers may vary. Gravity is the key link between the high density of a molecular cloud and its role as the site of star formation.
12.4. Answers may vary. A dense core in a molecular cloud begins to collapse under gravity; gravitational potential energy is transformed into energy of motion; as densities increase, so do both temperature and pressure. A protostar is formed when a sufficient amount of material collects at the center. Subsequent evolution consists of continued collapse

until hydrogen fusion begins and a main-sequence star is born.
12.5. Answers may vary. A protostar of less than 0.08 solar mass becomes a brown dwarf; in a protostar above 0.08 solar mass, P-P chain hydrogen fusion becomes possible; in a protostar above 1.3 solar masses, the CNO fusion cycle dominates.
12.6. Answers may vary. High-mass stars produce ultraviolet photons capable of ionizing hydrogen, heating the gas and causing it to expand rapidly, and thereby sculpting material in its surroundings.

Narrow It Down:
Multiple-Choice Questions

2. c
6. a
12. c

To the Point:
Qualitative and Discussion Questions

22. Ionization caused by weak diffuse background radiation.
26. Molecular hydrogen radiates weakly at low temperature.
27. They are unstable against gravitational collapse.

Going Further: Quantitative Questions

36. 1.5.
38. 1.375.
40. 0.01.

Chapter 13

Checkpoint Questions

13.1. Answers may vary. Eight solar masses.
13.2. Answers may vary. The nonburning core of helium contracts and heats up, temperature rises, the fusion rate in the hydrogen-fusing shell increases and expands outward, and the star becomes a red giant.
13.3. Answers may vary. As an AGB star expands, its own stellar winds strip away its atmosphere at an increasing rate, setting up shock waves that heat and compress the gas and make it visible as a planetary nebula.
13.4. Answers may vary. Since electron degeneracy pressure is generated without consuming any fuel, a white dwarf stellar remnant is indefinitely stable against further gravitational collapse.
13.5. Answers may vary. The triple-alpha process.

13.6. Answers may vary. **a.** Type Ia supernova. **b.** Once the mass of an accreting white dwarf exceeds the Chandrasekhar limit, carbon fusion becomes possible and the star explodes.
13.7. Answers may vary. No, because a pulsar can be detected only if a beam of its radiation sweeps across the Earth.

Narrow It Down:
Multiple-Choice Questions

4. b
12. c
16. d

To the Point:
Qualitative and Discussion Questions

24. It expands and cools while its luminosity increases.
26. The radius decreases as the mass increases.
28. They will become neutron stars or black holes.

Going Further: Quantitative Questions

36. 64 times larger, temperature.
40. 83 nm, ultraviolet.
42. Since the peak wavelength, 6.7×10^{-7}, is equivalent to 670 nm and is in the visible range, this star has a fairly cool temperature of 4,328 K.

Chapter 14

Checkpoint Questions

14.1. Answers may vary. **a.** Nuclear fusion. **b.** The cores of lower-mass stars are held up by degeneracy pressure, while the gravity of the highest-mass stars overwhelms even this effect.
14.2. Answers may vary. **a.** Driving in a car. **b.** The distance traveled by light can be different in two different reference frames, but speed = distance traveled ÷ elapsed time.
14.3. Answers may vary. As an object moves through a warped region of space-time, it behaves exactly as if acted upon by a force.
14.4. Answers may vary. Space-time is stretched more on the near side of an infalling object than on the far side.
14.5. Answers may vary. 1974, indirect detection of gravitational waves; 2015, direct detection.

Narrow It Down:
Multiple-Choice Questions

4. e
8. b
10. d

To the Point:
Qualitative and Discussion Questions

26. Someone tossing a ball to you from a moving car.

28. A thought experiment illustrating the effects of time dilation predicted by special relativity.

30. Gravitational redshift, gravitational lensing, spaghettification.

Going Further: Quantitative Questions

36. 37.7 km; 7,250 km.

40. 43 AU.

42. 80.3 min.

Chapter 15

Checkpoint Questions

15.1. Answers may vary. Interstellar dust can dim or obscure distant stars and clusters.

15.2. Answers may vary. Galactic disk: a flat expanse of hundreds of billions of Population I stars and billions of solar masses of gas and dust. Bulge: a centrally located spherical collection of densely packed stars, dust, and gas. Halo: a spherical shell of older globular clusters surrounding the disk.

15.3. Answers may vary. We would expect to see roughly one arm for each galactic revolution, or about 50, whereas the evidence suggests at most four arms.

15.4. Answers may vary. **a.** The rapid conversion of molecular gas into stars; this is the mechanism by which the majority of stars in the Universe are formed. **b.** High-energy emissions from and rapidly orbiting stars near the center of our galaxy.

15.5. Answers may vary. Flat galactic rotation curves.

15.6. Answers may vary. A collapsing gas cloud leaves behind a halo of dark matter; luminous matter continues to collapse, forming globular clusters in the stellar halo; a disk forms, via conservation of angular momentum; and the bulge and supermassive black hole form in the dense central region.

Narrow It Down:
Multiple-Choice Questions

4. b, d

14. b

16. c

To the Point:
Qualitative and Discussion Questions

22. Smaller. It did not account for interstellar dust.

25. Roughly 8 kpc from the center and 0.03 kpc above the midplane of the disk.

33. Newton's laws yield mass and size of object being orbited.

Going Further: Quantitative Questions

38. $1.1 \times 10^8 \, M_{Sun}$.

40. 193 km/s.

44. 415 km/s.

Chapter 16

Checkpoint Questions

16.1. Answers may vary. Shapley: Universe consists of a single enormous galactic system. Curtis: island universe theory.

16.2. Answers may vary. Two distinct categories (spiral and barred spiral) branch off from S0.

16.3. Answers may vary. If two methods yield conflicting distance measures, this signals a problem with at least one of the techniques.

16.4. Answers may vary. **a.** A SMBH surrounded by a thin accretion disk gives rise to AGN outflow. The accretion disk is further surrounded by a doughnut-shaped torus and spherical halo of smaller clouds. **b.** The various types of AGNs result from viewing a single type at different angles.

16.5. Answers may vary. All yield rotation curves that do not drop off with distance as predicted by Newton's laws.

Narrow It Down:
Multiple-Choice Questions

4. e

8. a, d

12. b

To the Point:
Qualitative and Discussion Questions

24. Elliptical and irregular.

30. Distance and galactic recessional velocity.

33. Compactness, high energies, relativistic jets, and short timescale variations.

Going Further: Quantitative Questions

36. 35 Mpc.

40. 350 Mpc.

44. 2.9×10^{39} kg.

Chapter 17

Checkpoint Questions

17.1. Answers may vary. The light that is currently being received here on Earth shows us how a star or a galaxy looked when the light was emitted.

17.2. Answers may vary. The density of galaxies is greater in clusters.

17.3. Answers may vary. A cosmic web of filamentary walls surrounding voids.

17.4. Answers may vary. **a.** In the era of quasars, young galaxies still contained an abundance of material in their galactic centers feeding supermassive black holes. The Dark Ages of cosmic history correspond to an earlier epoch in which neutral hydrogen had already formed but prior to the emergence of stars and the first large black holes. **b.** They enable astronomers to construct timelines describing galaxy formation.

17.5. Answers may vary. **a.** The smallest structures appear first and serve as building blocks for larger structures. **b.** Cold dark matter facilitates gravitational clumping, while faster-moving hot dark matter particles resist such collapse.

Narrow It Down:
Multiple-Choice Questions

6. d

12. a

20. True

To the Point:
Qualitative and Discussion Questions

30. Quasars occur at an early stage in galaxy formation.

32. Greater density equals greater gravitational attraction.

33. Dark matter that is not moving at relativistic speeds.

Going Further: Quantitative Questions

36. 1.76×10^{10} years.

40. $9.3 \times 10^{14} \, M_{Sun}$.

44. 1.2×10^5 km/s.

Chapter 18

Checkpoint Questions

18.1. Answers may vary. Big Bang: the Universe has a beginning in which all of the mass-energy

bursts forth. Due to expansion, distances between galaxies increase. Steady-state: the Universe has always been expanding and will always be, yet the overall density remains constant as matter is continually created.

18.2. Answers may vary. Planck era: temperatures are so high that the four fundamental forces are unified. GUTs era: begins when temperatures drop low enough for gravity to emerge as a separate force and lasts until the period of inflation (a hyperrapid expansion of space-time) begins. Electroweak era: strong nuclear force emerges. Particle era: splitting of the electromagnetic and weak forces and emergence of particles of the standard model. Big Bang nucleosynthesis: formation of nuclei of the lightest chemical elements via nuclear fusion. Recombination era: matter and photons decouple, rendering the Universe transparent.

18.3. Answers may vary. Cosmic expansion predicts a hot, dense early Universe; Big Bang nucleosynthesis accurately predicts baseline proportions of light elements in the early Universe; CMB radiation accounts for light produced during the era of recombination.

18.4. Answers may vary. **a.** Finite and unbounded, saddle-shaped, or flat. **b.** CMB temperature variations point to a flat geometry.

18.5. Answers may vary. Quantum gravity is needed to roll time back to $t = 0$ of the Big Bang.

Narrow It Down:
Multiple-Choice Questions

4. c

6. a

16. b, c, e

To the Point:
Qualitative and Discussion Questions

25. Quarks and leptons; leptons do not feel the strong nuclear force.

29. 75% hydrogen, 25% helium; temperatures were too low.

31. The expansion rate is accelerating.

Going Further: Quantitative Questions

38. A factor of 528.

40. 1.0 K.

42. 2.9×10^4 km/s.

GLOSSARY

A

abiotic synthesis The creation of life from nonlife.

absorption The capture of a photon of a specific wavelength, causing an electron to jump from a lower to a higher energy level.

absorption line A dark line in a continuous spectrum. Specific patterns of absorption lines indicate the presence of a specific chemical element in the source.

acceleration An object's change of velocity per unit time.

active galactic nucleus (AGN) A highly luminous, energetic galactic nucleus powered by a super-massive black hole.

active optics A telescope technology in which mirror segments are constantly adjusted to prevent de-formation by external factors.

adaptive optics A telescope technology in which mirror segments are rapidly adjusted to account for atmospheric turbulence and improve image resolution when needed.

AGB See *asymptotic giant branch*.

AGN See *active galactic nucleus*.

AGN outflow Material driven away from the central regions of an active galactic nucleus.

albedo The reflecting power of an object, measured as the ratio of light reflected to light received.

altitude The angular distance from a point on the sky to a point on the horizon directly below it.

Amazonian period The evolutionary period of Mars that began about 3 billion years ago, marked by low atmospheric mass and dry, cold surface conditions.

amplitude A measure of a wave's strength, equal to half of the wave's total height.

anaerobic Requiring the absence of free oxygen to survive.

analemma The figure-eight pattern formed by chart-ing the daily position of the Sun at the same time of day over the course of 1 year.

angular resolution A measure of a telescope's ability to separate or distinguish features in a distant object.

angular size Angular distance measured on a sphere (such as Earth or the celestial sphere) in degrees, arcminutes, and arcseconds.

annular eclipse A solar eclipse that occurs when the positions of Moon and Sun are such that the lunar disk does not fully block the solar disk.

antimatter A form of matter that annihilates on contact with normal matter. A particle and its anti-particle have the same mass but opposite charge.

aperture The diameter of a telescope's main light-collecting lens or mirror.

aphelion (pl. aphelia) The distance of farthest approach of an object orbiting the Sun.

apogee The distance of farthest approach of an object orbiting Earth.

arcminute A measure of angular size. There are 60 arcminutes in 1 degree.

asteroid A rocky object that orbits the Sun and has a diameter from hundreds of meters to 1,000 km.

asteroid belt The disk-shaped Solar System region 2.1 to 3.5 AU from the Sun where many asteroids orbit.

astrobiology The interdisciplinary study of life in its astronomical context.

astrology The belief that there is a relationship between astronomical phenomena and events in the human world.

astrometric binary or **visual binary** Co-orbiting stars that have been detected by the proper motions of the components.

astronomical unit (AU) The average distance from Earth to the Sun, 1.5×10^{11} meters; used as a stan-dard of measure for Solar System objects.

astronomy The study of celestial objects such as moons, planets, stars, nebulas, and galaxies.

asymptotic giant branch (AGB) The region of the HR diagram to which a low- or intermediate-mass star evolves when it has used up the hydrogen and helium in its core. The star then has a cool surface, large radius, and high luminosity.

atom 1. The fundamental building block of matter, consisting of a dense central nucleus surrounded by a cloud of negatively charged electrons. 2. The smallest unit of matter that retains the properties of a given chemical element.

AU See *astronomical unit*.

aurora The emission of light in the upper atmo-sphere, visible at higher latitudes, caused by charged particles from the solar wind flowing along Earth's magnetosphere.

axial tilt See *obliquity*.

B

Balmer series Emission or absorption lines of the hydrogen atom as an electron moves between the first excited state and higher energy states.

bar A thick, rectangular region of stars extending across the center of some spiral galaxies.

barred spiral galaxy A spiral galaxy with a central rectangular bar.

basalt A type of volcanic silicate rock that forms at the ocean floor.

BBN See *Big Bang nucleosynthesis*.

belt A dark, sinking cloud region that forms a horizontal band in Jupiter's atmosphere because of convection and the planet's rapid rotation.

Big Bang The event that initiated cosmic expansion.

Big Bang nucleosynthesis (BBN) The nuclear fusion of light elements in the first few minutes after the Big Bang.

Big Crunch A theoretical contraction of the Universe that would occur if the combined matter-energy exceeded critical density.

binary accretion The process by which the collision of two objects results in the formation of one larger object.

binary-accretion-disk-magnetic-field model A theory of planetary nebula formation that involves a binary system and accretion disks. Jets from the disk sculpt the shape of the planetary nebula.

binary star One of two stars in orbit around a com-mon center of mass. Binary stars are born together.

binding energy The energy needed to break up an object being held together by a given force. The same amount of energy is released when the object forms.

black hole A region of space-time whose strong gravitational distortion prevents anything, including light, from escaping.

blackbody radiation Electromagnetic radiation from a dense, opaque body, whose spectrum depends on only the body's temperature.

blazar A form of radio-loud AGN that shows rapid variability in radio, optical, and X-ray wavelengths but few emission lines.

blueshift A decrease in the wavelength of light that results when the light source moves toward the observer or the observer moves toward the source.

Bohr atom A model of the atom, devised by Niels Bohr, in which electrons in discrete orbits surround a positively charged nucleus. Electron "jumps" between the discrete orbits determine the emission or absorption of light.

Bok globule See *dark globule*.

bound orbit An object's orbital motion around a planet.

bow shock The boundary where a strong flow runs into an obstacle.

brown dwarf A failed star that has too little mass to drive core temperatures high enough to initiate hydrogen-burning nuclear fusion.

bulge The roughly spherical distribution of stars and interstellar material in the central region of a galaxy.

C

C-type asteroid An asteroid that is carbonaceous (rich in carbon) and is typically dark.

caldera The depression at the top of a volcano.

calibration The use of a known parameter value obtained by a proven method to validate a measure-ment of that parameter by an unproven method.

Cambrian explosion The period about 540–505 mil-lion years ago when the diversity of life expanded at an accelerated rate.

capture hypothesis A hypothesis of the Moon's origin stating that the Moon formed elsewhere and was captured intact by Earth's gravity.

carbonaceous meteorite The most primitive type of meteorite, which formed from the components that built the Solar System.

carbon-nitrogen-oxygen cycle See *CNO cycle*.

Cassini Division A 4,800-km-wide region between Saturn's A and B rings that has been shaped by Sat-urn's moon Mimas.

CDM See *cold dark matter*.

celestial equator The extension of Earth's equator onto the celestial sphere.

celestial sphere An imaginary transparent sphere surrounding Earth, on which positions of stars, planets, and other celestial bodies are projected from extensions of Earth's coordinate system.

Centaur An object that has a mix of asteroid-like and comet-like properties.

center of mass The point around which two orbiting bodies move. It lies closer to the more massive body.

centrifugal force The illusion of outward force when an object is moving on a curved path.

centripetal acceleration Acceleration involving a force that keeps an object moving on a circular path. The acceleration is directed toward the path's center of curvature.

Chandrasekhar limit The maximum mass ($1.4\ M_{Sun}$) of a white dwarf; the mass that electron degeneracy pressure can support.

chromatic aberration The failure of a lens to focus all colors to the same point.

chromosphere The lower layer of the Sun's atmosphere.

circumpolar constellation A constellation that, from the viewer's perspective, never rises or sets.

climate change A significant, long-term change in the statistical distribution of weather patterns.

CMB See *cosmic microwave background*.

CME See *coronal mass ejection*.

CNO cycle (carbon-nitrogen-oxygen cycle) A chain of nuclear fusion reactions, involving carbon, nitrogen, and oxygen, that produces helium; the mechanism for powering stars of more than 1.3 solar masses.

cold dark matter (CDM) Nonluminous matter moving at a slower-than-relativistic speed.

coma The nebulous envelope around the nucleus of a comet, formed when the comet passes close to the Sun.

comet A body of rock and ice ranging in size from hundreds of meters to 1,000 km and typically orbiting the Sun on a highly elliptical orbit.

comparative planetology The study of planets' characteristics based on comparison with other known planets.

compression wave A wave of increased density that passes through background material; the density returns to normal after the wave passes.

condensation theory A theory of planetary formation in which planets are created from a disk of gas and dust around a newly formed star.

conservation of angular momentum The principle whereby the rotational speed of an object that spins or orbits a central point increases as the distance from the center decreases, and vice versa.

conservation of energy The principle whereby total energy remains constant in a system or process in which energy is not added or removed, even though the energy can be converted from one form into another.

constant of nature A term that does not change from one situation or time to another. The value of a constant of nature must be measured to be determined.

constellation A group of stars that form a pattern, and the designated region of the sky surrounding them.

convection The transfer of heat from one place to another by fluid motion.

convective zone The layer of the Sun above the radiative zone; where energy is transported outward by the motion of gas (convection).

core 1. Earth's innermost layer, composed primarily of iron. The inner core is solid; the outer core is in a fluid state. 2. A star's central region, where high temperatures and densities allow thermonuclear fusion reactions to occur, releasing energy.

core accretion model A model of gas giant formation in which an icy terrestrial planet grows by binary accretion up to a critical point and then rapidly pulls gas in from the surrounding disk.

corona 1. A large circular or oval structure surrounding volcanoes on some Solar System bodies, formed by upwelling. 2. The extended region of extremely hot, low-density gas surrounding the Sun.

coronal gas The phase of the interstellar medium that consists of extremely hot, low-density gas resulting from explosions of high-mass stars (supernovas).

coronal hole A region of low density where the solar wind is generated.

coronal mass ejection (CME) A large-scale ejection of mass and magnetic energy from the Sun's corona.

cosmic censorship The conjecture that a singularity cannot exist without an event horizon.

cosmic distance ladder The sequence of methods for determining the distance to astronomical objects; arranged by increasing distance over which each method is effective.

cosmic expansion The expansion of space-time since the Big Bang.

cosmic microwave background (CMB) Blackbody radiation released during recombination, with a peak wavelength in the red part of the electromagnetic spectrum. The expansion of the Universe now puts the peak of CMB in the microwave range.

cosmic ray A high-speed subatomic particle (typically, an atomic nucleus) from space.

cosmic singularity The initial condition of the expanding Universe as a state of infinite density, temperature, and space curvature from which the Big Bang was initiated.

cosmic web Structure on the largest cosmic scales, appearing as a network of interconnected filaments (joined where dense clusters are found) and surrounding voids.

cosmological look-back time A measure of time elapsed between the moment photons are detected on Earth and the moment at which they were emitted by a distant object.

cosmological principle The idea that, on average, the conditions in one part of the Universe are the same as those in any other part.

cosmology The field of astronomy that studies the Universe as a whole.

crater An approximately circular depression made in the surface of an object when a smaller body impacts it at high speed.

cratering The change in a planet's surface features due to impacts with comets and asteroids.

critical density The matter-energy per unit volume that is just sufficient to slow cosmic expansion such that the expansion proceeds indefinitely at a speed that approaches zero more and more closely over time.

crust Earth's solid outer layer.

D

Dark Ages The early era of cosmic evolution, when neutral hydrogen filled space and absorbed all visible and ultraviolet photons.

dark energy Energy driving the acceleration of the expansion of space-time.

dark globule or **Bok globule** A small molecular cloud containing a few solar masses' worth of material whose density and dust content render it opaque to light in visible wavelengths.

dark matter Nonluminous mass that is evident only by its gravitational influence.

declination The position of an object on the celestial sphere, measured similarly to latitude on Earth.

degeneracy pressure Pressure exerted because of quantum mechanical motions of particles resulting from confinement to a small volume.

density perturbation A small region where the density of hydrogen gas is slightly higher or lower than that of the surrounding regions.

deoxyribonucleic acid (DNA) A double-stranded polymer that encodes genetic information and is the biological basis of life on Earth. Its two strands form a double-helix shape.

deuterium An isotope of hydrogen with one proton and one neutron.

differential force of gravity The different strengths of the gravitational force from a massive body that the different parts of an extended object "feel" because the force decreases as the inverse square of the distance between them.

differential rotation The rotation of different regions of an object (different latitudes and/or depths) at different rates.

differentiation The process in planet formation whereby, during a planet's molten stage, the denser materials sink to the center while less dense materials rise to the surface, forming distinct layers.

dipole field The magnetic-field configuration consisting of north and south magnetic poles connected by magnetic-field lines.

DNA See *deoxyribonucleic acid*.

Doppler shift A change in wavelength that results because either a wave source or an observer moves relative to the other.

droplet condensation The transformation of molecules from gaseous to liquid state in the form of small droplets in an atmosphere. It generally occurs where temperature is decreasing.

dust The smallest bits of solid matter in the Universe, composed of atoms such as carbon and silicon.

dust tail The tail of a comet that is composed of dust driven from the nucleus by solar radiation; it points away from the Sun as the dust particles move on their own orbits.

dwarf planet A body, smaller than a planet, that orbits a star and has enough mass to become spherical by self-gravity but that has not cleared its neighboring region of small objects.

E

Earth mass The mass of Earth, 5.97×10^{24} kg, used as a relative unit of measure for other objects.

Earth-crossing asteroid (ECA) An asteroid whose highly elliptical orbit crosses Earth's orbit and is therefore a potential risk for collision with Earth.

ECA See *Earth-crossing asteroid*.

eccentricity A measure of the roundness of an ellipse, calculated as a ratio: the distance from the ellipse's center to its foci, divided by the length of the semimajor axis.

eclipse The passing of one celestial body through the shadow of another.

eclipsing binary Co-orbiting stars in which one star is observed passing in front of the other.

ecliptic The path that the Sun appears to follow against the background stars, as defined by Earth's orbit around the Sun.

Einstein ring The ring-shaped appearance of a distant background object that is caused by a gravitational lens formed via symmetric distribution of dark matter in the foreground.

ejecta blanket A layer of pulverized material that is blown out of a crater on impact.

electromagnetic radiation Oscillating electric and magnetic fields traveling through space.

electromagnetic spectrum The full range of electromagnetic radiation at different wavelengths, from gamma rays (very short) to radio waves (very long).

electron A subatomic particle with negative electric charge.

elephant trunk A column of molecular gas that forms at the edge of an H II region and seems to point toward the ionizing star or stars.

ellipse The geometric shape of an oval, in which the sum of the distances from two points (foci) to every point on the ellipse is constant. A circle is a special type of ellipse.

elliptical galaxy A galaxy that appears elliptical on the sky. They are generally triaxial, meaning that the width is different in each dimension.

emission The release of a photon of a specific wavelength when an electron jumps from a higher to a lower energy level.

emission line A bright line at a particular wavelength in a spectrum. Specific patterns of emission lines indicate the presence of a specific chemical element in the source.

Encke Division A 325-kilometer-wide gap within Saturn's A ring that has been shaped by Saturn's moon Pan.

energy The ability to perform work. Energy comes in many forms, such as kinetic, thermal, gravitational, and electromagnetic.

energy level One of the certain discrete values of orbital energy possible for an atom.

envelope The region surrounding a protostar that is composed of free-falling gas.

epicycle A secondary orbit whose center point orbits Earth; devised by Ptolemy to explain retrograde motion.

equant A point displaced from the center of a planet's orbit around which the planet displays uniform motion.

equinox A day when the hours of daylight and darkness are equal. In the Northern Hemisphere, the autumn equinox occurs on approximately September 21; the spring equinox, approximately March 21.

erosion or **weathering** The process by which surface features are worn away by the action of wind and water (including ice).

escape velocity The velocity required to escape the pull of a massive body's gravity.

eukaryotic cell A cell that stores its DNA in a cell nucleus.

event horizon The boundary of a black hole, located one Schwarzschild radius away from the singularity.

excited state Any energy state that is higher than the ground state of an atom.

extrasolar planet or **exoplanet** A planet orbiting any star other than the Sun.

extremophile An organism that thrives under extremes of temperature, pressure, and other conditions.

eyepiece In a telescope, the small lens through which the observer looks.

F

Fermi paradox The conclusion that no intelligent life exists in our galaxy beyond Earth, since the billions of years following the Milky Way's origin would have allowed advanced civilizations to have spread across interstellar space and reached Earth, or that we have not yet received signals indicating the existence of other technological civilizations.

field reversal A reorientation of north and south poles within a magnetic field such that the two poles swap positions.

first law of planetary motion The principle, advanced by Johannes Kepler, stating that planets move on elliptical orbits with the Sun at one focus.

fission hypothesis A hypothesis of the Moon's origin stating that the Moon formed from material pulled into orbit from the still-molten Earth by a close encounter with another body.

flare An eruption that launches material from the solar surface.

flocculent spiral A spiral galaxy with poorly defined spiral arms.

focus (pl. foci) Either of the two points interior to an ellipse that define its shape. The Sun is always at one focus of a planet's elliptical orbit.

force An interaction between two bodies that, if unbalanced, can change their speed and/or direction of motion.

force boson A particle that carries one of the four fundamental forces.

force law The principle, advanced by Isaac Newton, stating that the change in an object's velocity due to an applied net force is in the same direction as, and directly proportional to, the force but inversely proportional to the object's mass.

fossil magnetism Magnetic fields that are imprinted in a planet but were created by an earlier active dynamo phase.

fragmentation The shattering of solid bodies, such as planetesimals, caused by collisions.

frame of reference An individual perspective from which observations are made.

free-fall Motion occurring solely due to gravitational force.

free-fall time The length of time required for a cloud to collapse under the force of its own gravity.

frequency The number of peaks (or troughs) of a wave that pass a point in space in a single second.

fundamental forces The four basic ways matter can interact: gravity, electromagnetism, the strong nuclear force, and the weak nuclear force.

G

galactic disk A flattened distribution of stars and interstellar material within a galaxy.

galaxy Millions to billions of gravitationally bound stars. Galaxies also include gas, dust, and dark matter.

galaxy cluster A gravitationally bound collection of galaxies containing thousands of members.

galaxy group A gravitationally bound collection of galaxies, usually containing fewer than 50 members.

galaxy supercluster A chain of clusters and groups that are being shaped by their mutual gravitation.

Galilean moons The four largest moons of Jupiter: Io, Ganymede, Callisto, and Europa.

gamma ray A high-energy electromagnetic wave with a wavelength shorter than 10^{-11} meter.

gamma-ray burst (GRB) A brief, intense outburst of gamma rays at relativistic speeds during a hypernova or kilonova.

gas giant A planet that consists mostly of hydrogen and helium and has a mass many times that of Earth.

gene A section of a DNA or an RNA molecule that is the code for a particular trait or function.

general theory of relativity Albert Einstein's second theory of relativity, which extends the special theory of relativity by incorporating gravity; it states that matter and energy dictate how space-time curves, while space-time dictates how matter and energy move.

genetic drift A change in a gene's frequency due to random events other than gene mutation.

geocentric model A model stating that Earth is the body around which all other Solar System objects orbit.

giant elliptical galaxy The largest, most massive type of elliptical galaxy.

giant molecular cloud (GMC) A large-scale region of gas in molecular form spanning tens to hundreds of parsecs and containing up to 10 million solar masses.

globular cluster A gravitationally bound cluster of 100,000 or more stars born at the same time.

GMC See *giant molecular cloud*.

GOE See *Great Oxygenation Event*.

grand design spiral galaxy A spiral galaxy with well-defined spiral arms that may be traced all the way from the bulge to the outer regions of the disk.

grand unified theory (GUT) A yet-to-be-discovered theory to explain how the strong and weak nuclear forces and the electromagnetic force emerged from a unified force moments after the Big Bang.

gravitational contraction The shrinking of an object's radius and accompanying increase in its density due to the force of its own gravity.

gravitational lensing A change in the path of light passing a massive object due to the curvature of space-time.

gravitational potential energy Energy that an object possesses because of its position in a gravitational field.

gravitational wave A ripple in the fabric of space-time, produced whenever matter moves.

gravity An attractive force between any two massive bodies that depends on the product of the bodies' masses and the inverse square of the distance between them.

GRB See *gamma-ray burst*.

Great Attractor An unseen concentration of mass whose gravity affects the motion of galaxies in a large volume of space that includes the Milky Way.

Great Oxygenation Event (GOE) The period in Earth's history when oxygen built up relatively rapidly in the atmosphere; about 2.3 billion years ago.

greenhouse effect The phenomenon by which the fraction of incoming solar energy that would have been radiated back into space is trapped by greenhouse gases, leading to an increase in the average temperature of the planet.

greenhouse forcing An increase in greenhouse gas concentrations that leads to a change in a planet's climate.

greenhouse gas An atmospheric gas that absorbs infrared radiation and keeps it from being reemitted into space.

ground state The lowest energy state of an atom.

GUT See *grand unified theory*.

H

H I clouds The phase of the interstellar medium that is composed of low-temperature, neutral hydrogen gas.

H II region The phase of the interstellar medium that contains ionized hydrogen gas.

habitable zone The region around a star where liquid water can exist on a planet's surface.

half-life The period of time over which the total number of nuclei of a radioactive isotope will drop by one-half.

HDF See *Hubble Deep Field*.

heat energy See *thermal energy*.

Heisenberg uncertainty principle The quantum mechanical principle stating that as the position of a particle is specified to higher accuracy, its speed becomes increasingly uncertain, and vice versa.

heliocentric model A model stating that the Sun is the body around which all other Solar System objects orbit.

heliopause The outer boundary of the heliosphere.

helioseismology The study of oscillations in the Sun and their propagation, used to develop precise models of solar structure.

heliosphere A cavity, created by the solar wind, in the interstellar gas surrounding the Solar System.

helium flash The rapid internal ignition of helium fusion in the core of an evolved star.

Herbig-Haro (HH) object A protostellar jet that is distinguished by knots in the jet beam.

Hesperian period The evolutionary period of Mars from about 3.5 to 3 billion years ago marked by the end of volcanic activity and the slow loss of much of Mars's atmosphere.

HH object See *Herbig-Haro object*.

hierarchical formation An ordered sequence of formation, from smallest to largest structures.

homogeneous Uniformly distributed. A homogeneous distribution of matter has the same average properties at every point in space.

horizon The horizontal circle defining the lower edge of the sky.

hot dark matter Nonluminous matter moving at a relativistic speed.

hot Jupiter A Jupiter-sized planet orbiting very close to its star, usually at a fraction of an astronomical unit.

HR diagram A graph of luminosity versus surface temperature (or spectral class) for a population of stars.

Hubble Deep Field (HDF) Data resulting from a week-long exposure in 1995 by the Hubble Space Telescope of a tiny, dark patch of sky revealing thousands of galaxies at great distances.

Hubble flow The pattern of large-scale galaxy motion associated with the expansion of space-time.

Hubble tuning-fork diagram Edwin Hubble's scheme for representing galaxy types along a progression from highly elliptical to spiral or barred spiral.

Hubble's constant The slope of Hubble's law relating recession velocity and distance; expressed in kilometers per second per megaparsec [(km/s)/Mpc].

Hubble's law The linear relationship between a galaxy's recession velocity and its distance.

hydrated Containing water.

hydrocarbon An organic compound consisting entirely of hydrogen and carbon.

hydrodynamic instability model A model of gas giant formation in which small regions collapse, forming planets within a gravitationally unstable disk.

hydrogen-fusing shell The layer outside a star's core where hydrogen continues to fuse to helium after such reactions have ended in the core.

hydrostatic equilibrium A balance between gravity and pressure.

hypothesis (pl. hypotheses) A proposed explanation of a phenomenon that must be rigorously tested by experiments.

I

ice giant A planet that consists mostly of ices (water, methane, carbon dioxide, and other compounds) and has a mass many times that of Earth.

IMF See *initial mass function*.

impact hypothesis The currently accepted hypothesis of the Moon's origin, stating that a large, Mars-sized body struck Earth and forced mantle material into orbit, which gravitationally coalesced into the Moon.

inertia An object's resistance to changes in velocity (speed or direction of motion).

inertial law The principle, advanced by Isaac Newton, stating that objects in motion at constant velocity along a straight line continue in that way unless acted on by a net force.

inflation A hyperrapid expansion of space-time that occurred extremely soon after the Big Bang.

infrared wave An electromagnetic wave with a wavelength of 700 nm (where 1 nanometer = 1×10^{-9} meter) to 1 mm, just longer than that of visible light.

initial mass function (IMF) The mathematical relationship describing the relative abundance of stars of different masses.

instability A complex pattern of gas flow resulting from a disruption in an originally smooth flow.

intensity The energy of light that an object emits— either in total or at a specific wavelength.

interferometry The technique of combining the signals from many smaller telescopes to achieve the resolving power of a larger one.

interstellar dust grain Tiny bit of solid matter found in diffuse clouds in the regions between the stars.

interstellar gas Ionic, atomic, or molecular gas found in diffuse clouds in the regions between the stars.

interstellar medium (ISM) The gas and dust spread among the stars in different phases that depend on temperature and density conditions.

intracluster medium Thin gas in the space between galaxies in a galaxy cluster.

inverse square law A relationship whereby a quantity (such as gravity) decreases in proportion to the square of a variable (such as distance) as the variable increases.

ion tail The tail of a comet that is composed of gas from the coma and carried backward by the solar wind and thus always points directly away from the Sun.

ion torus (pl. tori) A doughnut-shaped region of emission where ions spiral around the magnetic field.

ionosphere or **thermosphere** The fourth (and highest) layer of Earth's atmosphere, extending from about 85 to 1,000 km, where most of the atoms are ionized.

irregular galaxy An amorphous galaxy lacking common structures.

irregular satellite A satellite orbiting in the opposite direction from its planet and/or on a highly inclined orbit in relation to the planet's equator.

ISM See *interstellar medium*.

isotope One of several forms of an element with different numbers of neutrons in the nucleus.

isotropic Uniform in appearance when viewed from different angles. An isotropic distribution of matter looks the same in all directions.

J

jet A stream of a liquid, gas, or small solid particles shot outward from a comet nucleus or other object in a focused beam.

K

K See *kelvin*.

KBO See *Kuiper Belt object*.

kelvin (K) A unit of temperature in the Kelvin scale, where 0 means absolutely no motion. One kelvin has the same magnitude as 1°C.

kinetic energy Energy that is due to an object's bulk motion.

Kirkwood gap A band of orbits in the main asteroid belt where very few asteroids are found due to orbital resonances with Jupiter.

Kuiper Belt The disk-shaped Solar System region 30 to 50 AU from the Sun, outside Neptune's orbit, containing small bodies of ice and rock.

Kuiper Belt object (KBO) A small body made of ice and rock that orbits in the Kuiper Belt, beyond Neptune's orbit.

L

large-scale structure of the Universe The hierarchy of cosmic structures on ever-larger scales from galaxy groups up to superclusters and voids (and beyond).

Late Heavy Bombardment A hypothetical epoch, 4.1–3.8 billion years ago, during which a large number of asteroids collided with bodies in the inner Solar System.

lateral gene transfer The movement of genetic material that does not proceed from parent to offspring.

latitude A measure of location specifying north–south position on Earth running from 0° at the equator to 90° north (North Pole) and 90° south (South Pole).

lenticular galaxy A galaxy with a bulge, limited evidence of a disk, and no spiral arms.

lepton A class of fundamental particles of matter that includes electrons, muons, tau particles, and neutrinos.

life expectancy–mass relationship The mathematical connection between a star's life expectancy and mass: the greater the star's mass, the shorter its life expectancy.

light curve A graph of an astronomical object's light output versus time.

light-year (ly) The distance that light travels in a year, 9.5×10^{15} meters; used as a standard of measure for large distances.

line of nodes The line defined by the intersection of the Moon's orbital plane and Earth's orbital plane around the Sun.

lithosphere Earth's solid outer layers, consisting of the crust and the uppermost regions of the mantle.

Local Group A gravitationally bound collection of galaxies within which the Milky Way resides.

longitude A measure of location specifying east–west position on Earth.

long-period comet A comet that originated in the Oort Cloud and has a highly eccentric orbit and an orbital period longer than 200 years.

luminosity A light source's total energy output per time.

luminosity class A system for designating a star as a giant or dwarf based on absorption lines.

lunar highlands Old, heavily cratered portions of the Moon's surface lighter in color than the maria.

ly See *light-year*.

Lyman series Emission or absorption lines of the hydrogen atom as an electron moves between the ground state and excited states.

M

MACHO See *massive compact halo object*.

macromolecule See *polymer*.

magma flooding The creation of smooth lowlands via lava flows that fill in craters.

magnetic dynamo The flow of electrically charged fluid that can generate a large-scale magnetic field in a planet, a star, or any other astronomical object.

magnetosphere A region of space around a planet where the planet's magnetic field exerts a strong influence.

magnification A measure of a telescope's ability to enlarge appearances in an image.

magnitude scale A categorization of stellar brightness in which faint objects are given higher numbers than brighter objects.

main sequence The diagonal band on the HR diagram representing fusion of hydrogen in stars' cores. High-mass stars appear at upper left; low-mass stars, at lower right.

mantle The silicate-rock layer between Earth's core and crust. High pressures cause material there to deform and flow, except in the uppermost mantle.

maria (sing. mare) Large, dark plains on the Moon, formed by ancient volcanic eruptions.

Martian spherule A tiny pebble-like structure ("blueberry") that was found in Martian soil and may have crystallized out of a water-rich solution.

mass A quantity of matter that is related to an object's resistance to changes in velocity; measured in units of kilograms.

massive compact halo object (MACHO) A proposed form of dark matter that includes objects such as black holes and extremely faint neutron stars.

matter Any substance made up of particles with mass that occupies space.

Maunder Minimum The period in the mid to late 1600s when the number of sunspots remained extremely low from one sunspot cycle to another.

megabar A unit of pressure equal to a million times Earth's surface pressure.

megalith A large stone used as a monument or part of a monument.

megaparsec (Mpc) A unit of distance equal to 1 million parsecs.

meridian A circular arc crossing the celestial sphere and passing through the local zenith and celestial pole.

mesosphere The third layer of Earth's atmosphere, extending from about 50 to 85 km.

metallic hydrogen Hydrogen at a pressure and temperature high enough for electrons in closely spaced atoms to be able to move freely, as in a metal.

metallicity The abundance of elements heavier than helium in an object.

meteor A meteoroid that has entered Earth's atmosphere. Heat from friction causes radiation that may be seen as a very brief streak of light or as a fireball.

meteor shower The appearance of frequent meteors within a few hours or several days. The meteors seem to emanate from a common location on the sky as Earth passes through the debris deposited along a comet's orbit.

meteorite A rock that was a meteoroid, then a meteor, and has reached Earth's surface.

meteoroid A rock that ranges in size from a grain of sand to a boulder and orbits the Sun.

microwave The shortest-wavelength radio wave, with a wavelength of 1 millimeter (1×10^{-3} meter) to about 0.1 meter.

Milky Way The spiral galaxy in which our Sun is located.

mini-Neptune A planet with a mass less than 15 Earth masses but substantially greater than Earth's mass.

molecular clouds The low-temperature, high-density phase of the interstellar medium where molecules can form. Molecular clouds are the site of star formation.

molecular disk See *molecular ring*.

molecular outflow The bipolar swept-up shells of molecular material, which may be driven by protostellar jets.

molecular ring or **molecular disk** A dense ring of molecular material orbiting the galactic center.

momentum The mass of an object multiplied by its velocity. Only the application of a force can change an object's momentum.

monomer A small molecule that, when repeated, forms a polymer.

moon A natural satellite of a planet.

Mpc See *megaparsec*.

muon An elementary particle that is extremely short-lived and heavier than an electron.

mythology A collection of stories used to explain phenomena whose physical basis is not understood.

N

naked singularity A singularity that lacks an event horizon, which cannot exist if cosmic censorship is correct.

natural selection The process by which an organism with traits that make it well adapted to its environment survives and reproduces, passing on those traits, more successfully than organisms without those traits.

neap tide A tide that is lower than average because the Sun's gravity partially cancels out that of the Moon at the Moon's first and third quarter phases.

nebula An interstellar cloud of dust, hydrogen, helium, and other ionized gases.

nebulosity Clouds in space.

Neolithic period The period extending from about 10,000 BCE to between 4500 and 2000 BCE, when farming was introduced.

net force The sum of all forces acting on an object.

neutrino An electrically neutral, weakly interacting elementary subatomic particle often created in nuclear reactions.

neutron A subatomic particle with no net electric charge within an atomic nucleus.

neutron star A stellar remnant supported by neutron degeneracy pressure that results from the gravitational collapse of a massive star.

Noachian period The evolutionary period of Mars from about 4 to 3.5 billion years ago, marked by a thick atmosphere and warmer climate.

north celestial pole The extension of Earth's North Pole onto the celestial sphere.

north magnetic pole The point on Earth's Northern Hemisphere at which the planet's magnetic field points vertically downward.

nova A nonterminal explosive nuclear fusion occurring on the surface of a white dwarf after it has accreted matter from a companion star.

nuclear fission The splitting apart of a nucleus into two or more smaller nuclei.

nucleobase A nitrogen-containing base; a component of a nucleotide in DNA and RNA.

nucleotide A monomer built of a sugar molecule, a phosphate group, and a nucleobase. Sugars and phosphates link, forming the backbones of DNA and RNA.

nucleus (pl. nuclei) 1. The dense region at the center of an atom, consisting of protons and neutrons. 2. The main body of a comet, composed of ice and rock.

O

objective In a refractor, the large lens that gathers light from the object being observed and focuses the light rays to form an image.

oblate Roughly spherical with some bulging at the equator.

obliquity or **axial tilt** The angle between a planet's rotation axis and a line perpendicular to the planet's orbital plane.

Occam's razor A principle stating that among competing hypotheses, the simplest one should be selected.

omega (Ω) The ratio of the observed density of matter-energy in the Universe to the critical density.

Oort Cloud A spherical region at the outer limits of the Solar System that has a radius of about 50,000 AU and may contain trillions of comets.

opacity The degree to which light is absorbed when passing through a material.

open star cluster A loosely distributed group of stars that were born from the same cloud at the same time and remain bound for some period before dispersing.

orbit The cyclical path in space that one object makes around another object.

orbital period The time an object takes to complete one revolution of an orbit.

orbital resonance The synchronized, periodic gravitational influences that arise when objects orbiting a body have periods that are whole-number ratios of each other.

orbital velocity The velocity required to keep a body in orbit around another body.

order of magnitude The size of a physical quantity in powers of 10.

outflow channel An extended region of scoured ground on Mars that includes features indicating high rates of fluid flow.

ozone A molecule composed of three oxygen atoms: O_3.

ozone hole A region within the stratosphere above Earth's poles that shows a substantial decrease in ozone abundance.

P

P-P chain See *proton-proton chain*.

P-wave Pressure wave or primary wave: a seismic wave in which atoms and molecules oscillate back and forth in the direction of the wave's propagation.

Paleolithic period The period extending from 2.6 million years ago (characterized by the earliest known use of stone tools) to 10,000 years ago.

parallax A displacement or difference in the apparent position of an object viewed along two different lines of sight.

parsec (pc) A unit often used for very large distances, equal to approximately 3.26 light-years.

partial lunar eclipse An eclipse that occurs when the Moon passes partially through Earth's umbral shadow.

partial solar eclipse An eclipse that occurs when a region of Earth's surface passes under the Moon's penumbral shadow.

particle accelerator An experimental apparatus used to study the structure of matter by colliding atoms and subatomic particles together at very high speeds (high energies).

particle era A period in cosmic evolution when the particles described by the standard model of particle physics emerged.

pc See *parsec*.

peculiar motion 1. The motion of stars in the Milky Way disk that is a departure from pure revolution around the galactic center. 2. The motion of a galaxy through space that is due to gravitational attraction, distinct from motion due to the expansion of space-time.

penumbra The outer region of a shadow cast by an extended object.

perigee The distance of closest approach of an object orbiting Earth.

perihelion (pl. perihelia) The distance of closest approach of an object orbiting the Sun.

period-luminosity relationship The correlation between the luminosity and period of a Cepheid variable star, allowing a star to act as a standard candle for determining distance.

phase (of the Moon) As seen by an observer on Earth, the appearance of the illuminated portion of the Moon, which changes cyclically as the Moon orbits Earth.

phase transition A change in the state of matter based on changes in temperature (or another quantity).

photoevaporation The process by which gas is launched from the surface of an accretion disk or molecular cloud through heating by ionizing radiation from a young, massive star.

photoionization The stripping of an electron from an atom or ion via the absorption of a photon.

photon A particle of light.

photosphere The Sun's thin, "surface" layer, from which radiation escapes into space.

phylogeny Relationships among species over time.

Planck length The size scale at which space-time should begin showing its quantum nature, about 10^{-44} meter.

planet A body that orbits a star and is massive enough for self-gravity to have pulled it into a spherical shape and to have cleared smaller objects from the neighborhood of its orbit.

planet migration A change in a planet's orbit after the planet has formed.

planetary cooling The process by which a planet loses the heat acquired during its formation.

planetary differentiation The separation by gravity of a planet into different layers, from densest at the core to least dense near the surface.

planetary nebula In low- and intermediate-mass stars, the evolutionary stage (between the AGB and white dwarf phases) in which material is driven off the star in a stellar wind, forming a luminous cloud around the star.

planetary system A star and its orbiting planets; the Solar System is an example of a planetary system.

planetesimal An irregular rocky object, typical of those from which planets are believed to have formed by accretion.

plasma A gas composed of charged particles such as electrons, protons, and ions.

plate A large rigid block—crust plus solid upper-mantle material—that floats on the deformable part of the mantle.

plate tectonics The movement of Earth's lithospheric plates due to underlying mantle flow.

plume A rising column of molten rock from a planet's mantle that breaks through the crust.

Polaris The star that is found very near Earth's north celestial pole (hence called the North Star) and around which the sky of the Northern Hemisphere appears to turn.

polymer or **macromolecule** A large molecule containing many atoms in repeating subunits (monomers), often with a chain structure.

Population I A category of stars, found primarily in the disk or bulge, that have high metallicity and therefore must have been born relatively recently in cosmic history.

Population II Stars found primarily in the stellar halo, which have lower metallicity and therefore must have been born relatively early in cosmic history.

positron The antimatter version of the electron, having the mass of the electron but a positive charge.

prebiotic molecules Molecules that act as the building blocks of life—necessary for the origin of life.

precession The rotation of a planet's spin axis, similar to that of a wobbling top.

pressure wave See *P-wave*.

primitive meteorite A type of meteorite that formed, largely unchanged, from the disk that produced the planets of the Solar System.

primordial abundance of light elements The proportions of hydrogen, helium, deuterium, and other light elements created in the first few minutes after the Big Bang.

principle of equivalence The equivalence between the "force" experienced in an accelerating frame of reference and the force of gravity in an inertial (non-accelerating) frame.

processed meteorite A type of meteorite that formed when larger asteroids broke apart during violent collisions.

prominence A giant arc of magnetism anchored in the photosphere.

proper motion An object's motion that is on the plane of the sky relative to other stars and can be detected by observations taken over time.

protein A polymer, made of amino acid monomers, that performs and assists in cellular functions.

proton A subatomic particle with positive electric charge.

proton-proton chain or **P-P chain** The sequence of re-actions by which hydrogen nuclei are converted into helium nuclei in Sun-like stars.

protoplanetary disk The disk that surrounds a young star, from which planets may form.

protostar A dense central object, plus an accretion disk and envelope, that forms after cloud collapse occurs but before nuclear fusion begins.

protostellar jet A focused beam of plasma launched at high speed from a protostar.

pseudoscience A claim, belief, or practice that is presented as scientific but does not adhere to a valid scientific method, lacks supporting evidence or plau-sibility, and/or cannot be reliably tested.

pulsar A rotating, magnetized neutron star whose radio waves produce periodic emission patterns that are observed on Earth as energy bursts.

Q

quantum field The distribution of energy that fills space, within which fundamental particles appear as excited states.

quantum gravity A theory (yet to be fully worked out) that unifies general relativity and quantum mechan-ics, describing how space-time becomes granular ("quantized") at the smallest scales.

quantum mechanics or **quantum physics** A branch of physics dealing with physical phenomena at the level of molecular, atomic, and subatomic scales.

quark A class of fundamental particles of matter that make up other particles, such as protons and neutrons.

quasar A quasi-stellar radio source; an extremely luminous and distant active galactic nucleus.

R

radial velocity An object's motion that is perpen-dicular to the plane of the sky along the observer's line of sight and can be detected via Doppler shift measurements.

radial velocity method An exoplanet detection method that measures changes in a star's velocity as it orbits a common center of mass with one or more planets.

radiative energy transport The transfer of thermal energy via electromagnetic radiation (photons).

radiative zone The layer of the Sun above the core; where energy is transported outward by electro-magnetic radiation (photons scattering off particles of mass).

radio galaxy A galaxy whose nucleus is the origin of relativistic bipolar jets.

radio wave An electromagnetic wave with a wave-length longer than 0.1 meter.

radioactive decay The process by which some iso-topes of an element transform over time into other isotopes or elements, emitting subatomic particles and energy in the process.

radiometric dating A method used to date a sample of material by comparing relative amounts of radio-active atoms with the "daughter" atoms into which they decay.

reaction law The principle, advanced by Isaac Newton, stating that for every applied force, there is an equal and opposite force.

recombination era The era about 380,000 years after the Big Bang in which protons captured electrons, forming the first hydrogen atoms and thus de-coupling matter from cosmic blackbody radiation.

reconnection The breaking and re-joining of magnetic-field lines.

red giant A cool star with a large radius that has evolved off the main sequence because hydrogen fuel in its core is depleted.

redshift An increase in the wavelength of light that results when the light source moves away from the observer or the observer moves away from the source.

redshift survey Measurements of redshift (reces-sional velocity) for many galaxies across a region of the sky, from which astronomers produce three-dimensional maps of galaxies.

reflection nebula A dense, dusty interstellar cloud made visible by reflected starlight.

reflector An optical telescope that uses curved mir-rors to collect light and form an image.

reflex motion The movement of a star in response to the gravity of an orbiting planet.

refraction The bending of light as it passes from one transparent medium to another, such as from water to air.

refractor An optical telescope that uses lenses to col-lect light and form an image.

regular rich cluster A spherical, centrally condensed galaxy cluster with more than 1,000 galaxies.

regular satellite A satellite orbiting in the same direc-tion as its planet and with a low inclination to the planet's orbital plane.

residual ice cap A polar region of frozen material such as CO_2 or water that stays constant in size with seasonal changes.

retrograde motion Apparent motion of a planet in a direction opposite to its normal motion.

ribonucleic acid (RNA) A single-stranded polymer that transcribes the code of DNA, placing amino acids in the correct sequence to create proteins.

RNA See *ribonucleic acid*.

Roche limit The closest a satellite can get to its cen-tral object before that object's tidal forces pull the satellite apart, when the satellite's own gravity can no longer hold it together.

Roche lobe The region around a star within which orbiting material is gravitationally bound to that star.

rotation The spin of an object around an internal axis.

rotation curve The graph of rotation speed as a func-tion of orbital radius for objects orbiting a common center, such as stars orbiting a galaxy.

runaway greenhouse effect An unstable situation in which a planet's temperature rises, leading to an increased greenhouse effect, which raises the tem-perature higher still; this feedback drives planetary temperatures to very high levels.

runoff channel A long, meandering feature on Mars that resembles a river system on Earth.

S

S-type asteroid An asteroid that contains mostly silicate rock and is typically more reflective than a C-type asteroid.

S-wave Shear wave or secondary wave: a seismic wave in which atoms and molecules oscillate perpen-dicular to the direction of the wave's propagation.

satellite 1. Any object that orbits a larger object. 2. A small body orbiting a planet.

scatter Change direction; e.g., change a photon's direction of propagation.

scattered disk The disk-shaped region beyond the Kuiper Belt that contains many small bodies made of ice and rock scattered by the gravity of the planets to outer regions of the Solar System.

scattering Rearrangement of the orbits of planets due to mutual gravitational interactions.

Schwarzschild radius The distance from a black-hole singularity to points where the escape velocity equals the speed of light.

scientific method A body of techniques for acquiring new knowledge that includes systematic observa-tion, experimentation, and theorizing.

scientific model An idea or set of ideas used to create testable explanations.

scientific notation A way to write very large or very small numbers using the form $a \times 10^b$, where the coefficient a is any real number and the exponent b is an integer.

seafloor spreading The creation of new lithospheric material due to the upwelling of magma as tectonic plates spread apart.

seasonal ice cap A polar region of frozen material such as CO_2 or water that increases and decreases in area and thickness seasonally.

second law of planetary motion The principle, advanced by Johannes Kepler, stating that the line connecting a planet to the Sun will sweep out equal areas of its orbit in equal times.

seismic wave A propagating wave in Earth's interior caused by an earthquake (or other energetic event).

semimajor axis Half of the long axis of an ellipse.

semiminor axis Half of the short axis of an ellipse.

Seyfert galaxy A galaxy whose nucleus shows broad emission lines, indicating highly ionized material moving at great speed.

shear wave See *S-wave*.

shepherd moon A planetary satellite whose gravity confines a ring to a narrow band.

shield volcano A volcano that forms from a mantle plume and tends to have shallow, sloping sides.

short-period comet A comet that originated in the Kuiper Belt or scattered disk and has an orbital period of 200 years or less.

sidereal day The time taken by Earth to rotate on its axis relative to the stars.

sidereal month The amount of time it takes the Moon to make one complete orbit in relation to the background stars.

silicate A mineral compound containing silicon and oxygen.

singularity The central point of a black hole, where density and space-time curvature is infinite.

small-angle formula A mathematical relationship between the angular size of an object, as measured on the sky, and the object's distance and physical size.

SMBH See *supermassive black hole*.

snow line The distance from a star beyond which the temperature is cold enough for water to exist as ice.

solar day The time between successive crossings of the same meridian on the sky by the Sun.

solar granulation The pattern of bright interiors and dark edges on the photosphere; generated by convection cells.

solar magnetic cycle A large-scale reorientation of the Sun's magnetic field occurring every 22 years.

solar max The period of peak activity in the sunspot cycle.

solar neutrino problem The apparent discrepancy between the expected number of neutrinos being emitted from the Sun and the number measured experimentally; resolved by the discovery that neutrinos oscillate among three types.

solar wind 1. A flow of material from the surface of the Sun that sweeps through interplanetary space. 2. A stream of energetic particles flowing off the Sun.

south celestial pole The extension of Earth's South Pole onto the celestial sphere.

south magnetic pole The point on Earth's Southern Hemisphere at which the planet's magnetic field points vertically upward.

space weather Coronal mass ejections sweeping through interplanetary space, which produce auroras when they hit Earth and may pose threats to astronauts and technological systems.

space-time A unified, four-dimensional geometry of the Universe that incorporates both space and time.

special theory of relativity Albert Einstein's first theory of relativity; it states that the laws of physics are the same for all observers in inertial (nonaccelerating) frames of reference and that the speed of light in a vacuum is the same for all observers, regardless of their relative motion.

spectral classification A fundamental categorization of stars, based on observed spectral lines, into seven categories denoted by the letters O, B, A, F, G, K, and M.

spectrograph A device for separating light from a source into its component wavelengths and measuring the energy at each wavelength.

spectroscopic binary Co-orbiting stars that have been detected by the Doppler shift of their absorption lines.

spectroscopic parallax A method of determining distance to an astronomical object that relies on the width of absorption lines in the object's spectrum.

spectrum (pl. spectra) The distribution of light intensity versus wavelength (or frequency).

spectrum of perturbations The range and strength of variations in density imposed on cosmic matter during inflation.

speed An object's change in position per unit of time.

spiral arm A spiral-shaped distribution of stars and interstellar material, typically extending from the bulge outward.

spiral density wave A wavelike pattern of compression that moves through the physical material of a galaxy's disk. Spiral density waves are seen in spiral arms.

spiral galaxy A pinwheel-shaped galaxy that may have spiral arms.

spring tide A tide that is higher than average because it is reinforced by the linear alignment of Sun, Earth, and Moon at the new and full Moon phases.

standard candle An object with known luminosity that can be used to measure distance.

standard model of particle physics A description of subatomic particles and forces.

star A sphere of gas, such as the Sun, that produces energy via nuclear fusion in its core.

starburst A rapid conversion of an appreciable fraction of molecular gas into stars.

steady-state model A theory, advanced by Fred Hoyle, stating that although the Universe is expanding, new matter is slowly created between galaxies, maintaining a constant average density.

Stefan-Boltzmann law The mathematical relationship between a blackbody's luminosity and its surface area and temperature. Hotter, larger objects glow more brightly.

stellar ejecta Material expelled in a stellar explosion such as a supernova.

stellar evolutionary track A model-generated path on the HR diagram representing the evolution of a star through time.

stellar halo The spherical distribution of Population II stars that are found mainly in globular clusters.

stellar occultation The blocking of a star's light by an object between the star and an observer's view.

stratosphere The second layer of Earth's atmosphere, extending from about 17 km to about 50 km.

strong nuclear force One of the four fundamental forces of nature, it binds particles together in the nucleus of an atom.

subatomic particle Any particle smaller than an atom. Subatomic particles include protons, neutrons, and electrons.

subduction A geologic process in which one edge of a tectonic plate is forced below the edge of another.

summer solstice The day (approximately June 21 in the Northern Hemisphere) with the most hours of sunlight, when the Sun appears highest in the sky.

sunspot A large, strongly magnetized region of the Sun's surface that appears darker because it is lower in temperature than its surroundings.

supercluster A collection of galaxy clusters; the largest type of object in the Universe.

super-Earth A planet with a mass greater than Earth's but substantially less than 15 Earth masses.

superluminal motion The illusion that an object's speed exceeds the speed of light; created by the object's motion toward the observer.

supermassive black hole (SMBH) A black hole of millions or billions of solar masses that exists at the center of a galaxy.

supernova The highly energetic explosion of a massive star or a star in a binary system.

supernova remnant An expanding shock wave that contains ejected material and swept-up interstellar material resulting from the explosion of a star.

synodic month The amount of time it takes the Moon to cycle through its phases.

T

T Tauri star A variable star of less than 3 solar masses that is in a later stage of star formation and exhibits bright X-ray flares.

temperature A measure of the average kinetic energy of all particles in a substance.

terracing The steplike appearance of steep walls in an impact crater after the walls slump due to gravity.

terrestrial planet A planet consisting primarily of solid material.

theory An explanation for a phenomenon that has undergone extensive testing and is generally accepted to be accurate.

thermal energy or **heat energy** Energy that is due to random motions of atoms.

thermal history The evolutionary pattern of heating and cooling of a planet.

thermal inversion An increase in temperature that occurs with increasing altitude (whereas temperature usually decreases with increasing altitude).

thermal velocity The average velocity of particles in a gas, which depends on the gas's temperature.

thermonuclear fusion The transformation of lighter atomic particles into heavier ones, requiring high temperatures.

thermonuclear runaway Nuclear fusion that ignites and spreads catastrophically, consuming all available fuel.

thermosphere See *ionosphere*.

third law of planetary motion The principle, advanced by Johannes Kepler, stating that the square of a planet's orbital period (in years) equals its average orbital radius (in astronomical units) cubed.

tidal bulge A distortion from spherical symmetry that is caused by tidal force.

tidal force Force acting on an object that is due to differential gravity from a second object.

tidal locking Created by tidal effects, the match-up between the rotation and revolution periods of an orbiting object, such that the object always shows the same side to the body it orbits.

tidal synchronization The synchronization of the rotation and orbital periods of a satellite (such as a planet or moon) by the gravity of the body it orbits. (For example, the satellite completes two rotations for every three of its orbits.)

tide The effect of differential gravity on an extended object, leading to distortion of spherical mass distributions, among other effects.

time dilation The effect of relativity in which time runs slower for observers that are moving at relativistic speeds relative to that of distant observers.

total lunar eclipse An eclipse that occurs when the Moon passes through Earth's umbral shadow.

total solar eclipse An eclipse that occurs when a region of Earth's surface passes under the Moon's umbral shadow and the Sun's disk is fully blocked.

totality The moment or duration of total concealment of the Sun or Moon during an eclipse.

transit method An exoplanet detection method that measures changes in the brightness of a star as a planet passes in front of it.

transmission The passing of light through matter without being absorbed.

trans-Neptunian object A minor body in the Solar System that orbits the Sun at a greater average distance than Neptune, 30 AU.

triggered star formation Star formation that is initiated by the compression of a molecular cloud by a supernova blast wave.

Trojan asteroid An asteroid co-orbiting with Jupiter in one of two groups: one group preceding Jupiter and the other group following it, both at the same distance.

troposphere The lowest layer of Earth's atmosphere, reaching a maximum altitude of 17 km; the location of most of Earth's weather and clouds.

Tully-Fisher law A relationship of galactic rotation speed to galaxy mass that is used to infer the galaxy's luminosity.

turbulence The random jostling motions of moving air.

21-cm emission line A wavelength of radio light emitted by hydrogen; considered by some to be a natural communication channel between Earth and other civilizations.

Type Ia supernova The thermonuclear explosion of a white dwarf that has accreted mass from a binary companion, causing it to exceed the Chandrasekhar limit and experience thermonuclear runaway.

Type II supernova The rapid collapse and violent explosion of a massive star.

U

ultraviolet light An electromagnetic wave with a wavelength of 10 nm to 450 nm, just shorter than that of visible light.

umbra The inner region of a shadow cast by an extended object.

unbound orbit An object's motion that carries the object away from a planet permanently.

uncompressed density The density an object would have if the effect of gravity were excluded.

unified model of AGNs A consolidated depiction of the morphology of the various types of active galactic nuclei.

universal law of gravity The dependence of the force of gravity between two objects as the product of their masses times Newton's constant divided by their distance of separation squared.

Universe The totality of space, time, matter, and energy.

V

variable star A star with regular periodic changes in its temperature and luminosity. Periods tend to be on the order of days to years.

velocity Rate of change in an object's position and direction per unit of time.

velocity dispersion The characteristic range of velocities of bodies in an astronomical object, such as stars in an elliptical galaxy.

verification The use of a proven measurement method to validate a new measurement method.

visible light The form of light to which the human eye responds, with a wavelength ranging from 400 nm to 700 nm.

visual binary See *astrometric binary*.

void A region between superclusters, spanning tens to hundreds of megaparsecs, where galaxy density is much lower than average.

volatile Easily vaporized (transformed into a gaseous state) at relatively low temperatures; or, a volatile substance.

W

warm interstellar medium (WISM) The phase of the interstellar medium that is warmed by the galaxy's diffuse ionizing radiation (the sum of all of its ionizing stellar sources).

wavelength The distance between successive peaks or troughs of a wave.

weakly interacting massive particle (WIMP) A proposed form for dark matter consisting of particles that interact only via gravity and the weak nuclear force.

weathering See *erosion*.

weight The downward force that gravity produces on an object.

weightlessness The absence of the sensation of weight during free-fall.

white dwarf The compact remnant of a low- or intermediate-mass star supported by electron degeneracy pressure.

Wien's law The mathematical relationship between a blackbody's temperature and the wavelength at which it radiates with peak intensity.

WIMP See *weakly interacting massive particle*.

winter solstice The day (approximately December 21 in the Northern Hemisphere) with the fewest hours of daylight, when the Sun appears lowest in the sky.

WISM See *warm interstellar medium*.

X

X-ray An electromagnetic wave with a wavelength between 10^{-11} and 10^{-8} meter.

X-ray burster A binary system that shows intense X-ray outbursts caused by the pileup of material on the surface of a neutron star.

X-ray halo A large, quasi-spherical region of hot gas surrounding a galaxy or galaxy cluster.

Z

zenith The point on the sky directly overhead from an observer.

zodiac The 13 constellations that lie close to the plane of the ecliptic, which the Sun appears to pass through over the course of a year.

zone A white, rising cloud region that forms a horizontal band in Jupiter's atmosphere because of convection and the planet's rapid rotation.

CREDITS

Front Matter

Page vi: Peter Hoffman/Redux; Photo courtesy of Carol Latta; Dartmouth College/Eli Burakian; p. vii: Photo by Bachrach; Kris Snibbe/Harvard University; Pat Hartigan; David Jewitt; Caltech; Courtesy of Crystal Brogan; p. viii: Les Mentel; Courtesy of Cornell University Photography; J. Adam Fenster, University of Rochester; Photo by Carolyn Russo/NASM; Bob Paz/Keck Institute for Space Studies; Mark Halper; Bill Youngblood/Caltech; p. ix: Courtesy Christopher P. McKay, NASA; Dennis Wise, University of Washington; Falk Herwig; Lika Guhathakurta; University of Illinois College of Liberal Arts & Sciences; Lara Eakins, Department of Astronomy, University of Texas at Austin; Didier Saumon; p. x: Trippy Photography; Photo by Andrea Cipriani Mecchi; © Macquarie University; Stan Woosley; Bill Youngblood for Caltech; Photo by Sue Weisler, Rochester Institute of Technology; p. xi: UW-Whitewater photo/Craig Schreiner; Courtesy of Debra Elmegreen; Dunlap Institute for Astronomy & Astrophysics, University of Toronto; Paul Green; James Porto; Bryce Richter/University of Wisconsin–Madison; Scott Kenyon; Kris Snibbe/Harvard University; p. xii: Johns Hopkins University; Photo by Peter Zhaoyu Zhou/USC; p. xv (clockwise from top left): GL Archive/Alamy Stock Photo; NASA/C. Henze; Courtesy Caltech/MIT/LIGO Laboratory; Courtesy Caltech/MIT/LIGO Laboratory; Photo by Sue Weisler, Rochester Institute of Technology; Michael Kramer; Department of Special Collections and University Archives, W. E. B. DuBois Library, University of Massachusetts, Amherst; Emilio Segrè Visual Archives/American Institute of Physics/Science Source; p. xiii (top): Courtesy of Crystal Brogan; (bottom): Scott Kenyon; p. xxi: J. Adam Fenster, University of Rochester

Chapter 1

Page xxiv: Westend61 GmbH/Alamy Stock Photo; p. 2 (top): Design Pics Inc./Alamy Stock Photo; (bottom): Andrew Clegg, NSF; p. 3 (left to right): NASA; Guillem Lopez/Alamy Stock Photo; Gary Weathers/Tetra Images/Corbis; p. 6 (top): NASA/NOAA/GSFC/Suomi NPP/VIIRS/Norman Kuring; (bottom): NASA/SDO; p. 7 (top): Robert Gendler/Stocktrek Images/Corbis; (bottom): NASA, ESA, Jennifer Lotz and the HFF Team (STScI); p. 8 (top to bottom): Portrait Essentials/Alamy Stock Photo; Science History Images/Alamy Stock Photo; Science History Images/Alamy Stock Photo; Painters/Alamy Stock Photo; Patrick Guenette/Alamy Stock Vector; Miguel Pérez-Ayúcar/Manuel Castillo/Michel Breitfellner/Miguel Sánchez-Portal for CESAR/ESA; Journal du Voyage de l'abbé Chappe en Californie, Tome 3 (C3/22), Observatoire de Paris; p. 9 (top): ESA & Planck Collaboration/Rosat/Digitised Sky Survey; (bottom): NASA; p. 13: Peter Hoffman /Redux; p. 15: robertharding/Alamy Stock Photo.

Chapter 2

Page 20: George Pachantouris/Getty Images; p. 22 (top a): NASA; (top b): M. Rich, K. Mighell, and J. D. Neill (Columbia University), and W. Freedman (Carnegie Observatories), NASA; (top c): NASA, NOAO, ESA, the Hubble Helix Nebula Team, M. Meixner (STScI), and T.A. Rector (NRAO); (bottom a): NASA/DMSP; (bottom b): Chris Cook/Science Source; p. 25 (both): © David Malin Images; p. 27 (left): Frank Zullo/Science Source; (right): AA/USNO; p. 30 (left): A. Morton/Science Source; (right): Babak Tafreshi/National Geographic Creative; p. 32: Radius Images/Corbis; p. 33: NASA/JPL/Caltech; p. 35: Alberto Levy; p. 36: © 1999 Fred Espenak (MrEclipse.com); p. 37: Photo by Brian Oyer; p. 38 (left): Bob King/Stellarium; (right): Alan Dyer/Visuals Unlimited, Inc.; p. 39 (top): Cenk E. Tezel & Tunç Tezel; (bottom): Photo courtesy of Carol Latta; p. 40 (top): Ivy Close Images/Alamy Stock Photo; (bottom left): nagelestock.com/Alamy Stock Photo; (bottom right): 2/Steve Allen/Ocean/Corbis; p. 41 (top): Gianni Dagli Orti/Shutterstock; (bottom): Dartmouth College/Eli Burakian; p. 44 (clockwise from top left): World History Archive/Alamy Stock Photo; bilwissedition Ltd. & Co. KG/Alamy Stock Photo; Royal Astronomical Society/Science Source; Quagga Media/Alamy Stock Photo; ESA–D. Ducros, 2013; SSPL/Getty Images; Album/Alamy Stock Photo; Chronicle/Alamy Stock Photo; Georgios Kollidas/Alamy Stock Photo.

Chapter 3

Page 50: James Canning, Astronomy Section, Rochester Academy of Science; p. 52: Sonia Halliday Photo Library/Alamy Stock Photo; p. 53: Christophe Boisvieux/Getty Images; p. 55 (left): World History Archive/Alamy Stock Photo; (right): Pictures Now/Alamy Stock Photo; p. 56 (clockwise from top left): VintageCorner/Alamy Stock Photo; FALKENSTEINFOTO/Alamy Stock Photo; dpa picture alliance/Alamy Stock Photo; Photo by Bachrach; Photo by Anne DeMarinis; Wikimedia, pd; bilwissedition Ltd. & Co. KG/Alamy Stock Photo; p. 57 (top): Photo by Bachrach; (bottom): GL Archive/Alamy Stock Photo; p. 61 (left): FineArt/Alamy Stock Photo; (right): Gianni Tortoli/Science Source; p. 62: NASA/GSFC/Arizona State University; p. 63 (top): John Rummel; (bottom left): Chronicle/Alamy Stock Photo; (bottom right): GRANGER – All rights reserved; p. 64: Photo by Peter Blackwood, www.blackwoodphoto.com; p. 65: Science History Images/Alamy Stock Photo; p. 66 (top): Lebrecht Music & Arts/Alamy Stock Photo; (bottom): Bettmann/Getty Images; p. 69 (bottom): SpaceX.

Chapter 4

Page 78: NASA/ESA/Hubble Heritage Team; p. 80 (top): NAIC Arecibo Observatory, a facility of the NSF; (bottom, left to right): Science Source; Royal Astronomical Society/Science Source; S. Beckwith (STScI) Hubble Heritage Team, (STScI/AURA), ESA, NASA; p. 87 (left): Atlantide Photoravel/Getty Images; (right): Dr. Arthur Tucker/Science Photo Library/Getty Images; p. 85: Kris Snibbe/Harvard University; p. 88: Hubble Heritage Team (AURA/ STScI/ NASA); p. 89 (top): ESA/Hubble & NASA; (bottom): NOAO/AURA/NSF/Science Source; p. 91 (clockwise from top left): HUPSF Observatory (14), olvwork360662, Harvard University Archives; Chronicle/Alamy Stock Photo; GL Archive/Alamy Stock Photo; NOAO/AURA/NSF/Science Source; Science History Images/Alamy Stock Photo; Historic Collection/Alamy Stock Photo; NOAO/AURA/NSF/Science Source; Len Collection/Alamy Stock Photo; p. 92 (top): Pat Hartigan; (bottom): Reprinted Figure 3 with permission from A. S. Stodolna, A. Rouzée, F. Lépine, S. Cohen, F. Robicheaux, A. Gijsbertsen, J. H. Jungmann, C. Bordas, and M. J. J. Vrakking, "Hydrogen Atoms under Magnification: Direct Observation of the Nodal Structure of Stark States," Phys. Rev. Lett. Vol. 110, 213001, 2013. © 2019 by the American Physical Society; p. 94: John Chumack/Science Source; p. 97 (left): Maximilian-Vlad Teodorescu; (right): NASA, ESA, Andrew Fruchter (STScI), and the ERO team (STScI + ST-ECF); p. 98 (left): Enrico Sacchetti/Science Source; (right): Wikimedia, pd; p. 101: Science History Images/Alamy Stock Photo; p. 102 (left): James Vaughn/SpaceFlight Insider; (right): NASA, JPL-Caltech, SINGS Team (SSC); p. 103 (top left): STS-103, STScI, ESA, NASA; (top right): NASA, ESA, G. Illingworth, D. Magee, and P. Oesch (University of California, Santa Cruz), R. Bouwens (Leiden University), and the HUDF09 Team; (bottom): Alexandr Mitiuc/Alamy Stock Photo; p. 104 (left): NASA; (right): NASA/ESA/STScI.

Chapter 5

Page 108: ESA/NASA, ESO and Danny LaCrue; p. 110 (top): David Jewitt, UCLA; (bottom): ART Collection/Alamy Stock Photo; p. 113 (all): JPL/NASA; p. 115: NEAR Project, Galileo Project, NASA; p. 118: Wikimedia, pd; p. 119 (top): Al Russell, Astronomy Section, Rochester Academy of Science; (bottom): Science History Images/Alamy Stock Photo; p. 120 (left): NASA/JPL/Caltech/Ames Research Center/University of Arizona; (center): Halley Multicolor Camera Team/Giotto Project/ESA; (right): NASA/JPL/UMD; p. 121 (left): NASA/ESA; (right): Hubble Space Telescope Comet Team; p. 122: David Jewitt; p. 123 (top, left to right): Science History Images/Alamy Stock Photo; ESA/Rosetta/MPS for OSIRIS Team MPS/UPD/LAM/IAA/SSO/INTA/UPM/DASP/IDA; Science History Images/Alamy Stock Photo; (center): Al Russell, ASRAS; (bottom): AP Photo/AP Video; p. 124 (top): NASA/MSFC/D. Moser, NASA's Meteoroid Environment Office; (bottom, left to right): The Natural History Museum, London/Science Source; The Natural History Museum/Alamy Stock Photo; The Natural History Museum/Alamy Stock Photo; p. 125 (top): NASA Earth Observatory; (bottom left): NASA/JPL-Caltech/UCLA/MPS/DLR/IDA; (bottom right): NASA/JPL-Caltech/UCLA/MPS/DLR/IDA/PSI; p. 127 (a): NASA/Johns Hopkins University Applied Physics Laboratory/Southwest Research Institute/Alex Parker; (b): NASA/JHUAPL/SWRI; (c): Science History Images/Alamy Stock Photo; p. 129: Caltech; p. 132 (top): Science History Images/Alamy Stock Photo; (center): NASA/JPL-Caltech/Palomar Observatory; (bottom): Gemini Observatory/AURA; p. 133: PHL / UPR Arecibo; p. 134: STSci; p. 138: NASA and The Hubble Heritage Team (STScI/AURA); p. 140: Courtesy of Crystal Brogan; p. 143 (clockwise from top left): Portrait Essentials/Alamy Stock Photo; Georgios Kollidas/Alamy Stock Photo; Mark McCaughrean (Max-

(bottom): "Photo: Thomas Bresson"; p. 294 (left): NASA, ESA, and the Hubble Heritage Team (STScI/AURA); (right): NASA, ESA, and Z. Levay (STScI); p. 295: Schlesinger Library, Radcliffe Institute, Harvard University; p. 297 (top left): © 2011 Fred Espenak (AstroPixels.com); (top right): Professor Greg Parker; (bottom): A. Dupree (CfA), R. Gilliland (STScI), NASA, p. 298: University of Illinois College of Liberal Arts & Sciences; p. 300: Len Collection/Alamy Stock Photo; p. 301 (left): INTERFOTO/Alamy Stock Photo; (right): Omikron/Getty Images; p. 303: Lara Eakins, Department of Astronomy, University of Texas at Austin; p. 307: Hubble Heritage Team (AURA/ STScI/ NASA); p. 308 (clockwise from top left): Schlesinger Library, Radcliffe Institute, Harvard University; INTERFOTO/Alamy Stock Photo; GL Archive/Alamy Stock Photo; SSPL/Getty Images; Library of Congress; Didier Saumon; p. 309: Didier Saumon.

Chapter 12

Page 312: SC Observatory Team: Mike Selby, Andy Chatman, Stefan Schmidt; p. 314 (top, both): Courtesy NASA/JPL-Caltech; (bottom): LAMBDA/NASA; p. 315: NASA, ESA, and the Hubble Heritage Team (STScI/AURA) Acknowledgment: J. Hughes (Rutgers University); p. 316 (top): Copyright: Tony Hallas; (bottom): NASA/JPL/Caltech/Harvard-Smithsonian Center for Astrophysics; p. 322 (top left): European Southern Observatory/Science Source; (top right): NASA, ESA, and The Hubble Heritage Team (STScI/ AURA), P. McCullough (STScI/AURA); (center): NRAO; (bottom): Richard Klein, Lawrence Livermore National Laboratory; Pak Shing Li, University of California, Berkeley; Tim Sandstrom, NASA Ames Research Center; p. 323 (clockwise from top left): European Southern Observatory/Science Source; Science History Images/Alamy Stock Photo; Edward Emerson Barnard Papers, Box 43, Vanderbilt University Special Collections; MPIA/Markus Nielbock; NASA; NASA/ JPL-Caltech/STScI; NASA; Trippy Photography; p. 324: Trippy Photography; p. 328 (top): NASA and A. Watson (Institut de Astronomia, UNAM, Mexico (STScl-PRC00-32b); (bottom): NASA/ESA; p. 329 (left): NASA/JPL-Caltech/A. Noriega-Crespo (SSC/Caltech), H. Kline (JPL); (right): Photo by Andrea Cipriani Mecchi; p. 331: SONYC Team/Subaru Telescope; p. 332: Al Russell, Astronomy Section, Rochester Academy of Science; p. 333 (left): NASA, ESA, N. Smith (U. California, Berkeley) et al., and The Hubble Heritage Team (STScI/AURA); (right): NASA/JPL-Caltech/ University of Colorado; p. 334 (both): NASA, ESA/ Hubble and the Hubble Heritage Team; p. 335 (left): NASA, ESA, M. Robberto (Space Telescope Science Institute/ESA) and the Hubble Space Telescope Orion Treasury Project Team; (right): NASA and K. Luhman (Harvard-Smithsonian Center for Astrophysics), STScl-PRC00-19.

Chapter 13

Page 338: NASA, ESA, J. Hester and A. Loll (Arizona State University); p. 340 (both): © Anglo-Australian Observatory, photo by David Malin; p. 341: © Macquarie University; p. 345 (left): GL Archive/Alamy Stock Photo; (center): SPL/Science Source; (right, top): WIYN/NOAO/NSF/Science Source; (right, center):

J. P. Harrington and K. J. Borkowski (University of Maryland), and NASA; (right; bottom): X-ray/Optical Composite (X-ray: NASA/UIUC/Y. Chu et al., Optical: NASA/HST); p. 347 (left to right): Image processed by Donald P. Waid. Based on observations made with the NASA/ESA Hubble Space Telescope, and obtained from the Hubble Legacy Archive, which is a collaboration between the Space Telescope Science Institute (STScI/NASA), the Space Telescope European Coordinating Facility (ST-ECF/ESA), and the Canadian Astronomy Data Centre (CADC/NRC/CSA); ESA/ Hubble & NASA; NASA, ESA, C.R. O'Dell (Vanderbilt University), and M. Meixner, P. McCullough, and G. Bacon (Space Telescope Science Institute); ESA/ Hubble/NASA; p. 349: Bettmann/Getty Images; p. 350: Stan Woosley; p. 353 (top): "On the Impact of Three Dimensions in Simulations of Neutrino-Driven Core-Collapse Supernova Explosions." Sean M. Couch. *The Astrophysical Journal*, Vol. 775, Number 1 (August 30, 2013). © AAS. Reproduced with permission; (bottom left): Image copyright 2005–2015 R. Jay GaBany, Cosmotography.com; (bottom right): NASA, ESA, Zolt Levay (STScI); p. 354: K. Nishiyama and F. Kabashima/H. Maehara, Kyoto University; p. 357 (clockwise from top left): bilwissedition Ltd. & Co. KG/Alamy Stock Photo; MPIA/NASA/Calar Alto Observatory; NG Images/Alamy Stock Photo; NASA, ESA, R. Kirshner (Harvard-Smithsonian Center for Astrophysics and Gordon and Betty Moore Foundation), and M. Mutchler and R. Avila (STScI); NASA/ESA Hubble Space Telescope/Science Source; Stan Woosley; © Anglo-Australian Observatory, photo by David Malin; Chris Robart; World History Archive/Alamy Stock Photo; p. 360: Daily Herald Archive/SSPL/Getty Images (NRAO/AUI/NSF)/Science Source.

Chapter 14

Page 364: Mark A Garlick/National Geographic Creative; p. 366: NASA/C. Henze; p. 367: GL Archive/ Alamy Stock Photo; p. 368: Bill Youngblood for Caltech; p. 377: Photo by Sue Weisler, Rochester Institute of Technology; p. 378: Xinhua/Alamy Stock Photo; p. 380: NASA/Goddard Space Flight Center/ESA/Gaia/ DPAC/Science Source; p. 383: NRAO, Cal Tech, Walter Jaffe/Leiden Observatory, Holland Ford/JHU/STScI, and NASA; p. 384 (bottom): NASA/CXC/University of Hertfordshire/M. Hardcastle et al; p. 385 (clockwise from top left): GL Archive/Alamy Stock Photo; NASA/ C. Henze; Courtesy Caltech/MIT/LIGO Laboratory; Courtesy Caltech/MIT/LIGO Laboratory; Photo by Sue Weisler, Rochester Institute of Technology; Michael Kramer; Department of Special Collections and University Archives, W. E. B. Du Bois Library, University of Massachusetts, Amherst; Emilio Segrè Visual Archives/American Institute of Physics/Science Source; p. 386 (left): NASA; (right): Caltech/MIT/LIGO Lab.

Chapter 15

Page 390: NASA, ESA, the Hubble Heritage Team (STScI/AURA), and A. Riess (STScI); p. 392 (top left): Image composite by Ingrid Kallick of Possible Designs, Madison Wisconsin. The background Milky Way image is a drawing made at Lund Observatory. High-velocity clouds are from the survey done at Dwingeloo Observatory (Hulsbosch & Wakker, 1988); (top right):

ESA; (center): NASA/Hubblesite; (bottom): Courtesy of Nick Lamendola, Astronomy Section – Rochester Academy of Science; p. 393 (top): Chronicle/Alamy Stock Photo; (center): Victoria Art Gallery, Bath and North East Somerset Council/Bridgeman Images; (bottom): Royal Astronomical Society/Science Source; p. 394 (top): 916 collection/Alamy Stock Photo; (bottom): S. Kafka/K. Honeycutt/Indiana University/ WIYN/NOAO/NSF/Science Source; p. 397: UW–Whitewater photo/Craig Schreiner; p. 398: Gerard Lodriguss/Science Source; p. 399: NASA; p. 400: Courtesy of Debra Elmegreen; p. 401: © Anglo-Australian Observatory, photo by David Malin; p. 404 (top): Fred Baganoff (MIT), Mark Morris (UCLA), et al., CXC, NASA; (bottom): UCLA Galactic Center Group – W. M. Keck Observatory Laser Team; p. 408: Emilio Segrè Visual Archives/American Institute of Physics/ Science Source; p. 409: John B. Carnett/Bonnier Corporation via Getty Images; p. 410 (clockwise from top left): NASA/JPL-Caltech; Georgios Kollidas/ Shutterstock; GL Archive/Alamy Stock Photo; Anonymous/AP/Shutterstock; NASA/JPL-Caltech/ GSFC/SDSS; Robert Gendler/Stocktrek Images/Corbis; Emilio Segrè Visual Archives/American Institute of Physics/Science Source.

Chapter 16

Page 416: Calar Alto Observatory; p. 418 (top left): Science History Images/Alamy Stock Photo; (top right): SPL/Science Source; (bottom left): Science History Images/Alamy Stock Photo; (bottom right): ClassicStock/Alamy Stock Photo; p. 419 (top left): NASA/JPL-Caltech; (top right): Courtesy Carnegie Institution for Science/Edwin Hubble; (bottom): Giovanni Benintende, astrogb.com; p. 420 (top): The Hubble Heritage Team (AURA/STScI/NASA); (bottom): NASA, ESA, and The Hubble Heritage Team STScI/AURA; p. 421 (top): NASA, ESA, and The Hubble Heritage Team (STScI/AURA); (bottom): NASA/ CXC/SAO; p. 422 (left to right): Royal Observatory, Edinburgh/Science Photo Library/ Science Source; NASA, ESA and A. Nota (STScI/ESA); (right): NASA, ESA, and The Hubble Heritage Team (STScI/AURA); p. 423 (top to bottom): Giovanni Benintende, astrogb .com; NASA, ESA, and The Hubble Heritage Team STScI/AURA; NASA, ESA, and The Hubble Heritage Team (STScI/AURA); NASA, ESA, and The Hubble Heritage Team (STScI/AURA); p. 425: NASA, ESA, the Hubble Heritage Team (STScI/AURA) – ESA/Hubble Collaboration and A. Evans (University of Virginia, Charlottesville/NRAO/Stony Brook University), K. Noll (STScI), and J. Westphal (Caltech); p. 426: Dunlap Institute for Astronomy & Astrophysics, University of Toronto; p. 429 (top): BJ Fulton/LCOGT/ PTF; (bottom): Courtesy of the Archives, California Institute of Technology; p. 430: A relation between distance and radial velocity among extra-galactic nebulae Edwin Hubble, Proceedings of the National Academy of Sciences Mar 1929, 15 (3) 168–173; DOI: 10.1073/pnas.15.3.168; p. 432: NASA/CXC/SAO/ H.Marshall et al.; p. 433 (top): X-ray – NASA, CXC, R. Kraft (CfA), et al. Radio – NSF, VLA, M. Hardcastle (U Hertfordshire) et al.; Optical – ESO, M.Rejkuba (ESO–Garching) et al.; (bottom): Paul Green; p. 434: Roeland P. van der Marel (STScI), Frank C. van den Bosch (University of Washington), and NASA/ESA;

p. 436: James Porto; p. 437 (clockwise from top left): Giuseppe Donatiello; Emilio Segrè Visual Archives/ American Institute of Physics/Science Source; NASA and The Hubble Heritage Team (STScI/AURA); NASA/ ESA/STScI; "Unified Schemes for Radio-Loud Active Galactic Nuclei," C. Megan Urry and Paolo Padovani, *Publications of the Astronomical Society of the Pacific*, Vol. 107 (September 1995). Copyright Astronomical Society of the Pacific, reprinted with permission of the authors; James Porto; UCLA Galactic Center Group – W. M. Keck Observatory Laser Team; © Max Planck Institute for Extraterrestrial Physics; Elena Zhukova.

Chapter 17

Page 442: Two Micron All-Sky Survey; p. 444: NRAO/ AUI and C. O'Dea & F. Owen; p. 447: (left to right): Peter Horree/Alamy Stock Photo; Chronicle/Alamy Stock Photo; nicehelix/Shutterstock; p. 448 (left): Robert Gendler, Subaru Telescope (NAOJ); (right): Johannes Schedler/Panther Observatory; p. 449 (left): Rogelio Bernal Andreo; (right): NASA/JPL-Caltech/L. Jenkins (GSFC); p. 450 (top): Khosroshahi, Maughan, Ponman, Jones, ESA, ING; (bottom): NASA; p. 452: Andrew Fruchter (STScI) et al., WFPC2, HST, NASA; p. 453: Bryce Richter/University of Wisconsin–Madison; p. 454: Science History Images/Alamy Stock Photo; p. 455 (top): This representation was originally done by Lars Lindberg Christensen, University of Copenhagen, Denmark; (bottom): Scott Kenyon; p. 457 (top): "Cosmography of the Local Universe." Hélène M. Courtois, et al., *The Astronomical Journal*, Vol. 146, Number 3 (August 14, 2013). © AAS. Reproduced with permission; (bottom): M. Blanton and the Sloan Digital Sky Survey, www.sdss.org; p. 459: Gemini Observatory/ NSF/AURA; 460 (top): NASA; ESA; G. Illingworth, D. Magee, and P. Oesch, University of California, Santa Cruz; R. Bouwens, Leiden University; and the HUDF09 Team; (bottom): NASA, ESA, and L. Infante (Pontificia Universidad Católica de Chile); p. 461: Kris Snibbe/ Harvard University; p. 462 (clockwise from top left): Pictorial Press Ltd./Alamy Stock Photo; FLHC 96/ Alamy Stock Photo; The 2dF Galaxy Redshift Survey team, www.2dfgrs.net; This representation was originally done by Lars Lindberg Christensen, University of Copenhagen, Denmark; Scott Kenyon; p. 465: simulations were performed at the National Center for Supercomputer Applications by Andrey Kravtsov (The University of Chicago) and Anatoly Klypin (New Mexico State University). Visualizations by Andrey Kravtsov.

Chapter 18

Page 470: JPL/NASA; p. 472: NASA and A. Reiss (STScI); p. 473 (top): Emilio Segrè Visual Archives/American Institute of Physics/Science Source; (bottom): SPL/ Science Source; p. 483 (top): Johns Hopkins University; (bottom): © Roger Ressmeyer/Corbis/VCG/Getty Images; p. 484: NASA/WMAP Science Team; p. 486 (left to right): NASA; ESA and the Planck Collaboration; NASA/WMAP Science Team; p. 489: Photo by Peter Zhaoyu Zhou/USC; p. 492 (clockwise from top left): NASA; John Zich, Courtesy University of Chicago; The University of Oklahoma; NASA, ESA, The Hubble Key Project Team, and The High-Z Supernova Search Team; NASA, ESA and STScI; © The Nobel Foundation/photo: U Montan; ©The Nobel Foundation/photo: U Montan; © The Nobel Foundation/photo: U Montan; NASA, ESA, and L. Infante (Pontificia Universidad Católica de Chile). Acknowledgment: NASA, ESA, and J. Lotz (STScI) and the HFF team.

Back Matter

Pages APP-7–10: Reproduced by permission. © 2019, *Astronomy* magazine, Kalmbach Publishing Co.; p. APP-11: Photo by Andrea Cipriani Mecchi; Bob Paz/ Keck Institute for Space Studies; UW-Whitewater photo/Craig Schreiner; Courtesy of Crystal Brogan; p. APP-12: Photo by Sue Weisler, Rochester Institute of Technology; Bill Youngblood for Caltech; © Macquarie University; Lara Eakins, Department of Astronomy, University of Texas at Austin; Courtesy of Debra Elmegreen; p. APP-13: Scott Kenyon; Photo by Bachrach; Dartmouth College/Eli Burakian; Kris Snibbe/Harvard University; p. APP-14: Paul Green; Lika Guhathakurta; Pat Hartigan; Mark Halper; Falk Herwig; p. APP-15: David Jewitt; Photo by Peter Zhaoyu Zhou/USC; University of Illinois College of Liberal Arts & Sciences; Caltech; p. APP-16: Photo courtesy of Carol Latta; Dunlap Institute for Astronomy & Astrophysics, University of Toronto; Kris Snibbe/ Harvard University; Courtesy of Cornell University Photography; Courtesy Christopher P. McKay, NASA; p. APP-17: J. Adam Fenster, University of Rochester; Trippy Photography; Johns Hopkins University; Didier Saumon; Bill Youngblood/Caltech; p. APP-18: Dennis Wise, University of Washington; Les Mentel; James Porto; Peter Hoffman/Redux; p. APP-19: Bryce Richter/ University of Wisconsin–Madison; Stan Woosley; Photo by Carolyn Russo/NASM.

INDEX

Page numbers in *italics* refer to illustrations and tables.